2006

Information and Communications for Development

2006

Information and Communications for Development

Global Trends and Policies

THE WORLD BANK
Washington, D.C.

1818 H Street, NW
Washington, DC 20433
Telephone: 202-473-1000
Internet: www.worldbank.org
E-mail: feedback@worldbank.org

1 2 3 4 09 08 07 06

Cover photo credits (clockwise from top left): Bill Lyons; Eric Miller; Bruno Lanvin; Alan Gignoux.

ISBN-10: 0-8213-6346-8
ISBN-13: 978-0-8213-6346-1
eISBN: 0-8213-6347-6
DOI: 10.1596/978-0-8213-6346-1

Library of Congress Cataloging-in-Publication Data has been applied for.

Table of Contents

PART II

ANNEXES

BOXES

FIGURES

TABLES

Foreword

Information and communication technology (ICT) has a critical role to play in development efforts around the world. There was a time when the benefits of applying ICT in fighting poverty and promoting economic growth were not widely understood. Many in the development community questioned how high-tech (and often expensive) communication technology could be used to alleviate such dire challenges as starvation, homelessness, and lack of basic education and health services. Lately, however, this view has given way to an understanding of ICT as an essential component of broader efforts to harness the free flow of information to increase voice, accountability, and economic development.

In recent years, developing countries and the international development community have started taking concrete actions to incorporate ICT into their economic policies and development agendas. Many countries are preparing and implementing national e-strategies that emphasize the ubiquity of connectivity as well as new applications in areas such as e-government and e-business. The Millennium Development Goals (MDGs), drawn from the United Nations Millennium Declaration and adopted in September 2000, have several specific targets involving ICT as a tool for reducing poverty. Nevertheless, improving the identification and measurement of the actual effects of ICT in development remains an important challenge going forward, especially in light of the rapid pace of change in the sector and the dearth of concrete, long-term data across countries.

In this context, *Information and Communications for Development 2006: Global Trends and Policies* offers a realistic assessment of experiences, trends, and outlook on the ICT sector, with a focus on actual results and justified expectations. It attempts to track and analyze global ICT development trends and to provide empirical evidence of the benefits that ICT is providing in terms of economic growth and poverty reduction. Indicators for the MDG targets, among others, have been incorporated into the ICT At-a-Glance tables compiled for this report.

It is our hope that this report will contribute to the creation of a basis for more systematic monitoring and evaluation of the progress and impact of ICT, as well as provide useful insights on ICT for development in general.

Katherine Sierra
Vice President, Infrastructure
The World Bank

Preface

The international community is increasingly committed to the monitoring and evaluation (M&E) of development programs. The World Bank Group has sharpened its focus on results in its own strategies, operations, diagnostic work, and instruments. It has put a premium on better measurement of outputs and outcomes in order to assess progress toward well-defined goals, increase accountability, and better understand the Bank's contribution to sector performance—ultimately supporting policy advice and decision making.

The Global Information and Communication Technologies Department (GICT) is the World Bank Group department that focuses on the ICT sector. In response to the call for increased M&E efforts for ICT by the development community, most notably during the World Summits on the Information Society held in Geneva in 2003 and Tunis in 2005, GICT is undertaking several initiatives to advance methods of monitoring results in ICT for development projects.

GICT's results measurement agenda aims to strengthen the availability of ICT sector indicators at country, micro, and project levels. Consolidating data from many different sources, GICT has defined a core set of information-and-communication-related indicators to create tables that present a snapshot of country-specific ICT issues. These "At-a-Glance" tables, found in Part II of this volume, provide an overview of ICT sector growth as it relates to the general economic status of particular countries.

The purpose of *Information and Communications for Development 2006: Global Trends and Policies* is to consolidate all these M&E efforts and share the findings with the development community. In addition to ICT At-a-Glance tables for 144 economies, this report includes analytical work that applies these data to a range of topics: investment trends, principles and practical solutions to extending ICT services, the role of ICT in doing business, trends in national e-strategies, and approaches to tracking ICT globally.

Further efforts are needed to coordinate and make M&E analysis fully operational in the arena of ICT for development. The World Bank Group stands ready to work with countries and the international community to determine how best to use ICT for poverty reduction and economic growth in the developing world. It is hoped that this report will play a positive role in these efforts.

Mohsen Khalil
Director, Global Information and Communication Technologies Department
The World Bank Group

Acknowledgments

This report was prepared by the Global Information and Communication Technologies Department (GICT) and the Development Economics Data Group (DECDG) of the World Bank Group.

Preparation of this report was led by an Editorial Committee consisting of Pierre Guislain (Chair), Christine Zhen-Wei Qiang (Team Leader), Bruno Lanvin, Michael Minges, and Eric Swanson. The principal authors include Aref Adamali, David A. Cieslikowski, George Clarke, John Coffey, Pierre Guislain, Naomi Halewood, Kaoru Kimura, Michael Minges, William Prince, Christine Zhen-Wei Qiang, Zaid Safdar, and Björn Wellenius.

The ICT At-a-Glance tables were prepared by David A. Cieslikowski, Naomi Halewood, Kaoru Kimura, Michael Minges, William Prince, and Christine Zhen-Wei Qiang.

The systems development team included Reza Farivari, Saurabh Gupta, Shahin Outadi, and William Prince of DECDG.

Inputs, comments, guidance, and support at various stages of the report's preparation were received from Magdi Amin, Seth Ayers, Shaida Badiee, Antonio Estache, Richard Fix, Juan Manuel Galarza Tohen, Rachele Gianfranchi, Carlos Gomez, Clive Harris, Timothy Irwin, Ada Karina Izaguirre, Charles Kenny, Anat Lewin, Michel Maechler, Kayoko Shibata Medlin, Marisela Montoliu Munoz, Juan Navas-Sabater, Isabel Neto, Cecile Thioro Niang, Vincent Palmade, Marta Priftis, Anwar Ravat, Robert Stephens, David Townsend, Hien Thu Tu, Lee Tuthill, and Subramaniyam Venkatraman.

Debra Naylor, Hope Steele, Paul Holtz, and Mark Wahl provided overall design direction, editing, and layout. Special thanks are due to Nancy A. Lammers, Stephen McGroarty, Santiago Pombo-Bejarano, and Janet H. Sasser from the World Bank Office of the Publisher for supervision of the production and dissemination of the report.

Abbreviations and Acronyms

$	All dollar amounts are U.S. dollars unless otherwise indicated.
2G	second-generation (mobile communication technology)
3G	third-generation (mobile communication technology)
AAG	at-a-glance
APDIP	Asia-Pacific Development Information Programme
ATM	automated teller machine
B2B	business-to-business
BOPSY	Balance of Payments Statistics Yearbook
BOT	build-operate-transfer
BRIC	Brazil, Russia, India, and China
CAFTA	Central American Free Trade Agreement
CERT	computer emergency response team
CRM	customer resource management
CIO	chief information officer
DAC	Development Assistance Committee (OECD)
DECDG	Development Economics Data Group (World Bank)
DOI	Digital Opportunity Initiative
DOT Force	Digital Opportunity Task Force
DSL	digital subscriber line
ECOSOC	Economic and Social Council (UN)
ECTEL	Eastern Caribbean Telecommunications Authority
EDI	electronic data interchange
FDI	foreign direct investment
GATS	General Agreement on Trade in Services (WTO)
G2B	government-to-business
G2C	government-to-citizens
G2G	government-to-government
GICT	Global Information and Communication Technologies Department (World Bank Group)
GSM	Global System for Mobile Communications
HHI	Herfindahl-Hirschman Index
ICC	International Chamber of Commerce
ICDL	International Center for Distance Learning
ICS	Investment Climate Surveys
ICSID	International Center for Settlement of Investment Disputes
ICT	information and communication technology
IFC	International Finance Corporation
IMF	International Monetary Fund
IP	Internet protocol
IPR	intellectual property rights
ISP	Internet service provider
ITU	International Telecommunication Union
M&A	mergers and acquisitions
MDGs	Millennium Development Goals
M&E	monitoring and evaluation

MPDF	Mekong Project Development Facility
NPV	net present value
OECD	Organisation for Economic Co-operation and Development
OECS	Organization of Eastern Caribbean States
PIACs	public Internet access centers
PIAP	public Internet access point
PKI	public key infrastructure
PPI	Private Participation in Infrastructure
PTO	public telecommunications operator
SME	small and medium enterprise
TERENA	Trans-European Research and Education Networking Association
TFP	total factor productivity
TNI	Transnationality Index
UMTS	universal mobile telecommunications system
UN	United Nations
UNCITRAL	United Nations Commission on International Trade Law
UNCTAD	United Nations Conference on Trade and Development
UNDESA	United Nations Department of Economic and Social Affairs
UNECA	United Nations Economic Commission for Africa
UNPAN	United Nations Online Network in Public Administration and Finance
UPU	Universal Postal Union
VoIP	voice-over Internet protocol
WB	World Bank
WDI	World Development Indicators
WHO	World Health Organization
WiFi	wireless fidelity
WiMAX	worldwide interoperability microwave access
WITSA	World Information Technology and Services Alliance
WLL	Wireless Local Loop
WSIS	World Summit on the Information Society
WTO	World Trade Organization

Part I

Chapter 1

Overview

Pierre Guislain, Christine Zhen-Wei Qiang, Bruno Lanvin, Michael Minges, and Eric Swanson

In the past few decades, information and communication technology (ICT) has transformed the world. Its potential for reducing poverty and fostering growth in developing countries has increased rapidly. Mobile telephones provide market links for farmers and entrepreneurs. The Internet delivers vital knowledge to schools and hospitals. Computers improve public and private services, and increase productivity and participation. By connecting people and places, ICT has played a vital role in national, regional, and global development, and holds enormous promise for the future.

It has been over 20 years since the first telephone company was privatized, 10 since the World Wide Web emerged, and 5 since the telecommunications bubble burst. How has ICT driven—and evolved in response to—these and other events? What has been learned about ICT trends and the policies that shape an information society? And how can further advances be fostered and facilitated?

Information and Communications for Development 2006: Global Trends and Policies, prepared jointly by the World Bank Group's Global ICT Department and Development Economics Data Group, analyzes ICT in light of lessons from both developing and developed countries. It examines the roles of the public and private sectors, identifying the challenges involved in adopting and expanding ICT use as well as the benefits (direct and indirect) of doing so.

Part I of the report assesses topics essential to developing ICT. It contains chapters on investment (chapter 2), access (chapter 3), diffusion and use (chapter 4), country policies and strategies (chapter 5), and targets, monitoring, and evaluation (chapter 6). Each chapter provides a theoretical and qualitative framework supported by quantitative evidence. Where limited data impede comprehensive economic analysis, a case study approach is used.

Part II presents the new World Bank ICT At-a-Glance tables for 144 economies, which show the most recent national data on key indicators of ICT development, including access, quality, affordability, efficiency, sustainability, and applications. The data enable assessments and comparisons both over time and across economies, so they help gauge ICT capacity, performance, and progress, as well as suggest areas for future interventions. The tables also include detailed snapshots of the economic context and ICT sector structure in each economy.

Experience over the past decade has shown that a vibrant and competitive information and communication sector is a prerequisite for developing information societies. Lessons can be drawn that may be particularly valuable for countries launching new policies and programs to increase ICT's contribution to economic and social development and extend its reach to a greater part of the population. Additional and complementary efforts are also needed to ensure the success of such programs, particularly increased monitoring and evaluation, based on appropriate goals, targets, and indicators.

When tailored to needs, ICT has the potential to raise growth in businesses of any size and countries at any stage of development. Related, but even more important, is ICT's role in reducing poverty and inequality, both within and across countries. Thus it is crucial that ICT move closer to the mainstream of development economics and policies—nationally, regionally, and globally. Given ICT's far-reaching payoffs—and the many efforts required to achieve them—this report is aimed not only at ICT experts but also at the broader development community.

Recent Progress and Lessons

ICT plays a vital role in advancing economic growth and reducing poverty. Research in the 1960s and 1970s showed how telecommunications strengthens economic production and distribution, public service delivery, and government administration. In the 1980s information became recognized as a crucial factor of production, along with capital and labor. And in the 1990s globalization and the increasing information intensity of economic activity, coupled with rapid technological change and demand growth, made ICT critical to competitiveness and growth.

In recent years the world's policy makers have recognized that ICT provides key inputs for economic development, contributes to global integration, and enhances public sector effectiveness, efficiency, and transparency. There is also growing consensus that countries seeking to strengthen their investment climates (for foreign as well as domestic investors) should make it a priority to improve ICT access and quality. Moreover, country conditions that bolster ICT investment—including sound economic policies, strong property rights, liberalized markets, limited restrictions on entry and ownership, and predictable regulation—contribute to a healthy overall business environment and so to growth throughout the economy.

Firms that use ICT grow faster, invest more, and are more productive and profitable than those that do not (box 1.1). For example, sales growth is 3.4 percentage points higher and value added per employee $3,400 more among developing country firms that use e-mail to communicate with clients and suppliers (table 1.1). Profits are substantially higher among firms using ICT.

Box 1.1 Information and Communication Technology—Essential to Productivity Growth and Private Sector Development

A vibrant private sector—one where firms invest, create jobs, and increase their competitiveness—promotes growth and expands opportunities for poor people in developing countries. ICT is an essential part of national infrastructure and private sector potential. It can create business opportunities, especially for companies located far from urban centers, and improve links among firms, suppliers, and clients. When used well, ICT can also make management and operations more efficient.

The Internet can be especially valuable for firms in developing countries because it provides opportunities to connect to markets and participate in trade, domestic and foreign. A recent survey of 56 developed and developing countries found a significant link between Internet access and trade growth—with the greatest benefits accruing to developing countries with the weakest trade links.

As with other factors of production, such as capital and labor, ICT use differs based on businesses' size, ownership, and export orientation. In developing countries Web site and computer (though not necessarily e-mail) use are more common among service firms than firms engaged in manufacturing, agroindustry, and construction. Web site and e-mail use are especially high in the telecommunications, information technology, real estate, and hotel and restaurant industries, and among exporters and foreign-owned firms.

Among regions, firms in Central and Eastern Europe use such technology the most, reflecting its correlation with national income. But Web sites and e-mail are also widely used in some low-income countries—Bangladesh, Kenya, Moldova, and Tanzania—suggesting that ICT is not a luxury.

Table 1.1 Effect of ICT Use on Enterprise Performance in Developing Countries

Indicator	Enterprises that do not use ICT	Enterprises that use ICT	Difference
Sales growth (percent)	0.4	3.8	3.4
Employment growth (percent)	4.5	5.6	1.2
Profitability (percent)	4.2	9.3	5.1
Investment rate (percent)[a]	n.a.	n.a.	2.5
Re-investment rate (percent)[a]	n.a.	n.a.	6.0
Labor productivity (value added per worker, dollars)	5,288	8,712	3,423
Total factor productivity (percent)	78.2	79.2	1.0

Note: n.a. Not applicable.

a. Because the investment and re-investment rates are limited dependent variables (that is, they are truncated below at zero), their marginal changes are not equal to their coefficients. For that reason their average rates are not calculated. The unconditional mean for investment is 6 percent and for re-investment, 44 percent.

Source: Authors' analysis based on data from the World Bank Investment Climate Surveys 2000–2003.

ICT is also crucial to sustainable poverty reduction, because it makes a country's economy more efficient and globally competitive, improves health and education services, and creates new sources of income and employment for poor people. In addition, ICT enhances social inclusion and promotes more effective, accountable, democratic government, especially when combined with effective freedom of information and expression.

Over the past 25 years, developing countries have considerably increased ICT access, especially for telephone services. Developing countries accounted for more than 60 percent of the world's telephone lines (fixed and mobile) in 2005, up from less than 20 percent in 1980 (table 1.2). During this period, population increased by half and real GDP more than doubled in these countries—while the number of telephone subscribers rose 28-fold.

Recent growth has been especially striking. Although the proportion of subscribers to total population in developing countries did not even double between 1980 and 1990, over the next decade it nearly quintupled, from 27 to 129 per 1,000 people. And between 2000 and the end of 2005, it is estimated to have tripled, reaching almost 400 subscribers per 1,000 people.

Table 1.2 Population, GDP, and Telephone and Internet Access in Developing Countries, 1980–2005

Indicator	1980	1990	2000	2005[a]
Population (billions)	3.6 *81*	4.4 *83*	5.1 *84*	5.4 *84*
GDP (US$ trillions)[b]	3.1 *18*	4.2 *18*	5.9 *19*	7.5 *21*
Total telephones (per 1,000 people)	14 *17*	27 *22*	129 *38*	393 *61*
Fixed telephones (per 1,000 people)	14 *17*	27 *23*	83 *43*	135 *57*
Mobile telephones (per 1,000 people)	n.a.	0.09 *4*	46 *32*	258 *63*
Internet users (per 1,000 people)	n.a.	n.a.	15 *20*	67 *41*

Note: Numbers in italics are percentages of the world total. n.a. Not applicable.

a. Projected.
b. In constant 2000 U.S. dollars.

Source: Authors' analysis based on data from World Bank 2005 and ITU 2003.

Most of the recent growth has involved mobile phones, which now outnumber fixed ones. In Nigeria the number of mobile subscribers jumped from 370,000 in 2001 to 16.8 million in September 2005, making its mobile market the second largest in Africa. In the Philippines, which has had more mobile than fixed telephone subscribers since 2000, mobile subscribers continue to multiply. By the end of 2005, the country had about 40 million mobile subscribers—six times more than in 2000.

Mobile phones have an especially dramatic impact in developing countries—substituting for scarce fixed connections, increasing mobility, reducing transaction costs, broadening trade networks, and facilitating searches for employment. With prepaid services and calling cards, even poor households have been able to benefit from increased telephone access.

Telephone services now reach many small cities and towns, and by 2005 half of the world's households had telephones. Among developing regions the telephone subscription rate is highest in Europe and Central Asia, where between 2000 and 2004 it more than doubled to 730 per 1,000 people (figure 1.1). But growth was highest in Sub-Saharan Africa, with the rate tripling—albeit to a still-low 103 subscribers per 1,000 people.

Other types of ICT have also expanded rapidly in recent years. The latest estimates indicate that worldwide, Internet use more than quadrupled between 2000 and 2005. Again, Europe and Central Asia is in the lead among developing regions, with 117 Internet users per 1,000 people in 2004—four times as many as in 2000 and six to eight times as many as in South Asia and Sub-Saharan Africa. During this period the fastest growth, 370 percent, occurred in the Middle East and North Africa.

Liberalization and competition—and the resulting increase in private investment—have driven the development of telecommunications infrastructure and ICT in general. Capital is crucial to the development and expansion of robust telecommunications networks. Because developing countries often lack the capital—as well as the technology and managerial know-how—needed to develop such networks, many have turned to private investors, domestic and foreign.

By opening their telecommunications markets through well-designed reforms, governments can create competitive markets that grow faster, lower costs, facilitate innovation, and respond better to user needs. As a result, the traditional monopoly model of telecommunications services—based on extensive state control and protected national markets—has eroded, in concert with rapid technological advances in the sector and fundamental changes in economic policy in developing countries.

Over the past two decades telecommunications markets have undergone unprecedented liberalization in every region—though the pace and scale of reform have varied, and markets for fixed local and international telephone services remain closed or barely open in about half of developing countries. Effective competition between multiple providers helps expand access and results in cheaper, more modern services.

In 2003, 130 of 164 countries with available data had at least three competing providers of mobile services. The Democratic Republic of Congo has six competing mobile

Figure 1.1 Telephone and Internet Access by Region, 2000 and 2004
(per 1,000 people)

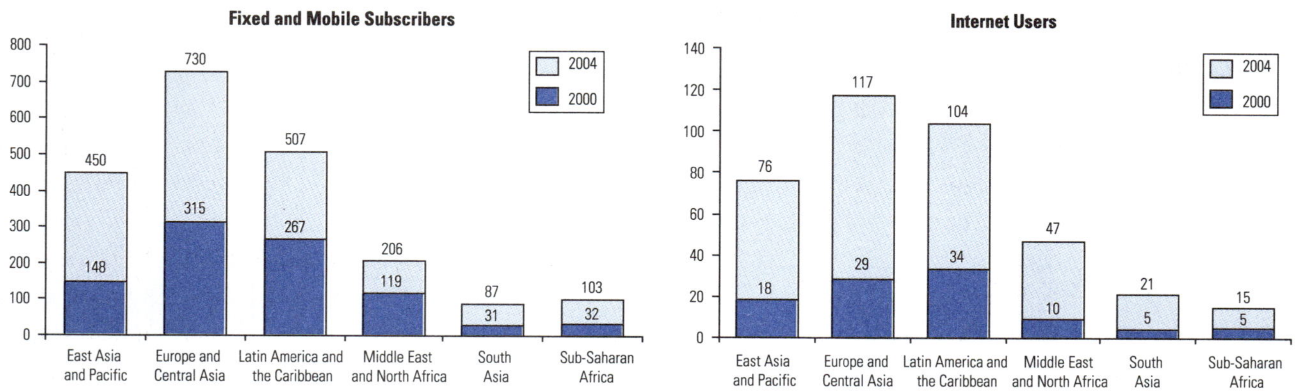

Source: Authors' analysis based on World Bank data.

telephone operators, giving it a mobile phone density 13 times that of Ethiopia—which has a similar income per capita but just one operator. In Algeria almost no one had a mobile subscription in 2000. But in 2003, after a second operator began providing services, nearly 5 percent of people did—and when a third operator entered the market in 2004, that share leapt to more than 15 percent by the end of the year and to 32 percent by September 2005. Similarly, Grenada issued new licenses in 2002, and between 2000 and 2004 the number of mobile subscribers soared from 45 to 860 per 1,000 people.

In markets for international telephone services, full competition leads to prices about half those in countries with limited competition. Among 30 African and Latin American countries that undertook telecommunications reforms in the 1980s and 1990s, those that introduced competition saw the sector grow and costs fall faster than those that delayed competition.

The Internet has also spurred a growing wave of innovation, ushering in new services and more cost-effective network solutions—especially in countries where service providers are allowed to build their own networks and gateways. New wireless technology is resulting in innovative business models and holds the promise of connecting poor users, extending competition to all market segments, and accelerating development of broadband infrastructure and access. Such technology is affordably priced and commercially viable in a number of countries, in both urban and rural areas. For example, a single broadband Internet connection in a village can provide access for numerous institutional programs (such as e-government and computers in schools) and private users.

Privatization and technological advances have boosted foreign direct investment (FDI)—a major source of ICT financing. In 1988 Chile privatized its incumbent operator, triggering the first wave of telecommunications-related FDI in developing countries, typically through divestitures of state companies to foreign investors. Since then more than 80 developing countries have privatized their incumbent telecommunications providers.

A second wave of telecommunications FDI started in the mid-1990s as governments, aiming to increase access to and revenue from communication services, awarded new licenses for mobile telephony and encouraged foreign investment. In 2000–03 mobile projects accounted for 51 percent of FDI in telecommunications, up from 7 percent a decade earlier.

During both waves, foreign investors were seeking new markets, higher returns, and diversified exposure. Many governments welcomed FDI as a way to expand networks, develop new services, and generate revenue through license fees. FDI also brought stronger, longer commitments than did other types of foreign investment, as well as new skills, technology, and management approaches. Between 1990 and 2003, 122 of 154 developing countries received foreign investment in telecommunications.

FDI in telecommunications jumped from $2 billion in 1990 to $33 billion in 1998—but gradually fell to about $13 billion in 2002 and 2003. Still, the decline in FDI has been smaller for telecommunications than for other infrastructure sectors. And although FDI to acquire government assets dropped significantly after 2000, flows for sector expansion stayed at the same level as during the boom years.

During 1990–2003, telecommunications projects accounted for 12 percent of FDI in developing countries. Latin America and the Caribbean attracted more than half of FDI in telecommunications, while Europe and Central Asia received about a quarter. These large shares reflect the prominence of middle-income countries in telecommunications FDI: during 1990-2003 low-income countries received just 6 percent of such investment.

Developing countries are home to a growing number of FDI providers. Although the largest foreign direct investors in telecommunications are multinational corporations from Europe and the United States, in recent years FDI originating in developing countries has become a fast-growing trend. By 2003 these South-South investments accounted for more than a quarter of telecommunications FDI in developing countries, up from a negligible share in the early 1990s (figure 1.2). Most such investment came from countries that were among the early liberalizers in their regions.

Some investors from developed countries have reduced FDI due to the bursting of the telecommunications bubble in 2000, compromised balance sheets following major investments or acquisitions, disappointing returns on some projects (both at home and abroad), and pessimism about emerging markets. For example, many global players invested in the developing markets of Latin America and East Asia during the 1990s, but have since withdrawn. Global operators have also pulled out of Sub-Saharan

Figure 1.2 Telecommunications FDI in Developing Countries, 1990–2003

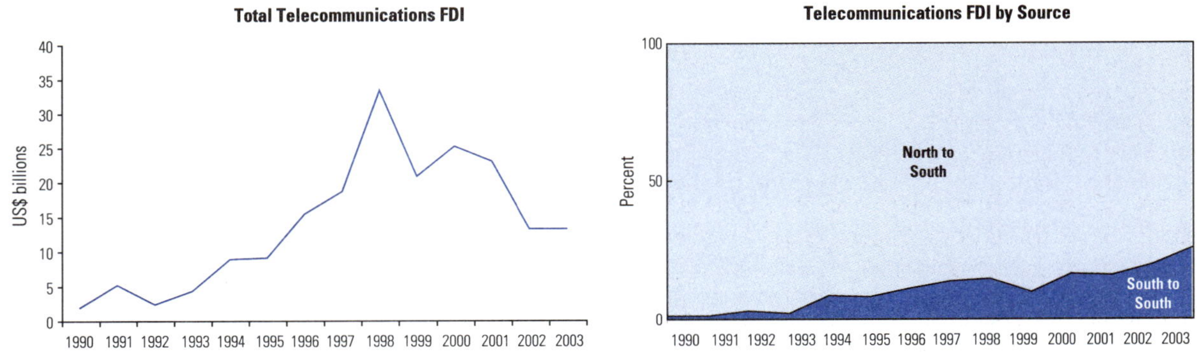

Source: Authors' analysis based on data from World Bank 2004.

Africa, creating a bigger role for financial and regional investors. The region's three largest mobile phone operators, accounting for nearly half of telecommunications FDI, are all regional firms.

But growing South-South investment is also due to growing wealth and capital account liberalization in some emerging market economies—trends that have increased the supply of capital in these countries and enabled their companies to invest abroad. By 2002, 4 of the 30 largest international telecommunications corporations were from developing countries. Other factors favoring South-South investment include geographic proximity and ethnic and cultural ties.

Most South-South telecommunications investors stick to their home or neighboring regions: during 1990–2003 more than 85 percent of such FDI stayed in the same region. Countries that avoid imposing unnecessary requirements that might exclude otherwise qualified bidders, and create a level playing field that provides fair opportunities to new entrants regardless of size or origin, are more likely to attract South-South and regional FDI.

Consistent, predictable, and transparent sector policies and regulation are essential to remove market impediments. Obstacles to well-functioning markets often remain even after extensive sector reforms. In Peru all segments of the telecommunications market have been open since 1995, but telephone services in provincial towns and marginal areas of big cities remain well below the levels achieved in other developing countries with comparably open markets. The challenge everywhere is to enable operators to tailor their service offerings and technical choices as effectively and efficiently as possible.

The regulatory improvements needed to achieve that goal often include opening markets to new entrants (including small domestic entrepreneurs), rebalancing retail tariffs, establishing an effective cost-based interconnection regime, securing reasonable access to existing infrastructure, and making radio spectrum available to a wider range of service providers (box 1.2). Consistent and transparent processes—for legal, regulatory, and administrative procedures and institutions—are the main requirements.

Some traditional regulatory provisions may stand in the way of new technologies, decentralized supply, and other innovations. In addition, high taxation can discourage investment by telecommunications operators and suppress demand from users. And as the cost of manufacturing cell phones continues to fall, government taxes and duties on their import, sale, and use remain a binding constraint on extending information and communication services to poor people.

Future Challenges—and Options for Addressing Them

Gaps in access to ICT remain large, as do differences in adoption of ICT applications. Access to ICT is highly unequal across and within countries. In particular, developing countries still have far to go to make ICT commonplace in governments, schools, and businesses. Moreover, fast growth in large emerging markets—especially Brazil, China, and India—masks slower development elsewhere. Although progress has been made reaching out to rural areas and the urban poor, in many countries these groups still lag behind. And the advanced information and communication services available through the Internet initially reach mainly better-off groups.

Box 1.2 Regulatory Measures That Help Markets Work Better

Open entry. Lifting restrictions on the entry of new service providers accelerates expansion to underserved and unserved areas. Replacing individual licensing with a regime of general authorization facilitates entry, eases the regulatory burden on companies and authorities, and enhances transparency.

Cost-based pricing. To minimize market distortions, retail prices must reflect differences in the costs of providing services. Geographically averaged prices discourage investment in high-cost areas. Tariff policy should be reviewed on a regular basis and prices rebalanced when necessary.

Cost-based interconnection. An effective interconnection regime is essential. High-cost areas should receive larger than average shares of revenues for completing outgoing calls and larger than average payments for incoming calls.

Access to infrastructure. New entrants must have reasonable access to the network infrastructure of incumbent operators. Unbundling the local loop, sharing physical infrastructure, and co-locating the equipment of different operators lowers barriers to competitive entry, increases revenues of incumbents, and reduces public inconvenience.

Access to the radio spectrum. Making radio spectrum available promptly and at low cost encourages the use of new wireless technology. Specific measures include increasing the allocation of spectrum for license-exempt use and mobile service.

A United Nations index measuring the availability and sophistication of e-government applications shows that developed countries score much higher than developing ones (UNPAN and UNDESA 2004). And whereas most developed nations have connected nearly all their primary and secondary schools to the Internet, just 38 percent of developing countries have done so—and less than 1 percent of many African countries (table 1.3). Differences in the number of secure Internet servers, a proxy for the availability of e-commerce, are similarly stark. While developed nations have more than 300 such servers per 1 million people, developing nations have fewer than 2. Canada has more secure servers than all developing countries combined.

The public and private sectors both have important roles. In market economies the private sector is primarily responsible for providing ICT services, and competitive, private sector–led markets go a long way toward making these services widely available. The public sector's main role is to provide a sound policy framework, regulate markets where they do not work well enough on their own, and support additional service provision where markets do not achieve economic and social objectives. The public sector is also an increasingly important user of ICT, particularly in the context of e-government, making it a major actor in fostering ICT uptake.

Table 1.3 Measures of Electronic Government, Education, and Commerce in Developing and Developed Countries, 2004

Country group	E-government index	E-education (% primary and secondary schools connected to Internet)	E-commerce (secure Internet servers per 1 million people)
Developing	0.27	38	1.9
Developed	0.68	94	319
Ratio of developed to developing	n.a.	2.5	165

Note: Based on incomplete data, with 142 countries for e-government, 68 for e-education, and 122 for e-commerce. n.a. Not applicable.

Source: World Bank 2004; Netcraft 2004; UNPAN and UNDESA 2004.

Targeted public intervention to expand services can maximize social returns. Governments must play a leading role in promoting the modernization and extension of ICT infrastructure and services. Even in well-performing markets, there is typically a divide between what service providers are willing or able to do on commercial grounds and what governments consider necessary from a development or broader economic perspective.

Thus, most developing countries continue to face ICT challenges, including service gaps among poor households and in rural areas (figure 1.3). Especially if such gaps persist after steps have been taken to make markets work, public resources can be used to narrow them and to develop sustainable markets for the private provision of the desired services. Many governments have made bridging gaps in access to ICT services an explicit public policy goal, as with universal service programs for telephones.

One successful approach has been to use competitive bidding to award subsidies to ICT providers. Subsidies can catalyze private investment by offsetting political constraints on prices, reflecting external benefits that would not contribute to operating revenues, and helping providers overcome entry barriers resulting from front-loaded investment requirements or financial market failures. Subsidies are transparent in that they have known costs, are subject to the discipline of public budgets, and can be paid as output objectives are met. In addition, subsidies can be targeted to desired beneficiaries. Capital contributions and risk guarantees may also be appropriate in some cases.

Demand aggregation can be used to increase access to ICT services, particularly advanced ones such as the Internet. For example, by combining the demand for connectivity among public agencies, schools, and hospitals, then seeking competitive bids, governments can help finance and expedite the expansion of broadband networks, particularly in rural areas. Government commitment to buying broadband capacity reduces commercial risks for investors asked to build infrastructure with high upfront costs and untested market demand. This approach also fosters competition in this market segment by creating opportunities for new entrants.

Determining proper levels of public support for fledgling markets can be difficult, and requires the examination of costs and benefits from the viewpoint of the economy as a whole. Such support involves tradeoffs between various development goals and limited public resources, and even with the best policies it can take years for a market to develop. Moreover, not all unprofitable services merit public support: only projects that require support to become viable and yield economic benefits at least equal to their costs should be eligible for it. In addition, using public resources to help develop new markets is not always justified, especially if services are not in high demand and widely and

Figure 1.3 Household Access to Telephones in Europe and Central Asia, by Country and Year

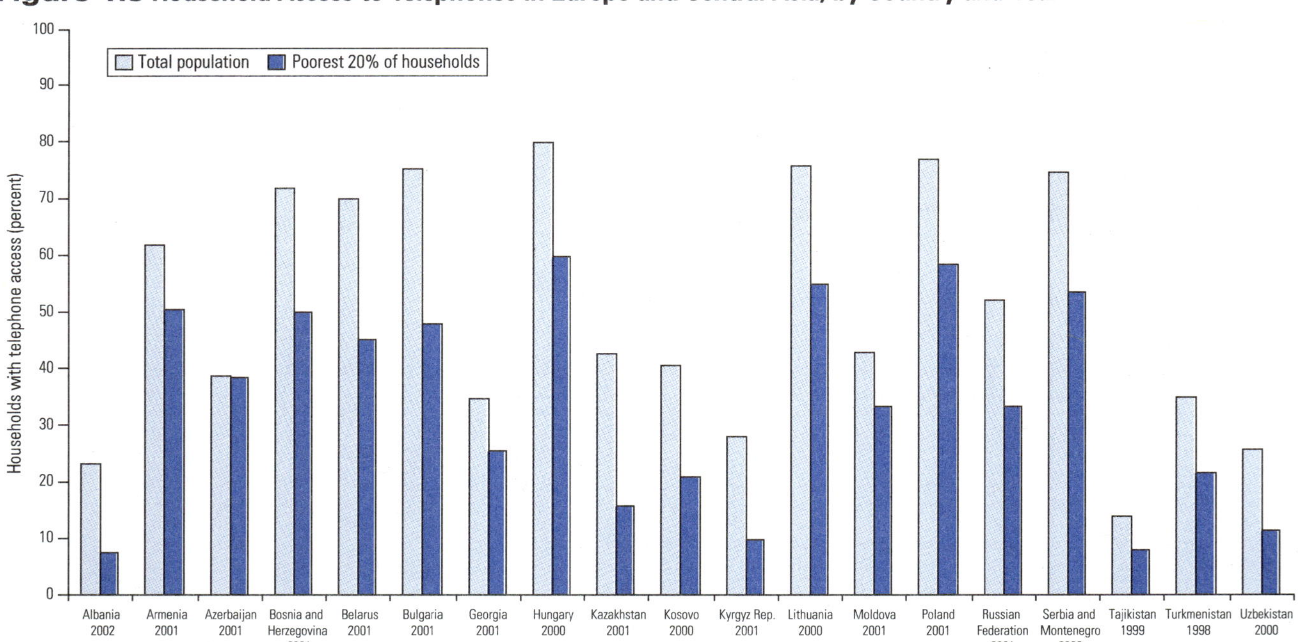

Source: World Bank staff estimates based on household surveys 1998–2002.

equitably shared. Any public support for private businesses should also have a clear exit strategy.

The private sector should play a critical role as an early adopter and significant beneficiary of ICT, yet businesses often face obstacles to ICT adoption. The potential benefits of ICT—for growth, investment, profitability, and productivity—provide a sound basis for promoting its use by businesses in developing countries. This is especially important because firms play a significant role as early adopters of new technology. But businesses in developing countries often face obstacles to adopting and using ICT applications, including the following:

- Slow, unreliable, insufficient, and expensive telecommunications services. Corruption is part of the problem: many firms have to pay bribes for service licenses or telephone connections.
- Limited incentives to change business models and operating structures when the costs of adopting ICT are significant and returns uncertain (figure 1.4). Many firms—especially those that are small, domestically owned, or nonexporting—do not recognize the value of ICT strategies or applications for their activities, and do not feel that they need ICT beyond a basic level of connectivity.
- Lack of trust in online business activities and related legal impediments.
- Shortages of workers skilled in ICT.

To foster investment and use of ICT, governments must implement policies and regulations that remove such barriers and help businesses identify and exploit its potential. Steps should be taken to provide a supportive legal and regulatory environment for e-business (including e-security policies and programs), enhance technological diffusion, and overcome market failures in areas such as demand aggregation (for example, by launching e-government services of interest to businesses) and skill development (by providing and supporting ICT training). Although income is a major constraint on demand, businesses and people everywhere are willing to spend a significant share of their income on information and communication if given the chance.

Another policy priority is to increase liberalization and competition in ICT markets—to stimulate investment in bandwidth, raise demand through lower prices, and boost efficiency and innovation in the provision of services. Policies should also promote neutrality among competing and emerging technologies to encourage interoperability, innovation, and choice among services.

E-strategies can prioritize and guide national ICT development. The World Summits on the Information Society (WSIS), held in Geneva in December 2003 and in Tunis in November 2005, encouraged developing countries to adopt e-strategies. Such strategies, tied to a country's overall development goals, are intended to promote and coordinate ICT investments and extend their benefits to larger segments of

Figure 1.4 The Most Significant Costs That Obstruct E-Business
(percentage of survey respondents)

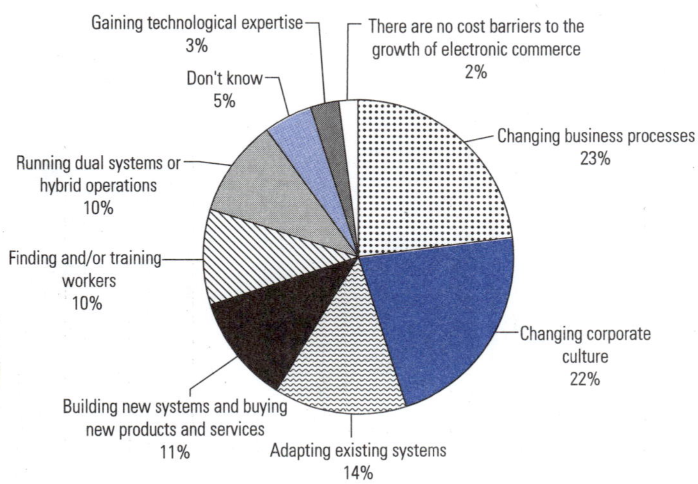

Source: WITSA 2000.

these countries' populations. E-strategies are being developed and implemented by governments around the world, in both developing and developed countries.

A review of 40 of these strategies conducted for this report finds that more than 85 percent aim to expand ICT use in governments and schools, expand telecommunications infrastructure, and provide an adequate legal and regulatory framework (figure 1.5). But the actions identified to achieve these goals vary considerably, partly because of income differences across countries. Other common themes include ICT industry development, information technology skill development, and e-business growth.

Although many countries have made significant progress on e-strategies, more work is needed to make them effective tools for development. E-strategies need to better address the obstacles identified above by providing stronger links to broader development goals—particularly in low-income countries—and increasing their focus on monitoring and evaluation. The likely development impact of an e-strategy can be assessed by evaluating its

- links to the country's overall development goals—political, economic, and social;
- use of indicators to benchmark analysis and formulate clear and realistic goals and targets;
- implementation mechanisms, including institutions, processes, and resources; and
- monitoring and evaluation mechanisms, especially the responsibilities and budgets for these efforts.

The e-strategies reviewed were strong in establishing development links (especially the strategies produced by middle-income countries) and providing implementation details, and many were effective at setting targets (figure 1.6). But the strategies were much weaker when it came to monitoring and evaluation, with most saying little or nothing about institutions or structures for doing so. When formulating e-strategies, governments should make plans for monitoring and evaluation and commit resources to implement them. Such plans would help make e-strategy design and implementation more effective and relevant—without monitoring and evaluation it is impossible to measure the results and assess the impact of ICT initiatives.

WSIS provided the first globally agreed targets for ICT development. These targets now need to be linked to action-oriented indicators. Measuring the impact of ICT on development and evaluating the outputs of e-strategy implementation are essential for making e-strategies relevant and holding governments accountable for their implementation. WSIS set 10 far-reaching targets for ICT and information society development, to be achieved by

Figure 1.5 Thematic Areas of Focus for E-Strategies, by Number of Countries

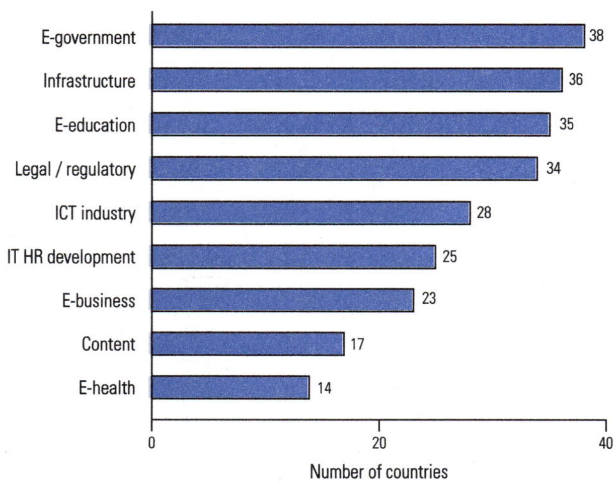

Note: HR = human resources

Source: World Bank staff analysis.

Figure 1.6 Strengths and Weaknesses of E-Strategies

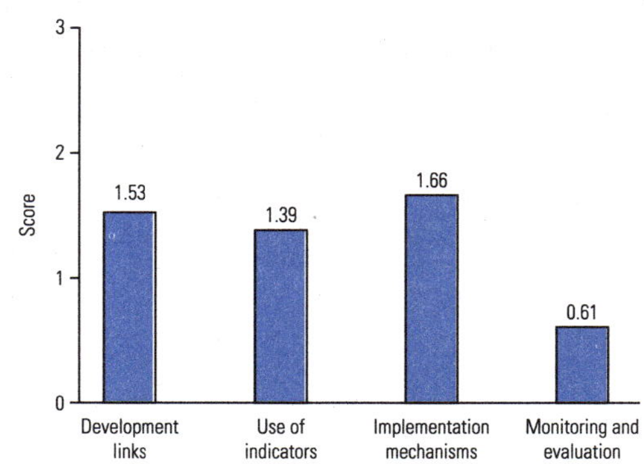

Note: Explanation of scoring—*development links*: none (0), mention general development goals (1), reference national programs (2), specify national and sectoral programs (3). *Use of indicators*: none (0), sporadic (1), common (2), extensive (3). *Implementation mechanisms* and *monitoring and evaluation*: none (0), vague (1), specific (2), no ambiguities (3).

Source: World Bank staff analysis based on a review of 40 e-strategies.

2015. These targets look beyond traditional measures of telecommunications development and include goals such as connectivity for governments, schools, hospitals, and rural areas.

Because these targets are broad—and, in some cases, poorly defined or difficult to measure or apply—this report proposes discrete, concrete, comparable indicators to help individual countries as well as the international community monitor progress and benchmark performance.

Although the targets provide a good starting point for quantifying ICT development, most countries will be unable or unwilling to pursue all 10 with equal vigor. Accordingly, monitoring and evaluation efforts should take into account the priorities that governments place on each target. A recent survey of policy makers found that connecting schools to ICT is the top international priority, followed by connecting scientific and research centers. Monitoring and evaluation should focus on targets that are clear, for which action can make a difference, and on which progress can be tracked between now and 2015. Finally, parallel activities are needed to identify policies that accelerate achievement of the targets, such as improving regulations and identifying appropriate financing and business models.

Countries should increase efforts to collect and disseminate ICT data. The availability of current data in most developing countries is limited, which does not allow for policy making based on timely and relevant benchmarking. Some developing countries compile measurable, comparable indicators to track their progress and benchmark their performance in developing an information society. But many do not. For example, many governments cite school connectivity as an important ICT goal. Yet data on ICT prevalence in schools are lacking in many countries, making it difficult to measure progress or evaluate ICT's contribution to education. A lot of the data needed to compile the indicators should not be difficult to collect, and can be gathered as part of normal administrative reporting.

Some countries with good monitoring data, such as Chile and Thailand, can demonstrate solid evidence of the success of their policies (figure 1.7). To foster progress toward the development of a global information society, all governments need to collect and maintain data for key ICT indicators.

The international community can facilitate more effective coordination of efforts to compile global information, establish a database for benchmarking, and provide technical assistance to interested countries. ICT data collected by national and regional authorities are often cumbersome to locate. These data need to be standardized and put in a central, global repository. In addition, agreement is needed on which data should be compiled and benchmarked internationally and which organizations should be responsible for various indicators. To support such efforts, this report compiles nearly 30 ICT indicators for 144 economies, which offer data on access, quality, affordability, efficiency and sustainability, and applications. They allow countries to compare themselves with other countries and to assess their progress in different ICT areas over a five-year period (2000 to 2004).

Figure 1.7 Internet Access in Thai Schools and Computers in Chilean Schools

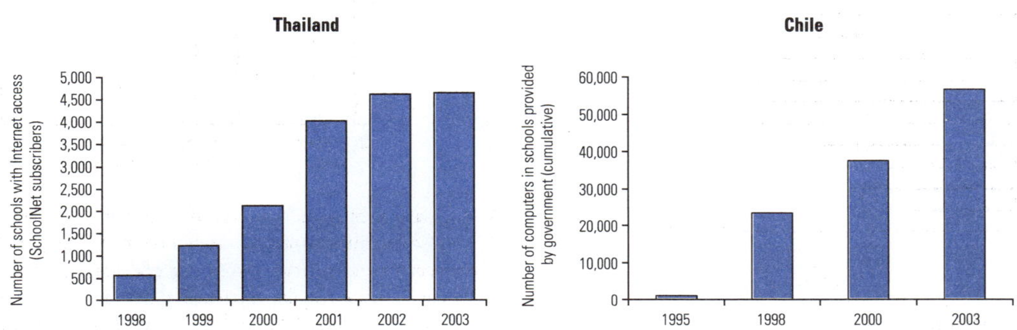

Source: Enlaces 2005; NECTEC 2003.

Recognizing—and Seizing—Opportunities

This report assesses what has worked well and what has not worked as well in developing ICT around the world. The digital divide between countries is not insurmountable, as shown by the phenomenal success of new technology—such as mobile telephony—in bridging the access gap, as well as the positive impact of efforts to stimulate competition and develop independent regulation of telecommunications markets. But many developing countries still require improvements in their ICT policies and strategies. Among the biggest challenges facing these countries are weak policy and implementation capacity, opposition from vested interests, and persistent obstacles to adoption of ICT. Many also lack adequate tools to monitor, evaluate, and guide investments in ICT and connectivity in underserved areas.

The World Bank hopes to make this report a regular publication. Doing so would contribute to continuity in benchmarking of ICT data and in monitoring and assessment of trends, themes, and programs supporting ICT in developing countries.

This report will have achieved its purpose if the views, analyses, data, and indicators it contains help stakeholders determine how their separate and collective efforts can yield the highest returns and contribute to inclusive information societies around the world.

References

Enlaces. 2005. Data from Enlaces Web site of Chile's Ministry of Education. http://www.enlaces.cl/index.php (accessed September 11, 2005).

ITU (International Telecommunication Union). 2003. *World Telecommunication Development Report.* Geneva: ITU.

NECTEC (National Electronics and Computer Technology Center). 2003. *Thailand ICT Indicators.* Bangkok, Thailand.

Netcraft. 2004. "Netcraft Secure Server Survey." Bath, United Kingdom. http://news.netcraft.com/archives/2003/04/09/netcraft_ssl_survey.html.

UNPAN (United Nations Online Network in Public Administration and Finance) and UNDESA (United Nations Department of Economic and Social Affairs). 2004. UN *Global E-Government Readiness Report 2004: Towards Access for Opportunity.* New York. http://www.unpan.org/egovernment4.asp.

WITSA (World Information Technology and Services Alliance). 2000. *International Survey of E-commerce 2000.* Arlington, Va.

World Bank. 2004. "Private Participation in Infrastructure (PPI) Project Database 2004." Washington, D.C. http://ppi.worldbank.org/.

———. Various years. "Investment Climate Surveys." Washington, D.C. http://rru.worldbank.org/InvestmentClimate/.

———. 2005. World Development Indicators database, World Bank.

Chapter 2

Foreign Direct Investment in Telecommunications in Developing Countries

Pierre Guislain and Christine Zhen-Wei Qiang

In an industry as capital-intensive as telecommunications, access to capital is key to ensuring the deployment and expansion of a robust network. Until the mid-1980s, telecommunications services were made available in most countries on a monopoly basis either by state-owned service providers or by heavily regulated private entities.[1] The traditional monopoly model involved extensive state control and protected national markets. As a result, foreign direct investment (FDI) in the sector was minimal when the first privatization and liberalization wave began. Indeed, by 1990, only $2 billion out of a total telecommunications investment of $20 billion in developing countries was FDI.

The public offerings of shares in British Telecommunications in the United Kingdom in 1984 and NTT (Nippon Telegraph and Telephone) in Japan in 1985 opened the sector to private investment and competition. The breakup of AT&T (American Telephone and Telegraph) in the United States in 1984 enabled competition and the growth of new companies. Led by these public offerings and the dissolution of the U.S. monopoly, the traditional model began to erode. This erosion coincided with rapid technological changes in the sector. Fundamental transformations in economic policy in developing countries began at the same time, leading to the ascendancy of free market principles in national economic policies and international trade.

The privatization of incumbent operators in Chile (1988) launched the first wave of FDI in telecommunications into developing countries, which needed not only capital but also management skills and technology transfers to accelerate the expansion and modernization of their telecommunications networks. Historically run as public monopolies, telecommunications companies were often burdened with organizational and operational inefficiencies. Heavy public debt levels and fiscal consolidation precluded injections of public funds. Domestic private investors had no experience in managing telecommunications enterprises and were not able to provide all the required investment capital by themselves (Guislain 1997).

The second wave of FDI in telecommunications occurred with the revolution in mobile telephone technology. The driving force of this revolution was the introduction of digital cellular services in the mid-1990s. The cost effectiveness of cellular networks and the generally more liberal attitude of governments toward this new technology created a boom in demand in developed and developing countries alike. The result was a renewed rush to invest, this time in an entirely new network infrastructure (UN 2004).

During both waves of FDI, strategic foreign investors were drawn into developing economies to seek new market opportunities, higher returns, and diversification of risk. Governments in countries receiving FDI saw it as an opportunity to expand existing telecommunications networks or install new telecommunications services. Through these new services, especially in mobile telephony, they extracted immediate revenues in the form of license fees. In addition, foreign investors tended to be the only ones with the expertise and finances needed to provide telecommunications services in markets that had hitherto been served only by the incumbent monopolist. FDI brought greater commitment and longer-term engagement by foreign investors as well as new management approaches, technology, and skills transfer to the host countries.

As FDI has typically been the driver of sector growth in liberalizing economies, FDI levels are an important indicator of sector development. A country that ends its sector monopoly and has an open investment regime will see the share of public investment shrink and foreign investment rise rapidly. As the market grows, becomes more competitive, and matures, private domestic investment follows and often overtakes FDI.

This chapter illustrates FDI trends in telecommunications in developing countries from 1990 through 2003. It also analyzes the factors that lead to success in attracting and managing FDI, with a focus on regulatory issues and sector policies.

Trends in FDI Flows

Understanding the trends of FDI flows into developing countries can help policy makers interested in attracting private and foreign investment in the telecommunications sector make good, effective decisions.

Measurement

To identify FDI trends in telecommunications, investment flows must be measured and data collected. But measuring FDI flows into developing countries is difficult because the data are usually limited and of poor quality. Annex 2A lists 10 data sources for FDI and private investment in the telecommunications infrastructure. Although each source has certain limitations, as summarized in the annex, these sources as a whole can be used to distill some generalized investment trends in the sector.

In this chapter, the World Bank Private Participation in Infrastructure (PPI) Project Database (http://ppi.worldbank.org/, 2004 update) is used as the main data source for analysis, primarily because it covers more developing countries than other sources reviewed. The PPI Project Database tracks information on more than 2,700 infrastructure projects with private investment participation in low- and middle-income countries, of which 25 percent are in the telecommunications sector. Box 2.1 provides key definitions and outlines some limitations of this database.

Investment Volume

Trends in the volume of FDI in telecommunications have fluctuated in line with the privatization and liberalization of the sector. FDI flows at first increased extremely rapidly, from a low base of about $2 billion in 1990 to a peak of $33 billion in 1998. Flows then hovered between $20 and $25 billion in the following three years, and dropped to about $13 billion per year in 2002 and 2003 (figure 2.1).

Between 1990 and 2003, 122 of 154 developing countries financed telecommunications infrastructure projects with foreign investment. Investments included expanding facilities and acquiring assets, as well as license fees paid to governments. During this period, foreign investors or consortia of foreign and domestic private investors participated in more than 460 telecommunications infrastructure projects, making foreign investment commitments of about

Figure 2.1 Total and Telecommunications FDI, 1990–2003

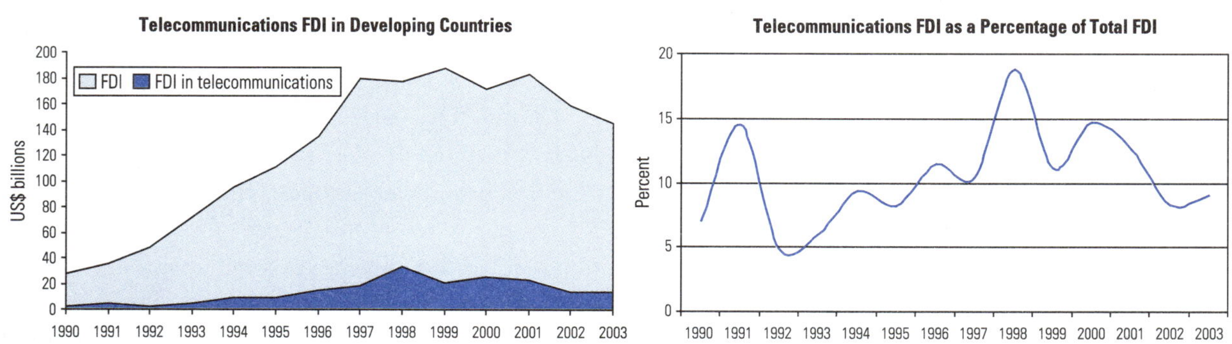

Source: Authors' analysis based on data from the World Bank PPI Project Database (2004 update).

Box 2.1 PPI Investment Data Definitions and Limitations

The World Bank PPI Project Database has figures that include investments in both facilities and government assets. Investments in facilities are either in greenfield facilities or in expanding or modernizing existing infrastructure. Investments in government assets are resources the investor spends on acquiring publicly owned property such as state-owned enterprises (divestiture proceeds), rights to provide services in a specific area (license proceeds), or the use of a specific radio spectrum (spectrum fees).

Because the PPI Project Database is not designed to track foreign investment, only the percentage of the private participation and the origin of the investor ("sponsor") in projects are available. We conducted further research to determine the type of investor based on the following criteria made in relation to the recipient country:

- If a project has either one sponsor that is foreign or multiple sponsors that are all foreign, then the sponsor type is taken to be purely *foreign*.
- If a project has multiple sponsors among which at least one is foreign and one is domestic, then the sponsor type is taken to be *consortium*.
- If a project does not have any foreign sponsor, then the sponsor type is taken to be *domestic*.
- *Public* refers to the public share in private consortia but does not include purely public-financed projects.

The FDI share of the project amount is calculated according to the type of sponsor and information on the composition of consortia from other sources. The analysis in this chapter is based on the 2004 update of the PPI Project Database and reflects the ownership structure and consortium arrangement as of mid-2004.

Several other data sources in annex 2A—including the ITU (International Telecommunication Union) Regulatory and Competition Database and the UNCTAD (United Nations Conference on Trade and Development) FDI Database, as well as the Business Monitor Online Database—have been reviewed to check consistency. Although major new investment projects of large operators were all included in the PPI Project Database, it did not capture some smaller projects or reinvested profits. We estimate that the data in this chapter cover approximately 75 percent of FDI projects and 85 percent of the total telecommunications FDI amount.

Source: Authors' analysis.

$194 billion—an amount that is 11.5 percent of the total FDI inflows of $1.7 trillion to developing countries, and 30 percent of the total investment of $650 billion in the telecommunications sector in developing countries.[2] If one excludes China—which received about 30 percent of total FDI to developing countries but no telecommunications FDI—the share of telecommunications in total FDI was over 16 percent.

Telecommunications FDI increased sharply during the first half of the 1990s, growing on average by 40 percent annually during 1990–5, and reached an average of $23 billion a year from 1996 to 2000 (this figure drops to $15 billion if one excludes the privatization of Telebras, the Brazilian incumbent, which constituted the single largest transaction during the period; see figure 2.2). After this boom, annual flows dropped. However, levels in 2001–3 ($16.5 billion per year on average) were still much higher than those in 1990–5 ($5.2 billion per year on average).

Purely foreign investment (without domestic partners) accounted for 42 percent of total private investment and funded 59 percent of projects. In addition, foreign investors' participation in consortium projects allowed domestic investors to gain exposure to the sector alongside seasoned international operators and investors. As a result, an

Figure 2.2 Telecommunications FDI in Developing Countries, 1990–2003

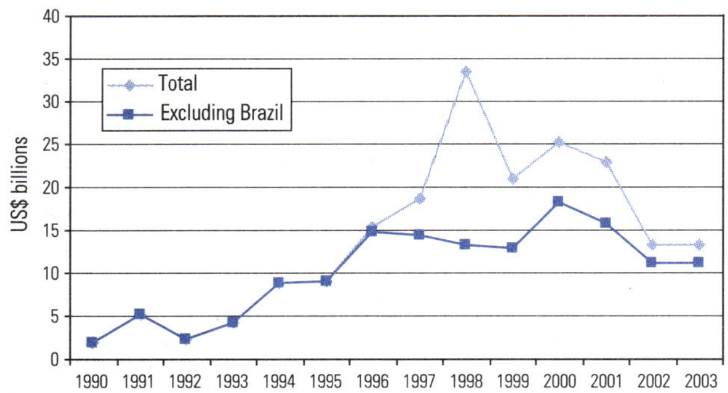

Note: The numbers in the figure refer to the annual total telecommunications FDI.

Source: Authors' analysis based on data from the World Bank PPI Project Database (2004 update).

Figure 2.3 Private Telecommunications Projects by Source of Funding, 1990–2003

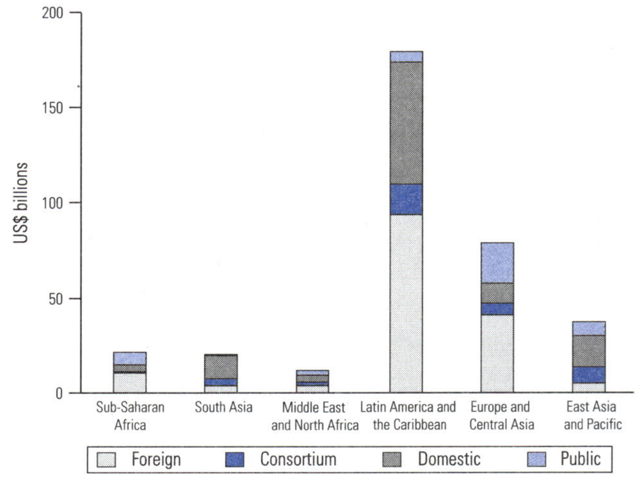

Source: Authors' analysis based on data from the World Bank PPI Project Database (2004 update).

additional 22 percent of total capital for private telecommunications projects, or $79.5 billion, was raised from public and domestic private investors through consortium arrangements with foreign investors.

Regional and Country Breakdown

A regional breakdown of private telecommunications projects by source of funding displays notable differences (figure 2.3).[3]

Latin America and the Caribbean and Europe and Central Asia are the two regions with the highest percentage of countries in which *private* capital has become a significant source of funding for the telecommunications sector.[4] Moreover, the dominant players were mainly foreign: these two regions together received about 80 percent of the overall worldwide FDI flows in telecommunications (figure 2.4).

In Latin America and the Caribbean, bidding requirements for privatization transactions, which represent the bulk of the Latin American telecommunications FDI volume, called for international operators or consortia including such operators. European and Central Asian countries were transitioning from planned economies, had little domestic private capital, and opened their doors to foreign investors. In contrast, in South Asia and in East Asia and Pacific, a substantial portion of telecommunications investments came from domestic investors, including large family groups that historically kept their investments within the region.

The large share of FDI received by Latin America and the Caribbean and by Europe and Central Asia also reflects the prominence of middle-income economies as recipients of FDI in telecommunications projects (figure 2.5). Of the $194 billion total telecommunications FDI flows to

Figure 2.4 Telecommunications FDI by Region, 1990–2003

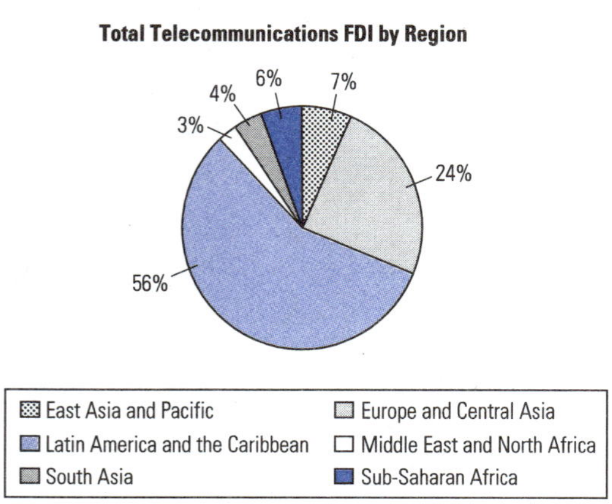

Region	Telecommunications FDI (US$ billions)	Telecommunications as a percentage of regional GDP
East Asia and Pacific	13.6	0.07
Europe and Central Asia	47.2	0.32
Latin America and the Caribbean	109.8	0.47
Middle East and North Africa	5.5	0.07
South Asia	7.4	0.10
Sub-Saharan Africa	10.8	0.24
Total	194	0.27

Source: Authors' analysis based on data from the World Bank PPI Project Database and the WDI Database (2004 update).

Figure 2.5 Telecommunications FDI by Income, 1990–2003

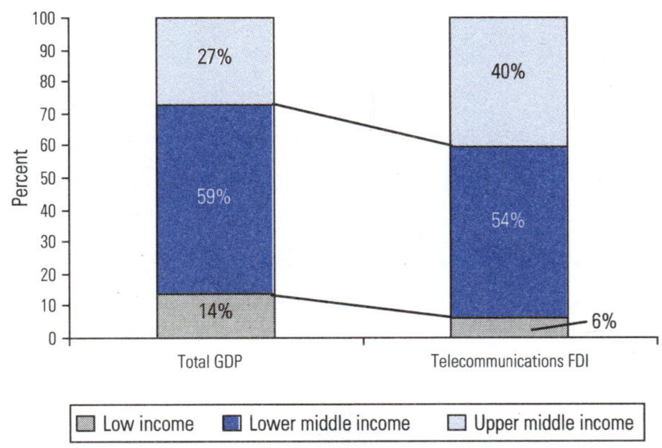

Source: Authors' analysis based on data from the World Bank PPI Project Database and the WDI Database (2004 update).

developing countries during 1990–2003, 54 percent went to 48 of the 55 lower-middle-income countries.[5] Upper-middle-income countries also had significant foreign participation, attracting 40 percent of total telecommunications FDI flows to developing countries. Only the remaining 6 percent went to low-income countries; of this, only 20 percent ($2 billion) went to Sub-Saharan Africa. However, although volumes are admittedly low, the number of low-income countries with foreign participation in telecommunications infrastructure grew from 20 in the first half of the 1990s to 51 (out of 66) by 2003.

The top 10 recipients of FDI in telecommunications projects accounted for about 70 percent of total foreign investment in the sector in 1990–2003. Inflows to Brazil alone amounted to $51 billion, making it the largest single recipient (accounting for 26 percent of the total). Several other large developing economies—such as Argentina, Poland, Indonesia, and Turkey (see table 2.1)—were also in the top 10.

Although the BRIC (Brazil, Russian Federation, India, and China) economies together attracted 42 percent of the total $1.7 trillion FDI to developing countries from 1990 through 2003, their share of telecommunications FDI was only 30 percent. Brazil, with $51 billion, was the top single recipient of telecommunications FDI; India (at $4.8 billion) and Russia (close to $2 billion) ranked 11th and 17th, respectively. China, the single largest recipient of overall FDI—attracting close to 30 percent of the total FDI flows from 1990 to 2003—had no foreign direct investments in telecommunications because of the restrictions it placed on FDI in this sector.

In per capita terms, however, small economies—such as the Seychelles, Estonia, Belize, and Grenada—attracted the highest levels of FDI (table 2.2). This is not entirely surprising, as small countries are not able to take advantage of economies of scale; they thus incur higher per capita investment costs. Small countries also represent the major exceptions to the absence of FDI in telecommunications prior to the late 1980s—in particular, small island economies in the Caribbean and the Pacific where Cable and Wireless of the United Kingdom was the (foreign) incumbent monopoly operator.

Types of Investment

FDI flows can be distinguished by regions, and also by type of project and composition. The PPI Project Database

Table 2.1 Top 10 Recipient Countries by Total Telecommunications FDI, 1990–2003

Rank	Country	Number of projects	Telecommunications FDI (US$ billions)	Percentage of global telecommunications FDI
1	Brazil	29	51.2	26
2	Argentina	7	22.8	12
3	Hungary	17	11.5	6
4	Venezuela, R.B.	6	8.9	5
5	Peru	8	8.0	4
6	Poland	10	8.0	4
7	Chile	20	6.2	3
8	Czech Republic	14	6.1	3
9	Indonesia	17	5.6	3
10	Turkey	3	5.4	3
Total		131	133.7	69

Source: Author's analysis based on data from the World Bank PPI Project Database (2004 update).

Table 2.2 Top 10 Recipient Countries by Telecommunications FDI per Capita, 1990–2003

Rank	Country	Total telecommunications FDI (US$ millions)	Population (millions)	FDI per capita (US$)
1	Hungary	11,506	10.2	1,131
2	Seychelles	71	0.08	884
3	Argentina	22,841	34.7	658
4	Czech Republic	6,073	10.3	590
5	Estonia	708	1.4	500
6	Belize	116	0.23	497
7	St. Kitts and Nevis	20	0.04	461
8	Chile	6,185	14.5	427
9	Grenada	41	0.1	424
10	Lithuania	1,439	3.6	405

Source: Authors' analysis based on data from the World Bank PPI Project Database (2004 update).

reports three main types of projects in telecommunications infrastructure:

- *divestitures* (privatization of state-owned telecommunications companies and sale of private entities),
- *greenfield projects* (operations involving new licenses and investments in new companies), and
- *concessions* (fixed-term management and operation contracts with major capital expenditure).

The composition of global telecommunications FDI flows has changed over time, reflecting the evolution of the sector (figure 2.6). In the first half of the 1990s, they were dominated by large privatizations; in the latter half of the decade, new license awards (mostly mobile telephony licenses) rose to prominence.

The First Wave: Privatization. The first wave of FDI flows to telecommunications in developing countries occurred in the early to mid-1990s and was characterized by divestitures, typically through sales of controlling stakes in the incumbent state-owned operator to foreign strategic investors. Since the first major privatization of an incumbent operator in a developing country (Chile) in 1988,[6] more than 80 developing countries have privatized their incumbents (Qiang and Guislain 2003). Privatization proceeds

Figure 2.6 Telecommunications FDI by Type of Project, 1990–2003

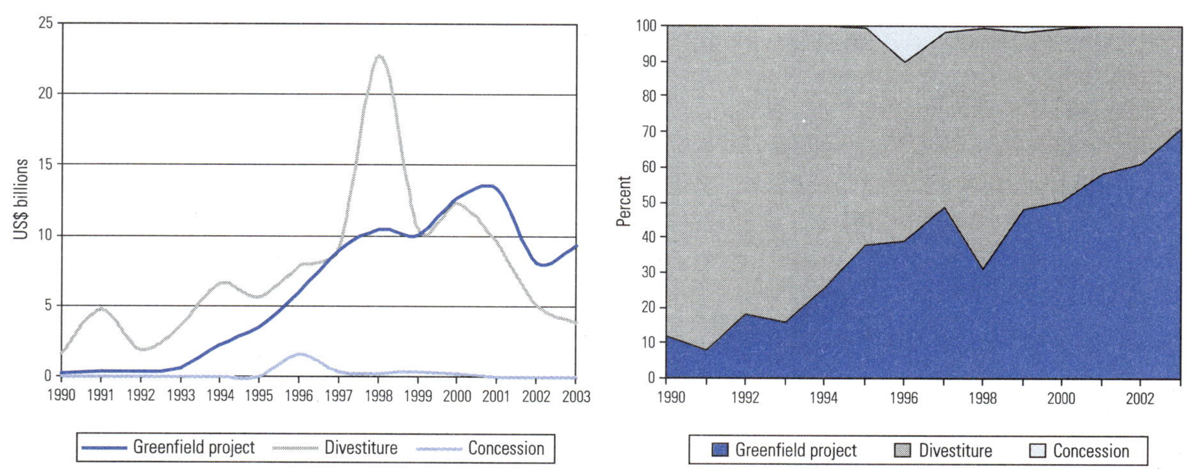

Source: Authors' analysis based on data from World Bank PPI Project Database (2004 update).

paid by foreign investors to the selling governments accounted for $57 billion; additional investments of $137 billion were made in the privatized facilities.

Privatizations accounted for 70 percent of total investment in projects with foreign participation in Latin America and the Caribbean, 48 percent in Europe and Central Asia, and 35 percent in Sub-Saharan Africa (figure 2.7). Large privatization transactions in Latin America and the Caribbean during that period included

- Argentina ($3.3 billion in 1990-92),
- Mexico ($7.5 billion between 1990 and 1994),
- República Bolivariana de Venezuela ($2.9 billion between 1991 and 1996),
- Cuba ($1.4 billion in 1994),
- Peru ($3.1 billion between 1994 and 1996), and
- Brazil ($34 billion in 1998 from the privatization of the Telebras system).

In the first half of the 1990s, governments—intent as they were on maximizing privatization proceeds—often granted the privatized companies 5-to-10-year monopoly rights in basic services. These exclusivities have, however, had an adverse impact on overall sector development and are increasingly running afoul of World Trade Organization (WTO) commitments on trade in telecommunications services. As a result, more recent privatizations have featured shorter monopoly or duopoly periods and come close to promoting full competition (Harris 2003).

Figure 2.7 Telecommunications FDI in Developing Countries by Region and Type of Project, 1990–2003

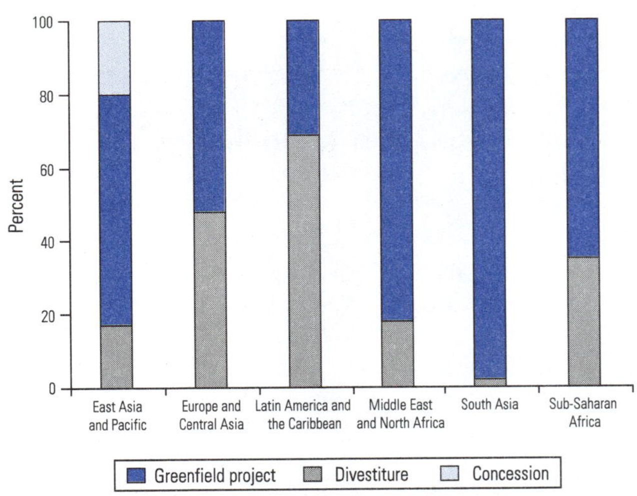

Source: Authors' analysis based on data from World Bank PPI Project Database (2004 update).

East Asia and Pacific, relative to the size of the region's economy, has attracted only modest FDI in telecommunications—amounting to merely 0.07 percent of the region's GDP (only one-seventh of the percentage in Latin America and the Caribbean, as shown in figure 2.4). This can be explained by three factors: the absence of telecommunications FDI in China, the low number of East Asian and Pacific countries that privatized their incumbent operators by selling controlling stakes to strategic investors, and the region's mobilization of domestic private capital.[7]

East Asia and Pacific was also the main region to use concessions.[8] Some of the biggest FDI in East Asia and Pacific went to Indonesia in 1995–6 as "joint operations schemes" (akin to 15-year concession arrangements) with international consortia that included PT Telkom, the state-owned incumbent operator, among their shareholders.[9] These concession schemes were, however, inherently unstable and ran into difficulties as a result of the 1997 East Asian crisis and the devaluation of major currencies in the region. PT Telkom ended up buying out its foreign partners.[10]

The Second Wave: Mobile Growth. The rising share of greenfield projects among FDI investments reflects the rising number of countries opening their telecommunications sector to competition and the shrinking number of assets to be privatized. But it is also a result of the impact of the revolution in mobile technology, which has led to the rapid deployment of mobile (or cellular) telecommunications networks.

Market enthusiasm for mobile communications can be seen in the increasing ratio of mobile FDI in total telecommunications FDI, which has risen from 7 percent on average in 1990–3 to 30 percent in 1994–9, and up to 51 percent in 2000–3. Over the same period, fixed-access and long-distance FDI remained at about 30 percent of total telecommunications FDI. More than half of all projects with foreign participation that were carried out between 1990 and 2003 were mobile projects (see figure 2.8); in the past two years, the share has been 95 percent .

The soaring number of mobile projects confirms the obvious: that mobile telephony has become the most dynamic telecommunications segment in developing countries. The experience of Nigeria demonstrates this well (see box 2.2). By requiring lower infrastructure investments than

Figure 2.8 Telecommunications FDI by Segment, 1990–2003

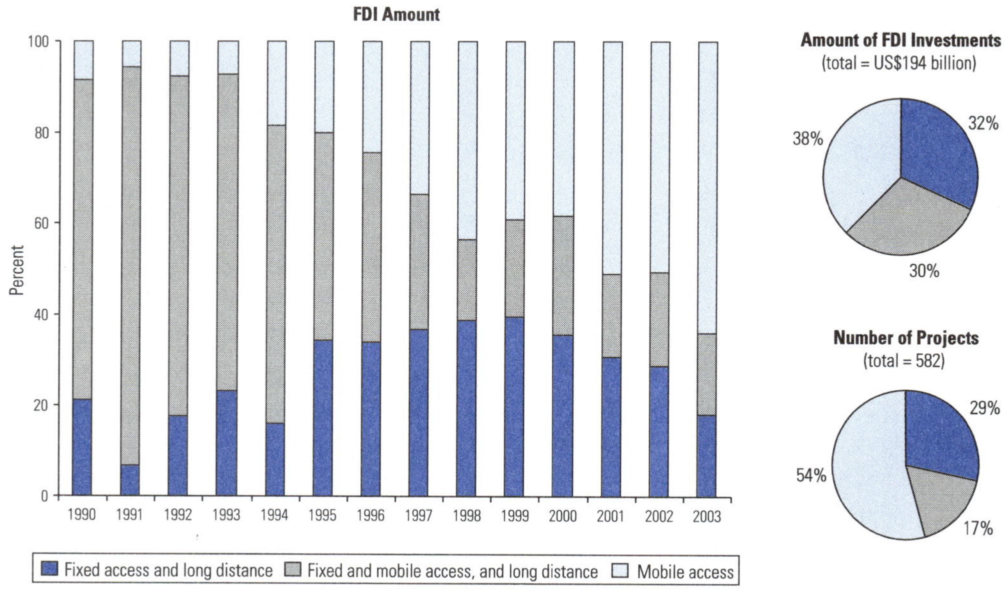

Source: Authors' analysis based on data from World Bank PPI Project Database (2004 update).

fixed-line networks, cellular systems offer a cheap, quick way to expand coverage. They also have the potential to reach remote areas, which is a crucial feature from a development point of view. From an institutional perspective, liberalization of the mobile market was easier as this was a new service with few established vested interests. Also, the potential of mobile telephony, both as a growth sector and a disruptive technology, was initially not widely recognized (Ure 2004).

Between 1990 and 2003, 213 foreign operators began providing mobile services—either on a stand-alone basis or along with basic fixed-line services—in more than 100 developing countries. By 2003, of the 164 countries with

Box 2.2 Mobile FDI in Nigeria

Following the adoption of the National Telecommunications Policy in 2000 and the award of three GSM (Global System for Mobile Communications) licenses to private operators in 2001, mobile investment and network rollout increased rapidly in Nigeria. The most dynamic new entrant was Mobile Telephone Networks (MTN) of South Africa, which achieved a 42 percent market share by March 2005. Globacom followed (100 percent held by local shareholders) with a 24 percent market share, V-Mobile (majority-owned by private investors, see box 2.7) with 24 percent, and then M-Tel (fully owned by state-owned incumbent NITEL) with 10 percent. New mobile subscriptions increased from about 28,250 per month during 2001 to more than 500,000 per month in 2004, raising the number of mobile subscribers from 370,000 in 2001 to about 11 million by March 2005. Mobile penetration rates rose from 0.3 percent to 8.2 percent over the same period. Industry reports estimate that foreign investment in the telecommunications sector had reached $3.5 billion by the end of 2004, making it the second biggest recipient of private investment in the country, behind only the oil and gas sector. Mobile telephony represented more than 70 percent of this investment, at $2.5 billion.

Source: Authors' analysis of World Cellular Networks Datapack 2005; Middle East and Africa Wireless Analyst 2004.

available data, 130 had three or more competing digital mobile operators (Qiang and Guislain 2003).

Increased Domestic Financing and Emerging Mergers and Acquisitions. The overall financing pattern of the telecommunications sector changed after the two major waves of FDI, and sector FDI flows dropped after the 1996–2000 boom. The decline was smaller than it was in other infrastructure sectors, however, suggesting the strength and resilience of this sector despite global economic downturn and sector-specific factors—including the bursting of the telecommunications bubble and the overextension of established operators in advanced economies (Izaguirre 2005; Palmade 2004).

It is worth noting that the decline in telecommunications FDI flows did not reflect a slowdown in sector expansion. In part, this was because of the increase in domestic private financing, which took care of some of the telecommunications capital needs. And, although FDI flows for the acquisition of government assets (divestiture proceeds and license fees) dropped significantly after 2000, FDI flows for sector expansion remained at the same annual level as during the boom (figure 2.9). The high FDI flows of the mid-1990s reflected large one-off privatizations that would have been difficult to sustain even in the most favorable environment.

Over the past decade, the bulk of telecommunications FDI has thus been directed to finance infrastructure rollout, a trend that can be expected to continue in the future. A recent paper by Fay and Delgado (2003)—which considers telecommunications to be the highest-growth infrastructure sector for low- and middle-income countries—projects the annual investment needs in developing countries from 2005 to 2010 at $40 billion for fixed lines and $56 billion for mobile ones.[11] Such needs, if met, imply an increase from the 2000 level in annual investment of 16.2 percent and 26.6 percent, respectively. Internal sources (such as retained earnings) and domestic private financing are likely to increase rapidly to finance this continued expansion, thus reducing the share of FDI in total new investment (World Bank 2005a).

This is not to say, however, that the privatization trend has died. Many developing-country governments still own the incumbent operator, or part thereof. The majority of those state-owned operators are under increasing stress, caught between growing pressure from private competition and the rigidities inherent in state ownership. The year 2005 may turn out to be another bumper year for privatization: it has witnessed major transactions in the Czech Republic (51.1 percent of Cesky was sold to Telefónica for $3.5 billion), Pakistan (26 percent of Pakistan Telecommunication Company Ltd. [PTCL] was sold to Orascom for $2.6 billion), and Turkey (55 percent of Turk Telecommunications was sold to a consortium led by Saudi Oger Telecommunications and Telecommunications Italia for $6.5 billion).[12]

Figure 2.9 Average Annual FDI in Telecommunications, 1990–2003

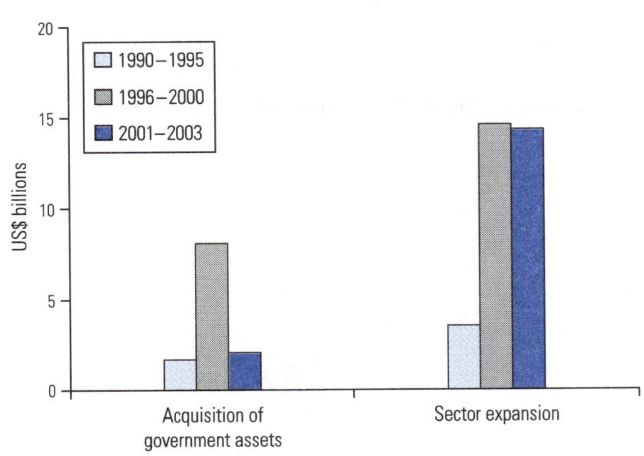

Source: Authors' analysis based on data from World Bank PPI Project Database (2004 update).

Investor Profile

The opening up of the telecommunications market, the new mobile technologies, the explosion of the Internet, and the financial crisis that hit many telecommunications investors from advanced economies in the early 2000s have all contributed to the evolution of the investor profile in the past 15 years.

Telecommunications: A Multiplayer Global Industry

Removing restrictions to competition and FDI since the 1990s has created new opportunities for foreign investors. The development of new technologies has further facilitated liberalization by reducing entry costs, introducing new services (mobile phone, paging, Internet, voice-over IP [Internet protocol]), and undermining service segmentation. Wireless transmission improvements and higher productivity in undersea fiber-optic cable systems have lowered the cost per circuit. Digital technologies now make it possible to distribute voice, data, and video on the same

Figure 2.10 **Fixed, Mobile, and Internet Subscribers in All Countries, 1990–2003**
(per 100 people)

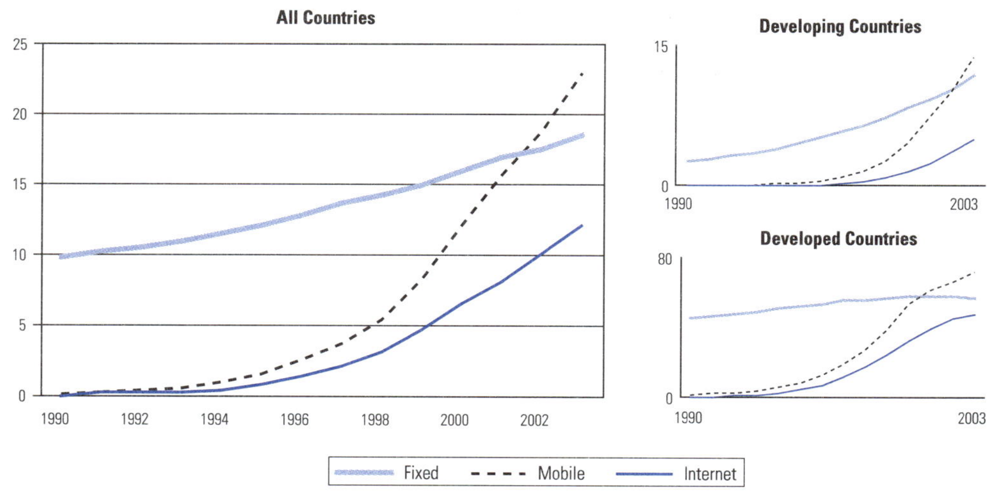

Source: Authors' analysis based on data from ITU 2004.

communications channel, thus transmitting more information per cable and portion of spectrum. Such developments have enabled investors to become multifaceted information providers and to expand across national borders. Global and regional telecommunications companies, competing outside their home bases, have emerged in this new environment.

As a result of the host of new opportunities, the telecommunications sector, once composed monopolistic domestic providers, has become a multiplayer global industry. International telephone traffic and revenues grew rapidly in the early 1990s. This growth, along with the explosion of mobile and Internet usage (see figure 2.10), encouraged the large telecommunications corporations to diversify out of their highly competitive home markets, which were characterized by relatively slow growth, into emerging markets with higher growth prospects. These diversified firms dominated the substantial investment in the industry in the 1990s, and they expanded very rapidly. In 2002, there were eight telecommunications firms listed among the world's 100 largest (by presence in host countries) multinational corporations—four times the number listed in 1992 (UNCTAD 2004). On average, they were each present in 30 host countries (see table 2.3).

Market opportunities and the comparative long-run profit potential generally determine the flow and direction of their foreign investments (Palmade 2004). The potential for such profit is a function of economic and political factors that include the size of the market, income per capita, and economic growth. Multinational corporations tend to invest in foreign markets where they have a "natural" cultural or historical affinity that arises from geographic proximity, common language, trade relationships, and political ties (for example, with former colonies). Examples are the investments by Telefónica of Spain as well as BellSouth and other U.S. operators in Latin America, Telstra (Australia) in Asia, and Deutsche Telekom in Central and Eastern Europe.

Between 1990 and 2003, the 10 largest foreign direct investors in telecommunications were all multinational corporations from Europe and the United States. They accounted for $110 billion, or 57 percent, of the total FDI in telecommunications in developing countries (see table 2.4).

Although telecommunications FDI flows to developing countries were generally high during the mid- and late-1990s, they decreased after 2000. Several factors contributed to this reduction:

- the bursting of the telecommunications bubble in advanced economies;
- the compromised balance sheet of leading global operators, following major investments or acquisitions and high bids for 3G (third-generation mobile communication technology) licenses; and
- pessimism about emerging markets following the East Asian, Russian, and Argentine economic crises.

Table 2.3 Telecommunications Multinational Corporations Ranked by Presence in Host Countries, 2002

Rank	Corporation	Home economy	Assets (US$ billions) Foreign	Assets (US$ billions) Total	Sales (US$ billions) Foreign	Sales (US$ billions) Total	TNI (percent)[a]	Presence in number of host countries[b]
1	France Télécom	France	73	112	20	49	50	42
2	Telecom Italia	Italy		85	6	30	20	41
3	Deutsche Telekom	Germany	37	121	17	51	32	28
4	AT&T	United States		55	2	38	4	28
5	Cable & Wireless	United Kingdom	4	26	5	8	37	27
6	Teliasonera AB	Sweden	18	24	1	7	75	26
7	BT Group PLC	United Kingdom	2	43	2	29	8	26
8	Tele2 AB	Sweden	5	6	4	5	77	23

Note: Blank cells indicate negligible values.

a. The Transnationality Index (TNI) is calculated as the average of the following three ratios: foreign assets to total assets, foreign sales to total sales, and foreign employment to total employment.

b. Presence of multinational corporations in host countries includes subsidiary enterprises, associate enterprises, and branches.

Source: UNCTAD 2004.

Additionally, some investors grew too fast and stretched too far; disappointing returns from some projects, both in their home countries and abroad, have also affected their appetite.

Moreover, macroeconomic and regulatory risks in host countries were causing even major international companies to reassess further investments, to slow down their expansion abroad, or even to withdraw from some markets where assets were underperforming or were considered nonstrategic. In 2001–3, France Télécom exited Argentina, El Salvador, and Indonesia and offered for sale assets in Brazil. Deutsche Telekom sold its shares in telecommunications operators in Malaysia, the Philippines, and Ukraine. Verizon left Argentina, the Czech Republic, and Mexico. Telia Sonera divested assets in Brazil, Hungary, and India (Izaguirre 2005).

Table 2.4 Top 10 Telecommunications Foreign Direct Investors, 1990–2003

Rank	Investor	Country of origin	Telecommunications FDI (US$ billions)	Destination regions (amounts in US$ billions): East Asia and Pacific	Europe and Central Asia	Latin America and the Caribbean	Middle East and North Africa	South Asia	Sub-Saharan Africa
1	Telefonica Internacional Holdings	Spain	34.0			33.8	0.2		
2	Telecom Italia	Italy	17.3		2.0	15.3			
3	France Télécom	France	13.5	0.55	5.7	5.9	0.4		1.0
4	Deutsche Telekom	Germany	11.4	0.66	10.7			0.03	
5	Verizon	United States	9.6	0.44	0.6	8.5		0.04	
6	Portugal Telecom	Portugal	7.7			7.4	0.23		0.15
7	MCI	United States	6.3			6.3			
8	BellSouth Corporation	United States	3.9			3.9			
9	SBC Communications	United States	3.8			1.7			2.1
10	TeliaSonera	Sweden	3.4	0.24	2.5	0.6		0.07	

Note: Blank cells indicate negligible values.

Source: Authors' analysis based on data from the World Bank PPI Project Database (2004 update).

The Rise of South-South and Regional Investors

The withdrawal of some developed-country investors has given investors from developing countries the chance to acquire assets at low prices and compete for new licenses. At the same time, increasing wealth and capital account liberalization in some emerging market economies have increased the supply of capital from the South and enabled their companies to invest abroad.

This is reflected in the composition of the list of the 30 largest telecommunications multinational corporations (see annex 2B). Though still dominated by companies from Europe and the United States, by 2002 the top-30 list included four companies from developing countries: Datatec (South Africa), América Móvil (Mexico), MTN Group (South Africa), and Telekom Malaysia. These investors tend to be operators from large developing countries investing within their own regions. They also tend to be from countries that reformed early: privatization and competition forced them to become more efficient. At the same time, their exposure to competition was limited as they were generally protected from full market liberalization. This combination allowed them to generate high margins and profits at home, available for investment abroad in new companies or acquisitions.

Developing-country investors are increasing their share of total telecommunications-related FDI (see figure 2.11). From 2001 to 2003, South-South FDI accounted for over 36 percent of total inflows and close to 20 percent of the total number of telecommunications projects, compared with only 23 percent and 11 percent, respectively, in 1990–9.

The rise in South-South FDI has been stimulated by structural, cyclical, and policy factors similar to those driving North-South FDI flows. These factors include maturing domestic markets and increased competition limiting profit opportunities (UNCTAD 2004). Specific factors—especially geographic proximity—seem to favor South-South investment. Since the cost of acquiring reliable information about foreign markets can be high for relatively small companies, they tend to invest in neighboring countries where they have gained familiarity through trade, cultural, or family links and thus comprehend the complexities of investing in these markets. They also tend to be more knowledgeable about local conditions than multinational operators and more tolerant of political risk.

In addition, when a multinational company considers a market too small or marginal to invest in directly, it may do so through a subsidiary. Investments by Vodacom of South Africa (partly owned by Vodafone of the United Kingdom) and Sonatel of Senegal (a subsidiary of France Télécom) are examples of multinational corporations investing through regional affiliates.

Over 85 percent of South-South FDI flows during 1990–2003 stayed within the same geographic region (see table 2.5). East Asia and Pacific, Europe and Central Asia, Latin America and the Caribbean, and the Middle East and North Africa received South-South FDI only from investors in their respective regions. This trend is even stronger when one considers investors from the Middle East and North Africa for whom Africa and South Asia are often perceived as part of their "natural" region. Moreover, nonregional investors are withdrawing: the largest Asian investor in Sub-Saharan Africa, Telekom Malaysia, for example, is gradually pulling out of the continent.

In Latin America and the Caribbean, early movers from the largest advanced economies—AT&T, BellSouth, France Télécom, MCI, and SBC—retreated after the burst of the

Figure 2.11 Telecommunications FDI, North-South versus South-South, 1990–2003

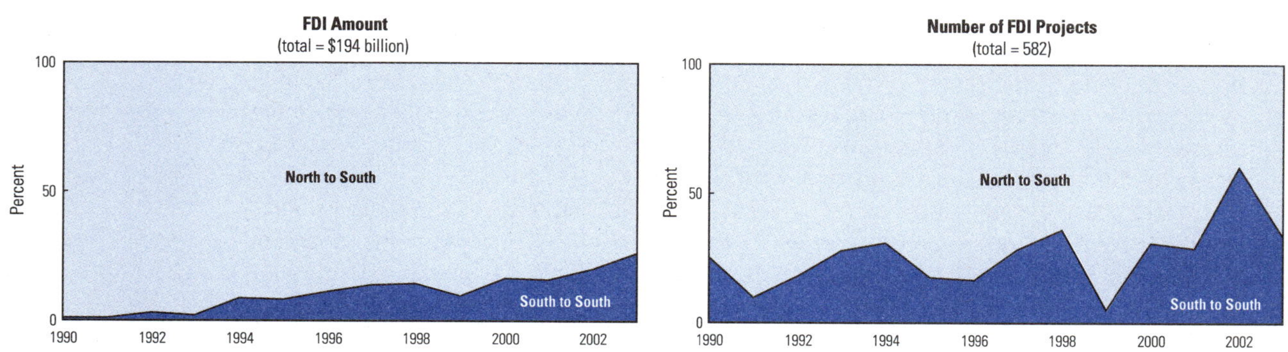

Source: Authors' analysis based on data from the World Bank PPI Project Database (2004 update).

Table 2.5 Intraregional South-South Telecommunications FDI, 1990–2003
(as a percentage of total South-South FDI)

Region of investor	Destination region: East Asia and Pacific	Europe and Central Asia	Latin America and the Caribbean	Middle East and North Africa	South Asia	Sub-Saharan Africa
North to South	72	93	90	52	75	51
South to South	28	7	10	48	25	49
East Asia and Pacific	100				24	50
Europe and Central Asia		100				
Latin America and the Caribbean			100			
Middle East and North Africa				100	36	5
South Asia					40	
Sub-Saharan Africa						45

Note: Based on the largest 75 investors in telecommunications, accounting for 95 percent of total telecommunications-related FDI in developing countries between 1990 and 2003.

Source: Authors' analysis based on data from the World Bank PPI Project Database (2004 update).

telecommunications bubble and the decline in demand for fixed-line services in which they had invested heavily.[13] This left as principal foreign operators Portugal Telecom and Spain's Telefónica, both counting on common cultural roots to convey strategic commercial advantages. It also created room for regional investors such as Telmex of Mexico and its spin-off, América Móvil, which is now the largest wireless provider in Latin America (box 2.3).

Many global operators invested in the developing markets of the East Asia and Pacific region in the 1990s, but most have since withdrawn. It is the regional companies that have kept or expanded their investments in this region,

Box 2.3 Expansion of Mexican Operators in Latin America

Once a state-owned monopoly, Teléfonos de México (Telmex) was privatized in 1990. It remains the largest telecommunications operator in Mexico, where it enjoyed a 78 percent market share in international long distance and 52 percent in Internet dial-up access in December 2004.

Telmex has expanded aggressively and has become a leading regional player in Latin America. In 2004, it acquired AT&T Latin America (with operations in Argentina, Brazil, Chile, Colombia, and Peru) and paid $113.5 million for an 80 percent stake in Techtel, Argentina's data and voice provider. In early 2005, it acquired MCI's equity stake in Brazilian long-distance operator Embratel. Today, Telmex has a direct presence in Argentina, Brazil, Chile, Colombia, Mexico, Peru, Puerto Rico, and the United States.

América Móvil was spun off from Telmex in 2000 and operates Mexico's largest mobile company, Telcel. It has expanded aggressively in Latin America in recent years, investing over $2 billion and purchasing assets from Bell Canada International, SBC, and AT&T, as these operators pulled out of Latin America. This acquisition strategy, funded by América Móvil's profits in Mexico, has opened up new markets for the operator: it now has (mostly fully owned) subsidiaries in Argentina, Brazil, Colombia, Ecuador, El Salvador, Guatemala, Honduras, Nicaragua, Paraguay, Peru, and Uruguay as well as in the United States.

América Móvil tripled its sales during the last two years, and is now the largest wireless provider in the region. With about 66 million customers by mid-2005, it is one of the 10 largest mobile operators in the world.

Source: Business Monitor International 2004 and América Móvil Web sites (accessed on September 16, 2005).

Table 2.6 Mobile Operators in Sub-Saharan Africa, 2004

Operator	Ownership	Subscribers (millions)	Subscribers excluding South Africa (millions)	Number of countries	Population under license	Fiscal year 04 revenue (US$ millions)
Celtel	Mobile Telecommunications Company K.S.C. (Kuwait)	6	6	13	234	614
MTN	Employees (18.7%); institutional and private investors	14.3	6.3	5	179	4,266
Vodacom	Telkom SA 50%, Vodafone 35%, VenFin Limited 15%	15.5	2.6	4	106	3,960

Note: Cable and Wireless (25%) and SBC Communications (15.5%) sold their shares in MTN in 1998. In 2002, employees acquired their shares through a management buy-out (Newshelf 664).

Source: World Bank, adapted from Celtel, MTN, and Vodacom, March 2005.

including SingTel (with a presence in Bangladesh, Indonesia, the Philippines, and Thailand), Telekom Malaysia (with operations in Bangladesh, Cambodia, Indonesia, Pakistan, Sri Lanka, and Thailand), and Shinawatra from Thailand (investing in Cambodia and the Lao People's Democratic Republic) (Ure 2004).

In Sub-Saharan Africa, global operators have also gradually phased out, to be replaced by regional players. Vivendi, for example, sold its equity stake in a Kenyan mobile operator in 2004 to Celtel as part of a debt reduction effort required after its ill-fated attempt to establish a global communications and media empire. France Télécom and its subsidiary Orange, although present in six African markets, have not shown strong strategic interest in the region in recent years. The three main mobile operators in Sub-Saharan Africa today are regional operators: South Africa's MTN and Vodacom, and Celtel, owned by the Mobile Telephone Corporation (MTC) of Kuwait (see table 2.6 and box 2.4). These operators are expanding their presence across Africa, having provided about $5 billion in FDI in telecommunications to other countries in the region. Together they represent about 47 percent of total FDI in Sub-Saharan Africa.

Box 2.4 Celtel

Celtel International, previously known as MSI, was founded in 1998 by a group of shareholders (including its chairman, a Sudanese investor) and international investment institutions such as the International Finance Corporation.

Celtel embarked on an aggressive investment strategy in Africa. Emphasizing smaller markets, it aimed to establish a regional brand and reap economies of scale. It issued $356 million in equity between 1998 and 2003 to finance African licenses (around $30 million) and network deployment. Investments in Africa in 2004 were $250 million, an increase of 140 percent over the previous year, resulting in a doubling of the number of subscribers from 2.5 million at the end of 2003 to 5.2 million a year later, and an additional 16 percent increase to 6 million subscribers by March 2005. By that time, the company had invested over $900 million in telecommunications.

Today, Celtel has equity stakes in operators in Burkina Faso, Chad, the Democratic Republic of Congo, the Republic of Congo, Gabon, Kenya, Malawi, Niger, Sierra Leone, Sudan, Tanzania, Uganda, and Zambia. The company is the market leader in all countries where it has a presence, with the exceptions of Tanzania and Uganda. In early 2005, Celtel was acquired by MTC of Kuwait for $3.34 billion.

Source: Authors' analysis.

As mentioned above, cross-regional South-South FDI flows have been limited. The biggest recipient has been Sub-Saharan Africa. About $2 billion (or one-fifth) of foreign investments in telecommunications in Africa came from East Asian and Pacific sponsors during 1990–2003, most of it in the mid-1990s. However, Telekom Malaysia, the most important Asian operator in Sub-Saharan Africa, has sold its 12.6 percent stake in Telkom South Africa and announced that it intends to divest its stakes in the Ghanaian and other African operators. The company has decided to refocus its overseas investments on countries closer to Malaysia. Interestingly, Chinese companies have made some inroads in Africa in recent years, though (so far) on a relatively modest scale.[14]

Middle East and North African FDI to Sub-Saharan Africa has included investments from Morocco and Tunisia to Mauritania, and from the Arab Republic of Egypt (through Orascom) to 12 Sub-Saharan African countries. Orascom, which invested in the late 1990s, exited from most of these countries as well as from Jordan and the Republic of Yemen during 2002–4 in an attempt to refocus its strategy on core operations in the Middle East. The most recent Middle East and North African FDI in Sub-Saharan Africa was the acquisition by MTC (Kuwait) of Celtel in March 2005 (see box 2.4).

South Asia has also benefited from South-South cross-regional FDI. In addition to the $455 million from within its own region, it received $1 billion of investment from East Asian sponsors and $2.9 billion from Middle East and North African ones—the latter inflows will almost double following the sale in March 2005 of 26 percent of PTC (Pakistan) to Orascom of Egypt for $2.6 billion.

Most of these FDI flows are categorized as cross-regional because of the World Bank regional classifications. In fact, however, investors and recipients are often located in neighboring regions. The limited South-South FDI across continents and the recent divestment by investors to refocus on regional investments underscores the importance of geographic proximity as well as links of culture and trade.

Investors: From Strategic to Financial

Financial and equity investors—such as investment banks, private equity, and mutual funds—became more active in the telecommunications sector in developing countries in the late 1990s, bringing an alternative source of private capital.[15] These investors are usually driven purely by financial returns on their investments. Among the top 20 telecommunications investors from developing countries, Isbank (Turkey) and Banco Opportunity, Banco Safra, and Techold (all from Brazil) are financial investors.

Financial investors typically look for companies with strong cash flows and good business fundamentals. The cash flows attract bank loans, syndicated for large borrowings, which provides the leverage for which the funds are looking. Having good fundamentals implies that the company is probably already a major force in the industry, either as a dominant player or a leading innovator.

The funds' typical investments are in the range of 10 to 25 percent stakes, but on occasion they can reach to 15 to 40 percent or even outright control. Unlike strategic investors, financial and equity investors look for shorter-term investments of about 3 to 5 years, with rates of return of 25 percent or better. In some cases these funds pair up with operators in order to mitigate company, market, and managerial risk. For example, Newbridge Capital partnered with ChinaNet to purchase Asia Global Crossing in November 2002.

During 2004–5, major private equity transactions in the telecommunications sector consisted mostly of buyouts of mature firms with strong fundamentals and an industry track record.[16] To take just one example, in June 2004 the private equity firm Advent International acquired a 65 percent stake in Bulgaria's fixed-line telecommunications operator, the Bulgarian Telecommunications Company (BTC), and agreed to invest $450 million in telecommunications infrastructure. One of the last European telecommunications operators to be privatized, BTC needed major investments to significantly upgrade its network. Advent, active in Central and Eastern Europe, had previously invested in Cesky Mobil in the Czech Republic and Connex in Romania.

This trend of growth in portfolio and financial investment is typical of a maturing sector that is moving away from the old public sector monopoly paradigm toward market-based financing mechanisms.

Factors for Success: Attracting Telecommunications FDI

A range of conditions affect the FDI-friendliness of a country's telecommunications sector. The basic contributing factors mirror what applies to other sectors of the economy.

Healthy General Business Environment

A business-friendly overall environment is an important ingredient for attracting and retaining FDI in telecommunications. Factors that favor FDI for any sector are sound macroeconomic policies, low political risk, ease of market entry, the protection of property and investor rights, and—more generally—reliable contract enforcement. Good infrastructure, a skilled work force, and favorable tax policies are additional factors.

The emergence of South-South and regional FDI adds new considerations to the business environment. Most South-South investors are operators with limited global experience. Many of them have smaller market shares and less experience in distributing products and services in other countries than multinational operators have.[17] They may have less bargaining power in negotiating for new contracts or resolving disputes in existing contracts. They may also have greater difficulty in meeting qualification requirements for tenders and lower creditworthiness with local banks, making it more difficult to raise financing. Unlike multinationals, they are often not well positioned to raise capital by issuing new bonds or stocks.

To maximize competition, countries wishing to attract FDI should be careful not to insist on unnecessary requirements that might exclude otherwise qualified bidders. They should focus instead on creating a level playing field that provides fair opportunities to new entrants irrespective of size or origin.

Liberalized Telecommunications Market

The first sector-specific step countries have taken to attract foreign investment has been to liberalize the telecommunications sector by opening it up to private and foreign investors and allowing competition. This is the most fundamental factor for attracting FDI because if there is no opportunity to invest in a country (other than purchasing the incumbent operator, where that option is offered), there can be no FDI. Governments have also realized that any restriction they place on investment (be it foreign or domestic) raises the cost of financing (and ultimately of services), thus making investment less likely.

The past two decades have witnessed an unparalleled opening and modernization of the telecommunications sector in all regions, although the pace and scale of reform have varied by region and country. Despite this progress, the 2005 ITU Regulatory and Competition Database suggests that as of 2004, 50 percent of developing countries retained monopolies on international telephony, 43 percent still had monopolies on fixed local services, and 13 percent on mobile services (see figure 2.12). A recent survey of 15 Asian economies notes that all but 1 had restrictions on telecommunications FDI, including maximum foreign share of 30 to 49 percent of total equity (Ure 2004).

A similar picture is found around the world (box 2.5), pointing to a considerable unfinished liberalization agenda and many missed opportunities for FDI. Ure reports that one foreign investor pulled out of a partnership investment in an Asian telecommunications company because "the regulations . . . made it very clear that we couldn't be in control—capped at 49 percent. That was the main reason we left" (Ure 2004, p. 32).

In most cases, foreign ownership restrictions limit takeover risk and hence management accountability, and reduce investment incentives, thereby inhibiting effective, profit-oriented management. Furthermore, limited ownership arrangements were more attractive when the investors had the security of a monopoly situation. With the gradual opening up of telecommunications markets, there is decreasing acceptance among foreign investors of these restrictions.

Policy makers in developing countries should further reduce foreign ownership restrictions in the telecommunications sector to reap the benefits FDI brings in terms of

Figure 2.12 Telecommunications Sector Liberalization in Developing Countries, 1998 Compared with 2004

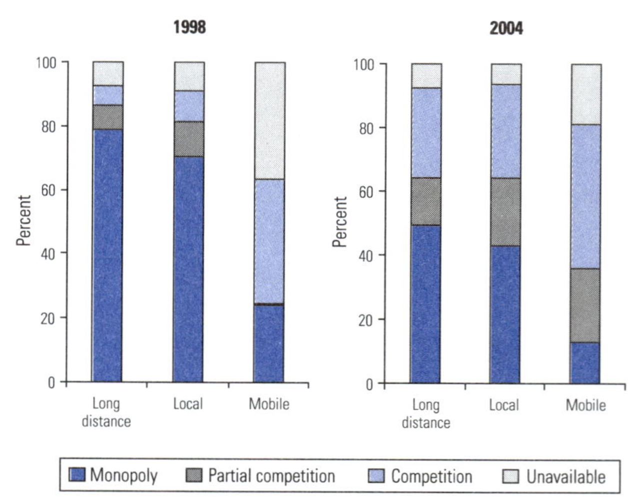

Source: Authors' analysis based on data from the ITU Regulatory and Competion Database 2005.

Box 2.5 Foreign Ownership Restrictions

Many countries are sensitive about foreign ownership of strategic or otherwise sensitive assets. Arguments in favor of restrictions are often linked to sovereignty or national security issues. Countries also express concern that foreign investment may crowd out local companies. Restrictions may be embodied in the country's constitution, laws, or regulations. The constitution had to be amended in Brazil and Mexico, for example, to allow the privatization and liberalization of the telecommunications sector (Guislain 1997).

In some countries, including the United States, foreign ownership rights are restricted to minority shares, typically 25–49 percent. Countries with such limits in telecommunications include Malaysia (30 percent); Kenya and the Philippines (40 percent); and India, Mexico, and Poland (49 percent, though in Poland the restriction does not apply to local services operators).

Proctor and Olivier (2002) analyzed the impact of FDI restrictions in Canada and India. They concluded that FDI restrictions led to an increase in the cost of capital in the telecommunications sector, slowing down investment and artificially prolonging the dominance of the incumbents. In India, for example, FDI is limited to 49 percent for most services, but it is allowed for up to 74 percent for Internet service with a gateway, certain infrastructure providers, and radio paging. Up to 100 percent FDI is permitted there for Internet service providers (ISPs) without gateways, and for providing e-mail and voice mail. The Indian government is now in the process of increasing the foreign investment ceiling from 49 percent to 74 percent in an effort to mobilize additional capital to fulfill its ambitious telephone penetration targets.

FDI may also be restricted to special arrangements under which foreign investors may purchase assets (typically as minority stakeholders) only in partnerships with domestic residents. In China and Vietnam, for example, until very recently, private and foreign involvement in basic telecommunications networks could occur only through special investment schemes. In these schemes, foreign companies finance capital investment and share in revenues, but they have no equity share and have only limited or no management control. In Indonesia, to take another example, the 1989 telecommunications law limited the award of basic telecommunications licenses to companies in which the incumbent public operator is a shareholder. And in yet another example, Turkish foreign investment law excluded foreign investment in concessions; it was modified in 1994 to pave the way for the privatization of Türk Telekom, which is to materialize more than 10 years later.

Source: Authors' analysis.

lower cost of capital (Procter and Olivier 2002) and higher productivity, coverage, and quality of services (Harris 2003; Palmade and Anayiotas 2004; Qiang and Guislain 2003).

Consistent Regulatory Framework

For most telecommunications operators and potential investors, "regulatory risk" is the principal factor in their investment strategies.[18] Foreign investors respond positively to a stable and predictable regulatory framework. They are willing to pay more for a telecommunications company if it operates in a well-established regulatory framework, as this stability reduces uncertainty and thereby risk (Kirkpatrick, Parker, and Zhang 2004). The independence of regulatory bodies from political influence and from capture by incumbent operators promotes regulatory credibility. Not surprisingly, countries that established independent regulatory authorities before they privatized and liberalized their telecommunications sector have seen more outside investments than countries that did not (Wallsten 2002).

The principal requirement for a regulatory framework is consistency—not only with respect to the law, regulations, and administrative procedures, but also with respect to how those are applied in practice.[19] Well-managed regulatory bodies operating under a clear legal framework lead to greater predictability in decision making; greater predictability reduces the cost, time, and risk to which potential investors and participants are exposed. This is particularly important for foreign investors who are less familiar with the local laws, business culture, and officials.

Inconsistency, delays, uncertainty, and policy confusion increase the cost of doing business and deter investments. For example, the Thai Telecommunications Act of 2001 reduced the foreign ownership ceiling from 49 percent to 25 percent. But the government, under intense industry pressure, quickly reversed the decision. Such policy reversals send the signal that either regulations are not well thought through, which is a capacity issue, or that special interests wield significant influence, which is an independence and integrity issue (Ure 2004). Either way, the inconsistency acts as a deterrent to FDI.

Dispute Resolution Mechanisms

Some countries have a relatively short FDI track record. The reliability of these countries' regulatory institutions or the strength of their governments' commitment to adhere to key contractual provisions may not be clear. Governments may consider providing guarantees to investors during their country's transition to full competition and regulatory independence. Guarantees could be in the form of government commitments to predefined regulatory frameworks, government payment of contractually obligated subsidies, and specifically agreed dispute resolution processes. For the additional comfort of potential investors, such guarantees may be backed by international financial institutions (Gupta et al. 2002).

However, privatization agreements governing the sale of a government's stake in the incumbent operator to foreign investors often included provisions that guaranteed the buyer certain exclusive rights or an increase in tariffs over a certain period. These guarantees have often created stumbling blocks for the continued growth of the sector. Similarly, inherently unstable agreements such as build-operate-transfer (BOT) agreements, concessions, and revenue- sharing arrangements have often outlived their usefulness before the lapse of their contractual term. In these cases, the agreements may be terminated early or converted into more stable mainstream licenses.

When rights given to investors are exorbitant or when technological or market conditions make them no longer reasonable, governments and legislators often end up modifying the concerned laws, regulations, or license conditions. In many countries, this has led to terminating the incumbent's monopoly early, to mandatory interconnection, to introducing new tariff structures, or to changing the terms and conditions of licenses. Any of these changes may alter initial contract terms significantly. Although license or concession agreements do not prevent the legislature from modifying the legal framework, they may provide redress (such as compensation rights) to aggrieved investors.[20] Governments need to carefully consider the tradeoff between maintaining the status quo (if that is an option at all), unilaterally modifying the initial terms of service provision, or reaching some form of agreement with the concerned service providers.

The resolution of this type of dispute is challenging for the government or regulator, caught between the rival objectives of introducing competition or other regulatory changes and of honoring contractual commitments to investors. Ending exclusivities, for instance, should normally lead to a more competitive market and greater foreign investment (see box 2.6).[21] In some cases, the parties are able to reach an amicable agreement. For example, Is-Tim (Aria)—a Turkish mobile operator 49 percent owned by Telecom Italia Mobile (TIM's bid to obtain this license in 2001, $2.5 billion, was the largest ever single FDI in Turkey)—claimed that the Turkish Telecommunications Authority did not enforce its roaming rights with other operators. This case was filed in Paris under International Chamber of Commerce (ICC) rules. The parties were later able to reach an amicable settlement—brokered in part by the Italian and Turkish prime ministers—which involved the merger in 2004 of Is-Tim and Aycell, the Türk Telekom mobile subsidiary.

In other instances, disputes between foreign investors and the host government have led to litigation. Most FDI contracts call for international arbitration to settle disputes because foreign investors are often not comfortable with the judicial system, practices, and traditions of the host country. Moreover, more countries have signed bilateral investment treaties giving their investors the right to go to international arbitration with respect to investments in the host country, even when their contract does not include an arbitration clause. This helps explain the large increase in FDI arbitration cases over the last decade (box 2.7).

Arbitration tends to be a last resort, however, to be used when all communication has broken down and agreement cannot be reached. It typically implies the end of the business relationship between the country and the investor.

Box 2.6 Ending the Exclusivity of Cable and Wireless in the Caribbean

Until the mid-1980s, Cable and Wireless (C&W) was one of the few foreign direct investors in telecommunications in developing countries. Its operations (typically with exclusivities) were mostly in the Commonwealth (see endnote 1). In the Caribbean, C&W was able to negotiate licenses with very long exclusivity periods, in some cases for 20 years or more. The common belief was that the island developing states that C&W operated in were too small to sustain competition. However, a growing perception that lack of competition was keeping communication costs high and hence limiting these countries' participation in the global economy led to a reevaluation of the C&W monopoly.

Five member countries of the Organization of Eastern Caribbean States (OECS)—Dominica, Grenada, St. Kitts and Nevis, St. Lucia, and St. Vincent and the Grenadines—joined together to negotiate with C&W for an early termination of its market exclusivities. In 2000 they established a regional telecommunications authority, the Eastern Caribbean Telecommunications Authority (ECTEL), to coordinate their sector policies and regulations, negotiations with C&W, and new licensing approaches.

The five countries worked with ECTEL to renegotiate existing exclusivity arrangements with the incumbent. In 2001, OECS member states reached an agreement with C&W on phasing out its exclusivity rights. New licenses were issued starting in 2002. Unlike previous termination agreements, such as those in Hong Kong (China) and Singapore, the termination agreement with C&W provided no compensation.

The end of the monopoly led to a significant increase in foreign investment in the OECS, especially in mobile telephony. Direct investment in the sector increased from $40 million in 2001 to more than $80 million in 2003 and over $90 million in 2004. Thanks to intense competition from new entrants, mobile penetration increased in Dominica from 1.6 percent in 2000 to 60 percent in 2004; in Grenada it increased from 4.5 percent to 86 percent. Meanwhile retail prices fell on average by about 50 percent.

Source: World Bank 2005b; World Bank and ITU 2004.

Arbitration is not suitable for working on the many day-to-day issues that may pit a foreign investor against the government or regulator of the host country. Other forms of settlement—such as mediation and technical expert intervention—are needed to deal with such disputes as they arise.

International Agreements

FDI in telecommunications has been bolstered by the WTO General Agreement on Trade in Services (GATS) and the commitments undertaken under its auspices, especially the 1997 Basic Telecommunications Agreement. These agreements open markets to foreign suppliers of telecommunications services. As of June 2005, 104 of 148 WTO member countries had made some commitments on telecommunications in their GATS schedules, including 96 governments (whose suppliers account for over 95 percent of the world's telecommunications revenues) that made commitments on basic telecommunications.

These commitments lay the foundation for improved market access and the liberalization of investment, both foreign and domestic. They strengthen investor confidence by demonstrating that a country intends to reform its telecommunications sector in a nonreversible way. They also provide recourse to foreign investors (through their government) to settle disputes under the WTO dispute resolution system. GATS commitments, in this way, enhance and underpin the domestic sector reform agenda of developing countries, providing investors with greater predictability.

Specific country commitments are classified along four modes. Mode 3, "commercial presence," opens the telecommunications sector to FDI. If explicit restrictions are introduced in the GATS schedule, it provides for a more limited opening. Such possible restrictions include a specific

Box 2.7 Selected FDI Disputes in Telecommunications

As of mid-2005, a number of telecommunications FDI disputes were at various stages of adjudication. The following paragraphs summarize some of the key issues raised by these cases.

Argentina: The government's decision to freeze utility tariffs following a 70 percent devaluation of the Argentine peso in 2002 has generated considerable litigation. One telecommunications dispute has been referred to the International Center for Settlement of Investment Disputes (ICSID). In July 2003, Telefónica of Spain, an investor in Telefónica de Argentina SA, claimed $3.8 billion in damages from the Argentine government as compensation. The tribunal was constituted in April 2004, but as of September 2005, no decision had yet been issued in this dispute.

Belize: In March 2005, a U.S. district court in Miami fined the government of Belize for defying its order to return control over Belize Telecommunications to a U.S. investor. The government installed a new board in the company after the investor failed to pay an installment on the purchase price of his shares. The investor argued that the government had failed to protect him from competition as agreed in the context of the sale.

Ghana: Telekom Malaysia acquired a 30 percent stake in Ghana Telekom following a privatization tender in 1997. Telekom Malaysia lent $50 million to the government in 2000 as advance on an additional 15 percent equity stake. Its management contract, which gave it control over the company, expired in 2002. Disagreements between shareholders ensued, and the government appointed a new board and took over control of the company. Telekom Malaysia filed an arbitration claim under United Nations Commission on International Trade Law (UNCITRAL) rules in London in 2003 pursuant to the Malaysia-Ghana bilateral investment treaty. The parties have reached an agreement for the government to reimburse the amount of the loan and buy back the shares.

Lebanon: The government decided in 2001 to terminate BOT or concession-type agreements with two mobile companies—France Télécom Mobile Liban, owned by France Télécom and local investors, and Libancell, owned by Sonera of Finland and local investors. The government alleged, among other things, that the companies offered services to more subscribers than authorized by their contract. The owners claimed expropriation and filed international arbitration proceedings in 2002. In February 2005, the arbitral tribunal ordered the government to pay France Télécom $266 million in indemnification. Libancell owners were awarded a similar amount.

Nigeria: A dispute opposes Econet of Zimbabwe and the main shareholders of V-Mobile, a private mobile operator, over the proposed sale of 51 percent of V-Mobile's shares to Vodacom (South Africa) and Virgin Mobile (United Kingdom). Econet, which has a 5 percent shareholding, claims it has preemptive rights. It also claims that it has a contract to manage V-Mobile's network. The dispute was referred to the permanent court of arbitration in The Hague, which appointed an arbitration panel. The panel found it had no jurisdiction—the case is now (September 2005) with the Nigerian federal high court.

Source: Authors' analysis based on INVEST-SD News Bulletin 2004; World Bank and ITU 2004.

ceiling on foreign ownership in the sector or the incumbent operator, or limiting the opening to certain market segments.

More than 93 governments also committed in GATS to telecommunications regulatory principles. Along with the overall GATS disciplines on regulation and transparency, these principles are considered to be best practices fundamental to ensuring minimum standards of good regulatory behavior, effective competition, and a stable climate for investors.

Countries wishing to accede to the WTO also have to make telecommunications commitments deemed reasonable by the major WTO member countries with whom accession is negotiated. This by and large implies more liberal commitments than those made by many of the countries that joined WTO before 1998. WTO accession is contributing to sector reform in countries such as Russia, Saudi Arabia, Ukraine, and Vietnam.

The first WTO dispute settlement case in telecommunications (and in services, more generally) was filed in 2002. The United States complained that Mexico had failed to meet several of its obligations under the GATS. The panel charged with settling this dispute concluded that the interconnection rates charged by Mexican operators to terminate international calls from the United States were not cost-oriented and that Mexico had failed to maintain appropriate measures to prevent anticompetitive practices, both required under the Reference Paper that Mexico had adopted as part of its GATS commitments. The panel also found that Mexico did not ensure access to public telecommunications networks in Mexico to U.S. telecommunications operating companies on reasonable terms, thus breaching provisions in the GATS Annex on Telecommunications. The panel, however, found that Mexico had not committed to allow international simple resale using cross-border leased circuits, as also claimed by the United States. The WTO dispute resolution body approved the panel's findings in June 2004. Mexico and the United States agreed on a timetable for Mexico to revise its international long-distance rules to allow competitive negotiation of interconnection charges and to issue regulations for companies without own facilities to provide international telecommunications services using the networks of Mexican concessionaires. These measures were satisfactorily implemented by August 2005 (Wellenius, Galarza, and Guermazi 2005).

Besides multilateral agreements, developing countries increasingly view bilateral and regional investment treaties as a way to enhance cooperation and to promote flows of FDI in the telecommunications sector. The bilateral trade agreement between the United States and Vietnam, for example, gives U.S. companies the right to invest in certain segments of the telecommunications sector even though domestic law has not yet opened these to competition and Vietnam has not yet joined the WTO. Similarly, the Central American Free Trade Agreement (CAFTA) calls for liberalizing the telecommunications sector in Central American countries and opening it up to foreign investment. CAFTA has already provided impetus for sector reform in Costa Rica and elsewhere. The focus of trade negotiation has thus moved beyond the cross-border movement of goods and services and now embraces the movement of capital as well. This has resulted in dramatic increases in trade linkages and cross-border capital flows.

Looking Forward

This chapter summarizes trends in telecommunications FDI in developing countries over the last two decades. As governments in developing countries privatized incumbent operators, liberalized foreign investment regimes and their telecommunications sectors, and established consistent regulatory frameworks, FDI levels in telecommunications grew about 10-fold from very low levels in the early 1990s to about $16.5 billion per year in the early years of this decade. The rapidly changing nature of the sector makes it difficult to predict what the future holds for telecommunications FDI. The following issues are likely to affect upcoming FDI trends in the sector:

- **Sustainability of the level of telecommunications FDI flows.** Will overall telecommunications FDI flows continue at their current levels? The likelihood of sustained or increased inflows may depend on many factors, including continued globalization of the economy, the recovery of capital markets and the global telecommunications industry, renewed appetite for privatization (as seen in early 2005), continued liberalization of markets, increasing demand for high-capacity broadband networks, and investor interest in providing access to rural areas. On the other hand, internal sources of financing and domestic private financing are likely to increase rapidly in many developing countries, reducing the need for FDI.

- **Profile of FDI investors.** Will the share of South-South telecommunications FDI continue to grow? Are there likely to be new Southern investors, such as those from the Gulf region who emerged in 2004–5? Will investment from China increase as its government promotes outward FDI by offering loans on preferential terms, tax rebates, and investment insurance? Will private equity investors increase their appetite for developing-country telecommunications investments? Are new technology developments likely to change investment requirements and lead to a different investor profile and destination of FDI? The fast-rising voice-over IP technology could prove disruptive in this respect.

M&As in the telecommunications sector of developing countries are beginning to increase. Privatized fixed-line operators as well as new mobile operators are merging or being taken over by other companies, domestic or foreign. The takeover of African mobile operator Celtel by MTC of Kuwait, the sale by Orascom of many of its mobile operations in Africa and the Middle East, and the acquisition spree of Telefónica, Telmex, and América Móvil in Latin America are all cases in point. Will a group of global or regional telecommunications operators emerge and take control of much of the global industry?

- **Asian impetus.** Will the rivalry between China and India boost global FDI further? In particular, will FDI be boosted by China's WTO commitments and India's ambitious teledensity targets? What will result from India's decision to raise the FDI ceiling in that country from 49 percent to 74 percent? China's reliance on competition among public operators and domestic financing has been successful, but it is idiosyncratic. Is this approach, which is already evolving as a result of China's recent accession to the WTO, sustainable? To what extent will China open its telecommunications sector to foreign investment?

- **WTO agenda.** Will telecommunications commitments and related sector reform by WTO accession candidates—including Algeria, Ethiopia, Iran, Russia, Saudi Arabia, and Vietnam—create new opportunities for FDI? Will the WTO Doha Development Round succeed and include a deepening of commitments in the telecommunications sector?

- **Regional integration.** Will small markets (such as those of African countries and Caribbean and Pacific islands) garner the political will to pursue economies of scale by integrating their telecommunications markets, bringing about a new wave of telecommunications FDI?

It is difficult to foresee how these issues will evolve since over the past two decades, the nature of both investors and investment has changed continuously. However, it is almost certain that, given the increasing trend toward globalization, FDI in telecommunications will continue. FDI has had a tremendous impact on building telecommunications networks in the developing world, and it will remain an essential ingredient for reducing the digital divide.

Annex 2A

Table 2A Data Sources of FDI in Telecommunications

Organization	Database	What it provides	How to access it
1. Business Monitor International	Business Monitor Online	This is a database of foreign companies and their investments in the telecommunications industry, along with detailed, periodic reports of the sector structure and performance within each country.	http://www.businessmonitor.com/ (registration required)
2. Informa Telecoms & Media	World Cellular Networks Datapack	This database provides information on cellular market share and penetration; operators, investors, and infrastructure; and technologies and licenses by country.	http://www.emc-database.com/ (registration required).
3. European Union (EU) Eurostat	EU Direct Investment	This database covers FDI data in EU member countries.	http://epp.eurostat.cec.eu.int/portal/page?_pageid=1073,1135280&_dad=portal&_schema=PORTAL&p_product_code=FDI_YBK
4. International Finance Corporation (IFC)	Investments in Telecommunications Projects	This database provides data on IFC investments in telecommunications projects worldwide.	http://ifcln001.worldbank.org/IFCExt/spiwebsite1.nsf/USUSSearch?openform

Table 2A *continued*

Organization	Database	What it provides	How to access it
5. International Monetary Fund (IMF)	International Financial Statistics Yearbook, or Balance of Payments Statistics Yearbook (BOPSY)	Embedded within Balance of Payment statistics, FDI statistics from IMF do not include geographic or industrial sector breakdown.	http://www.imf.org/external/pubs/cat/longres.cfm?sk=17592.0
6. International Telecommunication Union (ITU)	Regulatory and Competition Database	This database provides privatization data of telecommunications operators.	http://www.itu.int/ITU-D/treg/profiles/guide.asp?lang=en
7. International Telecommunication Union (ITU)	World Telecommunications Indicators Database	This provides data on telecommunications investments referring to the expenditure associated with acquiring the ownership of telecommunications infrastructure. These data include expenditure on initial installations and on additions to existing installations. The database does not have a breakdown of public and private investments.	http://www.itu.int/ITUD/ict/publications/world/world.html
8. Organisation for Economic Co-operation and Development (OECD)	International Direct Investment	This database provides inflow and outflow data in the telecommunications sector to and from OECD countries. It covers OECD countries only.	http://lysander.sourceoecd.org/vl=10914548/cl=73/nw=1/rpsv/statistic/s14_about.htm?jnlissn=16081080
9. World Bank	Private Participation in Infrastructure (PPI) Project Database	This database provides sector-specific data for private participation in infrastructure in developing countries. This is the main data source used in this report to approximate FDI in the telecommunications sector.	http://ppi.worldbank.org
10. United Nations Conference on Trade and Development (UNCTAD)	Foreign Direct Investment Database	Telecommunications FDI data are grouped with transport services.	http://www.unctad.org/Templates/Page.asp?intItemID=1923&lang=1

Source: Authors' compilation.

Annex 2B

Table 2B The 30 Largest Telecommunications Multinational Corporations in the World, 2002
(amounts in US$ billions)

Rank	Corporation	Home economy	Assets: Foreign	Assets: Total	Sales: Foreign	Sales: Total	TNI (%)[a]	Host countries[b]
1	France Télécom	France	73	112	20	49	50	42
2	Telekom Italia	Italy	..	85	6	30	20	41
3	Deutsche Telekom	Germany	37	121	17	51	32	28
4	AT&T	United States	..	55	2	38	4	28
5	Cable & Wireless	United Kingdom	4	26	5	8	37	27
6	Teliasonera AB	Sweden	18	24	1	7	75	26
7	BT Group PLC	United Kingdom	2	43	2	29	8	26
8	Tele2 AB	Sweden	5	6	4	5	77	23

(*Table continues on the following page.*)

Table 2B *continued*

Rank	Corporation	Home economy	Assets Foreign	Assets Total	Sales Foreign	Sales Total	TNI (%)[a]	Host countries[b]
9	Telefónica SA	Spain	36	71	11	27	50	19
10	NTT	Japan	..	158		99		19
11	Vodafone Group PLC	United Kingdom	207	233	34	42	85	19
12	KDDI Corp.	Japan	..	24	..	21	..	19
13	Colt Telecom Group PLC	United Kingdom	2	4	1	2	57	13
14	SBC Communications Inc.	United States	..	95	..	43	..	12
15	Datatec Limited	South Africa		1	2	2	96	12
16	Verizon Communications Inc.	United States	14	167	3	67	7	11
17	TDC A/S	Denmark	..	13	4	7	55	10
18	BellSouth Corp.	United States	3	49	2	22	8	10
19	KPN	Netherlands	..	26	3	12	24	9
20	Level 3 Communications Inc.	United States	1	9	1	3	16	9
21	Cellstar Corp.	United States	..	..	2	2	75	9
22	Swisscom R	Switzerland	2	12	3	11	23	8
23	Iberdrola	Spain	4	24	1	10	12	6
24	Sing Tel Ltd.	Singapore	16	19	3	6	62	6
25	América Móvil	Mexico	2	11	2	6	31	6
26	MTN Group Ltd.	South Africa	1	4	1	2	36	6
27	NTL Inc.	United States	13	13	3	3	100	5
28	Telstra Corporation	Australia	3	21	1	11	12	4
29	Telekom Malaysia Berhad	Malaysia	..	9	..	3	..	4
30	BCE Inc.	Canada	1	39	1	20	3	3

Note: .. Not available.

a. The Transnationality Index (TNI) is calculated as the average of the following three ratios: foreign assets to total assets, foreign sales to total sales, and foreign employment to total employment.

b. Presence of multinational corporations in host countries includes subsidiary enterprises, associate enterprises, and branches.

Source: UNCTAD 2004.

Endnotes

1. Notable exceptions to the public ownership model in the sector were the United States and the Philippines, where the sector was always private, as well as a number of Commonwealth countries where services were provided by Cable and Wireless, a U.K. company nationalized in 1946–7 and re-privatized following a 1981 law. The monopoly model was, however, near universal by the early 1980s, even though in the late nineteenth and early twentieth century competition had prevailed in many countries (Guislain 1997).
2. The total FDI figure is based on UNCTAD 2004. The total telecommunications investment figure is based on ITU 2004.
3. The breakdown of countries among regions used in this report follows the World Bank's classification. See http://www.worldbank.org/data/countryclass/classgroups.htm for details. Differences in absolute levels of regional FDI need to be read in light of the different size, population, and wealth of these regions.
4. According to Estache (2005), the percentage of countries with significant private financing in the telecommunications sector in 2002 by region was: Latin America and the Caribbean (67 percent of countries), Europe and Central Asia (58 percent), South Asia (50 percent), Sub-Saharan Africa (41 percent), East Asia and Pacific (38 percent), and, last, the Middle East and North Africa (23 percent).
5. The seven middle-income countries that did not receive FDI in telecommunications between 1990 and 2003 were Bosnia and

Herzegovina, Djibouti, the Federated States of Micronesia, the Marshall Islands, Suriname, Syrian Arab Republic, and Turkmenistan.

6. In 1988, Chile sold 49 percent of shares in incumbent local operator CTC to foreign investors for $270 million, and 45 percent of ENTEL, the long-distance operator, to a combination of Telefónica de Espana, Chase Manhattan Bank, employees, and pension funds.

7. Divestitures in East Asia and Pacific have commonly been carried out through public offerings of minority stakes on local or regional stock exchanges, with the government keeping majority control of the companies.

8. There were seven projects run as concessions in Indonesia, Lao People's Democratic Republic, and Vietnam during 1995–6, totaling about $2.8 billion. Thailand also used the concession model, but these transactions were not captured by the PPI Project Database.

9. These joint operations with foreign investors allowed Indonesia to almost double the number of fixed lines between 1996 and 2000.

10. Other examples of concessions in the telecommunications sector awarded to foreign investors include Turkey (later transformed into regular licenses after payment of a fee) and Lebanon. In the latter case, however, disputes between government and operators emerged, leading to the termination of the concessions and international arbitration (see box 2.7).

11. At the country level, in India, for instance, an estimated $60 billion would be needed for investment in the telecommunications sector from 2003 to 2010 to achieve the government's goal of a teledensity of 15 phones per 100 people.

12. On the other hand, in a few countries, such as Ghana, privatized incumbent operators are being taken over by governments as a result of failed privatizations, leading at least temporarily to foreign disinvestment.

13. BellSouth sold its assets to Telefónica in 2004 in order to finance Cingular's acquisition of AT&T Wireless.

14. ZTE, a Chinese vendor, runs a joint venture mobile operation in the Republic of Congo with the local operator and bought a 51 percent stake in Niger Telecommunications when the company was privatized. Distacom of Hong Kong became the strategic investor in Telecom Malagasy (Telma) in Madagascar, paying $12.6 million for a 68 percent stake and committing $165 million in additional investments over five years.

15. Another fund-raising channel is the public offering, with a growing number of telecommunications companies listing on stock exchanges, domestic as well as international.

16. This trend is taking place in developed countries as well. In December 2004, Cinven, a private equity firm, bought a 50.01 percent majority stake in the cable operating subsidiaries of France Télécom and Vivendi Universal for $716 million (AltAssets, accessed on September 16, 2005)

17. Telecommunications investors from the South involve a diverse range of companies from big and aggressive operators, such as América Móvil, Orascom, or MTN, to smaller ones such as Econet of Zimbabwe.

18. The aforementioned survey (see Ure 2004) in Asia included 10 risk factors: rate of return, scale of investment, direct control, country risk, currency risk, regulatory risk, other insurable risks, non-insurable risks, local taxation laws, and other issues.

19. Investors tend to emphasize consistency over transparency. Local partners are often in charge of dealing with local authorities, and choosing the right local partners who are influential but also commercially sound is of paramount importance.

20. A country's national sovereignty allows it to adopt new legislation and regulations or amend to existing legal texts, subject to constitutional and other legal requirements. Such normative changes can be made even if the government entered into a contractual commitment with an investor not to modify the legal environment that applies to the investment; in such cases, the investor may be entitled to compensation, based on the terms of the contract.

21. In most instances, the profits of incumbent monopolists have increased after the introduction of competition; although this may at first appear contradictory, the benefits of competition tend to outweigh the inefficiency of monopoly, even at company level. Indeed, the incumbent is forced to improve service delivery and cut costs to withstand new entry, while benefiting from the tremendous power of incumbency. Moreover, Wallsten (2003) found that exclusivity periods granted at privatization are correlated with a significant decrease in the level of the incumbent's investments.

References

AltAssets. "Private Equity News and Research." http://www.altassets.com/index.php (accessed on September 16, 2005).

América Móvil. http://www.americamovil.com (accessed on September 16, 2005).

Business Monitor Online. 2004. http://www.businessmonitor.com (accessed on September 16, 2005).

Estache, Antonio. 2005. "PPI Partnerships versus PPI Divorces in Developing Countries (Or Are We Switching from PPPI to PPDI?)" Policy Research Working Paper 3470, World Bank, Washington, DC.

Fay, Marianne, and Tito Yepes Delgado. 2003. "Investing in Infrastructure: What Is Needed from 2000 to 2010." Policy Research Working Paper 3102, World Bank, Washington, DC.

Guislain, Pierre. 1997. *The Privatization Challenge: A Strategic, Legal, and Institutional Analysis of International Experience.* Washington, DC: World Bank.

Guislain, Pierre, and Michel Kerf. 1995. "Concessions: The Way to Privatize Infrastructure Sector Monopolies." Finance and Private Development Note No. 59, World Bank, Washington, DC.

Gupta, Pankaj, Ranjit Lamech, Farida Mazhar, and Joseph Wright. 2002. "Mitigating Regulatory Risk for Distribution Privatization: The World Bank Partial Risk Guarantee." Energy & Mining Sector Board Discussion Paper Series No. 5, World Bank, Washington, DC.

Harris, Clive. 2003. "Private Participation in Infrastructure in Developing Countries: Trends, Impacts, and Policy Lessons." Working Paper 5, World Bank, Washington, DC.

ITU (International Telecommunication Union). 2004. World Telecommunications Indicators Database. ITU, Geneva. http://www.itu.int/ITU-D/ict/publications/ world/world.html.

———. 2005. Regulatory and Competition Database. ITU, Geneva. http://www.itu.int/ITUD/treg/profiles/guide.asp?lang=en (accessed on September 16, 2005).

INVEST-SD. 2004. *Investment Law and Policy Weekly News Bulletin* April 13. International Institute for Sustainable Development (IISD). http://www.iisd.org/investment.

Izaguirre, Ada Karina. 2005. "Private Telecommunications Projects: Private Activity Down 15 Percent in 2003." Viewpoint Note 288, World Bank, Washington, DC.

Kirkpatrick, Colin, David Parker, and Yin-Fang Zhang. 2004. "Foreign Direct Investment in Infrastructure in Developing Countries: Does Regulation Make a Difference?" Paper presented at the Asian Development Bank Institute (ADBI) Annual Conference, Tokyo, Japan, December 6.

Middle East and Africa Wireless Analyst. 2004. *The International Business Newsletter of the Middle East and Africa.* August 4, 2–4.

Palmade, Vincent. 2004. "FDI Trends: Foreign Direct Investment Flows to Top 20 Developing-Country Destinations." Unpublished, World Bank, Washington, DC.

Palmade, Vincent, and Andrea Anayiotas. 2004. "FDI Trends." Viewpoint Note 273, World Bank, Washington, DC.

Procter, Dean, and Simon-Pierre Olivier. 2002. "Capital Flows and Cost of Capital: The Importance of Liberalized Investment Rules for a Competitive Telecommunications Sector." Paper presented at the conference "Switching to High Growth: Issues in Policy and Regulation in the India Telecommunications Sector," New Delhi, October 19.

Qiang, Christine Zhen-Wei, and Pierre Guislain. 2003. "Foreign Direct Investments in the Telecommunications Sector." In *ICT for Development: A World Bank Perspective,* ed. Pierre Guislain, 28–33. Washington, DC: World Bank.

UN (United Nations). 2004. *The Report of the Task Force on Financial Mechanisms for ICT for Development: A Review of Trends and an Analysis of Gaps and Promising Practices.* New York: United Nations.

UNCTAD. 2004. *World Investment Report 2004: The Shift to Services.* New York and Geneva: United Nations.

Ure, John. 2004. "Infrastructure in East Asia and the Pacific: The Way Forward." Unpublished, World Bank, Washington, DC.

Wallsten, Scott. 2002. "Does Sequencing Matter? Regulation and Privatization in Telecommunications Reforms." Unpublished, World Bank, Washington, DC.

———. 2003. "Returning to Victorian Competition, Ownership, and Regulation: An Empirical Study of European Telecommunications at the Turn of the 20th Century." Paper presented at the 2003 Annual Bank Conference on Development Economics, Washington, DC, April 10.

Wellenius, Björn, Juan Galarza, and Boutheina Guermazi. 2005. "Telecommunications and the WTO: The Case of Mexico." Mimeo, World Bank, Washington, DC.

World Bank. 2005a. "Financing Telecommunications Needs in the Developing World." Unpublished, World Bank, Washington, DC.

———. 2005b. "The Impact of Reform on Telecommunications Prices and Services in the Countries of the OECS." Unpublished, World Bank, Washington, DC.

World Bank and ITU. 2004. "Dispute Resolution in the Telecommunications Sector: Current Practices and Future Directions." Unpublished, World Bank and ITU, Washington DC and Geneva.

Chapter 3

Extending Communication and Information Services: Principles and Practical Solutions

Björn Wellenius

Access to communication and information services exhibits major inequalities among and within countries. These disparities matter insofar as they constrain economic development and poverty reduction. This chapter highlights five principles that guide good practices for extending these services in changing technological and market environments:

- The first step is to allow markets to work. Competitive, private-led markets go a long way toward making communication and information services available to the entire population. Reforms in the telecommunications sector open the way.
- Extending services beyond their current reach in a reformed market environment starts with removing remaining obstacles. Regulatory frameworks may need to be revised to accommodate new technologies and business models.
- Public sector support can help extend services beyond the market, but this must be justified in terms of the economy as a whole. A case can often be made for narrowing gaps in established markets, but justifying public sector support for developing new markets is less straightforward.
- When justified in terms of the economy as a whole, the public sector can stimulate demand or jump-start supply in order to develop sustainable markets for the private provision of the desired services. Cash subsidy is the preferred instrument for jump-starting supply, but capital contributions and guarantees may be appropriate in some circumstances.
- Supply subsidies are preferably determined and allocated through the market. Competition among firms for subsidies is a proven tool. Success hinges on critical demand and supply factors and on the enabling environment.

What Gaps, and Do They Matter?

Access to and use of communication and information services presents major inequalities among and within countries. These inequalities have been of great concern for policy makers for at least four decades, but historically they tend to surge, decline, and be overtaken by new challenges.[1] In 1980, developing countries had 81 percent of the world's population but only 18 percent of GDP and 17 percent of all telephones. There were more phones in Manhattan than in all of Africa. In developing countries, the main cities typically had 10 times the number of telephones per inhabitant than the rest of the country. Many rural areas, where most of the world's poor people lived, had no service at all. Forging this "missing link" in the development chain became a rallying cry for the international development community (Maitland 1984).

The gap between developing countries as a group and developed countries has narrowed considerably. Between 1980 and 2005 the number of telephones (fixed and mobile) in developing countries has multiplied 40-fold (while population grew by one-half and real GDP more than doubled). Their share in the world's stock of telephones more than tripled. These countries now have six times as many phones (fixed and mobile) per dollar of GNP than high-income countries, compared with 1980 when both had roughly similar ratios (table 3.1).

Significant differences remain, nonetheless, among and within developing countries. Fast growth in large emerging markets—notably China, India, and Brazil—masks slower development in other economies. Progress has been made reaching out to rural areas and the urban poor, but in many countries these groups still lag in relative terms. Moreover, in the meantime more advanced communication and information services have become available through the Internet, but initially they are reaching mainly the better-off population groups. As services that require connecting to separate networks are becoming available through a single network, the Internet, broadband access to this network is now also at issue.[2] Bridging this new gap—the "digital divide"—among and within countries is a major theme of the World Summit on the Information Society (WSIS 2003).

These disparities matter insofar as they constrain economic development and poverty alleviation. Research in the 1960s and 1970s documented the importance of telecommunications for economic production and distribution, delivery of public services, and government administration (Saunders, Warford, and Wellenius 1983). In the 1980s information came to be recognized as a fundamental factor of production, along with capital and labor. During the 1990s globalization and the increasing information intensity of economic activity, coupled with rapid technological changes and demand growth, turned information and communication into critical factors of competitiveness and growth.

By the early 2000s a broad consensus had been reached among policy makers worldwide that information and communication provide key inputs for economic development, contribute to global integration while helping retain the identity of traditional societies, and enhance the effectiveness, efficiency, and transparency of the public

Table 3.1 World Population, GDP, and Telephones

Indicator	1980	1990	2000	2005[a]
Low- and middle-income countries				
Population (billions)	3.6	4.4	5.1	5.4
GDP (trillion US$)[b]	3.1	4.2	5.9	7.5
Fixed telephone lines (millions)	52	117	422	727
Mobile telephones (millions)	0	0.4	235	1,393
Total telephones, fixed + mobile (millions)	52	117	657	2,120
Total telephones (per US$ millions GDP)	17	28	111	283
Total telephones (per 1,000 people)	14	27	129	393
Total telephones (as percentage of world total)	17	22	38	61
High-income countries				
Population (billions)	0.84	0.90	0.97	1.00
GDP (trillion US$)[b]	14.5	19.7	25.6	28.5
Fixed main telephone lines (millions)	260	403	561	541
Mobile telephones (millions)	0	11	505	818
Total telephones, fixed + mobile (millions)	260	414	1,066	1,359
Total telephones (per US$ millions GDP)	18	21	42	48
Total telephones (per 1,000 people)	310	460	1,080	1,280
Total telephones (as percentage of world total)	83	78	62	39

a. Forecasts: population and GDP are World Bank projections, fixed and mobile telephones are ITU projections.
b. In constant 2000 U.S. dollars. A trillion is 1 million million (10^{12}).

Source: Author's analysis based on data from World Bank 2005 and ITU 2003.

sector (World Bank 2002). Access to information and communication is considered crucial to a sustainable agenda of poverty reduction because it increases the efficiency and global competitiveness of the economy, enables better delivery of health and education services, and creates new sources of income and employment for poor populations (Navas-Sabater, Dymond, and Juntunen 2002). It also enhances political participation and more effective and accountable democratic government (UNECA 1999).

The First Step: Letting the Market Work

The first step in extending services is to let the market work. Competitive, private-led markets go a long way toward making communication and information services available throughout the population. In market economies, responsibility for providing information and communication services rests primarily with the private sector. The main role of the public sector is to provide the policy framework, regulate the market to the extent that it does not work well enough by itself, and support additional service provision where the market alone does not suffice to meet economic and social objectives. The role of the public sector as user of communication and information technologies is also increasingly important, particularly in the context of e-government programs.

Sector Reforms Open the Way

Telecommunications sector reforms enable access to communication and information services. From the late 1980s, developing countries started to replace telecommunications monopolies (mostly state owned) by increasingly open, competitive, private-led market structures. Annual telecommunications investment in these countries increased from about $14 billion in the early 1980s to $30 billion in the early 1990s, and to $95 billion in the late 1990s—of which about one-half came from the private sector (Wellenius, Braga, and Qiang 2000).[3] Following a decline in the early 2000s, telecommunications investment is again rising. Increased investment has resulted in fast development of existing and new services. Annual growth of fixed telephone service, which had hovered around 7 to 10 percent in the 1970s and 1980s, shot up to 16 percent in the second half of the 1990s. Service was extended to many smaller cities and towns; by 2000 about one-third of all households had telephones, and the backlog of new connections dropped from several years to a few months or weeks.[4] Telephone kiosks and shops proliferated in Honduras, Indonesia, Senegal, and other countries. Additionally, in less than 10 years, mobile phones caught up with a century of fixed-telephone development, prepaid plans reached large numbers of formerly excluded low-income users,[5] and short text messaging became common as a low-priced communication option (ITU 2002 and Oestmann 2003). In 2002, 41 percent of Chilean rural households had a mobile phone whereas only 9 percent had a fixed phone (SUBTEL 2003). Fixed and mobile phones, taken together, expanded at 19 percent in the 1990s and 20 percent in the first half of the 2000s. A new wave of innovation is underway following the introduction of the Internet, ushering in new services as well as more cost-effective network solutions.

Competition Is the Cornerstone

Competition is the cornerstone of successful development of communication and information services. Yet about one-half of all low- and middle-income countries still have closed or barely open telecommunications markets (Rossotto, Sekkat, and Varoudakis 2003). Competitive markets grow faster, lower costs, facilitate innovation, and respond better to users' needs. The phone density in the Democratic Republic of Congo, with six competing mobile operators, is 13 times higher than in Ethiopia, which has a similar income per capita but only a single operator (*The Economist* 2005c; ITU 2004).[6] Limited competition in international services results in prices about twice those in countries with full competition (Rossotto et al. 2004). Among 30 African and Latin American countries that undertook reforms in the 1980s and 1990s, those that opened major market segments to competition before privatizing the incumbent or at the same time grew and reduced costs faster than those that privatized first and introduced competition later (Wallsten 1999).[7]

Similar conclusions result from an analysis of 86 developing countries worldwide (Fink, Mattoo, and Rathindran 2002). The lengthy exclusivity periods initially granted in Latin America at the time of privatizations in the early 1990s resulted in high prices for domestic and international telephone service and leased circuits; the adverse effects of exclusivity carried over to the emerging Internet service in those countries well into the late 1990s, even after competition was introduced (Wellenius 2000).

The Internet has taken off much faster in countries where service providers are allowed to build their own networks and gateways. Morocco's emerging information economy was boosted by dramatic telecommunications sector growth after introducing mobile competition in 1999. The country's leading position in the region and overall competitiveness was then compromised by delays in opening the rest of the telecommunications market (World Bank 2004).

Removing Remaining Obstacles

Extending services beyond their reach in reformed markets starts by identifying remaining constraints on the functioning of the markets and addressing them to the fullest extent possible.

Although income is an overriding constraint on demand, people everywhere are willing to spend a significant proportion of their income in communication and information if given the opportunity. Telecommunications typically accounts for 2–5 percent of GDP, averaging 3.4 percent in middle-income countries and 1.8 percent in low-income countries (ITU 2003).[8] The poorest 20 percent of households in Chile, for example, spend over 3 percent of their income on telecommunications, compared with 2 percent on water and 4 percent on electricity (Melo 2000). In Mauritius, the poorest 10 percent of households spend 2.9 percent of their income on communication, compared with 2.4 percent for all households and 2.7 percent for the 20 percent highest-income households; on average, rural households spend 78 percent of the amount spent on communication by urban households (CSO 2005).[9] Peruvians, taking all urban households together, spend about 1.5 percent of their income on telephone service (fixed and mobile); among households with service, the poorest 20 percent spend on telephone services about the same percentage of their income as the wealthiest 20 percent (OSIPTEL 2004).

The challenge is to enable operators to tailor their service offerings and technical choices to tap this revenue potential effectively. By removing market impediments and supplementing this with government support to reach specific goals beyond the market, policy makers can maximize the return to society from public resources and intervention. Opening markets to new entrants—including small local entrepreneurs—rebalancing tariffs, putting in place an effective cost-oriented interconnection regime, securing reasonable access to existing infrastructures, and making radio spectrum available are among key regulatory improvements often needed to achieve this result (see box 3.1).

Obstacles to well-functioning markets often remain even after economic and sector reforms have progressed quite far. In Peru, for example, all segments of the telecommunications market have been open since 1995, but telephone service in provincial towns and marginal areas of big cities remained well below the level achieved in other low- and middle-income countries with comparably open markets. A study of mostly low-income urban areas in 2004 estimated that there was an unmet demand of about 620,000 phone connections. Meeting this demand would increase the telephone penetration outside the capital city by about 40 percent—from 7.5 phones per 100 inhabitants to 10.6 phones per 100 inhabitants. The main factors preventing competitive entry to meet this demand were retail pricing distortions (including high fixed charges), outdated regulatory structure for retail pricing, uneconomic definition of local calling areas, and regulated call charges that discouraged investment. Other factors were high interconnection charges and price squeeze, and a lack of cost-based access to physical infrastructure (posts, masts, ducts, and rights of way) held by the incumbent and other utilities. Overcoming these regulatory issues, which also affected the performance of the telecommunications sector overall, would go a significant way toward extending services throughout low-income urban and peri-urban areas at no cost in public funding (OSIPTEL 2004).

Facilitating Innovation

It may be necessary to revisit the regulatory framework from the viewpoint of its ability to support innovation. Some regulatory provisions, although appropriate for service supply by large companies using traditional technologies, now may stand in the way of new technologies and decentralized supply through a wider range of business models. The trend is toward replacing individual licensing of operators by a regime of general authorization, and making more spectrum available for shared and license-exempt use. New fixed wireless technologies are resulting in innovative business models and hold the promise of extending competition to all market segments, accelerating development of broadband infrastructures, and connecting the rural and urban poor. At the heart of these new technologies is the application of advanced digital processing techniques, which

Box 3.1 Regulatory Measures Help the Market Work Better

Open entry. Removing restrictions on entry accelerates expansion to unserved areas. New entrants may be better placed than incumbents to provide cost-effective service and bring additional capital and management resources. They are often also smaller and nimbler than incumbents, faster to adopt new lower-cost technologies, more focused on customer needs and marketing, better attuned to local conditions and business opportunities, and better able to assess and deal with local risks. Replacing individual licensing by a regime of general authorization facilitates entry, reduces the regulatory burden on companies and authorities, and reduces the opportunity for corruption.

Cost-oriented pricing. Retail prices must reflect cost differences among services and across the country. Retail tariffs may need to be rebalanced, raising fixed and local call charges as monopoly rents (especially in international services) are competed away. Temporary measures, however, may be needed to support low-income households during the transition to fully commercial prices. Geographically averaged prices discourage investment in high-cost areas and lead to loss of market share in contested low-cost areas.

Cost-oriented interconnection. An effective interconnection regime is essential for competition to develop. Interconnection charges should recognize that calls to and from high-cost areas cost more than those between low-cost areas. High-cost areas should retain a larger-than-average share of revenues generated by outgoing calls and receive a larger-than-average payment for completing incoming calls. Failure to properly pay for the termination of incoming calls undermines the business case for reaching out to unserved areas.

Access to infrastructure. New entrants must have reasonable access to the network infrastructures of incumbent operators. Unbundling the local loop helps develop competition in new services, especially broadband Internet. Sharing physical infrastructure and co-locating equipment of different operators lowers barriers to competitive entry, increases revenues of incumbents, and contains environmental damage and public inconvenience. Domestic mobile roaming is necessary for customers of operators with limited networks to have service countrywide.

Access to the radio spectrum. Making spectrum available promptly and at low cost encourages the use of new wireless technologies. Specific measures include increasing the allocation of spectrum for license-exempt use and for mobile service. Spectrum is a prime resource that can be used to accelerate network development and reach unserved areas and population groups at lower cost than conventional radio and cable solutions. Spectrum should not be rationed by artificially high fees or by restricting availability below demand.

Source: Author's analysis.

enable users to share spectrum without causing insurmountable interference to each other (*The Economist* 2004).

In the context of improved regulatory frameworks, cost-effective strategies can be devised to roll out public broadband access to the Internet. These strategies would be aimed at meeting both private user needs and the requirements of institutional programs, such as e-government and computers in schools. For example, using wireless technologies such as WiFi and WiMAX, a single broadband Internet point in a village or even an entire region could potentially be sufficient to provide access for numerous prospective users, both institutional and private. Thus the business case for creating a commercially viable rural Internet service, in conjunction with a public institutional network strategy, may be increasingly realistic. Franchising services could mobilize local entrepreneurship while providing the benefits of a larger organization that contributes technical assistance, training, quality standards, name branding, payment vehicles, bulk procurement, and opportunity to share experiences.

Other Constraints

Markets may also be constrained by factors that are largely outside the control of the sectoral authorities. Governments often extract huge license fees from operators, which raises the cost of doing business, creates disincentives for entry, and even drives active competitors out of the market. The cumulative burden of uncoordinated taxation often results in high costs that suppress demand and discourage investment.

Telecommunications operators and users in most African countries, for example, are subject to a wide range of taxes, including excise, value-added, and import taxes; local taxes; and in-kind contributions. These are imposed and changed at various times by government departments (such as finance), local authorities (such as municipalities), and the sector regulator (through license and spectrum fees) with little coordination or regard for the impact on business viability, service provision, and use (Stoeder 2005). As the cost of manufacturing cell phones continues to decline, taxes and duties levied by governments on importing, selling, and using these devices is becoming the binding constraint on extending communication and information services among the poor (*The Economist* 2005b).

Consistent, predictable, and transparent rules of the game are essential. Some of the main obstacles experienced by operating companies in Sub-Saharan Africa relate to uncertainty on government policy objectives, business incentives (such as early termination of tax holidays), terms of engagement (such as license renewals and new service obligations), fees (such as license fees, spectrum user charges, and contributions to universal service funding), and market structures (such as unclear timetable and rules for increased competition). These uncertainties are compounded—in other regions of the world as well as Sub-Saharan Africa—by lack of transparency of the regulatory process, unclear procedures for decisions, and uneven treatment of competitors. Weak or nonexistent competition law and authority place the burden of establishing and enforcing competitive safeguards on the emerging telecommunications regulator. Financial market failures, such as lack of long-term debt financing in local currency needed by small and medium operators, restrict market entry.

Extending Access Beyond the Market

Even in well-performing markets, gaps typically remain between what service providers are prepared to do solely on commercial grounds and what governments consider necessary from broader development perspectives. Many rural areas remain without service because of high costs or low revenue potential. Service packages are sometimes mandated by the authorities or undertaken voluntarily by the operating companies to keep services affordable for low-income urban households as tariffs are rebalanced, and these packages may lose money.[10] Incomplete infrastructure networks (roads and electricity as well as telecommunications) can prevent service providers from expanding service coverage and local entrepreneurs from entering the market. Externalities, such as the benefits to a whole village resulting from Internet use by only a few people, are not reflected in the service provider's operating revenues, and rollout of new services is slower than would be optimal from the viewpoint of the whole economy.

Public Sector Support

Public sector resources can be used to narrow critical gaps that remain in the long run despite taking the necessary steps to make the market work well. Striking the right balance between enabling the market and providing public sector support can, however, be quite difficult. Even with the best policies, it may take years for the market to play out. Although it may be hard to foresee with precision how far the market will ultimately go, that should not rule out some market-supporting mechanisms from being introduced even as market efficiency measures are given precedence.

The decision to extend a service beyond the market, however, must be supported by thorough examination from the viewpoint of the whole economy. Government intervention in the market is essentially a political decision beyond the narrow province of telecommunications policy and regulation, for it involves trading off different development goals that accrue across society and impose conflicting demands on limited public resources. But not all unprofitable services deserve public support. Several questions should be answered before the government intervenes:

- Are the benefits to the economy likely to more than offset the costs?
- Will the services effectively reach and be used by the target population?
- Are the support instruments well suited to address specific obstacles?
- Are they consistent with overall economic policies?

- Is the cost in public resources kept at a minimum for given results?
- Is the amount of public resources to be used for this purpose justified?

Economic analysis provides a framework for thinking through (and to some extent quantifying) key aspects of public sector support. The costs of providing the additional services are compared with the benefits of so doing, from the viewpoints of both the companies providing the service and the economy at large. The analysis from the company's viewpoint indicates how much support would be needed to render the provision of the additional services commercially viable. The analysis from the economy's viewpoint indicates whether the benefits to society at large (not only to the immediate participants) equal or exceed the cost of the resources used. Only projects that need support to become viable *and* yield benefits to the economy at least equal to their costs should be eligible for public sector support. In practice, public sector budgetary constraints further limit the extent and pace at which this set of projects can be supported. Cost-benefit analysis points to priorities among the different projects as well as choices between the projects and alternative uses of public resources in other sectors (Belli et al. 2001). Estimating demand by end users, government programs, and nongovernment organizations lies at the center of economic justification, which must assess the risk that these demands may not materialize and examine options to contain these risks.

Public Sector Support for Narrowing Gaps in Established Markets

Public sector support to narrow gaps in established markets is often justified. A strong case can be made for supporting a service that (1) has been taken up on commercial terms by a substantial part of the country's population to which the service is available (that is, the value of the service has been demonstrated though the market) *and* (2) people deprived of the service are deemed to suffer serious economic or social disadvantage. This is the classic case for universal service programs. Fixed and mobile voice communication have by now passed the market test in many countries. The economic and social effects of the fixed telephone have been well known for decades. Low- and middle-income countries have focused on supporting the provision of communal facilities, especially public telephones in villages and low-income urban areas lacking service,[11] while high-income countries have supported the provision of telephone service in every household.

There is a compelling argument for extending access to voice communication beyond the market in developing countries. In recent years, mobile phones have served this purpose exceptionally well. Where available, growth rates have been phenomenal. On aggregate there are now more mobile than fixed phones, and about 70 percent of the developing world's population (over 50 percent in Sub-Saharan Africa) live within the footprint of mobile phone service.[12] Mobile phones both substitute for scarce fixed connections and add mobility. The economic impact is dramatic, especially in terms of reducing transaction costs, broadening trade networks, and facilitating the search for work, and it is much larger in low- and middle-income countries than in high-income countries (Waverman, Meschi, and Fuss 2005). Prepaid service is taken up massively by otherwise excluded low-income households. Mobile phones are widely shared and rented out by the call by local entrepreneurs, serving as *de facto* public telephones. In the villages of Bangladesh, which are among the poorest in the world, women entrepreneurs provide payphone service at a profit using mobile phones (Lawson and Meyenn 2000). Farmers and fishermen use mobile phones to figure out where they can get the best prices for their products, small businesses shop around for supplies, and in some countries (such as Zambia) they can make cashless payments (*The Economist* 2005a). Operators and users creatively exploit features of mobile technology to implement pro-poor approaches.[13] Whether extending access to mobile services requires public sector support or merely needs the remaining obstacles to the market to be removed is, however, a matter to be assessed in each situation.

Public Sector Support for Launching New Markets

Justifying the use of public sector resources to help develop new markets is less straightforward. There are strong arguments, and a growing body of empirical evidence, in favor of early adoption of advanced communication and information services as tools of economic development and poverty reduction. These advanced services, however, have not yet passed the market test in any developing country.[14] Moreover, these services cut a new divide between people who have use for them and those who do not (Kenny

2003). Use is constrained by income, education, and occupation—that is, by poverty itself—as well as by social and cultural factors (including gender bias) and relevance of the information content that can be accessed. Initial deployment of the Internet thus arguably increases inequality, although in the longer run the benefits may trickle down and all groups gain. Success is likely to depend on a number of demand and supply factors that are subject to great uncertainty and difficult to coordinate.

There is thus a growing contrarian view that government support should be limited to services for which demand can materialize quickly once service is made available, as is demonstrably the case of mobile phones, and let the Internet and related services advance at their own pace through the market (Fink and Kenny 2003; *The Economist* 2005a). The policy options are less clear cut, however, since some services in high demand among poor people, such as live television, may generate significant revenues but require broadband networks (Townsend 2002). New solutions to deliver old services at more affordable prices, such as voice over Internet protocol (VoIP), also require broadband networks. And people living on one dollar a day may be unable to help themselves to the Internet but can benefit from advanced services intermediated through radio and television, human facilitation, and appropriate interfaces.

Providing Public Sector Support

Public sector support to extend communication and information services seeks to develop sustainable markets for the private provision of the desired services. Support is designed to turn socially desirable investments that are not profitable by themselves into commercially viable undertakings by stimulating demand, jump-starting supply, or both. Projects that are not demonstrably good for society at large, or that are unlikely to ever stand on their own generally, should not be supported. This principle applies broadly to infrastructure services (such as electricity, water supply and sanitation, and transport, as well as telecommunications). Services that exhibit large externalities and public good properties (such as education, health, and legal systems) are normally supported on a recurrent basis.[15] The principle reflects a position on economic policy as much as it reflects the practical observation that projects that must rely on continued support are unduly vulnerable to the availability of public funds or aid.

Stimulating Demand

Demand can be stimulated by making established services more affordable and new services more relevant. In some countries (for example, Ireland and Kazakhstan), government agencies help eligible households (such as low-income, elderly, or disabled persons living alone) pay for phone service together with other basic services, such as rent, food, or electricity. In the United States, households that qualify under low-income support programs receive discounts on their telephone bills; the discount is paid to the companies providing the service. Demand can also be stimulated by supplying selected services at prices below commercial levels. In Chile, for instance, rural public phone calls are priced below full cost recovery, and initial investment is subsidized to offset the suppliers' losses. In the United Kingdom, optional lifeline phone service aimed at low-income households offered, for a low fixed quarterly price, unlimited incoming and emergency calls and flexible payment plans for additional outgoing calls.[16] Improving customer service and accelerating take-up of computers can also help. Telephone customers in Egypt dial up their choice of Internet service provider (ISP), paying only the price of the phone call without having to subscribe to the ISP. A personal computer and one year of Internet access are available to residential telephone customers in Morocco for less than retail prices. Ensuring access to relevant content is a key success factor in rolling out new services.

Governments can stimulate demand for new services—the Internet in particular—by launching applications of interest to the public, such as online tax filing or access to birth certificates and land records. Aggregating the demand of various government branches and agencies and inviting competitive tenders rather than each branch developing its own can be a powerful driver of private rollout of new services. Government commitment to purchase broadband capacity can be a practical tool to reduce commercial risk if investors are asked to build facilities before their value has been fully tested in the market. The procurement choices that national and local authorities make, as well as the information content they develop and make available to the public through these technologies, can have a strong effect on the viability of competition and opportunities for new entrants, including small entrepreneurs.

Jump-Starting Supply

Cash subsidy is a proven public sector instrument for jump-starting supply. Cash subsidies can be effective to catalyze additional private investment in several ways. They can help service suppliers overcome entry barriers that result from front-loaded investment requirements or financial market failures, offset temporary political constraints on prices, or reflect external benefits that otherwise would not contribute to operating revenues.[17] Cash subsidies are transparent, in the sense that they have known costs and are subject to the discipline of public sector budgets. They can also be accurately targeted to the desired beneficiaries. Subsidies are paid as the output objectives are met.[18] In Peru, for example, private operators that committed to install payphones in villages with no telephones received part of the subsidy on the award of their contract, part once the equipment was installed, and the rest in semi-annual installments for several years, contingent on compliance with performance standards (Cannock 2001).

Capital contributions and risk guarantees may also be appropriate in particular circumstances. Public sector investment in private-led businesses with the expectation of future returns on investment can help overcome financial market failures. Because in a market economy the primary responsibility for providing services rests with the private sector, a clear exit strategy for the public sector is essential. Guarantees against risks that are under the government's control can mitigate the impact of political and regulatory uncertainty.[19] In contrast, tax breaks, concessional financing, in-kind grants, and guarantees of risks not under the government's control are less transparent and accurate, and often go against good taxation and other public policies (Irwin 2003).[20]

Good Subsidy Practice

Good subsidy practice commits all participants to contribute to financing the provision of services:

- Service providers invest and risk their own resources to set up the facilities and provide the services during a given time under specified conditions.

- Government subsidies help service providers meet some investment and start-up costs. Subsidies are designed to reduce access barriers—such as initial connection, equipment, or installation charges—to which low-income groups are especially sensitive.

- Customers typically pay for the use of services, at least as much as is needed to meet operating and maintenance costs. Where domestic installations are involved, customers are also required to pay part of the investment cost, as a confirmation of economic demand for service and commitment to pay for service use.

Subsidies must be neutral about competition among service providers, service alternatives, and technologies. Subsidies must also be explicit and transparent, flowing externally to the sources and beneficiaries. Cross-subsidies within operating companies to meet uneconomic service obligations are neither desirable nor readily sustainable in an increasingly competitive market environment.

Subsidies can be channeled through various points in the service supply chain. As the provision of communication and information services is increasingly being separated from the provision of networks carrying these services, subsidies can be applied to either. In Chile, for example, companies received subsidies to extend the public telecommunications network and install and operate public phones in designated villages under agreed terms and conditions (Wellenius 2002). In Bolivia and Nicaragua, subsidies are being provided to extend mobile networks to areas having no telephone service. In Uganda, companies are being subsidized to extend the network and install Internet points of presence in provincial towns; actually providing Internet services to the public is left to competing companies, including local entrepreneurs. A joint venture between several states and private companies is being developed in eastern Africa to bring broadband domestic and international access throughout the region (Guislain et al. 2005).

Financing Subsidies

Specialized telecommunications development funds are often used to collect and disburse subsidies.[21] From the viewpoint of overall economic efficiency and equity, subsidies should be financed from general revenues, as in Chile and Nepal. Subsidies, however, are often financed by levies on telecommunications revenues—for example, 1 percent of gross operating revenues in Peru and Uganda, and 2 percent in the Dominican Republic and Nicaragua (Navas-Sabater 2005).[22] One argument supporting this option is that the net benefits of the subsidies will, in many cases, flow back to the industry contributors through opportunities to compete for these subsidies, increased traffic, and overall growth, and will also result in external benefits to current users. A practical argument is that funding from general revenues is subject to the uncertainties of annual budgetary cycles. Arguably, however, it is inequitable to require users of mature services (mainly fixed and mobile phones), including growing numbers of low-income families, to subsidize early adopters of new technology, who tend to be among the better-off.

Development funds have a mixed record in financing telecommunications subsidies. They have performed well in some countries but in others the monies collected disbursed only slowly, some funds were raided by government treasuries, and private companies are sometimes reluctant to channel money through a public sector agency.[23] Moreover, because mobile services are quickly reaching virtually the whole population, special funds may no longer be needed to close any remaining gaps in access to voice services. Several countries (for example, Bolivia, Chile, Colombia, Peru, and Uganda) have broadened the scope of their funds to help develop more advanced services and facilities.

Competing for Subsidies

Subsidies are preferably determined and allocated through the market. Competition among firms for subsidies—initially developed for rural payphone programs in middle-income countries—is increasingly found in telecommunications in lower-income countries. Competition is also being used for other rural infrastructure services (electricity, water and sanitation, and transport). Competition among firms for subsidies results in lower subsidies, more effective mobilization of private investment, and greater transparency than occurs in traditional public sector funding of these investments (Wellenius, Foster, and Calvo 2004).

Competition among Firms for Subsidies

Competition among firms for subsidies comprises the following main steps:

- The government defines the broad objectives, target population, and levels of funding of the subsidy program. It also establishes key service conditions, such as types of services to be provided, quality standards, maximum retail prices, and duration of service commitments.
- Specific service requirements are primarily identified by prospective beneficiaries and communities. Economic and technical analysis is used to aggregate these service requirements into projects, select and prioritize projects that are likely to be desirable from the viewpoint of the economy at large but not commercially viable on their own, and determine the maximum subsidy justified for each project.[24]
- Private firms submit competitive bids for these projects. Subject to meeting service conditions and complying with rules that apply to all providers, bidders are free to develop their business strategies and their choice of technology.
- Subsidies can be awarded to the bidders that require the lowest amounts. Alternatively, bids can be invited for fixed subsidies and awarded against other quantifiable service measures, such as the lowest price to end users or the fastest service rollout.
- Subsidies are paid in full or in installments that are linked to implementation of investments and provision of service.
- Service providers own the facilities and bear all construction and commercial risks. No additional subsidies are available downstream for the same services.
- The government monitors and enforces service quality and pricing standards, protects users against arbitrary changes of service, and provides investors with stable rules of the game. It also evaluates the effectiveness and impact of the subsidies, and draws lessons that can be applied to subsequent projects.

Success Factors

The success of competition among firms for subsidies hinges on critical demand and supply factors and on the enabling environment:

- **Demand side.** The ultimate constraint on the demand side is the users' willingness and ability to pay at least as much as is needed to keep the service going after initial installation and startup. This probably is the case in even the poorest countries and rural areas as far as sustaining communal voice services, but it limits how quickly higher levels of service may be provided. Government programs, such as e-government and computers in schools, should also pay for the communication and information services needed to reach target population groups. The combined revenues from meeting public and private demands can go a long way toward accelerating rollout of new services.

- **Supply side.** The primary concern on the supply side is to ensure effective competition for the subsidies. Existing operating companies, equipment providers, and new firms set up for this purpose are likely to participate only in countries where the local telecommunications market is sufficiently developed. Creating opportunities for new entrepreneurs to join the market, on a single-community basis in many cases, can promote diversity, decentralization, and broader competition. But effective competition for subsidies may decline as the market consolidates and firms and entrepreneurs establish positions on geographical or market niche bases.

- **Enabling environment.** A clear, stable, and credible legal and regulatory sectoral and general business framework is needed for prospective service providers to make reasonable estimates of costs and revenues and to assess the risks they are being asked to assume. Significant institutional capacity needs to be in place to design and manage the process and funding of the subsidies. The WSIS Task Force on Financial Mechanisms has identified government and regulatory capacity as arguably the single most important constraint on the development of communication and information service (WSIS 2004).

Alternative Modalities of Competition for Subsidies

Competition among firms, as outlined above, is a proven tool to determine and allocate subsidies to carry out predetermined projects. Alternative modalities could be considered as well, however. Other forms of competition for subsidies include competition among projects proposed by communities or firms, competition among regional governments for central funds, and competition among sectors for a share of these funds. Implicit in all modalities is competition among technologies and business models for delivering the services.

Some of these alternatives, such as competition among projects initiated by suppliers or communities, have been used in other infrastructure service sectors (such as electricity) and could be tried for communication and information services. Or bidders could be given more freedom to configure the projects, subject to meeting prescribed aggregate service targets. This would further reduce the extent to which the authorities second-guess the market. It also might be simpler to prepare when a large number of places or smaller operators are involved, and it would allow suppliers greater flexibility to respond to demand and tailor their business strategies. The design of competition for subsidies could also be adapted to support micro enterprises, such as individual local entrepreneurs, perhaps by developing standard screening and financing packages or by applying solutions devised for supporting small and medium enterprises generally.

Conclusion

Public sector support for extending services beyond the market is no substitute for undertaking broad sector reforms and following up by removing remaining impediments to well-working markets. Accelerated rollout of new services and networks should start by cleaning up the regulatory framework to accommodate new technologies and business models. It is probably unrealistic to expect policy makers to wait several years for improved rules to fully work their way through the markets before undertaking public sector support programs. But the onus is on the proponents of such programs to show convincingly that all that is reasonably possible to improve market conditions is already being done. The default should be no public sector support.

That a particular service outcome is desirable from sectoral, development, or political viewpoints is not a sufficient reason to commit public sector resources to its pursuit. There are many such desirable endeavors that seek funding. It is up to the proponents to demonstrate—in terms of the economy as a whole, not only in terms of the telecommunications sector—that the public resources required are worth using to this particular end rather than to meet competing demands. Analytical tools designed and used for many years for similar purposes can be applied consistently. These tools have limitations, but within defined boundaries they provide essential inputs to what is ultimately a political decision.

Efforts to extend services beyond the market must be commensurate with economic development overall. Inadequate access to communication and information is as much a result of poverty as it is a contributing factor. The ultimate constraint on the pace at which it is possible and meaningful to extend these services is the private and public users' ability to use the services productively and their willingness to contribute to paying for them. Primary reliance on the private sector for providing services depends on having in place enlightened and capable public sector players to make the programs work. Strengthening this capacity, especially at the regulatory level, is often necessary.

Endnotes

1. In 1965 the International Telecommunication Union (ITU) established a specialized group (GAS-5) to look into telecommunications in developing countries.
2. Examples of services traditionally provided through separate networks are voice communication through fixed and mobile telephone networks, television and radio through terrestrial and satellite broadcasting and cable networks, and broadband Internet through improved telephone and cable networks and WiFi hotspots.
3. These figures overstate somewhat the volume of new resources made available to the sector, since they include payments for transferring the ownership of existing assets (privatization of state enterprises) and acquiring new licenses.
4. In 2000 about 30 percent of households in developing countries had fixed telephones (author's estimates using ITU data). Today the proportion of households with either fixed or mobile phones is likely to be considerably higher. Although some fixed phones have been replaced by mobiles, and some homes with fixed phones now also have mobile phones, many mobile phones are likely to be in households that earlier had no phones at all. These averages mask major differences by income groups. ITU data for 2002 show that about 8 percent of households in low-income countries had fixed phones, compared with 49 percent and 59 percent in lower- and upper-middle income countries, respectively (ITU 2003). These numbers have only indicative value, since individual country data are often inconsistent and subject to errors.
5. In 2003, about 80 percent of Africa's 34 million mobile phones were prepaid, as were over 90 percent of South Africa's new customers. Low-income users had been largely excluded from conventional fixed and mobile service because they were not creditworthy, were unwilling to commit to fixed payments, or needed more effective control over call expenditure.
6. Based on recent World Development Indicators data. The original story as reported in the references showed the phone density in the Democratic Republic of Congo as being 15 times that in Ethiopia.
7. See also chapter 2 for the role of foreign direct investment in advancing competition.
8. These figures understate willingness to pay because they include countries where demand for services is not being met.
9. Communication comprises postal services as well as telecommunications services and equipment. Data are for 2001–2.
10. Preliminary calculations in Bulgaria and Latvia, for example, suggested in 2000 that tariff rebalancing might cause as many as one-third of all customers to drop their service unless they received some support.
11. The main exception has been Colombia, where a system has been in place for many years to provide telephones in the less affluent urban households cross-subsidized by taxing high-income households and businesses. The system eventually has proven unsustainable and is being replaced by an urban payphones program. Moreover, an increasing number of households are returning their subsidized residential phones and taking up mobile service at fully commercial prices.
12. Actual take-up is much lower, though, as the high price of making calls discourages many potential users.
13. For example, a pre-paid user with low or zero credit available can call and end the call prior to connection, expecting the called party to call back. Credit transfers between subscribers allows sharing of pre-paid cards or reselling credit in smaller amounts. Text messages addressed to special servers can be used to obtain specialized information (for example, send the word "coffee" to a predefined number and receive the price of coffee in the capital city markets).
14. Internet use is growing rapidly, and in some developing countries advanced services might pass the market test in the foreseeable future. In Chile, Jamaica, and Malaysia, for example, it is estimated that between one-fourth and one-third of the populations used the Internet in 2003, albeit mainly through communal facilities such as cyber-cafes (the proportion of connected households is much lower). Advanced services are being taken up quickly in the more developed countries. In the European Union, one-half of individuals and nine out of ten enterprises used the Internet in 2004. The gaps between high- and middle-income EU countries, however, remain substantial. See EU (2005).
15. Some communication and information projects, such as national emergency calling numbers, may be deemed of sufficiently high social value to be supported even if they can never be commercially viable on their own.
16. Prepaid mobile service, however, overtook fixed lifeline options as they were being developed.
17. High initial investment and start-up costs followed by low recurrent costs means that the prices that make investments privately profitable in the early stages of market development can be much higher than the marginal cost of production (or may even not exist), thus leading to investment levels well below socially optimal levels (Serra 2000). One example of a financial sector market failure is the lack of access to long-term investment capital. An example of temporary political constraint on prices is the gradual phasing out of cross-subsidies for low-income telephone customers as prices are rebalanced toward costs. Gains to the whole village as some inhabitants

learn to use new services is an example of benefits that do not accrue fully as operating revenues.

18. Whenever possible, subsidies within a single budget cycle are preferred because they minimize uncertainty about subsidy amounts and availability. When governments commit to provide subsidies over longer periods, they assume liabilities resembling those created by taking on debt, and both the cost of these commitments and the new fiscal risks need to be considered. See Boyle and Irwin (2005).

19. Regulatory risk falls under this category, for example, the risk that the authorities do not allow operators to change prices as prescribed in the license or regulations.

20. Tax breaks include exemptions from import duties. In-kind contributions would include, for example, selling state assets at below-market prices, and giving free use of scarce highly valued spectrum or rights of way. (To the extent that high market prices of the spectrum do not reflect real scarcities but rather result from obsolete government policies, granting spectrum rights would have no real cost to the economy.) The grant element of concessional financing should be treated as a capital contribution. Risks that are not under the government's control include commercial risks, such as those resulting from uncertainty of future demand for new services. To the extent that the government is a major source of demand (for example, of broadband for e-government and distance education programs), capacity purchasing contracts can be practical tools to reduce commercial risk by investors being asked to roll out facilities before their value has been tested fully in the market. For a general discussion of public sector support, see Irwin (2003).

21. For a concise review of the experience of telecommunications development funds, including key factors of success, see Dymond and Oestmann (2002). Wellenius (2002) discusses in detail the success factors and problems encountered in implementing the Chilean development fund, which pioneered competition among firms for rural subsidies.

22. Funding (from various sources) to extend telecommunications services beyond the market ranges between 0.3 percent of sector revenues spent in Chile between 1995 and 2000 to provide payphones in rural areas, to over 5 percent in the United States to support service in high-cost areas and low-income households. Typical funding is 1 to 2 percent of sector revenues.

23. An alternative to conventional telecommunications development funds could be a virtual fund, in which payments are made directly among operating companies to settle differences between mandatory contributions to development funding and expenses incurred to extend services beyond the market. This may be an attractive option where government agencies are deemed ineffective or corrupt.

24. A standard technique is to identify projects with positive social NPV (present value of benefits net of costs) and negative private NPV. Projects meeting these conditions can be ranked by social NPV per unit of maximum subsidy to be made available. Ranking this way maximizes the social return per dollar of subsidy, but it helps the worst-off localities only after all the others.

References

Belli, Pedro, Jock R. Anderson, Howard N. Barnum, John A. Dixon, and Jee-Peng Tan. 2001. *Economic Analysis of Investment Operations: Analytical Tools and Practical Applications.* Washington, DC: World Bank Institute Development Studies, World Bank.

Boyle, Glenn, and Timothy Irwin. 2005. "Estimating the Fiscal Risks and Costs of Output-Based Payments." *OBApproaches* 08 (July), Global Partnership on Output-Based Aid, World Bank, Washington, DC. http://www.gpoba.org/newsletters/OBApproaches _Measuring_Risk.pdf.

Cannock, Geoffrey. 2001. "Expanding Rural Telephony: Output Based Contracts for Pay Phones in Peru." In *Contracting for Public Services: Output-Based Aid and Its Applications*, ed. Penelope Brook and Suzanne Smith, 15–21. Washington, DC: World Bank.

CSO (Central Statistics Office). 2005. *Household Budget Survey 2001/02 Analytical Report.* Central Statistics Office, Ministry of Finance and Corporate Affairs, Port Louis. http://statsmauritius.gov.mu/report/natacc/hbsrep02/hbs.pdf

Dymond, Andrew, and Sonja Oestmann. 2002. "Universal Access and Rural Communication Development Funds: Success Factors Worldwide and Practical Insights from Uganda." Paper prepared for Intelecon Research and Consultancy Ltd., Vancouver. http://www.inteleconresearch.com/pdf/intelecon%20paper%20-%20ua%20&%20rur.pdf.

The Economist. 2004. "On the Same Wavelength." Special Report Spectrum Policy, *The Economist* 372 (8388): 61–3.

———. 2005a. "Behind the Digital Divide." *The Economist* 374 (8417): 22.

———. 2005b. "Mobile Phones and Development: Less Is More." *The Economist* 376 (8434):11.

———. 2005c. "The Real Digital Divide." *The Economist* 374 (8417): 11.

EU (European Union). 2005. *Internet Usage by Individuals and Enterprises 2004.* Eurostat, Statistics in Focus, report KS-NP-05-018. http://epp.eurostat.cec.eu.int/portal/page?_pageid=1073,46587259&_dad=portal&_schema=PORTAL&p_product_code=KS-NP-05-018

Fink, Carsten, and Charles Kenny. 2003. "W(h)ither the Digital Divide?" *info* 5 (6): 15–24.

Fink, Carsten, Aaditya Mattoo, and Randeep Rathindran. 2002. "An Assessment of Telecommunications Reform in Developing Countries." Policy Research Working Paper 2909, World Bank, Washington, DC. http://wdsbeta.worldbank.org/external/

default/WDSContentServer/IW3P/IB/2002/11/11/000094946_02102904035023/Rendered/PDF/multi0page.pdf.

Guislain, Pierre, Mavis Ampah, Laurent Besancon, Cecile Niang, and Alexandre Serot. 2005. "Connecting Sub-Saharan Africa." Working Paper 51, World Bank, Washington, DC.

Irwin, Timothy 2003. "Public Money for Private Infrastructure." Working Paper 10, World Bank, Washington, DC.

ITU (International Telecommunication Union). 2002. *World Telecommunications Development Report*, section 2. Geneva: International Telecommunication Union.

———. 2003. *World Telecommunications Development Report*. Geneva: International Telecommunication Union.

———. 2004. *African Telecommunications Indicators*, section 4. Geneva: International Telecommunication Union.

Kenny, Charles. 2003. "The Internet and Economic Growth in Less-Developed Countries: A Case for Managing Expectations?" *Oxford Development Studies*. 31 (1): 100–13.

Lawson, Cina, and Natalie Meyenn. 2000. "Bringing Cellular Phone Service to Rural Areas: Grameen Telecom and Village Payphones in Bangladesh." Viewpoint 205, World Bank, Washington, DC. http://www.worldbank.org/viewpoint/html-notes/205/205lawson.pdf

Maitland, Donald. 1984. *The Missing Link: Report of the Independent Commission for World-Wide Telecommunications Development*. International Telecommunication Union, Geneva. http://www.itu.int/osg/spu/sfo/missinglink/The_Missing_Link_A4-E.pdf.

Melo, Jose. 2000. "Telecommunications and the Poor." Unpublished, World Bank, Washington, DC.

Navas-Sabater, Juan. 2005. "Universal Access and Output-Based Aid in Telecommunications and ICT." Dissemination Note Global Information and Communications Technologies Department, World Bank, Washington, DC. http://lnweb18.worldbank.org/ict/resources.nsf/a693f575e01ba5f385256b500062af05/f9845b5620512a29852570310056bfba/$FILE/obaNoteFinal.pdf.

Navas-Sabater, Juan, Andrew Dymond, and Niina Juntunen. 2002. "Telecommunications and Information Services for the Poor." World Bank Discussion Paper No. 432, World Bank, Washington, DC.

Oestmann, Sonja. 2003. "Mobile Operators: Their Contribution to Universal Service and Public Access." Vancouver: Intelecon Research and Consultancy Ltd. http://www.inteleconrsearch.com/pdf/mobile%20%26%20us%20-%20for%20rru.pdf

OSIPTEL. 2004. "Marginal-Urban and Rural Telecommunications Development." In *OSIPTEL Telecommunications Capacity Building Technical Assistance Project: Report for Tasks 1-6*, Volume 1. Clifford Chance LLP, Washington, DC. http://rru.worldbank.org/documents/osiptel.pdf

Rossotto, Carlo Maria, Khalid Sekkat, and Aristomene Varoudakis. 2003. "Opening up Telecommunications to Competition and MENA Integration in the World Economy." Working Paper Series No. 33, World Bank, Washington, DC.

Rossotto, Carlo Maria, Björn Wellenius, Anat Levin, and Carlos R. Gomez. 2004. "Competition in International Voice Communications." Working Paper 42, World Bank, Washington, DC.

Saunders, Robert, Jeremy Warford, and Björn Wellenius. 1983. *Telecommunications and Economic Development*. Baltimore and London: The Johns Hopkins University Press.

Serra, Pablo. 2000. "Subsidies in Chilean Public Utilities." Policy Research Working Paper 2445, World Bank, Washington, DC.

Stoeder, Margo. 2005. "Introduction: Regulatory Landscape in Africa." African Mobile Operators Workshop, April 13–14, World Bank, Washington, DC.

SUBTEL. 2003. *Análisis de estadísticas por hogar del sector telecomunicaciones según censo de población y vivienda Chile 2002*. Informe Estadístico No. 7, Subsecretaría de Telecomunicaciones, Santiago. http://www.subtel.cl.

Townsend, David N. 2002. "The World Cup and Communications Development: A New Vision?" Paper presented at the Universal Access and Rural Connectivity Regional Workshop for Southern and Eastern Africa, Dar-es-Salaam. http://www.dntownsend.com/dnta/Reports/DNT%20WCup%20Com.doc

UNECA (United Nations Economic Commission for Africa). 1999. "Information and Communication Technologies for Improved Governance in Africa." Paper presented at the United Nations Economic Commission for Africa "Africa Development Forum," Addis Ababa, October 24–28. http://www.uneca.org/adf99/governance.htm#democratisation.

Wallsten, Scott. 1999. "An Empirical Analysis of Competition, Privatization, and Regulation in Telecommunications Markets in Africa and Latin America." Policy Research Working Paper 2136, World Bank, Washington, DC. http://econ.worldbank.org/external/default/main?pagePK=64165259&theSitePK=469372&piPK=64165421&menuPK=64166093&entityID=000094946_99072206440581

Waverman, Leonard, Meloria Meschi, and Melvyn Fuss. 2005. "The Impact of Telecoms on Economic Growth in Developing Countries. Africa: The Impact of Mobile Phones." The Vodafone Public Policy Papers Series No. 2: 10–23. http://www.vodafone.com/assets/files/en/AIMP_17032005.pdf

Wellenius, Björn. 2000. "Regulating the Telecommunications Sector: The Experience of Latin America." In *Regulatory Policy in Latin America: Post-Privatization Experience*, ed. Luigi Manzetti, 189–242. Miami: North-South Center Press.

———. 2002. "Closing the Gap in Access to Rural Communications: Chile 1995–2002." World Bank Discussion Paper 430, World Bank, Washington, DC.

Wellenius Björn, Carlos Braga, and Christine Zhen-Wei Qiang. 2000. "Investment and Growth of the Information Infrastructure: Summary Results of a Global Survey." *Telecommunications Policy* 24 (2000): 639–43.

Wellenius, Björn, Vivien Foster, and Christina Malmberg Calvo. 2004. "Private Provision of Rural Infrastructure Services: Competing for Subsidies." Policy Research Working Paper 3365, World Bank, Washington, DC.

World Bank. 2002. *Information and Communication Technologies: A World Bank Group Strategy.* Washington, DC: World Bank.

———. 2004. *Morocco: Developing Competition in Telecommunications.* Washington, DC: GICT Policy Division, World Bank.

———. 2005. World Development Indicators database, country classification as of July 2005.

WSIS (World Summit on the Information Society). 2003. "Declaration of Principles: Building the Information Society: A Global Challenge in the New Millennium." World Summit on the Information Society, Geneva. http://www.itu.int/wsis/docs/geneva/official/dop.html. A summary of the main points of the Declaration is accessible at http://www.itu.int/wsis/basic/key.html

———. 2004. "The Report of the Task Force on Financial Mechanisms for ICT for Development." World Summit on the Information Society, Geneva. http://www.itu.int/wsis/tffm/final-report.pdf

Chapter 4

The Role of ICT in Doing Business

Christine Zhen-Wei Qiang, George R. Clarke, and Naomi Halewood

A vibrant private sector—where firms invest, create jobs, and improve their productivity—promotes growth and increases opportunities for poor people (World Bank 2004). There is a growing consensus in the development community that in order to improve the investment climate in their countries, governments should place a high level of priority on improving access to information and communication technology (ICT) as well as its quality.

A good communication and information infrastructure is an essential part of overall infrastructure; it can improve the connectivity between firms, suppliers, and clients as well as provide business opportunities, especially for companies that are physically distant from urban centers. Barr (2000) suggests that entrepreneurs with wider, more diverse contacts have more productive enterprises and are more likely to have access to supplier credit. A network of entrepreneurs—either by exchanging information through an industry-specific online bulletin or by establishing relationships with distant clients—can feed back into the traditional face-to-face network and accelerate endogenous firm growth.

Besides increasing firms' networks, ICT also reduces the historical advantages of long-established firms. This allows firms from developing countries greater opportunities to participate in international trade and become connected to foreign markets. Clarke and Wallsten (2004), in a study of 27 high-income countries and 66 developing countries, found that a 1 percentage point increase in the number of Internet users boosts total exports by 4.3 percentage points, and increases exports from low-income countries to high-income countries by 3.8 percentage points.[1] Freund and Weinhold (2004), in a study that included 56 developed and developing countries, also found a significant link between access to the Internet and trade growth for the period 1997–9. Specifically, they found that developing countries with the fewest established trade links benefit the most from using the Internet.

Furthermore, the internalization of ICT applications affects business operations directly. The ability to transfer information seamlessly through shared electronic files and networked computers improves the efficiency of business processes such as documentation, data processing, and other back-office functions (for example, organizing incoming orders and preparing invoices). Increasingly sophisticated ICT applications—such as customer resource management (CRM) and electronic data interchange (EDI)—allow firms to store, share, and use their acquired knowledge.[2] All this can reduce inefficiencies in the use of capital and labor and can lower operational and transaction costs among economic agents, thus improving the productivity and profitability of firms.

A growing number of firm- and industry-level studies have recently been carried out on the links between ICT and growth. One advantage of microeconomic studies is that they provide specific information on why and how firms adopt ICT applications and which aspects of the ICT adoption process are especially useful. Examination of firm-level data also helps to reveal the effects experienced by different types of firms. The results can then be used to formulate targeted public policies and reforms. For example, in a country with high potential for agricultural exports, expanding a basic service such as e-mail may be the most beneficial policy because it would allow producers to communicate electronically with their agribusiness constituents. In another country, the garment industry may require more sophisticated corporate data services to provide the edge it needs to compete in overseas markets. For these reasons, we will also use firm-level data.

This chapter is organized as follows. In the next two sections we describe the firm surveys in more detail. We present the pattern of firm behavior related to ICT use along with the effect of ICT use on enterprise performance based on our empirical analysis. The following two sections identify the key obstacles to ICT adoption. We conclude by suggesting a number of policy issues that should be addressed to remove the obstacles identified.

Firm Behavior Related to ICT Use

The data used in this study come from Investment Climate Surveys (ICS) conducted by the Private Sector Development Network and Development Economics Research Group at the World Bank, in collaboration with local partners in the countries involved.[3] These surveys were conducted through face-to-face interviews between 1999 and 2003, covering 20,000 firms from 26 sectors in 56 low- and middle-income countries (see table 4.1) in East Asia and Pacific, Europe and Central Asia, Latin America and the Caribbean, the Middle East and North Africa, South Asia, and Sub-Saharan Africa.[4] The surveys include core questions, which are identical across countries. They also include country- and region-specific questions that are intended to provide greater information on specific investment climate issues that highlight country or regional characteristics.

ICS has the following three indicators for ICT use:

1. percentage of a firm's workforce regularly using computers in their jobs
2. whether a firm uses e-mail to interact with its clients and suppliers
3. whether a firm uses a Web site to interact with its clients and suppliers.

These indicators show which firms are currently using ICT in the 56 developing countries surveyed (see annex 4B for country-level summary statistics of these three and other ICT-related indicators).

Rates of E-Mail and Web Site Use

Firms in both manufacturing and service sectors use e-mail at a similar rate, but the use of Web sites and computers is much higher for service firms than for manufacturing ones. Although over 55 percent of firms in both sectors

Table 4.1 Countries in the ICS Sample

Region	Country
East Asia and Pacific	Cambodia (503), China 2002 (1,548), China 2003 (2,401), Indonesia (713), Philippines (716)
Europe and Central Asia	Albania (170), Armenia (171), Azerbaijan (170), Belarus (250), Bosnia and Herzegovina (182), Bulgaria (250), Croatia (187), Czech Republic (268), Estonia (170), Georgia (174), Hungary (250), Kazakhstan (250), Kosovo (329), Kyrgyz Republic 2002 (173), Kyrgyz Republic 2003 (102), Latvia (176), Lithuania (200), Macedonia, FYR 2002 (170), Moldova 2002 (174), Moldova 2003 (103), Poland 2002 (500), Poland 2003 (108), Romania (255), Russian Federation (506), Serbia 2001 (402), Serbia 2003 (408) and Montenegro (100), Slovak Republic (170), Slovenia (188), Tajikistan 2002 (176), Tajikistan 2003 (107), Turkey (514), Ukraine (463), Uzbekistan 2002 (260), Uzbekistan 2003 (100), the former Yugoslavia (250)
Latin America and the Caribbean	Bolivia (671), Brazil (1,642), Ecuador (453), Guatemala (455), Honduras (450), Nicaragua (232), Peru (583)
Middle East and North Africa	Algeria (557), Morocco (859)
South Asia	Bangladesh (1,001), India 2000 (895), India 2002 (1,827), Pakistan (965)
Sub-Saharan Africa	Eritrea (78), Ethiopia (427), Kenya (284), Mozambique (1,940), Nigeria (232), Tanzania (276), Uganda (300), Zambia (207)

Note: Numbers in parentheses are the number of firms surveyed in each country.

Source: Authors' compilation.

Figure 4.1 E-Mail, Web Site, and Computer Use by Sector

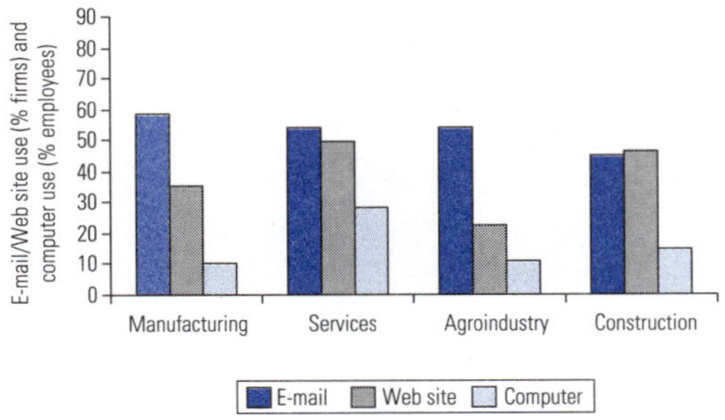

Source: Authors' analysis based on data from the World Bank Investment Climate Surveys 2000–2003.

reported using e-mail to interact with their clients and suppliers, 50 percent of service firms and only 35 percent of manufacturing firms use Web sites for the same purpose (see figure 4.1).

Exports of the manufacturing sector are twice the level of service sector exports, and the sector is likely to be more mature than service, finance, and retail sectors in most developing economies. In its markets, networks of clients and suppliers are already established, thus e-mail alone would suffice for many manufacturing firms to keep in contact with existing clients. Service firms, on the other hand, need to differentiate their products more than traditional manufacturers do, so they may want to use Web sites to disseminate information and cast a wider net to recruit new clients.

Computer use differs from e-mail use in that it does not necessarily infer communication. It suggests rather that the firm has invested in automation or processes business information electronically. A much higher rate of employees (40 percent more) use computers in service firms than in manufacturing firms. Service firms have a higher proportion of desk workers and may require many of them to work on individual computers, whereas manufacturing firm employees can share computers, or a single computer can run a series of processes.

Top ICT Users from Service Sector

Firms in the IT and telecommunications industries appear to be the heaviest users of both Web sites and e-mail (Cluster 1 in figure 4.2). This is expected because these firms themselves are providing these services. In addition, close to 70 percent of firms use ICT applications to interact with clients and suppliers in both the real estate and the hotel and restaurant sectors. Although data on e-mail and Web site use are not available for the accounting and finance sector, this subsector has the highest percentage (67 percent) of employees using computers at work.

Figure 4.2 Web Site and E-Mail Use by Subsector

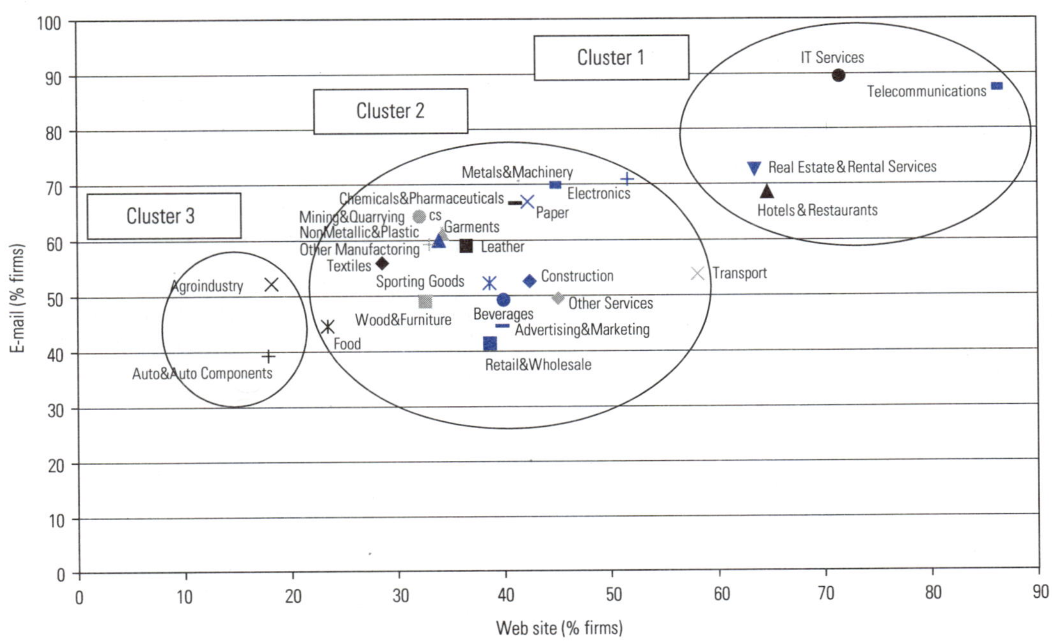

Source: Authors' analysis based on data from the World Bank Investment Climate Surveys 2000–2003.

Figure 4.3 E-Mail and Web Site Use by Size, Export Orientation, and Ownership

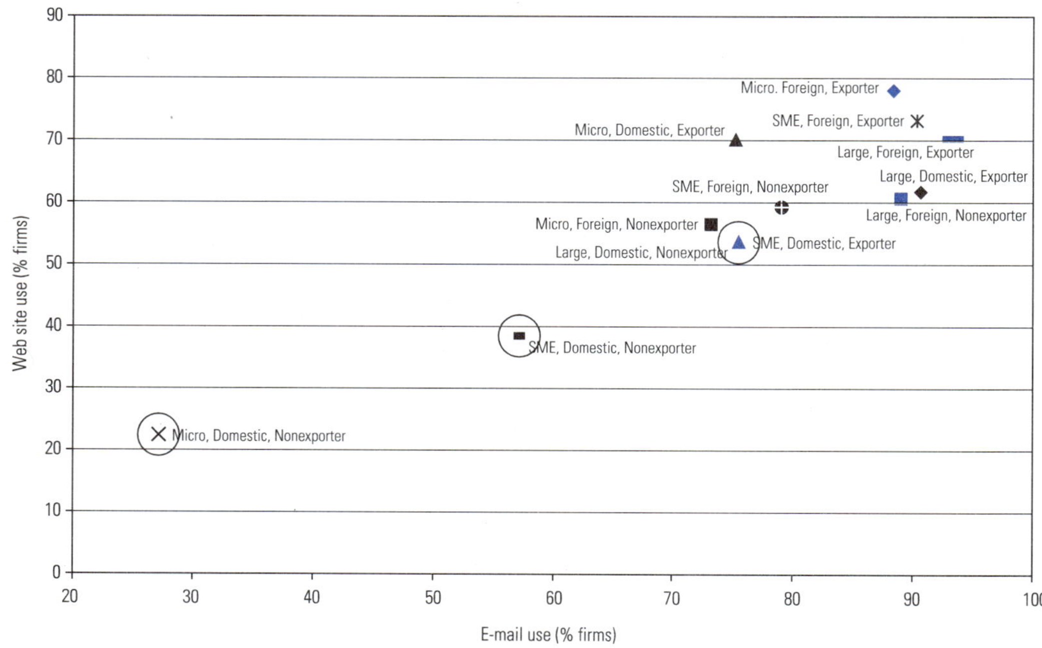

Note: SME - small and medium enterprise.

Source: Authors' analysis based on data from the World Bank Investment Climate Surveys 2000–2003.

The second cluster in figure 4.2 is composed mostly of manufacturing firms. These firms' usage rates range from 30 to 50 percent for Web sites and from 50 to 70 percent for e-mail. The traditional sectors that have driven many developing economies forward, such as the agroindustry and automotive industry sectors, seem to be lagging in ICT use and fall into the third cluster. In most developing economies, an expanding service sector with information as an integral part could propel a shift toward ICT adoption.

Exporters and Foreign Firms More Likely to Use ICT

The upper-right cluster in figure 4.3 shows that exporters and foreign-owned firms rely significantly on e-mail and Web sites. Also, firms that are either exporters *or* have foreign ownership are relatively heavy ICT users regardless of the size of the firm.[5] In contrast, the size of a firm becomes a critical factor when the firm is both a nonexporter and domestically owned. Among the *micro nonexporting domestic firms*, only 27 percent use e-mail and 22 percent use Web sites to interact with clients and suppliers. If computer use affects firm productivity and IT expands networking within sectors and industries, the micro nonexporting domestic firms may not be benefiting from these externalities.

ICT Use Correlated with Income Level

ICT use is highly correlated with income level. Countries with high ICT usage (that is, countries where between 60 and 90 percent of firms use e-mail or Web sites) have an average per capita GDP of $12,500. The top 10 e-mail and Web site users for client interactions are mostly in transition countries in Central and Eastern Europe. Brazil and Ecuador are the exceptions, both geographically and in their much lower income levels ($7,767 and $3,684, respectively).

Several low-income countries, such as Bangladesh, Kenya, Moldova, and Tanzania, have an average rate of e-mail and Web site use of between 50 and 60 percent, suggesting that ICT use is not a luxury (figure 4.4).

Does ICT Use Improve Enterprise Performance?

Increasing access to ICT is not a goal in itself. The real question is whether ICT use improves the performance of an enterprise. In this section, ICS data are used to test whether this is the case by looking at several measures of enterprise performance, including sales and employment growth, labor productivity, total factor productivity, and investment

Figure 4.4 E-Mail and Web Site Use by Country

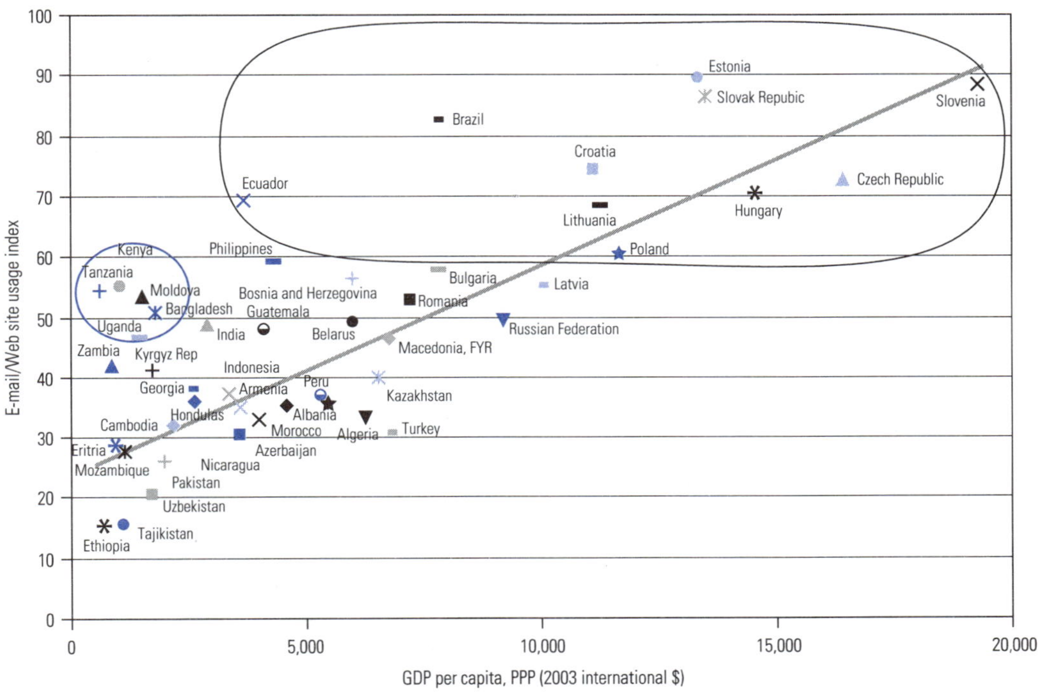

Source: Authors' calculations based on data from the World Bank Investment Climate Surveys 2000–2003.

(see table 4.2). These performance measures are described in greater detail in annex 4A.

To reduce the possibility of finding a spurious correlation between enterprise performance and technology use, the impact of technology use on enterprise performance is estimated in a regression framework. This controls for other factors that might affect both enterprise performance and technology use.[6] These factors include firm size, export status, foreign ownership, age of the firm, country characteristics, and sector characteristics.[7] The empirical results are summarized in table 4.3. The detailed empirical findings are presented in annex 4C, tables 4C.1 to 4C.3.

Enterprise growth. Two measures of enterprise growth—sales and employment growth—are included in the analysis. After controlling for other enterprise characteristics (ownership, export status, size, and age) and for unobserved country and sector characteristics, we find that enterprises that used ICT more intensively tended to grow faster. Enterprises that used e-mail to communicate with their clients and suppliers grew 3.4 percentage points faster per year in terms of sales and 1.2 percentage points faster in terms of employment than those that did not. This is relatively high, given that the average annual growth rates for these variables were 3 percent and 5 percent, respectively.

Table 4.2 Enterprise Performance Measures

Variable	Measure
Sales growth	Average real annual sales growth for past three years
Employment growth	Average annual growth in employment for past three years
Profitability	Profits divided by sales
Investment rate	Total new investment over book value of current assets
Re-investment rate	Percentage of net profits re-invested in the enterprise
Labor productivity	Value added per worker in U.S. dollars
Total factor productivity	Technical efficiency

Source: Authors' analysis.

Table 4.3 Effect of ICT Use on Enterprise Performance in Developing Countries

Performance Indicator	Enterprises that do not use ICT	Enterprises that use ICT	Improvement
Enterprise growth			
Sales growth (percent)	0.4	3.8	3.4
Employment growth (percent)	4.5	5.6	1.2
Profitability (percent)	4.2	9.3	5.1
Investment[a]			
Investment rate (percent)	n.a	n.a	2.5
Re-investment rate (percent)	n.a	n.a	6.0
Productivity			
Labor productivity (value added per worker, dollars)	$5,288	$8,712	$3,423
Total factor productivity (percent)	78.2	79.2	1.0

Note: n.a. Not applicable.

a. Because the investment rates and re-investment rates are limited dependent variables (that is, they are truncated below at zero), the marginal improvements are not equal to the coefficients. For this reason, the authors do not calculate the average rates. The unconditional means for the two variables are 6 percent and 44 percent, respectively.

Source: Authors' analysis based on data from the World Bank Investment Climate Surveys 2000–2003.

Profitability. Enterprises that used ICT more intensively were also more profitable. Profits as a percentage of sales were about 5 percentage points higher on average for enterprises that used the Internet to communicate with clients and suppliers than profits for those that did not, or more than 60 percent higher than that of the average enterprise in the sample, the profits of which were equal to about 7 percent of sales.

Investment. The empirical analysis includes two measures of investment: investment as a share of assets and the re-investment rate (investment as a share of profits). In general, enterprises that used ICT more intensively tended to invest more than other enterprises, although the impact of ICT on investment appeared to be less than its impact on growth and profitability. Enterprises that used e-mail to communicate with clients and suppliers re-invested about 6 percent more of their profits than other enterprises. This was only 15 percent higher than the average reinvestment rate of about 44 percent of profits. However, since enterprises that used ICT also tended to be more profitable, the impact of the higher investment rate is magnified.

Productivity. Finally, both labor and total factor productivity are higher for enterprises that use ICT more intensively. The coefficients for the three ICT variables are positive and statistically significant in both regressions. On average, technical efficiency is about 1 percentage point higher for enterprises that communicate with their clients and suppliers using e-mail (that is, they produce about 1 percent more output with the same inputs). Value-added for these enterprises is about $3,400 higher per employee (the average level is about $7,000 for enterprises in the sample). The larger relative impact on labor productivity is probably because enterprises that use information technology more intensively are more capital intensive overall.

In summary, enterprises that use ICT more intensively are more productive, grow faster, invest more, and are more profitable. These results are robust across different measures of ICT use and across different measures of enterprise performance.

Impact of Telecommunications Infrastructure on ICT Adoption

If, as shown above, ICT applications offer "digital opportunities" to firms in developing countries, the natural assumption is that all firms would invest in these technologies and adopt them for doing business. This, however, has not happened. There continue to be significant differences in ICT adoption rates depending on income levels, firm size, ownership structure (foreign or domestic), and export orientation.

Box 4.1 Impact of Telecommunications Infrastructure on ICT Use

Impact of Slow Connectivity on Internet Use

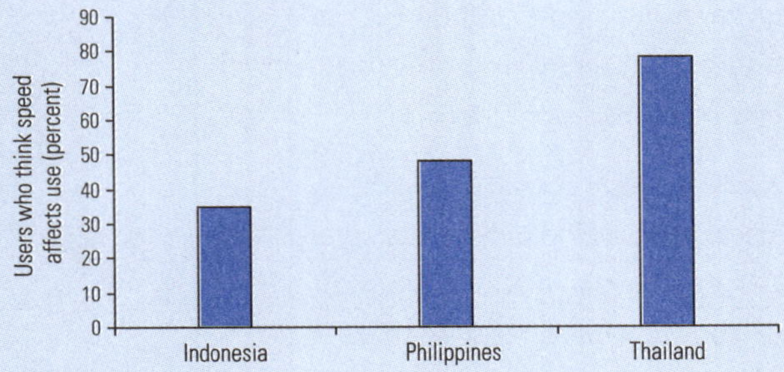

The lack of available telephone lines and Internet service providers (ISPs) has been cited as a major barrier to ICT use among small and medium enterprises (SMEs) in Indonesia, the Philippines, Sri Lanka, and Thailand. Availability was seen as being particularly problematic outside major cities. Speed of connectivity was also a major concern (see figure). In Indonesia, for instance, 66 percent of Internet users found service slow or very slow, with some businesses avoiding online research or downloading documents because of the slow speed and frequent disconnections. In Thailand, speed of connection was considered the primary criterion for choosing an ISP.

Source: Asia Foundation 2002.

Among the obstacles to adopting ICT that can explain this phenomenon, insufficient availability, affordability, and poor quality of the existing telecommunications services are likely to be initial concerns for enterprises in many developing countries. One major constraint is bandwidth for data traffic. Slow and unreliable access limits ICT use (see box 4.1).[8]

Telecommunications services are also much more expensive in developing countries than in developed countries, even without taking income level into account. Prices for broadband (defined by the ITU as a dedicated connection to the Internet of 256 kbs or faster) varied significantly between income groups and regions (see figure 4.5). In 2004, the monthly charge for a 256 kbs speed connection for high- and middle-income countries was on average $180. For low-income countries (with a GNI per capita of $825 or less), the charge was $1,976.

ICS Data on Telecommunications

The ICS surveys contain two types of questions on telecommunications:

1. Objective questions on sector performance. Examples of questions include how many days telephone service was interrupted during the previous year, how much telecommunications service interruptions cost the firm (as a percentage of sales), how long it takes to get a

Figure 4.5 Monthly Broadband Charges by Income Group and Region, 2003

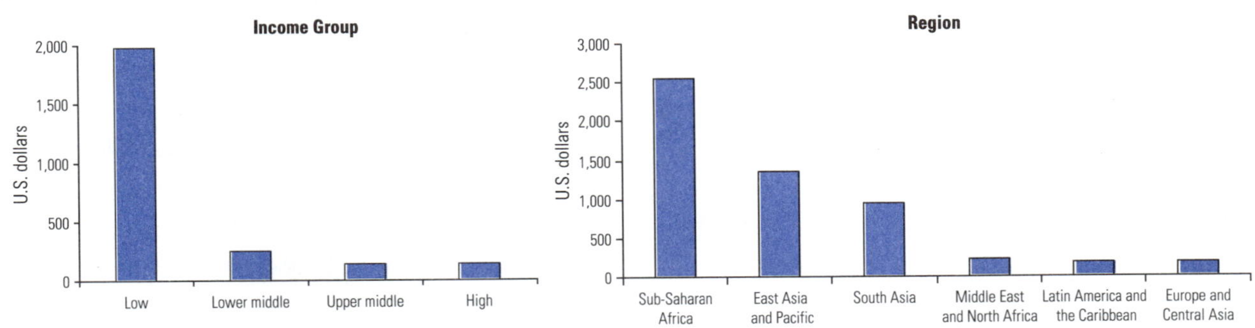

Note: Broadband means a connection speed of more than 256 kbs. Categories calculated by authors using 2003 World Bank income and region classifications.

Source: Authors' analysis based on data from ITU 2003.

fixed-line connection, and whether a bribe needs to be paid to get a fixed-line connection.

2. Subjective questions on enterprises' perceptions about telecommunications service. These questions most notably include asking managers how significant an obstacle poor telecommunications service is for the operations and growth of their business.[9]

There are large differences in wait days for telephone connections across countries and regions. For example, businesses in the Middle East and North Africa region wait, on average, 132 days for a telephone line connection. This is almost eight times the wait in either the East Asia and Pacific region or the Europe and Central Asia region. Businesses in the Middle East and North Africa and in Sub-Saharan Africa are more than five times as likely to experience telephone service interruptions than firms in Latin America and the Caribbean or in South Asia (see figure 4.6).

Firms in many developing countries also face regulatory and bureaucratic delays. Depending on the nature of the business and its dependence on ICT, an individual firm may need to offer a bribe to speed up transactions or to acquire a particular telecommunications service. Out of the 22 countries where the question was asked, respondents reported paying bribes in 20 of them (the only countries where bribes were not reported were Poland and Uzbekistan). Firms in over a dozen countries indicated the amounts of bribes (see figure 4.7). In Honduras, firms paid an average bribe close to 1,400 percent of the telephone connection charge in order to get the connection.

Our analysis also shows that a bribe was paid by 10 percent of the firms (out of the total of 3,837 that responded) to get a fixed-line connection. Only 6 percent of firms in IT services (16 firms) and none of the firms in telecommunications paid bribes. Furthermore, both telecommunications and IT firms indicated relatively few wait days: 4.5 and 15.3, respectively, compared with 36 days on average. Firms in the telecommunications and IT services industries are likely to have easier access to regulators and telecommunications authorities.

Figure 4.6 Quality of Telephone Service by Region

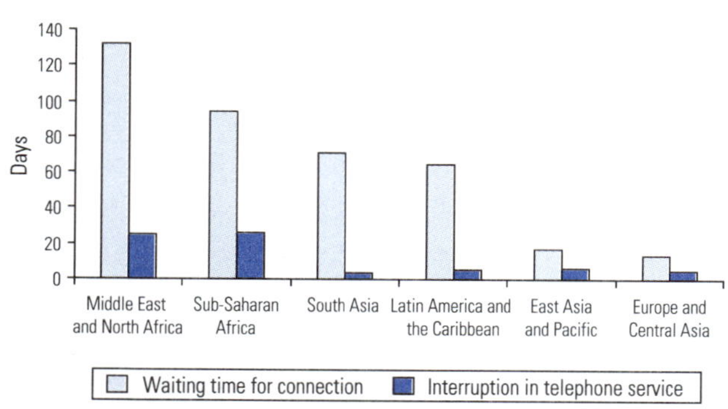

Source: Authors' analysis based on data from the World Bank Investment Climate Surveys 2000–2003.

Figure 4.7 Bribe Value as a Percentage of Telephone Connection Charge

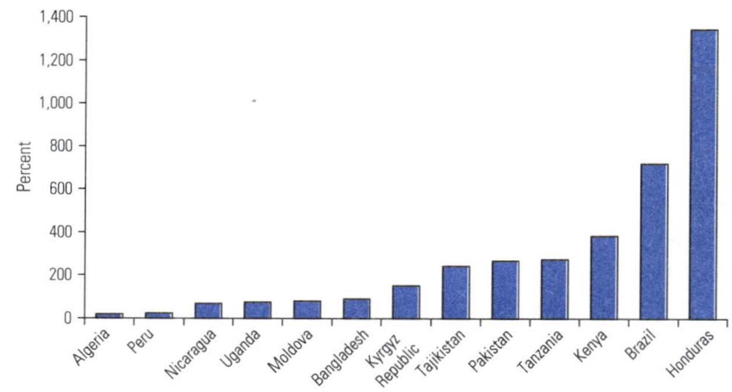

Source: Authors' analysis based on data from the World Bank Investment Climate Surveys 2000–2003.

Impact on ICT Adoption

To investigate more fully the impact of telecommunications services on the probability that the firm uses ICT, multivariate Probit analysis was used to estimate how much the quality of telecommunications infrastructure affects e-mail, Web, and computer use (see annex 4C for the details of this analysis).

The number of days it takes to get a fixed-line telephone is used as a proxy for the quality of telecommunications service. This measure was chosen because it is available for more countries than other measures. The negative and statistically significant coefficients for this variable (see table 4C.4 in annex 4C) suggest that, when the quality of telecommunications service is poor, firms are less likely to use e-mail or the Web to communicate with clients, and are less likely to use computers.

The effect of the quality of telecommunications service, however, appears relatively small. For example, raising the service quality from the level observed at the 20th percentile of quality to the level observed at the 80th percentile would increase the average probability that a firm in the sample would use e-mail to communicate with clients and suppliers by 3.5 percentage points (a 5.4 percent increase). Similarly, higher-quality service would increase the probability that a firm would use the Web to communicate

Table 4.4 Effect on IT Use of Moving from Poor to Good Telecommunications Service

Quality of telecommunications service	Increase in probability of using e-mail to communicate with clients (percentage points)	Increase in probability of using Web to communicate with clients (percentage points)	Increase in number of workers using computers (percentage points)
Days waiting for a connection	3.5	1.8	2.2

Note: Probabilities are calculated for all enterprises in the sample and then averaged over all enterprises to find the average probability.

Source: Authors' analysis based on data from the World Bank Investment Climate Surveys 2000–2003.

with clients and suppliers by 1.8 percentage points (a 5 percent increase). It would also increase the number of workers using computers by 2.2 percentage points (a 12.6 percent increase) (see table 4.4).

Although the results for e-mail and Web use are robust for the different measures of the quality of telecommunications service, the results for the percentage of workers using computers are not robust.[10] This is not surprising because computer use alone does not rely directly on telecommunications service.

Other Factors Affecting ICT Use

Because the ICA surveys were conducted to provide a general analysis of investment climate, the coverage of ICT questions was limited. The questions mainly required "yes" or "no" responses on ICT use, but they did not provide further insights about

1. how often a firm uses ICT,
2. for what purpose a firm uses ICT (such as information search, monitoring the market/prices, banking and financial services, and so on),
3. e-commerce (for example, purchase and sales via EDI or Internet), or
4. barriers to the use of ICT.

In order to identify the obstacles that hinder the adoption and use of ICT by firms in developing countries, we reviewed five other firm-level surveys that included specific questions about reasons for not adopting ICT.

- **London School of Economics (LSE) and Institute of Development Studies (IDS) E-Commerce Survey (2003).** More than 180 e-marketplace sites were surveyed to examine their role in supporting firms seeking to participate in international trade. Seventy-four managers from exporting firms in the garments and horticulture sectors in Bangladesh, Kenya, and South Africa, along with 37 other key informants in these countries and Europe were interviewed about their experiences with the uptake of business-to-business (B2B) e-commerce applications (Humphrey et al. 2003).
- **Mekong Project Development Facility (MPDF) Survey of Current IT Usage at SMEs in Vietnam (2003).** Manufacturing and trading firms located in Hanoi and Ho Chi Minh City were interviewed to assess how ICT is applied in small and medium enterprises (SMEs) in Vietnam, and to identify the pressing issues of ICT adoption in those firms (MPDF 2003).
- **CRITO Global E-Commerce Survey (2002).** This survey included 2,139 firms in 10 economies (Brazil, China, Denmark, France, Germany, Japan, Mexico, Singapore, Taiwan [China], and the United States). Questions in the survey covered the use of e-commerce technologies, drivers for Internet use and barriers to conducting business on the Internet, benefits from e-commerce use, prevalence of online sales, and online service offering (CRITO 2002).
- **South Africa E-Commerce Survey (2002).** The empirical evidence is drawn from 120 firm-level interviews and 31 personal interviews with industry experts in the South African manufacturing sector regarding firm perception and experiences in e-mail and Internet use for conducting business operations (Moodley 2002).
- **World Information Technology and Services Alliance (WITSA) International Survey of E-Commerce (2000).** Twenty-eight of the WITSA members from both developed and developing countries were interviewed about their views on the best way to encourage the growth of e-commerce (WITSA 2000).

According to these surveys, there is a range of reasons that firms in developing countries hesitate to use ICT (table 4.5).

Table 4.5 Barriers to ICT Adoption in Selected Countries

Barrier	Brazil	China	Singapore	Global
Need for face-to-face interaction	..	..	38	34
Concern about privacy of data or security issues	49	45	48	44
Customers do not use the technology	48	33	..	31
Finding staff with e-commerce expertise	34	..	..	..
Costs of implementing an e-commerce site	34	..	45	34
Making needed organizational changes	..	..	38	..
Level of ability to use the Internet as part of business strategy	..	31	..	..
Business laws do not support e-commerce	..	41	..	..
Inadequate legal protection for Internet purchase	41	55	44	34

Note: The numbers refer to the percentage of firms citing obstacles (top five for each country). .. Not available.

Source: CRITO 2002, quoted in Wong and Ho 2004.

These reasons vary considerably among sectors and countries.[11] They also change over time and depend on a country's progress along the ICT adoption ladder.

Besides the limitations of the existing telecommunications infrastructure in developing countries (as already revealed through the ICS analysis), the principal constraints in adopting and using ICT applications can be summarized as follows:

1. lack of applicability and little incentive to change business models when returns are not clear;
2. lack of trust in e-business processes and legal protection including privacy of data, online transactions, authentication, and security; and
3. shortage of ICT-skilled labor.

Lack of Applicability and Little Incentive to Change Business Models When Returns Are Not Clear

Both the LSE/IDS survey and the WITSA survey found that many firms in developing countries are not fully aware of the opportunities and benefits that can be derived from an ICT-enabled business environment. Clearly, for firms to adopt ICT strategies and applications, the potential returns must outweigh investment and maintenance costs. Beyond a certain level of connectivity (PC, Internet access, and online information or marketing), many firms will stay with traditional business processes if they believe ICT is unsuitable for their business, or if expected returns are small.

But the decision of whether to adopt ICT often depends on the perceived benefits to the firm's existing business operations and clientele networks. This is contrary to the notion that the Internet will enable access to larger markets and therefore create the opportunities for firms' business to grow. For example, if a firm's existing clients and suppliers are not using the Internet and do not have the necessary ICT capabilities, the firm may not see any direct benefit of using ICT in its business operations. Moodley (2002), in his survey of South African firms, found that even those firms that do use e-mail and Web sites do so with existing clients and suppliers, and do not seem to be expanding their markets.

Although direct benefits are sometimes not clear, there is increasing evidence that e-business is a complex and costly undertaking that calls for substantial investments and institutional changes (figure 4.8). Adopting ICT could affect all aspects of a company's operations and business processes, especially its distribution, supply chain, customer service, marketing, and finance. This involves costs that range from building new systems and buying new products (such as computers) and services to running dual systems or hybrid operations (incorporating old business processes) and gaining technical expertise (such as hiring skilled workers to build a Web site).[12] Moreover, the ICS surveys also show that the ability to absorb ICT costs differs according to firm characteristics (box 4.2). There are, however, no systematic methodologies to help firms assess the costs of adopting ICT and pursuing online activities.[13]

Even when a firm is convinced that it is necessary to integrate ICT in its business operations, a typical developing-country firm may still face the following obstacles:

Figure 4.8 The Most Significant Costs That Obstruct E-Business
(percentage of survey respondents)

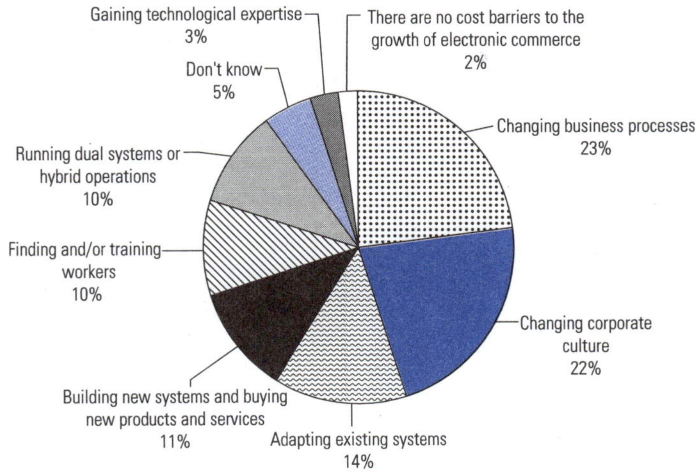

Source: WITSA 2000.

1. lack of understanding how ICT can help to conduct business differently (see box 4.3);
2. lack of accepted, coherent, and industry-specific e-business models in practice;
3. inertia and internal reluctance to change existing corporate culture and operating structures; and
4. logistical challenges of integrating into a real-time environment (for example, inadequate transport infrastructure to deliver the firm's products).

Lack of Trust in E-Processes and Legal Protection

All five firm-level surveys found that the lack of trust in e-mail or Internet-based activities is one of the major

Box 4.2 Perception of ICT Affordability

The ICS data analysis shows that the ability to absorb ICT costs differs according to firm size. Smaller firms find it more difficult (see figure). Although 63 percent of micro firms said ICT services were not affordable, only 25 percent of very large firms felt the same way. The figure also shows that foreign, export-oriented firms tend to have more positive perceptions about ICT affordability. One explanation for this might be that foreign-owned firms are more likely to receive technical support and resources from their parent companies, and are also richer on average than domestic firms, so their affordability is higher. Export companies may understand the benefits of ICT use more easily because of the nature of their business. Another explanation might be that service affordability is more of a concern and the benefits harder to imagine for domestic, nonexporting firms.

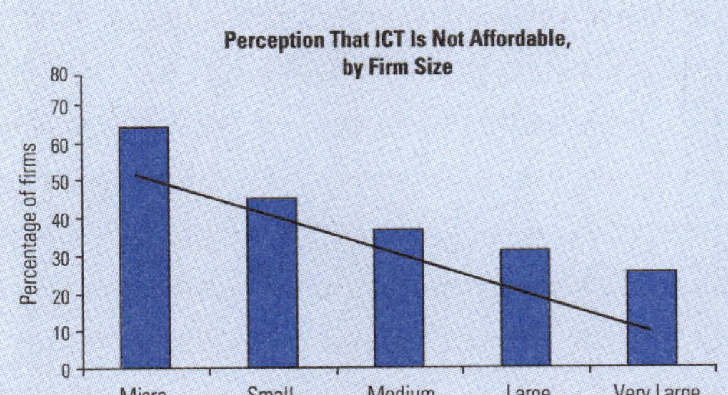

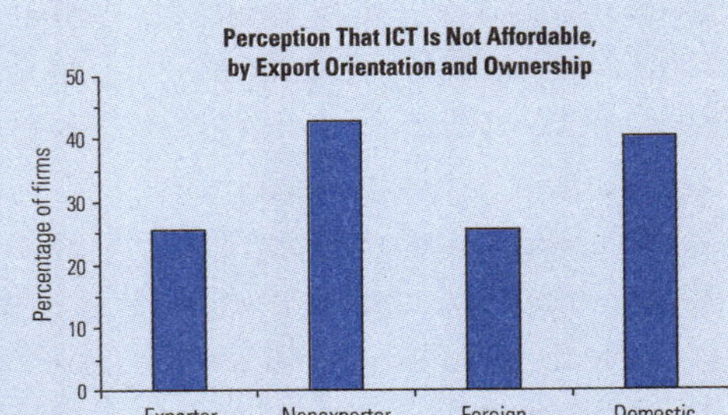

Source: Authors' analysis based on data from the World Bank Investment Climate Surveys 2000–2003.

Box 4.3 Internet: No More than a Bulletin Board?

Business dropped dramatically at a small hotel in Senggigi, Indonesia, after a riot, with occupancy rates falling to less than 10 percent. To boost business, in 1999 the hotel invested in developing a Web site, hoping to attract more customers. However, the managers lacked basic understanding of how to use the Web site and the Internet. After the hotel paid a one-time fee to the Web designer, it never updated or improved the site, nor did it check to see whether the site had been registered on any one of the major search engines. The hotel had five computers, but only one was connected to the Internet. That computer was located in the director's office and only the manager was allowed to check e-mails, and that only once a month. Not surprisingly, the company was dissatisfied with its "e-business" experiment. Such cases are not uncommon in developing countries where firms expect ICT to transform their business without analyzing *how* ICT helps them to conduct business differently. Furthermore, a Web site is a live communication and transaction medium requiring constant monitoring for it to add business value.

Source: Asia Foundation 2002.

constraints for uptake of ICT. From a firm's perspective, trust in e-processes implies confidence that

- online information and communication are accountable for the quality, reliability, and legality of products and services;
- personal information and finances are secure;
- electronic purchases, fund transfers, and business deals are considered equal to paper-based transactions in terms of validity.

Both the WITSA and the LSE/IDS surveys highlighted the firms' fear that the technology platform is not robust enough to protect online transactions, information privacy, and data integrity. Uncertainty about the identity of communicating parties could further complicate these concerns (see box 4.4 for an illustration of how a trading company deals with electronic business inquiries). Internet security software such as firewalls, encryption technology, antivirus packages, authentication routines, and security administration systems have been developed to mitigate these concerns.

On top of the technical dimension of adopting ICT in business practices, shifting from traditional interactions to electronic transactions has led to a number of legal and regulatory concerns. The lack of a satisfactory redress mechanism when things go wrong online may strongly discourage online transactions. Moreover, in the case of cross-border transactions, often there is no harmonized legal framework with rules that pertain to the determination of jurisdiction and applicable law, nor are there mechanisms that ensure the cross-border enforcement of legal rulings. Businesses can risk being sued in multiple jurisdictions under a number of inconsistent laws.

Particularly in developing countries, access to reliable advice on appropriate security measures is often at a premium. In these countries, the cost of keeping abreast of rapid developments in the complex technological and legal issues involved in online transactions may be too high for many firms.

Shortage of ICT-Skilled Labor

Another barrier is the availability of ICT competencies within the firm. The WITSA survey listed workforce concerns in order of importance (see figure 4.9).

Skilled labor plays an important role because it affects the absorption rate of ICT applications within a country. Some nations depend on a small set of skilled local or foreign ICT workers, which limits the development of e-business. For example, the Bhutanese government, in order to encourage ICT use by businesses, recently began to incorporate ICT programs at the university level to supplement its existing ICT-skilled labor force, which was composed mostly of Indian migrant workers (Bhutan, UNDP, and World Bank 2002).

However, this does not necessarily mean that all countries require a highly ICT-skilled workforce to integrate the newest technologies available. The labor force should,

Box 4.4 A Fruit and Vegetable Trader's Experience in the E-Marketplace

A small fruit and vegetable trading company in Nairobi registered with open e-marketplaces and was interacting with potential buyers who, through e-mails, inquired about macadamia nuts from Switzerland, carrots from Romania, and oranges from the Ukraine. The owner of the firm outlined his procedures for dealing with an e-mail inquiry:

- Ignore any inquiry from a firm with a hotmail or Yahoo e-mail address.
- Check to see if the company making the inquiry has a Web site, and possibly consult Web-based "Yellow Page" business directories.
- Consider the product requested. It is better to deal with robust products (for example, fruit rather than vegetables) because the logistical requirements are simpler.
- Check on the Internet to see who the competitors are in the market being supplied, in particular to see whether and when there are supply gaps in the market.
- Follow up the inquiry with a quote based on the cost of local sourcing, packaging, freight costs, and the exporter's markup. Contact with the customer is predominantly by e-mail.

Although the trading firm was thorough in its online dealings, after some unfortunate experiences of nonpayment, the exporter was demanding a 50 percent down payment and the remainder on proof of shipment.

Source: Humphrey et al. 2003.

ideally, adjust to ICT technology and applications that are used in the country and within industries. For example, a study of Philippine workers made it evident that Internet-based businesses do not necessarily require highly skilled labor (De Vera 2002). Rather, there is a need for semi-skilled ICT labor in certain industries, such as tourism, electrical communications equipment, and telecommunication services (for example, in call centers).

Apart from the lack of ICT-skilled and semi-skilled labor, a significant number of the firms also lack managerial understanding and skills for successfully integrating ICT applications. Many firms rely on internal ICT enthusiasts as their main source of ICT skill and knowledge. Business executives who run the firms may not know what the options are for ICT solutions; in many ways, they do not know what they need to know to make decisions about the

Figure 4.9 Main Constraints on ICT-Skilled Workforce

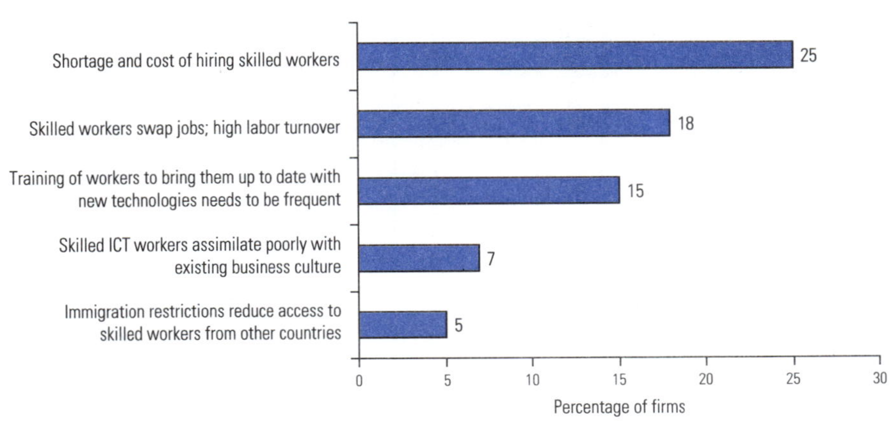

Source: WITSA 2000

allocation of resources and effort regarding ICT equipment and human resources.

Appropriate management of ICT adoption requires a firm to have personnel who can combine both business and technology disciplines to do the following:

- select from a broad range of ICT applications (from basic use of e-mail to the much broader concept of data exchange or supply chain management) that would save time and resources for business processes and strengthen the firm's core competence;
- provide guidelines on which types of ICT technologies and telecommunications services would be most appropriate for their firm (or industry);
- avoid incompatible systems and networks between firms, clients, and suppliers (or regions); and
- maintain and upgrade ICT applications to fit business needs and technology changes.

The professional advice of ICT consultants can help, but such consulting is both scarce and costly in most developing countries.

Public Policies

This analysis and other microeconomic studies have identified a number of barriers that hinder firms' adoption of ICT. The potential benefits of ICT applications in businesses are more likely to come to fruition if policy makers focus on removing these obstacles. Specifically, governments can

- create appropriate environments for ICT uptake. This includes liberalizing markets to expand and improve network infrastructure, providing a supportive legal and regulatory environment for e-business, and taking steps to enhance technological diffusion.
- target programs to overcome market failures. Often, such programs are needed in particular areas such as demand aggregation (for example, by developing e-government services and encouraging firms to use them) and skill formation (for example, by emphasizing ICT in curriculum).

Easing Bottlenecks in the Telecommunications Infrastructure

The availability of a wide range of high-quality communications services at competitive prices is important: this allows firms to choose technologies and services appropriate to their business needs. Experience in OECD countries shows that the availability of affordable access to a high-speed telecommunications infrastructure is closely linked with a firm's migration to e-business (OECD 2004). In developing countries, analog modems are still the dominant way to access the Internet, and speed of access remains an issue for business use. New broadband technologies, such as cable modem and digital subscriber line (DSL), offer faster connections. Governments will need to play a leading role in promoting the modernization and extension of the national information infrastructure.

A priority for policy is therefore to continue to emphasize liberalizing the telecommunications market and promoting effective competition (see box 4.5). This will stimulate new investment in additional bandwidth, increase demand for communication services through falling prices, and promote greater efficiency and innovation in the provision of infrastructure and services. Policy initiatives to promote technology neutrality among competing and developing technologies are also important for encouraging interoperability, innovation, and choice among services.

E-Trust

The government's support of a regulatory framework for trust, security, enforcing authentication mechanisms, and combating cyber-crime—combined with internationally accepted privacy and consumer protection protocols—is essential in encouraging business use of ICT applications. Of particular relevance for firms are low-cost online dispute resolution mechanisms both among firms and between firms and consumers.

Countries are developing e-security policies and programs, including public key infrastructure (PKI) and computer emergency response teams (CERT). These policies and programs are often a combination of top-down, government-led initiatives together with bottom-up, industry-led programs. PKI supports digital signatures and other public security services, and is necessary in laying the foundation for e-applications such as e-signature, e-notarization, and e-certification. The CERT program, a multi-agency information-sharing framework, aims at assessing and managing e-security risks, providing training and emergency alert and response services, and offering backup to ensure the continuity of network systems and applications.

Box 4.5 Pro-competitive Regulatory Framework to Ease Telecommunications Bottlenecks

Based on the ICS analysis, firms in countries with private operators or a separate telecommunications regulatory authority experience fewer wait days for telephone connections and interruptions in telephone service (top pair of figures). Among the top 10 countries experiencing many wait days for telephone connections, only two have a competitive market (bottom pair of figures).

Waiting Time for Telephone Connection and Interruption in Telephone Service by Type of Operator Ownership

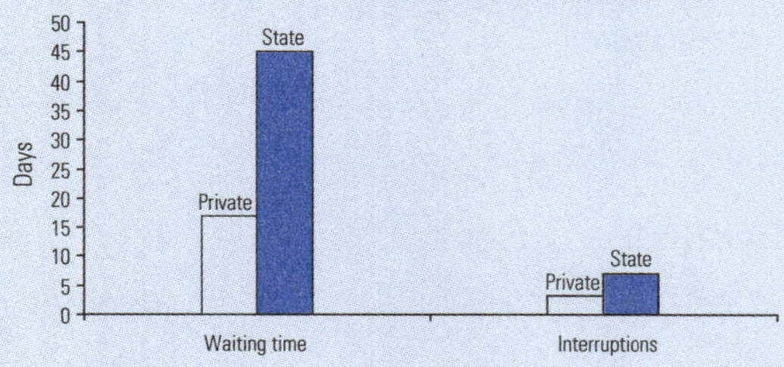

Waiting Time for Telephone Connection and Interruption in Telephone Service by Presence of Regulator

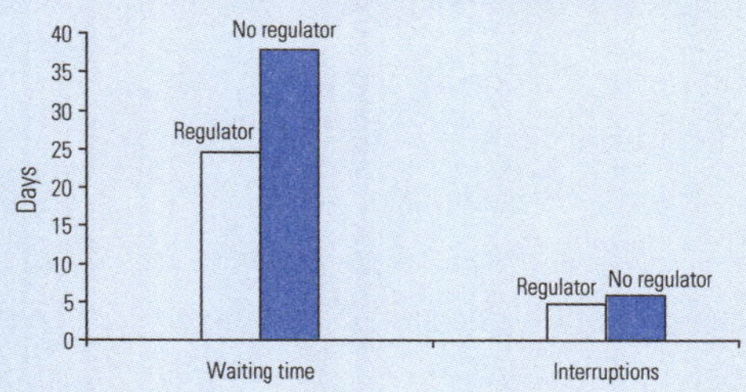

Source: Authors' analysis based on data from the World Bank Investment Climate Surveys 2000–2003.

Waiting Time

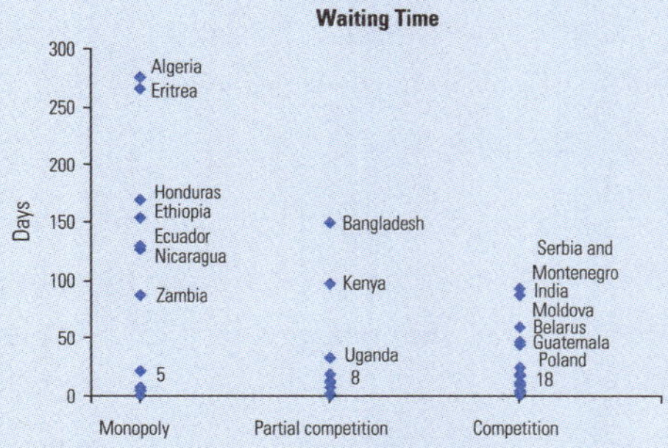

Internet Monthly Price Basket

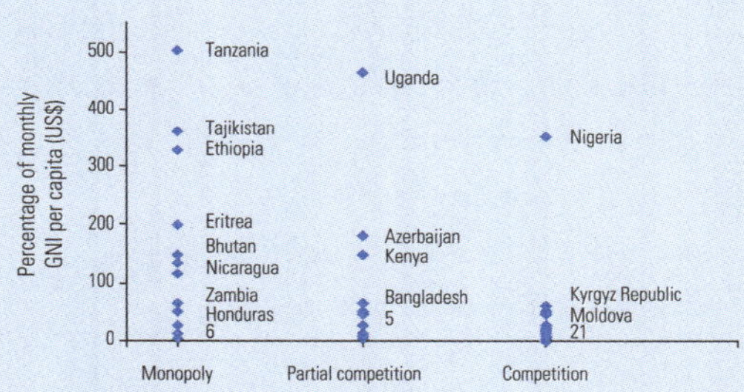

Source: Authors' analysis based on data from ITU 2003 (price and sector structure) and from World Bank (waiting time and GNI per capita).

The government's role in providing a pro-competitive regulatory framework is clearly critical to ease the bottlenecks hampering the development of a telecommunications infrastructure.

Demand Aggregation

Firms will adopt ICT when they are convinced of its applicability and potential returns. They need to make realistic assessments of e-business opportunities, benefits, and costs. Policy measures that encourage business associations or other groups to provide awareness-raising and consultation services can be beneficial. A measure that could mitigate some of the costs associated with ICT use, open source software has been proposed as an economically viable solution for establishing and upgrading ICT systems. Open source software can foster wider compatibility between different applications and operating systems, and it can lower the cost of ICT adoption, especially for SMEs.

In addition, the provision of information and services online by the government itself can demonstrate the effects of ICT on businesses by spreading awareness of the potential for online delivery and interaction, and offer incentives for the adoption of ICT. Governments, as model users, can also act as standard-setters for ICT adoption by firms. In order to ensure access to public services and obligatory requirements for business purposes, firms would adjust their choice of systems and software to maintain interoperability with these online government services. Public e-procurement provides such an example. Government demand aggregation to provide services can also contribute to promoting trust and security in online transactions.

Building Human Resources

For ICT investment to be fully effective, one has to take advantage of "network externalities." This requires that a large enough number of firms make the investment and become connected to the network. Moreover, network externalities are fully exploited only when those using ICT communicate among themselves effectively, rapidly, and efficiently. This requires that users have certain levels of technical, analytical, and problem-solving skills.

Lack of ICT and business skills is a widespread impediment to effective uptake once adoption decisions have been made. Governments have major roles to play in providing basic ICT skills in compulsory schooling, and an important role in conjunction with educational institutions, businesses, and individuals in providing the framework to encourage ICT skill formation at higher levels as well.

Governments might also help firms to confront the difficulties of technological learning in a developing-country context. It can assist companies by putting in place a set of complementary institutions and associations that provide advice on technology choice, identify and disseminate information of best practices, and assess the available e-commerce technologies and approaches. This can be especially effective through the support of training and skills development.

Conclusion

Results from firm-level survey analysis suggest that ICT is widespread in businesses across sectors and income levels. Also, ICT is playing an important role in allowing businesses to grow faster and become more productive—this alone suggests that creating an appropriate environment to exploit ICT is important. Governments can liberalize the telecommunications market to allow affordable access to network infrastructure, provide a supportive legal and regulatory environment for e-business, and overcome market failures in areas such as demand aggregation and skill formation. But it is clear that the relevance of ICT varies across industries and income levels, implying the need for realism and a tailored response. Beyond a certain level of connectivity (computer use, Internet access, online information, or marketing), many firms in developing countries may choose to stay with traditional business processes either because the more sophisticated ICT applications are unsuitable for their business, or because expected returns from business reengineering are small. Any supporting role played by ICT is crucially dependent on business processes within firms and the broader economic environment beyond them.

Annex 4A: Measures of Firm Performance

This annex describes the construction of the performance variables used in this study:

- sales growth
- employment growth
- investment rate
- re-investment rate
- profitability
- labor productivity
- total factor productivity (TFP).

Descriptions of the formulas used to calculate the various productivity measures, other than TFP, are described in table 4A.1. The construction of the measure for TFP is explained in detail below.

Data

The data used in this study come from Investment Climate Surveys (ICS), which were filled out by trained enumerators during face-to-face interviews that lasted about three hours. Most questions were answered by the owner or managing director of the firm. Exceptions were the specialized questions on employment and company accounts; these were directed toward the relevant professionals (for example, human resource managers or accountants).

The surveys, conducted between 1999 and 2003, cover firms from 26 sectors in about 56 low- and middle-income countries in all regions. The surveys include both core questions and country- and region-specific questions. Although, in principle, the core survey should be identical across countries, some questions have been added or dropped over time and not all questions have been asked in all countries. For instance, balance sheet and income statement data were not collected for some countries, meaning that most productivity measures cannot be calculated for the firms in those countries. Table 4A.2 contains information on which

Table 4A.1 Productivity Measures

Variable	Measure
Sales growth	**Average real annual sales growth for past three years** If data were not available for all three years, then the average was calculated over available years. Nominal values are converted to real values using GDP deflator (that is, sector-specific deflators were not available).
Employment growth	**Average annual growth in employment for past three years** If data were not available for all three years, then average was calculated over available years. Workers include both temporary and permanent workers.
Profitability	**Profits divided by sales** Profits are sales less expenditures on intermediate inputs, energy, interest expenditures, and overhead costs.
Investment rate	**Total new investment over book value of current assets**
Re-investment rate	**Percent of net profits re-invested in establishment** Question was asked directly of managers during survey (it is not calculated). Data are missing for firms with negative net profits.
Labor productivity	**Value added per worker** Value added is sales less expenditures on intermediate inputs and energy (electricity, fuel, and other). Workers include both temporary and permanent workers. Converted to U.S. dollars using the average exchange rate for the year of the survey.
Total factor productivity	**Technical efficiency (see text for calculations)**

Source: Authors' analysis.

surveys contained enough information to calculate each of the productivity measures.

The surveys were conducted in a uniform way across countries. The sampling frames were stratified across location within each country, subsector, and enterprise size. When recent census data were available, the random samples were constructed using census data. If recent census data were not available, the lists were constructed using lists of enterprises from government agencies (for example, from the National Bureau of Statistics in Tanzania). As a result, the surveys ultimately cover the formal sector—firms that need to be registered are included. When firms could not be located or refused to participate in the survey, they were replaced with new firms with characteristics as similar (in terms of size, sector, and location) as possible.

Table 4A.2 Countries with Enough Data to Compute Productivity Measures

	Year	Employment growth	Sales growth	Profitability	Investment rate	Re-investment rate	Labor productivity	Total factor productivity
Albania	2002	X						
Algeria	2002	X	X					
Armenia	2002	X						
Azerbaijan	2002	X						
Bangladesh	2002	X	X	X	X		X	X
Belarus	2002	X						
Bosnia and Herzegovina	2002	X						
Bolivia	2000		X	X				
Brazil	2003	X	X		X	X		
Bulgaria	2002	X						
Cambodia	2003	X	X	X	X	X	X	X

(*Table continues on the following page.*)

Table 4A.2 *continued*

	Year	Employment growth	Sales growth	Profitability	Investment rate	Re-investment rate	Labor productivity	Total factor productivity
China	2002	X	X	X			X	X
China	2003	X	X	X		X	X	X
Croatia	2002	X						
Czech Republic	2002	X						
Ecuador	2003	X	X	X		X	X	X
Eritrea	2002	X	X		X		X	X
Estonia	2002	X			X			
Ethiopia	2001	X	X	X			X	X
Macedonia, FYR	2002	X						
Georgia	2002	X						
Guatemala	2003	X	X			X		
Honduras	2003	X	X			X		
Hungary	2002	X						
India	2000	X	X	X	X		X	X
India	2002	X	X	X	X		X	X
Indonesia	2003	X	X		X	X		
Kazakhstan	2002	X						
Kenya	2003	X	X	X		X	X	
Kosovo	2003	X	X			X		
Kyrgyz Republic	2002	X						
Kyrgyz Republic	2003	X	X	X		X	X	X
Latvia	2002	X						
Lithuania	2002	X						
Moldova	2002	X						
Moldova	2003	X	X	X		X	X	X
Montenegro	2003	X	X			X		
Morocco	2000	X	X	X			X	X
Mozambique	2002	X	X	X			X	
Nicaragua	2003	X	X			X		
Pakistan	2002		X	X	X		X	X
Peru	2002	X	X	X		X	X	X
Philippines	2003	X	X	X	X	X	X	X
Poland	2002	X						
Poland	2003	X	X	X		X	X	X
Romania	2002	X						
Russian Federation	2002	X						
Serbia	2001	X	X					
Serbia	2003	X	X			X		
Slovak Republic	2002	X						
Slovenia	2002	X						

Table 4A.2 *continued*

	Year	Employment growth	Sales growth	Profitability	Investment rate	Re-investment rate	Labor productivity	Total factor productivity
Tajikistan	2002	X						
Tajikistan	2003	X	X	X		X	X	X
Tanzania	2003	X	X	X	X	X	X	X
Turkey	2002	X						
Uganda	2003	X	X	X		X	X	
Ukraine	2002	X						
Uzbekistan	2002	X						
Uzbekistan	2003	X	X	X		X	X	X
Yugoslavia	2002	X						
Zambia	2002	X	X	X	X	X	X	X
Total		59	35	24	13	23	24	21

Source: Authors' analysis.

Estimating Total Factor Productivity

TFP is estimated using a stochastic frontier approach and assuming a Cobb-Douglas production function. This essentially estimates how far firms fall short of the amount that they could produce given the amount of capital and labor they would use if they were fully efficient (that is, as efficient as the most efficient enterprises in the sample). Formally, the equation estimated is

$$\ln S_{ik} = \sum_j D_{ijk} (\beta_j + \alpha_{jL} \ln L_{ik} + \alpha_{jK} \ln K_{ik} + \alpha_{jI} \ln I_{ik}) + \lambda_k - \mu_{ik} + \varepsilon_{ik} \qquad (4A.1)$$

where S_i is sales for firm i in the year of the survey, L_i is the number of full-time employees (permanent and temporary), K_i is the net book value of fixed assets for firm i in the year of the survey, and I_i is the value of intermediate inputs. All monetary variables (for example, sales, intermediate inputs, and capital) were collected in local currency units, which were converted to U.S. dollars using average exchange rates. To control for the possibility that the annual exchange rates might not be long-run exchange rates, country dummies (I_k) are included in the base regression. Since the monetary variables are in natural logs, the country dummies effectively control for the possibility that the results might be affected by exchange rates.[14]

In addition to labor, capital, and intermediate inputs, equation 4A.1 also contains a series of 26 sector dummies, D_{ijk}, that are equal to 1 if firm i in country k is affiliated with sector j.[15] The sector dummies also interact with labor, capital, and intermediate inputs to allow different sectors to use different production technologies—that is, labor, capital, and intermediate input intensities are not assumed to be the same across sectors. Technically, this is done by multiplying the sector dummies by the variables representing capital, labor, and intermediate inputs.

The error term is assumed to have two components: ε_i, which represents random statistical noise (for example, noise from measurement error or individual firm-level shocks), and μ_i, which represents the firm's technical efficiency (that is, how far the firm is from the production possibilities frontier). The first term, ε_i, is a two-sided error term, which is assumed to have a normal distribution. The second term, μ_i, is assumed to be nonnegative (that is, greater than or equal to zero) and to follow an exponential distribution.[16] The two error terms are assumed to be independently and identically distributed. The technical efficiency of firm i is calculated using the following formula:

$$TE_i = E\{\text{Exp}\ (\mu_i | \varepsilon_i)\}. \qquad (4A.2)$$

Annex 4B: ICT-Related Investment Climate Survey Data by Country

Table 4B.1 Telecommunications Results

Country	Survey year	Number of firms	Fixed line[a]	Telecom major or severe obstacle to business operations and growth (% firms)	Unavailable main line telephone service (days)	Unavailable main line telephone service (hours)	Unavailable main line telephone service (total % sales lost)	Workforce regularly using computers in their jobs (percent)	E-mail use for interacting with clients and suppliers (% firms)
Albania	2002	170	C	18.24	8.79	..	..	..	38.24
Algeria	2002	557	M	15.67	25.20	..	..	..	41.72
Armenia	2002	171	M	11.11	7.90	..	..	..	39.77
Azerbaijan	2002	170	P	3.64	3.21	..	..	..	32.35
Bangladesh	2002	1001	C	24.65	..	..	..	..	70.30
Belarus	2002	250	C	2.83	1.75	..	..	..	53.60
Bolivia	2001	671	M	..	..	..	..	..	..
Bosnia and Herzegovina	2002	182	M	3.41	4.43	..	..	..	58.24
Brazil	2003	1642	C	6.16	2.06	13.00	1.23	17.53	92.02
Bulgaria	2002	250	C	6.85	1.72	..	..	..	61.60
Cambodia	2003	503	P	3.21	4.55	4.85	2.62	21.03	40.76
China	2002	1548	P	23.53	..	..	..	33.33	..
China	2003	2400	P	..	..	..	..	35.97	..
Croatia	2002	187	C	1.08	1.18	..	..	..	76.47
Czech Republic	2002	268	C	2.26	0.83	..	..	..	75.37
Ecuador	2003	453	M	18.14	12.38	41.61	7.62	22.25	83.22
Eritrea	2002	78	M	14.29	12.16	..	..	..	47.44
Estonia	2002	170	C	5.36	1.94	..	..	..	89.41
Ethiopia	2002	427	M	29.40	..	..	..	..	24.65
Georgia	2002	174	P	6.32	13.87	..	..	..	39.08
Guatemala	2003	455	C	6.61	2.56	6.80	4.37	12.75	66.37
Honduras	2003	450	M	18.26	6.47	5.50	7.23	7.68	50.22
Hungary	2002	250	C	1.61	1.60	..	..	..	74.40
India	2000	895	C	..	..	..	..	..	44.39
India	2002	1827	C	5.33	..	..	..	17.03	62.07
Indonesia	2003	713	P	9.12	1.81	..	2.64	13.21	49.93
Kazakhstan	2002	250	C	2.87	5.63	..	..	..	43.20
Kenya	2003	284	P	44.12	35.83	37.06	..	13.78	78.42
Kosovo	2003	329		17.19	4.90	4.58	3.04	17.99	29.35
Kyrgyz Republic	2002	173	C	1.04	15.91	..	..	..	28.90
Kyrgyz Republic	2003	102	C	3.53	2.21	19.50	0.23	..	40.20
Latvia	2002	176	C	3.41	0.75	..	..	..	56.25
Lithuania	2002	200	C	4.00	1.38	..	..	..	69.00

Table 4B.1 *continued*

Country	Survey year	Number of firms	Fixed line[a]	Telecom major or severe obstacle to business operations and growth (% firms)	Unavailable main line telephone service (days)	Unavailable main line telephone service (hours)	Unavailable main line telephone service (total % sales lost)	Workforce regularly using computers in their jobs (percent)	E-mail use for interacting with clients and suppliers (% firms)
Macedonia, FYR	2002	170		3.70	3.29	..	..	..	45.88
Moldova	2002	174	C	4.85	1.73	11.80	..	..	37.93
Moldova	2003	103	C	2.30	4.14	..	0.16	..	53.40
Montenegro	2003	100	C	..	..	..	..	..	..
Morocco	2000	859	M	..	..	..	..	..	49.27
Mozambique	2002	194	M	20.65	13.59	..	..	6.87	36.31
Nicaragua	2003	452	M	12.39	6.96	5.79	9.51	7.93	37.39
Nigeria	2001	232	C	59.39	..	..	..	..	..
Pakistan	2003	965	C	9.24	2.66	..	..	10.65	33.58
Peru	2002	583	C	4.03	6.68	89.19	..	27.90	50.27
Philippines	2003	716	C	11.34	1.88	9.46	..	13.57	48.76
Poland	2002	500	C	3.70	1.43	..	..	..	66.40
Poland	2003	108	C	4.83	0.53	10.44	0.06	..	66.67
Romania	2002	255	C	7.11	1.98	..	..	..	57.25
Russian Federation	2002	506	D	4.29	7.79	..	..	..	53.56
Serbia	2001	402	C	13.93	7.90	..	..	27.67	45.27
Serbia	2003	408	C	..	..	..	..	..	..
Serbia and Montenegro	2003	508	C	..	2.21	7.14	0.54	25.52	53.94
Slovak Republic	2002	170	C	1.81	1.40	..	..	..	83.53
Slovenia	2002	188	M	1.06	1.12	..	..	..	87.23
Tajikistan	2002	176	M	6.33	43.36	9.22	0.41	..	10.80
Tajikistan	2003	107	M	4.09	4.99	..	..	..	7.62
Tanzania	2003	276	M	11.76	49.56	10.81	..	10.29	58.43
Turkey	2002	514	P	10.94	0.82	..	..	..	51.17
Uganda	2003	300	P	5.19	17.83	30.29	..	7.07	38.67
Ukraine	2002	463	D	5.36	3.41	..	..	..	57.02
Uzbekistan	2002	260	D	2.08	23.58	..	5.00	..	21.15
Uzbekistan	2003	100	D	5.65	1.80	..	..	..	14.00
Yugoslavia, the former	2002	250		10.33	13.35	..	..	..	70.00
Zambia	2002	207	M	32.85	40.05	13 48	..	10 03	83 50

Note: .. Not available. C = competition; D = duopoly; LCU = local currency unit; M = monopoly; P = partial competition.

a. Level of competition for fixed-line local services in 2004.

Source: Authors' analysis based on data from the World Bank Investment Climate Surveys 2000–2003.

Table 4B.2 ICT Results

Country	Survey year	Number of firms	Fixed line[a]	Web site use for interacting with clients and suppliers (% firms)	ICT services are NOT affordable (% firms)	ICT services poor in quality (% firms)	Waiting time for main line telephone connection (days)	Gift/ payment expected for connection (% firms answering yes)	Gift/payment expected for connection (value in LCU)
Albania	2002	170	C	32.35	18.36	..	18.10	..	..
Algeria	2002	557	M	24.74	..	19.71	276.53	4.67	719.71
Armenia	2002	171	M	30.41	..	..	5.43	..	..
Azerbaijan	2002	170	P	28.24	..	..	2.17	..	..
Bangladesh	2002	1001	C	31.12	20.10	16.94	150.43	..	147,809.93
Belarus	2002	250	C	44.80	..	..	59.92	..	..
Bolivia	2001	671	M	..	..	..	..	10.91	..
Bosnia and Herzegovina	2002	182	M	54.95	..	..	1.93	..	..
Brazil	2003	1642	C	73.14	..	21.60	18.22	2.10	301.94
Bulgaria	2002	250	C	54.00	..	..	20.05	..	..
Cambodia	2003	503	P	23.06	71.97	17.24	4.15	..	50.63
China	2002	1548	P	..	..	85.73	12.11	5.51	..
China	2003	2400	P	..	..	14.94	6.05	..	1.07
Croatia	2002	187	C	72.19	..	..	5.00	..	..
Czech Republic	2002	268	C	69.78	..	..	1.86	..	..
Ecuador	2003	453	M	55.41	8.05	12.79	129.68	27.36	176.59
Eritrea	2002	78	M	10.26	17.78	42.50	266.46	..	..
Estonia	2002	170	C	89.41	..	..	2.36	..	..
Ethiopia	2002	427	M	6.34	48.56	44.96	154.90	..	..
Georgia	2002	174	P	36.78	..	..	8.85	..	..
Guatemala	2003	455	C	29.23	22.96	9.27	47.68	3.92	560.00
Honduras	2003	450	M	21.78	32.97	14.05	170.11	14.36	11,419.05
Hungary	2002	250	C	66.00	..	..	4.41	..	..
India	2000	895	C	..	..	..	..	..	..
India	2002	1827	C	35.90	..	..	86.68	..	..
Indonesia	2003	713	P	24.33	60.87	9.71	26.59	4.25	36,564.14
Kazakhstan	2002	250	C	36.80	..	..	5.98	..	..
Kenya	2003	284	P	32.00	..	..	98.82	55.49	8,824.36
Kosovo	2003	329		35.99	3.23	6.67	26.96	15.79	3.68
Kyrgyz Republic	2002	173	C	28.90	..	..	21.11	..	2,500.00
Kyrgyz Republic	2003	102	C	42.57	..	..	11.57	30.43	..
Latvia	2002	176	C	53.98	..	..	2.83	..	..
Lithuania	2002	200	C	67.50	..	..	2.23	..	..
Macedonia, FYR	2002	170		47.06	..	..	8.19	..	..
Moldova	2002	174	C	29.89	..	..	7.83	..	..
Moldova	2003	103	C	53.40	..	..	88.20	20.00	900.00
Montenegro	2003	100	C	..	..	..	..	..	..
Morocco	2000	859	M	16.56	..	..	..	..	..
Mozambique	2002	194	M	19.16	48.00	32.50	21.20	..	..
Nicaragua	2003	452	M	16.19	37.89	14.29	127.60	11.71	2,765.64

Table 4B.2 *continued*

Country	Survey year	Number of firms	Fixed line[a]	Web site use for interacting with clients and suppliers (% firms)	ICT services are NOT affordable (% firms)	ICT services poor in quality (% firms)	Waiting time for main line telephone connection (days)	Gift/ payment expected for connection (% firms answering yes)	Gift/payment expected for connection (value in LCU)
Nigeria	2001	232	C	..	..	..	..	..	..
Pakistan	2003	965	C	18.34	47.60	22.04	25.29	..	4,945.90
Peru	2002	583	C	..	..	4.38	9.93	..	141.90
Philippines	2003	716	C	24.42	29.31	21.12	13.21	2.40	13.20
Poland	2002	500	C	65.00	..	..	7.61	..	..
Poland	2003	108	C	69.44	..	..	44.57	..	..
Romania	2002	255	C	53.73	..	..	6.03	..	..
Russian Federation	2002	506	D	48.42			14.42		
Serbia	2001	402	C			22.84	63.51		
Serbia	2003	408	C	33.08					
Serbia and Montenegro	2003	508	C	45.67	7.49	24.55	88.08	19.05	59.64
Slovak Republic	2002	170	C	80.59	..	..	2.85	..	..
Slovenia	2002	188	M	89.36	..	..	7.76	..	..
Tajikistan	2002	176	M	6.25	..	..	17.41	..	180.00
Tajikistan	2003	107	M	2.88	..	..	5.96	33.33	..
Tanzania	2003	276	M	23.48	29.96	18.44	23.07	18.89	110,000.00
Turkey	2002	514	P	50.78	..	..	1.52	..	..
Uganda	2003	300	P	10.00	..	..	33.41	18.32	83,000.00
Ukraine	2002	463	D	54.21	..	..	13.82	..	..
Uzbekistan	2002	260	D	16.15	..	..	20.00	..	..
Uzbekistan	2003	100	D	14.00	..	..	8.93	..	..
Yugoslavia, the former	2002	250		66.80	..	..	29.69	..	..
Zambia	2002	207	M	27.18	16.56	28.48	88.55	..	0.63

Note: .. Not available. C = competition; D = duopoly; LCU = local currency unit; M = monopoly; P = partial competition.

a. Level of competition for fixed-line local services in 2004.

Source: Authors' analysis based on data from the World Bank Investment Climate Surveys 2000–2003.

Annex 4C: Regression Results

Effect of ICT Use on Enterprise Performance

In the empirical analysis of the effects of ICT use on enterprise performance, the following model is used:

$$performance_{ijk} = \alpha + \beta \mathrm{ICT}_{ijk} + \gamma X_{ijk} + \lambda_k + \nu_j + \omega_t + \varepsilon_{ijk} \qquad (4C.1)$$

where $performance_{ijk}$ is the performance of enterprise i in sector j in country k. The performance measures are listed in annex 4A. ICT_{ijk} is an indicator reflecting enterprises' use of ICT. It is based on the three ICT indicators discussed above: a dummy variable indicating that the enterprise communicates with clients and suppliers using e-mail; a dummy variable indicating that the enterprise communicates with clients and suppliers using the Web (other than through e-mail); and the percentage of employees who regularly use computers. These three indicators are used to test the robustness of results. They are typically available for different countries and they capture different aspects of ICT use.[17]

In addition to the ICT variables, the model also includes X_{ijk}, a vector of enterprise level characteristics. These include enterprise size, age, exports as a percentage of sales, and dummy variables representing whether the enterprise is state- or foreign-owned.

Finally, the model includes a series of dummy variables to control for country (λ_k), sector of operations (v_j), and survey year (ω_t). The sector dummies control for systematic differences between enterprises operating in different sectors that might affect enterprise performance. Similarly, the country dummies control for differences between countries—such as macroeconomic stability, other government policies, and educational achievement—that might affect both technology use and enterprise performance. That is, the differences in productivity are interpreted as differences between enterprises within the same country. Finally, the dummies for the survey year control for systematic differences in the global economy that might affect enterprise performance and for changes in technology that might affect technology use over time.

Tables 4C.1 through 4C.3 present the regression results of the effect of ICT use on enterprise performance.

Impact of the Quality of Telecommunications Services on the Probability of Firms Using ICT

In order to investigate the impact of telecommunications services on the probability that a firm uses ICT, the following equation was used to estimate how much the quality of telecommunications infrastructure affects e-mail, Web, and computer use:

Table 4C.1 Effect of Computer Use on Different Measures of Enterprise Performance (regression results)

	(1)	(2)	(3)	(4)	(5)	(6)	(7)
Indicator	Sales growth	Employment growth	Value added per worker (US$ thousands)	Investment (share of assets)	Re-investment rate	Profitability	Technical efficiency
Observations	8,177	8,710	4,911	3,628	2,089	4,714	4,331
Country and year dummies	Yes	Yes	Yes	Yes	Yes	Yes	Yes
Sector dummies	Yes	Yes	Yes	Yes	Yes	Yes	Yes
Computer use	0.0001	−0.0001	0.1712***	0.0003	0.1465***	0.0007*	0.0004***
(Percentage of workers who use computers)	(0.64)	(0.92)	(16.78)	(1.53)	(4.41)	(1.78)	(7.69)
Worker	0.0144***	0.0249***	0.2455	0.0350***	4.1367***	0.0112*	−0.0005
(Natural log)	(4.03)	(11.84)	(1.33)	(9.24)	(7.27)	(1.68)	(0.47)
Age	−0.0746***	−0.0275***	−0.0079	−0.0322***	−5.4862***	−0.0164	−0.0035*
(Natural log)	(11.81)	(7.52)	(0.23)	(4.83)	(−5.75)	(1.33)	(1.88)
Exports	0.0006***	−0.0006***	−0.0019*	0.0000	0.0144	−0.0008**	−0.0000
(Percentage of sales)	(3.20)	(5.38)	(1.95)	(0.23)	(0.47)	(2.30)	(0.36)
Enterprise is majority foreign owned	0.0004	−0.0169*	8.4137***	−0.0041	−12.1374***	0.0119	0.0145***
(Dummy)	(0.02)	(1.82)	(10.39)	(0.27)	(−4.72)	(0.41)	(3.37)
Enterprise is state owned	−0.0267	−0.0237**	−2.4230***	−0.0061	−16.1606***	−0.0273	−0.0113***
(Dummy)	(1.56)	(2.30)	(2.95)	(0.33)	(−4.97)	(0.91)	(2.65)
Constant	−0.0437	0.1033*	1.7257	0.0856	74.1805***	−0.5379	0.6959***
(Dummy)	(0.27)	(1.93)	(0.15)	(0.49)	(6.11)	(1.28)	(11.80)
R^2	0.04	0.12	0.17	..	..	0.05	0.04

Note: .. Not applicable. The numbers in parentheses are *t*-statistics.

***, **, * denote statistical significance at the 1 percent, 5 percent, and 10 percent levels, respectively. All regressions include sector and year dummies.

Source: Authors' analysis based on data from the World Bank Investment Climate Surveys 2000–2003.

Table 4C.2 Effect of E-Mail Use on Different Measures of Enterprise Performance (regression results)

Indicator	(1) Sales growth	(2) Employment growth	(3) Value added per worker (US$ thousands)	(4) Investment (share of assets)	(5) Re-investment rate	(6) Profitability	(7) Technical efficiency
Observations	9,504	13,285	6,076	3,907	4,878	5,953	5,350
Country and year dummies	Yes	Yes	Yes	Yes	Yes	Yes	Yes
Sector dummies	Yes	Yes	Yes	Yes	Yes	Yes	Yes
Uses e-mail to communicate with clients and suppliers	0.0335***	0.0119*	3.4233***	0.0547***	6.8826***	0.0506***	0.0099***
(Dummy)	(3.82)	(1.90)	(8.66)	(4.35)	(3.93)	(2.80)	(3.87)
Worker	0.0091***	0.0207***	–0.2013	0.0257***	3.2737***	–0.0103	–0.0044***
(Natural log)	(2.85)	(9.40)	(1.40)	(5.87)	(5.53)	(1.56)	(4.69)
Age	–0.0750***	–0.0576***	1.0012***	–0.0438***	–4.0490***	0.0012	0.0040**
(Natural log)	(14.57)	(15.72)	(4.20)	(6.37)	(–4.32)	(0.11)	(2.52)
Exports	–0.0000	–0.0003***	0.0048	–0.0004**	–0.0254	0.0002	0.0001
(Percentage of sales)	(0.34)	(2.79)	(0.91)	(2.47)	(–1.02)	(0.75)	(1.57)
Enterprise is majority foreign owned	0.0029	0.0008	6.1198***	0.0263	–4.2040*	0.0338	0.0075*
(Dummy)	(0.21)	(0.09)	(9.93)	(1.52)	(–1.84)	(1.19)	(1.85)
Enterprise is state owned	–0.0632**	–0.1069***	4.5120***	–0.0256	–28.8728***	–0.1108*	0.0200**
(Dummy)	(2.22)	(8.32)	(3.35)	(0.72)	(–3.55)	(1.79)	(2.21)
Constant	0.1022	0.1018**	30.4289***	0.2186	–2.500	–0.1302	0.7523***
(Dummy)	(0.42)	(2.55)	(8.82)	(0.76)	(–0.10)	(0.31)	(28.13)
R^2	0.05	0.10	0.19	..	..	0.05	0.03

Note: .. Not applicable. The numbers in parentheses are *t*-statistics.

***, **, * denote statistical significance at the 1 percent, 5 percent, and 10 percent levels, respectively. All regressions include sector and year dummies.

Source: Authors' analysis based on data from the World Bank Investment Climate Surveys 2000–2003.

Table 4C.3 Effect of Web Use on Different Measures of Enterprise Performance (regression results)

Indicator	(1) Sales growth	(2) Employment growth	(3) Value added per worker (US$ thousands)	(4) Investment (share of assets)	(5) Re-investment rate	(6) Profitability	(7) Technical efficiency
Observations	8,926	12,699	5,527	3,710	4,825	5,422	4,900
Country and year dummies	Yes	Yes	Yes	Yes	Yes	Yes	Yes
Sector dummies	Yes	Yes	Yes	Yes	Yes	Yes	Yes
Uses Web to communicate with clients and suppliers	0.0203**	0.0178***	2.3402***	0.0210	3.6031**	0.0393**	0.0041
(Dummy)	(2.24)	(2.76)	(5.66)	(1.63)	(2.24)	(1.99)	(1.53)
Worker	0.0135***	0.0204***	–0.0064	0.0319***	3.7100***	–0.0032	–0.0032***
(Natural log)	(4.09)	(8.96)	(0.44)	(6.92)	(6.35)	(0.45)	(3.31)
Age	–0.0777***	–0.0603***	1.0376***	–0.0464***	–4.3009***	0.0017	0.0026
(Natural log)	(14.60)	(15.87)	(4.31)	(6.42)	(–4.54)	(0.15)	(1.61)

(*Table continues on the following page.*)

Table 4C.3 *continued*

Indicator	(1) Sales growth	(2) Employment growth	(3) Value added per worker (US$ thousands)	(4) Investment (share of assets)	(5) Re-investment rate	(6) Profitability	(7) Technical efficiency
Exports	–0.0000	–0.0003***	0.0049	–0.0004**	–0.01675	0.0001	0.0000
(Percentage of sales)	(0.26)	(2.65)	(0.89)	(2.52)	(–0.67)	(0.48)	(1.40)
Enterprise is majority foreign owned	0.0040	0.0025	6.3262***	0.0344*	–4.0198*	0.0346	0.0108***
(Dummy)	(0.28)	(0.27)	(10.37)	(1.91)	(–1.74)	(1.19)	(2.66)
Enterprise is state owned	–0.0670**	–0.1039***	3.1836**	–0.0292	–29.4981***	–0.1277**	0.0168*
(Dummy)	(2.28)	(7.89)	(2.35)	(0.77)	(–3.62)	(1.97)	(1.81)
Constant	–0.0168	0.1082***	2.8770	0.0020	..	–0.1506	0.7878***
(Dummy)	(0.03)	(2.66)	(1.11)	(0.01)	..	(0.36)	(8.88)
R^2	0.05	0.11	0.18	..	..	0.05	0.02

Note: .. Not applicable. The numbers in parentheses are *t*-statistics.

***, **, * denote statistical significance at the 1 percent, 5 percent, and 10 percent levels, respectively. All regressions include sector and year dummies.

Source: Authors' analysis based on data from the World Bank Investment Climate Surveys 2000–2003.

$$\text{ICT Use}_{ijk} = \beta_1 + \beta_2 \text{ Telecom}_{ijk} + \beta_3 \text{ IT Quality}_{ijk} + \beta_4 \text{ X}_{ijk} + \beta_5 C_k + \omega_t + \varepsilon. \quad (4C.2)$$

The dependent variables are the ICT-use variables by firm *i* in sector *j* in country *k*, the same ones used in the previous section: dummy variables indicating that the company uses e-mail and the Web to communicate with clients, and a continuous variable, truncated at zero, indicating the percentage of employees who use computers in their jobs. For the two regressions with dummy variables as the dependent variables, Probit analysis is used. Since the percentage of employees who use computers in their jobs is truncated above at 100 percent (that is, all employees use computers) and below at 0 percent (that is, no employees do), Tobit

Table 4C.4 Effect of Telecommunications Service on IT Use
(regression results)

Indicator	(1) Uses e-mail to communicate with clients and suppliers (dummy)	(2) Uses Web to communicate with clients and suppliers (dummy)	(3) Percentage of employees using computers
Observations	7,286	7,194	7,603
Technology use			
Quality of information technology services	0.2254***	0.1387***	3.6647***
(Average for firms in sector/region)	(5.21)	(2.76)	(3.95)
Telecommunications sector performance			
Days to get telephone	–0.0009***	–0.0004***	–0.0465***
(Average for firms in sector/region)	(6.68)	(2.99)	(4.43)
Enterprise characteristics			
Workers	0.1679***	0.1226***	3.0550***
(Natural log)	(31.03)	(24.53)	(13.49)
Age of firm	–0.0348***	–0.0010	–2.5098***
(Natural log)	(4.25)	(0.13)	(6.16)
Exports	0.0009***	–0.0007***	–0.0136

Table 4C.4 *continued*			
Indicator	(1) Uses e-mail to communicate with clients and suppliers (dummy)	(2) Uses web to communicate with clients and suppliers (dummy)	(3) Percent of employees using computers
(As a percentage of sales)	(4.15)	(3.52)	(1.16)
Enterprise is foreign owned	0.0884***	0.0077	9.1440***
(Dummy)	(3.58)	(0.37)	(8.94)
Enterprise is state owned	–0.2882***	–0.1940***	–8.0229***
(Dummy)	(6.18)	(4.65)	(7.35)
Country characteristics			
Per capita GDP	0.0939***	0.1764***	7.3547***
(Natural log)	(9.37)	(16.53)	(10.97)

Note: The numbers in parentheses are *t*-statistics.

***, **, * denote statistical significance at the 1 percent, 5 percent, and 10 percent levels, respectively. All regressions include sector and year dummies.

Source: Authors' analysis based on data from the World Bank Investment Climate Surveys 2000–2003..

estimation is used for that variable. In all cases the error term, ε_{ijk}, is assumed to have a normal distribution.

The number of days it takes to get a fixed-line telephone is used as a proxy for the quality of telecommunications service as the main independent variable. This measure is used because it is available for more countries than alternative measures. Because of concerns about endogeneity (for example, firms' perceptions about quality might be affected by the use of information technology), average values of the quality variables for firms in the same sector and region of the country are used. Because an individual firm's use of information technology is unlikely to affect other firms' perceptions, this seems a reasonable proxy for quality of service.

The regressions include several additional variables of country- as well as enterprise-level characteristics, such as enterprise size, age, exports as a percentage of sales, and dummy variables representing whether the enterprise is state- or foreign-owned.

Table 4C.4 shows the regression results of the effect of telecommunications.

Endnotes

1. The results are robust to controlling for the possibility that Internet use is endogenous (that is, that causation also runs in the opposite direction).
2. For example, customer databases with a history of client-specific correspondence help managers and employees respond more effectively to customers.
3. See http://rru.worldbank.org/InvestmentClimate/ for a full explication of the survey.
4. The sectors covered in the ICS are Textiles; Leather; Garments; Agro-Industry; Food; Beverages; Metals and Machinery; Electronics; Chemicals and Pharmaceuticals; Construction Materials; Wood and Furniture; Non-metallic and Plastic Materials; Paper, Printing and Publishing; Sports; Information Technology; Other Manufacturing; Telecommunications; Accounting and Financial Products; Advertising and Marketing; Other Services; Retail and Wholesale Trade; Hotels and Restaurants; Transportation; Real Estate and Rental Services; Mining and Quarrying; and Autos and Auto Parts. Although this study covers 56 countries, some countries had more than one survey conducted in different years. Therefore, as shown in annex 4A, some of the regressions include close to 60 surveys.
5. The size of a firm is defined as the sum of the number of permanent and temporary workers. Large firms have more than 100 employees. SMEs range from 10 to 100 employees. Micro firms have fewer than 10 workers.
6. For example, because of the high cost of making international calls in many developing countries, and because of the need of foreign-owned enterprises to keep in touch with their head offices, these enterprises might be more likely to use the Internet than domestic enterprises (Clarke 2004). But foreign-owned enterprises also tend to be more technically efficient.
7. Although, ideally, it would be desirable to control for the educational attainment of the firms' workforce, comparable data were not easily available for the firm surveys used in this study. One concern is that firms with highly educated workers might be more likely to use ICT services and to exhibit higher productivity. Omitting a variable to control for the educational status of workers could bias the coefficients for ICT use upward in the empirical analysis. Another single-country study, however, has found—after controlling for worker and manager education—that total factor productivity (technical efficiency)

was higher in firms in South Africa that use computers more intensively (World Bank 2005).

8. A firm's propensity to invest in new ICT technologies also depends on the effect that other infrastructure services, such as electricity and transport, have on its production process.

9. A common criticism of perception-based measures of the investment climate is that they might not reflect reality. Idiosyncratic factors (for example, the manager's temperament or expectations about sector performance) might determine perceptions about sector performance rather than demonstrating the actual performance of the sector. In this case, however, the perception-based indexes appear to be consistent with objective measures of sector performance even after controlling for other factors that affect perceptions about telecommunications service. Based on a regression analysis, findings showed that managers were more likely to rate telecommunications as a serious problem in countries where it takes a long time to get a fixed-line connection, where service interruptions are more common, and where bribes are needed to get a new connection.

10. The coefficient for bribes is negative but is statistically insignificant as to the regression for Web use.

11. SMEs (with fewer than 250 employees, according to the OECD definition) and micro firms (with fewer than 10 employees) in some developed countries experience similar constraints in ICT adoption, as reported by the OECD survey *ICT, E-Business and SMEs* (2004).

12. According to the ITU (2001), the average cost to build a fully electronic commerce–enabled site was $250,000 in 2001. This figure can range from $500,000 to $2 million for larger companies. WITSA (2000) estimated that the average time to build a fully functioning site was five months, but this often stretched to a year.

13. Costs, however, of both technology and the skills to implement it, are beginning to decline as electronic features are incorporated into existing products and skills become more widely available.

14. Since the variables are in natural logs, the country dummies will remove any exchange rate effects. That is, once country dummies are included, the coefficients on labor, capital, and intermediate inputs will be identical whatever exchange rate is used. In practice, however, the estimates of technical efficiency are virtually identical whether country dummies are included or not. The correlation between the estimates with and without country dummies is about 0.97.

15. In practice, there are too few data to estimate production functions for retail and wholesale trade, hotels and restaurants, transportation, real estate and rental services, and mining and quarrying. Therefore, production functions are estimated for only 21 sectors out of the original 26 sectors listed in the ICS.

16. In practice, results are very similar under different distributional assumptions. The simple correlation between this measure of technical efficiency and estimates assuming a half-normal distribution is 0.97.

17. Because these variables are highly correlated and because they are available for different countries, they are included one at a time. Their coefficients therefore should be interpreted with care. Including all variables simultaneously would result in the sample size being reduced considerably, making it more difficult to estimate the impact of the indicators. As a result, the effects of the different variables on enterprise performance should not be summed. Rather, they should be seen as general indicators of information technology use and interpreted in this light.

References

Asia Foundation. 2002. "SME and eCommerce." http://www.asiafoundation.org/ICT/surveys.html. (accessed September 16, 2005)

Barr, Abigail. 2000. "Social Capital and Technical Information Flows in the Ghanaian Manufacturing Sector." *Oxford Economic Papers* 52 (3): 539–59.

Bhutan (the government), UNDP (United Nations Development Programme), and World Bank. 2002. "Bhutan Private Sector Survey." Draft report.

Clarke, George. 2004. "The Effect of Enterprise Ownership and Foreign Competition on Internet Diffusion in the Transition Economies." *Comparative Economic Studies* 46 (2): 341–70.

Clarke, George, and Scott Wallsten. 2004. "Has the Internet Increased Trade? Evidence from Industrial and Developing Countries." World Bank Policy Research Working Paper 3215, World Bank, Washington, DC.

CRITO (The Center for Research on Information Technology and Organizations). 2002. "Global E-Commerce Survey." Draft paper. The Center for Research on Information Technology and Organizations at the University of California, Irvine. http://www.crito.uci.edu/.

De Vera, Roberto. 2002. "The Employment Impact of Business to Consumer E-Commerce on the Philippine Workers." Philippine APEC Study Center Network Discussion Paper No. 2002–09, Makati City, Philippines.

Freund, Caroline L., and Weinhold, Diana. 2004. "The Effect of the Internet on International Trade." *Journal of International Economics* 62 (1): 171–89.

Humphrey, John, Robin Mansell, Daniel Paré, and Hubert Schmitz. 2003. *The Reality of E-Commerce with Developing Countries.* London: London School of Economics and Institute of Development Studies.

ITU (International Telecommunication Union). 2001. "Internet and Electronic Commerce." http://www.itu.int/newsarchive/wtd/2001/FeatureE-commerce.html.

———. 2003a. *Birth of Broadband: ITU Internet Reports.* Geneva: ITU.

———. 2003b. *World Telecommunication Development Report.* Geneva: ITU.

———. 2004. *ITU Internet Reports: The Portable Internet.* Geneva: ITU.

Moodley, Sagren. 2002. "Competing in the Digital Economy? The Dynamics and Impacts of B2B E-commerce on the South African Manufacturing Sector." Discussion Paper 2002/79, WIDER United Nations University, Helsinki.

MPDF (Mekong Project Development Facility). 2003. *Current IT Usage at SMEs.* Mekong Project Development Facility, Ho Chi Minh City.

OECD (Organisation for Economic Co-operation and Development). 2004. *ICT, E-Business and SMEs.* Paris: OECD.

WITSA (World Information Technology and Services Alliance). 2000. *International Survey of E-commerce 2000.* Arlington, VA: World Information Technology and Services Alliance.

Wong, Poh-Kam, and Yuen-Ping Ho. 2004. "E-Commerce in Singapore: Impetus and Impact of Globalization." GEC Project, CRITO, Irvine University. http://www.crito.uci.edu.

World Bank. Various years. "Investment Climate Surveys."-Washington, DC. http://rru.worldbank.org/InvestmentClimate/

———. 2004. *Doing Business in 2004: Understanding Regulation.* Washington, DC: World Bank.

———. 2005. "South Africa: An Assessment of the Investment Climate." Unpublished, World Bank, Washington, DC.

Chapter 5

Trends in National E-Strategies: A Review of 40 Countries

Aref Adamali, John Oliver Coffey, and Zaid Safdar

E-strategies have been on the international development agenda in recent years. The Group of Eight (G-8) and the United Nations (UN), among others, have advocated the need for developing countries to establish information and communication technology (ICT) programs to better use ICT for development. In 2000, the Digital Opportunity Task Force (DOT Force) was launched under the auspices of the G-8. In 2001, the UN Secretary General—with the recommendation of the Economic and Social Council (ECOSOC)—created the ongoing ICT Task Force. In 2003, world leaders met for the first phase of the World Summit for the Information Society (WSIS) and adopted a Plan of Action encouraging that national e-strategies be developed by the time the second phase of WSIS convenes in November 2005.[1]

Some countries, mostly developed ones, initiated e-strategies on their own. They recognized the potential ICT has for their economies and societies. They championed plans and actions that included ICT as an important part of their respective national strategies. Countries that launched e-strategies early on and followed through are reaping benefits today; these countries are regarded as forerunners in ICT development. For instance, Singapore began its ICT program in 1991, the United States did in 1993, and Canada, Japan, and most European nations started shortly thereafter (UNCTAD 2003, page 64).

The estimated number of national e-strategies worldwide ranged between 70 and 90 until 2003.[2] Before that year, some countries had not begun formulating e-strategies. Others had already written their ICT plans and were in the process of implementing them. Still others, such as Singapore, the Republic of Korea, and Mauritius, had completed several cycles of e-strategies and were further ahead in their ICT development.

Developing a national e-strategy is a daunting task. It requires an understanding of the social and development priorities of a country. It requires vision and leadership from the highest levels of government. It requires rationalizing how individual ICT objectives are to be carried out, both in assigning responsibility to individual government agencies and in committing financial resources. It also requires government emphasis on measuring results so outcomes can be assessed and future directives can be planned based on real data and concrete information (see annex 5A for a selected list of resources for ICT policies and e-strategies).

This chapter gives an overview of how countries have been formulating national e-strategies and what they have been focusing on. It examines written e-strategy plans of 40 countries (annex 5B), both developed and developing, taken from a wide spectrum of income and region groups (figure 5.1).[3] It evaluates the strength, scope, and direction of these countries' ICT programs based on the goals stated in their e-strategy documents. It produces lessons learned that developing countries can review and use to strengthen their ICT programs. The chapter is a time-invariant study of what

Figure 5.1 E-Strategy Review Sample: Country Composition by Region and Income Level (percent)

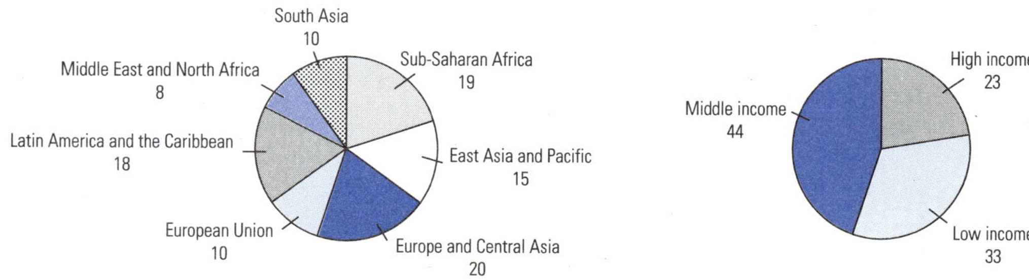

Source: Authors' analysis.

e-strategy programs say; it therefore does not deal with the actual implementation of particular e-strategies or results they may have produced.

This chapter has three parts. The first explains the review methodology used in analyzing the e-strategies. The methodology illustrates not only how the review is done but also the parameters on which e-strategy designs should theoretically be based. The second part presents the results of the analyses. The results look at how e-strategies are conceptualized and on what objectives and interventions they focus. And the third gives recommendations to national ICT strategists and policy makers on formulating effective e-strategies that could prove to be successful and help them achieve their stated goals.

What Is an E-Strategy?

At the national level, *e-strategy* refers to a plan of action—typically a strategy document written by state leaders—illustrating how ICTs are to be developed and used to achieve the economic, social, and development objectives of a country. E-strategy thus guides and focuses government priorities in ICT development. It explains how institutions interact with one another and how they share resources and responsibilities for ICT development. It specifies a multisector activity that involves leaders from government, the private sector, academia, and civil society.

This chapter deals exclusively with e-strategy at the national level. It uses the terms *e-strategy, national e-strategy, national ICT plan,* and *national ICT strategy* interchangeably.

E-Strategy and Development

E-strategy objectives are tied to the country's overall development objectives, which include topics such as education, health, government, business, and industry. ICT development is intended not as an end in itself but as a means to fulfilling the larger development needs of a country. Linking e-strategy to a country's development strategy also gives credibility to the ICT program and confers wider acceptance to it outside ICT circles.

Connecting e-strategy to development requires coordination and sequencing across governmental agencies. For example, if a country wants to introduce distance education, it should tie its initiative not only to e-strategy objectives (such as promoting e-literacy or enhancing the use of ICT in education) but also to "d-strategy" and more generic policy objectives (such as developing ICT usage or improving education delivery in general). The latter may involve the diversification of its economy from traditional to newer sectors (World Bank 2005).

E-Strategy Life Cycle

E-strategies move through several stages of life cycle, as shown in the diagram in figure 5.2.

The e-strategy life cycle can be broken down into three parts. At the beginning, a national ICT vision is developed. This vision takes into account the current ICT availability, development objectives, and input from various stakeholders. Next, responsible institutions and organizations are identified to carry out the assigned tasks. Finally, the

Figure 5.2 A Typical E-Strategy Life Cycle

Source: Authors' analysis.

e-strategy is monitored to assess the level of progress achieved in the country's ICT capability.

The e-strategy life cycle thus explained does not show how monitoring and evaluation (M&E) is integral to e-strategy evolution and how it feeds back into e-strategy formulation. But it does denote one basic idea: e-strategy formulation drives both implementation and M&E. Hence, the e-strategy plan has to look forward—not only in *what* it hopes to achieve but also *how* it intends to do so.

At times, e-strategies evolve along with the country's development needs and implementation capacities. In such cases, they undergo multiple life cycles. For example, Singapore is today implementing its fourth e-strategy. From 1980 to 1985, it conducted an ICT program to overcome shortages of national resources and to exploit its human resource base. From 1986 to 1991, its "National IT Plan" aimed to spread network connections to the private sector and professional groups through trade, legal, and health services. In 1992, it embarked upon an "IT 2000 – Intelligent Island" plan. In 2000, it started an "Infocomm 21 – Infocomm Capital" program (Tipson and Fritelli 2003).

E-Strategy Review Methodology

Looking at the e-strategies from the 40 selected countries, two forms of analysis have been carried out:

- *Analysis I: How are e-strategies formulated?* The e-strategies are evaluated for how they link to a country's development goals, how they indicate institutional and budgetary support for implementation, and how they incorporate M&E mechanisms. This chapter uses an analytical framework designed to assess basic elements of e-strategies and to allow comparisons across country, income, and region groups.
- *Analysis II: What do e-strategies focus on?* The main themes of each e-strategy, such as e-government or ICT infrastructure, are examined. For each theme, prominent objectives and interventions are identified. For example, within e-government, government-to-government (G2G) applications may be seen by some countries as a top objective and process reform a primary means to develop G2G.

Framework for E-Strategy Analysis I

Four criteria are used to evaluate national e-strategies in Analysis I (figure 5.3). Mapped to the e-strategy life cycle discussed earlier, the four review criteria are

- development linkages,
- use of indicators,
- implementation mechanisms, and
- M&E mechanisms.

Each criterion is scored on a scale of 0 to 3, where 0 is low and 3 is high. Scores are assigned on a normative basis based on how countries perform relatively. Annex 5C shows details of the scoring scale and a summary of scorecards from Analysis I. Descriptions of each review criterion and the rationale for using it follow.

Development Linkages. This criterion determines how tightly e-strategy is linked to the country's larger political, economic, and social development goals. To evaluate development linkages, e-strategies are scored based on closeness of the e-strategy to its stated objectives and to the country's other goals.

Figure 5.3 Framework for E-Strategy Analysis I

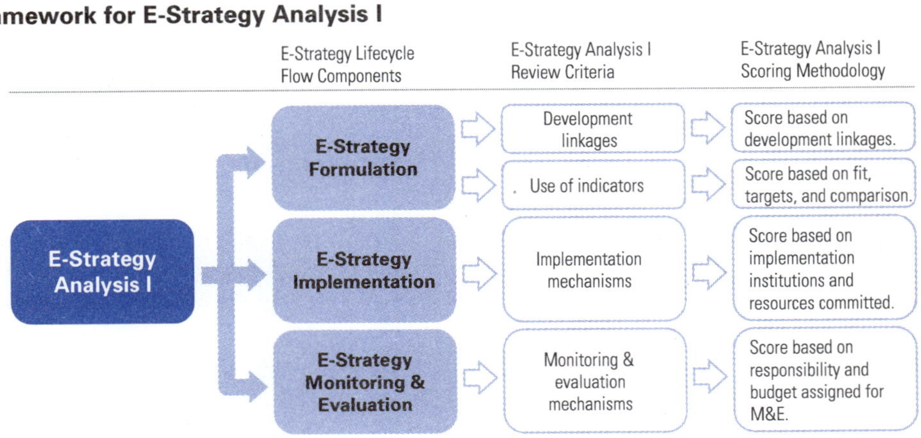

Source: Authors' analysis.

Use of Indicators. This criterion gauges the use of data indicators in e-strategies (see also chapter 6 for more discussion of the importance of indicators). Using indicators is essential for accurately benchmarking baseline analysis, for formulating targets, and for M&E (which is considered separately in the analysis). Benchmarking is useful for assessing the country's current level of ICT development.

The e-strategies are assessed for the degree to which they do cross-country comparisons in baseline assessments and target setting. Cross-country comparisons help provide the context in which countries can understand their current level of development. Incorporating benchmarks relative to other countries also helps identify areas of potential comparative advantage.

To evaluate the use of indicators, e-strategies are scored on four points:

1. Fit to goals: the extent to which data are selected or customized to fit the main thematic areas of the e-strategy and the initiatives they are intended to advance.
2. Baseline data: the extent to which baseline data are used in understanding the country's current state of ICT development.
3. Target setting: the extent to which quantitative and qualitative targets are established to achieve the main objectives of the e-strategy.
4. Cross-country comparison: the extent to which cross-country information is integrated into baseline analysis and used in establishing credible targets.

Implementation Mechanisms. This criterion evaluates the types of institutions designated to manage e-strategy implementation. The e-strategies are assessed for the degree of clarity with which they address implementation mechanisms and related roles and responsibilities. E-strategies must be explicit about implementation roles if they are to move from being conceptual plans to practical tools that can lead a country's ICT development efforts.

To evaluate implementation, e-strategies are scored on two factors :

1. Institutional structure and responsibility: whether e-strategies are specific about what institutions would lead implementation of key components and whether e-strategies clarify responsibility and report operational mechanisms.
2. Budget: whether specific details are given about budgetary requirements and about potential funding sources to implement key initiatives.

Monitoring and Evaluation Mechanisms. This criterion assesses whether M&E is an explicit part of the e-strategy and whether there is a clear plan as to how M&E will be conducted. The starting assumption is that M&E is integral to the design and implementation of effective e-strategies. Incorporating M&E ensures that e-strategies are explicit and realistic in what they aim to achieve. It also ensures that their implementation is regularly assessed and realigned so that scarce public resources are properly used. The credibility of e-strategies depends upon a solid and realistic M&E foundation. E-strategies

Figure 5.4 Framework for E-Strategy Analysis II

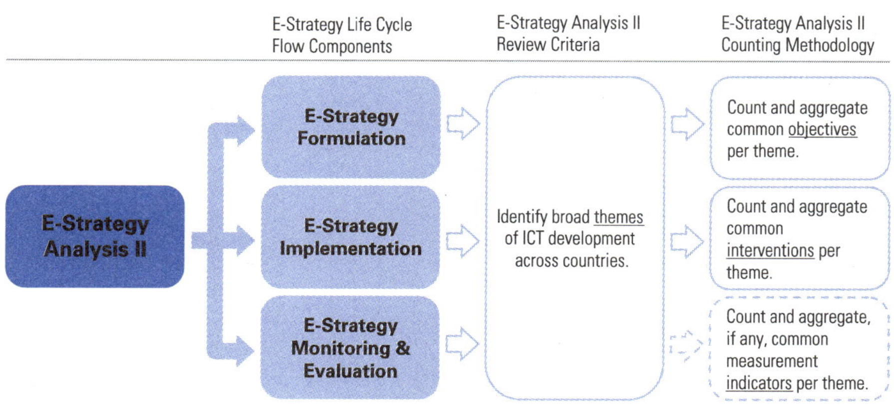

Source: Authors' analysis.

are scored on two factors in evaluating M&E: structure and responsibility, and budget.

Framework for E-Strategy Analysis II

Two steps are taken to evaluate e-strategies in Analysis II (figure 5.4). First, a broad set of ICT development themes are identified across all 40 countries. Second, a common set of objectives and interventions for each theme are aggregated.

As opposed to normative scoring, the themes, objectives, and interventions are now counted and their relative frequency noted among given countries. Annex 5D shows the detailed tabulated results from Analysis II for the 40 countries.

Measurement indicators shown in the diagram in figure 5.4 can help gauge e-strategy performance in a specific thematic or application area. In our review, such indicators were not seen in e-strategies. Incorporating data indicators to benchmark and measure progress based on objectives is an area countries could improve in their ICT strategies.

Analysis I: Trends in National E-Strategies

This section presents results of *how* e-strategies are formulated, highlighting their strengths and weaknesses.

Overview

Overall, e-strategies perform marginally in their designs based on the analysis (figure 5.5). On a scale of 0 to 3, they fall short of midway, at 1.3. This indicates that although countries have made significant progress in setting up e-strategies for ICT development, they need to do more. The ICT strategies show better results in providing implementation details and forming development linkages, but they are weak in incorporating M&E. E-strategies from middle-income economies and the Middle East and North Africa region score relatively higher than those from other income or regional groups. E-strategies from Mozambique, Rwanda, Trinidad and Tobago, and Ukraine have overall best scores in the four analytical categories.

Development Linkages

The e-strategies score relatively well on development linkages, meaning that ICTs are fairly strongly tied to overall development objectives (figure 5.6). Of the four main categories for which e-strategies are assessed, development linkages score second to implementation mechanisms.

Overall, middle-income countries score highest on this measure. Mauritius, Trinidad and Tobago, Tunisia, and Ukraine—all middle-income economies—receive good scores on development linkages. ICT may be seen by middle-income countries to be a relevant tool for tackling what they may regard as the "next level" of development challenges. For example, Trinidad and Tobago has articulated an e-strategy that is clear in describing how ICT development fits in with its nontechnology activities. The country's ICT strategy explicitly states its intention to contribute to the National Development Plan ("Vision 2020") by creating greater social equity through providing universal access to ICT. The e-strategy establishes a timetable and a methodology to determine how the expansion of ICT in the country can be leveraged to support economic, social, and environmental policy objectives.

Figure 5.5 Analysis I Review, Average Scores

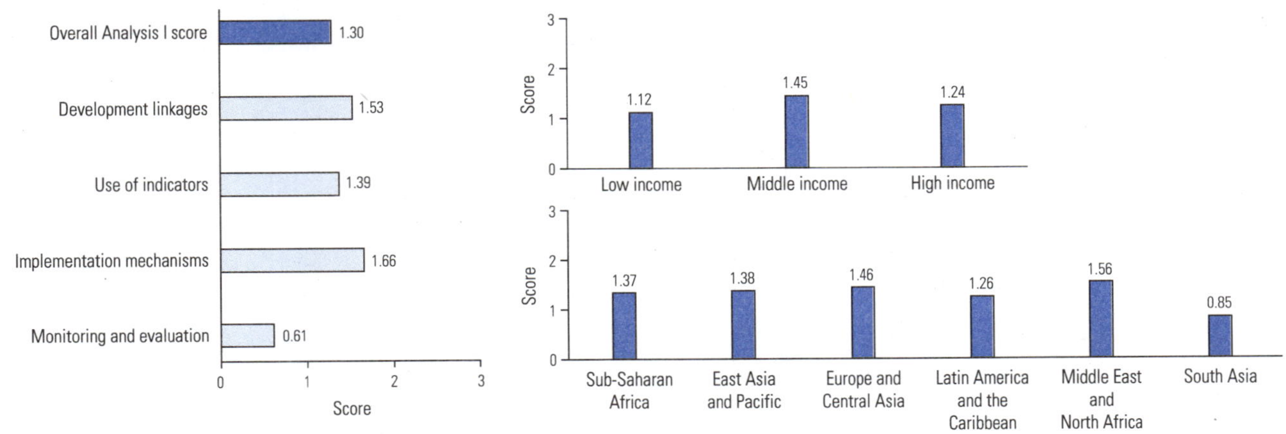

Source: Authors' analysis.

Figure 5.6 Development Linkages by Income Group, Average Scores

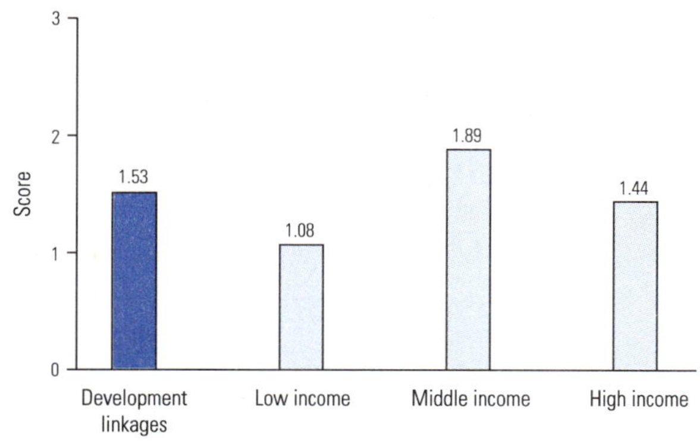

Source: Authors' analysis.

High-income countries, on the other hand, score low in development linkages because they presumably do not find it necessary to draw close and unambiguous linkages between their development initiatives and the role of ICT; the linkages may be sufficiently obvious to them. Low-income countries likewise may find it difficult to conceive of and communicate the linkage between ICT and their many daunting development challenges in a cohesive e-strategy document. They may also assign higher priorities to basic necessities such as food and health than to developing ICT.

Use of Indicators

On average, the e-strategies score worse on the use of indicators than they do on development linkages, meaning that the countries reviewed use little or no data in formulating their current state analysis or developing future targets (figure 5.7).

As many as two-thirds of the e-strategies perform weakly on the use of cross-country comparison. Such poor performance indicates that strategy formulators are not crafting e-strategies to take into account where a country stands vis-à-vis other countries in ICT development. This is noteworthy because cross-country comparison is commonly used in much of the ICT-for-development literature, and because countries are often presented in terms of their relative e-readiness or e-development rankings on a number of indexes.

Though many e-strategies score low on how well they use data to fit with goals or how well they compare their own ICT development with that of other countries, many of the e-strategies score high on their use of baseline data and targets. A greater number of countries score high in their use of targets—that is, targets are embedded throughout their e-strategies—than they do in the other indicator categories. This is understandable because e-strategies are forward-looking documents, charting out new territory for development and establishing targets by which to guide this process.

Low-income countries—especially from Europe and Central Asia, Latin America and the Caribbean, and South Asia—show weak performance in the use of indicators (figure 5.8). Middle-income nations, however, specifically Jordan and Tunisia, score high on the use of indicators. Jordan's "Reach" e-strategy, for example, incorporates indicators and targets throughout its plan. It draws extensively on comparisons with other countries with which Jordan either competes or that it seeks to emulate in similar levels of success in ICT development (box 5.1).

Figure 5.7 Categories for Use of Indicators by Income Group, Average Scores

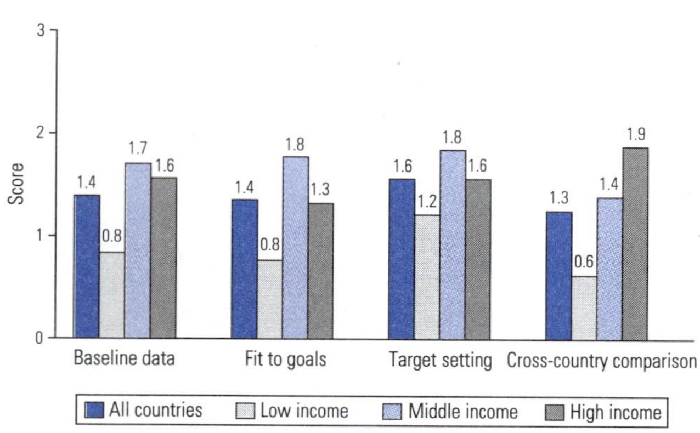

Source: Authors' analysis.

Figure 5.8 Use of Indicators by Income Group, Average Scores

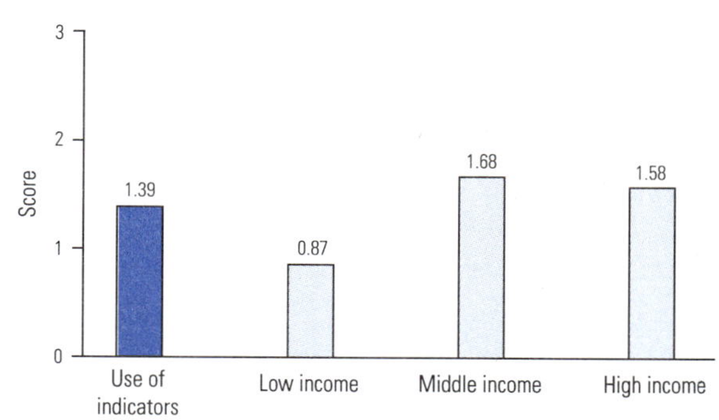

Source: Authors' analysis.

Box 5.1. Jordan's E-Strategy and Its Challenges

Jordan launched an e-strategy in 2000 and focused it on developing its ICT sector. It used baseline data extensively and provided a current status of the ICT sector. It drew data from other countries—mainly the Arab Republic of Egypt, India, Ireland, and Israel—to make cross-country comparisons. It looked at

- the number of software and IT services firms,
- the number of people employed in the sector,
- sector revenues,
- annual growth rates,
- the number of employees per firm,
- the revenue per employee,
- the total value of sector exports,
- the value of exports per employee,
- major products of the sector,
- the sources of foreign direct investment (FDI), and
- leading private investors or partners participating in the sector.

Based on these data, the e-strategy reviewed Jordan's comparative strengths and weaknesses and established three high-level goals:

- create 30,000 IT-related jobs by 2004,
- generate $550 million in annual exports by 2004, and
- ensure $150 million in cumulative FDI by 2004.

The e-strategy was effective in laying out specific actions and deliverables for individual initiatives, grouped in six key focus areas to achieve its overarching targets. But the e-strategy was weak in implementation and M&E, and it fell short of meeting its goals by the year 2004.

An updated e-strategy was developed four years after the original launch, in 2004. It reviewed progress made against both the three high-level targets and the deliverables from each individual initiative. Taking into account the slower growth from preceding years, the revised e-strategy lowered the initial targets substantially. The deadline for reaching the targets was extended from 2004 to 2006, the annual export target from the ICT sector was reduced from $550 million to $100 million, a new target of $550 million in domestic revenues was added, and the FDI target was increased by an additional $20 million.

Jordan's case shows how all elements of e-strategy design are important. Laying out an ICT plan using real data and indicators is necessary. But emphasizing implementation and measuring results is also essential for realistically achieving stated goals and targets.

Source: World Bank 2005.

Implementation Mechanisms

Implementation mechanisms score highest among the four e-strategy formulation criteria for countries reviewed. The two categories of implementation mechanisms are shown in figure 5.9. The majority of e-strategies are clearer in the detail they provide on institutions to lead the implementation of e-strategy than they are on any other criteria on which they are assessed.

Though many e-strategies score high on implementation detail, considerably fewer are equally specific about how to finance the implementation. Most e-strategies score lower on budget details than they do on institutional structure; two-fifths provide no budget information at all (despite being explicit about institutional structures).

Slovenia is one exception. In Slovenia, the implementation plan is set out in a detailed matrix that includes policy

Figure 5.9 Categories of Implementation Mechanisms, Average Scores

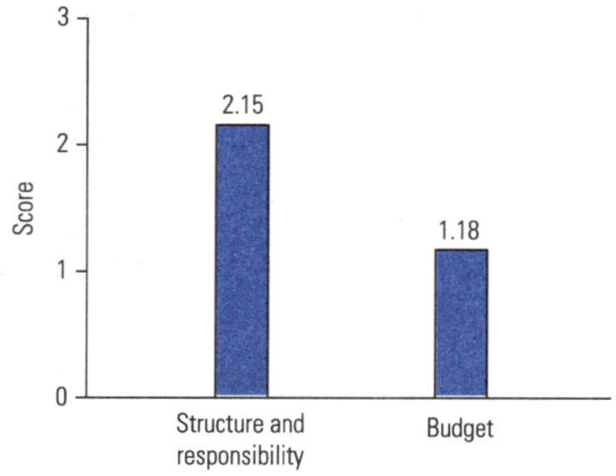

Source: Authors' analysis.

objectives, supporting interventions, the status of each intervention, a measure describing the risks and dependencies of each intervention, and the government agency responsible for each intervention. Linked to this description of implementation and responsibility is a detailed explanation of the funding sources and how those funds will be deployed. Three sources of financing are identified:

1. *Existing ministry budgets.* The financing of the e-strategy will be arranged to be consistent with the decentralized nature of the implementation responsibility of the e-strategy. In other words, each government ministry and agency will use existing budget money to execute their respective e-strategy responsibilities; financing will not in this instance be drawn from a central fund.

2. *European Union subsidies.* The central e-strategy coordinating body will disburse EU subsidies according to their priority in the National Development Plan, subject to the creation of cost appraisals.

3. *International Financial Institution funds.* These resources are regarded as an "additional" source of funding to be used to realize objectives that could not be completed by using funds described above. A special government resolution and needs analysis is required to obtain this line of credit.[4]

High-scoring e-strategies adhere to two general models concerning the types of structures that are responsible for implementing the e-strategy. The first model, where the implementation of the e-strategy is fully centralized, has the central government taking full responsibility for defining and implementing its elements. The second model has a decentralized implementation structure. In this model, different government ministries, agencies, and other stakeholders (such as the private sector) are responsible for defining and implementing parts of the e-strategy. These different entities answer to a central government oversight and coordination body.

Nigeria is an example of the first, centralized model. The role assigned to the National Information Technology Development Agency (NITDA) is to implement, monitor, evaluate, regulate, and verify ICT activities on an ongoing basis. NITDA acts under the supervision and coordination of the Federal Ministry of Science and Technology. National programs to foster the development and growth of ICT in Nigeria are operated and directly controlled by NITDA by consulting—in some cases collaborating—with key stakeholders. Chief Information Technology Officers are appointed in all federal agencies to advise NITDA, but they are not responsible for implementing programs per se.

Mozambique exemplifies the second, decentralized model. It emphasizes the fact that successful ICT strategy depends on the active participation of all sectors of society and the economy, including the beneficiaries. At the highest level, a National Consultative Forum is made up of diverse stakeholders from academia, the development sector, the public sector, the private sector, and civil society. Implementation partnerships are formed at the provincial level, where ICT commissions are responsible for implementing the ICT strategy. For example, the Professional ICT Curriculum and Certification program is implemented jointly by the Provincial Digital Resource Centers, the ICT policy commission, and private sector companies. A central ICT Policy Implementation Technical Unit is responsible primarily to support and advise the regional implementation bodies.

Monitoring and Evaluation

Countries from all income and region groups perform poorly in their use of M&E (figure 5.10). The vast majority of e-strategies say little or nothing about institutions or structures to monitor and evaluate their progress. Of the few e-strategies that are more specific about M&E, even fewer provide budgetary details about how to finance it.

M&E is a critical area of focus for ICT policy makers. Country leaders should, when they are formulating their e-strategy, plan to set up M&E and should commit specific financial resources to it. Doing this would help make e-strategy design and implementation effective and relevant.

Figure 5.10 M&E by Region, Average Scores

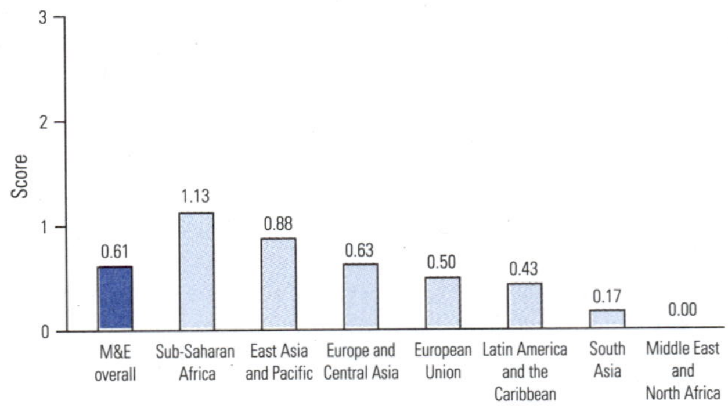

Source: Authors' analysis.

Without M&E, it is impossible to measure results and assess the impact of ICT initiatives.

There are three countries that are exceptions: Mozambique, Rwanda, and Nigeria. Countries in Sub-Saharan Africa overall perform slightly better in both structure/responsibility and budgetary aspects of M&E among countries studied. For example, Rwanda's e-strategy lays out in explicit detail institutional responsibilities for M&E and how M&E activities are to be integrated in the implementation machinery and timeline (box 5.2). Approaching it differently than Rwanda does, Mozambique has created projects that focus on data gathering and analysis as stand-alone initiatives of its larger strategy.

Analysis II: Thematic Areas of Focus

This section presents results of *what* e-strategies focus on, illustrating differences in priority objectives and intervention tactics across countries.

Overview

E-strategies vary in their objectives and initiatives to achieve ICT development. There are nine significant thematic areas on which the ICT strategies focus (figure 5.11).[5]

In aggregate terms, four of the nine themes occur in over 85 percent of the e-strategies. These are the following:

- *E-government:* providing services and information via the Internet by the government to companies, citizens, and other sections of government.
- *Infrastructure:* constructing physical components such as fiber-optic backbone and wired and wireless networks over which electronic communications are transmitted and received.

Figure 5.11 Thematic Areas of Focus for E-Strategies, by Number of Countries

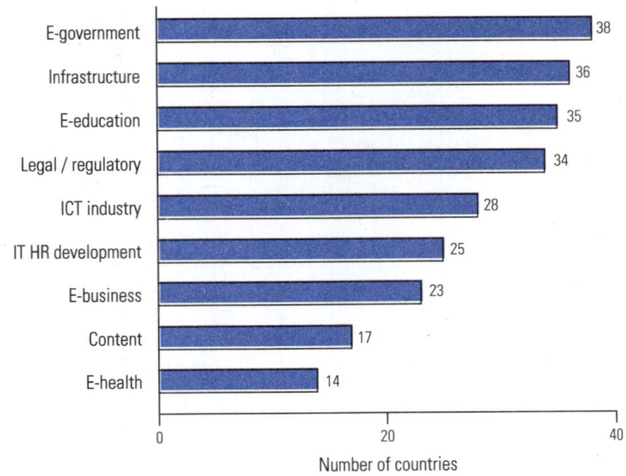

Note: HR = human resources.

Source: Authors' analysis.

- *E-education:* using ICT in education to improve teaching and school administration and to provide basic e-literacy to all levels of school system and to adult learners.
- *Legal/Regulatory:* creating and modifying legal and regulatory mechanisms to enable and support ICT adoption in business and government and to safeguard users of ICT.

The remaining five themes occur in at least 40 percent of e-strategies:

- *ICT industry:* creating or expanding domestic ICT production of hardware, software, and services for local or foreign markets.
- *IT HR development:* developing human resources with ICT skills to support domestic ICT industry and attract foreign business operations.
- *E-business:* using ICT in traditional or new e-commerce businesses to reduce costs, improve competitiveness, and increase market reach.
- *Content:* creating locally relevant multimedia content to encourage ICT use. Also considered is using ICT to store cultural and historical media.
- *E-health:* using ICT in the administration and provision of health services and health information.

Box 5.2 Embedding M&E into Implementation Structures in Rwanda

Rwanda's e-strategy shows details of M&E in its ICT program, which starts with the country president, who chairs the annual e-strategy review meeting, and ends with the teams on the ground that implement the e-strategy.

The e-strategy implementation is carried out at three levels. First, the National Information Technology Commission (NITC), headed by the president, gives oversight and guidance to the overall mission. Second, the Rwanda International and Technology Authority (RITA), reporting to the NITC, serves in a hands-on advisory and consultancy role on ICT issues for different government agencies. Third, numerous Plan Execution and Coordination Committees (PECCs), spread throughout government agencies and private sector institutions and coordinated and supported by RITA, take responsibility for actually implementing individual initiatives within the e-strategy.

M&E is implemented by PECCs and managed by RITA; ultimately M&E is managed by the NITC. PECCs have to be intimately familiar with e-strategy targets to implement actions and report progress. Each PECC submits a biannual report to a special M&E unit of RITA. The reports from different PECCs are consolidated by RITA and given to an NITC committee biannually. The committee submits an annual implementation report, including M&E results, to the president and the cabinet.

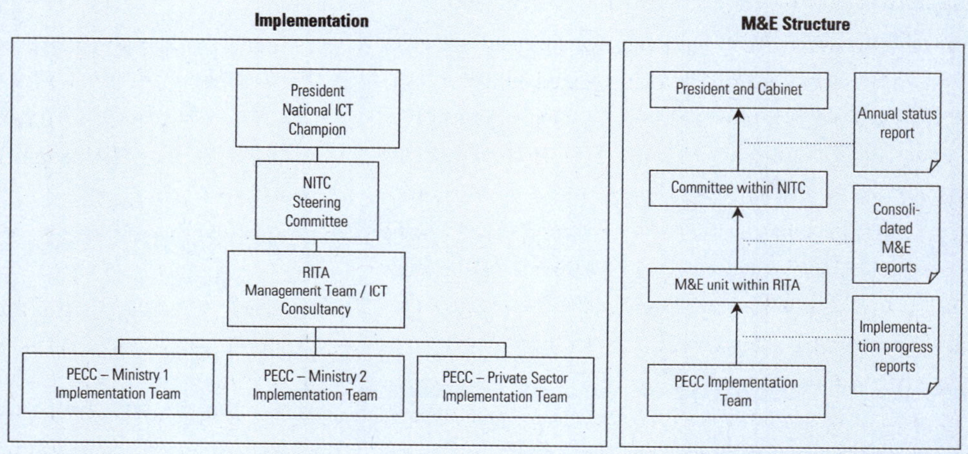

The Rwanda model combines implementation-level ownership of M&E with executive oversight. It attempts to balance the involvement of M&E by teams at the implementation level with M&E expertise at the management and steering committee level. The information and analysis produced from the multistage process forms the basis of the e-strategy and is reviewed by the highest levels of government.

The payoffs of the organization, structure, and M&E focus of Rwanda's e-strategy will be seen in coming years.

Source: World Bank 2005.

In general, the objectives identified in national e-strategies converge, particularly for e-government, infrastructure, e-education, and legal and regulatory reform themes. This suggests that these objectives are fundamental to the creation of an information society and provide the foundation for more specialized applications in, for example, the ICT sector, e-health, e-business, and content development.

In contrast, the types of actions cited to implement the commonly sought objectives by countries diverge. This is partly due to income differences. ICT maturity is related to wealth; consequently, countries employ different methods to achieve their objectives. For example, Rwanda and Hong Kong (China) both focus on government-to-citizen e-services, yet they approach the same objective differently. In Rwanda, public access points are aimed as a key intervention to disseminate government information, whereas in Hong Kong, interactive televisions are desired as a new medium to offer government information and services.

E-Government

E-government is defined in the e-strategies as the provision of services and information by electronic means between different sections of government (G2G), between government and business (G2B), and between government and citizens (G2C). E-government is the most commonly occurring component across the surveyed e-strategies (more than 95 percent of e-strategies included this component). It is seen as a strong tool for improving government productivity, administrative effectiveness, and cost savings.

In most e-strategies, the e-government component refers to a "single window" approach to integrated online public services. This often manifests itself as an e-government portal that serves as the conduit for online services. Examples include land and property registration or records maintenance (G2C), e-procurement (G2B), and centralized census and population data (G2G). There is an even spread of focus to develop G2G, G2B, and G2C across the e-strategies that contain e-government (figure 5.12). Each objective occurs in about 60 percent of the e-strategies.

Common interventions in implementing e-government applications include reforming government processes (more than 55 percent), computerizing and networking government agencies (more than 50 percent), and developing standards and protocols for ICT system interoperability (also more than 50 percent). Government process reform, the intervention most cited, typically refers to changes in internal business processes brought about by automation that complements or replaces labor-intensive methods and systems. For example, Jamaica plans to replace its paper-based customs processing system with a paperless one to improve efficiency and reduce transaction costs.

Developing countries emphasize reforming existing bureaucratic processes and expanding the capacity of government networks to realize e-government objectives (figure 5.13). Developed nations, however, stress the importance of streamlining IT standards that facilitate interoperability and overcome integration issues of electronic systems.

E-strategies in low- and middle-income nations not only call for developing e-government applications but also focus on generating demand for online services. They try to create awareness among citizens and businesses of the benefits of online transactions.

Some approaches to e-government are worth highlighting. Mozambique's plan for e-government, while ambitious, focuses initially on institutional and systems-level computerization rather than G2C services. To this end, Mozambique will conduct a survey of the state of ICTs in public institutions before undertaking online public services. Poland develops this approach further by focusing

Figure 5.12 **E-Government Objectives by Percentage of Countries in Income Group**

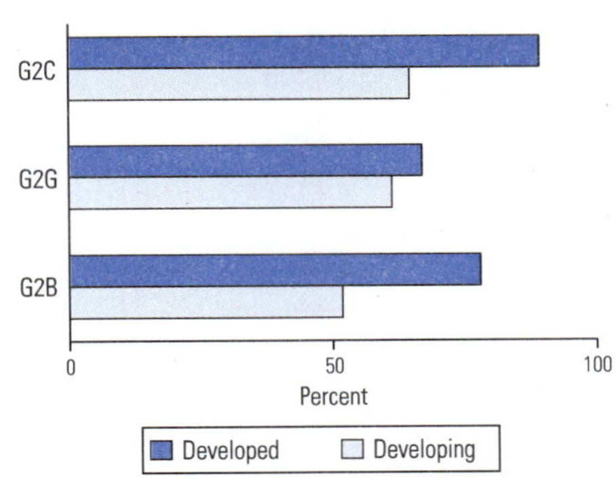

Note: G2B = government-to-business; G2C = government-to-citizens; G2G = government-to-government.

Source: Authors' analysis.

Figure 5.13 **E-Government Interventions by Percentage of Countries in Income Group**

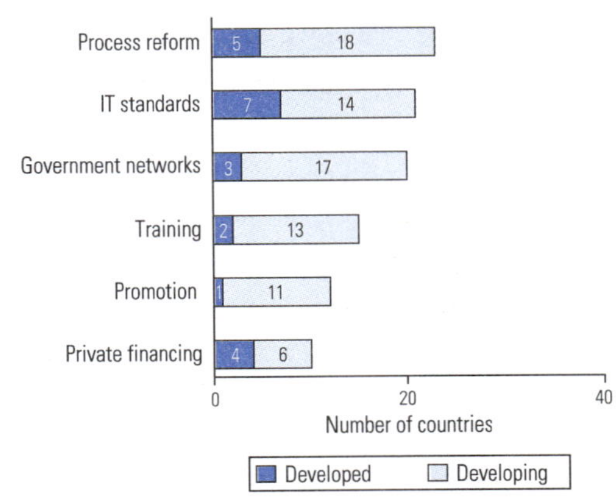

Source: Authors' analysis.

on a prioritized list of e-government projects with due consideration to productivity gains. Thailand will begin its e-government project with pilot projects in ministries as a way to identify common and shareable data as well as to identify ministries and government agencies that are ready for computerization and process reform. These approaches contrast with many (if not most) e-strategies that merely list services to be automated or put online—thus the high count for G2C services—with little consideration to prioritization or return on investment.

Infrastructure

Telecommunications infrastructure is crucial and fundamental to using ICT for development. Without the proper infrastructure, e-strategy is less likely to succeed, and projects may stagnate or never get off the ground. This is why most of the ICT strategies surveyed (90 percent) specify telecommunications infrastructure as an area of focus.

"Universal access" is the most prevalent focus within the infrastructure component, occurring in more than 65 percent of e-strategies. To provide universal access entails providing equal access to voice and data communications networks across the country, in rural as well as underprivileged urban areas. It also often includes an emphasis on financing access through specially earmarked universal access funds.

The second most common objective within the infrastructure theme is broadband development (more than 50 percent of the e-strategies surveyed identified this objective). This is frequently viewed as a way to generate consumer demand for online services and thus spur the development of such services by private businesses and government agencies. The next most significant focus is providing "telecenters" (specified by more than 50 percent of e-strategies surveyed). This focus encompasses the creation of Internet access nodes (for example, Internet kiosks and Internet Automated Teller Machines, also known as ATMs) for public use in regions where "last mile" access of ICT services to homes and businesses is not widely available.

It is worth noting that low-income countries are seeking universal access and they are the only ones focusing on creating, extending, and upgrading backbone networks (figure 5.14). This suggests that high- and middle-income countries have already implemented appropriate backbone telecommunications infrastructure. Middle-income countries look instead for Internet kiosks, ATMs, and other delivery mechanisms to extend the reach of existing networks. The lack of focus on telecenters by high-income economies suggests that such services are in less demand in these countries because affordable, basic "last mile" local telecommunications infrastructure is already available. E-strategies from high-income countries focus on the deployment of ubiquitous broadband to households.

Figure 5.14 Infrastructure Objectives by Percentage of Countries in Income Group

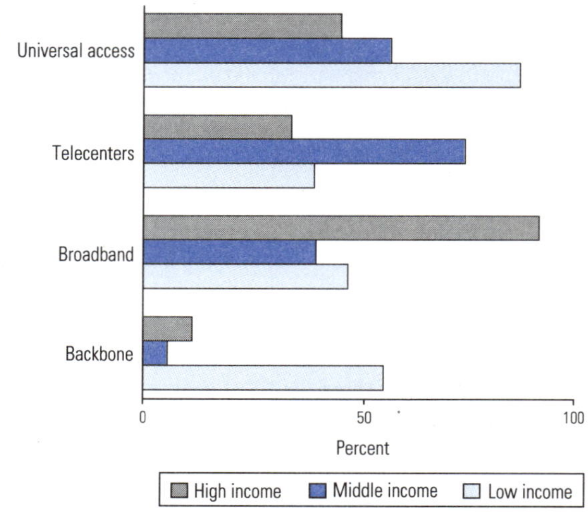

Source: Authors' analysis.

To achieve the above objectives, the majority of e-strategies encourage the development of regulatory structures and supervisory agencies to manage a competitive, market-driven modern telecommunications infrastructure sector. Governments assume the role of facilitators, using regulation to allow other participants to get involved and ensure fair competition. This allows the other participants to help fund the implementation of the e-strategy, rather than putting scarce government funds into infrastructure development (see also chapter 2 on the role of FDI). Regulatory agencies thus oversee the introduction and ongoing management of private competition in telecommunications, a management that includes supervision of interoperability and interconnection issues among different service providers.[6]

E-Education

E-education is a focus area in 88 percent of the national ICT strategies surveyed. The principal objective of this focus is e-literacy (that is, basic computer and application skills such as using spreadsheets and surfing the Web) in the formal and informal education system (see also chapter 6). There is a fairly even spread of focus across primary, secondary, and tertiary institutions as well as adult and community training centers, although income levels partly dictate the e-education priorities of countries.[7]

Several different interventions to advance e-education are addressed in e-strategies. Teacher training, school and center connectivity, institutional capacity development, and distance learning are cited in over 50 percent of the e-strategies—Poland's e-education strategy is a case in point (box 5.3). Curriculum development and quality assurance follow, cited in over 30 percent of the e-strategies.

Interestingly, only low- and middle-income countries address distance learning and quality assurance (figure 5.15). The ICT strategies that address distance learning usually aim to increase the reach of the education system to areas that do not have formal schools. In this way, they also provide individuals or groups with a curriculum designed by the education ministry. They complement traditional higher education facilities, which tend to be fewer in developing countries. In quality assurance, countries assert the need for nationally and internationally recognized standards of e-literacy. For example, Mozambique plans to use a program similar to the International Center for Distance Learning (ICDL) to meet the needs of the public and private sector for professionals with appropriate technical skills.

Legal and Regulatory Reform

Legal and regulatory components feature in 85 percent of the national e-strategies. The principal focus of this e-strategy component is to revise existing legal and regulatory structures concerned with ICT and to create new laws that facilitate ICT-related activities (see also chapter 2 on the importance of a consistent regulatory framework). A broad range of reforms are cited in the e-strategies. Among the reforms identified are rules to govern trade (for example, intellectual property rights, taxes, tariffs), safeguard personal privacy (for example, data protection), and facilitate e-commerce (for example, e-contracts, digital signatures, and e-payment systems).

Box 5.3 E-Education in Poland

Poland's program for e-literacy articulates a clear understanding of the status of education in Poland in relation to the demands of the modern economy. It is also clear on the potential that ICT has to help modernize the education system to meet those demands. Poland's e-education plan incorporates all of the most common interventions and objectives found in the national e-strategies of middle-income countries:

- providing ICT equipment and connectivity to schools,
- training teachers,
- educating adults as well as students,
- expanding teaching resources,
- accrediting institutes and certifying programs, and
- developing curricula.

The e-strategy emphasizes the need for ICT-qualified teachers as a first logical step toward the e-education objectives.

More generally, Poland's e-strategy proposes a new educational paradigm to prepare students for employment in the global and emerging local information-based economy. Poland looks to the experience of other industrialized countries: those countries have created competitive economies by increasing the technical competence of the workforce to achieve economic and social progress. Poland's e-strategy emphasizes the importance of training (and retraining) workers with market-oriented technical and language skills. It highlights the need to create a system of continuous learning so that workers and students can acquire new skills relevant to the fast-changing demands of the modern workplace throughout their working careers.

Source: Authors' analysis.

Figure 5.15 E-Education Objectives by Percentage of Countries in Income Group

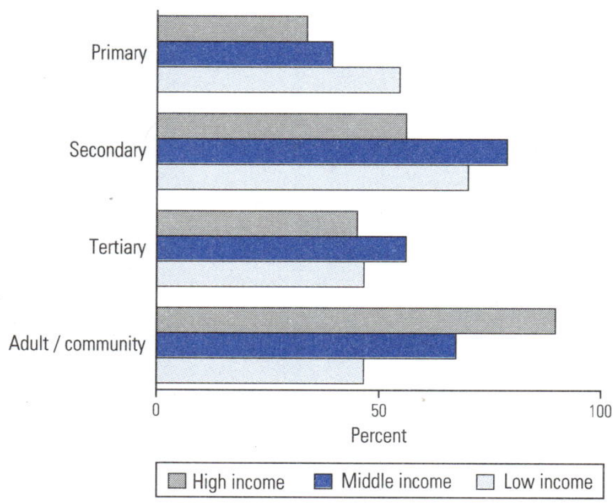

Source: Authors' analysis.

There is a strong emphasis in the e-strategies on legal interventions that focus on the business sector (more than 70 percent; see also chapter 4). This is followed by interventions that center on the government (more than 40 percent) and civil society (40 percent), which involves protecting user information and prohibiting illegal activity (figure 5.16). The commercial orientation of e-strategies is evident in their support of the business sector (for reasons of economic growth) while safeguarding the rights and interests of users and consumers in transactions with government and businesses, and in personal communications.

For trust and confidence measures, the e-strategies stress legal interventions to combat cyber-crime (80 percent of the e-strategies) nearly twice as often as they do online privacy (over 40 percent). For interventions for a stable and active ICT and business environment, the e-strategies include provisions for protection of intellectual property rights (more than 70 percent) and stimuli for the commercial sector such as tax incentives and reduced tariffs (more than 50 percent) (figure 5.17).

Figure 5.16 Regulatory Objectives by Percentage of Countries in Income Group

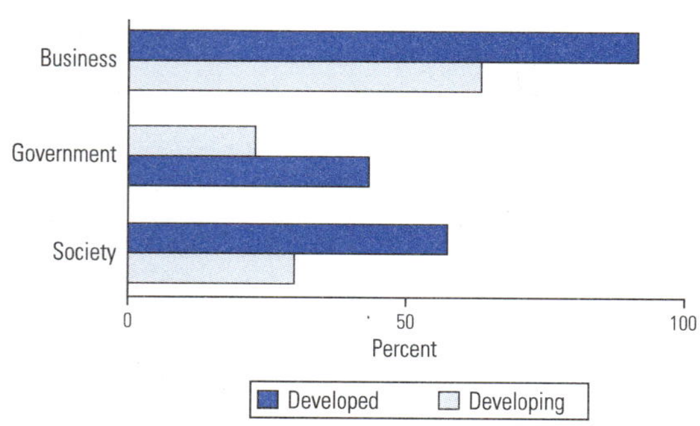

Source: Authors' analysis.

It is worth noting that a number of countries (almost 25 percent)—including Bangladesh, Indonesia, Jamaica, and Romania—stipulate adherence to regional and international legal frameworks. Korea is seeking a leadership role in the effort to create such frameworks. This suggests that compliance with legal and regulatory conventions is perceived as a necessary precondition for integration into the global ICT environment.

ICT Industry Development

Developing the ICT production and service sectors is identified in more than two-thirds of the ICT strategies. Over 90 percent of the e-strategies for this theme center on producing software and hardware and providing IT services, such as outsourced development of back office systems for the export market (table 5.1).[8] The distribution of objectives for product type is shared quite uniformly among software (more than 50 percent), IT services (30 percent), and hardware (more than 25 percent). Twice as many e-strategies center on export markets for their ICT sectors than on their domestic markets. The attention to foreign markets is seen as important for attracting foreign investment and encourages joint ventures between domestic and foreign companies (see also chapter 2 for discussion about attracting foreign investors and joint ventures).

The overriding motivation for the focus on foreign markets is the wish to increase flows of foreign currency to

Figure 5.17 Regulatory Interventions by Percentage of Countries

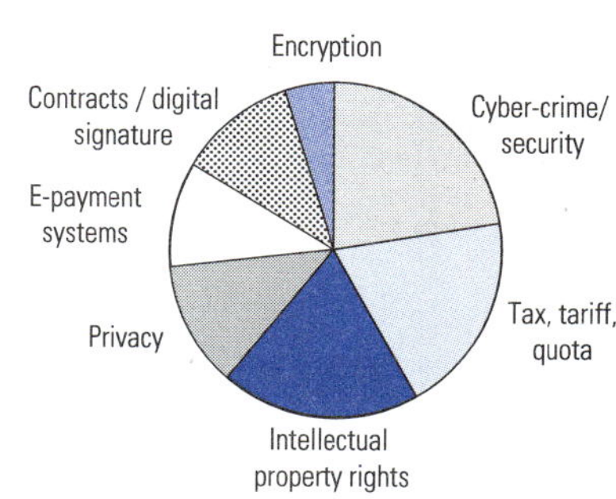

Source: Authors' analysis.

Table 5.1 Economy Distribution by ICT Industry Segment

Market	Software	Hardware	IT services
Domestic	Angola, Bangladesh, Hong Kong (China), Rwanda, Thailand, Tunisia, Ukraine, Vietnam	Angola, Bangladesh, Rwanda, Ukraine, Vietnam	Bangladesh, Dominican Republic, Hong Kong (China), Rwanda, Tunisia, Ukraine
International	Angola, Azerbaijan, Bangladesh, Colombia, Egypt (Arab Rep. of), Hong Kong (China), India, Jordan, Korea (Rep. of), Rwanda, Thailand, Tunisia, Ukraine, Vietnam	Angola, Bangladesh, India, Korea (Rep. of), Rwanda, Ukraine, Vietnam	Dominican Republic, Hong Kong (China), India, Jordan, Rwanda, Tunisia, Ukraine

Source: Authors' analysis.

the producing country. These findings are consistent with current trends in outsourcing that show how the production of hardware, software, and IT services is increasingly being relocated to countries with a skilled workforce and comparatively lower labor costs.

The data also suggest that, in general, the success of ICT sector development depends on key enabling interventions that *directly* support the commercialization of technology innovations. Among these interventions are technology company incubators (more than 80 percent), support for R&D (also more than 80 percent), and the promotion of ICT products and services domestically and internationally (more than 60 percent). The e-strategies also seek to nurture the ICT sector using less direct interventions—though to a lesser extent—such as support to business associations and quality assurance (both appear in more than 30 percent of the e-strategies).

Jordan is an example of a country with an e-strategy that contains a broad cross-section of objectives and interventions to support the ICT sector. In addition to the above-mentioned focus areas and interventions, Jordan plans to promote company collaboration for joint marketing and training. Furthermore, the government of Jordan will provide technical and financial assistance to build capabilities in operations, marketing, and management. Jordan aims to help Jordanian companies float their stocks on the Amman Stock Exchange. In part, this is to help companies stem the outflow of ICT-related expertise by allowing companies to issue stocks as incentives to retain their employees.

E-Business

E-business, occurring in over 50 percent of the strategies, is not one of the most common themes in e-strategies. Typically the e-strategies define e-business as the use of ICT in business to reduce transaction costs, to broaden market reach, and to increase the productivity and speed of doing business (see also chapter 4 on the role of ICT in doing business). Some e-strategies cite e-business as a catalyst for modernizing the private sector in general (that is, not the ICT sector per se); other e-strategies, such as that of Trinidad and Tobago, seek to encourage e-business with a view to increase demand for domestically produced ICT products. To this end, the government of Trinidad and Tobago plans to co-develop, with the private sector, an integrated e-business application for local companies to conduct online sales. Furthermore, the government has offered to help identify comparative advantages for local e-businesses and will collaborate in the development of "Skillnet," a service that provides recruitment, learning, and career information.

By far the most common target for e-business initiatives in the e-strategies are small and medium enterprises (over 70 percent of the e-strategies), which would benefit most from ICT development. This contrasts with medium and large enterprises, which are the target beneficiaries of less than 5 percent of the e-strategies that focus on e-business. These businesses are typically well ahead in the use of ICT, often even further than government, and hence require less external support.

The predominant approach to support e-business adoption entails general promotion and education efforts that aim to demonstrate the benefits of e-business applications to non-ICT related businesses (this is identified in over 90 percent of e-strategies that address e-business) (figure 5.18). For example, many ICT strategies plan to run workshops to demonstrate productivity tools (such as spreadsheets and word processors) and e-commerce transactions (such as e-payments and contracts). Outreach and public relations—rather than direct subsidies or financial

Figure 5.18 E-Business Interventions by Number of Countries

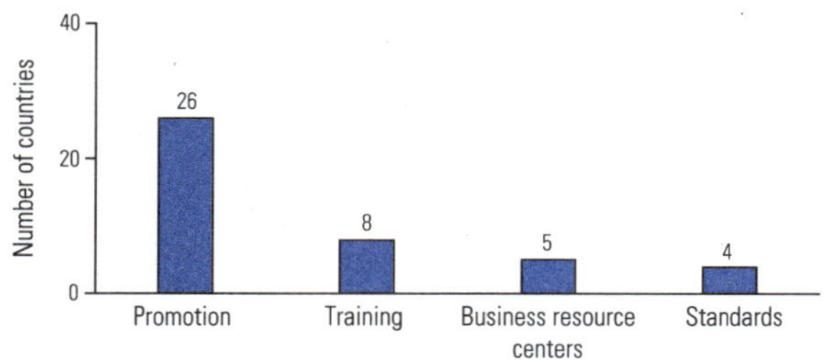

Source: Authors' analysis.

incentives—are among more common approaches used to increase awareness of the benefits of e-business in the business community as well as among the general public.

The large difference between promotion and the next major intervention, training, suggests the rather limited scope government has for supporting e-business. It is noteworthy that only middle-income countries explicitly plan training initiatives for e-business.[9]

Information Technology Human Resources Development

Although almost all ICT strategies surveyed consider human resources to be central to developing the use of ICT in the economy and society, just over half of those dedicate particular attention to developing professional IT expertise. These e-strategies uniformly view a workforce of technologists—including programmers, network administrators, and designers—to be fundamental to the ICT sector.[10]

Of those e-strategies that focus on information technology human resources development (IT HRD), the vast majority center their initiatives on developing technology professionals to meet the needs of the ICT sector (more than 90 percent), and to a lesser extent on the non-ICT business sector (almost 50 percent) and government (over 30 percent). This concentration on the needs of the ICT sector can be attributed to the commonly held view that this sector has the greatest potential for driving immediate and long-term economic growth.

The leading interventions for this component of e-strategies all have to do with creating or expanding teaching capacity. Chief among these are initiatives to build new institutional capacity (that is, funding and other support to build technical training centers) to train IT professionals, an intervention identified in over 70 percent of e-strategies within this theme. Next come interventions to enhance existing IT training capacity (specified in over 60 percent of the strategies) and the development of IT curricula (in more than 50 percent of the strategies). Enhancing existing IT training capacity is intended to increase the number of qualified IT graduates by increasing the number of IT instructors, enlarging class sizes, and enhancing the availability of other educational resources.

Over 50 percent of e-strategies include initiatives that involve the private sector in a variety of capacities. Special consideration is given to the private sector as advisers to the government on IT HRD. The private sector is also often cited as a source of internships and work placements, and as a way to align skills provision to demand. A few e-strategies, including those from Chile and the United Kingdom, provide a basis for their IT HRD initiatives by surveying current technology trends and working with the ICT sector to predict future labor needs.[11]

Content

Over 40 percent of the e-strategies focus on content by either establishing a multimedia production industry or digitizing national heritage and cultural content in local languages for domestic use.

The e-strategies consider the content creation industry in many instances to be a new market opportunity. But they also consider that as a duty to preserve national identity and cultural works. Poland, for instance, plans to create new content and digitize existing content (for example, works from the national library) that is of interest, and of possible use, to the public, including tourists and foreign investors. Similarly, Ireland has established a "digital hub" to produce digital content that includes national archives, national art collections, digital maps, and works from the national library. Tunisia plans to integrate digital content creation into university curricula and create several multimedia educational institutes that have institutional relations with foreign multimedia academies.

Focus on content is more prevalent in e-strategies from middle- and high-income countries. This suggests that it may not constitute a priority first step for ICT development in lower-income countries.

E-Health

An e-health strategy component occurs in only 35 percent of the surveyed national ICT strategies. This focus encompasses the use of ICT in the administration of health care organizations, the delivery of clinical services, and the creation of awareness of health issues in the general public.

In those e-strategies that focus on e-health, a relatively equal distribution of emphasis between two leading objectives exists. These objectives are the use of ICT in the administration of health care organizations and online access to health education. This is followed by the delivery of clinical services, which is cited in over 40 percent of the e-strategies.

The e-strategies contain several initiatives to support and advance the use of ICT in the health care sector. Some of these initiatives are health center connectivity, instituting technical standards and protocols for integrating systems and exchanging data, training personnel, referring patients, and using online billing systems. Of these interventions, connecting health care centers to ICTs—meaning both network connectivity within the health care centers and connectivity to external networks such as regional hospitals and clinics—occurs by far the most frequently. This is cited in more than 50 percent of the e-strategies with an e-health focus. Developing technical standards for interoperability follows. In the e-strategies, connectivity is most often considered for delivering clinical services and consulting via telemedicine. However, connectivity is also viewed as benefiting health care administration by enabling improved communications, sharing resources, collecting data, and providing health information services within and among health care institutions.

Many e-strategies with an e-health focus state the need to become more cost-efficient by creating, sharing, and integrating their systems and technologies to create economies of scale. They also cite the need to increase productivity and efficiency. As might be expected, variations exist in focus areas across country, region, and income categories. For example, many Sub-Saharan African countries (such as Tanzania) plan to establish systems for essential humanitarian services (such as HIV information and nutritional surveillance). More advanced countries focus on improving and extending existing advanced services such as single patient records (Finland) and smart cards for integrated patient information systems (Czech Republic).

Recommendations

This chapter offers the following five recommendations for formulating effective e-strategies. They are all based on lessons learned from the review of 40 e-strategies.

1. Establish a Limited Number of Objectives and Clear Interventions to Achieve Them

Establishing ICT development priorities is one of the main reasons to develop a national e-strategy. It entails making tradeoffs between what will be implemented in the near term and what will be carried out later through subsequent e-strategies. This is necessary to channel limited resources to priority objectives and to establish a logical sequencing of initiatives (such as focusing initially more on infrastructure development than on ICT applications). Once key objectives have been established, explicit initiatives or projects that will be implemented to achieve them should be developed and budgeted for.

The number of major objectives that the reviewed e-strategies tackle ranges from three to nine. Many e-strategies share common objectives: e-government, infrastructure, e-education, and legal and regulatory reform appear in over 85 percent of them. The methods proposed in the e-strategies to achieve these objectives vary considerably, however. There are usually few parallels between the types of interventions to be undertaken and income category. There are some exceptions to this rule—such as basic infrastructure development being a main focus of low-income countries and broadband development a focus of middle- and high-income countries.

2. Link E-Development Objectives to National Economic and Social Development Goals

E-strategies are to a certain degree "sales pitches" for allocating often-scarce national financial resources to ICT development. This requires making a clear case for how ICT development supports the economic and social development of a country. Such linkages should be not only at a conceptual level—important as this may be to establish the foundations for the e-strategy in general—but should also drive through to all of the key objectives and related initiatives that the e-strategy advances.

On the whole, middle-income countries tend to develop such linkages, suggesting that they can more easily conceive of the importance of ICTs to a range of development challenges as they move beyond dealing with basic development needs.

3. Develop Clear, Realistic, Data-Driven Targets

Developing a credible e-strategy begins with outlining the country's current state of ICT development. Based on the analysis, future goals can then be developed, with clear interim and final targets. This should be done both at an aggregate level for the e-strategy as a whole and for theme-specific or sector-specific initiatives.

The e-strategies all score relatively poorly in their use of data. This is somewhat to be expected, as meaningful ICT-related data are not readily accessible. It is also a worrying trend. If e-strategies are to serve as road maps for ICT development, they will not be credible without clear data and analysis. A number of countries are developing ICT goals and objectives without clear benchmarks for measuring the degree to which they have achieved their goals. Where they do establish clear benchmarks, they too often do so without sufficient reference to their current state of ICT readiness, and so are unable to gauge the level of effort and resources required to achieve the objectives. This can lead to unrealistic goals and targets.

4. Clarify Implementation Responsibility and Budgetary Commitments

Establishing a framework to govern the implementation of an e-strategy is key to ensuring that it moves from being a theoretical blueprint to a document that guides actual change. This framework should clarify implementation responsibility for all of the e-strategy's core components, including the funding of each component's implementation.

In this regard, the reviewed e-strategies score very well, advancing a range of both centralized and decentralized structures. However, the e-strategies say considerably less about how implementation is to be financed. Although the level of clarity on implementation structures and responsibility is a good trend, it rings hollow if it is not accompanied with equal clarity on related financing.

5. Ensure that Monitoring and Evaluation Forms a Key Component of the E-Strategy

Monitoring and evaluation (M&E) has to form a central pillar of formulating the e-strategy and the subsequent process of implementing it. Developing M&E components of e-strategies is a way to ensure that ICT strategies are explicit and realistic about what they aim to achieve. M&E also checks that the implementation of those ICT strategies is regularly assessed and realigned to ensure the efficient use of often-scarce resources. Additionally, results from M&E should be used to establish the needs and direction of current and future e-strategies.

The e-strategies that were reviewed scored poorest on this measure, with few exceptions. The exceptions tended to be e-strategies that were stronger on all other measures. Good, clear M&E is essential not only to ensuring an e-strategy's effective implementation and efficient use of resources but also to improving the manner in which the e-strategy itself is to be formulated. Thinking through M&E elements of the e-strategy forces policy makers to clarify objectives, set targets, and outline how the targets will be met by implementing specific initiatives.

Annex 5A

Table 5A Resources and Links for ICT Policies and E-Strategies

Organization	Type	Links
WSIS	International Summit	World Summit on the Information Society, http://www.itu.int/wsis
ITU E-strategies	International Organization	E-Strategies Homepage, http://www.itu.int/ITU-D/e-strategy
UN ICT Task Force	International Organization	UN ICT Task Force, http://www.unicttaskforce.org
UNCTAD	International Organization	UNCTAD, E-Commerce Branch, http://r0.unctad.org/ecommerce
World Bank	International Organization	Global Information and Communication Technologies, Policy Division, http://info.worldbank.org/ict/policy.cfm
SDNP, UNDP	International Organization, Regional Mission	Sustainable Development Networking Programme, Regional Mission http://www.sdnp.undp.org
APDIP, UNDP	International Organization, Regional Mission	Asia-Pacific Development Information Programme, Regional Mission http://www.apdip.net
APC	Nongovernmental	Association for Progressive Communications, http://www.apc.org/english/index.shtml
Bridges.org	Nongovernmental	http://www.bridges.org
Development Gateway E-strategies	Not-for-profit	E-Strategies and Action Plans, http://topics.developmentgateway.org/ict/rc/BrowseContent.do~source=RCContentUser~folderId=3041
ePol-Net	Nongovernmental	Global E-Policy Resource Network, http://195.218.115.39/pport/web/Sitemap/3
Markle Foundation	Nongovernmental	http://www.markle.org

Source: Authors' compilation.

Annex 5B

Table 5B Countries Covered in E-Strategy Review

Number	Economy reviewed	Link to country's published e-strategy[a]
1	Albania	http://www.undp.org.al/?elib,428
2	Angola	http://www.uneca.org/aisi/nici/Angola/angola.htm
3	Azerbaijan	http://www.nicts.az:8101
4	Bangladesh	http://www.bccbd.org/html/itpolicy.htm
5	Bhutan	http://www.dit.gov.bt/guideline.php?id=7
6	Bolivia	http://www.aladi.org/nsfaladi/ecomerc.nsf/0/E8147919B55D97A403256BEA004D2EDA/$File/lineamientos.pdf?OpenElement
7	Chile	http://www.agendadigital.cl/agenda_digital/agendadigital.nsf/vwDocumentosWebLink/27363116E8E6631704256E5800549FE3?OpenDocument
8	Hong Kong, China	http://www.info.gov.hk/digital21/eng/strategy2004/strategy_main.html
9	Colombia	http://www.agenda.gov.co/
10	Czech Republic	http://www.micr.cz/scripts/detail.php?id=1288
11	Dominican Republic	http://www.edominicana.gov.do/
12	Egypt, Arab Rep. of	http://www.uneca.org/aisi/nici/Egypt/egypt.htm
13	Finland	http://www.tietoyhteiskuntaohjelma.fi/esittely/en_GB/introduction/
14	Ghana	http://www.uneca.org/aisi/nici/Ghana/ghana.htm
15	India	
	Andhra Pradesh	http://www.gipi.org.in/state_policy/andhra.pdf
	Delhi	http://delhigovt.nic.in/icetpolicy.pdf
	Haryana	http://www.gipi.org.in/state_policy/haryana.pdf
	National	http://www.gipi.org.in/ITPolicyInIndia.php
16	Indonesia	http://www.sdnbd.org/sdi/issues/IT-computer/policy/indonesia.pdf

(*Table continues on the following page.*)

Table 5B *continued*

Number	Economy reviewed	Link to country's published e-strategy[a]
17	Ireland	http://www.taoiseach.gov.ie/index.asp?locID=181&docID=1773
18	Jamaica	http://unpan1.un.org/intradoc/groups/public/documents/CARICAD/UNPAN009931.pdf
19	Japan	http://www.kantei.go.jp/foreign/policy/it/index_e.html
20	Jordan	http://www.reach.jo
21	Korea, Rep. of	http://www.ipc.go.kr/ipceng/public/public_view.jsp?num=2007&fn=&req=&pgno=3
22	Mauritius	http://ncb.intnet.mu/ncb/downloads/index.htm
23	Mozambique	http://www.markle.org/downloadable_assets/mz_final_ict_strategy.pdf
24	Nigeria	http://www.uneca.org/aisi/nici/Documents/IT%20policy%20for%20Nigeria.pdf
25	Norway	http://odin.dep.no/nhd/engelsk/publ/rapporter/bn.html
26	Poland	http://www.mnii.gov.pl/mniien/index.jsp?place=Lead08&news_cat_id=110&news_id= 891&layout=2&forum_id=136&page=text
27	Romania	http://unpan1.un.org/intradoc/groups/public/documents/UNTC/UNPAN016044.pdf
28	Russian Federation	http://www.e-rus.ru/eng
29	Rwanda	http://www.uneca.org/aisi/nici/Documents/rwanpap2.htm
30	Singapore	http://www.ida.gov.sg/idaweb/aboutida/infopage.jsp?infopagecategory=&infopageid= I226&versionid=2
31	Slovenia	http://unpan1.un.org/intradoc/groups/public/documents/UNTC/UNPAN015723.pdf
32	South Africa	http://www.tsicanada.com/documents/Strategy.pdf
33	Tanzania	http://www.tanzania.go.tz/pdf/ictpolicy.pdf
34	Thailand	http://www.nectec.or.th/intro/e_nationalpolicy.php
35	Trinidad and Tobago	http://www.gov.tt/nict/
36	Tunisia	Hard copy only
37	Ukraine	http://www.e-ukraine.com.ua
38	United Kingdom	http://www.cabinetoffice.gov.uk/e-government/
39	Venezuela	http://www.mct.gov.ve
40	Vietnam	http://mpt.gov.vn/english/introduction/?thucdon=in

a. Last accessed July 2005.

Source: Authors' compilation.

Annex 5C: E-Strategy Analysis I Scorecards

Table 5C.1 E-Strategy Analysis I Summary by Region

Region	Development linkages[a]	Use of indicators: Fit to goals[b]	Use of indicators: Baseline data[c]	Use of indicators: Target setting[d]	Use of indicators: Cross-country comparison[e]	Implementation: Institutional structure and responsibility[f]	Implementation: Budget[g]	M&E: Structure and responsibility[h]	M&E: Budget[i]
East Asia and Pacific	1.50	1.25	1.25	2.00	1.00	2.75	0.75	1.50	0.25
Europe and Central Asia	1.63	1.88	1.63	1.50	1.63	2.13	1.75	1.13	0.13
European Union	1.29	1.00	1.43	1.43	1.86	2.00	0.43	0.86	0.14
Latin America and the Caribbean	1.57	0.86	1.57	1.43	1.14	2.43	1.14	0.86	0.00
Middle East and North Africa	2.00	2.67	2.33	2.67	2.00	2.67	1.00	0.00	0.00
South Asia	0.67	0.33	0.67	1.67	0.33	2.00	1.67	0.33	0.00
Sub-Saharan Africa	1.75	1.50	1.00	1.25	0.63	1.63	1.38	1.38	0.88

a. Scale: None (0), mention general development goals (1), reference national programs (2), specify national and sectoral programs (3).

b. Scale: None (0), vague (1), specific (2), no ambiguities (3).

c. Scale: None (0), sporadic (1), common (2), extensive (3).

d. Scale: None (0), sporadic (1), common (2), extensive (3).

e. Scale: None (0), sporadic (1), common (2), extensive (3).

f. Scale: None (0), vague (1), specific (2), no ambiguities (3).

g. Scale: None (0), vague (1), specific (2), no ambiguities (3).

h. Scale: None (0), vague (1), specific (2), no ambiguities (3).

i. Scale: None (0), vague (1), specific (2), no ambiguities (3).

Source: Authors' analysis.

Table 5C.2 E-Strategy Analysis I Summary by Income Level

Income level	Development linkages[a]	Use of indicators: Fit to goals[b]	Use of indicators: Baseline data[c]	Use of indicators: Target setting[d]	Use of indicators: Cross-country comparison[e]	Implementation: Institutional structure and responsibility[f]	Implementation: Budget[g]	M&E: Structure and responsibility[h]	M&E: Budget[i]
High	1.44	1.33	1.56	1.56	1.89	2.11	0.67	1.00	0.11
Middle	1.89	1.78	1.72	1.83	1.39	2.22	1.22	1.00	0.06
Low	1.08	0.77	0.85	1.23	0.62	2.08	1.46	0.92	0.62

a. Scale: None (0), mention general development goals (1), reference national programs (2), specify national and sectoral programs (3).

b. Scale: None (0), vague (1), specific (2), no ambiguities (3).

c. Scale: None (0), sporadic (1), common (2), extensive (3).

d. Scale: None (0), sporadic (1), common (2), extensive (3).

e. Scale: None (0), sporadic (1), common (2), extensive (3).

f. Scale: None (0), vague (1), specific (2), no ambiguities (3).

g. Scale: None (0), vague (1), specific (2), no ambiguities (3).

h. Scale: None (0), vague (1), specific (2), no ambiguities (3).

i. Scale: None (0), vague (1), specific (2), no ambiguities (3).

Source: Authors' analysis.

Annex 5D: E-Strategy Analysis II Scorecards

Figure 5D Analysis II Summary by Region, Top Four Focus Areas
(percentage of countries that identify the topic as a focus area)

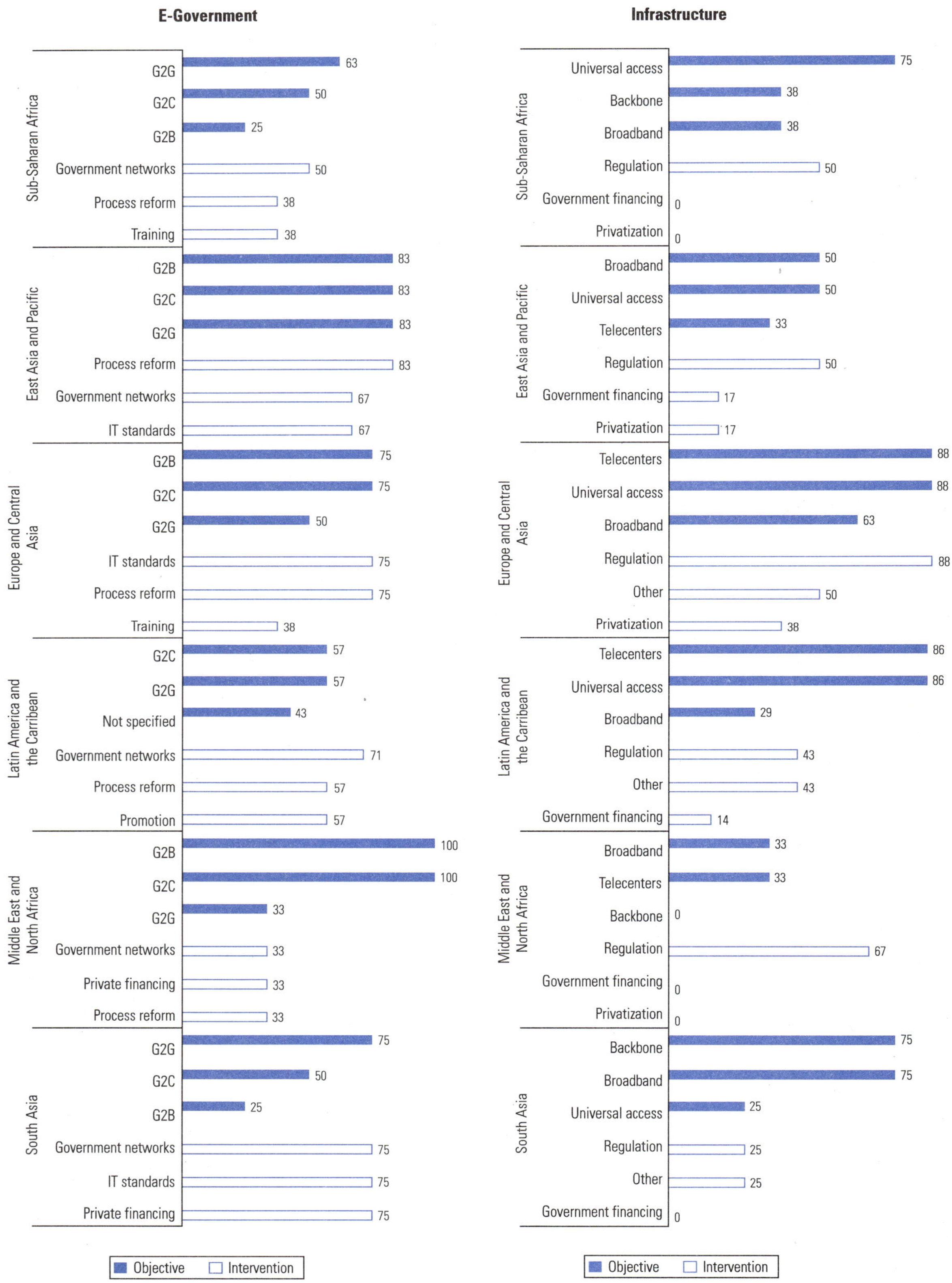

Figure 5D *continued*

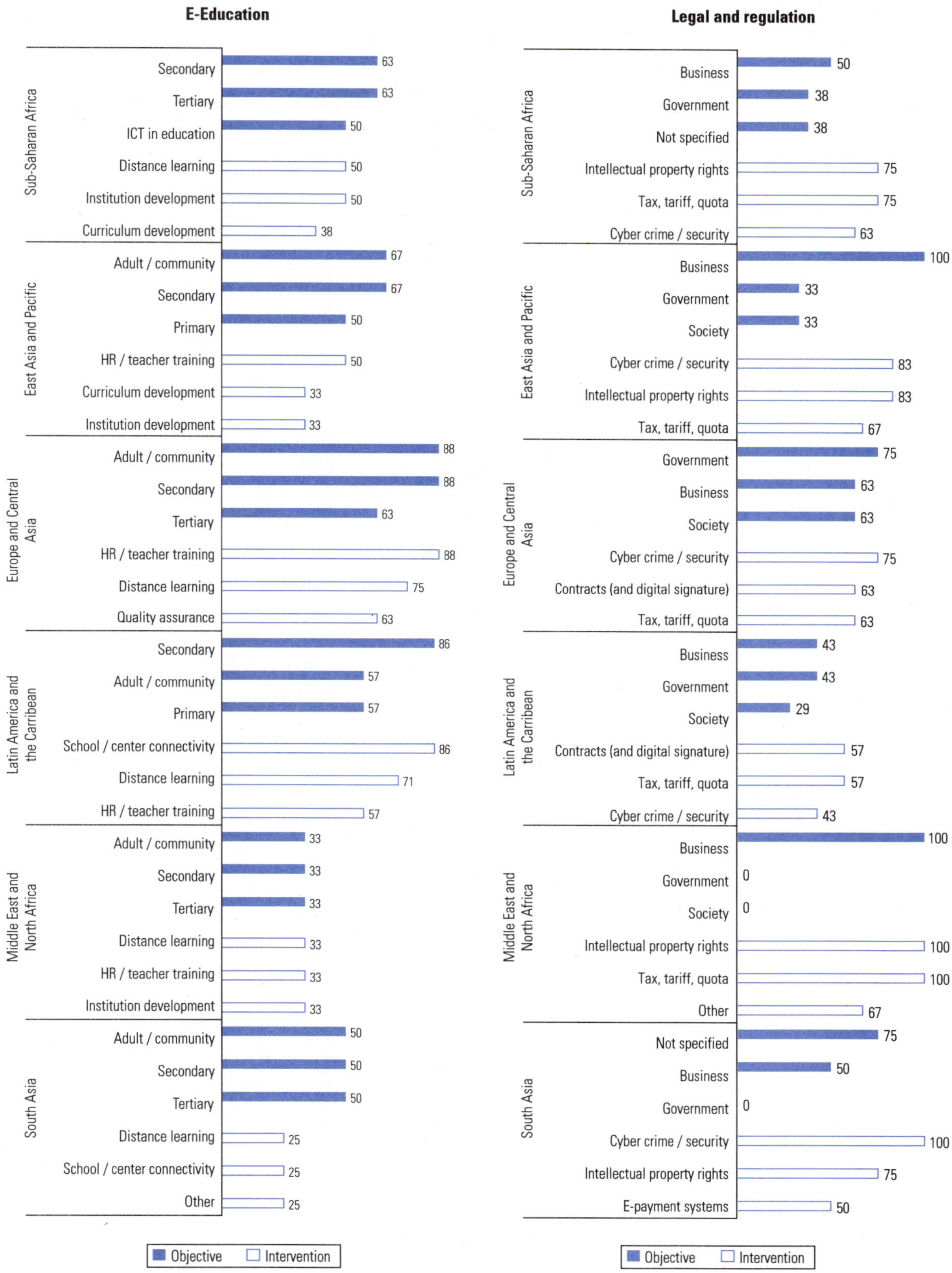

Note: G2B = government-to-business; G2C = government-to-citizens; G2G = government-to-government; HR = human resources.
Source: Authors' analysis.

Table 5D.1 E-Strategy Focus Area: E-Government

Economy	E-government										
	Objective				Intervention						
	G2G	G2B	G2C	Not specified	Government networks	IT standards	Training	Process reform	Private sector participation	Promotion	Other
Albania	x	x	x		x		x	x		x	
Angola	x						x			x	
Azerbaijan				x							
Bangladesh	x	x	x		x					x	
Bhutan				x		x	x		x	x	
Bolivia	x		x		x	x				x	
Chile				x	x	x		x			
Colombia	x		x								
Czech Republic	x	x	x			x		x			
Dominican Rep.				x	x			x		x	
Egypt, Arab Rep. of	x	x	x		x		x	x			
Finland		x	x				x	x	x		
Ghana				x	x						x
Hong Kong, China	x	x	x			x		x	x		x
India	x		x		x	x	x	x	x		
Indonesia	x	x	x		x	x		x	x		
Ireland	x	x	x		x	x		x			
Jamaica	x	x	x				x	x		x	
Japan	x				x	x	x		x		
Jordan		x	x								
Korea, Rep. of	x	x	x		x	x		x			
Mauritius	x		x			x					
Mozambique		x			x			x			x
Nigeria	x		x		x	x	x	x			
Norway	x	x	x			x					
Poland		x	x		x	x	x	x	x	x	
Romania	x	x	x			x	x	x			x
Russian Federation						x		x			
Rwanda	x	x	x		x		x	x		x	x
Singapore			x							x	
Slovenia		x	x			x					
South Africa											
Tanzania	x		x								
Thailand	x	x			x	x	x	x	x		x
Trinidad and Tobago	x	x	x		x	x	x	x		x	
Tunisia		x	x						x	x	
United Kingdom	x	x	x			x		x	x		x
Ukraine	x	x	x			x		x			
Venezuela, R. B. de				x	x						
Vietnam	x	x	x		x		x	x			

Note: G2B = government-to-business; G2C = government-to-citizens; G2G = government-to-government.

Source: Authors' analysis.

Table 5D.2 E-Strategy Focus Area: Infrastructure

	Infrastructure								
	Objective					Intervention			
Economy	Universal access	Info/tele centers	Backbone	Broadband	Regional internet exchanges	Government financing	Regulation/ licensing	Privatization	Other
Albania	x	x					x	x	
Angola	x						x		
Azerbaijan	x						x		
Bangladesh	x		x		x				x
Bhutan		x	x	x					
Bolivia	x	x							
Chile		x		x				x	
Colombia	x	x	x			x	x		
Czech Republic		x		x			x		x
Dominican Rep.	x								x
Egypt, Arab. Rep. of							x		
Finland		x		x					x
Ghana	x						x		
Hong Kong, China				x			x		
India			x	x					
Indonesia	x	x	x						
Ireland	x		x	x		x	x		x
Jamaica	x	x					x		
Japan				x			x		
Jordan				x					
Korea, Rep. of	x			x		x			x
Mauritius									
Mozambique	x	x	x	x	x				
Nigeria	x		x	x			x		
Norway				x					
Poland	x	x		x			x		
Romania	x	x		x			x		x
Russian Federation	x	x				x		x	x
Rwanda	x	x	x	x			x		
Singapore									
Slovenia	x	x		x		x	x		
South Africa									
Tanzania	x				x				
Thailand		x					x	x	
Trinidad and Tobago	x	x		x			x		x
Tunisia		x					x		
United Kingdom	x	x		x			x		
Ukraine	x	x		x			x	x	x
Venezuela, R. B. de	x	x							
Vietnam	x			x			x		

Source: Authors' analysis.

Table 5D.3 E-Strategy Focus Area: E-Education/E-Literacy

	E-education/e-literacy													
	Objective					Intervention								
Economy	Primary	Secondary	Tertiary	Adult/ community	Not specified	Use of ICTs in education	School/ center connectivity	Computer labs	Curriculum development	Learning institution capacity development	HR/ teacher training	Distance learning	Quality assurance	Other
Albania	x	x	x	x			x	x	x		x	x	x	
Angola				x		x				x				
Azerbaijan				x								x		
Bangladesh														
Bhutan		x		x										
Bolivia	x	x				x				x	x		x	
Chile		x		x			x		x	x	x		x	x
Colombia	x	x	x	x			x	x	x			x		x
Czech Republic	x	x	x	x			x			x	x		x	x
Dominican Rep.	x	x		x			x		x	x		x		
Egypt, Arab. Rep. of				x					x	x				
Finland				x		x				x			x	x
Ghana	x	x	x						x	x	x	x		x
Hong Kong, China		x	x	x							x			x
India		x	x				x							x
Indonesia	x	x	x	x						x		x		
Ireland	x	x		x			x			x				x

Jamaica	x	x	x			x	x	x			x	x	x	
Japan			x	x								x		
Jordan														
Korea, Rep. of				x		x	x	x	x	x				
Mauritius		x												
Mozambique					x		x			x		x	x	
Nigeria	x	x	x			x			x		x	x		
Norway		x		x			x		x	x	x			x
Poland		x		x			x		x	x	x	x	x	x
Romania		x		x				x	x		x		x	
Russian Federation	x	x	x								x	x		
Rwanda	x	x	x	x		x	x	x	x	x	x	x	x	
Singapore					x									
Slovenia	x	x	x	x			x		x	x	x	x		
South Africa			x	x										
Tanzania	x	x	x	x		x							x	
Thailand	x	x	x	x							x	x		x
Trinidad and Tobago		x	x	x			x				x			x
Tunisia		x	x								x	x		
United Kingdom	x	x	x	x		x				x				
Ukraine		x	x	x						x	x	x	x	
Venezuela, R. B. de					x							x		
Vietnam	x	x				x			x		x			

Source: Authors' analysis.

Table 5D.4 E-Strategy Focus Area: Legal/Regulation

	Legal/regulation											
	Objective				Intervention							
Economy	Government	Business	Society	Not specified	Intellectual property rights	Tax, tariff, quota	Privacy (personal information)	Contracts (and digital signature)	Cyber-crime security	Encryption	E-payment systems	Other
Albania					x		x	x		x	x	x
Angola	x	x			x	x						
Azerbaijan	x		x			x			x			
Bangladesh		x			x				x	x	x	x
Bhutan		x		x			x	x	x			x
Bolivia	x					x						
Chile	x	x	x		x	x		x	x		x	x
Colombia				x					x		x	
Czech Repubic		x	x						x			
Dominican Rep.												
Egypt, Arab Rep. of		x			x	x	x		x			x
Finland		x	x				x		x			x
Ghana				x	x	x			x			x
Hong Kong, China		x			x				x	x		
India				x	x				x	x	x	
Indonesia	x	x	x		x	x	x	x	x		x	x
Ireland	x	x	x		x	x	x	x	x		x	x

Jamaica		x			x	x	x	x				x
Japan				x	x		x		x			
Jordan		x			x	x						x
Korea, Rep. of		x	x				x	x	x		x	x
Mauritius												
Mozambique				x	x	x	x		x			
Nigeria	x	x	x		x	x	x		x			x
Norway		x	x			x		x	x			x
Poland	x			x		x		x	x			
Romania	x	x	x			x	x		x	x	x	
Russian Federation	x	x	x					x				
Rwanda		x			x	x			x		x	
Singapore		x			x	x						
Slovenia	x	x	x			x		x	x			x
South Africa				x	x							
Tanzania	x	x	x			x			x			x
Thailand	x	x			x	x	x	x	x		x	
Trinidad and Tobago	x	x	x		x		x	x	x	x		x
Tunisia		x			x	x					x	
United Kingdom		x		x	x		x		x			
Ukraine	x	x			x	x	x	x	x			x
Venezuela, R. B. de				x		x		x				
Vietnam		x		x	x	x			x			x

Source: Authors' analysis.

Table 5D.5 E-Strategy Focus Area: ICT Industry

	ICT industry													
	Objective							Intervention						
	Product				Target market									
Economy	Hardware	Software	IT services	Not specified	Domestic	International	Not specified	Cyber park/ incubation	R&D	Access to finance	Promotion	Business networks/ clusters	Quality	Other
Albania				x	x			x	x					
Angola	x	x			x	x				x				
Azerbaijan		x				x		x	x		x			
Bangladesh	x	x	x		x	x		x	x		x	x		x
Bhutan				x		x				x	x		x	
Bolivia														
Chile						x			x	x	x	x		
Colombia		x				x		x		x			x	
Czech Republic														
Dominican Rep.			x		x	x					x			
Egypt, Arab Rep. of		x				x		x	x		x			
Finland														
Ghana			x					x		x	x			x
Hong Kong, China		x	x		x	x		x	x		x	x		
India	x	x	x			x		x	x	x			x	
Indonesia				x					x					
Ireland				x			x	x	x	x	x	x		

Jamaica				x		x				x	x			
Japan									x					
Jordan		x	x			x		x		x	x		x	x
Korea, Rep. of	x	x				x		x	x	x	x		x	x
Mauritius				x	x	x		x			x			
Mozambique				x			x	x						
Nigeria				x				x	x	x				
Norway				x			x	x	x		x			
Poland														
Romania					x			x	x					x
Russian Federation				x		x				x	x	x		
Rwanda	x	x	x		x	x		x	x	x			x	
Singapore				x			x	x	x			x		
Slovenia														
South Africa				x		x		x	x	x	x	x		
Tanzania				x		x			x					
Thailand		x			x	x		x	x	x	x		x	
Trinidad and Tobago				x	x	x			x			x		x
Tunisia		x	x		x	x		x	x	x		x		
United Kingdom														
Ukraine	x	x	x		x	x		x	x	x	x		x	
Venezuela, R.B. de				x			x	x	x					
Vietnam	x	x			x	x		x	x	x	x		x	x

Source: Authors' analysis.

Table 5D.6 E-Strategy Focus Area: IT Human Resources Development

Economy	IT human resources development									
	Objective				Intervention					
	Government	Business	IT sector	Not specified	Enhance existing capacity	Build new capacity	Curriculum development	Quality assurance	Private sector collaboration	Other
Albania		x	x			x	x			
Angola										
Azerbaijan										
Bangladesh			x				x		x	
Bhutan										
Bolivia										
Chile			x			x			x	
Colombia										
Czech Republic										
Dominican Rep.							x			
Egypt, Arab Rep. of	x		x		x	x	x			x
Finland										
Ghana	x	x	x		x				x	
Hong Kong, China		x	x		x	x	x	x	x	
India			x		x	x	x		x	x
Indonesia	x		x			x	x		x	
Ireland			x			x			x	

Jamaica			x		x	x				
Japan			x		x			x	x	
Jordan	x	x	x				x	x	x	
Korea, Rep. of			x		x	x		x	x	x
Mauritius	x		x		x					
Mozambique	x	x	x		x		x			
Nigeria			x							
Norway		x	x	x						
Poland										
Romania										
Russian Federation	x		x	x		x				
Rwanda	x	x	x		x	x	x			x
Singapore				x		x				
Slovenia		x			x	x				
South Africa		x	x			x	x		x	x
Tanzania				x	x	x				
Thailand		x	x					x	x	x
Trinidad and Tobago	x	x	x		x	x	x	x	x	
Tunisia			x		x			x	x	
United Kingdom		x			x					
Ukraine			x		x					
Venezuela, R.B. de										
Vietnam						x				x

Source: Authors' analysis.

Table 5D.7 E-Strategy Focus Area: E-Business

	E-business							
	Objective: firm size			Intervention				
Economy	SMME and SME	Medium and large	Not specified	Training	Promotion	Standards	Business resource centers	Other
Albania	x			x	x			x
Angola						x		
Azerbaijan	x		x		x			
Bangladesh			x		x			
Bhutan			x		x			
Bolivia			x					
Chile	x				x			
Colombia								
Czech Republic			x					
Dominican Rep.			x		x			
Egypt, Arab Rep. of			x	x	x			
Finland	x			x	x			
Ghana	x				x			
Hong Kong, China			x		x			
India			x					
Indonesia	x				x			
Ireland	x			x			x	x

Jamaica			x	x	x		x	
Japan	x							x
Jordan								
Korea, Rep. of	x	x			x	x		
Mauritius	x			x	x		x	
Mozambique			x		x			
Nigeria			x		x			
Norway								
Poland						x		
Romania	x				x		x	
Russian Federation								
Rwanda	x				x		x	
Singapore	x		x		x			x
Slovenia	x				x			x
South Africa								
Tanzania	x	x			x			
Thailand	x	x			x	x		x
Trinidad and Tobago	x			x	x			
Tunisia	x				x			x
United Kingdom	x				x			
Ukraine								
Venezuela, R. B. de								
Vietnam			x	x	x			

Note: SME = small and medium enterprises; SMME = small, medium, and micro enterprises.

Source: Authors' analysis.

Table 5D.8 E-Strategy Focus Area: E-Health

Economy	E-health							
	Objective			Intervention				
	Clinical service delivery	Administration	Health education	Other (disease, emergencies)	Not specified	Health center connectivity	Standards	Other
Albania	x	x		x			x	
Angola								
Azerbaijan								
Bangladesh	x	x	x			x		
Bhutan								
Bolivia								
Chile		x				x		
Colombia								
Czech Republic			x	x		x		
Dominican Rep.								
Egypt, Arab Rep. of		x	x					
Finland		x		x		x	x	x
Ghana	x							
Hong Kong, China								
India								
Indonesia	x	x						
Ireland		x	x	x				x

Jamaica							
Japan		x		x		x	x
Jordan							
Korea, Rep. of		x		x	x		
Mauritius							
Mozambique		x	x				
Nigeria		x	x		x		x
Norway					x		
Poland	x						
Romania			x		x		
Russian Federation							
Rwanda	x	x	x	x	x		x
Singapore							
Slovenia		x	x				
South Africa							
Tanzania		x	x				x
Thailand							
Trinidad and Tobago							
Tunisia							
United Kingdom							
Ukraine							
Venezuela, R. B. de			x		x		
Vietnam							

Source: Authors' analysis.

Endnotes

1. Article 8 of the WSIS Plan of Action states that "development of national e-strategies ... should be encouraged by all countries by 2005."
2. The United Nations Development Programme estimates the number to be 90; McConnell International puts it at 70.
3. The sample is based on the availability of past and current e-strategy documents from countries representing all income and region groups.
4. The relation between the funding and attainment of strategic objectives is further described in the strategy by showing how the funding will be deployed through direct investments (development and use of ICT), indirect investments (in products and services), and new ICT-based business models.
5. A focus area is considered "significant" if it appears in 5 or more of the 40 e-strategies that were surveyed.
6. Many e-strategies (for example, those of Albania and the Czech Republic) stipulate the need to adhere to regional and international regulatory standards and agreements.
7. Secondary schools are the most common (more than 70 percent) followed closely by adult and community training centers (also more than 70 percent) and tertiary educational institutions (more than 50 percent).
8. IT services are a subset of ICT services and include call centers, data processing, customer support, sales, operations, and project management. Many countries, including Bhutan, Hong Kong (China), Japan, and Rwanda, plan to outsource the development and/or administration of e-government services to local ICT companies.
9. Training, according to the strategies, encompasses capacity building in management skills as well as e-literacy.
10. For the purposes of the e-strategy analysis, the distinction is made between adult e-literacy training (basic skills in computers and computer applications) and professional training in information technology, which is an applied set of skills aimed at supplying the ICT sector with technology professionals.
11. For the ratio of income levels and number of countries selected, it is worth noting that no high-income countries cited the provision of IT professionals to government agencies as a component of their strategies. Other than this exception, no interventions or focus areas vary significantly from the ratio of income levels and number of countries selected.

References

The following are useful documents dealing with formulation or implementation of e-strategies.

APC (Association for Progressive Communications). 2004. "E-Strategies and the World Summit on the Information Society." May 2004. http://rights.apc.org/documents/estrategies. pdf (accessed September 2005).

G-8 DOI (Digital Opportunity Initiative). 2001. "Creating a Development Dynamic: Final Report of the Digital Opportunity Initiative." July, 2001. http://www.opt-init.org/framework.html (accessed September 2005).

Tipson, Frederick, and Claudia Fritteli. 2003. *Global Digital Opportunities: National Strategies of 'ICT for Development.'* New York: Markle Foundation.

UN ICT Task Force. 2003. Tools for Development: *Using ICT to Achieve the Millennium Development Goals.* December.

—-. *2004. Second Annual Report of the Information and Communication Technologies Task Force,* 04-34245 (E), New York, May 2004.

UNCTAD (United Nations Conference on Trade and Development). 2003. *E-Commerce and Development.* UNCTAD/SIDTE/ECB/2003/1. New York and Geneva: United Nations.

UNDP-APDIP (United Nations Development Programme-Asia-Pacific Development Information Programme). 2004. *ICT Policies and e-Strategies in the Asia-Pacific.* New Delhi: Elsevier.

World Bank. 2005. *Monitoring and Evaluation Toolkit for E-strategies Results.* Global Information and Communication Technologies Department, World Bank: Washington, DC.

Chapter 6

Tracking ICTs: World Summit on the Information Society Targets

Michael Minges

In December 2003, the World Summit on the Information Society (WSIS) adopted a Plan of Action for promoting the development of information and communication technology (ICT) (WSIS 2003). The plan includes 10 targets for ICT connectivity to be achieved by 2015 (box 6.1). These targets are distinctive for several reasons. They are the first globally agreed targets for measuring ICT development.[1] They also look beyond the traditional telecommunications sector by including, for example, connectivity in government, schools, and hospitals.

Governments, multilateral organizations, researchers, and others may find it useful to evaluate progress toward the WSIS targets. These targets provide benchmarks that governments can incorporate into e-strategies to measure their own progress and that of other countries (see chapter 5 of this volume). For example, countries such as Lebanon and Uganda have elaborated ICT development goals (figure 6.1) that can be benchmarked to the WSIS targets. Multilateral and bilateral assistance agencies can use the targets to monitor projects with an ICT component. Development plans and poverty reduction strategies can include the WSIS targets to set standards in relevant areas. There is evidence that ICT applications are contributing to progress in critical development areas such as health and education. Researchers may wish to measure the impact of ICTs on progress toward social and economic development targets of the UN Millennium Development Goals (MDGs).[2]

Because the targets of the WSIS Plan of Action have so many uses, it is particularly important to translate them into measurable indicators that can be monitored on an ongoing basis. The WSIS targets are broadly defined (see box 6.1).[3] They suggest specific indicators or quantitative benchmarks, but these are not identified. Therefore the search for indicators must begin by distinguishing what is measurable and what is not. The selected indicators should be internationally comparable and pertinent to a wide set of countries. They should also be sufficiently robust to provide reliable evidence of trends and levels, and not so complex as to limit their construction and use. They should be understandable and accessible to ordinary citizens as well as researchers and policy makers.

How Important Are the WSIS Targets for ICT Policy Makers?

Because there are many targets and because some countries might find it difficult to pursue them all simultaneously, it is useful to prioritize them. An online survey conducted by the International Telecommunication Union (ITU) asked respondents to rank the WSIS targets by importance (figure 6.2) (ITU 2004). Responses are grouped into three levels of relevance: targets above 80 percent are considered *high priority*, those scoring between 60 and 70 percent are considered *medium*

Box 6.1 WSIS Targets

Based on internationally agreed development goals, including those in the Millennium Declaration, which are premised on international cooperation, indicative targets may serve as global references for improving connectivity and access in the use of ICTs in promoting the objectives of the Plan of Action, to be achieved by 2015. These targets may be taken into account in the establishment of the national targets, considering the different national circumstances:

A. to connect villages with ICTs and establish community access points;

B. to connect universities, colleges, secondary schools, and primary schools with ICTs;

C. to connect scientific and research centers with ICTs;

D. to connect public libraries, cultural centers, museums, post offices, and archives with ICTs;

E. to connect health centers and hospitals with ICTs;

F. to connect all local and central government departments and establish Web sites and e-mail addresses;

G. to adapt all primary and secondary school curricula to meet the challenges of the Information Society, taking into account national circumstances;

H. to ensure that all of the world's population has access to television and radio services;

I. to encourage the development of content and to put in place technical conditions in order to facilitate the presence and use of all world languages on the Internet;

J. to ensure that more than half the world's inhabitants have access to ICTs within their reach.

Source: WSIS 2003.

priority, and the others are considered to be of *less* importance. Some targets may be ranked lower than others because they are already close to being achieved on a global level. For example, terrestrial radio broadcasting covers 95 percent of the world's population and television covers 85 percent; therefore, ensuring that the world's population has access to radio and television services (target H) would have a lower priority, as not much effort is required to meet this target.

Supplementary indicators are proposed for countries that have achieved the basic targets. Many of the targets are supply oriented, premised on providing ICT infrastructure to various institutions and populations. This chapter compares some of the targets with demand indicators, which can demonstrate levels of use. For example, connecting public libraries is a laudable goal, but if few people are using them, then the target might merit adjustment. Demand indicators are also a useful check on the validity of the data used to measure the targets, many of which are currently compiled from administrative records.

Target A: Connect Villages with ICTs and Establish Community Access Points

Basic indicators
- **Percentage of villages (localities) with a telephone**
- **Percentage of rural population covered by a mobile telephone network signal**
- **Number of public Internet access points per 100 inhabitants**

Supplementary indicator
- **Percentage of villages (localities) with Internet access**

This target has two parts: connecting villages with ICTs and establishing community access points. In the first part, the phrase "to connect villages" is clear, but "with ICTs" is vague—do we mean radios and TVs, telephone service,

Figure 6.1 ICT Targets in Uganda and Lebanon

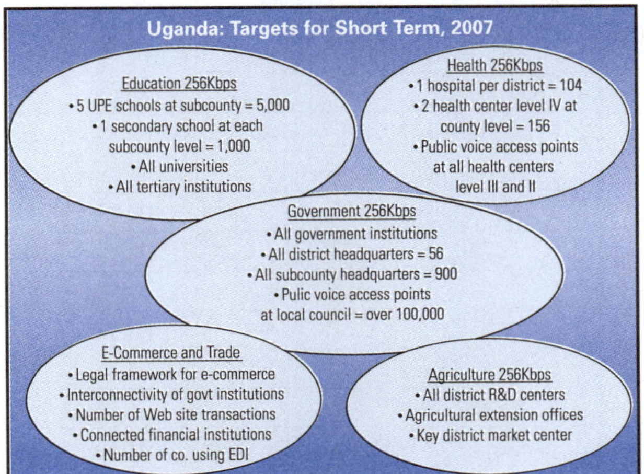

Some of Lebanon's ICT Targets

- Villages are already connected, 30 community access points by 2004
- Connecting public institutions (government agencies, schools, universities, libraries, hospitals) by 2005
- Secondary school curriculum revised in 2002. All students to reach ICT competency level as part of the formal education by 2010
- All universities & research centers to be connected by 2005, all secondary schools by 2008, and all primary schools by 2010
- All public libraries and museums are to be connected by 2008
- All hospitals are to be connected by 2006 and health/social centers by 2010
- Internet penetration rate to reach 18 percent by 2006 and 25 percent by 2010
- Broadband connectivity services to be offered by 2005
- Building awareness of the use of ICTs to all segments of the society by 2010

Source: Adapted from Uganda Communications Commission 2005 and American University of Beirut. 2003.

Figure 6.2 Importance of WSIS Targets

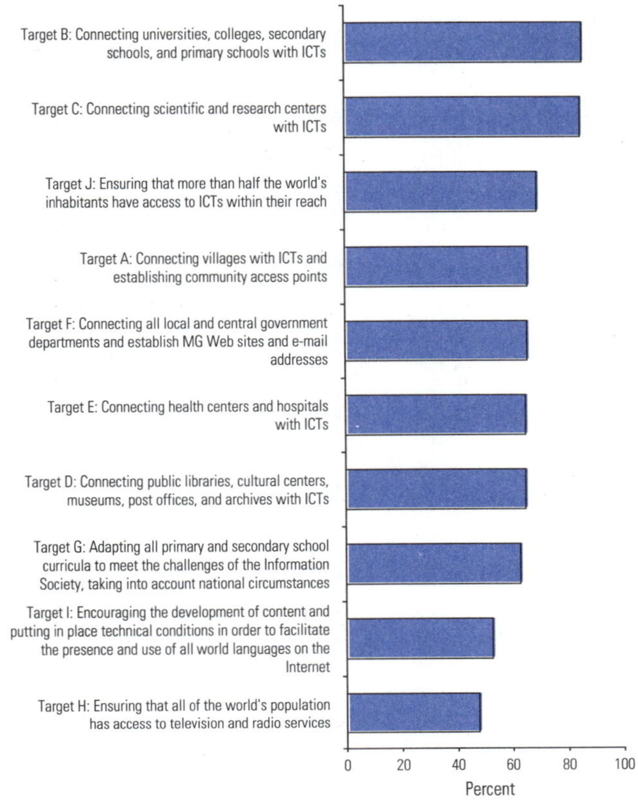

Note: Based on the results of 1,250 respondents from an online survey conducted between April 10 and May 10, 2004.

Source: Adapted from ITU 2004.

Internet access? Broadcasting is dealt with in target H, but Internet access is related to the second part of this target. Therefore connecting villages with ICTs might focus on providing telephone service, with the indicator being the *percentage of villages with telephone service.*[4] The necessary data should normally be available from the administrative records of telephone operators.

Connecting Villages with ICTs. The target of connecting villages with ICTs was rated as having medium importance in the ITU survey. As just over half the world's population was living in rural areas in 2002, providing ICT access in villages is an important goal. However, there are methodological complications in measuring village connectivity. The definition of *village* varies among countries in terms of population size and administrative divisions. Some countries do not explicitly use the word *village* in their statistics but refer instead to *localities.* And no international data set is available with the number of villages per country. An extrapolation of data for the top 10 developing countries with the largest rural populations suggests that there may be around 2.7 million villages in the world, of which some three-quarters already have telephone service (table 6.1). One reason for this surprisingly high level is the progress that China and India—the two countries with the largest rural populations in the world—have made in connecting villages. In 2002, 87.9 percent of Chinese "administrative villages" had a telephone; in India more than 90 percent of its some 600,000 villages had a so-called Village Public Telephone.

Though wider in scope, an alternative measure could be the *percentage of the population covered by a mobile telephone network signal.* This measure is possibly more relevant than telephone availability because mobile phones are the prevalent mode of communications in most developing countries.[5] It also avoids some of the methodological complications of measuring village connectivity and is available for many countries. Mobile population coverage could be adapted to reflect the proportion of the rural population

Table 6.1 Village Connectivity: Top 10 Developing Countries by Rural Population

Rank	Country	Population (millions) 2002	Rural population (percent) 2002	Rural population (millions) 2002	Number of villages (thousands)	Average village population	Number of villages with fixed telephone (thousands)	Villages with fixed telephone (percent)	Mobile population coverage (percent) 2003 Total	Mobile population coverage (percent) 2003 Estimated rural
1	China	1,295	62	803	930	867	772	83	73	56
2	India	1,050	72	756	607	1,242	468	77	41	19
3	Indonesia	217	56	122	69	1,745	27	39	85	73
4	Bangladesh	144	76	109	86	1,272	68	79	50	34
5	Pakistan	150	66	99	125	795	12	10	41	11
6	Nigeria	121	54	65	90	727	..	..	43	..
7	Vietnam	80	75	60	9	6,780	9	98	29	6
8	Ethiopia	69	85	59	10	5,837	1	11	..	..
9	Thailand	62	68	42	69	614	55	80	92	88
10	Egypt, Arab Rep. of	71	58	41	5	8,826	1	21	98	97
	Top 10	3,259	66	2,156	2,000	1,079	1,412	71	58	..

Note: Estimated mobile rural population coverage assumes that all inhabitants in urban areas are already covered. It is derived by subtracting the percentage of the population in urban areas from the overall percentage of the population covered by mobile telephony. .. Not available.

Sources: World Bank (population, mobile population coverage), UNDP (rural population), and the following national sources: *Bangladesh*: Bari 2004, Minges and Simkhada 2002; *China*: NBS 2003; *Egypt*: American Chamber 2005, Egypt State 1995; *Ethiopia*: Library of Congress 1991. Villages with telephone derived from rural call stations: *India*: Minges and Simkhada 2002; *Indonesia*: DGT 2002; *Pakistan*: Minges and Simkhada 2002; *Nigeria*: CDC 1991; *Thailand*: NSO 1999; *Vietnam*: Anh 2002, VNPT 2004.

covered. Indeed, in several countries, such as the Arab Republic of Egypt and Indonesia, the percentage of people with rural mobile coverage is estimated to be higher than the percentage of people in villages with a fixed telephone (table 6.1).

As table 6.1 shows, there are sometimes big differences between measures of village fixed-line and rural mobile coverage. Some countries may want to calculate either or both of these indicators. What constitutes a village varies widely, ranging from 614 people in Thailand to 8,826 in Egypt. The village data are also from diverse sources and some almost 25 years old. Both definitions and data availability and timeliness all need to be strengthened before accurate comparisons can be made about village ICT connectivity.

Establishing Community Access Points. Although mobile coverage reflects the *possibility* of access, it does not indicate that people are in fact connected. Providing access to a range of ICTs through a central facility is a more concrete manifestation of connectivity.[6] The definition of a *community access point*, in the second part of the target, is vague. It could refer to a particular type of government-sponsored facility, to any kind of public venue (for example, post offices, libraries, or schools), or even to privately operated Internet cafes. The ITU has organized several workshops on defining *community access* and proposed the indicator *percentage of localities with public Internet access centers (PIACs)* (ITU and Secretaría de Comunicaciones y Transportes 2004).

Because this indicator is not presently widely available, an interim measure might be the *number of public Internet access points (PIAP) per 1,000 inhabitants*, which the European Union included in its original e-Europe benchmarking indicators (figure 6.3). The target of establishing community access points is similar to some of the other targets. Targets B and D call for connecting schools and locations such as post offices and libraries, which provide public Internet access in some countries. There could be some overlap with this target. Also, public access points may not be as relevant in countries with a high level of household Internet access. Therefore it might be useful to contrast public Internet access penetration with the percentage of homes that have (private) Internet access. On the demand side, it would be interesting to contrast public Internet access penetration with the *percentage of the population that utilizes the access points*, a question often asked in Internet user surveys.

Figure 6.3 Number of Public Internet Access Points, 2003
(per 1,000 people)

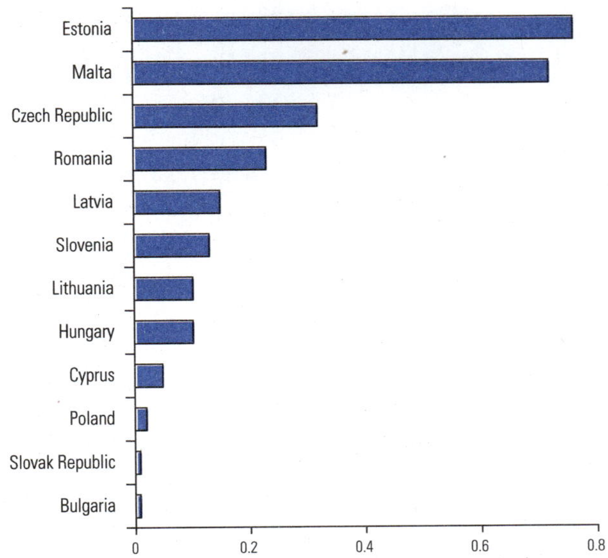

Source: eEurope+ 2003.

Target B: Connect Universities, Colleges, Secondary Schools, and Primary Schools with ICTs

Basic indicators
- **Percentage of schools with a computer**
- **Percentage of schools with Internet access**

Supplementary indicators
- **Students per computer**
- **Percentage of schools with broadband Internet access**

This target was the highest ranked in the ITU survey, attesting to the benefits that many see from connecting schools. Providing educational institutions with ICTs is attractive because young people are quick to grasp new skills and thus form the basis for a cyber-savvy nation. Developing countries have relatively young populations, so reaching this target would have a strong impact in those countries. It is, therefore, perhaps even more relevant for developing countries than for developed ones. Connecting schools also provides access to huge amounts of information and allows teachers to use resources on the Internet that expand and enrich the curriculum.

The target does not identify which ICTs schools are to be connected with, but available data generally cover computers and Internet access. Basic indicators would include the *percentage of universities, colleges, secondary, and primary schools with a computer* as well as the *percentage of universities, colleges, secondary, and primary schools with Internet access.*

Although there are a growing number of data on ICT infrastructure in schools, they are not available from a central source. Their scattered nature adds to the complexity of locating and comparing ICT school data. Comparability is often difficult both because countries use different indicators to measure school connectivity and because data for all countries are not always available for the same year. Statistics for developing countries are not widely available. It should, however, be fairly straightforward to collect the data because the indicators are clear and most statistical offices or ministries responsible for education regularly compile information on the number of schools and students.[7]

Providing computers and Internet access to most schools has been achieved by most developed countries as well as some middle-income developing ones.[8] For example, 92 percent of primary and secondary schools in the European Union already had Internet access by 2002 (Gallup Europe 2002). When countries have met the target of networking their schools, a higher level of connectivity—such as broadband access—might also be monitored.[9] More precise indicators—such as the *number of computers available per student* and the *number of computers connected to the Internet*—would allow a higher level of benchmarking. It would also exclude computers and Internet access used only for administrative purposes.

Although connecting educational institutions to the Internet is a laudable goal, the reality is that in some countries students have a higher level of access outside of their school. Although developing ICT skills is an impetus for connecting schools, it is appropriate to ask if students can obtain those skills elsewhere. The Program for International Student Assessment (PISA) surveyed 15-year-old students about the availability of a computer in their school and other locations (OECD 2003).

The results are interesting for a couple of reasons:

- They show that in some countries, there is a higher level of access to computers outside of schools than inside.
- For some countries, the data show a lower level of computer availability than the official statistics show.

This suggests that although schools may have computers and Internet access, they are not available to all students.

Evidence suggests that realistic plans backed with adequate funding can succeed in connecting schools rapidly. Countries such as Singapore have provided ICTs in schools through careful planning with clear objectives and financial support. Singapore's five-year $1.2 billion Master Plan for IT in Education, launched in 1997, connected all schools to the Internet by 2002—quite an impressive achievement, as only around one-third of schools were connected in 1998 (Ministry of Education [Singapore] 1997).

As a policy in some countries, telecommunications operators are encouraged to provide discounted Internet access to educational institutions. Specific projects in some countries have also helped to dramatically increase ICT infrastructure availability in schools. One such example in a developed country is the *e-rate* in the United States. Enacted as part of the U.S. Telecommunications Act of 1996, the e-rate called for operators to provide Internet access to schools at affordable prices. No doubt this has helped boost Internet connectivity in U.S. public primary and secondary schools from 65 percent in 1996 to 100 percent in 2003 (National Center for Education Statistics 2005).

Special programs targeting ICTs in schools have been successful in some developing countries. The government-funded Enlaces program in Chile, launched in 1995, has connected 4,677 schools (almost 80 percent of them with broadband) and provided 57,000 computers.[10] In Estonia, the Tiger Leap program began in 1996. By 2000, all schools in Estonia had a computer and 75 percent had access to the Internet.[11] Thailand's SchoolNet, which started in 1995, had connected 4,654 schools to the Internet by 2003.[12]

A number of countries are incorporating ICT indicators and targets in their education plans. For example, South Africa's White Paper on e-Education (Department of Education [South Africa] 2003) calls for tracking indicators such as the ratio of learners to computers and levels of Internet connectivity on an annual basis. It also calls for boosting the level of connectivity in schools—from the 27 percent that had a computer for teaching in 2002 to 100 percent having access to a networked computer facility by 2013.

Target C: Connect Scientific and Research Centers with ICTs

Basic indicator • **Availability of a national educational and research computer network**

Supplementary indicator • **Bandwidth of national educational and research network**

This target is the second-highest ranking among respondents of the ITU survey. It is also one of the hardest to measure. The concept of scientific and research centers may be intuitively clear, but definitions for statistical purposes are lacking. The boundary between scientific and research centers associated with academic institutions and those sponsored by governments is often blurred; the former might also be included in target B (connecting universities). Moreover, in some countries, significant research is carried out by private organizations. Even if private research institutions are included in the target, they can be difficult to identify. As a result, many countries do not explicitly have statistics on the number of scientific and research centers in their countries.

Because determining the denominator for this target is so difficult, an alternative is to examine the *characteristics of research networks that can be compared between countries.* This assumes that it is desirable for scientific and research centers to be connected to an academic and research network, and that these centers would connect to such networks if the facilities were available. The Trans-European Research and Education Networking Association (TERENA) provides a directory of national research networks for some 50 countries in the European, Central Asian, and Mediterranean region on its Web site, including details about the number of institutions connected in each country and bandwidth (table 6.2) (TERENA 2005).

Target D: Connect Public Libraries, Cultural Centers, Museums, Post Offices, and Archives with ICTs

There are two connectivity aspects for these types of institutions. Because the overall aim of the WSIS targets is to enhance access, and these are public institutions, measurement might be focused on ICT access provided at these locations to the general public. Alternatively, some of these institutions are repositories of useful information and it would be beneficial to have that information online. This calls for a measurement such as the percentage of these institutions with Web sites. Much of the supply-side data for these areas could be collected through administrative records.

Public libraries

Basic indicators • **Percentage of public libraries providing access to the Internet**
• **Percentage of public libraries with a Web site**

UNESCO has published data on the number of public libraries by country.[13] Because data are lacking for a number of countries, it is not possible to derive regional or world

Table 6.2 Selected Research and Education Networks, 2005

Country	Network	Research institutions connected	Backbone bandwidth	Network map
Algeria	CERIST	11	155 Mb/s =< 622 Mb/s	http://www.arn.dz/carte.htm
Azerbaijan	AzNET	5	622 Mb/s =< 1.2 Gb/s	http://www.aznet.org
Kazakhstan	KazRENA	9	34 Mb/s =< 155 Mb/s	http://www.kazrena.kz
Morocco	MARWAN	2	34 Mb/s =< 155 Mb/s	http://www.marwan.ma/docs/topology.pdf
Syrian Arab Republic	SHERN	3	..	http://www.shern.net
Turkey	ULAKBIM	14	34 Mb/s =< 155 Mb/s	http://www.ulakbim.gov.tr
Uzbekistan	UzSciNet	24	2 Mb/s or below	www.uzsci.net/info

Note: .. Not available. Gb/s = gigabytes per second; Mb/s = megabytes per second.

Source: World Bank, adapted from TERENA.

totals. Some developed countries have data on ICTs in libraries, but the number of public libraries in many developing countries is often difficult to determine.

Some countries have identified public libraries as attractive locations for providing Internet access. Countries where statistics on ICTs in libraries are collected tend to focus on this public service aspect of their institutions. Australia measures the *number of Internet workstations per public library* as well as a demand side indicator—the *percentage of individuals who access the Internet from public libraries.*[14] The U.S. National Center for Education Statistics publishes an annual report that includes the *percentage of public libraries providing access to the Internet,* the *number of users of electronic resources per 1,000 population per typical week,* and the *number of Internet terminals in public libraries for public use per inhabitant and per library* (National Center for Education Statistics 2005).

An online directory of library holdings is a valuable service. The European Union includes the *availability of electronic card catalogues in public libraries* as one of the 20 policy indicators composing its e-government indicator. The *percentage of public libraries with a Web site* could be a proxy for this indicator, because it would then be possible to have the catalogue online.

Cultural Centers

Basic indicators
- **Percentage of cultural centers with Internet access**
- **Percentage of cultural centers with a Web site**

The concept of *cultural center* is not statistically well defined. Although a few countries publish statistics on the number of cultural centers, most do not. Apart from the vagueness of what a cultural center is, the purpose of connecting them with ICTs also needs to be established. One possibility is that because cultural centers are typically places where the public gathers, they are good candidates for providing public access to ICTs. In that case, they might fall under the category of community access points (target A). Until a robust definition of cultural center is created, it will be difficult to collect and compile data for this subtarget.

Museums

Basic indicator
- **Percentage of museums with a Web site**

Most people go to museums to see an exhibit, not for access to ICTs. Thus, the main reason for connecting museums is to provide online access to information about their collections. A suitable indicator would be the *percentage of museums with a Web site.* This indicator has been compiled for regional studies and could possibly be derived from portal sites. According to one source, there are 41,000 museums in 199 countries.[15] There are a number of portals providing links to museums with Web sites, with one claiming links to 37,000 museums around the world (not all of which are online).[16] The information is not summarized by country, nor is there an international data set on the number of museums in a country—let alone data on how many are connected to the Internet or have a Web site. At the national level, there are scant data regarding ICTs in museums. It is difficult to reach a consensus about what type of ICT information might be useful or—for the purposes of WSIS—what it means for a museum to be connected with ICTs.[17]

Post Offices

Basic indicators • **Postal service has a Web site**
• **Percentage of post offices with public Internet access**

Connecting post offices could significantly expand access to ICTs for the public at large. According to the Universal Postal Union (UPU), 96 percent of the world's inhabitants were served by some 660,000 postal establishments in 2003 (UPU (2004). If they were all connected to the Internet and provided public access, then almost the whole world would, theoretically, have access to the Internet from post offices.

The UPU collects several ICT-related indicators including online services available in post offices (for example, Internet postage service, Internet bill payment, e-mail service, Internet goods ordering service, and so on), and *post offices offering public Internet access* (figure 6.4).

One major limitation with the UPU data is that they refer to the simple existence of a service at some post office. A country could have 1,000 post office locations, but if only one out of the 1,000 provides Internet access, then that country would be considered to provide access through its post office. It is analytically more useful to know the *percentage of post offices that provide Internet access.*

A number of countries are promoting the use of post offices to achieve universal access goals for ICTs. As part of the Community Access Point Program in Mauritius, for example, 96 post offices are to be equipped with at least two computers and Internet access; by April 2005, 24 interactive kiosks were in operation. In Bhutan, some 38 post offices will offer e-post services allowing users to send and receive e-mail.

Archives

Basic indicator • **National archive has a Web site presence**

Many countries have established national archives to store important records. Some countries also have regional and local archives. The International Council on Archives, a UNESCO partner, has a directory of its institutional members with address information and, if available, a link to their Web site.[18] There is no data set of the number of archives by country or the percentage online, but this could probably be compiled from the available information. An initial indicator could be *whether the national archive has a Web site* (table 6.3).

Figure 6.4 Post Offices and ICT, 2003

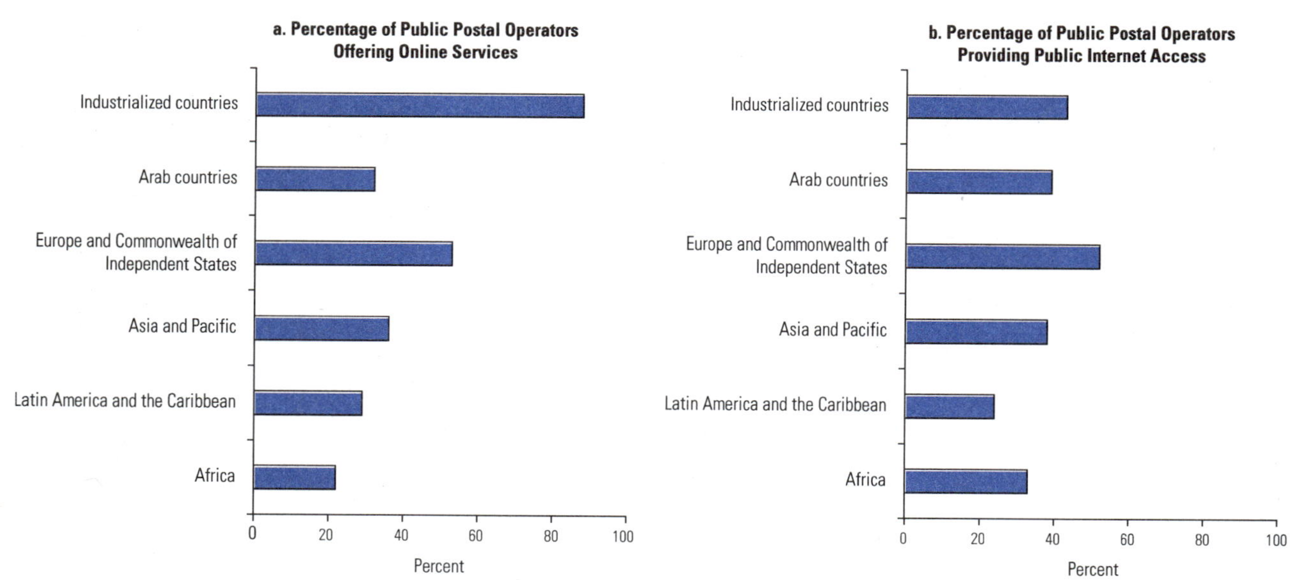

Note: Figure 6.4a refers to the post office having online services described in the text; figure 6.4b refers to the availability of Internet access for the public from at least one post office.

Source: UPU 2004.

Table 6.3 National Archives Online in Southeast Asia

Country	National archive	Web site
Brunei	Museums and Archives Department	http://www.museums.gov.bn/
Cambodia	National Archives	http://www.camnet.com.kh/archives.cambodia/
Indonesia	National Archives	http://www.anri.go.id/
Malaysia	National Archives	http://www.arkib.gov.my/
Singapore	National Archives	http://www.nhb.gov.sg/NAS/
Thailand	National Archives	http://www.culture.go.th/archives
Vietnam	State Records and Archives Department	http://www.luutruvn.gov.vn/

Source: Adapted from International Council on Archives (http://www.ica.org).

Target E: Connect Health Centers and Hospitals with ICTs

Basic indicators
- **Percentage of hospitals/health centers with Internet access**
- **Percentage of hospitals/health centers with a Web site**

Connecting health centers and hospitals with ICTs offers many benefits including improving internal administrative processes, allowing health staff online access to medical information, and providing the possibility for remote diagnosis. ICTs in a health care setting can also help achieve several of the UN Millennium Development Goals related to health.

Measuring this target on a global level is complicated by the fact that there is no source for official data on the number of hospitals or health centers in the world. The World Health Organization (WHO) does not regularly collect information on the number of hospitals, nor does it compile data on ICT connections in health institutions. The International Hospital Federation lists over 40,000 hospitals around the world; however, these data are based only on countries that have reported the data.[19] A number of national statistical offices or ministries of health publish statistics on the number of hospitals and sometimes on the number of health centers. There have been one-off reports, such as a regional publication for Latin America that featured a table showing the *percentage of hospitals with information systems* and the *percentage of hospitals with a computer.*[20] An EU project collected data on the *percentage of hospitals and health centers with a Web site* for about a dozen Mediterranean countries (EU 2001).

Although this target aims at the connectivity of hospitals and health centers, connectivity for doctors (and for patients to communicate electronically with them) might be equally important. In the first round of its e-Europe indicators, the European Union used the *percentage of general practitioners with a computer in their office* and the *percentage with Internet access in their office* for measuring e-health. The current EU e-health indicators are the *percentage of general practitioners using electronic patient records* and, on the demand side, the *percentage of population (aged 16 and over) using Internet to seek health information whether for themselves or others.*

Target F: Connect all Local and Central Government Departments and Establish Web Sites and E-Mail Addresses

Basic indicator
- **Central government Web site presence**

Supplementary indicator
- **Percentage of basic public services available online**

Unlike many of the other targets, which are focused on measuring access or use of public facilities to increase access, this target has multiple purposes. Connecting all local and central government departments can enhance internal processes, which may improve governance. Establishing Web sites, however, has a public service aspect. Establishing e-mail addresses can help government employees communicate with each other more effectively, but it can also benefit the public by providing them with an online contact.

This target is difficult to compare across countries because the structure of governments differs so widely around the world. Public services in a city state such as Singapore are essentially provided by the central government; in the Swiss federalist system, many services are carried out at the local level. The way a government is presented online also varies—some countries opt to centralize information at a ministry level and others have Web sites for different administrative units.

Figure 6.5 The State of E-Government, 2004

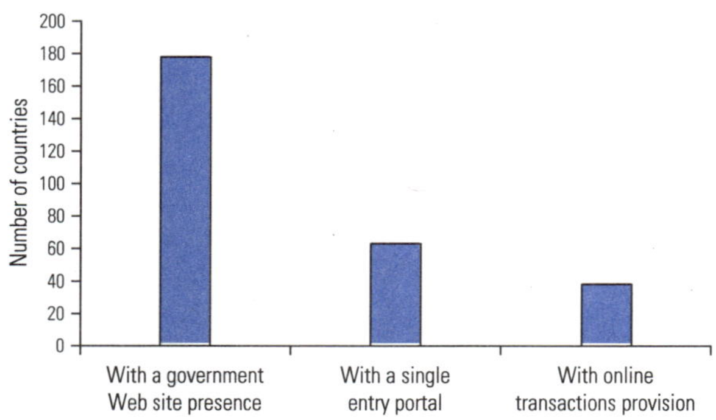

Source: World Bank, adapted from UN 2004.

The target of establishing government Web sites is relatively easy to meet, at least at a central government level. According to the United Nations, out of 191 UN Member States, 178 had a *central government Web site presence* (UN 2004). Because Web site presence can range from a few simple pages providing descriptive information to highly interactive systems that allow citizens to complete and dispatch government forms online, more sophisticated measures of e-government are needed (figure 6.5).

The United Nations has compiled an *index that calculates the degree of sophistication of central government Web sites.*

Table 6.4 The 20 Most Desirable Online Public Services

For citizens	For businesses
Income taxes	Social contribution for employees
Job search	Corporate tax
Social security benefits	Value-added tax
Personal documents	Registration of a new company
Car registration	Submission of data to the statistical office
Application for building permission	Customs declaration
Declaration to the police	Environment-related permits
Public libraries	Public procurement
Birth and marriage certificates	
Enrollment in higher education	
Announcement of moving	
Health-related services	

Note: *Social security benefits* includes unemployment benefits, child allowances, medical costs, and student grants. *Personal documents* includes passports and driver's licenses.

Source: Capgemini 2004.

Another approach is to derive an indicator based on how well governments are performing in providing commonly used public services applications online. The European Union has identified 20 such applications (see table 6.4), which range from car registrations to customs declarations. It has also created a composite indicator—*online availability of public services*—for tracking not only how many of these applications are online but also their degree of sophistication (for example, the provision of simple information, ability to download forms, possibility of completing transactions online).[21]

Another approach is to look at the demand side. For example, how many people are actually using government Web sites? Whether the public finds the available online government information useful is an important aspect of this indicator. The European Union uses a supplementary indicator for measuring this: *the percentage of individuals using the Internet for interacting with public authorities.*

Target G: Adapt all Primary and Secondary School Curricula to Meet the Challenges of the Information Society, Taking into Account National Circumstances

Basic indicator • **Country includes ICT in school curricula**

Target G is the second target related to education. Target B calls for connecting educational establishments; this one seeks to ensure that students are provided with the training needed to use ICTs in school and as part of everyday life. The target can be tracked by the simple indicator of *whether a country includes ICT courses in its curriculum.* Data are available for Europe about the availability of ICT in the curricula as a separate subject (figure 6.6) (Eurydice 2004). A related indicator would be the *number of teachers trained in ICT* because trained staff is essential to provide instruction. These data are provided for both primary and secondary schools. It would be useful to measure the quality of learning, but such measures are rare for any subject matter.

Target H: Ensure that All of the World's Population Has Access to Television and Radio Services

Basic indicators
- **Percentage of population covered by radio/television signal**
- **Percentage of households with a radio/television**

Figure 6.6 ICTs in the Secondary School Curriculum in Europe

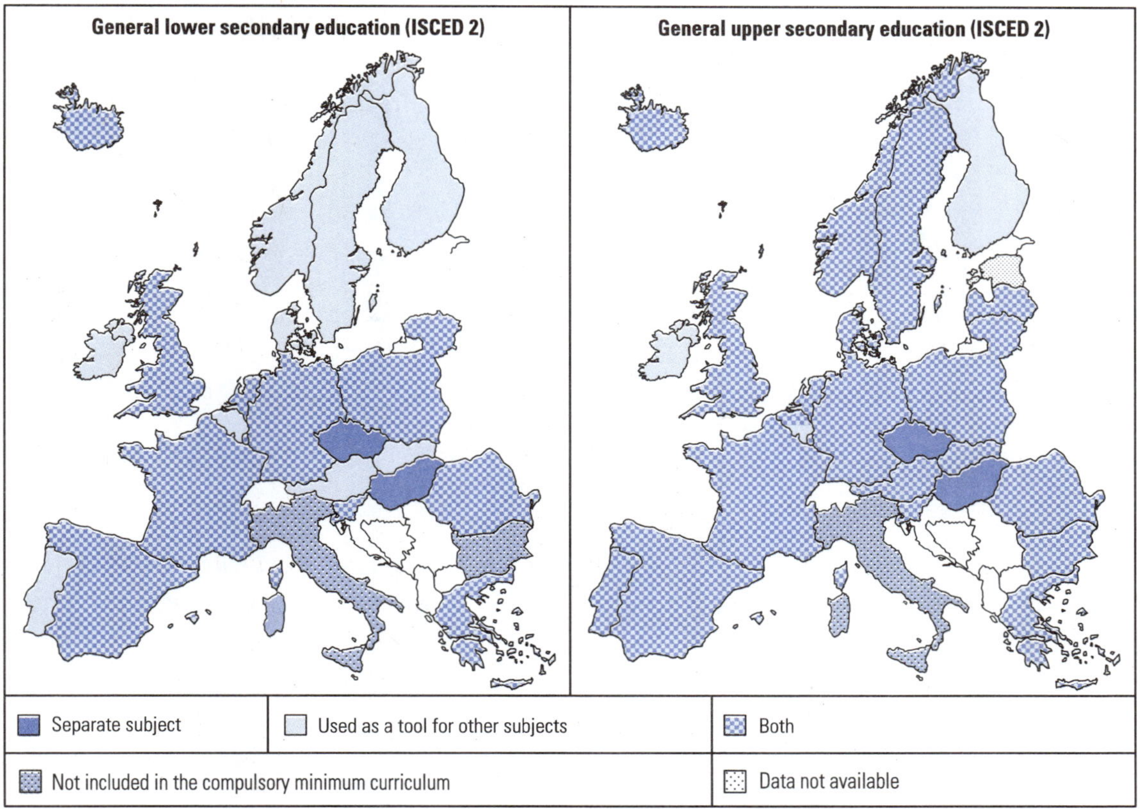

Source: Eurydice 2004.

Supplementary indicator • **Percentage of households with multichannel/digital television**

This is the easiest of all the WSIS targets to measure, because it is the most specific. Unlike most of the other targets, which contain the vague term *ICTs*, this one explicitly calls for access to radio and television services. The measurement is conceptually clear and data are generally available (although reliability can be a concern depending on the indicator selected).

Radio and television sets per 100 inhabitants have traditionally been used as indicators to measure broadcasting penetration.[22] These data are rarely compiled from demand-side data using household surveys, which ask how many broadcast sets are available in a home. Instead the indicators are typically derived from supply estimates. Although the data may have been reliable for some countries, where broadcast license schemes required broadcast set owners to register them, they have, over time, declined in utility. In certain countries the data are less useful because of significant evasion of set registration to avoid paying the license fee. The number of radio or television sets could also be estimated from shipment records. This requires good sales data as well as valid assumptions about replacement and resale rates. But perhaps most tellingly, the number of sets in a country does not give a clear picture of access: some families can have many sets, while others have few or none. From an access perspective, it is not the number of sets that matters but rather the availability of at least one set for a family.

One of the most popular indicators for measuring access is the *percentage of households with a radio or television set.* This is widely asked in household surveys. A less specific indicator of access is the *percentage of population covered by broadcasting service.* This measures theoretical access regardless of whether people actually have a set.

Coverage is fairly widespread according to ITU statistics for 2002 (figure 6.7). Some 69 percent of households had a radio set, while 75 percent had a television in that year.

Figure 6.7 Broadcast Access, 2002

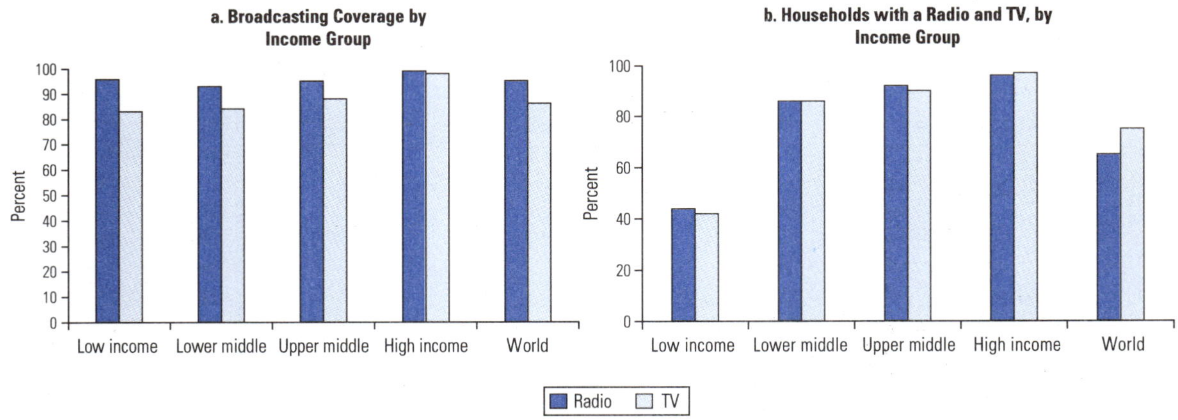

Source: World Bank, adapted from UN 2004.

Global population coverage for radio was 95 percent, and 86 percent for television. Clearly, this target is close to being achieved, which may be why respondents to the ITU survey showed less interest in it (fewer than half thought this was an important target).[23] Another reason for the respondents' lackluster response to the target may be, as some policy makers believe, that new ICTs are to be favored over traditional ones.[24] Given widespread availability of basic over-the-air broadcasting in many countries, more advanced targets might be proposed—such as the percentage of households with cable or satellite service or with a digital television.

There are two points worth noting regarding the broadcasting targets. First, statistically, there are fewer households with radios than there are with televisions in a number of countries. This seems counterintuitive because radios are less expensive than televisions, can operate on battery power, and appear to be ubiquitous (they are built into cars, hi-fi systems, alarm clocks, and so forth). The most plausible explanation is the questionnaire design of household surveys. They typically ask if a household has a stand-alone radio rather than asking about a radio as part of another device. Unless household surveys are modified to reflect this possibility, the world may, statistically at least, never reach the target of achieving complete radio access. A second point is that there is a very close relationship between electricity and television. In general, unless electricity is available, it is difficult to operate a television set. Therefore, achieving global access to television at a household level will depend on the availability of electricity.

Target I: Encourage the Development of Content and Put in Place Technical Conditions in Order to Facilitate the Presence and Use of all World Languages on the Internet

Basic indicator • **Share of country's Internet hosts in the world in relation to population**

Supplementary indicator • **Percentage of top 50 Web sites in a country that are local**

This target, like target G, differs from the others in that it is not about connectivity. Instead it refers to the information that ICTs deliver. That may be one reason that it was not ranked high in importance in the ITU survey. Another may be that national language content may already be available for many people: 96 percent of the world's people speak 4 percent of the languages.

The first part of the target is concerned with developing content. As content is available for most of the world's widely spoken languages, one might take this target to mean developing locally relevant content and content in languages that are poorly represented on the Internet. The European Union developed an e-Content indicator that measured the *percentage of the top 50 Web sites in a country that were local* (figure 6.8a).[25] To derive this indicator, surveys are needed about what Web sites users actually access.[26] Although these types of surveys are carried out for many developed countries, they are rare for most developing ones. One decision to be made in working with this indicator is whether to include major portals and Web services that have developed localized versions as national content.[27]

Figure 6.8 Measuring Content

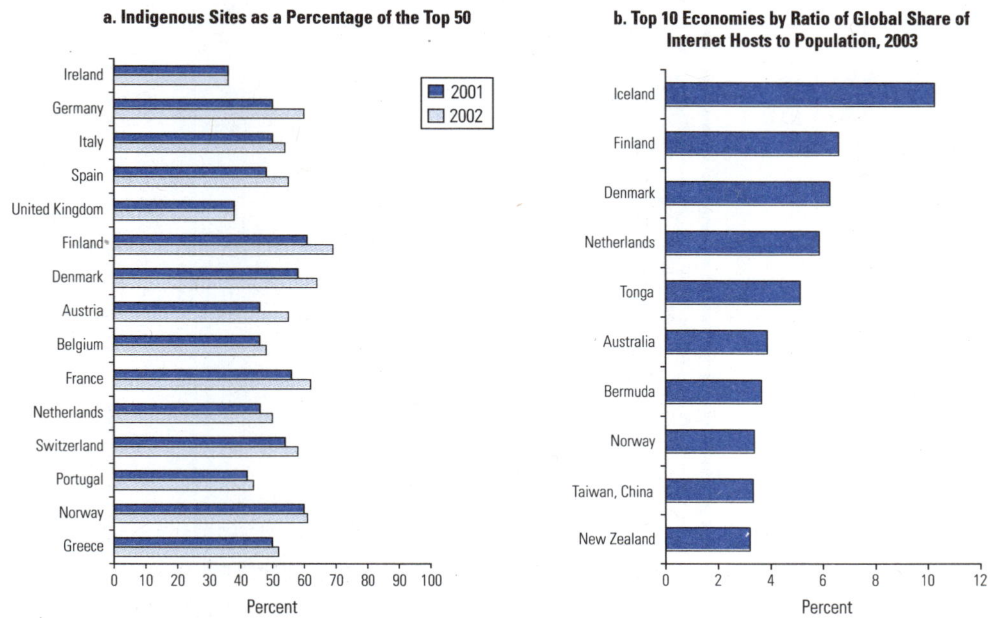

Note: Figure 6.8a shows indigenous sites as a percentage of the top 50 (including portals and Internet service providers); figure 6.8b is derived as follows: (number of Internet hosts in country / number of Internet hosts in the world) / (country's population / world population).

Source: Adapted from RIPE and ISC.

UNESCO has commissioned background reports for the second WSIS summit that will propose methods for measuring local content.[28] One indicator is the *share of a country's global Internet hosts in relation to its population.* This indicator is attractive because it is widely available. The number of Internet hosts is surveyed for all countries on a regular basis by several organizations. If the share of a country's Internet hosts (that is, hosts using that country's two-letter domain name) in the world is equal to or greater than its share of world population, then one could assume that there is a reasonable amount of content available for that country. By this measure, countries from the Nordic region lead the world in having a relatively large amount of content (figure 6.8b).

Using this alternate indicator does have some limitations. Not all content on Internet hosts using a country's domain might be local. For example, some countries such as Tonga (.TO) expressly promote their catchy domain name for worldwide use. Conversely, businesses and individuals in some countries use other domains (such as .COM) to host large volumes of content.

Achieving the second part of the target is less about "technical conditions" and more about financial support and a private-public partnership. Because audiovisual material can be developed for unwritten languages, it would even be possible to include content based on unwritten languages, although accessing that information would still require some computer literacy. Because there are over 6,000 languages in the world, many of which do not have a written alphabet and are spoken by small groups of people, efforts may need to focus on written languages.[29] What is critical is to ensure that all national and widely spoken languages have no barriers to being present in cyberspace. Here, the major barrier appears to be financial rather than technical.[30]

Target J: Ensure that More than Half the World's Inhabitants Have Access to ICTs within Their Reach

Basic indicators
- **Percentage of population covered by mobile telephony**

Advance indicators
- **Percentage of population covered by third-generation mobile telephony**
- **Percentage of households with Internet access**

This target too presents a challenge to measure. Along with target H on broadcasting, this target refers to people. Thus, some kind of penetration indicator is needed, so an appropriate numerator needs to be selected. One approach

Box 6.2 Differences between Access, Subscription, and Use

One difficulty often confronted when comparing ICT statistics is that measurement terms are not used precisely. For example, a clear distinction exists between *use*, *access*, and *ownership* or *subscription*, but the terms are sometimes used interchangeably. *Access* means that a person could utilize an ICT because it is available, but may not necessarily do so. *Use* means that a person is actually utilizing an ICT. *Ownership* or *subscription* means that the person possesses an ICT device or subscribes to an ICT service. Another point of confusion is that some surveys ask households whether they have access to an ICT service, rather than asking whether the service is available from the home. For example, a household would be counted as having Internet access even if access was not available from the home, but the head of household had access from work.

People collecting data should try to be specific about what they mean and use the most appropriate term. Ideally, statistics should be compiled on all three concepts: access, use, and ownership or subscription. Comparing access, use, and ownership helps identify barriers that may have important policy implications. For example, if the level of usage does not match the level of accessibility, this suggests that there are other barriers besides infrastructure affecting the take-up of ICTs. The level of ownership, measured through purchase or subscription to an ICT good or service, can reflect how convenient it is to use ICTs.

Source: ITU 2003.

would be to track a number of ICTs (for example, radio, television, fixed telephone, mobile telephone, computers, Internet), but some of these are covered in other targets. Selecting too many indicators would complicate tracking the target. It might also cause some countries to focus on a particular ICT to the detriment of others. Ideally, a single indicator would be best.

It is noteworthy that this target specifies *access* and not *use*. Existing indicators are based on supply or usage (either subscriptions or users) rather than access. The target also specifies that ICTs should be "within their reach." The idea seems to be that ICTs be easily available either publicly or privately, and that they be affordable and actually used. The concept of availability itself has multiple dimensions (see box 6.2). Even though infrastructure may be available, people may lack the necessary terminal equipment, may not be able to afford the needed service, or may have no desire to use it. Extending the concept of access into these other areas makes the task of measurement more complex, because additional indicators would be needed to measure access to terminal equipment, affordability, and motivation.

One indicator for measuring the target of accessibililty is the *percentage of the population covered by a mobile network*. This indicator relates directly to the unit of measurement in the target (the world's inhabitants) and is widely published by mobile operators. Today's second-generation (2G) mobile networks provide voice service; most also offer text messaging capability. They can also be used for data transmission (albeit relatively slow at 9.6 kbps). A mobile network thus offers the potential for access to several ICTs.[31] Third-generation (3G) mobile networks offer broadband speeds with the ability to deliver more complex ICT applications. This target might have two measures of access: 2G and 3G mobile coverage.

More than 80 percent of the world's population was covered by a 2G mobile network in 2002 (figure 6.9a). In 2004, there were some 86 million 3G mobile network subscribers in the 36 countries that had launched such networks, or 9 percent of all mobile subscribers (figure 6.9b).[32] The world is not far from achieving global connectivity for basic access from mobile telephone networks. A more challenging target may be needed for some countries. The European Union, for instance, uses the *percentage of households with Internet access* as its accessibility indicator.

Conclusion

The targets included in the WSIS Plan of Action are the first globally endorsed goals for ICT development. As such, they offer the international community a standard for

Figure 6.9 Mobile Population Coverage

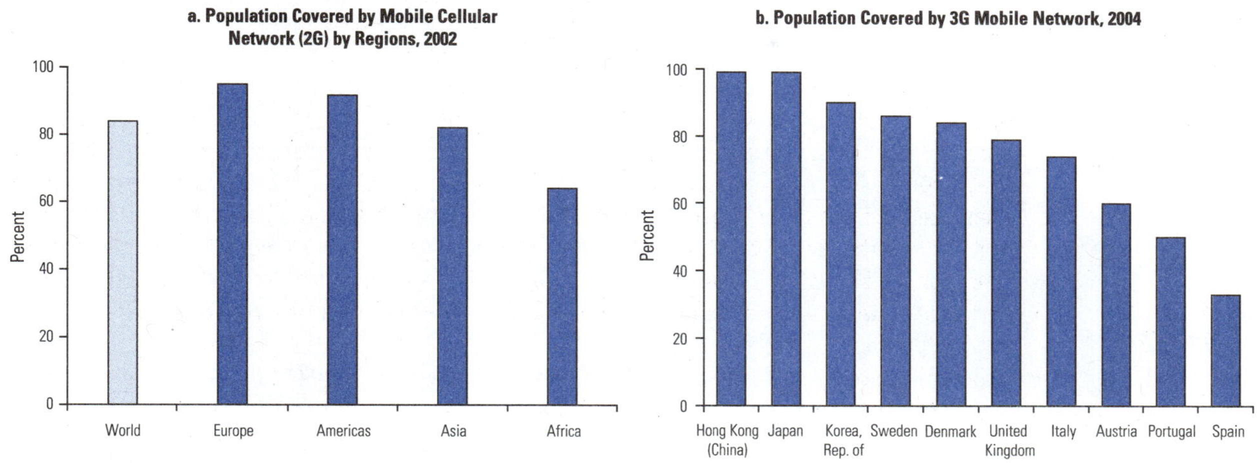

Sources: Adapted from ITU 2003; World Bank estimates.

benchmarking progress toward a global information society. However, the targets suffer from some shortcomings in their definition, scope, and relevance. A number are impractical, insignificant, or ambiguous. In some cases this is because of ambiguity in the referenced population—for example, "village" or "science and research center," which have no internationally comparable definitions—rather than the ICT component of the target. Many of the targets focus on the ends without identifying why they are important. Such identification would have helped determine relevance and priorities.

Applying the targets to existing data offers only an incomplete picture of global information society development. This could be improved by

- reaching a consensus on the exact indicators to be used to monitor the WSIS targets,
- encouraging countries not already collecting information society statistics to start doing so,
- agreeing which international organizations should compile global information,
- establishing a repository of the data to be used for benchmarking, and
- developing parallel activities such as identifying policies to accelerate achievement of the targets (for example, improving regulatory environments, identifying financing and business models, and so on).

Thirty-two indicators have been proposed to monitor the targets (table 6.5). Where possible, these indicators are based on the text of the target as well as what regional organizations and national authorities currently collect. However, not all the targets deserve equal treatment. One possibility might be to distinguish poorly defined targets (for example, bringing all languages to the Internet) from those that are clear, for which action can make a difference, and for which progress can be tracked between now and 2015.

Table 6.5 Indicators for Monitoring the WSIS Targets

Target / basic indicator	Note
A. Connect villages with ICTs and establish community access points	
• Percentage of villages with telephone service	ITU asks for this in questionnaire but does not regularly disseminate.
• Percentage of rural population covered by mobile signal	Could be derived from existing indicator on national mobile population coverage.
• Public Internet access points per 100 inhabitants	EU had this as an indicator.
• Percentage of localities with public Internet access centers	New ITU indicator.

(*Table continues on the following page.*)

Table 6.5 *continued*

Target / basic indicator	Note
B. Connect universities, colleges, secondary schools, and primary schools with ICTs	
• Percentage of schools with computers • Percentage of schools with Internet access • Students per computer	UNESCO would be the logical coordinator for these indicators. EU has *Number of pupils per computer with Internet connection (broadband/ non-broadband)* for its e-learning indicator.
• Percentage of schools with broadband access	
C. Connect scientific and research centers with ICTs	
• Availability of a national educational and research network • Bandwidth of national education and research networks	EU used bandwidth indicator as part of original set of eEurope indicators. The Trans-European Research and Education Networking Association (TERENA, http://www.terena.nl) compiles many data on this subject.
D. Connect public libraries, cultural centers, museums, post offices, and archives with ICTs	
• Percentage of public libraries providing access to the Internet • Percentage of public libraries with a Web site	UNESCO has collected data on the number of public libraries. Some countries collect data on ICTs in public libraries.
• Percentage of cultural centers with Internet access	Difficult to compile because of the lack of definition for "cultural center"
• Percentage of museums with a Web site	There are a number of portals that have links to museum Web sites. Estimated number of museums in the world is around 40,000.
• Post office is online • Percentage of post offices with Internet access	Some 660,000 post offices in the world. UPU collects several statistics on the overall availability of ICTs in a country's postal system. These need to be disaggregated at a unit level to derive the percentage with Internet access.
• National archive is online	There are portals with links to archives.
E. Connect health centers and hospitals with ICTs	
• Percentage of hospitals/health centers with Internet access	These data are not currently available from a central source.
• Percentage of hospitals/health centers with a Web site	
F. Connect all local and central government departments and establish Web sites and e-mail addresses	
• Central government Web site presence • Number of basic public services fully available online	UNPAN publishes in annual e-government report; 178 of 191 UN members have central government Web site.
G. Adapt all primary and secondary school curricula to meet the challenges of the Information Society, taking into account national circumstances	
• Country includes ICT in school curricula	Eurydice (http://www.eurydice.org/) collects this information for Europe.
H. Ensure that all of the world's population has access to television and radio services	
• Percentage of population covered by radio signal • Percentage of population covered by TV signal • Percentage of households with a radio • Percentage of households with a TV • Percentage of households with multi-channel TV	Some of these data are generally available. Radio household data tend to be collected in developing nations but very few developed ones. Note that statistically, there are sometimes more TV than radio households; this anomaly is due to the fact that surveys often ask only about the existence of a stand-alone radio set and do not include radios in alarm clocks, stereos, or cars.
I. Encourage the development of content and to put in place technical conditions in order to facilitate the presence and use of all world languages on the Internet	
• Percentage share of Internet hosts	Internet host data are available from several sources.
• Percentage of local sites in top 50 Web sites	EU had used this as an e-content indicator.
J. Ensure that more than half the world's inhabitants have access to ICTs within their reach	
• Percentage of population covered by 2G mobile network	ITU collects these data.
• Percentage of population covered by 3G mobile network	
• Percentage of households with Internet access	Asked for in some household surveys.

Note: EU = European Union; ITU = International Telecommunication Union; UNESCO = UN Educational, Social, and Cultural Organization; UNPAN = United Nations Online Network in Public Administration and Finance; UPU = Universal Postal Union; WHO = World Health Organization.

Source: Author's compilation.

Although some countries collect data on some of the indicators that have been identified, many do not. Countries should be encouraged to undertake data collection as soon as possible, and provided with support for this, because the deadline for achieving the targets—2015—is only a decade away. Most of the MDGs take 1990 as a base year from which to measure progress. For the WSIS targets, a common base year such as 2000 could be used. Many of the data should not be difficult to collect; many of the underlying data are already collected as part of normal administrative reporting. For example, most ministries of education around the world compile statistics on the number of students and schools. It should be fairly easy to extend this to include the number of schools with computers or connected to the Internet. See annex 6A for an example of data collection for the WSIS targets.

The international community can help by providing technical assistance and forging public-private partnerships. Existing information society statistics collected by national or regional authorities are often cumbersome to locate; they need to be linked to a central portal. Maintaining such a portal might be a task for the partnership of international agencies that has already been active in identifying a core list of information society indicators to take on.

A successful model showing how WSIS target tracking could move forward are the MDGs. A group of international agencies collects the data necessary for monitoring the different MDGs. This group meets regularly and reviews progress. A similar process could be envisioned for the WSIS targets, with the goal of agreeing on a comparable set of indicators by 2006 and a mid-term review around 2010 to gauge progress and possibly refine the indicators.

Annex 6A: A Benchmarking Example of the WSIS Targets—Estonia

In order to test their feasibility, the indicators proposed above are applied to Estonia (table 6A.1). Estonia was chosen for this exercise because the data are available. The Statistical Office of Estonia compiles many of the needed base statistics, and the Estonian Informatics Centre compiles numerous ICT-related indicators. It should be noted that Estonia is a good example—because so many data are available—but it is not representative of many developing countries. Estonia has also been successful at applying ICTs throughout different sectors of the economy, so it might be interesting to benchmark other countries relative to Estonia. Its recent EU membership has also meant that the country applies standard definitions for a number of the information society statistics it collects.

- Regarding village connectivity (WSIS target A), Estonia's 4,000 villages, by law, must be connected to the telephone network if the technical means for doing so exist. With mobile telephony population coverage of 99 percent, virtually all villages would have access to at least voice telephone service. For community access, Estonia had 0.76 public Internet access points (PIAPs) per 1,000 inhabitants in 2003, ranking it first among the new EU member countries. Government policy provides for free Internet access from PIAPs.
- With regard to educational connectivity (WSIS target B), Estonia's Tiger Leap program, launched in 1996, has been instrumental in providing schools with information technology facilities.[33] All schools had computers by the year 2000; at the beginning of 2003, 98 percent had Internet connections.
- The Estonian Educational and Research Network (EENet), a state agency administered by the Ministry of Education and Research, was created in 1993 to manage the science and education computer network (WSIS target C). EENet extends to most Estonian counties. At the end of 2003, over 200,000 researchers, students, and teachers used the network, and 455 institutions had a permanent connection (about 60 percent of all customers). Most of the institutions (85 percent) are educational. The category of educational institutions includes vocational schools and schools of general education, colleges, and universities, as well as public Internet access points, libraries, and archives. Research institutions make up 4 percent of the institutions having permanent connection; cultural institutions make up 7 percent; and other organizations in the field of education, research, and culture make up 4 percent.
- With regard to connecting public facilities (target D), Estonia compiles cultural statistics including data on the number of public libraries and museums. Data are available on the number of public libraries offering public Internet access facilities. There are no published data on the number of museums with a Web site, but this could be derived from the link to online museums on the

Table 6A.1 Benchmarking Estonia

Target / indicator	Value	Date	Note / source
A. Connect villages with ICTs and establish community access points			
Number of villages	4,432	2003	SOE
Percentage of villages with telephone serviceBy law, must be connected to the telephone network if the technical means for doing so exist			
Number of PIAPs	487	Jun-02	RIA
Number of PIAPs per 1,000 inhabitants	0.36	Jun-02	RIA
B. Connect universities, colleges, secondary schools, and primary schools with ICTs			
Percentage of schools with a computer	100	2004	Tiger Leap. Since 2000.
Number of computers per 100 pupils at primary level	4	2003	RIA
Number of computers per 100 pupils at secondary level	4	2003	RIA
Number of computers per 100 pupils at primary level	88	2004	Tiger Leap
C. Connect scientific and research centers with ICTs			
Has academic and research networkhttp://www.eenet.ee/englishEENet/usersurvey2003.html			
D. Connect public libraries, cultural centers, museums, post offices, and archives with ICTs			
Number of public libraries	573	2003	SOE
Number of PIAPs in libraries	288	Jun-02	RIA
Percentage of libraries providing Internet access to the public	33	2001	RIA
Number of museums	200	2003	Including rural municipality museums. SOE.
Number of museums with Web sites	30	2005	http://www.kul.ee/index.php?path=0x294x871
Number of post offices	548	2004	"The administration provides public Internet access points" UPU
Eesti Post Web site	http://www.post.ee		
National Archives Web site	http://www.ra.ee/?topic=25		
E. Connect health centers and hospitals with ICTs			
Number of hospitals	51	2002	SOE
Percentage of general practitioners with PC in the consulting room	99	Jun-03	RIA
Percentage of general practitioners with PC with Internet access	88	Jun-03	RIA
F. Connect all local and central government departments and establish Web sites and e-mail addresses			
Number of public administration agencies	64	Jul-04	RIA
Government portal	http://www.riik.ee		
Percentage of PCs connected to the Internet	95.1	2003	RIA
G. Adapt all primary and secondary school curricula to meet the challenges of the Information Society, taking into account national circumstances			
ICT included in primary and secondary curricula			Tiger Leap
H. Ensure that all of the world's population has access to television and radio services			
Population coverage of radio (percent)	100	2005	"Can be listened to all over Estonia" Eesti Radio
Population coverage of television (percent)	99.96	2005	Eesti TV
Households with a radio (percent)	87	2003	SOE
Households with a color TV (percent)	93	2003	SOE
I. Encourage the development of content and to put in place technical conditions in order to facilitate the presence and use of all world languages on the Internet			
Percentage of population Estonian mother tongue	67	2000	Official language
Number of EE hosts	237,461	Jan-05	Internet Software Consortium
As a percentage of total hosts	0.07	Jan-05	Calculated
Estonian population as a percentage of world	0.02	2004	Calculated

Table 6A.1 *continued*

Target / indicator	Value	Date	Note / source
J. Ensure that more than half the world's inhabitants have access to ICTs within their reach			
Mobile population coverage	99	2004	EMT GSM network. Source: Telia Sonera.
Households with Internet access (percent)	13.9	2002	RIA

Note: PIAPs = public Internet access points; RIA = Estonian Informatics Center; SOA = Statistical Office of Estonia; UPU = Universal Postal Union.

Source: Adapted from SOE (http://www.stat.ee) and RIA (http://www.ria.ee).

Ministry of Culture Web site. Estonia does not use the statistical concept of cultural center or make reference to it in Ministry of Culture documents. The National Archives of Estonia is online. Eesti Post has a Web site and is engaged in a number of ICT projects, including an innovative e-ticket system that allows users to purchase and store public transportation tickets using their national ID card.

- According to the Estonian Information Centre, practically all general practitioner doctors had a computer in their offices (99 percent in June 2003) and 88 percent had a connection to the Internet (target E).
- The Estonian government consists of 64 public agencies staffed by 21,400 people (in 2003). Practically all of the staff who require a personal computer have one (15,600), and 95 percent of personal computers are connected to the Internet (WSIS target F).
- ICTs have been incorporated into the Estonian educational curricula as both a subject and a tool for teaching other subjects (WSIS target G). Over 100 software packages have been created in Estonian covering language, history, culture, and nature. ICT competency exams are under development.
- Virtually the whole country (99 percent) is covered by radio and television broadcasting; 93 percent of households have a color TV and 87 percent have a radio (WSIS target H).
- According to the 2000 census, 67 percent of the population speaks Estonian as their mother tongue. (Another 30 percent speak Russian as a first language.) Most government Web sites are bi- or trilingual (Estonian and Russian and/or English). Estonia had 237,461 Internet hosts in January 2005 or 0.07 percent of the world total. This is 3.5 times greater than the share of the Estonian population in the world, suggesting that there is considerable online content available in the official language (WSIS target I).
- Regarding access to ICTs within people's reach (WSIS target J), 99 percent of the Estonian population is covered by a 2.5G mobile network. That means that virtually all of Estonia's inhabitants have theoretical access to voice service as well as data service at speeds equivalent to dial-up (26–53 kbps as measured by one of Estonia's mobile operators). EDGE mobile technology, which provides average data speeds in Estonia of 115 kbps, was launched in 2004 in the capital Tallinn with coverage to expand throughout the country. Third-generation universal mobile telecommunications system (UMTS) licenses were issued in 2003, which calls for 30 percent population coverage within seven years of launch. Data transmission speeds must be at least 144 kb/s in towns and 64 kb/s elsewhere; in practice, average throughput with such networks tends to be higher, in the range of 384 kb/s.

Endnotes

1. Although a few ICT-related indicators are included in the Millennium Development Goals (MDGs) of the United Nations, they form part of a broader development framework. The WSIS targets are the first targets entirely devoted to ICT; unlike the MDGs, they propose specific levels of achievement to be reached by 2015.
2. The MDGs set targets for reductions in poverty, improvements in health and education, and protection of the environment. See http://ddp-ext.worldbank.org/ext/MDG/home.do (accessed June 14, 2005).
3. The WSIS targets have been explored in several documents. See ITU (2003, chapter 4) and OSILAC (2005).
4. It would be analytically useful to further identify whether the service is fixed, mobile, or both.

5. The use of mobile networks to provide telephone service in villages has proven successful in some countries. The case of Grameen Village Phone in Bangladesh is widely documented. Pan-African mobile operator MTN (Mobile Telephone Networks) has launched a Village Phone project in several countries that aims to get around the lack of electricity by using solar-powered equipment.

6. The availability of a range of ICTs at a center also implies the availability of electricity, often a significant barrier to connectivity.

7. There are data difficulties with educational statistics. It is sometimes difficult to determine the precise number of schools, as teachers and pupils move between classes. It is also difficult to determine whether informal educational institutions should be included. Also, access to ICT is sometimes restricted to administrative use, an eventuality that should be taken into account.

8. Tunisia, a lower-middle-income developing country and host of the second WSIS summit, reports that all of its secondary and tertiary institutions have Internet access. All primary schools are to be connected by 2006. These data are reported on the Agence Tunisienne d'Internet Web site at http://www.ati.tn/ (accessed March 24, 2005).

9. Statistics Canada carried out a detailed survey on ICTs in primary and secondary schools that contains a variety of indicators, including the type of Internet connection (Plante and Beattie 2004).

10. See the Ministry of Education (Chile) at www.enlaces.cl.

11. See Tiger Leap Foundation. "Tiger under Magnifying Glass." http://www.tiigrihype.ee/eng/publikatsioonid/tiigerluup_eng/tiigerluup_eng.html (accessed September 18, 2005).

12. See NECTEC (National Electronics and Computer Technology Center) at http://www.school.net.th/articles/schoolnetpaper.html (accessed September 20, 2005).

13. See UNESCO "Public Libraries," http://stats.uis.unesco.org/ TableViewer/tableView.aspx?ReportId=14 (accessed March 28, 2005).

14. This information came from the Web site of the Australian Bureau of Statistics entitled "Measures of a knowledge-based economy and society, Australia." Unfortunately, the link announces that this has been discontinued: http://www.abs.gov.au/Ausstats/abs@.nsf/94713ad445ff1425ca25682000192af2/4f377c757da4394fca256d97002c1a68!OpenDocument (accessed September 18, 2005).

15. See the American Association of Museums Web site http://www.aam-us.org/aboutmuseums/abc.cfm (accessed September 18, 2005).

16. See http://www.musee-online.org. Another portal is the World Wide Web (WWW) Virtual Library museums pages (VLmp) http://icom.museum/vlmp supported by International Council of Museums (ICOM), a UNESCO-affiliated organization.

17. One of the few examples of data on ICTs in museums is Shahzad Mohammad. October 2004. *ICT Baseline Research for Museums and Archives.* http://www.ymlac.org.uk/uploads/ ICT percent20 Baseline percent20Research.pdf (accessed September 18, 2005).

18. See the International Council on Archives (ICA) online membership directory at http://www.ica.org/members.php?pcat=A&plangue=eng and the UNESCO Archives Portal at http://portal.unesco.org/ci/en/ev.php-URL_ID=5761&URL_DO=DO_TOPIC&URL_SECTION=20phabet and dictionary for all of the existing nonwritten languages at half a billion euros.

19. See the International Hospital Federation database of hospitals at http://www.hospitalmanagement.net/ihf/publication_5.html (accessed September 18, 2005).

20. Unfortunately, the data are almost a decade old. They are from an extract of a report at http://www.virtual.epm.br/material/healthcare/spanish/E0404.pdf.

21. There can be large benefits from making administrative forms available online. An e-government project in the Maldives has estimated that the there will be a 70 percent reduction in time-delaying visits to public agencies by allowing property owners the ability to register and renew vehicle and vessel registrations online. See http://www.itu.int/ITU-D/ict/cs/maldives/material/CS_MDV.pdf (accessed September 18, 2005).

22. UNESCO had collected data on radio and television sets through 1999. The ITU also collects those data as well as other broadcasting-related indicators.

23. The ability to reach this target depends more on the availability of electricity than on a lack of ICTs. There is a striking relationship between household television penetration and the percentage of households with electricity.

24. This is unfortunate since there are many examples of the development benefit of radio and television.

25. The measurement is based on the number of hits to Web sites in a month. See Technopolis Group (2003).

26. See the Nielsen//NetRatings "Usage Metrics" at http://direct.www.nielsen-netratings.com/news.jsp?section =dat_to.

27. For example, MSN (Microsoft Network) has over 40 local versions, and the popular search engine Google has interfaces available for more than a hundred languages.

28. The reports on multilingualism on the Internet have been commissioned to Indiana University and FUNREDES. The Indiana University research aims to create an index to make comparisons of linguistic diversity on the Internet. See http://www.informatics.indiana.edu/colloquia/default.asp?archive=true&id=322&st=&year=2003.

29. The figure cited is sourced to Ethnologue but is the subject of some controversy; the actual figure is estimated by others to be much lower. See http://www.ethnologue.com/ethno_docs/ distribution.asp. Only some 1,500 languages have a written alphabet. To put the figures in perspective, 96 percent of the world speaks 4 percent of the languages. One estimate puts the cost of creating an alphabet and dictionary for all of the existing nonwritten languages at half a billion euros.

30. *See Mèthodes pour informatiser des langues et des groupes de langues peu dotèes.* http://portal.unesco.org/ci/fr/ev.php-URL_ID=16735&URL_DO=DO_TOPIC&URL_SECTION=201.html (accessed September 18, 2005).

31. Mobile phones are also more attractive as terminal devices than computers for many in developing countries: "We can learn lessons from why the mobile phone has been successful in Asia ... it is relatively robust, relatively small, you don't need a desk, you don't need to be in a particular place. And you don't have to be literate to use them or speak English. These are all constraints when it comes to operating a computer" (BBC News 2003).

32. Although an operator may install a 3G network, it does not mean that all of its existing subscribers will immediately be transferred to it. Subscriber conversion often takes many years during which parallel networks are kept in operation.

33. Estonia's Tiger Leap Foundation Web site is http://www.tiigrihype.ee/eng/index.php.

References

Agence Tunisienne d'Internet. http://www.ati.tn/ (accessed March 24, 2005).

American Association of Museums. http://www.aam-us.org/aboutmuseums/abc.cfm (accessed September 18, 2005).

American Chamber of Commerce in Egypt. "Doing Business in Egypt" online edition from http://www.amcham.org.eg/dbe/ (accessed September 20, 2005).

American University of Beirut. 2003. "R&D Potentials in ICT: Case of Lebanon," by Hassan Diab. Paper presented at "The Role of Foresight in Implementing the EU Wider Neighborhood Policy, and in the Structuring of a Euro-Mediterranean Research and Innovation Area" meeting, Malta, November 13–14.

Anh, Mai Thi Phuong. 2002. "Vietnam." *Proceedings of Multi-country Study Mission on Integrated Local Community Development to the Republic of China.* Asia Productivity Organization. http://www.apo-tokyo.org/icd/papers/E-Publications/02. IntegLocCommDev/02.IntegLocCommDev.htm (accessed September 20, 2005).

Bari, Zain. 2004. "Grameen Bank's Microcredit Powers Village Phone Program." *Grameen Dialogue.* July. http://www.grameen-info.org/dialogue/commonElements/htmls/backissues.html (accessed September 20, 2005).

BBC News. 2003. "Asia Plays with Hi-Tech Visions." October 10, 2003. http://news.bbc.co.uk/1/hi/technology/3177348.stm.

Capgemini. 2004. "Online Availability of Public Services: How Is Europe Progressing? Web-Based Survey on Electronic Public Services." Report prepared for the European Commission Directorate General for Information Society and Media. October.

CDC (Centers for Disease Control and Prevention). 1991. "Dracunculiasis Eradication: Ghana and Nigeria, 1990." *International Notes Update.* April 19. http://wonder.cdc.gov/wonder/prevguid/m0001960/m0001960.asp (accessed September 20, 2005).

Commission of the European Communities. 2002. *eEurope 2005: Benchmarking Indicators.* http://europa.eu.int/idabc/servlets/Doc?id=18431 (accessed June 14, 2005).

DGT (Directorate General of Posts and Telecommunications). 2002. "The Implementation of ICT Development in Indonesia." Paper presented at the Second Meeting on Asia-Pacific Initiative for the Information Societies, Negara Brunei Darussalam, August 5–9. http://unpan1.un.org/intradoc/groups/public/documents/ APCITY/UNPAN006373.pdf (accessed September 20, 2005).

eEurope+ 2003. "Progress Report." February 2004.

Department of Education (South Africa). September 2003. *White Paper on e-Education.* Department of Education: Cape Town. http://www.info.gov.za/whitepapers/2003/e-education.pdf.

Egypt State Information Service. 1995. "Arab Republic of Egypt Yearbook." Online edition available from http://www.\sis.gov.eg/public/yearbook/YEARBOOK/CH01SC11.HTM (accessed September 20, 2005).

EU (European Union). 2001. "WWW Indicators, Mediterranean Countries, Synthesis 1999–2000." Final ESIS project report. http://www.eu-esis.org/esis2www/synthMED7.htm (accessed September 22, 2005).

Eurydice. 2004. *Key Data on Information and Communication Technology in Schools in Europe.* Eurydice: Brussels. http://www.eurydice.org/Documents/KDICT/en/FrameSet.htm (accessed September 18, 2005).

Gallup Europe. 2002. Eurobarometer, Flash 118. http://www.eosgallupeurope.com/webreports/Report20FL%20118%20Headteacher%20&%20Internet%203.pdf (accessed June 14, 2005).

ITU (International Telecommunication Union). 2003. *World Telecommunication Development Report 2003.* Geneva: ITU.

———. 2004. "'Global' Support for Information Society Targets." Press Release, May 17. http://www.itu.int/newsarchive/press_releases/2004/12.html (accessed June 14, 2005).

ITU (International Telecommunication Union) and Secretaría de Comunicaciones y Transportes (Mexico). 2004. Recommendation of the Global Indicators Workshop on Community Access to ICTs, Mexico City, November 16-19. http://www.itu.int/ITU-D/ict/mexico04/recom/Recomendations_04_e.pdf (accessed June 14, 2005).

Library of Congress. 1991. *Ethiopia: A Country Study.* Online edition available from http://countrystudies.us/ethiopia/ 122.htm (accessed September 20, 2005).

Minges, Michael, and Pratikshya Simkhada. 2002. "A Closer Look at South Asia." *ITU News.* December. http://www.itu.int/itunews/issue/2002/10/southasia.html (accessed September 20, 2005).

Ministry of Education (Chile). "Resultados de una dÉcada." http://www.enlaces.cl/index.php (accessed September 11, 2005).

Ministry of Education (Singapore). 1997. "Master Plan for IT in Education." http://www.moe.gov.sg/edumall/mpite/index.html (accessed September 15, 2005).

National Center for Education Statistics (U.S.) 2005. *Public Libraries in the United States: Fiscal Year 2002.* http://nces.ed.gov/pubsearch/pubsinfo.asp?pubid=2005356. (accessed September 15, 2005).

NSO (National Statistical Office), Thailand. 1999. "Basic Statistics of Village by Region." http://www.nso.go.th/portal/page?_pageid=133, 1&_dad=portal&_schema=PORTAL (accessed September 20, 2005).

NBS (National Bureau of Statistics) of China. 2003. *China Statistical Yearbook.* NBS: Beijing. http://www.stats.gov.cn/english/index.htm (accessed September 20, 2005).

NECTEC (National Electronics and Computer Technology Center). "SchoolNet Thailand: An Information Infrastructure for the Future of Thailand." http://www.nectec.or.th/english/ (accessed September 20, 2005)

OECD (Organisation for Economic Co-operation and Development). 2003. Program for International Student Assessment (PISA) 2003 database. http://pisaweb.acer.edu.au/oecd_2003/oecd_pisa_data.html (accessed September 18, 2005).

OSILAC (Observatory for the Information Society in Latin America and the Caribbean). 2005. *Benchmarking the Plan of Action of the World Summit on the Information Society (WSIS) in Latin America and the Caribbean.* Santiago: UN. http://unpan1.un.org/intradoc/groups/public/documents/UN/UNPAN020039.pdf (accessed September 18, 2005).

Parsad, Basmat, and Jennifer Jones. 2005. *Internet Access in U.S. Public Schools and Classrooms: 1994-2003.* U.S. Department of Education (NCES 2005-015). Washington, DC: National Center for Education Statistics. http://nces.ed.gov/pubsearch/ pubsinfo.asp?pubid=2005015.

Plante, Johanne, and David Beattie. 2004. *Connectivity and ICT Integration in Canadian Elementary and Secondary Schools: First Results from the Information and Communications Technologies in Schools Survey, 2003–2004.* Ottawa: Statistics Canada. http://www.statcan.ca/english/research/81-595-MIE/81-595-MIE2004017.pdf.

Technopolis Group. 2003. "Indicators for European Digital Content for the Global Networks: Final Report for the Second Measurement for DG Information Society." Technopolis Group: Vienna. ftp://ftp.cordis.lu/pub/econtent/docs/final_report_content_indicators2.pdf (accessed September 18, 2005).

TERENA (Trans-European Research and Education Networking Association). "Compendium Information 2005." http://www.terena.nl/compendium/ (accessed September 18, 2005).

Uganda Communications Commission. 2005. "Uganda's Experience on Building ICT Indicators," by Ann Rita Ssemboga. Paper presented at the World Telecommunications/ICT indicators (WTI) Meeting, Geneva, February 11–12.

UN (United Nations). 2004. Division for Public Administration and Development Management. *Global E-Government Readiness Report 2004.* http://www.unpan.org/egovernment4.asp (accessed March 15, 2005).

UNCTAD (United Nations Conference for Trade and Development). "Partnership Project Document." http://measuring-ict.unctad.org/QuickPlace/measuring-ict/Main.nsf/h_Index/FBFD3BDF8A2AC11EC1256EF5005C2CC5/?OpenDocument (accessed June 14, 2005).

UNESCO (United Nations Educational, Scientific and Cultural Organisation) Institute of Statistics. Database extract "Public Libraries." http://stats.uis.unesco.org/TableViewer/tableView.aspx?ReportId=14 (accessed March 28, 2005).

UPU (Universal Postal Union). 2004. *Postal Market 2004.* UPU: Berne. http://www.upu.int/statistics/en/postal_market_2004_review_and_outlook_en.pdf.

VNPT (Vietnam Posts and Telecommunications Corporation). "Plan of VNPT in 2004." http://www.vnpt.com.vn/home 181203.asp?ID=47&dataID=4645 (accessed September 20, 2005).

WSIS (World Summit on the Information Society). 2003. *Plan of Action.* http://www.itu.int/wsis/docs/geneva/official/poa.html (accessed June 14, 2005).

Part II

User's Guide to ICT At-a-Glance Country Tables

The World Bank ICT At-a-Glance country tables present the most recent country-specific ICT data from many sources available in one place. They offer a snapshot of the economic and social context and the structure and performance of the ICT sector in each of the 144 countries covered in the report.

Tables

Economies are presented alphabetically. Data are shown for 144 economies with populations of more than 1 million for which timely and reliable information exists. The table of Key ICT Indicators for Other Economies presents 64 additional economies—those with sparse data, smaller economies with populations of between 30,000 and 1 million, and others that are members of the International Bank for Reconstruction and Development, or as it is commonly known, the World Bank.

The data in the tables are categorized into three sections:

- *Economic and social context* provides a snapshot of the country's macroeconomic and social environment.
- *ICT sector structure* provides an overview of regulatory and policy status in the telecommunications sector.
- *ICT sector performance* provides statistical data on the ICT sector with indicators for access, quality, affordability, institutional efficiency and sustainability, and ICT applications.

Aggregate Measures for Income Groups and Regions

The aggregate measures for income groups include 208 economies (those economies listed in the At-a-Glance country tables plus those in the Other Economies table) wherever data are available.

The aggregate measures for regions include only low- and middle-income economies (note that these measures include developing economies with populations of less than 1 million, including those listed in the Other Economies table). The country composition of regions is based on the World Bank's analytical regions and may differ from common geographic usage (see the section on the classification of economies at the end of the user's guide).

Values for the indicators under *ICT sector structure* that are nonnumerical cannot be aggregated into income and regional groups.

Charts

The GNI per Capita chart shows that indicator from 2000 to 2004 for the country and the region average.

The ICT MDG Indicators chart is based on the three ICT indicators selected for measuring the Millennium

Development Goals (MDGs). The three ICT indicators are *fixed and mobile subscribers (per 1,000 people),* based on MDG indicator 47; *personal computers (per 1,000 people),* based on MDG indicator 48a; and *Internet users (per 1,000 people),* based on MDG indicator 48b. For more information, visit the World Bank's MDG Web site: http://ddp-ext.worldbank.org/ext/ MDG/home.do.

The Price of Call to the United States chart shows the trend of the price of a three-minute, peak-rate telephone call (in U.S. dollars) to the United States between 2000 and 2004 for the country and the region average.

The Total Telecommunications Revenue chart shows the trend of telecommunications revenue as a percentage of GDP from 2000 to 2004 for the country and the region average.

Statistics

Statistics are the backbone of any analysis. Especially in the ICT sector, because the technologies are changing so quickly, acquiring reliable statistics and ensuring their relevance requires continual review of current trends and data.

Data Consistency and Reliability

Considerable effort has been made to standardize the data collected. However, full comparability of data among countries cannot be ensured and care must be taken in interpreting the indicators.

Many factors affect data availability, comparability, and reliability. Among these are the fact that statistical systems in developing countries are still weak; statistical methods, coverage, practices, and definitions differ widely among countries; and cross-country and intertemporal comparisons involve complex technical and conceptual problems that cannot be unequivocally resolved. Data coverage may not be complete because of special circumstances or because economies are experiencing problems (such as those stemming from conflicts) that affect the collection and reporting of data. For these reasons, although data are drawn from the sources thought to be most authoritative, they should be construed only as indicating trends and characterizing major differences among economies rather than offering precise quantitative measures of those differences.

Data Sources

Data are drawn from Global Insight, the International Telecommunication Union (ITU), Netcraft, the United Nations Department of Economic and Social Affairs, the United Nations Development Programme (UNDP), the United Nations Online Network in Public Administration and Finance (UNPAN), the World Economic Forum (WEF), the World Information Technology and Services Alliance (WITSA), and the World Bank.

Classification of Economies

For operational and analytical purposes, the World Bank's main criterion for classifying economies is GNI per capita. Every economy is classified as low income, middle income (these are subdivided into lower middle and upper middle), or high income. Note that classification by income does not necessarily reflect development status. Because GNI per capita changes over time, the country composition of income groups may change each year.

Low-income economies are those with a GNI per capita of $825 or less in 2004. Middle-income economies are those with a GNI per capita of more than $826 but less than $10,066. Lower-middle-income and upper-middle-income economies are separated at a GNI per capita of $3,255. High-income economies are those with a GNI per capita of $10,066 or more.

For more information on these classifications, see the World Bank's country classification Web site: http://www.worldbank.org/data/countryclass/countryclass. html.

Symbols

The following symbols are used throughout the At-a-Glance tables:

••

The symbol •• means that data are not available or that aggregates cannot be calculated because of missing data in the year shown.

0 or 0.0s

0 or 0.0 means zero or less than half the unit shown.

Albania

	Albania 2000	Albania 2004	Lower-middle-income group 2004	Europe & Central Asia Region 2004
Economic and social context				
Population, total (millions)	3	3	2,430	472
Urban population (% total population)	42	44	49	64
Poverty (% population below US$1 per day)	*<2*	*<2*	..	*3.6*
GNI per capita, Atlas method (current US$)	1,160	2,080	1,580	3,290
GDP growth, 1995–2000 and 2000–4 (%)	5.2	6.0	5.7	5.0
Adult literacy rate (% ages 15 and over)	*99*	..	90	97
Primary, secondary, tertiary school enrollment (% gross)	69	*69*	*72*	*81*
ICT sector structure				
Separate telecommunications regulator	Yes	Yes		
Status of main fixed-line operator	*Public*	Public		
Level of competition: international long distance	*M*	M		
Level of competition: mobile	*C*	*C*		
Level of competition: Internet service provider	*C*	C		
Government prioritization of ICT (scale 1–7)	..	..	3.8	3.8
ICT sector performance				
Access				
Telephone main lines (per 1,000 people)	49	88	192	*243*
International voice traffic (minutes per person)[a]	72	*155*	..	*28*
Mobile subscribers (per 1,000 people)	10	395	255	487
Population covered by mobile telephony (%)	*84*	*90*	*76*	*82*
Internet users (per 1,000 people)	1	*9*	70	*115*
Personal computers (per 1,000 people)	8	*11*	38	*73*
Households with television (%)	*90*	*90*	*89*	*92*
Quality				
Telephone faults (per 100 main lines per year)	70.2	*57.2*	..	*30.4*
Broadband subscribers (per 1,000 people)	0.0	*0.0*	12.6	*2.4*
International Internet bandwidth (bits per person)	1	4	58	148
Affordability				
Price basket for fixed line (US$ per month, residential)	3.0	5.6	*5.5*	*3.5*
Price basket for mobile (US$ per month)	..	24.3	8.9	10.3
Price basket for Internet (US$ per month)	..	*28.6*	*25.3*	*19.8*
Price of call to United States (US$ per 3 minutes)	4.59	1.34	1.45	1.06
Institutional efficiency and sustainability				
Total telecommunications revenue (% GDP)	1.8	*5.6*	3.6	3.5
Total telephone subscribers per employee	40	*319*	*195*	*150*
Total telecommunications investment (% revenue)	28.4	*12.8*	*25.5*	*19.0*
ICT applications				
ICT expenditure (% GDP)	..	..	*5.1*	*5.1*
E-government readiness index (scale 0–1)	..	0.16	0.26	0.39
Secure Internet servers (per 1 million people)	*0.3*	0.6	1.6	6.4
Schools connected to the Internet (%)	..	..	..	*65*

GNI per Capita, Atlas Method, 2000–4

Current US$

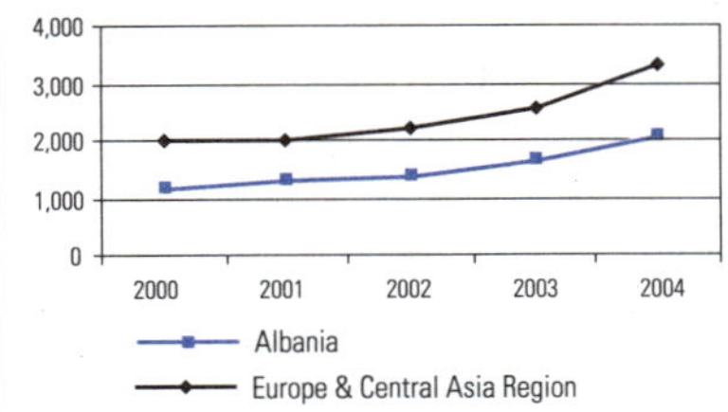

ICT MDG[b] Indicators, 2000–4

Number per 1,000 people

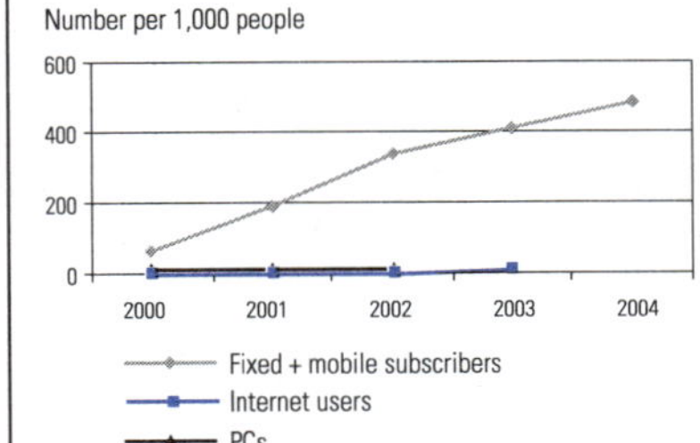

Price of Call to the United States, 2000–4

US$ per 3 minutes

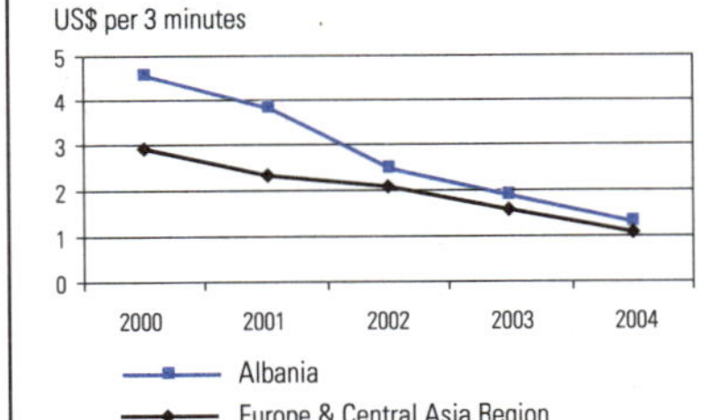

Total Telecommunications Revenue, 2000–4

Percentage of GDP

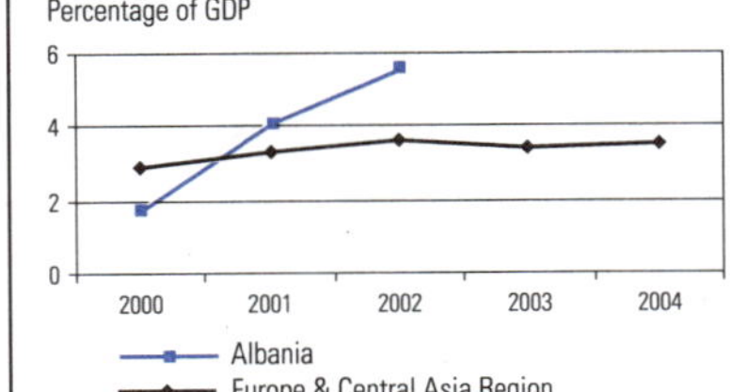

Notes: Figures in italics are for years other than those specified. .. indicates data are not available. C = competition; GDP = gross domestic product; GNI = gross national income; ICT = information and communication technology; M = monopoly; MDG = Millennium Development Goal; P = partial competition; and PCs = personal computers.

a. Outgoing and incoming.

b. Millennium Development Goal indicators 47, 48a, and 48b.

Sources: Economic and social context: UNDP and World Bank; ICT sector structure: ITU, WEF; ICT sector performance: Global Insight/WITSA, ITU, Netcraft, UNDESA, UNPAN, and World Bank. Produced by the Global Information and Communication Technologies Department and the Development Economics Data Group. For complete information, see Definitions and Data Sources.

Algeria

	Algeria 2000	Algeria 2004	Lower-middle-income group 2004	Middle East & North Africa Region 2004
Economic and social context				
Population, total (millions)	30	32	2,430	294
Urban population (% total population)	57	59	49	56
Poverty (% population below US$1 per day)	*<2*	..	..	*2.4*
GNI per capita, Atlas method (current US$)	1,580	2,280	1,580	2,000
GDP growth, 1995–2000 and 2000–4 (%)	3.2	4.8	5.7	4.5
Adult literacy rate (% ages 15 and over)	..	*70*	90	..
Primary, secondary, tertiary school enrollment (% gross)	71	*74*	*72*	*68*
ICT sector structure				
Separate telecommunications regulator	No	Yes		
Status of main fixed-line operator	*Public*	Public		
Level of competition: international long distance	*M*	M		
Level of competition: mobile	*M*	C		
Level of competition: Internet service provider	..	C		
Government prioritization of ICT (scale 1–7)	..	3.8	3.8	..
ICT sector performance				
Access				
Telephone main lines (per 1,000 people)	58	99	192	*118*
International voice traffic (minutes per person)[a]	16	..	..	*20*
Mobile subscribers (per 1,000 people)	3	151	255	*88*
Population covered by mobile telephony (%)	60	84	*76*	..
Internet users (per 1,000 people)	5	46	70	*47*
Personal computers (per 1,000 people)	7	*9*	38	*30*
Households with television (%)	79	98	*89*	*88*
Quality				
Telephone faults (per 100 main lines per year)	12.0	..	..	..
Broadband subscribers (per 1,000 people)	0.0	1.1	12.6	*0.2*
International Internet bandwidth (bits per person)	0	*4*	58	*15*
Affordability				
Price basket for fixed line (US$ per month, residential)	4.5	*5.1*	*5.5*	*4.9*
Price basket for mobile (US$ per month)	..	10.2	8.9	8.1
Price basket for Internet (US$ per month)	..	*17.8*	*25.3*	*24.5*
Price of call to United States (US$ per 3 minutes)	3.67	2.08	1.45	1.64
Institutional efficiency and sustainability				
Total telecommunications revenue (% GDP)	0.6	2.5	3.6	2.8
Total telephone subscribers per employee	103	..	*195*	..
Total telecommunications investment (% revenue)	*26.7*	..	*25.5*	*27.8*
ICT applications				
ICT expenditure (% GDP)	..	..	*5.1*	..
E-government readiness index (scale 0–1)	..	0.25	0.26	0.16
Secure Internet servers (per 1 million people)	..	0.1	1.6	0.6
Schools connected to the Internet (%)	..	53	..	..

GNI per Capita, Atlas Method, 2000–4
Current US$

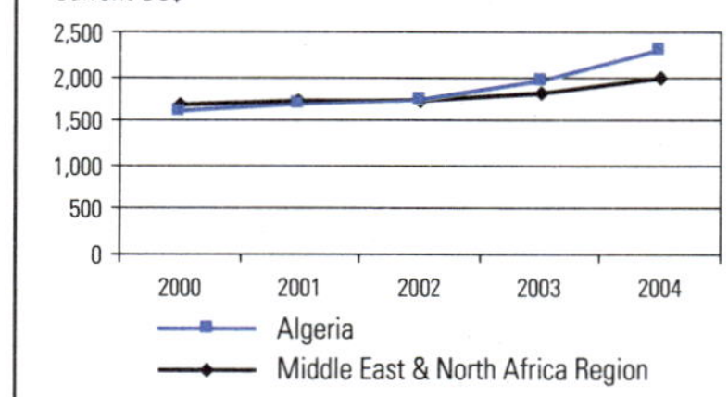

ICT MDG[b] Indicators, 2000–4
Number per 1,000 people

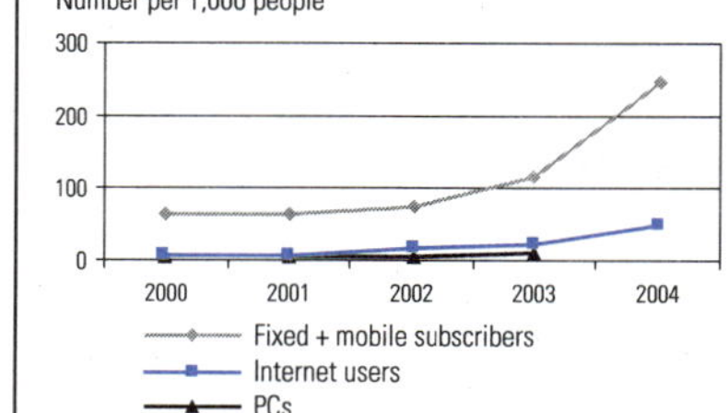

Price of Call to the United States, 2000–4
US$ per 3 minutes

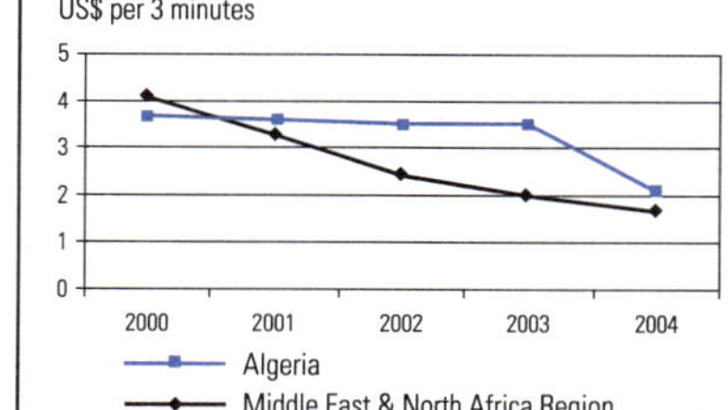

Total Telecommunications Revenue, 2000–4
Percentage of GDP

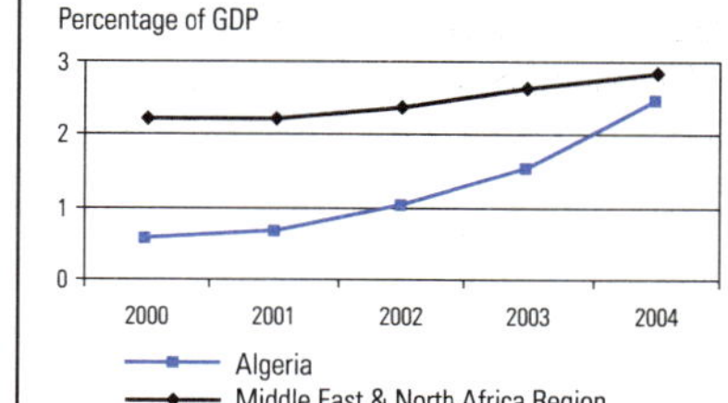

Notes: Figures in italics are for years other than those specified. .. indicates data are not available. C = competition; GDP = gross domestic product; GNI = gross national income; ICT = information and communication technology; M = monopoly; MDG = Millennium Development Goal; P = partial competition; and PCs = personal computers.

a. Outgoing and incoming.

b. Millennium Development Goal indicators 47, 48a, and 48b.

Sources: Economic and social context: UNDP and World Bank; ICT sector structure: ITU, WEF; ICT sector performance: Global Insight/WITSA, ITU, Netcraft, UNDESA, UNPAN, and World Bank. Produced by the Global Information and Communication Technologies Department and the Development Economics Data Group. For complete information, see Definitions and Data Sources.

Angola

	Angola 2000	Angola 2004	Lower-middle-income group 2004	Sub-Saharan Africa Region 2004
Economic and social context				
Population, total (millions)	12	14	2,430	719
Urban population (% total population)	33	36	49	37
Poverty (% population below US$1 per day)	..	..	..	*46.4*
GNI per capita, Atlas method (current US$)	480	1,030	1,580	600
GDP growth, 1995–2000 and 2000–4 (%)	6.3	8.1	5.7	3.9
Adult literacy rate (% ages 15 and over)	*67*	..	90	59
Primary, secondary, tertiary school enrollment (% gross)	*30*	..	*72*	*52*
ICT sector structure				
Separate telecommunications regulator	Yes	Yes		
Status of main fixed-line operator	*Public*	*Public*		
Level of competition: international long distance	*M*	C		
Level of competition: mobile	*P*	P		
Level of competition: Internet service provider	*P*	C		
Government prioritization of ICT (scale 1–7)	..	3.5	3.8	*4.4*
ICT sector performance				
Access				
Telephone main lines (per 1,000 people)	6	7	192	17
International voice traffic (minutes per person)[a]	6	*7*	..	..
Mobile subscribers (per 1,000 people)	2	53	255	86
Population covered by mobile telephony (%)	..	..	*76*	..
Internet users (per 1,000 people)	1	14	70	15
Personal computers (per 1,000 people)	1	4	38	*12*
Households with television (%)	9	*9*	*89*	*15*
Quality				
Telephone faults (per 100 main lines per year)	53.0	..	..	..
Broadband subscribers (per 1,000 people)	0.0	*0.0*	12.6	*0.1*
International Internet bandwidth (bits per person)	0	*1*	58	*4*
Affordability				
Price basket for fixed line (US$ per month, residential)	..	*11.9*	*5.5*	*8.5*
Price basket for mobile (US$ per month)	..	11.2	8.9	13.5
Price basket for Internet (US$ per month)	..	*78.8*	*25.3*	*54.8*
Price of call to United States (US$ per 3 minutes)	9.32	3.23	1.45	2.43
Institutional efficiency and sustainability				
Total telecommunications revenue (% GDP)	1.2	1.9	3.6	5.0
Total telephone subscribers per employee	45	*96*	*195*	*144*
Total telecommunications investment (% revenue)	..	*10.7*	*25.5*	*27.8*
ICT applications				
ICT expenditure (% GDP)	..	..	*5.1*	..
E-government readiness index (scale 0–1)	..	0.21	0.26	0.11
Secure Internet servers (per 1 million people)	..	0.2	1.6	1.9
Schools connected to the Internet (%)	..	..	..	..

GNI per Capita, Atlas Method, 2000–4

Current US$

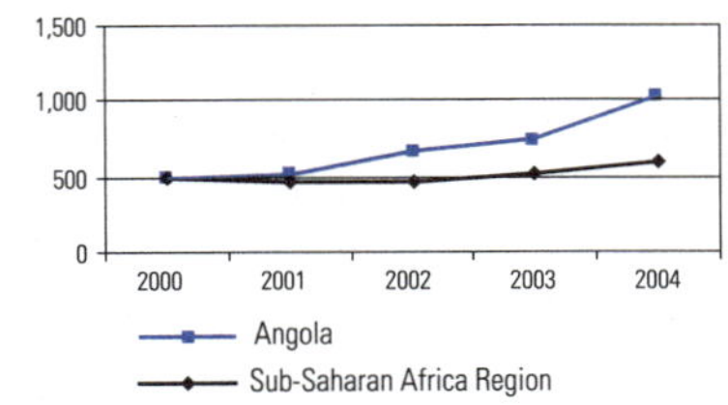

ICT MDG[b] Indicators, 2000–4

Number per 1,000 people

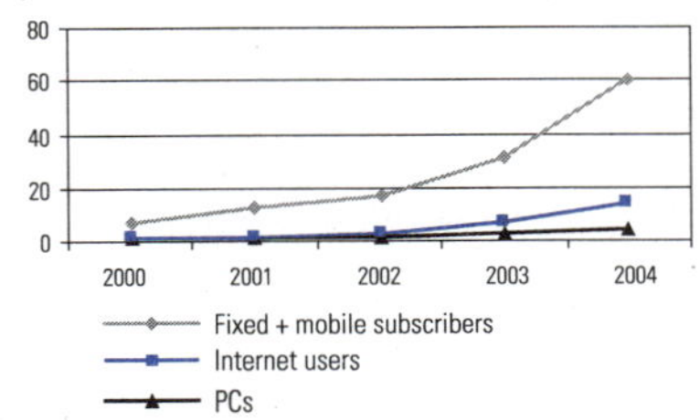

Price of Call to the United States, 2000–4

US$ per 3 minutes

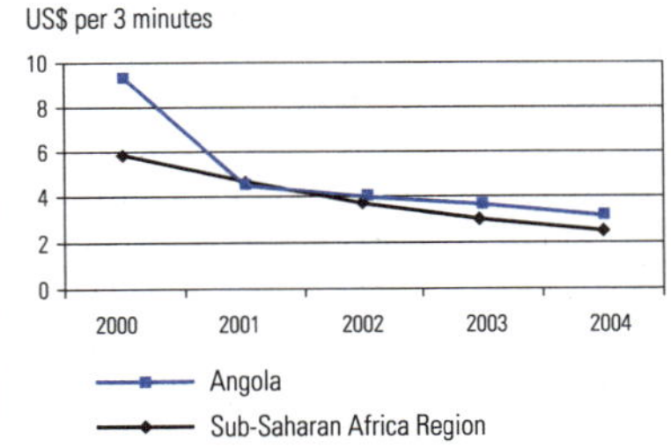

Total Telecommunications Revenue, 2000–4

Percentage of GDP

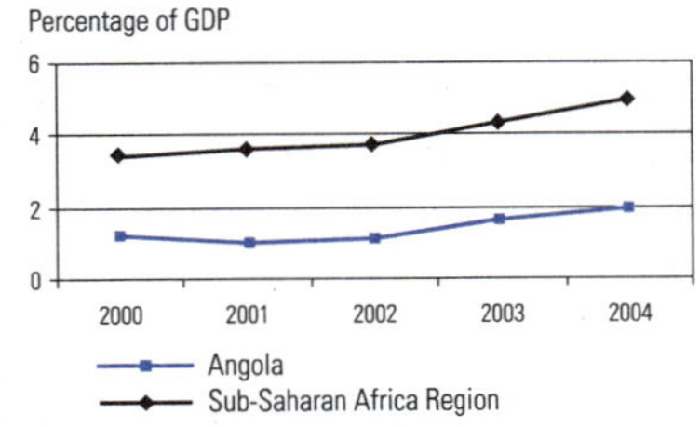

Notes: Figures in italics are for years other than those specified. .. indicates data are not available. C = competition; GDP = gross domestic product; GNI = gross national income; ICT = information and communication technology; M = monopoly; MDG = Millennium Development Goal; P = partial competition; and PCs = personal computers.

a. Outgoing and incoming.

b. Millennium Development Goal indicators 47, 48a, and 48b.

Sources: Economic and social context: UNDP and World Bank; ICT sector structure: ITU, WEF; ICT sector performance: Global Insight/WITSA, ITU, Netcraft, UNDESA, UNPAN, and World Bank. Produced by the Global Information and Communication Technologies Department and the Development Economics Data Group. For complete information, see Definitions and Data Sources.

Argentina

	Argentina 2000	Argentina 2004	Upper-middle-income group 2004	Latin America & the Caribbean Region 2004
Economic and social context				
Population, total (millions)	37	38	576	541
Urban population (% total population)	90	90	72	77
Poverty (% population below US$1 per day)	*7.7*	*3.3*	..	*9.5*
GNI per capita, Atlas method (current US$)	7,490	3,720	4,770	3,600
GDP growth, 1995–2000 and 2000–4 (%)	2.7	–0.1	2.7	1.5
Adult literacy rate (% ages 15 and over)	*97*	..	94	89
Primary, secondary, tertiary school enrollment (% gross)	92	*95*	*80*	*82*
ICT sector structure				
Separate telecommunications regulator	Yes	Yes		
Status of main fixed-line operator	*Private*	Private		
Level of competition: international long distance	*P*	C		
Level of competition: mobile	*P*	C		
Level of competition: Internet service provider	*C*	C		
Government prioritization of ICT (scale 1–7)	..	2.7	4.1	3.5
ICT sector performance				
Access				
Telephone main lines (per 1,000 people)	215	210	*220*	181
International voice traffic (minutes per person)[a]	25	40	*39*	..
Mobile subscribers (per 1,000 people)	164	353	490	324
Population covered by mobile telephony (%)	..	95	*84*	*76*
Internet users (per 1,000 people)	71	196	*133*	*104*
Personal computers (per 1,000 people)	70	97	*99*	*75*
Households with television (%)	95	97	*92*	*88*
Quality				
Telephone faults (per 100 main lines per year)	..	..	*20.3*	..
Broadband subscribers (per 1,000 people)	1.0	13.5	*3.7*	*5.2*
International Internet bandwidth (bits per person)	12	320	176	165
Affordability				
Price basket for fixed line (US$ per month, residential)	20.0	6.6	*13.9*	*9.0*
Price basket for mobile (US$ per month)	..	8.3	11.1	9.1
Price basket for Internet (US$ per month)	..	*13.3*	*20.8*	*31.5*
Price of call to United States (US$ per 3 minutes)	*2.77*	..	1.03	0.90
Institutional efficiency and sustainability				
Total telecommunications revenue (% GDP)	2.8	2.4	3.4	3.3
Total telephone subscribers per employee	448	*593*	*402*	..
Total telecommunications investment (% revenue)	23.9	11.9	*18.6*	..
ICT applications				
ICT expenditure (% GDP)	4.3	*5.7*	*5.0*	*5.3*
E-government readiness index (scale 0–1)	..	0.64	0.49	0.39
Secure Internet servers (per 1 million people)	*6.4*	10.1	10.7	8.6
Schools connected to the Internet (%)	..	..	*60*	..

GNI per Capita, Atlas Method, 2000–4

Current US$

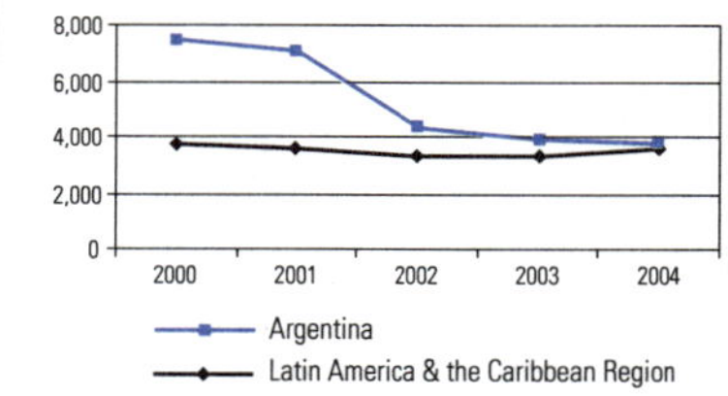

ICT MDG[b] Indicators, 2000–4

Number per 1,000 people

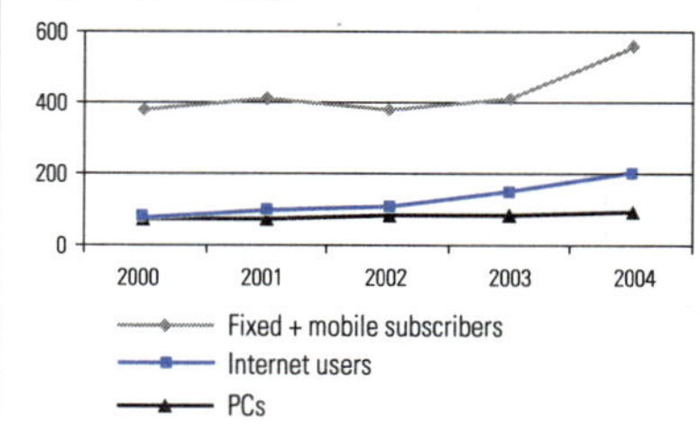

Price of Call to the United States, 2000–4

US$ per 3 minutes

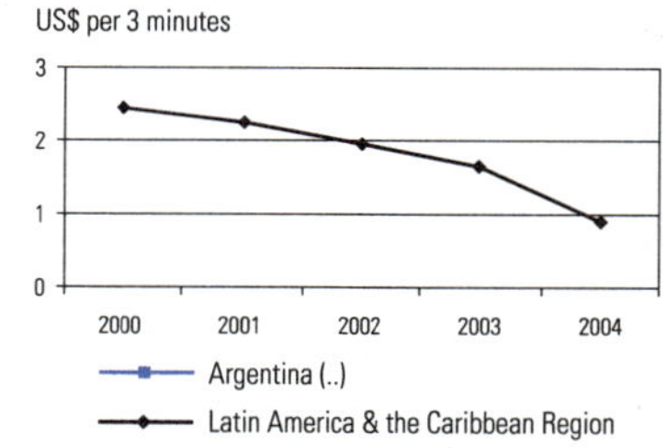

Total Telecommunications Revenue, 2000–4

Percentage of GDP

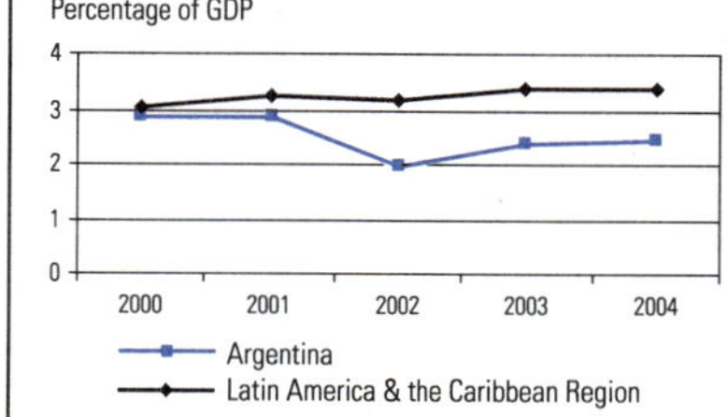

Notes: Figures in italics are for years other than those specified. .. indicates data are not available. C = competition; GDP = gross domestic product; GNI = gross national income; ICT = information and communication technology; M = monopoly; MDG = Millennium Development Goal; P = partial competition; and PCs = personal computers.

a. Outgoing and incoming.

b. Millennium Development Goal indicators 47, 48a, and 48b.

Sources: Economic and social context: UNDP and World Bank; ICT sector structure: ITU, WEF; ICT sector performance: Global Insight/WITSA, ITU, Netcraft, UNDESA, UNPAN, and World Bank. Produced by the Global Information and Communication Technologies Department and the Development Economics Data Group. For complete information, see Definitions and Data Sources.

Armenia

	Armenia 2000	Armenia 2004	Lower-middle-income group 2004	Europe & Central Asia Region 2004
Economic and social context				
Population, total (millions)	3	3	2,430	472
Urban population (% total population)	65	64	49	64
Poverty (% population below US$1 per day)	*12.8*	..	..	*3.6*
GNI per capita, Atlas method (current US$)	650	1,120	1,580	3,290
GDP growth, 1995–2000 and 2000–4 (%)	5.1	12.0	5.7	5.0
Adult literacy rate (% ages 15 and over)	*99*	..	90	97
Primary, secondary, tertiary school enrollment (% gross)	72	*72*	*72*	*81*
ICT sector structure				
Separate telecommunications regulator	..	..		
Status of main fixed-line operator	*Mixed*	Mixed		
Level of competition: international long distance	*M*	M		
Level of competition: mobile	*M*	M		
Level of competition: Internet service provider	*C*	C		
Government prioritization of ICT (scale 1–7)	..	..	3.8	3.8
ICT sector performance				
Access				
Telephone main lines (per 1,000 people)	171	190	192	*243*
International voice traffic (minutes per person)[a]	40	67	..	*28*
Mobile subscribers (per 1,000 people)	6	67	255	487
Population covered by mobile telephony (%)	*38*	81	*76*	*82*
Internet users (per 1,000 people)	13	*65*	70	*115*
Personal computers (per 1,000 people)	8	*20*	38	*73*
Households with television (%)	89	*91*	*89*	*92*
Quality				
Telephone faults (per 100 main lines per year)	55.5	*52.9*	..	*30.4*
Broadband subscribers (per 1,000 people)	0.0	*0.0*	12.6	*2.4*
International Internet bandwidth (bits per person)	..	*3*	58	148
Affordability				
Price basket for fixed line (US$ per month, residential)	8.1	*3.1*	*5.5*	*3.5*
Price basket for mobile (US$ per month)	..	12.2	8.9	10.3
Price basket for Internet (US$ per month)	..	*44.8*	*25.3*	*19.8*
Price of call to United States (US$ per 3 minutes)	*4.86*	2.42	1.45	1.06
Institutional efficiency and sustainability				
Total telecommunications revenue (% GDP)	3.0	3.0	3.6	3.5
Total telephone subscribers per employee	71	145	*195*	*150*
Total telecommunications investment (% revenue)	46.3	34.8	*25.5*	*19.0*
ICT applications				
ICT expenditure (% GDP)	..	..	*5.1*	*5.1*
E-government readiness index (scale 0–1)	..	0.25	0.26	0.39
Secure Internet servers (per 1 million people)	*0.3*	1.3	1.6	6.4
Schools connected to the Internet (%)	..	..	..	*65*

GNI per Capita, Atlas Method, 2000–4

Current US$

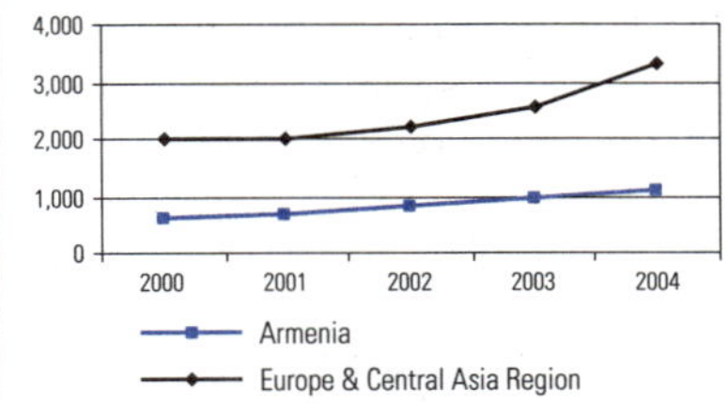

ICT MDG[b] Indicators, 2000–4

Number per 1,000 people

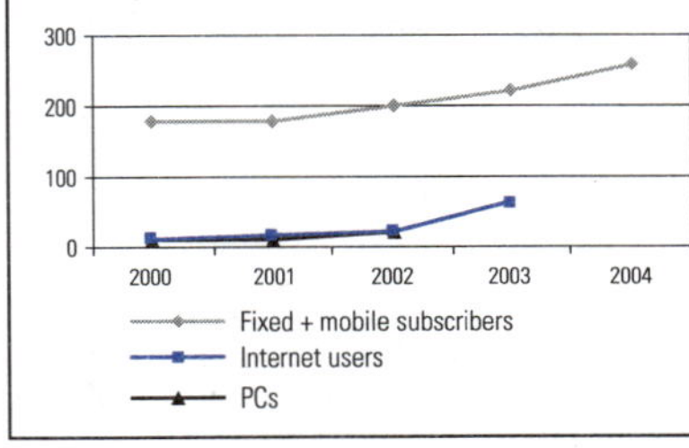

Price of Call to the United States, 2000–4

US$ per 3 minutes

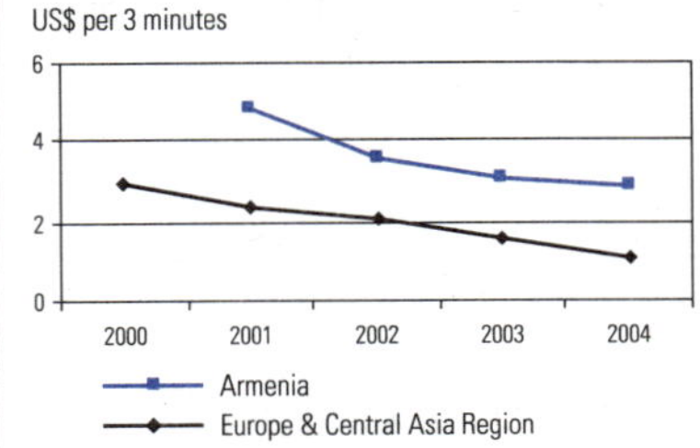

Total Telecommunications Revenue, 2000–4

Percentage of GDP

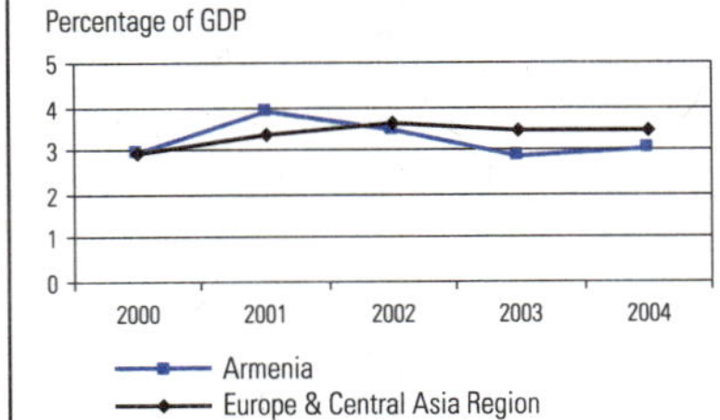

Notes: Figures in italics are for years other than those specified. .. indicates data are not available. C = competition; GDP = gross domestic product; GNI = gross national income; ICT = information and communication technology; M = monopoly; MDG = Millennium Development Goal; P = partial competition; and PCs = personal computers.

a. Outgoing and incoming.

b. Millennium Development Goal indicators 47, 48a, and 48b.

Sources: Economic and social context: UNDP and World Bank; ICT sector structure: ITU, WEF; ICT sector performance: Global Insight/WITSA, ITU, Netcraft, UNDESA, UNPAN, and World Bank. Produced by the Global Information and Communication Technologies Department and the Development Economics Data Group. For complete information, see Definitions and Data Sources.

Australia

	Australia 2000	Australia 2004	High-income group 2004
Economic and social context			
Population, total (millions)	19	20	1,001
Urban population (% total population)	91	92	77
Poverty (% population below US$1 per day)	..	..	..
GNI per capita, Atlas method (current US$)	20,090	26,900	32,040
GDP growth, 1995–2000 and 2000–4 (%)	4.1	3.3	2.0
Adult literacy rate (% ages 15 and over)	..	..	..
Primary, secondary, tertiary school enrollment (% gross)	114	*116*	*93*
ICT sector structure			
Separate telecommunications regulator	Yes	Yes	
Status of main fixed-line operator	*Mixed*	Mixed	
Level of competition: international long distance	*C*	C	
Level of competition: mobile	*C*	C	
Level of competition: Internet service provider	*C*	C	
Government prioritization of ICT (scale 1–7)	..	4.9	4.9
ICT sector performance			
Access			
Telephone main lines (per 1,000 people)	540	542	558
International voice traffic (minutes per person)[a]	224	..	..
Mobile subscribers (per 1,000 people)	446	887	767
Population covered by mobile telephony (%)	96	97	98
Internet users (per 1,000 people)	344	497	*480*
Personal computers (per 1,000 people)	469	616	*504*
Households with television (%)	97	96	*98*
Quality			
Telephone faults (per 100 main lines per year)	*8.3*	*8.0*	..
Broadband subscribers (per 1,000 people)	3.9	77.0	126.2
International Internet bandwidth (bits per person)	128	1,096	4,718
Affordability			
Price basket for fixed line (US$ per month, residential)	17.4	29.4	25.8
Price basket for mobile (US$ per month)	..	17.8	17.8
Price basket for Internet (US$ per month)	..	*18.1*	*20.9*
Price of call to United States (US$ per 3 minutes)	0.67	*0.68*	0.77
Institutional efficiency and sustainability			
Total telecommunications revenue (% GDP)	3.6	3.5	2.9
Total telephone subscribers per employee	252	*388*	*485*
Total telecommunications investment (% revenue)	26.4	*21.6*	*12.3*
ICT applications			
ICT expenditure (% GDP)	6.8	*5.9*	*7.1*
E-government readiness index (scale 0–1)	..	0.83	0.70
Secure Internet servers (per 1 million people)	*176.3*	408.7	311.4
Schools connected to the Internet (%)	..	*97*	*99*

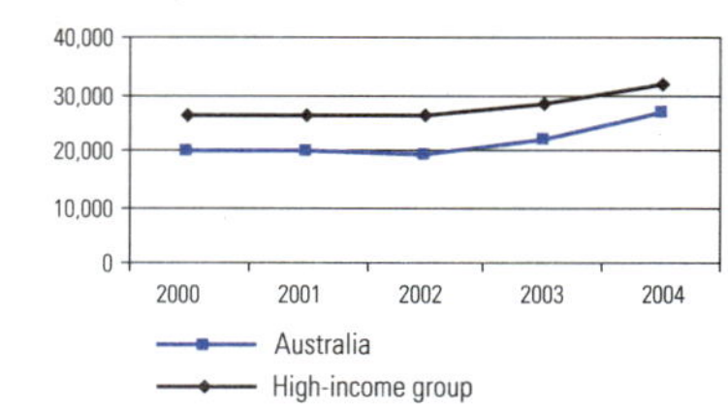

GNI per Capita, Atlas Method, 2000–4

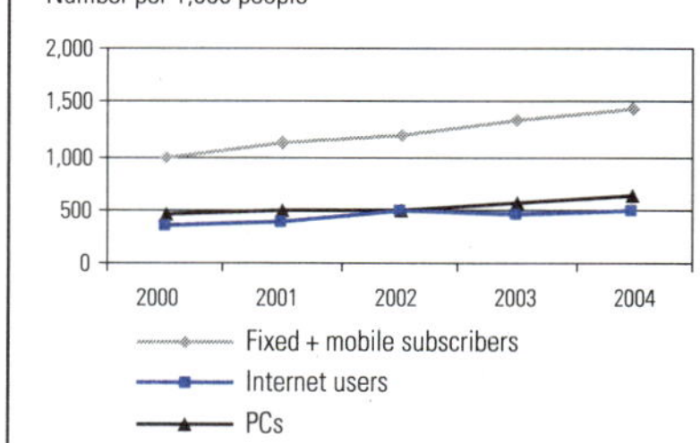

ICT MDG[b] Indicators, 2000–4

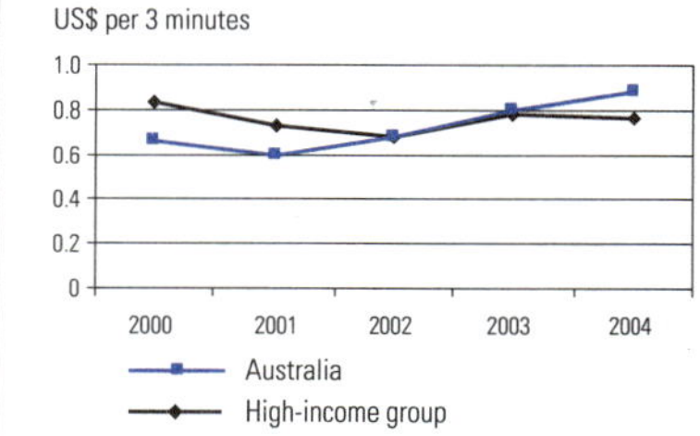

Price of Call to the United States, 2000–4

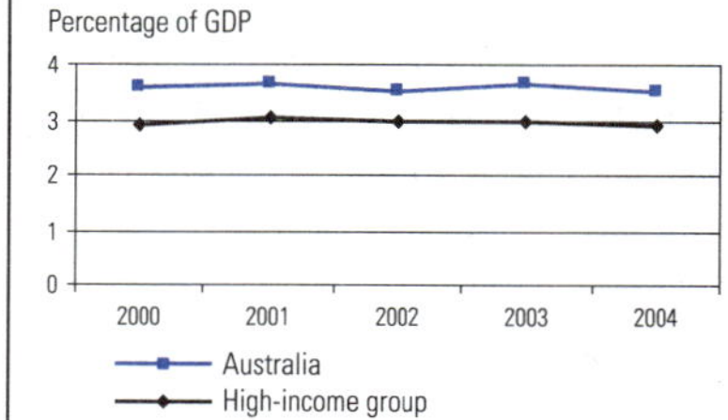

Total Telecommunications Revenue, 2000–4

Notes: Figures in italics are for years other than those specified. .. indicates data are not available. C = competition; GDP = gross domestic product; GNI = gross national income; ICT = information and communication technology; M = monopoly; MDG = Millennium Development Goal; P = partial competition; and PCs = personal computers.

a. Outgoing and incoming.

b. Millennium Development Goal indicators 47, 48a, and 48b.

Sources: Economic and social context: UNDP and World Bank; ICT sector structure: ITU, WEF; ICT sector performance: Global Insight/WITSA, ITU, Netcraft, UNDESA, UNPAN, and World Bank. Produced by the Global Information and Communication Technologies Department and the Development Economics Data Group. For complete information, see Definitions and Data Sources.

Austria

	Austria 2000	Austria 2004	High-income group 2004
Economic and social context			
Population, total (millions)	8	8	1,001
Urban population (% total population)	66	66	77
Poverty (% population below US$1 per day)	..	..	..
GNI per capita, Atlas method (current US$)	25,700	32,300	32,040
GDP growth, 1995–2000 and 2000–4 (%)	2.8	1.2	2.0
Adult literacy rate (% ages 15 and over)	..	..	..
Primary, secondary, tertiary school enrollment (% gross)	91	*89*	*93*
ICT sector structure			
Separate telecommunications regulator	Yes	Yes	
Status of main fixed-line operator	*Mixed*	Mixed	
Level of competition: international long distance	*C*	C	
Level of competition: mobile	*P*	C	
Level of competition: Internet service provider	*C*	C	
Government prioritization of ICT (scale 1–7)	..	4.9	4.9
ICT sector performance			
Access			
Telephone main lines (per 1,000 people)	499	440	558
International voice traffic (minutes per person)[a]	306	*256*	..
Mobile subscribers (per 1,000 people)	763	956	767
Population covered by mobile telephony (%)	98	98	98
Internet users (per 1,000 people)	337	397	*480*
Personal computers (per 1,000 people)	282	*396*	*504*
Households with television (%)	97	97	*98*
Quality			
Telephone faults (per 100 main lines per year)	5.4	*5.4*	..
Broadband subscribers (per 1,000 people)	23.8	102.0	126.2
International Internet bandwidth (bits per person)	1,034	6,730	4,718
Affordability			
Price basket for fixed line (US$ per month, residential)	23.3	30.6	25.8
Price basket for mobile (US$ per month)	..	26.4	17.8
Price basket for Internet (US$ per month)	..	*32.9*	*20.9*
Price of call to United States (US$ per 3 minutes)	1.19	0.71	0.77
Institutional efficiency and sustainability			
Total telecommunications revenue (% GDP)	2.4	*2.5*	*2.9*
Total telephone subscribers per employee	545	..	*485*
Total telecommunications investment (% revenue)	18.8	..	*12.3*
ICT applications			
ICT expenditure (% GDP)	6.0	*5.3*	*7.1*
E-government readiness index (scale 0–1)	..	0.70	0.70
Secure Internet servers (per 1 million people)	*83.3*	195.4	311.4
Schools connected to the Internet (%)	..	*94*	*99*

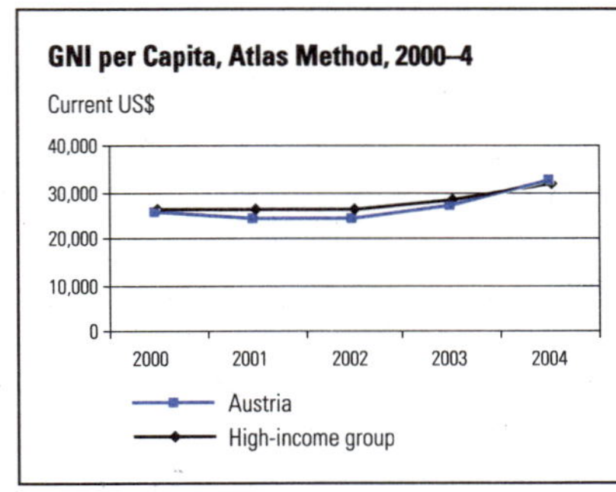

GNI per Capita, Atlas Method, 2000–4

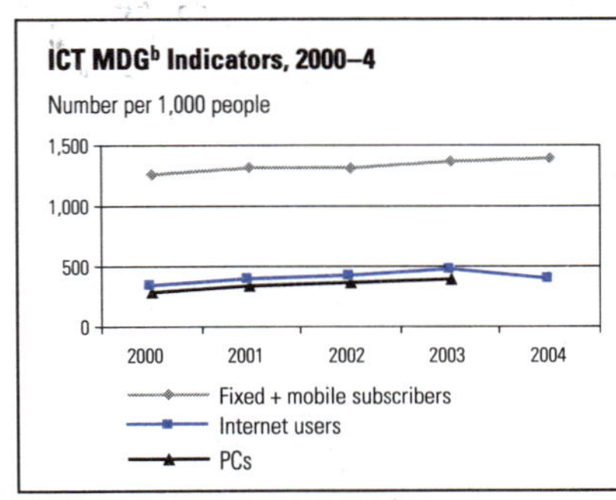

ICT MDG[b] Indicators, 2000–4

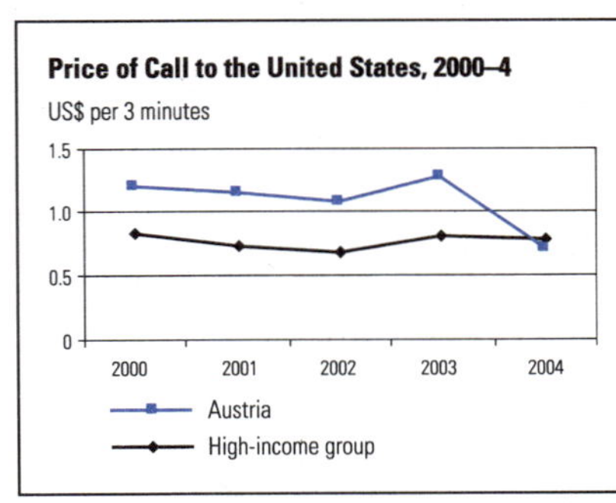

Price of Call to the United States, 2000–4

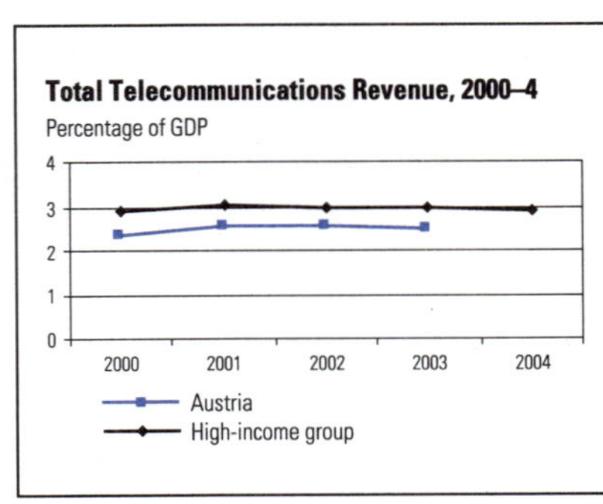

Total Telecommunications Revenue, 2000–4

Notes: Figures in italics are for years other than those specified. .. indicates data are not available. C = competition; GDP = gross domestic product; GNI = gross national income; ICT = information and communication technology; M = monopoly; MDG = Millennium Development Goal; P = partial competition; and PCs = personal computers.

a. Outgoing and incoming.

b. Millennium Development Goal indicators 47, 48a, and 48b.

Sources: Economic and social context: UNDP and World Bank; ICT sector structure: ITU, WEF; ICT sector performance: Global Insight/WITSA, ITU, Netcraft, UNDESA, UNPAN, and World Bank. Produced by the Global Information and Communication Technologies Department and the Development Economics Data Group. For complete information, see Definitions and Data Sources.

Azerbaijan

	Azerbaijan 2000	Azerbaijan 2004	Lower-middle-income group 2004	Europe & Central Asia Region 2004
Economic and social context				
Population, total (millions)	8	8	2,430	472
Urban population (% total population)	51	50	49	64
Poverty (% population below US$1 per day)	*10.9*	*3.7*	..	*3.6*
GNI per capita, Atlas method (current US$)	610	950	1,580	3,290
GDP growth, 1995–2000 and 2000–4 (%)	7.3	10.7	5.7	5.0
Adult literacy rate (% ages 15 and over)	*99*	..	90	97
Primary, secondary, tertiary school enrollment (% gross)	69	*69*	*72*	*81*
ICT sector structure				
Separate telecommunications regulator	..	..		
Status of main fixed-line operator	*Public*	Public		
Level of competition: international long distance	*M*	M		
Level of competition: mobile	*C*	C		
Level of competition: Internet service provider	..	..		
Government prioritization of ICT (scale 1–7)	..	..	3.8	3.8
ICT sector performance				
Access				
Telephone main lines (per 1,000 people)	100	123	192	*243*
International voice traffic (minutes per person)[a]	12	*14*	..	*28*
Mobile subscribers (per 1,000 people)	52	183	255	487
Population covered by mobile telephony (%)	94	*96*	*76*	*82*
Internet users (per 1,000 people)	1	54	70	*115*
Personal computers (per 1,000 people)	..	*33*	38	*73*
Households with television (%)	..	*99*	*89*	*92*
Quality				
Telephone faults (per 100 main lines per year)	52.0	*54.0*	..	*30.4*
Broadband subscribers (per 1,000 people)	0.0	*0.0*	12.6	*2.4*
International Internet bandwidth (bits per person)	..	*0*	58	148
Affordability				
Price basket for fixed line (US$ per month, residential)	17.6	*1.6*	*5.5*	*3.5*
Price basket for mobile (US$ per month)	..	10.5	8.9	10.3
Price basket for Internet (US$ per month)	..	*108.3*	*25.3*	*19.8*
Price of call to United States (US$ per 3 minutes)	7.10	4.18	1.45	1.06
Institutional efficiency and sustainability				
Total telecommunications revenue (% GDP)	1.6	1.7	3.6	3.5
Total telephone subscribers per employee	123	*229*	*195*	*150*
Total telecommunications investment (% revenue)	27.9	*12.0*	*25.5*	*19.0*
ICT applications				
ICT expenditure (% GDP)	..	..	*5.1*	*5.1*
E-government readiness index (scale 0–1)	..	0.20	0.26	0.39
Secure Internet servers (per 1 million people)	*0.1*	0.4	1.6	6.4
Schools connected to the Internet (%)	..	..	..	*65*

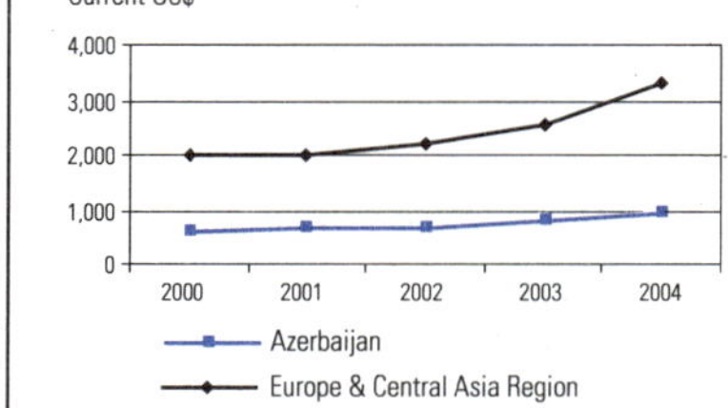

GNI per Capita, Atlas Method, 2000–4

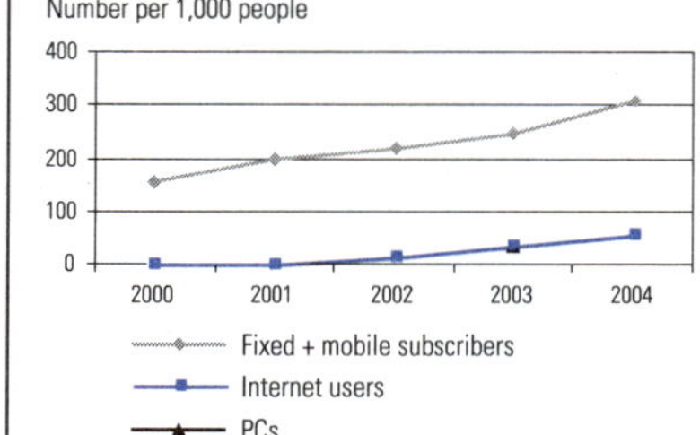

ICT MDG[b] Indicators, 2000–4

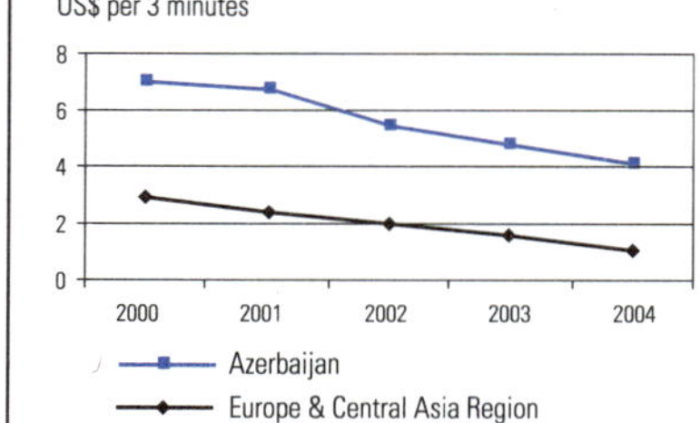

Price of Call to the United States, 2000–4

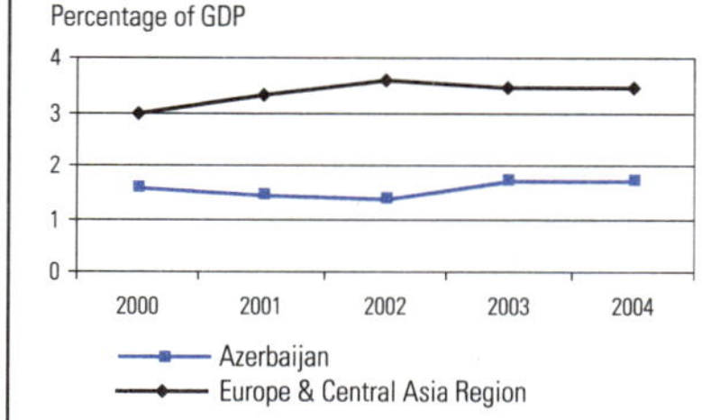

Total Telecommunications Revenue, 2000–4

Notes: Figures in italics are for years other than those specified. .. indicates data are not available. C = competition; GDP = gross domestic product; GNI = gross national income; ICT = information and communication technology; M = monopoly; MDG = Millennium Development Goal; P = partial competition; and PCs = personal compsuters.

a. Outgoing and incoming.

b. Millennium Development Goal indicators 47, 48a, and 48b.

Sources: Economic and social context: UNDP and World Bank; ICT sector structure: ITU, WEF; ICT sector performance: Global Insight/WITSA, ITU, Netcraft, UNDESA, UNPAN, and World Bank. Produced by the Global Information and Communication Technologies Department and the Development Economics Data Group. For complete information, see Definitions and Data Sources.

Bangladesh

	Bangladesh 2000	Bangladesh 2004	Low-income group 2004	South Asia Region 2004
Economic and social context				
Population, total (millions)	131	140	2,338	1,448
Urban population (% total population)	23	25	31	28
Poverty (% population below US$1 per day)	36.0	..	..	*31.3*
GNI per capita, Atlas method (current US$)	390	440	510	590
GDP growth, 1995–2000 and 2000–4 (%)	5.2	5.1	5.4	5.8
Adult literacy rate (% ages 15 and over)	..	*41*	61	58
Primary, secondary, tertiary school enrollment (% gross)	54	*53*	*55*	*56*
ICT sector structure				
Separate telecommunications regulator	No	Yes		
Status of main fixed-line operator	*Public*	Public		
Level of competition: international long distance	*M*	M		
Level of competition: mobile	*C*	C		
Level of competition: Internet service provider	..	C		
Government prioritization of ICT (scale 1–7)	..	4.4	..	5.3
ICT sector performance				
Access				
Telephone main lines (per 1,000 people)	4	*5*	33	41
International voice traffic (minutes per person)[a]	2	*3*	*4*	*4*
Mobile subscribers (per 1,000 people)	2	27	48	47
Population covered by mobile telephony (%)	..	*50*	*43*	*43*
Internet users (per 1,000 people)	1	*2*	20	21
Personal computers (per 1,000 people)	2	*4*	*8*	11
Households with television (%)	20	*29*	*16*	*32*
Quality				
Telephone faults (per 100 main lines per year)	..	..	..	*88.1*
Broadband subscribers (per 1,000 people)	*0.0*	*0.0*	0.5	0.6
International Internet bandwidth (bits per person)	0	1	*3*	4
Affordability				
Price basket for fixed line (US$ per month, residential)	10.7	*7.2*	*6.6*	3.2
Price basket for mobile (US$ per month)	..	3.7	11.6	3.2
Price basket for Internet (US$ per month)	..	*20.0*	*45.5*	*15.1*
Price of call to United States (US$ per 3 minutes)	4.14	1.21	1.95	1.21
Institutional efficiency and sustainability				
Total telecommunications revenue (% GDP)	0.8	1.3	2.3	1.9
Total telephone subscribers per employee	48	..	*89*	*89*
Total telecommunications investment (% revenue)	25.1	*15.3*	*27.8*	*15.3*
ICT applications				
ICT expenditure (% GDP)	2.1	*2.7*	*4.1*	*4.1*
E-government readiness index (scale 0–1)	..	0.08	0.12	0.34
Secure Internet servers (per 1 million people)	*0.0*	0.0	0.3	0.4
Schools connected to the Internet (%)	..	..	..	..

GNI per Capita, Atlas Method, 2000–4

Current US$

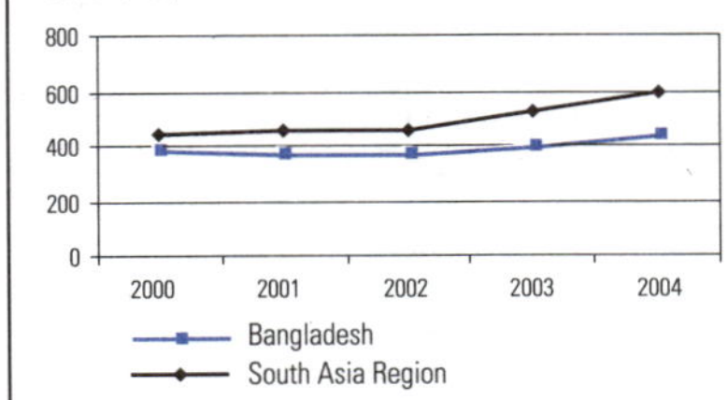

ICT MDG[b] Indicators, 2000–4

Number per 1,000 people

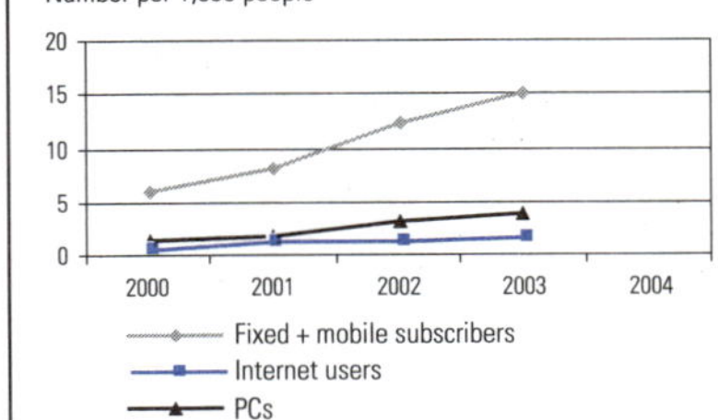

Price of Call to the United States, 2000–4

US$ per 3 minutes

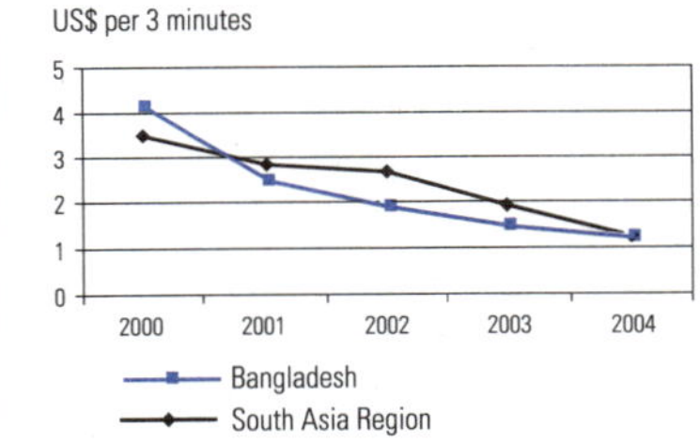

Total Telecommunications Revenue, 2000–4

Percentage of GDP

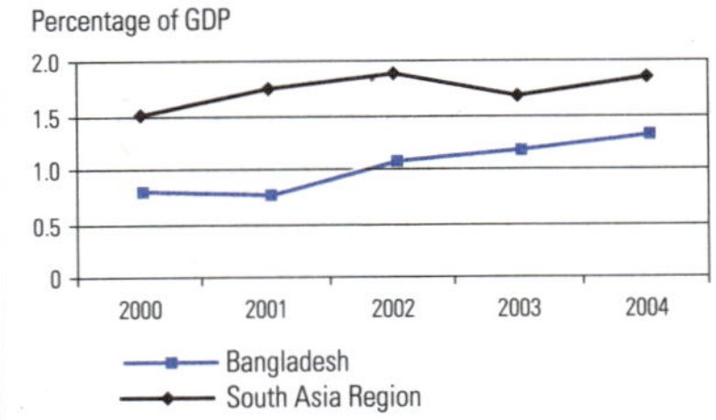

Notes: Figures in italics are for years other than those specified. .. indicates data are not available. C = competition; GDP = gross domestic product; GNI = gross national income; ICT = information and communication technology; M = monopoly; MDG = Millennium Development Goal; P = partial competition; and PCs = personal computers.

a. Outgoing and incoming.

b. Millennium Development Goal indicators 47, 48a, and 48b.

Sources: Economic and social context: UNDP and World Bank; ICT sector structure: ITU, WEF; ICT sector performance: Global Insight/WITSA, ITU, Netcraft, UNDESA, UNPAN, and World Bank. Produced by the Global Information and Communication Technologies Department and the Development Economics Data Group. For complete information, see Definitions and Data Sources.

Belarus

	Belarus 2000	Belarus 2004	Lower-middle-income group 2004	Europe & Central Asia Region 2004
Economic and social context				
Population, total (millions)	10	10	2,430	472
Urban population (% total population)	70	71	49	64
Poverty (% population below US$1 per day)	<2	..	..	*3.6*
GNI per capita, Atlas method (current US$)	1,280	2,120	1,580	3,290
GDP growth, 1995–2000 and 2000–4 (%)	6.7	6.7	5.7	5.0
Adult literacy rate (% ages 15 and over)	*100*	..	90	97
Primary, secondary, tertiary school enrollment (% gross)	88	*88*	*72*	*81*
ICT sector structure				
Separate telecommunications regulator	..	..		
Status of main fixed-line operator	*Public*	Public		
Level of competition: international long distance	*M*	M		
Level of competition: mobile	*C*	P		
Level of competition: Internet service provider	..	C		
Government prioritization of ICT (scale 1–7)	..	..	3.8	3.8
ICT sector performance				
Access				
Telephone main lines (per 1,000 people)	275	329	192	*243*
International voice traffic (minutes per person)[a]	39	*50*	..	*28*
Mobile subscribers (per 1,000 people)	5	249	255	487
Population covered by mobile telephony (%)	..	87	*76*	*82*
Internet users (per 1,000 people)	19	203	70	*115*
Personal computers (per 1,000 people)	..	*18*	38	*73*
Households with television (%)	87	91	*89*	*92*
Quality				
Telephone faults (per 100 main lines per year)	28.3	*24.8*	..	*30.4*
Broadband subscribers (per 1,000 people)	0.0	*0.0*	12.6	*2.4*
International Internet bandwidth (bits per person)	2	36	58	148
Affordability				
Price basket for fixed line (US$ per month, residential)	1.2	*2.0*	*5.5*	*3.5*
Price basket for mobile (US$ per month)	..	7.5	8.9	10.3
Price basket for Internet (US$ per month)	..	*12.8*	*25.3*	*19.8*
Price of call to United States (US$ per 3 minutes)	3.28	*2.25*	1.45	1.06
Institutional efficiency and sustainability				
Total telecommunications revenue (% GDP)	1.5	3.0	3.6	3.5
Total telephone subscribers per employee	105	*159*	*195*	*150*
Total telecommunications investment (% revenue)	28.8	*29.1*	*25.5*	*19.0*
ICT applications				
ICT expenditure (% GDP)	..	..	*5.1*	*5.1*
E-government readiness index (scale 0–1)	..	0.38	0.26	0.39
Secure Internet servers (per 1 million people)	*0.4*	0.4	1.6	6.4
Schools connected to the Internet (%)	..	..	..	*65*

GNI per Capita, Atlas Method, 2000–4
Current US$

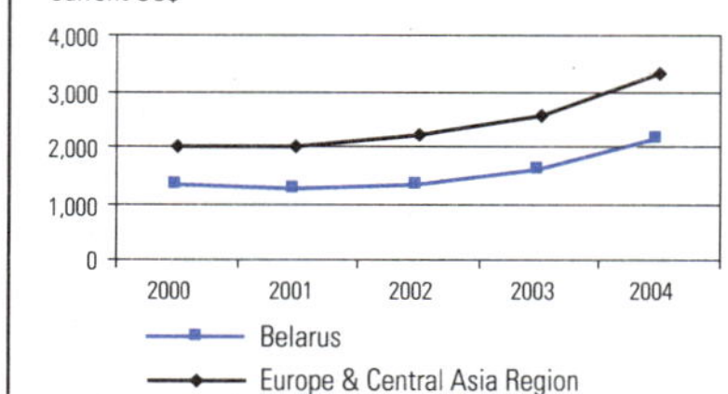

ICT MDG[b] Indicators, 2000–4
Number per 1,000 people

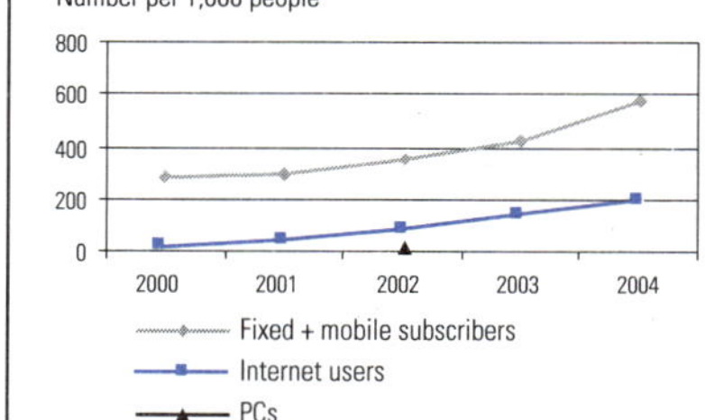

Price of Call to the United States, 2000–4
US$ per 3 minutes

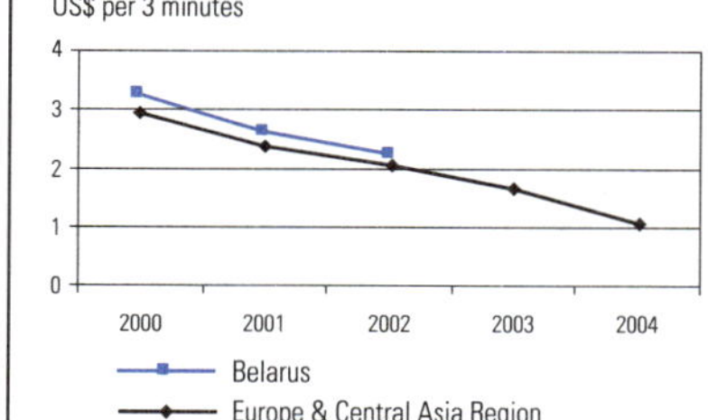

Total Telecommunications Revenue, 2000–4
Percentage of GDP

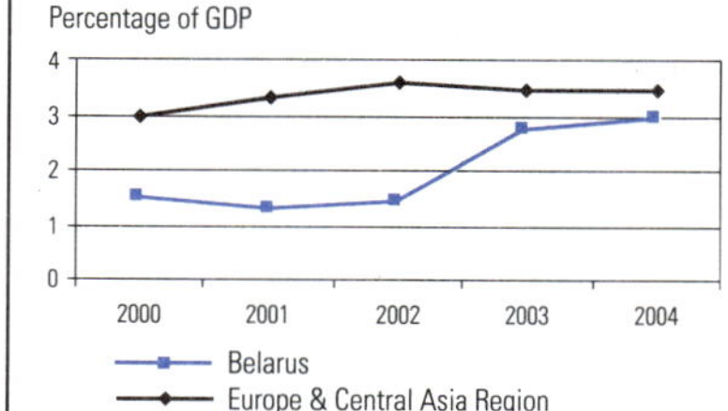

Notes: Figures in italics are for years other than those specified. .. indicates data are not available. C = competition; GDP = gross domestic product; GNI = gross national income; ICT = information and communication technology; M = monopoly; MDG = Millennium Development Goal; P = partial competition; and PCs = personal computers.

a. Outgoing and incoming.

b. Millennium Development Goal indicators 47, 48a, and 48b.

Sources: Economic and social context: UNDP and World Bank; ICT sector structure: ITU, WEF; ICT sector performance: Global Insight/WITSA, ITU, Netcraft, UNDESA, UNPAN, and World Bank. Produced by the Global Information and Communication Technologies Department and the Development Economics Data Group. For complete information, see Definitions and Data Sources.

Belgium

	Belgium 2000	Belgium 2004	High-income group 2004
Economic and social context			
Population, total (millions)	10	10	1,001
Urban population (% total population)	97	97	77
Poverty (% population below US$1 per day)	..	..	..
GNI per capita, Atlas method (current US$)	24,890	31,030	32,040
GDP growth, 1995–2000 and 2000–4 (%)	2.8	1.2	2.0
Adult literacy rate (% ages 15 and over)	..	..	..
Primary, secondary, tertiary school enrollment (% gross)	111	*114*	*93*
ICT sector structure			
Separate telecommunications regulator	Yes	Yes	
Status of main fixed-line operator	*Mixed*	Mixed	
Level of competition: international long distance	*C*	C	
Level of competition: mobile	*P*	P	
Level of competition: Internet service provider	*C*	C	
Government prioritization of ICT (scale 1–7)	..	4.3	4.9
ICT sector performance			
Access			
Telephone main lines (per 1,000 people)	517	462	558
International voice traffic (minutes per person)[a]	316	..	..
Mobile subscribers (per 1,000 people)	549	807	767
Population covered by mobile telephony (%)	99	99	98
Internet users (per 1,000 people)	163	*307*	*480*
Personal computers (per 1,000 people)	224	*251*	*504*
Households with television (%)	94	97	*98*
Quality			
Telephone faults (per 100 main lines per year)	3.5	*5.6*	..
Broadband subscribers (per 1,000 people)	12.4	155.7	126.2
International Internet bandwidth (bits per person)	1,829	11,296	4,718
Affordability			
Price basket for fixed line (US$ per month, residential)	22.0	34.6	25.8
Price basket for mobile (US$ per month)	..	24.9	17.8
Price basket for Internet (US$ per month)	..	*28.7*	*20.9*
Price of call to United States (US$ per 3 minutes)	1.67	0.75	0.77
Institutional efficiency and sustainability			
Total telecommunications revenue (% GDP)	2.3	2.6	2.9
Total telephone subscribers per employee	384	721	*485*
Total telecommunications investment (% revenue)	27.1	13.6	*12.3*
ICT applications			
ICT expenditure (% GDP)	6.1	*5.5*	*7.1*
E-government readiness index (scale 0–1)	..	0.77	0.70
Secure Internet servers (per 1 million people)	*33.2*	90.9	311.4
Schools connected to the Internet (%)	..	*93*	*99*

GNI per Capita, Atlas Method, 2000–4

Current US$

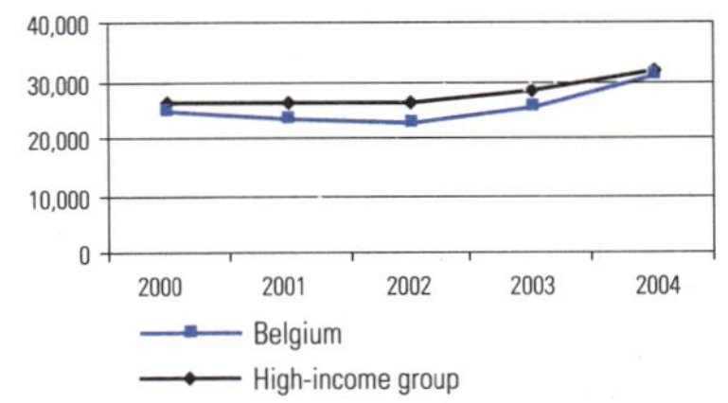

ICT MDG[b] Indicators, 2000–4

Number per 1,000 people

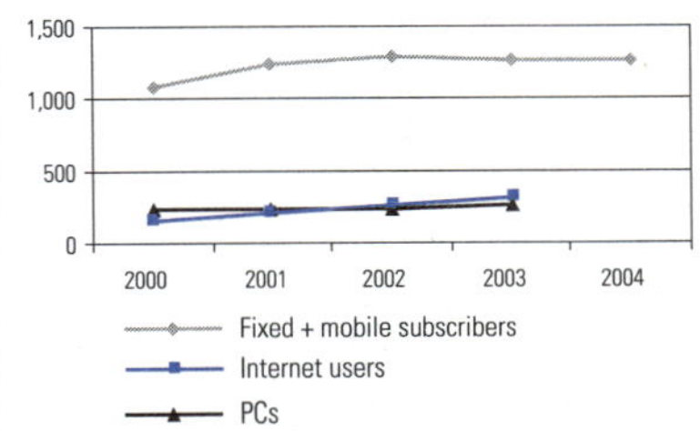

Price of Call to the United States, 2000–4

US$ per 3 minutes

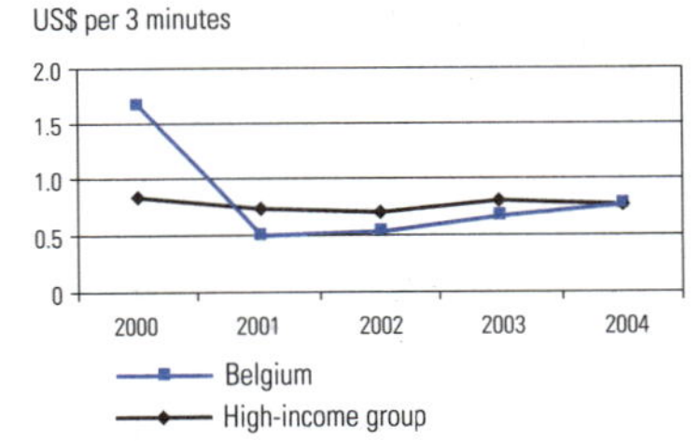

Total Telecommunications Revenue, 2000–4

Percentage of GDP

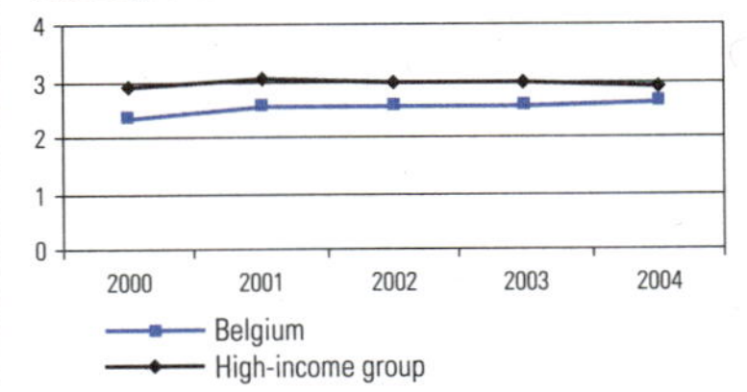

Notes: Figures in italics are for years other than those specified. .. indicates data are not available. C = competition; GDP = gross domestic product; GNI = gross national income; ICT = information and communication technology; M = monopoly; MDG = Millennium Development Goal; P = partial competition; and PCs = personal computers.

a. Outgoing and incoming.

b. Millennium Development Goal indicators 47, 48a, and 48b.

Sources: Economic and social context: UNDP and World Bank; ICT sector structure: ITU, WEF; ICT sector performance: Global Insight/WITSA, ITU, Netcraft, UNDESA, UNPAN, and World Bank. Produced by the Global Information and Communication Technologies Department and the Development Economics Data Group. For complete information, see Definitions and Data Sources.

Benin

	Benin 2000	Benin 2004	Low-income group 2004	Sub-Saharan Africa Region 2004
Economic and social context				
Population, total (millions)	6	7	2,338	719
Urban population (% total population)	42	45	31	37
Poverty (% population below US$1 per day)	..	..	..	*46.4*
GNI per capita, Atlas method (current US$)	390	530	510	600
GDP growth, 1995–2000 and 2000–4 (%)	5.3	4.5	5.4	3.9
Adult literacy rate (% ages 15 and over)	..	*34*	61	59
Primary, secondary, tertiary school enrollment (% gross)	49	*55*	*55*	*52*
ICT sector structure				
Separate telecommunications regulator	No	Yes		
Status of main fixed-line operator	*Public*	Public		
Level of competition: international long distance	*M*	M		
Level of competition: mobile	*M*	C		
Level of competition: Internet service provider	..	..		
Government prioritization of ICT (scale 1–7)	..	..	..	*4.4*
ICT sector performance				
Access				
Telephone main lines (per 1,000 people)	8	*10*	33	17
International voice traffic (minutes per person)[a]	6	..	*4*	..
Mobile subscribers (per 1,000 people)	9	*35*	48	86
Population covered by mobile telephony (%)	*23*	..	*43*	..
Internet users (per 1,000 people)	2	*10*	20	15
Personal computers (per 1,000 people)	2	*4*	*8*	*12*
Households with television (%)	18	*20*	*16*	*15*
Quality				
Telephone faults (per 100 main lines per year)	8.0	..	..	..
Broadband subscribers (per 1,000 people)	*0.0*	*0.0*	0.5	*0.1*
International Internet bandwidth (bits per person)	0	*7*	*3*	*4*
Affordability				
Price basket for fixed line (US$ per month, residential)	11.3	*12.2*	*6.6*	*8.5*
Price basket for mobile (US$ per month)	..	15.5	11.6	13.5
Price basket for Internet (US$ per month)	..	*46.4*	*45.5*	*54.8*
Price of call to United States (US$ per 3 minutes)	5.93	4.80	1.95	2.43
Institutional efficiency and sustainability				
Total telecommunications revenue (% GDP)	2.4	..	2.3	5.0
Total telephone subscribers per employee	87	..	*89*	*144*
Total telecommunications investment (% revenue)	54.2	..	*27.8*	*27.8*
ICT applications				
ICT expenditure (% GDP)	..	..	*4.1*	..
E-government readiness index (scale 0–1)	..	0.23	0.12	0.11
Secure Internet servers (per 1 million people)	*0.2*	..	0.3	1.9
Schools connected to the Internet (%)	..	..	..	..

GNI per Capita, Atlas Method, 2000–4

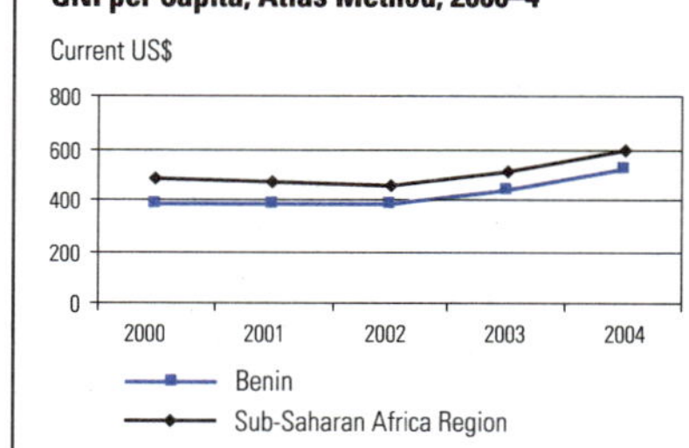

ICT MDG[b] Indicators, 2000–4

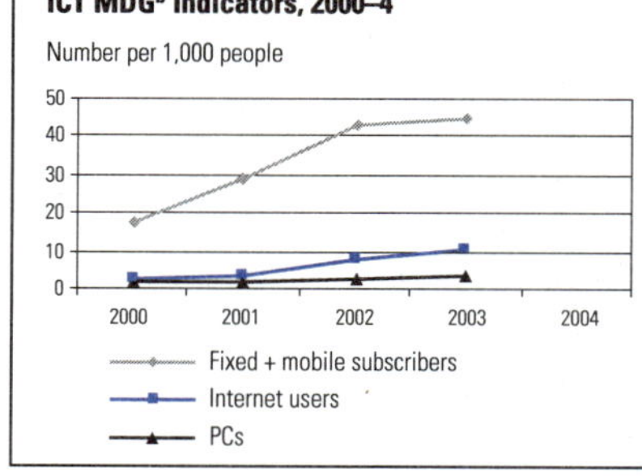

Price of Call to the United States, 2000–4

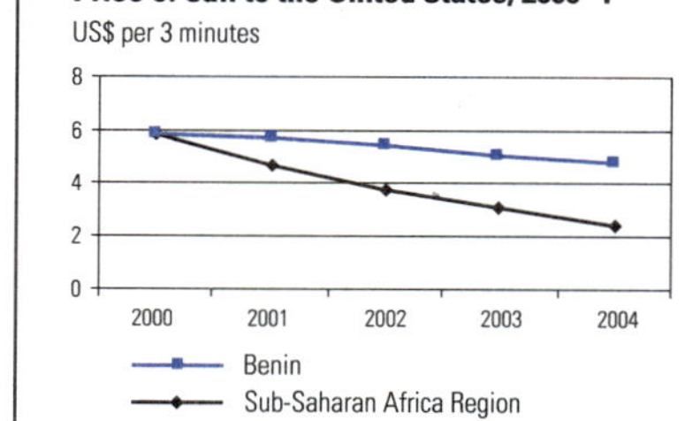

Total Telecommunications Revenue, 2000–4

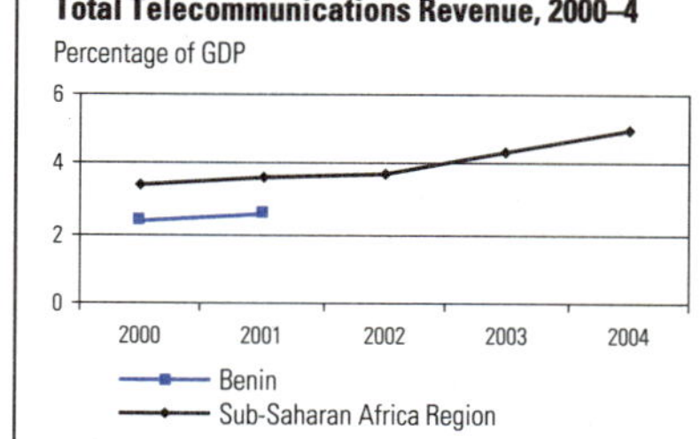

Notes: Figures in italics are for years other than those specified. .. indicates data are not available. C = competition; GDP = gross domestic product; GNI = gross national income; ICT = information and communication technology; M = monopoly; MDG = Millennium Development Goal; P = partial competition; and PCs = personal computers.

a. Outgoing and incoming.

b. Millennium Development Goal indicators 47, 48a, and 48b.

Sources: Economic and social context: UNDP and World Bank; ICT sector structure: ITU, WEF; ICT sector performance: Global Insight/WITSA, ITU, Netcraft, UNDESA, UNPAN, and World Bank. Produced by the Global Information and Communication Technologies Department and the Development Economics Data Group. For complete information, see Definitions and Data Sources.

Bolivia

	Bolivia 2000	Bolivia 2004	Lower-middle-income group 2004	Latin America & the Caribbean Region 2004
Economic and social context				
Population, total (millions)	8	9	2,430	541
Urban population (% total population)	62	64	49	77
Poverty (% population below US$1 per day)	*14.4*	..	..	*9.5*
GNI per capita, Atlas method (current US$)	1,000	960	1,580	3,600
GDP growth, 1995–2000 and 2000–4 (%)	3.5	2.6	5.7	1.5
Adult literacy rate (% ages 15 and over)	*87*	..	90	89
Primary, secondary, tertiary school enrollment (% gross)	84	*87*	*72*	*82*
ICT sector structure				
Separate telecommunications regulator	Yes	Yes		
Status of main fixed-line operator	*Mixed*	Private		
Level of competition: international long distance	*M*	M		
Level of competition: mobile	*P*	C		
Level of competition: Internet service provider	*C*	C		
Government prioritization of ICT (scale 1–7)	..	3.1	3.8	3.5
ICT sector performance				
Access				
Telephone main lines (per 1,000 people)	61	70	192	181
International voice traffic (minutes per person)[a]	14	24	..	..
Mobile subscribers (per 1,000 people)	70	200	255	324
Population covered by mobile telephony (%)	..	*60*	*76*	*76*
Internet users (per 1,000 people)	14	*35*	70	*104*
Personal computers (per 1,000 people)	17	36	38	*75*
Households with television (%)	46	..	*89*	*88*
Quality				
Telephone faults (per 100 main lines per year)	..	..	..	..
Broadband subscribers (per 1,000 people)	0.0	*0.0*	12.6	*5.2*
International Internet bandwidth (bits per person)	2	44	58	165
Affordability				
Price basket for fixed line (US$ per month, residential)	7.9	..	*5.5*	*9.0*
Price basket for mobile (US$ per month)	..	6.3	8.9	9.1
Price basket for Internet (US$ per month)	..	*22.3*	*25.3*	*31.5*
Price of call to United States (US$ per 3 minutes)	2.43	*1.89*	1.45	0.90
Institutional efficiency and sustainability				
Total telecommunications revenue (% GDP)	4.7	*3.8*	3.6	3.3
Total telephone subscribers per employee	321	..	*195*	..
Total telecommunications investment (% revenue)	33.3	..	*25.5*	..
ICT applications				
ICT expenditure (% GDP)	4.4	*5.6*	*5.1*	*5.3*
E-government readiness index (scale 0–1)	..	0.26	0.26	0.39
Secure Internet servers (per 1 million people)	*0.6*	1.8	1.6	8.6
Schools connected to the Internet (%)	..	..	..	..

GNI per Capita, Atlas Method, 2000–4

Current US$

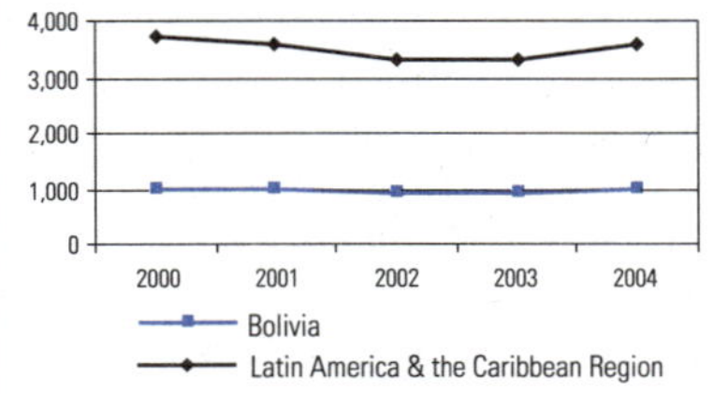

ICT MDG[b] Indicators, 2000–4

Number per 1,000 people

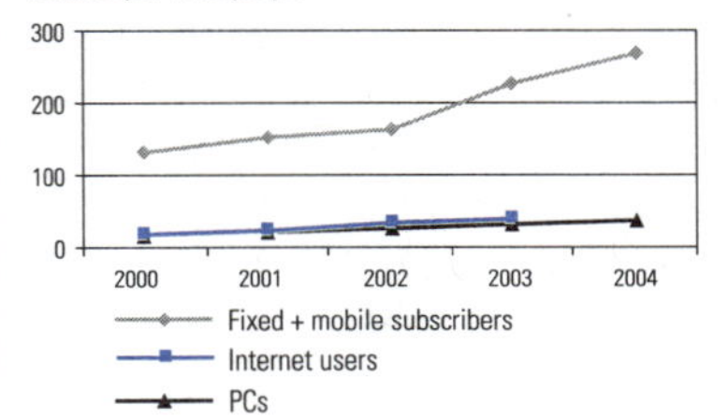

Price of Call to the United States, 2000–4

US$ per 3 minutes

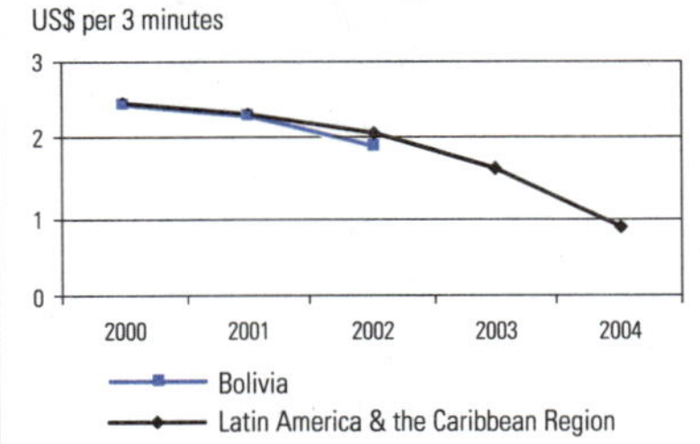

Total Telecommunications Revenue, 2000–4

Percentage of GDP

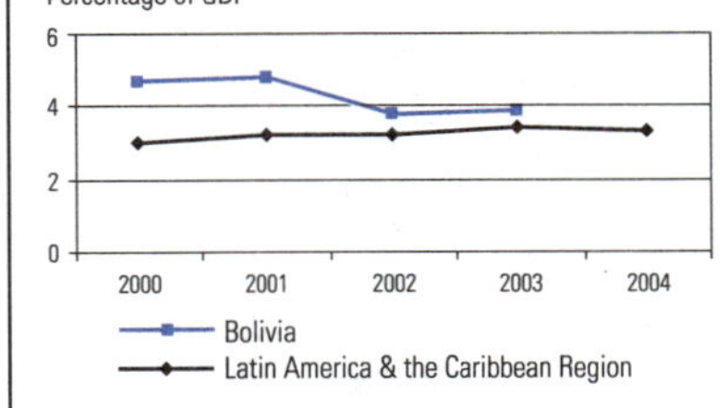

Notes: Figures in italics are for years other than those specified. .. indicates data are not available. C = competition; GDP = gross domestic product; GNI = gross national income; ICT = information and communication technology; M = monopoly; MDG = Millennium Development Goal; P = partial competition; and PCs = personal computers.

a. Outgoing and incoming.

b. Millennium Development Goal indicators 47, 48a, and 48b.

Sources: Economic and social context: UNDP and World Bank; ICT sector structure: ITU, WEF; ICT sector performance: Global Insight/WITSA, ITU, Netcraft, UNDESA, UNPAN, and World Bank. Produced by the Global Information and Communication Technologies Department and the Development Economics Data Group. For complete information, see Definitions and Data Sources.

Bosnia and Herzegovina

	Bosnia and Herzegovina 2000	Bosnia and Herzegovina 2004	Lower-middle-income group 2004	Europe & Central Asia Region 2004
Economic and social context				
Population, total (millions)	4	4	2,430	472
Urban population (% total population)	43	45	49	64
Poverty (% population below US$1 per day)	..	..	..	*3.6*
GNI per capita, Atlas method (current US$)	1,330	2,040	1,580	3,290
GDP growth, 1995–2000 and 2000–4 (%)	25.3	3.8	5.7	5.0
Adult literacy rate (% ages 15 and over)	95	..	90	97
Primary, secondary, tertiary school enrollment (% gross)	..	..	*72*	*81*
ICT sector structure				
Separate telecommunications regulator	Yes	Yes		
Status of main fixed-line operator	*Public*	Public		
Level of competition: international long distance	..	M		
Level of competition: mobile	..	P		
Level of competition: Internet service provider	..	C		
Government prioritization of ICT (scale 1–7)	..	3.8	3.8	3.8
ICT sector performance				
Access				
Telephone main lines (per 1,000 people)	206	*245*	192	*243*
International voice traffic (minutes per person)[a]	90	..	..	*28*
Mobile subscribers (per 1,000 people)	25	*274*	255	487
Population covered by mobile telephony (%)	60	*90*	*76*	*82*
Internet users (per 1,000 people)	11	*39*	70	*115*
Personal computers (per 1,000 people)	..	..	38	*73*
Households with television (%)	..	*87*	*89*	*92*
Quality				
Telephone faults (per 100 main lines per year)	..	..	..	*30.4*
Broadband subscribers (per 1,000 people)	0.0	*0.1*	12.6	*2.4*
International Internet bandwidth (bits per person)	7	78	58	148
Affordability				
Price basket for fixed line (US$ per month, residential)	4.6	*5.1*	*5.5*	*3.5*
Price basket for mobile (US$ per month)	..	9.1	8.9	10.3
Price basket for Internet (US$ per month)	..	*7.3*	*25.3*	*19.8*
Price of call to United States (US$ per 3 minutes)	2.96	3.62	1.45	1.06
Institutional efficiency and sustainability				
Total telecommunications revenue (% GDP)	4.2	..	3.6	3.5
Total telephone subscribers per employee	480	*238*	*195*	*150*
Total telecommunications investment (% revenue)	27.5	..	*25.5*	*19.0*
ICT applications				
ICT expenditure (% GDP)	..	..	*5.1*	*5.1*
E-government readiness index (scale 0–1)	..	0.22	0.26	0.39
Secure Internet servers (per 1 million people)	..	3.9	1.6	6.4
Schools connected to the Internet (%)	..	..	..	*65*

GNI per Capita, Atlas Method, 2000–4

Current US$

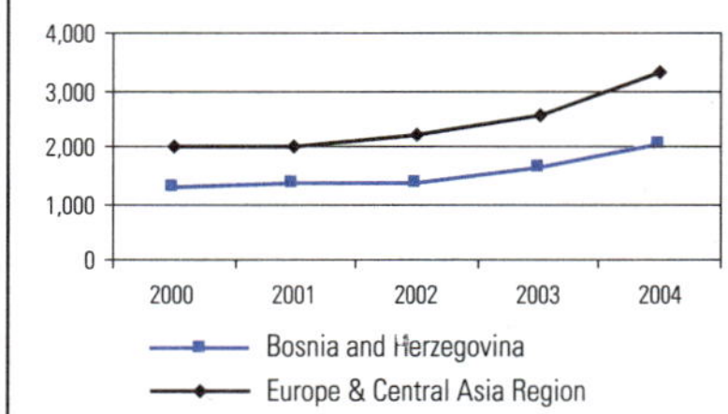

ICT MDG[b] Indicators, 2000–4

Number per 1,000 people

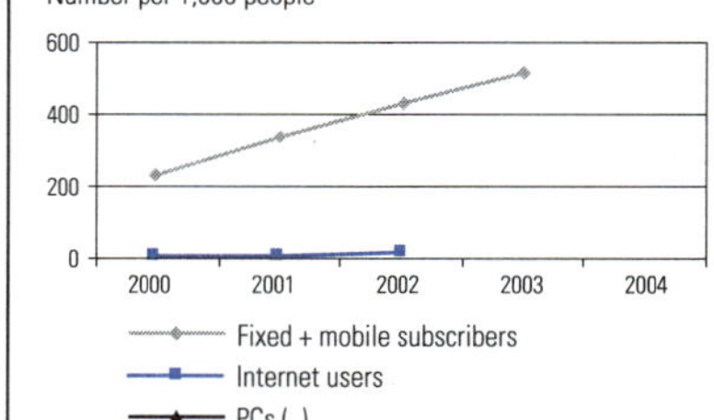

Price of Call to the United States, 2000–4

US$ per 3 minutes

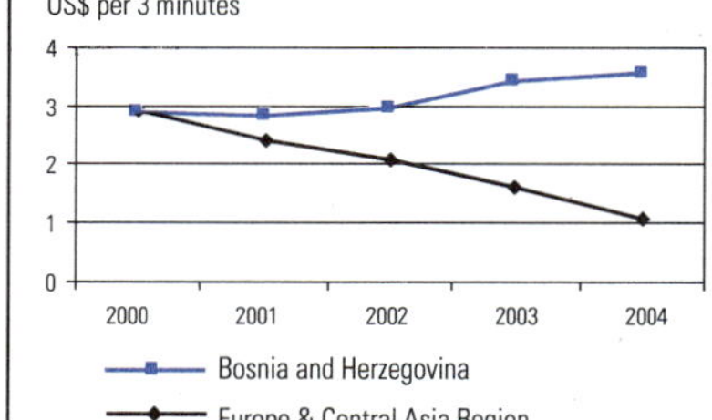

Total Telecommunications Revenue, 2000–4

Percentage of GDP

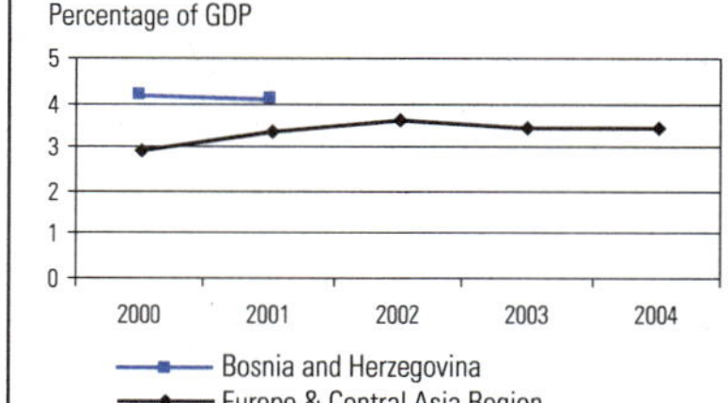

Notes: Figures in italics are for years other than those specified. .. indicates data are not available. C = competition; GDP = gross domestic product; GNI = gross national income; ICT = information and communication technology; M = monopoly; MDG = Millennium Development Goal; P = partial competition; and PCs = personal computers.

a. Outgoing and incoming.

b. Millennium Development Goal indicators 47, 48a, and 48b.

Sources: Economic and social context: UNDP and World Bank; ICT sector structure: ITU, WEF; ICT sector performance: Global Insight/WITSA, ITU, Netcraft, UNDESA, UNPAN, and World Bank. Produced by the Global Information and Communication Technologies Department and the Development Economics Data Group. For complete information, see Definitions and Data Sources.

Botswana

	Botswana 2000	Botswana 2004	Upper-middle-income group 2004	Sub-Saharan Africa Region 2004
Economic and social context				
Population, total (millions)	2	2	576	719
Urban population (% total population)	50	52	72	37
Poverty (% population below US$1 per day)	..	..	..	*46.4*
GNI per capita, Atlas method (current US$)	3,010	4,340	4,770	600
GDP growth, 1995–2000 and 2000–4 (%)	6.2	4.9	2.7	3.9
Adult literacy rate (% ages 15 and over)	..	*79*	94	59
Primary, secondary, tertiary school enrollment (% gross)	70	*70*	*80*	*52*
ICT sector structure				
Separate telecommunications regulator	Yes	Yes		
Status of main fixed-line operator	*Public*	Public		
Level of competition: international long distance	*M*	M		
Level of competition: mobile	*C*	P		
Level of competition: Internet service provider	*C*	C		
Government prioritization of ICT (scale 1–7)	..	4.9	4.1	*4.4*
ICT sector performance				
Access				
Telephone main lines (per 1,000 people)	81	79	*220*	17
International voice traffic (minutes per person)[a]	43	*64*	*39*	..
Mobile subscribers (per 1,000 people)	119	348	490	86
Population covered by mobile telephony (%)	..	85	*84*	..
Internet users (per 1,000 people)	15	25	*133*	15
Personal computers (per 1,000 people)	36	*41*	*99*	*12*
Households with television (%)	15	*15*	*92*	*15*
Quality				
Telephone faults (per 100 main lines per year)	..	..	*20.3*	..
Broadband subscribers (per 1,000 people)	0.0	*0.0*	*3.7*	*0.1*
International Internet bandwidth (bits per person)	4	23	176	*4*
Affordability				
Price basket for fixed line (US$ per month, residential)	4.8	11.3	*13.9*	*8.5*
Price basket for mobile (US$ per month)	..	11.1	11.1	13.5
Price basket for Internet (US$ per month)	..	*27.0*	*20.8*	*54.8*
Price of call to United States (US$ per 3 minutes)	*3.64*	2.88	1.03	2.43
Institutional efficiency and sustainability				
Total telecommunications revenue (% GDP)	3.5	*3.1*	3.4	5.0
Total telephone subscribers per employee	190	*341*	*402*	*144*
Total telecommunications investment (% revenue)	13.9	*6.8*	*18.6*	*27.8*
ICT applications				
ICT expenditure (% GDP)	..	..	*5.0*	..
E-government readiness index (scale 0–1)	..	0.29	0.49	0.11
Secure Internet servers (per 1 million people)	..	0.6	10.7	1.9
Schools connected to the Internet (%)	..	..	*60*	..

GNI per Capita, Atlas Method, 2000–4

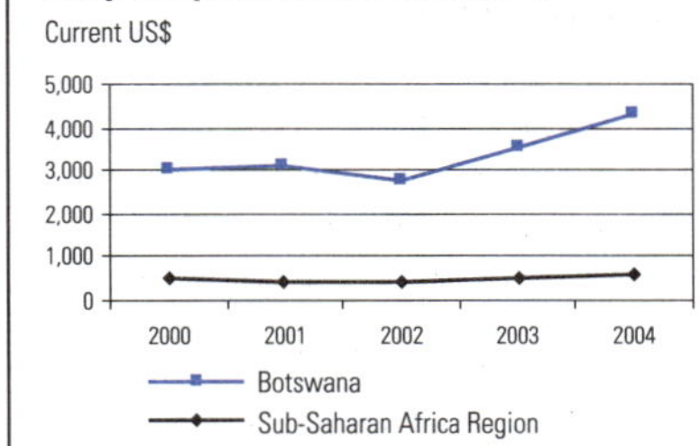

ICT MDG[b] Indicators, 2000–4

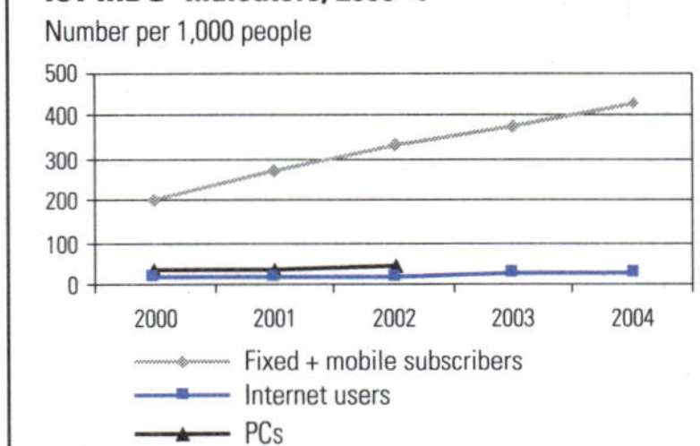

Price of Call to the United States, 2000–4

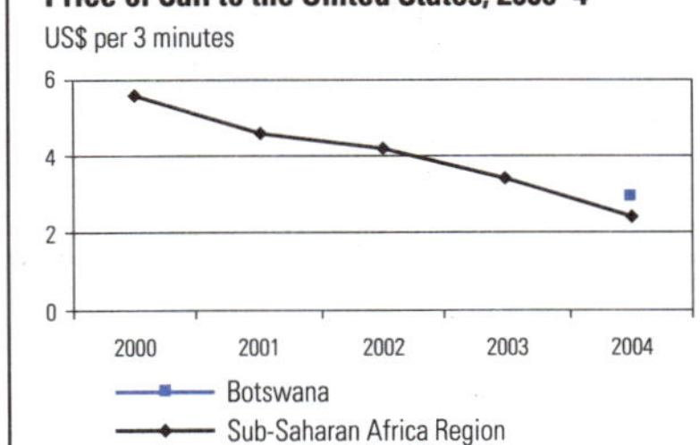

Total Telecommunications Revenue, 2000–4

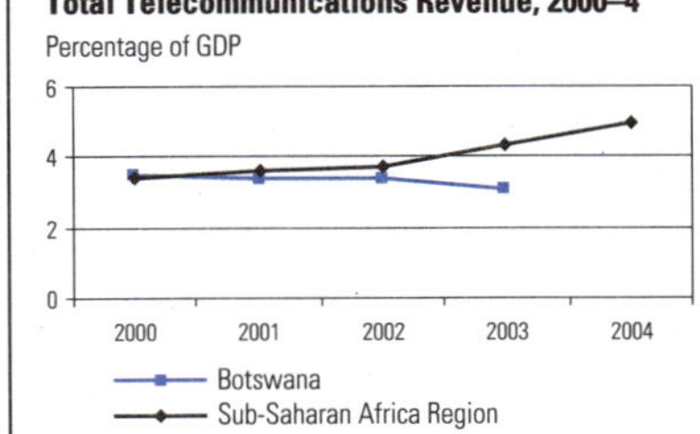

Notes: Figures in italics are for years other than those specified. .. indicates data are not available. C = competition; GDP = gross domestic product; GNI = gross national income; ICT = information and communication technology; M = monopoly; MDG = Millennium Development Goal; P = partial competition; and PCs = personal computers.

a. Outgoing and incoming.

b. Millennium Development Goal indicators 47, 48a, and 48b.

Sources: Economic and social context: UNDP and World Bank; ICT sector structure: ITU, WEF; ICT sector performance: Global Insight/WITSA, ITU, Netcraft, UNDESA, UNPAN, and World Bank. Produced by the Global Information and Communication Technologies Department and the Development Economics Data Group. For complete information, see Definitions and Data Sources.

Brazil

	Brazil 2000	Brazil 2004	Lower-middle-income group 2004	Latin America & the Caribbean Region 2004
Economic and social context				
Population, total (millions)	170	179	2,430	541
Urban population (% total population)	81	84	49	77
Poverty (% population below US$1 per day)	*9.9*	*8.2*	..	*9.5*
GNI per capita, Atlas method (current US$)	3,650	3,090	1,580	3,600
GDP growth, 1995–2000 and 2000–4 (%)	2.0	2.0	5.7	1.5
Adult literacy rate (% ages 15 and over)	..	*88*	90	89
Primary, secondary, tertiary school enrollment (% gross)	91	*91*	*72*	*82*
ICT sector structure				
Separate telecommunications regulator	Yes	Yes		
Status of main fixed-line operator	*Private*	Private		
Level of competition: international long distance	*P*	C		
Level of competition: mobile	*P*	C		
Level of competition: Internet service provider	*C*	C		
Government prioritization of ICT (scale 1–7)	..	4.4	3.8	3.5
ICT sector performance				
Access				
Telephone main lines (per 1,000 people)	182	237	192	181
International voice traffic (minutes per person)[a]	11	..	..	..
Mobile subscribers (per 1,000 people)	136	367	255	324
Population covered by mobile telephony (%)	..	*68*	*76*	*76*
Internet users (per 1,000 people)	29	*109*	70	*104*
Personal computers (per 1,000 people)	50	*86*	38	*75*
Households with television (%)	87	*90*	*89*	*88*
Quality				
Telephone faults (per 100 main lines per year)	3.1	*1.7*	..	..
Broadband subscribers (per 1,000 people)	0.6	12.8	12.6	*5.2*
International Internet bandwidth (bits per person)	5	154	58	165
Affordability				
Price basket for fixed line (US$ per month, residential)	11.4	*7.4*	*5.5*	*9.0*
Price basket for mobile (US$ per month)	..	18.9	8.9	9.1
Price basket for Internet (US$ per month)	..	*28.0*	*25.3*	*31.5*
Price of call to United States (US$ per 3 minutes)	1.15	*0.71*	1.45	0.90
Institutional efficiency and sustainability				
Total telecommunications revenue (% GDP)	3.7	4.0	3.6	3.3
Total telephone subscribers per employee	516	..	*195*	..
Total telecommunications investment (% revenue)	39.8	*30.8*	*25.5*	..
ICT applications				
ICT expenditure (% GDP)	5.6	*6.7*	*5.1*	*5.3*
E-government readiness index (scale 0–1)	..	0.64	0.26	0.39
Secure Internet servers (per 1 million people)	*6.0*	11.2	1.6	8.6
Schools connected to the Internet (%)	..	*50*	..	..

GNI per Capita, Atlas Method, 2000–4

Current US$

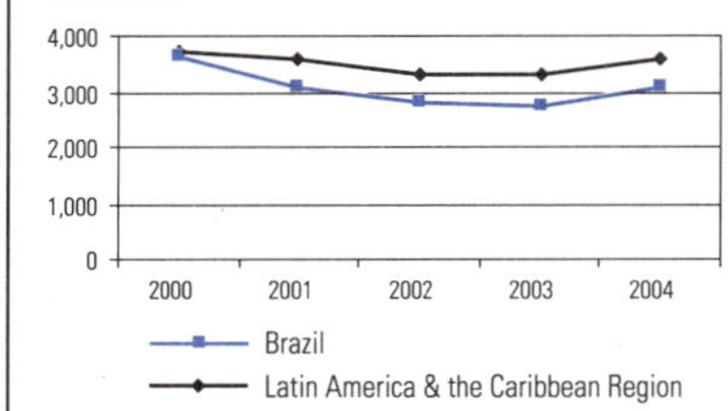

ICT MDG[b] Indicators, 2000–4

Number per 1,000 people

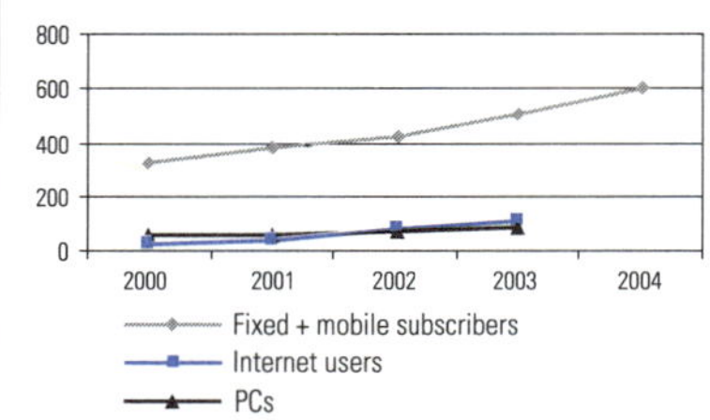

Price of Call to the United States, 2000–4

US$ per 3 minutes

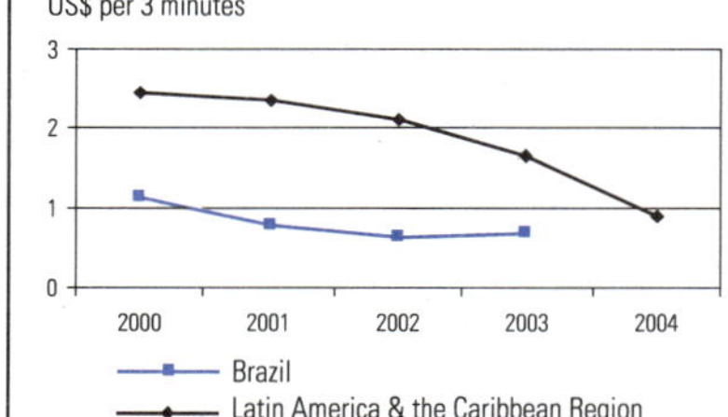

Total Telecommunications Revenue, 2000–4

Percentage of GDP

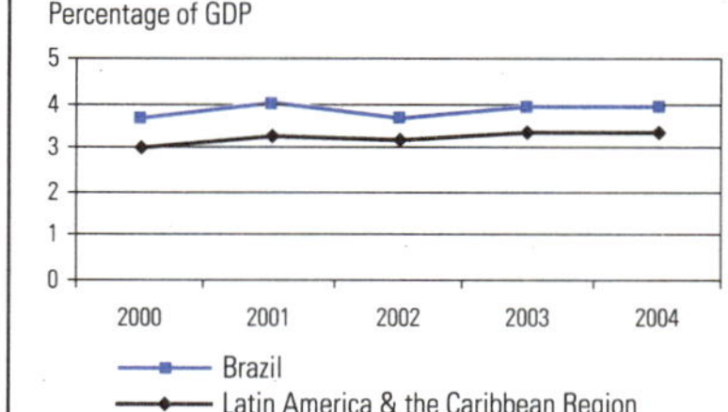

Notes: Figures in italics are for years other than those specified. .. indicates data are not available. C = competition; GDP = gross domestic product; GNI = gross national income; ICT = information and communication technology; M = monopoly; MDG = Millennium Development Goal; P = partial competition; and PCs = personal computers.

a. Outgoing and incoming.

b. Millennium Development Goal indicators 47, 48a, and 48b.

Sources: Economic and social context: UNDP and World Bank; ICT sector structure: ITU, WEF; ICT sector performance: Global Insight/WITSA, ITU, Netcraft, UNDESA, UNPAN, and World Bank. Produced by the Global Information and Communication Technologies Department and the Development Economics Data Group. For complete information, see Definitions and Data Sources.

Bulgaria

	Bulgaria 2000	Bulgaria 2004	Lower-middle-income group 2004	Europe & Central Asia Region 2004
Economic and social context				
Population, total (millions)	8	8	2,430	472
Urban population (% total population)	69	70	49	64
Poverty (% population below US$1 per day)	*<2*	*4.7*	..	*3.6*
GNI per capita, Atlas method (current US$)	1,600	2,740	1,580	3,290
GDP growth, 1995–2000 and 2000–4 (%)	–0.5	4.7	5.7	5.0
Adult literacy rate (% ages 15 and over)	*98*	..	90	97
Primary, secondary, tertiary school enrollment (% gross)	76	*78*	*72*	*81*
ICT sector structure				
Separate telecommunications regulator	Yes	Yes		
Status of main fixed-line operator	*Public*	Mixed		
Level of competition: international long distance	*M*	C		
Level of competition: mobile	*C*	C		
Level of competition: Internet service provider	*C*	*C*		
Government prioritization of ICT (scale 1–7)	..	3.7	3.8	3.8
ICT sector performance				
Access				
Telephone main lines (per 1,000 people)	358	356	192	*243*
International voice traffic (minutes per person)[a]	40	*44*	..	*28*
Mobile subscribers (per 1,000 people)	92	549	255	487
Population covered by mobile telephony (%)	..	98	*76*	*82*
Internet users (per 1,000 people)	53	219	70	*115*
Personal computers (per 1,000 people)	45	*51*	38	*73*
Households with television (%)	92	97	*89*	*92*
Quality				
Telephone faults (per 100 main lines per year)	4.8	*2.6*	..	*30.4*
Broadband subscribers (per 1,000 people)	0.0	*5.6*	12.6	*2.4*
International Internet bandwidth (bits per person)	5	80	58	148
Affordability				
Price basket for fixed line (US$ per month, residential)	*3.1*	*8.5*	*5.5*	*3.5*
Price basket for mobile (US$ per month)	..	17.3	8.9	10.3
Price basket for Internet (US$ per month)	..	*12.5*	*25.3*	*19.8*
Price of call to United States (US$ per 3 minutes)	2.55	0.57	1.45	1.06
Institutional efficiency and sustainability				
Total telecommunications revenue (% GDP)	3.1	*5.9*	3.6	3.5
Total telephone subscribers per employee	140	*199*	*195*	*150*
Total telecommunications investment (% revenue)	14.0	*44.7*	*25.5*	*19.0*
ICT applications				
ICT expenditure (% GDP)	3.9	*3.9*	*5.1*	*5.1*
E-government readiness index (scale 0–1)	..	0.51	0.26	0.39
Secure Internet servers (per 1 million people)	*2.3*	5.9	1.6	6.4
Schools connected to the Internet (%)	..	*60*	..	*65*

GNI per Capita, Atlas Method, 2000–4

Current US$

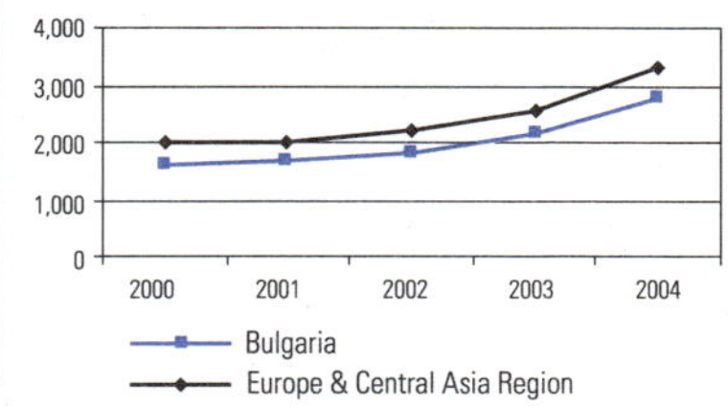

ICT MDG[b] Indicators, 2000–4

Number per 1,000 people

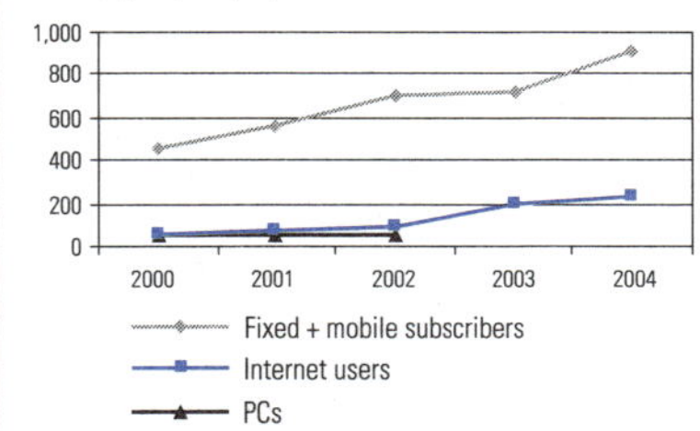

Price of Call to the United States, 2000–4

US$ per 3 minutes

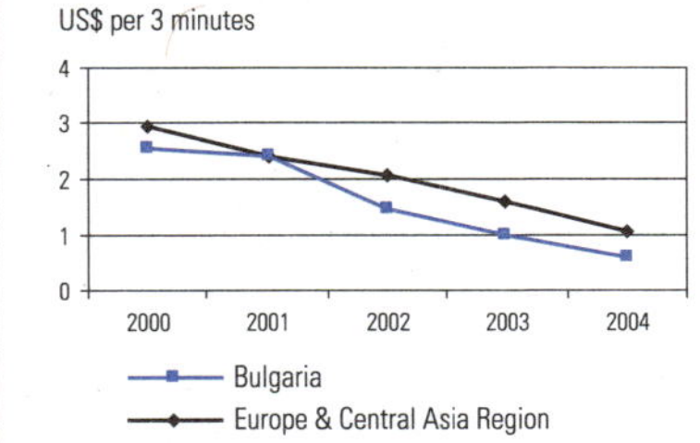

Total Telecommunications Revenue, 2000–4

Percentage of GDP

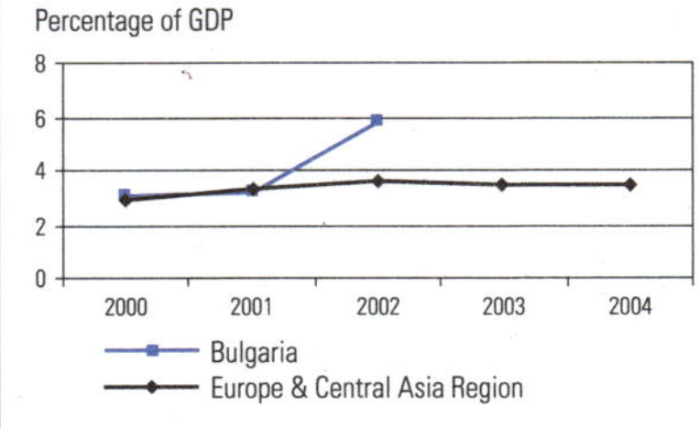

Notes: Figures in italics are for years other than those specified. .. indicates data are not available. C = competition; GDP = gross domestic product; GNI = gross national income; ICT = information and communication technology; M = monopoly; MDG = Millennium Development Goal; P = partial competition; and PCs = personal computers.

a. Outgoing and incoming.

b. Millennium Development Goal indicators 47, 48a, and 48b.

Sources: Economic and social context: UNDP and World Bank; ICT sector structure: ITU, WEF; ICT sector performance: Global Insight/WITSA, ITU, Netcraft, UNDESA, UNPAN, and World Bank. Produced by the Global Information and Communication Technologies Department and the Development Economics Data Group. For complete information, see Definitions and Data Sources.

Burkina Faso

	Burkina Faso 2000	Burkina Faso 2004	Low-income group 2004	Sub-Saharan Africa Region 2004
Economic and social context				
Population, total (millions)	11	12	2,338	719
Urban population (% total population)	17	18	31	37
Poverty (% population below US$1 per day)	*44.9*	..	..	*46.4*
GNI per capita, Atlas method (current US$)	250	360	510	600
GDP growth, 1995–2000 and 2000–4 (%)	4.2	5.2	5.4	3.9
Adult literacy rate (% ages 15 and over)	..	..	61	59
Primary, secondary, tertiary school enrollment (% gross)	22	*24*	*55*	*52*
ICT sector structure				
Separate telecommunications regulator	Yes	Yes		
Status of main fixed-line operator	*Public*	Public		
Level of competition: international long distance	*M*	M		
Level of competition: mobile	*C*	C		
Level of competition: Internet service provider	*C*	C		
Government prioritization of ICT (scale 1–7)	..	..	..	*4.4*
ICT sector performance				
Access				
Telephone main lines (per 1,000 people)	5	*5*	33	17
International voice traffic (minutes per person)[a]	*3*	*4*	*4*	..
Mobile subscribers (per 1,000 people)	2	*19*	48	86
Population covered by mobile telephony (%)	..	*60*	*43*	..
Internet users (per 1,000 people)	1	*4*	20	15
Personal computers (per 1,000 people)	1	*2*	*8*	*12*
Households with television (%)	7	*7*	*16*	*15*
Quality				
Telephone faults (per 100 main lines per year)	59.3	*19.7*	..	..
Broadband subscribers (per 1,000 people)	*0.0*	*0.0*	0.5	*0.1*
International Internet bandwidth (bits per person)	0	*1*	*3*	*4*
Affordability				
Price basket for fixed line (US$ per month, residential)	11.6	*11.9*	*6.6*	*8.5*
Price basket for mobile (US$ per month)	..	15.4	11.6	13.5
Price basket for Internet (US$ per month)	..	*45.4*	*45.5*	*54.8*
Price of call to United States (US$ per 3 minutes)	3.16	1.14	1.95	2.43
Institutional efficiency and sustainability				
Total telecommunications revenue (% GDP)	2.0	*2.0*	2.3	5.0
Total telephone subscribers per employee	62	*138*	*89*	*144*
Total telecommunications investment (% revenue)	38.0	*38.0*	*27.8*	*27.8*
ICT applications				
ICT expenditure (% GDP)	..	..	*4.1*	..
E-government readiness index (scale 0–1)	..	0.29	0.12	0.11
Secure Internet servers (per 1 million people)	..	0.2	0.3	1.9
Schools connected to the Internet (%)	..	..	..	..

GNI per Capita, Atlas Method, 2000–4
Current US$

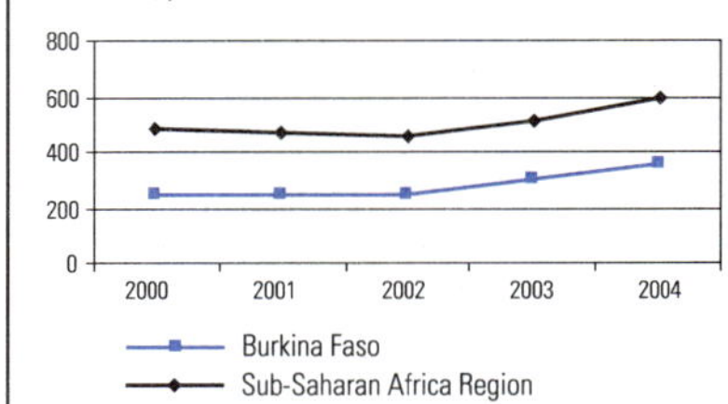

ICT MDG[b] Indicators, 2000–4
Number per 1,000 people

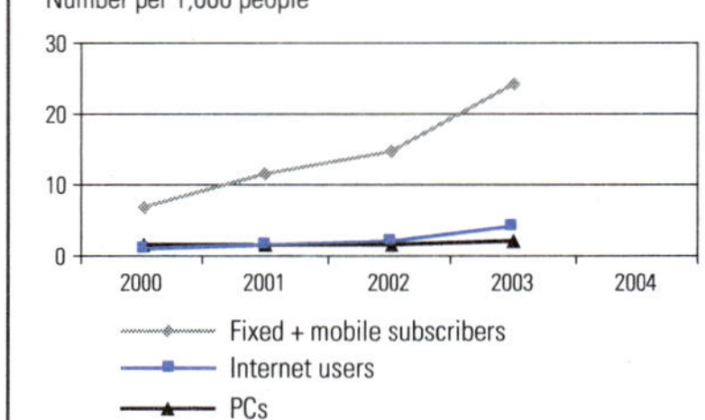

Price of Call to the United States, 2000–4
US$ per 3 minutes

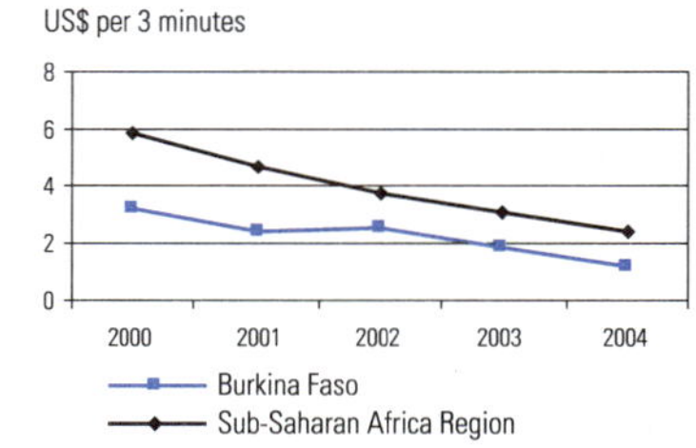

Total Telecommunications Revenue, 2000–4
Percentage of GDP

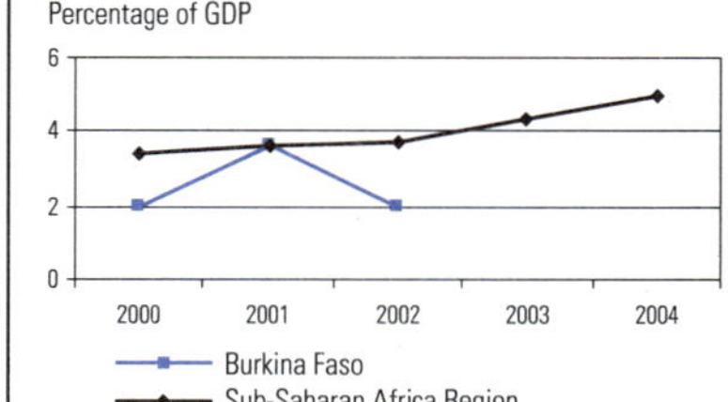

Notes: Figures in italics are for years other than those specified. .. indicates data are not available. C = competition; GDP = gross domestic product; GNI = gross national income; ICT = information and communication technology; M = monopoly; MDG = Millennium Development Goal; P = partial competition; and PCs = personal computers.

a. Outgoing and incoming.

b. Millennium Development Goal indicators 47, 48a, and 48b.

Sources: Economic and social context: UNDP and World Bank; ICT sector structure: ITU, WEF; ICT sector performance: Global Insight/WITSA, ITU, Netcraft, UNDESA, UNPAN, and World Bank. Produced by the Global Information and Communication Technologies Department and the Development Economics Data Group. For complete information, see Definitions and Data Sources.

Burundi

	Burundi 2000	Burundi 2004	Low-income group 2004	Sub-Saharan Africa Region 2004
Economic and social context				
Population, total (millions)	7	7	2,338	719
Urban population (% total population)	9	10	31	37
Poverty (% population below US$1 per day)	*54.6*	..	..	*46.4*
GNI per capita, Atlas method (current US$)	110	90	510	600
GDP growth, 1995–2000 and 2000–4 (%)	–0.3	2.7	5.4	3.9
Adult literacy rate (% ages 15 and over)	59	..	61	59
Primary, secondary, tertiary school enrollment (% gross)	32	*35*	*55*	*52*
ICT sector structure				
Separate telecommunications regulator	Yes	Yes		
Status of main fixed-line operator	*Public*	Public		
Level of competition: international long distance	*M*	C		
Level of competition: mobile	*C*	C		
Level of competition: Internet service provider	*C*	C		
Government prioritization of ICT (scale 1–7)	..	..	..	*4.4*
ICT sector performance				
Access				
Telephone main lines (per 1,000 people)	3	*3*	33	17
International voice traffic (minutes per person)[a]	1	..	*4*	..
Mobile subscribers (per 1,000 people)	2	*9*	48	86
Population covered by mobile telephony (%)	..	82	*43*	..
Internet users (per 1,000 people)	1	*2*	20	15
Personal computers (per 1,000 people)	1	*2*	*8*	*12*
Households with television (%)	11	*14*	*16*	*15*
Quality				
Telephone faults (per 100 main lines per year)	..	..	..	..
Broadband subscribers (per 1,000 people)	*0.0*	*0.0*	0.5	*0.1*
International Internet bandwidth (bits per person)	0	*1*	*3*	*4*
Affordability				
Price basket for fixed line (US$ per month, residential)	1.9	*4.5*	*6.6*	*8.5*
Price basket for mobile (US$ per month)	..	11.6	11.6	13.5
Price basket for Internet (US$ per month)	..	*58.6*	*45.5*	*54.8*
Price of call to United States (US$ per 3 minutes)	*7.35*	2.45	1.95	2.43
Institutional efficiency and sustainability				
Total telecommunications revenue (% GDP)	2.0	..	2.3	5.0
Total telephone subscribers per employee	50	*98*	*89*	*144*
Total telecommunications investment (% revenue)	179.3	..	*27.8*	*27.8*
ICT applications				
ICT expenditure (% GDP)	..	..	*4.1*	..
E-government readiness index (scale 0–1)	..	0.04	0.12	0.11
Secure Internet servers (per 1 million people)	..	0.3	0.3	1.9
Schools connected to the Internet (%)	..	..	..	..

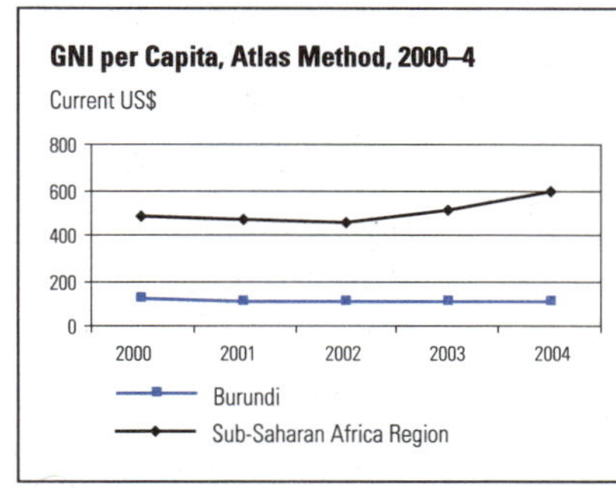

GNI per Capita, Atlas Method, 2000–4

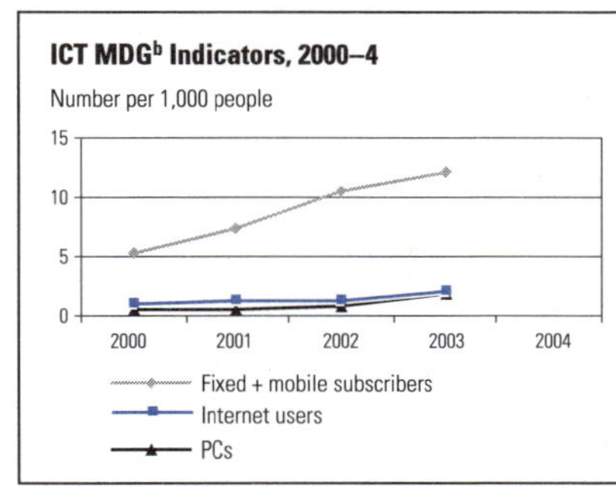

ICT MDG[b] Indicators, 2000–4

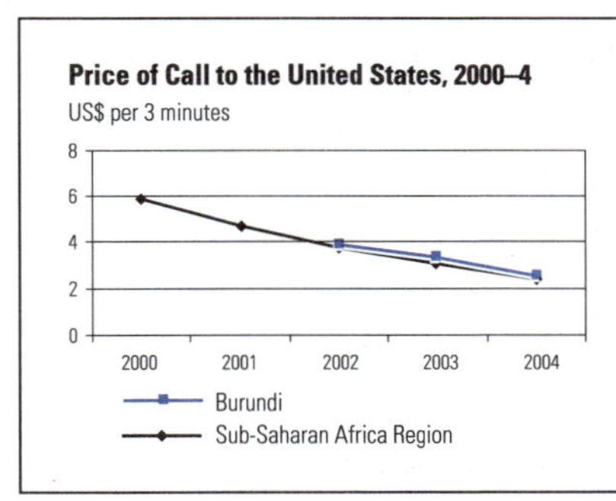

Price of Call to the United States, 2000–4

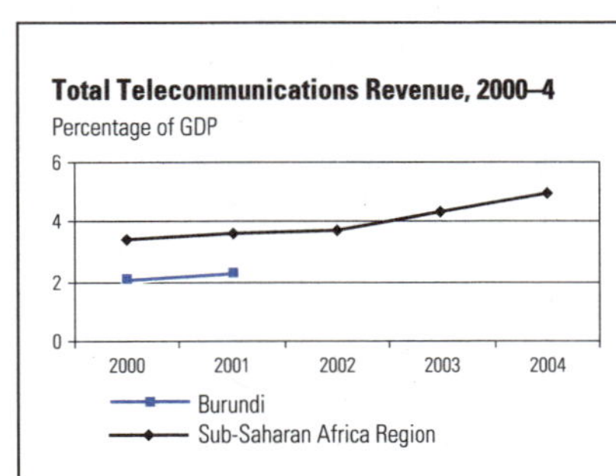

Total Telecommunications Revenue, 2000–4

Notes: Figures in italics are for years other than those specified. .. indicates data are not available. C = competition; GDP = gross domestic product; GNI = gross national income; ICT = information and communication technology; M = monopoly; MDG = Millennium Development Goal; P = partial competition; and PCs = personal computers.

a. Outgoing and incoming.

b. Millennium Development Goal indicators 47, 48a, and 48b.

Sources: Economic and social context: UNDP and World Bank; ICT sector structure: ITU, WEF; ICT sector performance: Global Insight/WITSA, ITU, Netcraft, UNDESA, UNPAN, and World Bank. Produced by the Global Information and Communication Technologies Department and the Development Economics Data Group. For complete information, see Definitions and Data Sources.

Cambodia

	Cambodia 2000	Cambodia 2004	Low-income group 2004	East Asia & Pacific Region 2004
Economic and social context				
Population, total (millions)	13	14	2,338	1,870
Urban population (% total population)	17	19	31	41
Poverty (% population below US$1 per day)	*34.1*	..	..	*14.9*
GNI per capita, Atlas method (current US$)	280	320	510	1,280
GDP growth, 1995–2000 and 2000–4 (%)	6.7	5.6	5.4	7.5
Adult literacy rate (% ages 15 and over)	..	74	61	90
Primary, secondary, tertiary school enrollment (% gross)	54	*59*	*55*	*68*
ICT sector structure				
Separate telecommunications regulator	..	..		
Status of main fixed-line operator	*Public*	Public		
Level of competition: international long distance	*P*	P		
Level of competition: mobile	*P*	P		
Level of competition: Internet service provider	*P*	P		
Government prioritization of ICT (scale 1–7)	..	..	..	4.8
ICT sector performance				
Access				
Telephone main lines (per 1,000 people)	2	3	33	194
International voice traffic (minutes per person)[a]	3	*3*	*4*	..
Mobile subscribers (per 1,000 people)	10	63	48	248
Population covered by mobile telephony (%)	80	*87*	*43*	*73*
Internet users (per 1,000 people)	0	*3*	20	75
Personal computers (per 1,000 people)	1	*2*	*8*	37
Households with television (%)	*23*	..	*16*	*80*
Quality				
Telephone faults (per 100 main lines per year)	*7.2*	..	..	..
Broadband subscribers (per 1,000 people)	0.0	*0.0*	0.5	13.4
International Internet bandwidth (bits per person)	0	*2*	*3*	52
Affordability				
Price basket for fixed line (US$ per month, residential)	16.7	*9.3*	*6.6*	*4.5*
Price basket for mobile (US$ per month)	..	4.0	11.6	5.1
Price basket for Internet (US$ per month)	..	*49.7*	*45.5*	*19.9*
Price of call to United States (US$ per 3 minutes)	6.00	*2.94*	1.95	1.20
Institutional efficiency and sustainability				
Total telecommunications revenue (% GDP)	0.6	2.7	2.3	3.6
Total telephone subscribers per employee	241	..	*89*	..
Total telecommunications investment (% revenue)	..	*29.1*	*27.8*	*31.0*
ICT applications				
ICT expenditure (% GDP)	..	..	*4.1*	*5.0*
E-government readiness index (scale 0–1)	..	0.21	0.12	0.21
Secure Internet servers (per 1 million people)	*0.2*	0.1	0.3	0.6
Schools connected to the Internet (%)	..	..	..	..

GNI per Capita, Atlas Method, 2000–4
Current US$

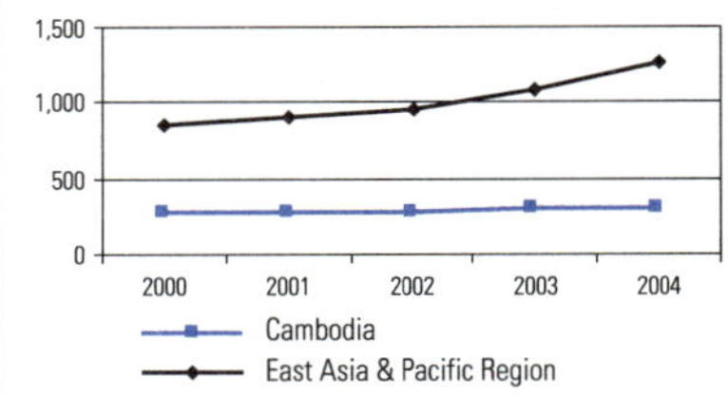

ICT MDG[b] Indicators, 2000–4
Number per 1,000 people

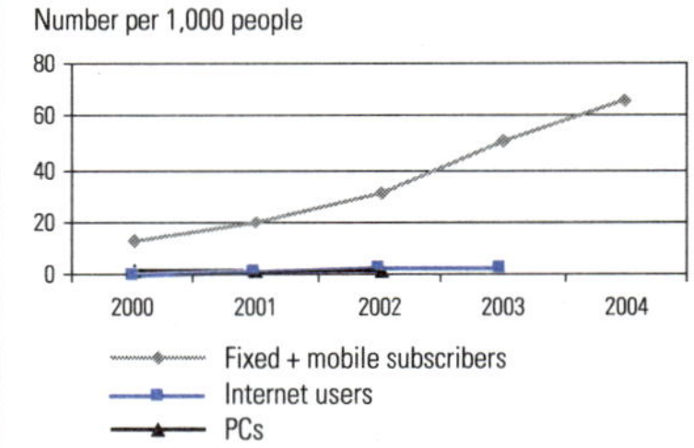

Price of Call to the United States, 2000–4
US$ per 3 minutes

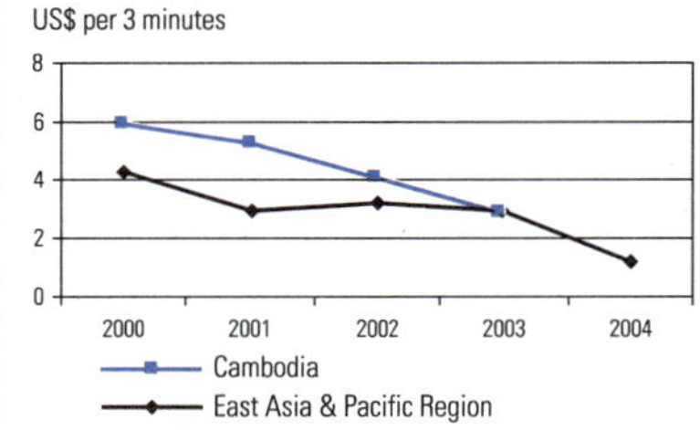

Total Telecommunications Revenue, 2000–4
Percentage of GDP

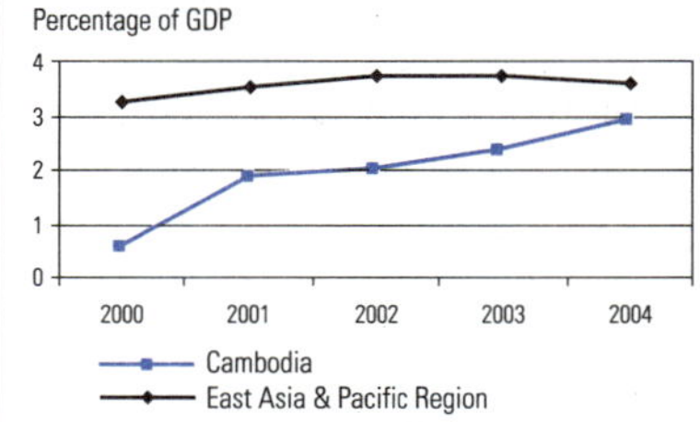

Notes: Figures in italics are for years other than those specified. .. indicates data are not available. C = competition; GDP = gross domestic product; GNI = gross national income; ICT = information and communication technology; M = monopoly; MDG = Millennium Development Goal; P = partial competition; and PCs = personal computers.

a. Outgoing and incoming.
b. Millennium Development Goal indicators 47, 48a, and 48b.

Sources: Economic and social context: UNDP and World Bank; ICT sector structure: ITU, WEF; ICT sector performance: Global Insight/WITSA, ITU, Netcraft, UNDESA, UNPAN, and World Bank. Produced by the Global Information and Communication Technologies Department and the Development Economics Data Group. For complete information, see Definitions and Data Sources.

Cameroon

	Cameroon 2000	Cameroon 2004	Low-income group 2004	Sub-Saharan Africa Region 2004
Economic and social context				
Population, total (millions)	15	16	2,338	719
Urban population (% total population)	49	52	31	37
Poverty (% population below US$1 per day)	*32.5*	*17.1*	..	*46.4*
GNI per capita, Atlas method (current US$)	570	800	510	600
GDP growth, 1995–2000 and 2000–4 (%)	4.8	4.6	5.4	3.9
Adult literacy rate (% ages 15 and over)	*68*	..	61	59
Primary, secondary, tertiary school enrollment (% gross)	56	*55*	*55*	*52*
ICT sector structure				
Separate telecommunications regulator	Yes	Yes		
Status of main fixed-line operator	*Public*	Public		
Level of competition: international long distance	*M*	M		
Level of competition: mobile	*P*	C		
Level of competition: Internet service provider	*C*	C		
Government prioritization of ICT (scale 1–7)	..	*3.0*	..	*4.4*
ICT sector performance				
Access				
Telephone main lines (per 1,000 people)	6	7	33	17
International voice traffic (minutes per person)[a]	6	..	*4*	..
Mobile subscribers (per 1,000 people)	7	93	48	86
Population covered by mobile telephony (%)	30	70	*43*	..
Internet users (per 1,000 people)	3	9	20	15
Personal computers (per 1,000 people)	3	9	*8*	*12*
Households with television (%)	17	*18*	*16*	*15*
Quality				
Telephone faults (per 100 main lines per year)	..	..	..	..
Broadband subscribers (per 1,000 people)	0.0	0.0	0.5	*0.1*
International Internet bandwidth (bits per person)	*0*	3	*3*	*4*
Affordability				
Price basket for fixed line (US$ per month, residential)	6.5	*6.7*	*6.6*	*8.5*
Price basket for mobile (US$ per month)	..	16.6	11.6	13.5
Price basket for Internet (US$ per month)	..	*51.7*	*45.5*	*54.8*
Price of call to United States (US$ per 3 minutes)	*3.25*	..	1.95	2.43
Institutional efficiency and sustainability				
Total telecommunications revenue (% GDP)	2.7	3.2	2.3	5.0
Total telephone subscribers per employee	90	..	*89*	*144*
Total telecommunications investment (% revenue)	38.3	*19.5*	*27.8*	*27.8*
ICT applications				
ICT expenditure (% GDP)	4.7	*4.9*	*4.1*	..
E-government readiness index (scale 0–1)	..	0.12	0.12	0.11
Secure Internet servers (per 1 million people)	..	0.2	0.3	1.9
Schools connected to the Internet (%)	..	..	..	..

GNI per Capita, Atlas Method, 2000–4
Current US$

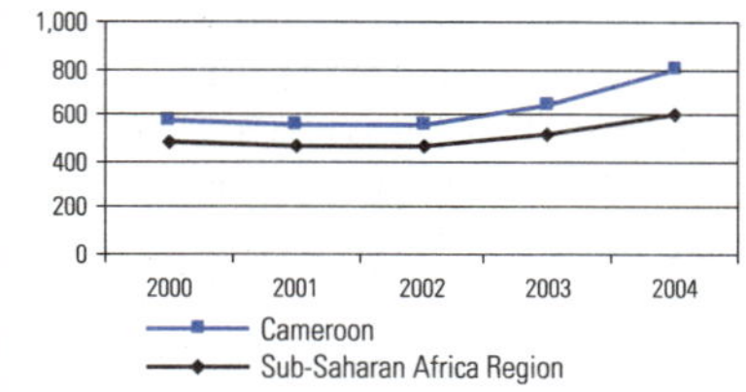

ICT MDG[b] Indicators, 2000–4
Number per 1,000 people

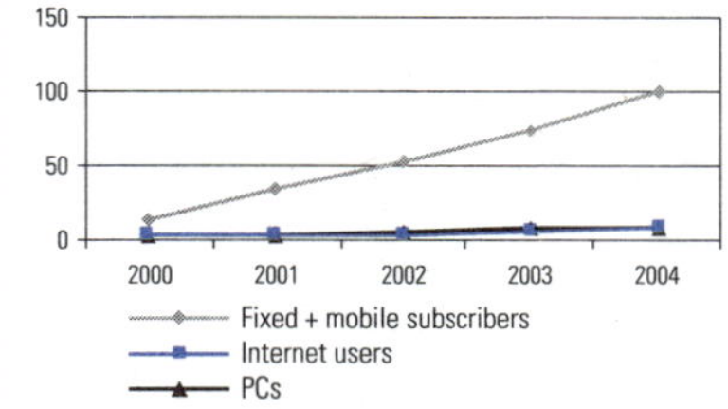

Price of Call to the United States, 2000–4
US$ per 3 minutes

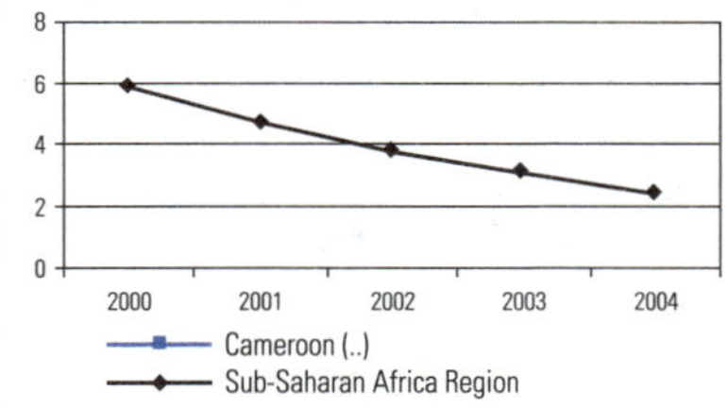

Total Telecommunications Revenue, 2000–4
Percentage of GDP

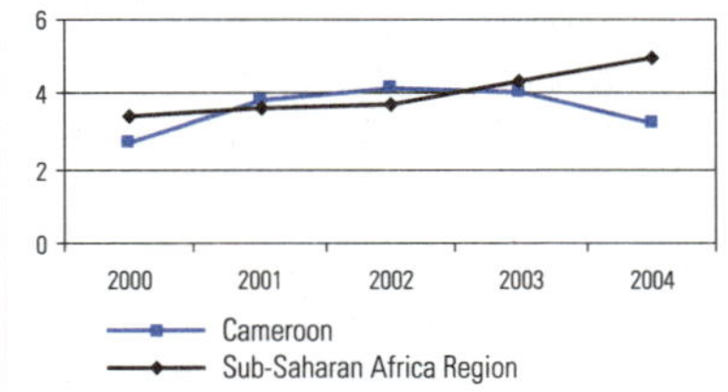

Notes: Figures in italics are for years other than those specified. .. indicates data are not available. C = competition; GDP = gross domestic product; GNI = gross national income; ICT = information and communication technology; M = monopoly; MDG = Millennium Development Goal; P = partial competition; and PCs = personal computers.

a. Outgoing and incoming.

b. Millennium Development Goal indicators 47, 48a, and 48b.

Sources: Economic and social context: UNDP and World Bank; ICT sector structure: ITU, WEF; ICT sector performance: Global Insight/WITSA, ITU, Netcraft, UNDESA, UNPAN, and World Bank. Produced by the Global Information and Communication Technologies Department and the Development Economics Data Group. For complete information, see Definitions and Data Sources.

Canada

	Canada 2000	Canada 2004	High-income group 2004
Economic and social context			
Population, total (millions)	31	32	1,001
Urban population (% total population)	79	81	77
Poverty (% population below US$1 per day)	..	..	..
GNI per capita, Atlas method (current US$)	21,820	28,390	32,040
GDP growth, 1995–2000 and 2000–4 (%)	4.3	2.5	2.0
Adult literacy rate (% ages 15 and over)	..	..	..
Primary, secondary, tertiary school enrollment (% gross)	95	..	*93*
ICT sector structure			
Separate telecommunications regulator	Yes	Yes	
Status of main fixed-line operator	*Private*	*Private*	
Level of competition: international long distance	*C*	C	
Level of competition: mobile	*P*	C	
Level of competition: Internet service provider	*C*	C	
Government prioritization of ICT (scale 1–7)	..	4.8	4.9
ICT sector performance			
Access			
Telephone main lines (per 1,000 people)	661	615	558
International voice traffic (minutes per person)[a]	439	..	..
Mobile subscribers (per 1,000 people)	284	451	767
Population covered by mobile telephony (%)	90	*93*	98
Internet users (per 1,000 people)	422	*544*	*480*
Personal computers (per 1,000 people)	419	*588*	*504*
Households with television (%)	99	*99*	*98*
Quality			
Telephone faults (per 100 main lines per year)	2.0	1.0	..
Broadband subscribers (per 1,000 people)	45.9	164.7	126.2
International Internet bandwidth (bits per person)	1,133	6,818	4,718
Affordability			
Price basket for fixed line (US$ per month, residential)	13.6	16.1	25.8
Price basket for mobile (US$ per month)	..	6.7	17.8
Price basket for Internet (US$ per month)	..	*12.7*	*20.9*
Price of call to United States (US$ per 3 minutes)	..	..	0.77
Institutional efficiency and sustainability			
Total telecommunications revenue (% GDP)	2.9	2.7	2.9
Total telephone subscribers per employee	280	*299*	*485*
Total telecommunications investment (% revenue)	23.9	16.1	*12.3*
ICT applications			
ICT expenditure (% GDP)	6.2	*5.8*	*7.1*
E-government readiness index (scale 0–1)	..	0.87	0.70
Secure Internet servers (per 1 million people)	*162.6*	484.0	311.4
Schools connected to the Internet (%)	..	*98*	*99*

GNI per Capita, Atlas Method, 2000–4
Current US$

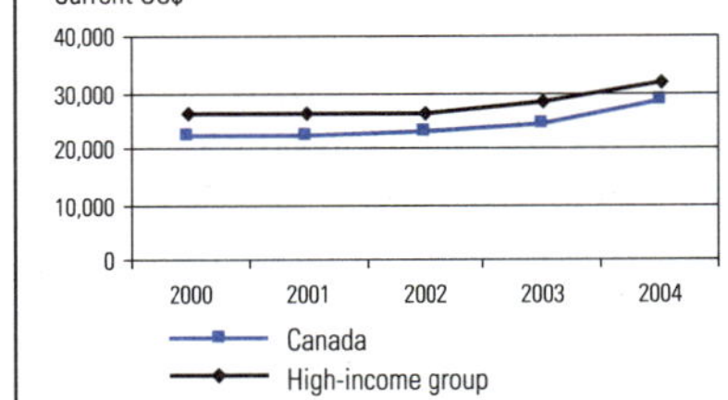

ICT MDG[b] Indicators, 2000–4
Number per 1,000 people

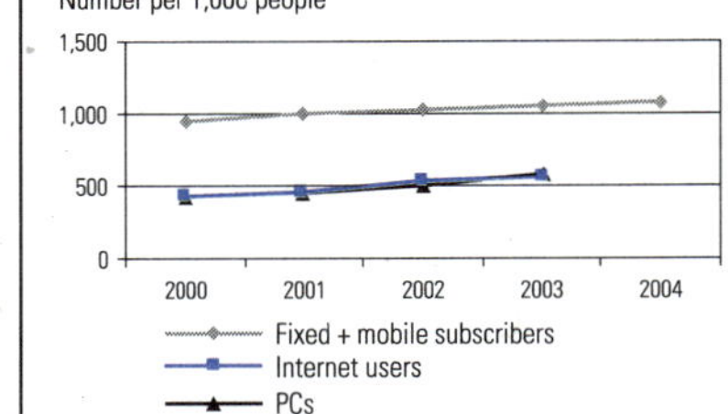

Price of Call to the United States, 2000–4
US$ per 3 minutes

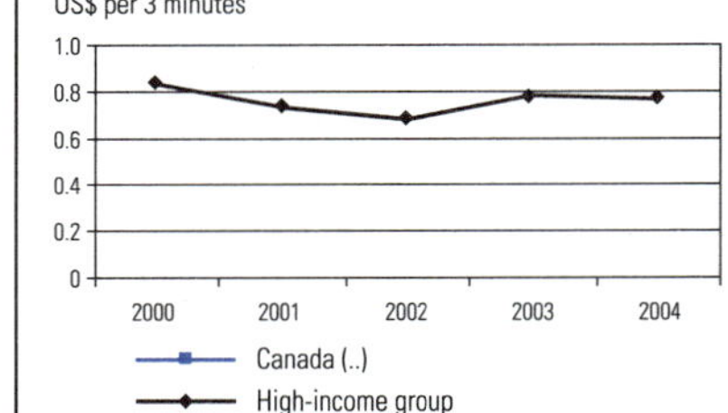

Total Telecommunications Revenue, 2000–4
Percentage of GDP

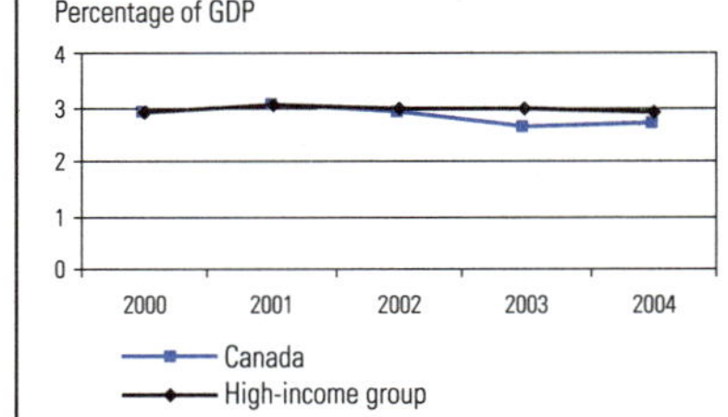

Notes: Figures in italics are for years other than those specified. .. indicates data are not available. C = competition; GDP = gross domestic product; GNI = gross national income; ICT = information and communication technology; M = monopoly; MDG = Millennium Development Goal; P = partial competition; and PCs = personal computers.

a. Outgoing and incoming.

b. Millennium Development Goal indicators 47, 48a, and 48b.

Sources: Economic and social context: UNDP and World Bank; ICT sector structure: ITU, WEF; ICT sector performance: Global Insight/WITSA, ITU, Netcraft, UNDESA, UNPAN, and World Bank. Produced by the Global Information and Communication Technologies Department and the Development Economics Data Group. For complete information, see Definitions and Data Sources.

Central African Republic

	Central African Republic 2000	Central African Republic 2004	Low-income group 2004	Sub-Saharan Africa Region 2004
Economic and social context				
Population, total (millions)	4	4	2,338	719
Urban population (% total population)	41	43	31	37
Poverty (% population below US$1 per day)	..	..	..	*46.4*
GNI per capita, Atlas method (current US$)	280	310	510	600
GDP growth, 1995–2000 and 2000–4 (%)	3.0	–1.4	5.4	3.9
Adult literacy rate (% ages 15 and over)	49	..	61	59
Primary, secondary, tertiary school enrollment (% gross)	35	*31*	*55*	*52*
ICT sector structure				
Separate telecommunications regulator	Yes	Yes		
Status of main fixed-line operator	*Mixed*	Mixed		
Level of competition: international long distance	*M*	M		
Level of competition: mobile	*C*	C		
Level of competition: Internet service provider	..	..		
Government prioritization of ICT (scale 1–7)	..	..	..	*4.4*
ICT sector performance				
Access				
Telephone main lines (per 1,000 people)	3	*2*	33	17
International voice traffic (minutes per person)[a]	2	..	*4*	..
Mobile subscribers (per 1,000 people)	1	*10*	48	86
Population covered by mobile telephony (%)	..	..	*43*	..
Internet users (per 1,000 people)	1	*2*	20	15
Personal computers (per 1,000 people)	2	*2*	*8*	*12*
Households with television (%)	2	*2*	*16*	*15*
Quality				
Telephone faults (per 100 main lines per year)	..	..	..	..
Broadband subscribers (per 1,000 people)	*0.0*	*0.0*	0.5	*0.1*
International Internet bandwidth (bits per person)	..	*0*	*3*	*4*
Affordability				
Price basket for fixed line (US$ per month, residential)	32.1	*32.8*	*6.6*	*8.5*
Price basket for mobile (US$ per month)	..	12.7	11.6	13.5
Price basket for Internet (US$ per month)	..	*175.0*	*45.5*	*54.8*
Price of call to United States (US$ per 3 minutes)	13.31	1.99	1.95	2.43
Institutional efficiency and sustainability				
Total telecommunications revenue (% GDP)	1.1	*1.0*	2.3	5.0
Total telephone subscribers per employee	35	*54*	*89*	*144*
Total telecommunications investment (% revenue)	1.0	*1.1*	*27.8*	*27.8*
ICT applications				
ICT expenditure (% GDP)	..	..	*4.1*	..
E-government readiness index (scale 0–1)	..	0.00	0.12	0.11
Secure Internet servers (per 1 million people)	..	..	0.3	1.9
Schools connected to the Internet (%)	..	..	..	..

GNI per Capita, Atlas Method, 2000–4
Current US$

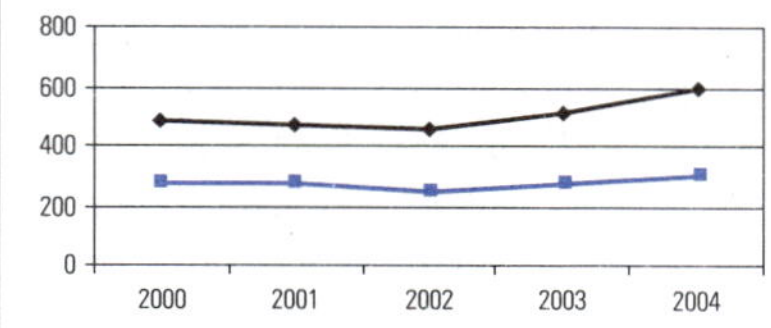

ICT MDG[b] Indicators, 2000–4
Number per 1,000 people

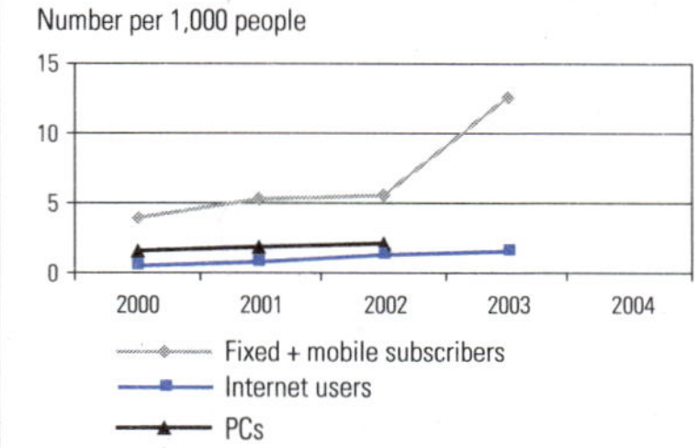

Price of Call to the United States, 2000–4
US$ per 3 minutes

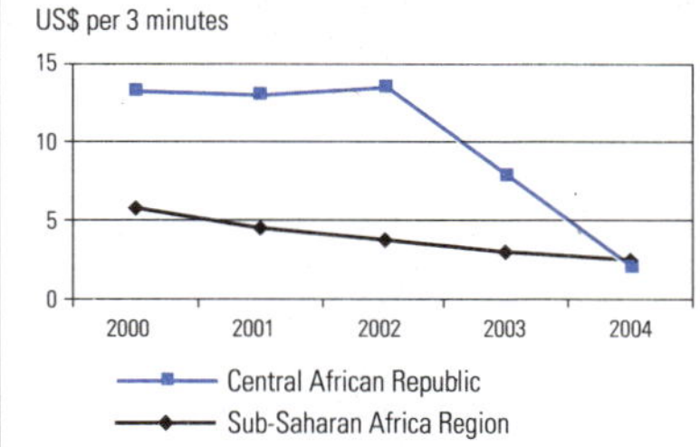

Total Telecommunications Revenue, 2000–4
Percentage of GDP

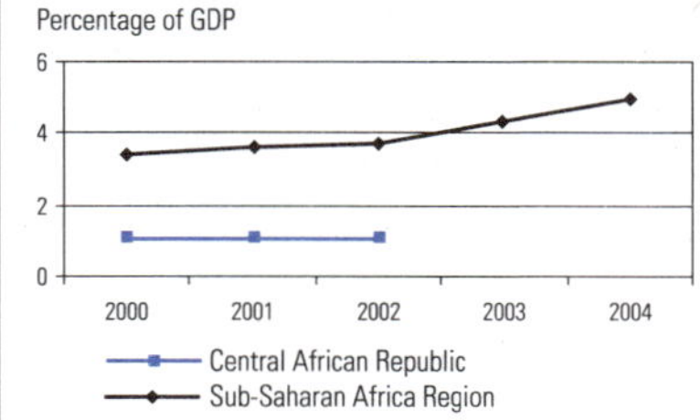

Notes: Figures in italics are for years other than those specified. .. indicates data are not available. C = competition; GDP = gross domestic product; GNI = gross national income; ICT = information and communication technology; M = monopoly; MDG = Millennium Development Goal; P = partial competition; and PCs = personal computers.

a. Outgoing and incoming.
b. Millennium Development Goal indicators 47, 48a, and 48b.

Sources: Economic and social context: UNDP and World Bank; ICT sector structure: ITU, WEF; ICT sector performance: Global Insight/WITSA, ITU, Netcraft, UNDESA, UNPAN, and World Bank. Produced by the Global Information and Communication Technologies Department and the Development Economics Data Group. For complete information, see Definitions and Data Sources.

Chad

	Chad 2000	Chad 2004	Low-income group 2004	Sub-Saharan Africa Region 2004
Economic and social context				
Population, total (millions)	8	9	2,338	719
Urban population (% total population)	24	25	31	37
Poverty (% population below US$1 per day)	..	..	..	*46.4*
GNI per capita, Atlas method (current US$)	190	260	510	600
GDP growth, 1995–2000 and 2000–4 (%)	2.6	14.3	5.4	3.9
Adult literacy rate (% ages 15 and over)	26	..	61	59
Primary, secondary, tertiary school enrollment (% gross)	35	*38*	*55*	*52*
ICT sector structure				
Separate telecommunications regulator	Yes	Yes		
Status of main fixed-line operator	*Public*	Public		
Level of competition: international long distance	*M*	M		
Level of competition: mobile	*P*	*C*		
Level of competition: Internet service provider	*M*	C		
Government prioritization of ICT (scale 1–7)	..	3.1	..	*4.4*
ICT sector performance				
Access				
Telephone main lines (per 1,000 people)	1	*2*	33	17
International voice traffic (minutes per person)[a]	1	..	*4*	..
Mobile subscribers (per 1,000 people)	1	*8*	48	86
Population covered by mobile telephony (%)	..	*8*	*43*	..
Internet users (per 1,000 people)	0	3	20	15
Personal computers (per 1,000 people)	1	*2*	*8*	*12*
Households with television (%)	2	*2*	*16*	*15*
Quality				
Telephone faults (per 100 main lines per year)	52.0	..	..	..
Broadband subscribers (per 1,000 people)	0.0	0.0	0.5	*0.1*
International Internet bandwidth (bits per person)	0	0	*3*	*4*
Affordability				
Price basket for fixed line (US$ per month, residential)	13.2	12.8	*6.6*	*8.5*
Price basket for mobile (US$ per month)	..	27.7	11.6	13.5
Price basket for Internet (US$ per month)	..	*68.9*	*45.5*	*54.8*
Price of call to United States (US$ per 3 minutes)	12.50	..	1.95	2.43
Institutional efficiency and sustainability				
Total telecommunications revenue (% GDP)	..	*4.0*	2.3	5.0
Total telephone subscribers per employee	31	..	*89*	*144*
Total telecommunications investment (% revenue)	..	..	*27.8*	*27.8*
ICT applications				
ICT expenditure (% GDP)	..	..	*4.1*	..
E-government readiness index (scale 0–1)	..	0.01	0.12	0.11
Secure Internet servers (per 1 million people)	..	..	0.3	1.9
Schools connected to the Internet (%)	..	..	..	..

GNI per Capita, Atlas Method, 2000–4

Current US$

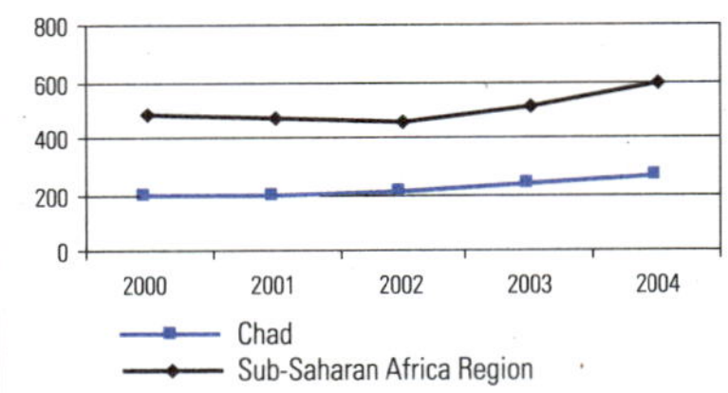

ICT MDG[b] Indicators, 2000–4

Number per 1,000 people

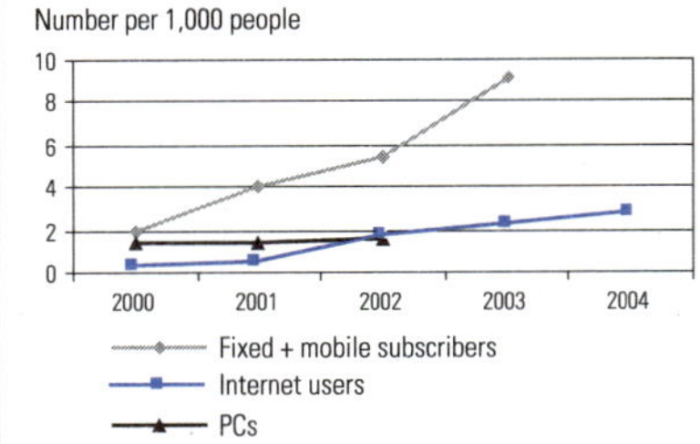

Price of Call to the United States, 2000–4

US$ per 3 minutes

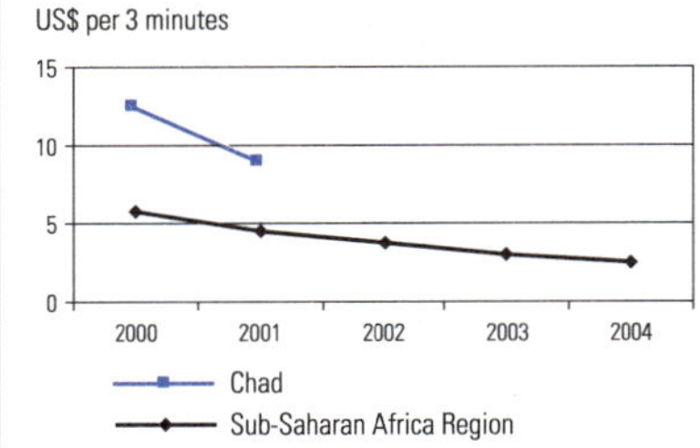

Total Telecommunications Revenue, 2000–4

Percentage of GDP

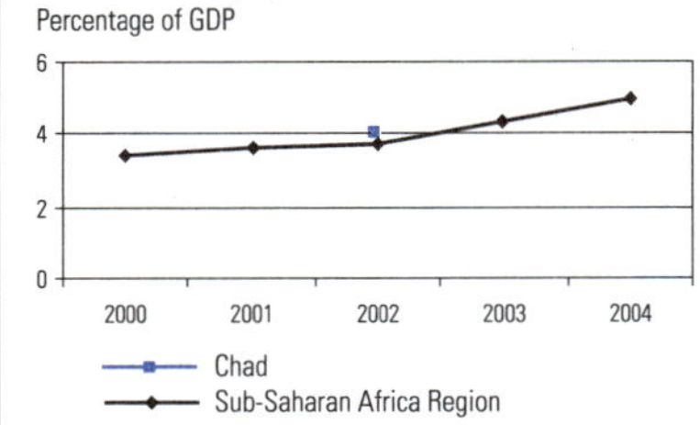

Notes: Figures in italics are for years other than those specified. .. indicates data are not available. C = competition; GDP = gross domestic product; GNI = gross national income; ICT = information and communication technology; M = monopoly; MDG = Millennium Development Goal; P = partial competition; and PCs = personal computers.

a. Outgoing and incoming.

b. Millennium Development Goal indicators 47, 48a, and 48b.

Sources: Economic and social context: UNDP and World Bank; ICT sector structure: ITU, WEF; ICT sector performance: Global Insight/WITSA, ITU, Netcraft, UNDESA, UNPAN, and World Bank. Produced by the Global Information and Communication Technologies Department and the Development Economics Data Group. For complete information, see Definitions and Data Sources.

Chile

	Chile 2000	Chile 2004	Upper-middle-income group 2004	Latin America & the Caribbean Region 2004
Economic and social context				
Population, total (millions)	15	16	576	541
Urban population (% total population)	86	87	72	77
Poverty (% population below US$1 per day)	<2	..	..	*9.5*
GNI per capita, Atlas method (current US$)	4,780	4,910	4,770	3,600
GDP growth, 1995–2000 and 2000–4 (%)	4.1	3.4	2.7	1.5
Adult literacy rate (% ages 15 and over)	..	*96*	94	89
Primary, secondary, tertiary school enrollment (% gross)	79	*82*	*80*	*82*
ICT sector structure				
Separate telecommunications regulator	..	..		
Status of main fixed-line operator	*Private*	Private		
Level of competition: international long distance	..	C		
Level of competition: mobile	..	C		
Level of competition: Internet service provider	..	C		
Government prioritization of ICT (scale 1–7)	..	4.8	4.1	3.5
ICT sector performance				
Access				
Telephone main lines (per 1,000 people)	217	208	*220*	181
International voice traffic (minutes per person)[a]	35	58	*39*	..
Mobile subscribers (per 1,000 people)	224	600	490	324
Population covered by mobile telephony (%)	99	99	*84*	*76*
Internet users (per 1,000 people)	167	263	*133*	*104*
Personal computers (per 1,000 people)	93	155	*99*	*75*
Households with television (%)	95	95	*92*	*88*
Quality				
Telephone faults (per 100 main lines per year)	25.0	40.0	*20.3*	..
Broadband subscribers (per 1,000 people)	0.5	30.0	*3.7*	*5.2*
International Internet bandwidth (bits per person)	12	796	176	165
Affordability				
Price basket for fixed line (US$ per month, residential)	15.7	16.4	*13.9*	*9.0*
Price basket for mobile (US$ per month)	..	17.0	11.1	9.1
Price basket for Internet (US$ per month)	..	*21.8*	*20.8*	*31.5*
Price of call to United States (US$ per 3 minutes)	2.45	..	1.03	0.90
Institutional efficiency and sustainability				
Total telecommunications revenue (% GDP)	3.5	3.8	3.4	3.3
Total telephone subscribers per employee	315	*567*	*402*	..
Total telecommunications investment (% revenue)	42.3	*23.2*	*18.6*	..
ICT applications				
ICT expenditure (% GDP)	6.0	*6.7*	*5.0*	*5.3*
E-government readiness index (scale 0–1)	..	0.88	0.49	0.39
Secure Internet servers (per 1 million people)	*9.2*	17.2	10.7	8.6
Schools connected to the Internet (%)	..	*62*	*60*	..

GNI per Capita, Atlas Method, 2000–4
Current US$

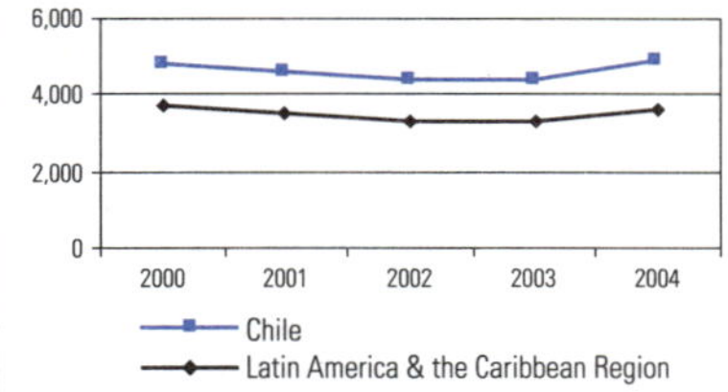

ICT MDG[b] Indicators, 2000–4
Number per 1,000 people

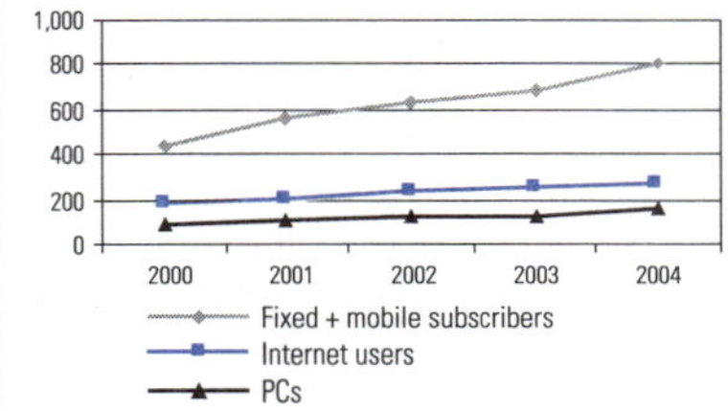

Price of Call to the United States, 2000–4
US$ per 3 minutes

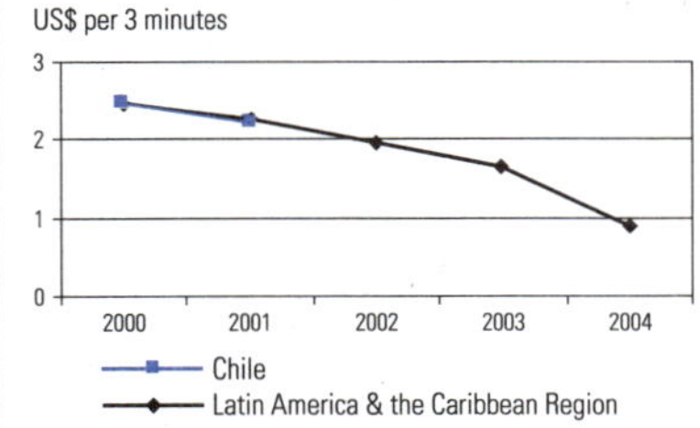

Total Telecommunications Revenue, 2000–4
Percentage of GDP

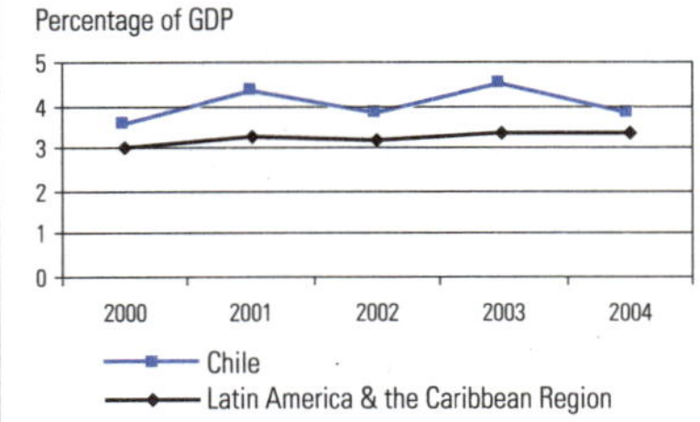

Notes: Figures in italics are for years other than those specified. .. indicates data are not available. C = competition; GDP = gross domestic product; GNI = gross national income; ICT = information and communication technology; M = monopoly; MDG = Millennium Development Goal; P = partial competition; and PCs = personal computers.

a. Outgoing and incoming.

b. Millennium Development Goal indicators 47, 48a, and 48b.

Sources: Economic and social context: UNDP and World Bank; ICT sector structure: ITU, WEF; ICT sector performance: Global Insight/WITSA, ITU, Netcraft, UNDESA, UNPAN, and World Bank. Produced by the Global Information and Communication Technologies Department and the Development Economics Data Group. For complete information, see Definitions and Data Sources.

China

	China 2000	China 2004	Lower-middle-income group 2004	East Asia & Pacific Region 2004
Economic and social context				
Population, total (millions)	1,263	1,296	2,430	1,870
Urban population (% total population)	36	40	49	41
Poverty (% population below US$1 per day)	*17.8*	*16.6*	..	*14.9*
GNI per capita, Atlas method (current US$)	840	1,290	1,580	1,280
GDP growth, 1995–2000 and 2000–4 (%)	8.2	8.7	5.7	7.5
Adult literacy rate (% ages 15 and over)	91	..	90	90
Primary, secondary, tertiary school enrollment (% gross)	67	*69*	*72*	*68*
ICT sector structure				
Separate telecommunications regulator	..	..		
Status of main fixed-line operator	*Public*	Mixed		
Level of competition: international long distance	*P*	P		
Level of competition: mobile	*P*	P		
Level of competition: Internet service provider	*C*	*C*		
Government prioritization of ICT (scale 1–7)	..	4.7	3.8	4.8
ICT sector performance				
Access				
Telephone main lines (per 1,000 people)	115	241	192	194
International voice traffic (minutes per person)[a]	4	..	..	..
Mobile subscribers (per 1,000 people)	68	258	255	248
Population covered by mobile telephony (%)	..	*73*	*76*	*73*
Internet users (per 1,000 people)	18	73	70	75
Personal computers (per 1,000 people)	16	40	38	37
Households with television (%)	86	91	*89*	*80*
Quality				
Telephone faults (per 100 main lines per year)	..	..	..	..
Broadband subscribers (per 1,000 people)	0.0	16.5	12.6	13.4
International Internet bandwidth (bits per person)	2	57	58	52
Affordability				
Price basket for fixed line (US$ per month, residential)	11.9	3.6	*5.5*	*4.5*
Price basket for mobile (US$ per month)	..	3.7	8.9	5.1
Price basket for Internet (US$ per month)	..	*10.1*	*25.3*	*19.9*
Price of call to United States (US$ per 3 minutes)	6.67	2.90	1.45	1.20
Institutional efficiency and sustainability				
Total telecommunications revenue (% GDP)	3.6	3.7	3.6	3.6
Total telephone subscribers per employee	..	*655*	*195*	..
Total telecommunications investment (% revenue)	69.8	33.3	*25.5*	*31.0*
ICT applications				
ICT expenditure (% GDP)	4.1	*5.3*	*5.1*	*5.0*
E-government readiness index (scale 0–1)	..	0.41	0.26	0.21
Secure Internet servers (per 1 million people)	*0.1*	0.2	1.6	0.6
Schools connected to the Internet (%)	..	..	..	..

GNI per Capita, Atlas Method, 2000–4
Current US$

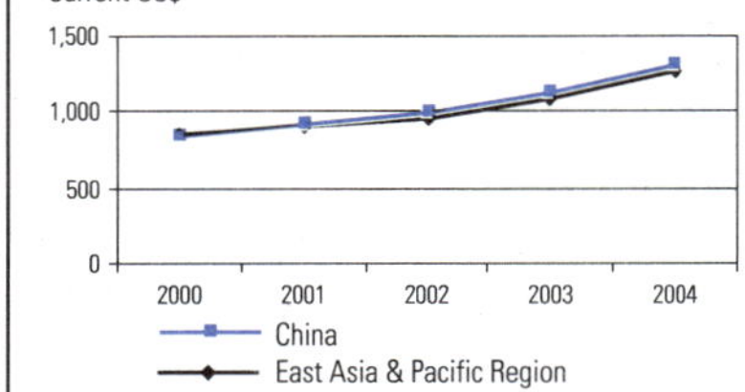

ICT MDG[b] Indicators, 2000–4
Number per 1,000 people

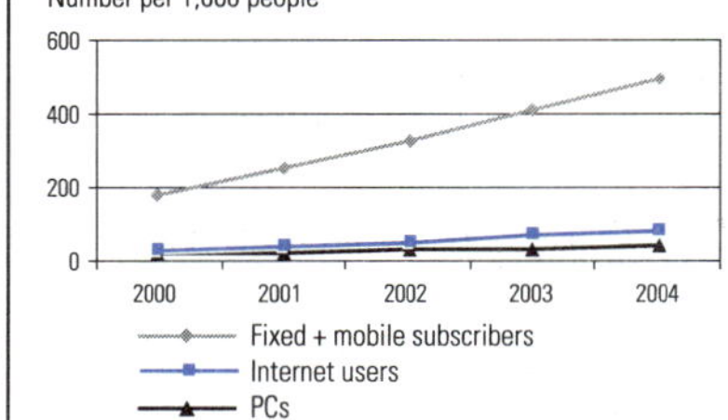

Price of Call to the United States, 2000–4
US$ per 3 minutes

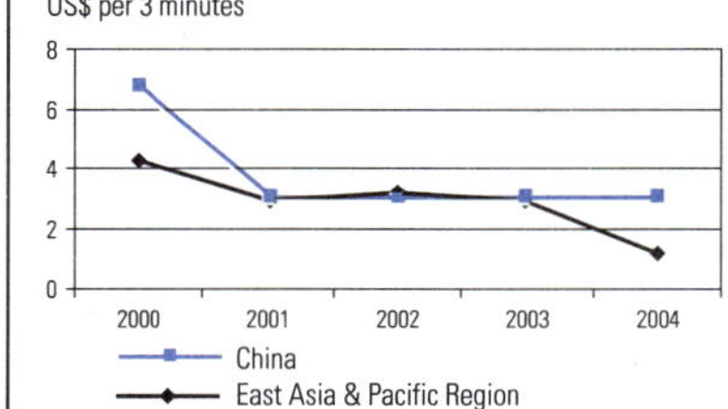

Total Telecommunications Revenue, 2000–4
Percentage of GDP

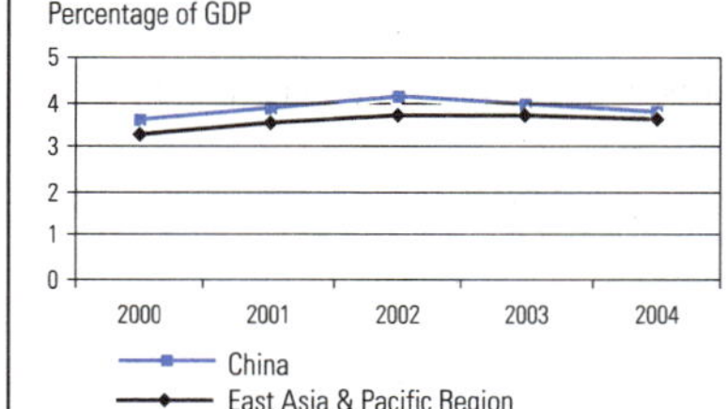

Notes: Figures in italics are for years other than those specified. .. indicates data are not available. C = competition; GDP = gross domestic product; GNI = gross national income; ICT = information and communication technology; M = monopoly; MDG = Millennium Development Goal; P = partial competition; and PCs = personal computers.

a. Outgoing and incoming.

b. Millennium Development Goal indicators 47, 48a, and 48b.

Sources: Economic and social context: UNDP and World Bank; ICT sector structure: ITU, WEF; ICT sector performance: Global Insight/WITSA, ITU, Netcraft, UNDESA, UNPAN, and World Bank. Produced by the Global Information and Communication Technologies Department and the Development Economics Data Group. For complete information, see Definitions and Data Sources.

Colombia

	Colombia 2000	Colombia 2004	Lower-middle-income group 2004	Latin America & the Caribbean Region 2004
Economic and social context				
Population, total (millions)	42	45	2,430	541
Urban population (% total population)	75	77	49	77
Poverty (% population below US$1 per day)	*8.2*	..	..	*9.5*
GNI per capita, Atlas method (current US$)	2,050	2,000	1,580	3,600
GDP growth, 1995–2000 and 2000–4 (%)	0.6	2.9	5.7	1.5
Adult literacy rate (% ages 15 and over)	..	*94*	90	89
Primary, secondary, tertiary school enrollment (% gross)	71	*71*	*72*	*82*
ICT sector structure				
Separate telecommunications regulator	Yes	Yes		
Status of main fixed-line operator	*Public*	Public		
Level of competition: international long distance	*C*	C		
Level of competition: mobile	*P*	P		
Level of competition: Internet service provider	*C*	C		
Government prioritization of ICT (scale 1–7)	..	4.0	3.8	3.5
ICT sector performance				
Access				
Telephone main lines (per 1,000 people)	170	170	192	181
International voice traffic (minutes per person)[a]	*8*	..	..	..
Mobile subscribers (per 1,000 people)	53	230	255	324
Population covered by mobile telephony (%)	..	*74*	*76*	*76*
Internet users (per 1,000 people)	21	84	70	*104*
Personal computers (per 1,000 people)	32	*41*	38	*75*
Households with television (%)	94	*92*	*89*	*88*
Quality				
Telephone faults (per 100 main lines per year)	44.3	*33.0*	..	..
Broadband subscribers (per 1,000 people)	0.2	2.8	12.6	*5.2*
International Internet bandwidth (bits per person)	15	*81*	58	165
Affordability				
Price basket for fixed line (US$ per month, residential)	*6.6*	*5.8*	*5.5*	*9.0*
Price basket for mobile (US$ per month)	..	9.1	8.9	9.1
Price basket for Internet (US$ per month)	..	*18.6*	*25.3*	*31.5*
Price of call to United States (US$ per 3 minutes)	*2.00*	..	1.45	0.90
Institutional efficiency and sustainability				
Total telecommunications revenue (% GDP)	2.7	4.9	3.6	3.3
Total telephone subscribers per employee	242	..	*195*	..
Total telecommunications investment (% revenue)	63.5	..	*25.5*	..
ICT applications				
ICT expenditure (% GDP)	8.5	*8.9*	*5.1*	*5.3*
E-government readiness index (scale 0–1)	..	0.64	0.26	0.39
Secure Internet servers (per 1 million people)	*1.6*	3.5	1.6	8.6
Schools connected to the Internet (%)	..	*50*	..	..

GNI per Capita, Atlas Method, 2000–4
Current US$

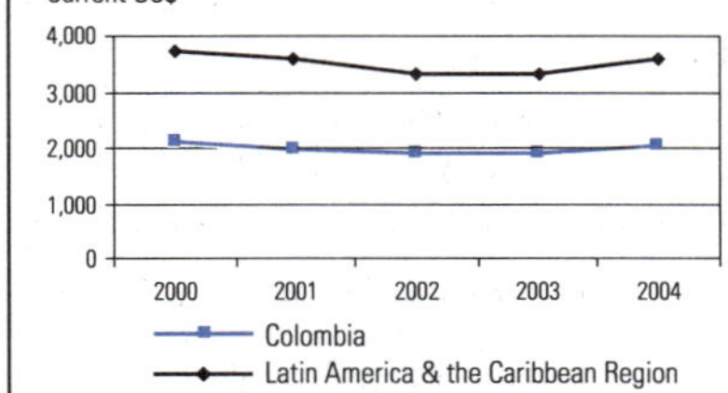

ICT MDG[b] Indicators, 2000–4
Number per 1,000 people

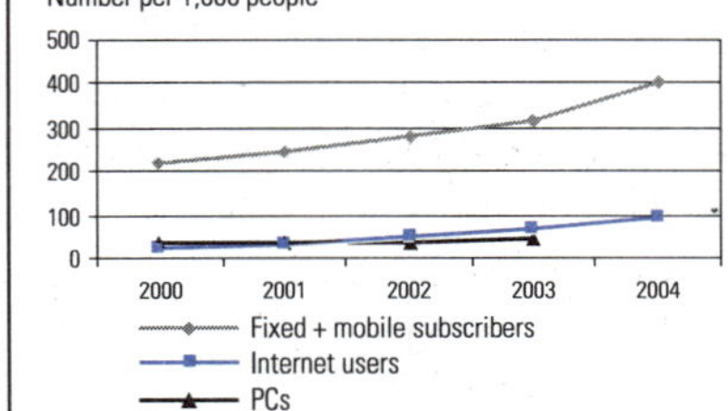

Price of Call to the United States, 2000–4
US$ per 3 minutes

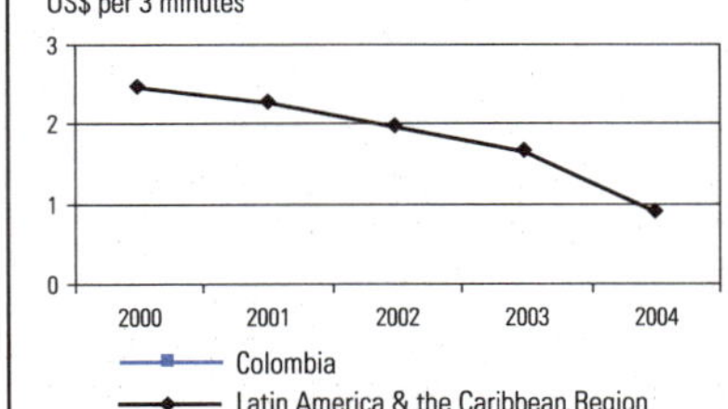

Total Telecommunications Revenue, 2000–4
Percentage of GDP

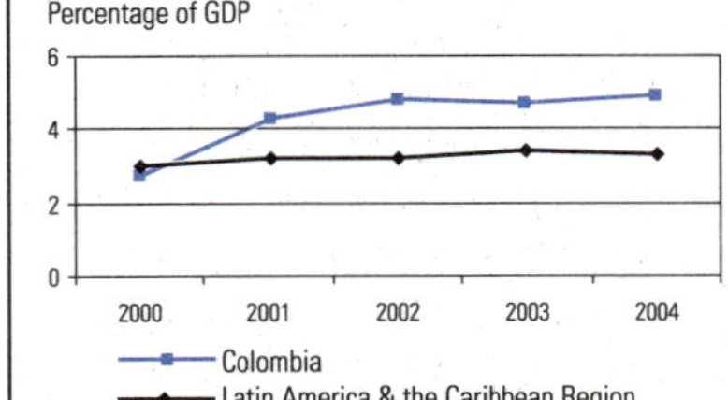

Notes: Figures in italics are for years other than those specified. .. indicates data are not available. C = competition; GDP = gross domestic product; GNI = gross national income; ICT = information and communication technology; M = monopoly; MDG = Millennium Development Goal; P = partial competition; and PCs = personal computers.

a. Outgoing and incoming.
b. Millennium Development Goal indicators 47, 48a, and 48b.

Sources: Economic and social context: UNDP and World Bank; ICT sector structure: ITU, WEF; ICT sector performance: Global Insight/WITSA, ITU, Netcraft, UNDESA, UNPAN, and World Bank. Produced by the Global Information and Communication Technologies Department and the Development Economics Data Group. For complete information, see Definitions and Data Sources.

Congo, Democratic Republic of

	Congo, Dem. Rep.of 2000	Congo, Dem. Rep.of 2004	Low-income group 2004	Sub-Saharan Africa Region 2004
Economic and social context				
Population, total (millions)	49	55	2,338	719
Urban population (% total population)	30	32	31	37
Poverty (% population below US$1 per day)	..	..	..	*46.4*
GNI per capita, Atlas method (current US$)	90	120	510	600
GDP growth, 1995–2000 and 2000–4 (%)	–3.9	3.5	5.4	3.9
Adult literacy rate (% ages 15 and over)	*65*	..	61	59
Primary, secondary, tertiary school enrollment (% gross)	*28*	..	*55*	*52*
ICT sector structure				
Separate telecommunications regulator	No	Yes		
Status of main fixed-line operator	*Public*	Public		
Level of competition: international long distance	*C*	P		
Level of competition: mobile	*C*	C		
Level of competition: Internet service provider	*C*	C		
Government prioritization of ICT (scale 1–7)	..	..	..	*4.4*
ICT sector performance				
Access				
Telephone main lines (per 1,000 people)	0	0	33	17
International voice traffic (minutes per person)[a]	..	..	*4*	..
Mobile subscribers (per 1,000 people)	0	37	48	86
Population covered by mobile telephony (%)	..	*55*	*43*	..
Internet users (per 1,000 people)	0	2	20	15
Personal computers (per 1,000 people)	..	..	*8*	*12*
Households with television (%)	1	*2*	*16*	*15*
Quality				
Telephone faults (per 100 main lines per year)	..	..	..	..
Broadband subscribers (per 1,000 people)	0.0	0.0	0.5	*0.1*
International Internet bandwidth (bits per person)	0	*0*	*3*	*4*
Affordability				
Price basket for fixed line (US$ per month, residential)	..	..	*6.6*	*8.5*
Price basket for mobile (US$ per month)	..	10.4	11.6	13.5
Price basket for Internet (US$ per month)	..	*74.0*	*45.5*	*54.8*
Price of call to United States (US$ per 3 minutes)	..	..	1.95	2.43
Institutional efficiency and sustainability				
Total telecommunications revenue (% GDP)	..	*4.6*	2.3	5.0
Total telephone subscribers per employee	..	..	*89*	*144*
Total telecommunications investment (% revenue)	..	..	*27.8*	*27.8*
ICT applications				
ICT expenditure (% GDP)	..	..	*4.1*	..
E-government readiness index (scale 0–1)	..	0.05	0.12	0.11
Secure Internet servers (per 1 million people)	..	..	0.3	1.9
Schools connected to the Internet (%)	..	..	..	..

GNI per Capita, Atlas Method, 2000–4

Current US$

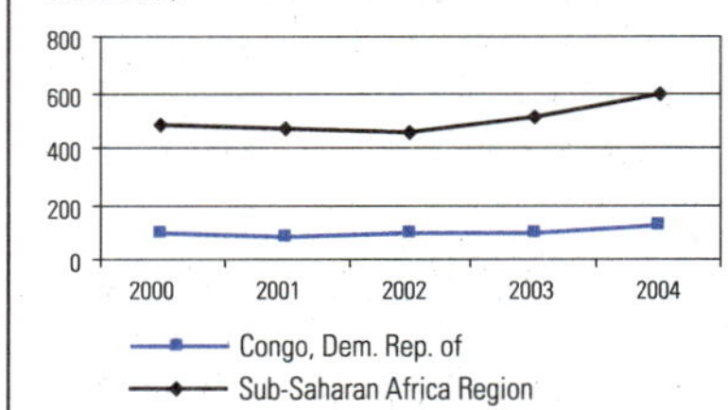

ICT MDG[b] Indicators, 2000–4

Number per 1,000 people

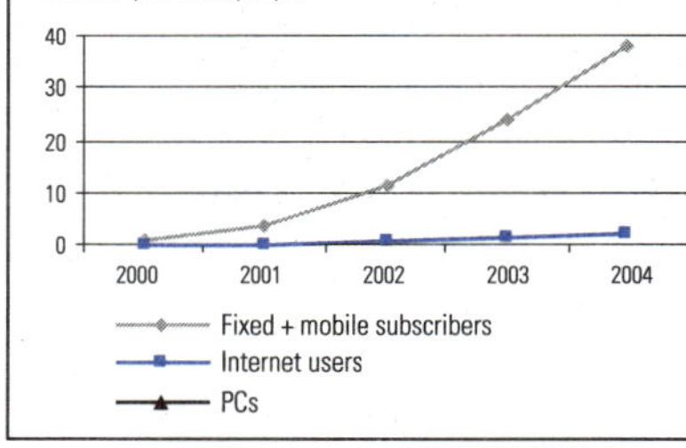

Price of Call to the United States, 2000–4

US$ per 3 minutes

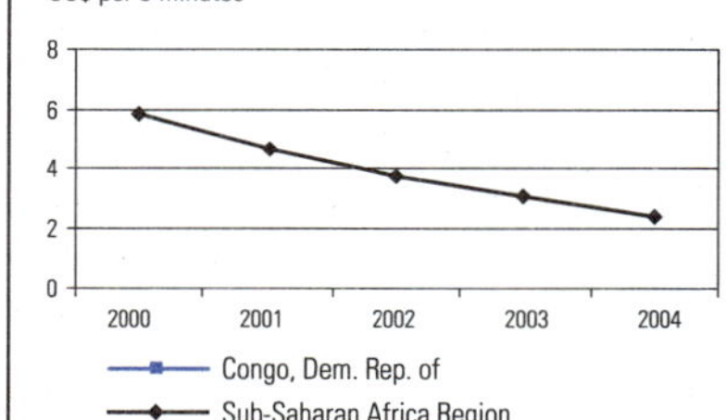

Total Telecommunications Revenue, 2000–4

Percentage of GDP

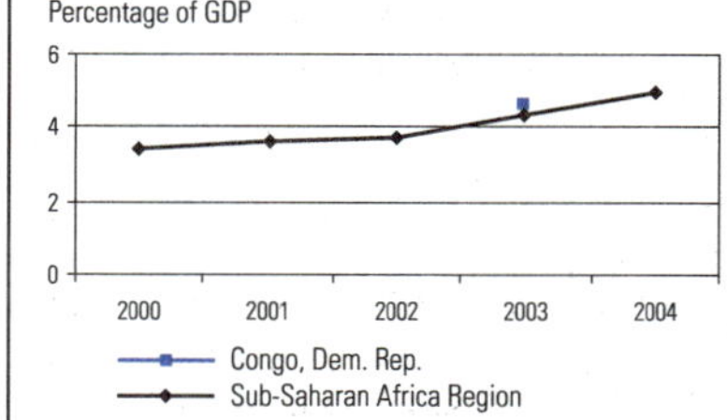

Notes: Figures in italics are for years other than those specified. .. indicates data are not available. C = competition; GDP = gross domestic product; GNI = gross national income; ICT = information and communication technology; M = monopoly; MDG = Millennium Development Goal; P = partial competition; and PCs = personal computers.

a. Outgoing and incoming.

b. Millennium Development Goal indicators 47, 48a, and 48b.

Sources: Economic and social context: UNDP and World Bank; ICT sector structure: ITU, WEF; ICT sector performance: Global Insight/WITSA, ITU, Netcraft, UNDESA, UNPAN, and World Bank. Produced by the Global Information and Communication Technologies Department and the Development Economics Data Group. For complete information, see Definitions and Data Sources.

Congo, Republic of

	Congo, Rep. of 2000	Congo, Rep. of 2004	Low-income group 2004	Sub-Saharan Africa Region 2004
Economic and social context				
Population, total (millions)	3	4	2,338	719
Urban population (% total population)	52	54	31	37
Poverty (% population below US$1 per day)	..	..	..	*46.4*
GNI per capita, Atlas method (current US$)	510	770	510	600
GDP growth, 1995–2000 and 2000–4 (%)	1.8	3.4	5.4	3.9
Adult literacy rate (% ages 15 and over)	..	*83*	61	59
Primary, secondary, tertiary school enrollment (% gross)	49	*47*	*55*	*52*
ICT sector structure				
Separate telecommunications regulator	..	..		
Status of main fixed-line operator	*Public*	Public		
Level of competition: international long distance	*C*	C		
Level of competition: mobile	*C*	C		
Level of competition: Internet service provider	..	..		
Government prioritization of ICT (scale 1–7)	..	..	..	*4.4*
ICT sector performance				
Access				
Telephone main lines (per 1,000 people)	6	*2*	33	17
International voice traffic (minutes per person)[a]	..	..	*4*	..
Mobile subscribers (per 1,000 people)	20	115	48	86
Population covered by mobile telephony (%)	..	*65*	*43*	..
Internet users (per 1,000 people)	0	*4*	20	15
Personal computers (per 1,000 people)	3	*4*	*8*	*12*
Households with television (%)	6	*6*	*16*	*15*
Quality				
Telephone faults (per 100 main lines per year)	..	..	..	..
Broadband subscribers (per 1,000 people)	0.0	*0.0*	0.5	*0.1*
International Internet bandwidth (bits per person)	*0*	*0*	*3*	*4*
Affordability				
Price basket for fixed line (US$ per month, residential)	..	..	*6.6*	*8.5*
Price basket for mobile (US$ per month)	..	17.5	11.6	13.5
Price basket for Internet (US$ per month)	..	*121.2*	*45.5*	*54.8*
Price of call to United States (US$ per 3 minutes)	..	5.39	1.95	2.43
Institutional efficiency and sustainability				
Total telecommunications revenue (% GDP)	..	..	2.3	5.0
Total telephone subscribers per employee	61	..	*89*	*144*
Total telecommunications investment (% revenue)	..	..	*27.8*	*27.8*
ICT applications				
ICT expenditure (% GDP)	..	..	*4.1*	..
E-government readiness index (scale 0–1)	..	0.15	0.12	0.11
Secure Internet servers (per 1 million people)	..	..	0.3	1.9
Schools connected to the Internet (%)	..	..	..	..

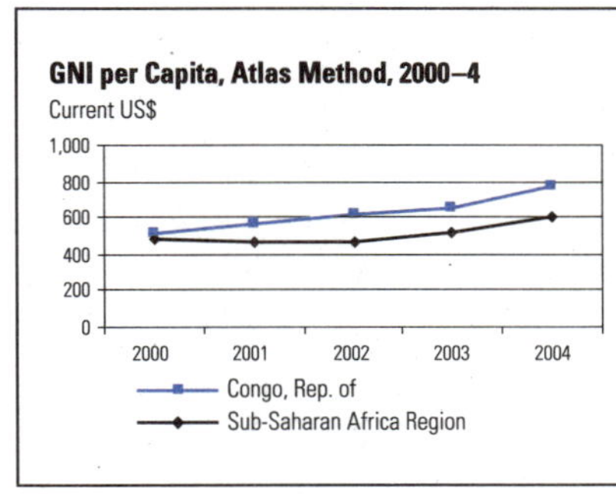

GNI per Capita, Atlas Method, 2000–4

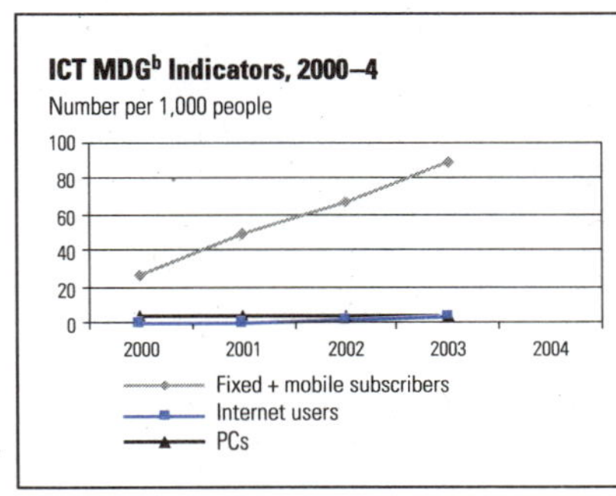

ICT MDG[b] Indicators, 2000–4

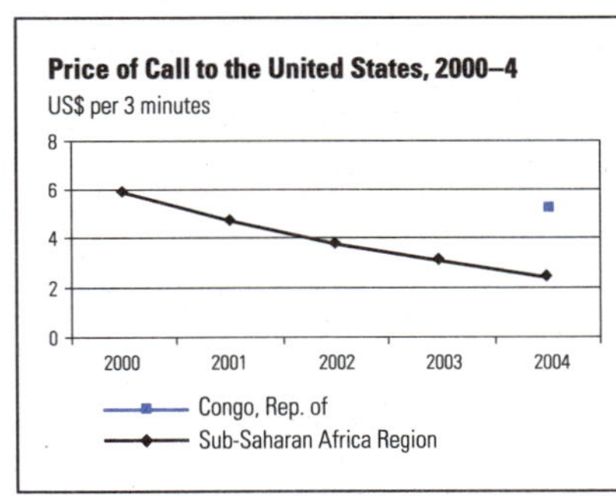

Price of Call to the United States, 2000–4

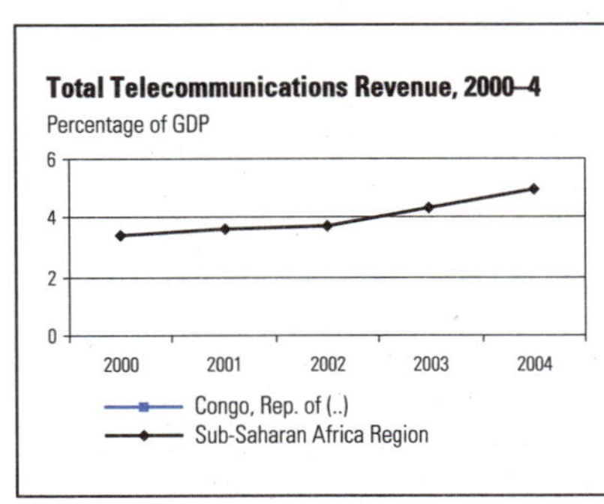

Total Telecommunications Revenue, 2000–4

Notes: Figures in italics are for years other than those specified. .. indicates data are not available. C = competition; GDP = gross domestic product; GNI = gross national income; ICT = information and communication technology; M = monopoly; MDG = Millennium Development Goal; P = partial competition; and PCs = personal computers.

a. Outgoing and incoming.

b. Millennium Development Goal indicators 47, 48a, and 48b.

Sources: Economic and social context: UNDP and World Bank; ICT sector structure: ITU, WEF; ICT sector performance: Global Insight/WITSA, ITU, Netcraft, UNDESA, UNPAN, and World Bank. Produced by the Global Information and Communication Technologies Department and the Development Economics Data Group. For complete information, see Definitions and Data Sources.

Costa Rica

	Costa Rica 2000	Costa Rica 2004	Upper-middle-income group 2004	Latin America & the Caribbean Region 2004
Economic and social context				
Population, total (millions)	4	4	576	541
Urban population (% total population)	59	61	72	77
Poverty (% population below US$1 per day)	2.0	..	..	*9.5*
GNI per capita, Atlas method (current US$)	3,820	4,670	4,770	3,600
GDP growth, 1995–2000 and 2000–4 (%)	5.7	3.9	2.7	1.5
Adult literacy rate (% ages 15 and over)	..	*96*	94	89
Primary, secondary, tertiary school enrollment (% gross)	67	*68*	*80*	*82*
ICT sector structure				
Separate telecommunications regulator	Yes	Yes		
Status of main fixed-line operator	*Public*	Public		
Level of competition: international long distance	*M*	M		
Level of competition: mobile	*M*	M		
Level of competition: Internet service provider	*M*	M		
Government prioritization of ICT (scale 1–7)	..	3.9	4.1	3.5
ICT sector performance				
Access				
Telephone main lines (per 1,000 people)	236	327	*220*	181
International voice traffic (minutes per person)[a]	54	..	*39*	..
Mobile subscribers (per 1,000 people)	56	227	490	324
Population covered by mobile telephony (%)	..	..	*84*	*76*
Internet users (per 1,000 people)	60	295	*133*	*104*
Personal computers (per 1,000 people)	157	*225*	*99*	*75*
Households with television (%)	85	91	*92*	*88*
Quality				
Telephone faults (per 100 main lines per year)	5.0	*4.2*	*20.3*	..
Broadband subscribers (per 1,000 people)	0.0	*0.1*	*3.7*	*5.2*
International Internet bandwidth (bits per person)	22	*125*	176	165
Affordability				
Price basket for fixed line (US$ per month, residential)	7.0	*5.9*	*13.9*	*9.0*
Price basket for mobile (US$ per month)	..	4.2	11.1	9.1
Price basket for Internet (US$ per month)	..	*25.8*	*20.8*	*31.5*
Price of call to United States (US$ per 3 minutes)	1.93	..	1.03	0.90
Institutional efficiency and sustainability				
Total telecommunications revenue (% GDP)	1.7	2.5	3.4	3.3
Total telephone subscribers per employee	231	*316*	*402*	..
Total telecommunications investment (% revenue)	44.4	*68.4*	*18.6*	..
ICT applications				
ICT expenditure (% GDP)	6.9	*7.5*	*5.0*	*5.3*
E-government readiness index (scale 0–1)	..	0.17	0.49	0.39
Secure Internet servers (per 1 million people)	*14.5*	53.2	10.7	8.6
Schools connected to the Internet (%)	..	*15*	*60*	..

GNI per Capita, Atlas Method, 2000–4
Current US$

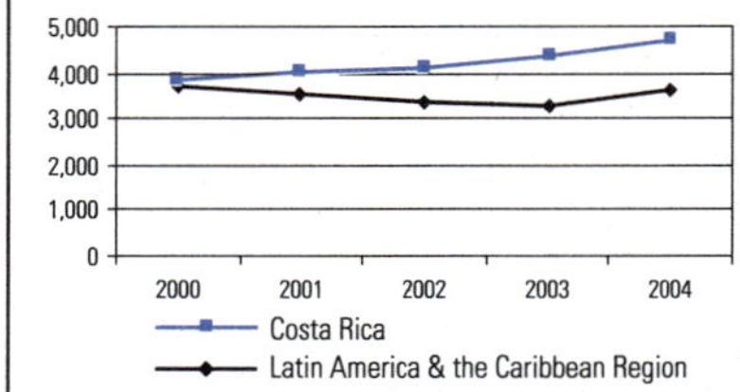

ICT MDG[b] Indicators, 2000–4
Number per 1,000 people

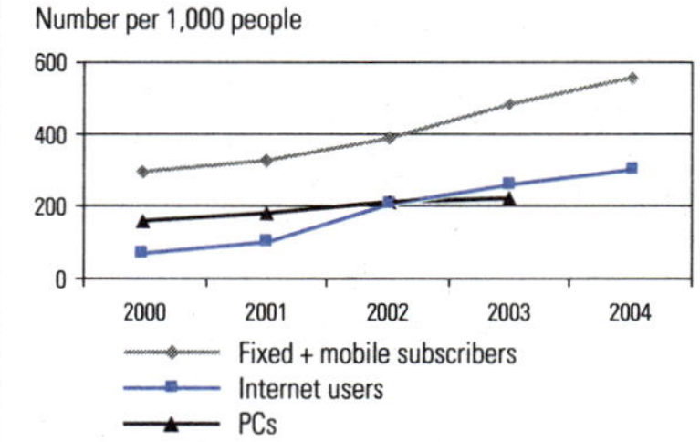

Price of Call to the United States, 2000–4
US$ per 3 minutes

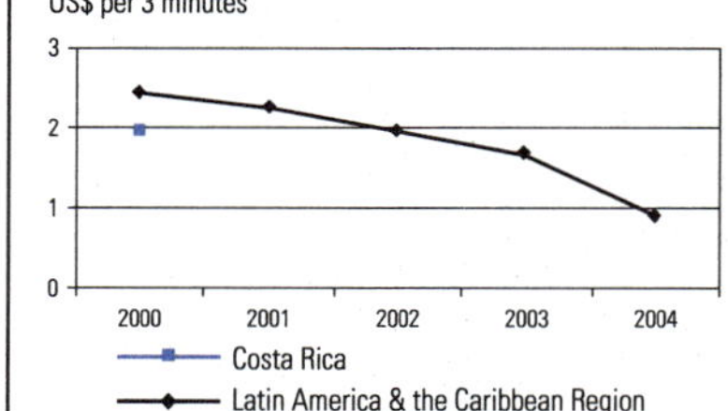

Total Telecommunications Revenue, 2000–4
Percentage of GDP

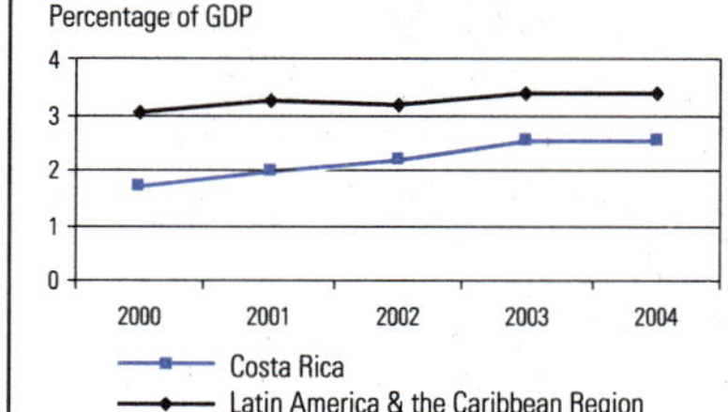

Notes: Figures in italics are for years other than those specified. .. indicates data are not available. C = competition; GDP = gross domestic product; GNI = gross national income; ICT = information and communication technology; M = monopoly; MDG = Millennium Development Goal; P = partial competition; and PCs = personal computers.

a. Outgoing and incoming.

b. Millennium Development Goal indicators 47, 48a, and 48b.

Sources: Economic and social context: UNDP and World Bank; ICT sector structure: ITU, WEF; ICT sector performance: Global Insight/WITSA, ITU, Netcraft, UNDESA, UNPAN, and World Bank. Produced by the Global Information and Communication Technologies Department and the Development Economics Data Group. For complete information, see Definitions and Data Sources.

Côte d'Ivoire

	Côte d'Ivoire 2000	Côte d'Ivoire 2004	Low-income group 2004	Sub-Saharan Africa Region 2004
Economic and social context				
Population, total (millions)	16	17	2,338	719
Urban population (% total population)	44	45	31	37
Poverty (% population below US$1 per day)	*15.5*	*10.8*	..	*46.4*
GNI per capita, Atlas method (current US$)	690	770	510	600
GDP growth, 1995–2000 and 2000–4 (%)	3.6	–1.5	5.4	3.9
Adult literacy rate (% ages 15 and over)	48	..	61	59
Primary, secondary, tertiary school enrollment (% gross)	41	..	*55*	*52*
ICT sector structure				
Separate telecommunications regulator	Yes	Yes		
Status of main fixed-line operator	*Mixed*	Mixed		
Level of competition: international long distance	*M*	P		
Level of competition: mobile	*P*	P		
Level of competition: Internet service provider	*C*	C		
Government prioritization of ICT (scale 1–7)	..	..	..	*4.4*
ICT sector performance				
Access				
Telephone main lines (per 1,000 people)	17	13	33	17
International voice traffic (minutes per person)[a]	9	*11*	*4*	..
Mobile subscribers (per 1,000 people)	30	98	48	86
Population covered by mobile telephony (%)	23	55	*43*	..
Internet users (per 1,000 people)	3	20	20	15
Personal computers (per 1,000 people)	6	*10*	*8*	*12*
Households with television (%)	32	*35*	*16*	*15*
Quality				
Telephone faults (per 100 main lines per year)	100.0	*81.0*	..	..
Broadband subscribers (per 1,000 people)	0.0	*0.0*	0.5	*0.1*
International Internet bandwidth (bits per person)	0	*2*	*3*	*4*
Affordability				
Price basket for fixed line (US$ per month, residential)	8.5	28.2	*6.6*	*8.5*
Price basket for mobile (US$ per month)	..	23.9	11.6	13.5
Price basket for Internet (US$ per month)	..	*67.2*	*45.5*	*54.8*
Price of call to United States (US$ per 3 minutes)	6.07	2.25	1.95	2.43
Institutional efficiency and sustainability				
Total telecommunications revenue (% GDP)	3.4	*3.7*	2.3	5.0
Total telephone subscribers per employee	189	..	*89*	*144*
Total telecommunications investment (% revenue)	23.7	*14.6*	*27.8*	*27.8*
ICT applications				
ICT expenditure (% GDP)	..	..	*4.1*	..
E-government readiness index (scale 0–1)	..	0.04	0.12	0.11
Secure Internet servers (per 1 million people)	..	0.2	0.3	1.9
Schools connected to the Internet (%)	..	..	..	..

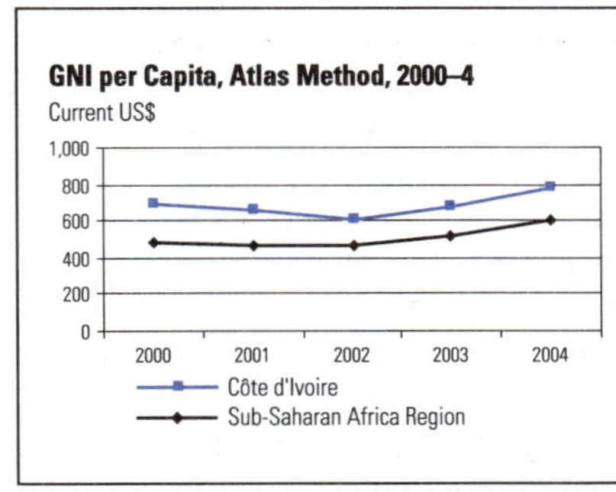

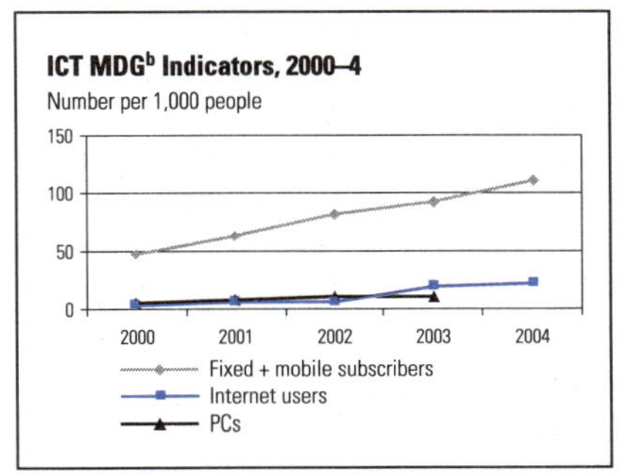

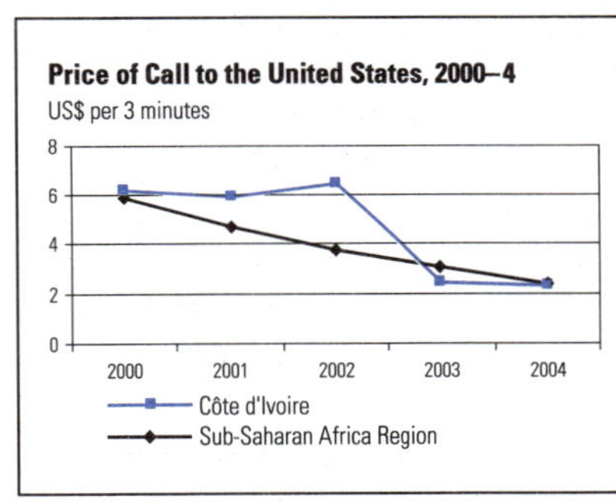

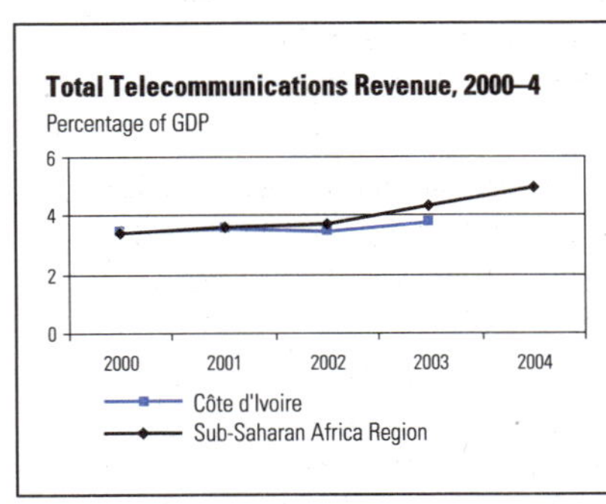

Notes: Figures in italics are for years other than those specified. .. indicates data are not available. C = competition; GDP = gross domestic product; GNI = gross national income; ICT = information and communication technology; M = monopoly; MDG = Millennium Development Goal; P = partial competition; and PCs = personal computers.

a. Outgoing and incoming.

b. Millennium Development Goal indicators 47, 48a, and 48b.

Sources: Economic and social context: UNDP and World Bank; ICT sector structure: ITU, WEF; ICT sector performance: Global Insight/WITSA, ITU, Netcraft, UNDESA, UNPAN, and World Bank. Produced by the Global Information and Communication Technologies Department and the Development Economics Data Group. For complete information, see Definitions and Data Sources.

Croatia

	Croatia 2000	Croatia 2004	Upper-middle-income group 2004	Europe & Central Asia Region 2004
Economic and social context				
Population, total (millions)	4	5	576	472
Urban population (% total population)	58	59	72	64
Poverty (% population below US$1 per day)	*<2*	*<2*	..	*3.6*
GNI per capita, Atlas method (current US$)	4,500	6,590	4,770	3,290
GDP growth, 1995–2000 and 2000–4 (%)	3.2	4.5	2.7	5.0
Adult literacy rate (% ages 15 and over)	*98*	..	94	97
Primary, secondary, tertiary school enrollment (% gross)	72	*75*	*80*	*81*
ICT sector structure				
Separate telecommunications regulator	Yes	Yes		
Status of main fixed-line operator	*Mixed*	Mixed		
Level of competition: international long distance	*M*	C		
Level of competition: mobile	*C*	C		
Level of competition: Internet service provider	*C*	C		
Government prioritization of ICT (scale 1–7)	..	*3.4*	4.1	3.8
ICT sector performance				
Access				
Telephone main lines (per 1,000 people)	393	419	*220*	*243*
International voice traffic (minutes per person)[a]	168	168	*39*	*28*
Mobile subscribers (per 1,000 people)	236	630	490	487
Population covered by mobile telephony (%)	98	98	*84*	*82*
Internet users (per 1,000 people)	68	*228*	*133*	*115*
Personal computers (per 1,000 people)	114	*171*	*99*	*73*
Households with television (%)	94	*93*	*92*	*92*
Quality				
Telephone faults (per 100 main lines per year)	..	*12.0*	*20.3*	*30.4*
Broadband subscribers (per 1,000 people)	0.0	5.0	*3.7*	*2.4*
International Internet bandwidth (bits per person)	..	313	176	148
Affordability				
Price basket for fixed line (US$ per month, residential)	*11.9*	*14.7*	*13.9*	*3.5*
Price basket for mobile (US$ per month)	..	14.4	11.1	10.3
Price basket for Internet (US$ per month)	..	*17.2*	*20.8*	*19.8*
Price of call to United States (US$ per 3 minutes)	..	..	1.03	1.06
Institutional efficiency and sustainability				
Total telecommunications revenue (% GDP)	4.5	5.4	3.4	3.5
Total telephone subscribers per employee	..	478	*402*	*150*
Total telecommunications investment (% revenue)	..	14.8	*18.6*	*19.0*
ICT applications				
ICT expenditure (% GDP)	..	..	*5.0*	*5.1*
E-government readiness index (scale 0–1)	..	0.39	0.49	0.39
Secure Internet servers (per 1 million people)	*13.7*	32.4	10.7	6.4
Schools connected to the Internet (%)	..	..	*60*	*65*

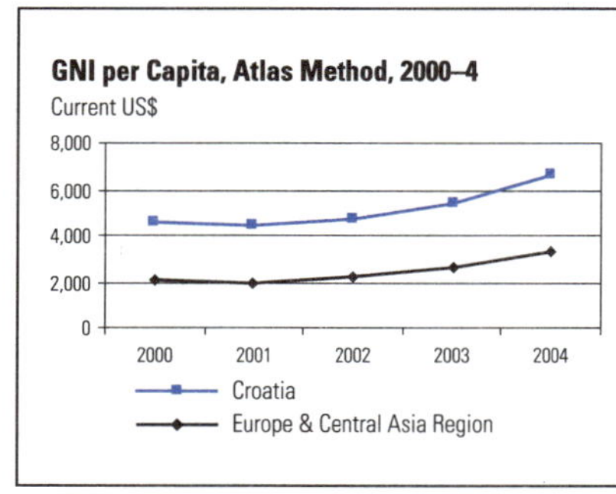

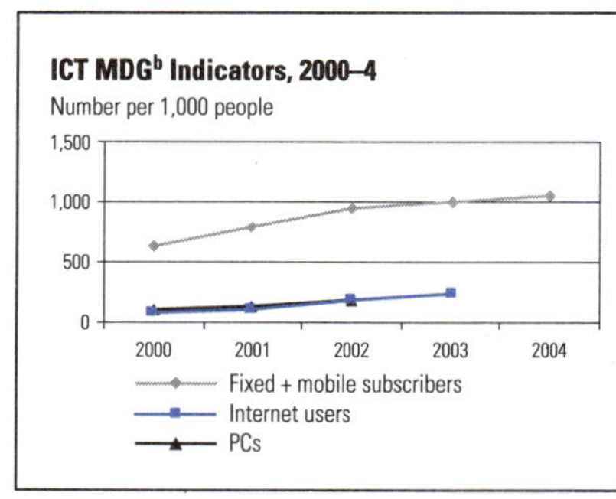

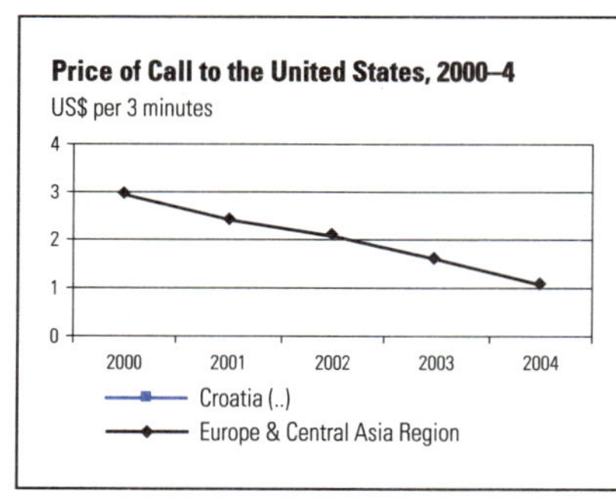

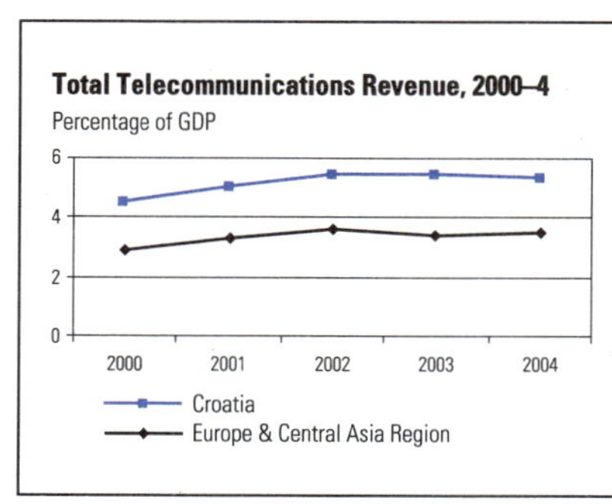

Notes: Figures in italics are for years other than those specified. .. indicates data are not available. C = competition; GDP = gross domestic product; GNI = gross national income; ICT = information and communication technology; M = monopoly; MDG = Millennium Development Goal; P = partial competition; and PCs = personal computers.

a. Outgoing and incoming.

b. Millennium Development Goal indicators 47, 48a, and 48b.

Sources: Economic and social context: UNDP and World Bank; ICT sector structure: ITU, WEF; ICT sector performance: Global Insight/WITSA, ITU, Netcraft, UNDESA, UNPAN, and World Bank. Produced by the Global Information and Communication Technologies Department and the Development Economics Data Group. For complete information, see Definitions and Data Sources.

Cuba

	Cuba 2000	Cuba 2004	Lower-middle-income group 2004	Latin America & the Caribbean Region 2004
Economic and social context				
Population, total (millions)	11	11	2,430	541
Urban population (% total population)	75	76	49	77
Poverty (% population below US$1 per day)	..	..	..	*9.5*
GNI per capita, Atlas method (current US$)	..	..	1,580	3,600
GDP growth, 1995–2000 and 2000–4 (%)	4.2	*2.0*	5.7	1.5
Adult literacy rate (% ages 15 and over)	..	*97*	90	89
Primary, secondary, tertiary school enrollment (% gross)	76	*80*	*72*	*82*
ICT sector structure				
Separate telecommunications regulator	..	..		
Status of main fixed-line operator	*Mixed*	Mixed		
Level of competition: international long distance	*M*	M		
Level of competition: mobile	..	P		
Level of competition: Internet service provider	*C*	C		
Government prioritization of ICT (scale 1–7)	..	..	3.8	3.5
ICT sector performance				
Access				
Telephone main lines (per 1,000 people)	44	*61*	192	181
International voice traffic (minutes per person)[a]	29	..	..	..
Mobile subscribers (per 1,000 people)	1	*4*	255	324
Population covered by mobile telephony (%)	41	..	*76*	*76*
Internet users (per 1,000 people)	5	*9*	70	*104*
Personal computers (per 1,000 people)	12	*24*	38	*75*
Households with television (%)	*81*	..	*89*	*88*
Quality				
Telephone faults (per 100 main lines per year)	10.0	..	..	..
Broadband subscribers (per 1,000 people)	0.0	*0.0*	12.6	*5.2*
International Internet bandwidth (bits per person)	2	..	58	165
Affordability				
Price basket for fixed line (US$ per month, residential)	12.4	..	*5.5*	*9.0*
Price basket for mobile (US$ per month)	..	20.0	8.9	9.1
Price basket for Internet (US$ per month)	..	*53.5*	*25.3*	*31.5*
Price of call to United States (US$ per 3 minutes)	7.35	..	1.45	0.90
Institutional efficiency and sustainability				
Total telecommunications revenue (% GDP)	2.6	..	3.6	3.3
Total telephone subscribers per employee	30	..	*195*	..
Total telecommunications investment (% revenue)	16.0	..	*25.5*	..
ICT applications				
ICT expenditure (% GDP)	..	..	*5.1*	*5.3*
E-government readiness index (scale 0–1)	..	0.09	0.26	0.39
Secure Internet servers (per 1 million people)	*0.2*	0.1	1.6	8.6
Schools connected to the Internet (%)	..	..	..	..

GNI per Capita, Atlas Method, 2000–4
Current US$

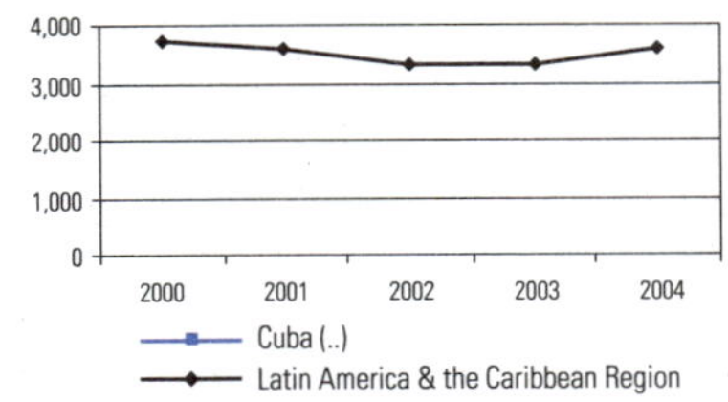

ICT MDG[b] Indicators, 2000–4
Number per 1,000 people

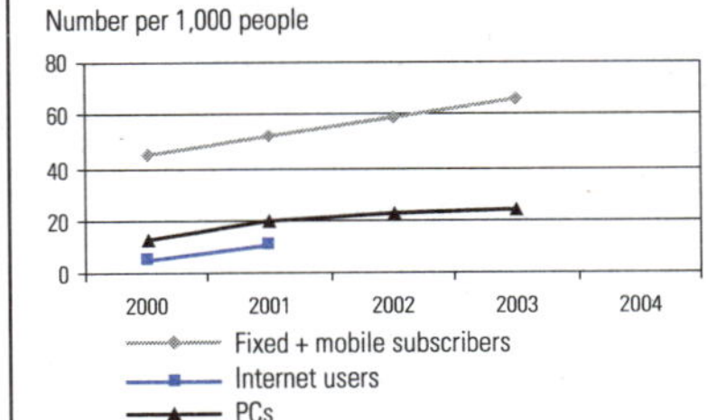

Price of Call to the United States, 2000–4
US$ per 3 minutes

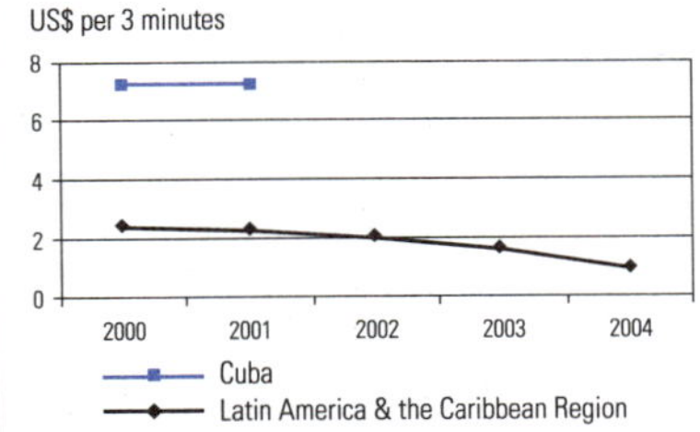

Total Telecommunications Revenue, 2000–4
Percentage of GDP

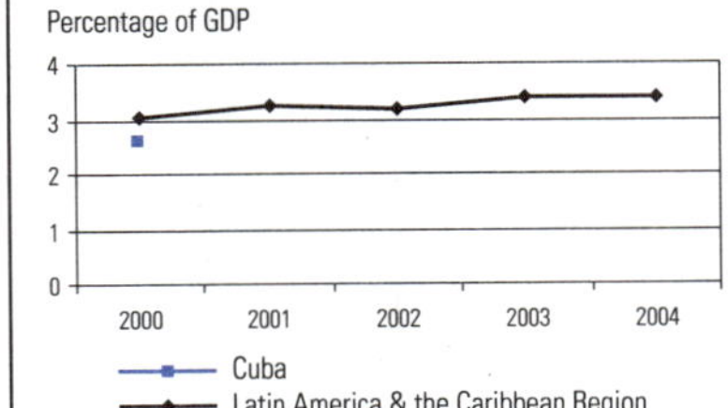

Notes: Figures in italics are for years other than those specified. .. indicates data are not available. C = competition; GDP = gross domestic product; GNI = gross national income; ICT = information and communication technology; M = monopoly; MDG = Millennium Development Goal; P = partial competition; and PCs = personal computers.

a. Outgoing and incoming.

b. Millennium Development Goal indicators 47, 48a, and 48b.

Sources: Economic and social context: UNDP and World Bank; ICT sector structure: ITU, WEF; ICT sector performance: Global Insight/WITSA, ITU, Netcraft, UNDESA, UNPAN, and World Bank. Produced by the Global Information and Communication Technologies Department and the Development Economics Data Group. For complete information, see Definitions and Data Sources.

Czech Republic

	Czech Republic 2000	Czech Republic 2004	Upper-middle-income group 2004	Europe & Central Asia Region 2004
Economic and social context				
Population, total (millions)	10	10	576	472
Urban population (% total population)	74	74	72	64
Poverty (% population below US$1 per day)	*<2*	..	..	*3.6*
GNI per capita, Atlas method (current US$)	5,690	9,150	4,770	3,290
GDP growth, 1995–2000 and 2000–4 (%)	0.9	2.9	2.7	5.0
Adult literacy rate (% ages 15 and over)	..	..	94	97
Primary, secondary, tertiary school enrollment (% gross)	76	*80*	*80*	*81*
ICT sector structure				
Separate telecommunications regulator	Yes	Yes		
Status of main fixed-line operator	*Mixed*	Mixed		
Level of competition: international long distance	*M*	C		
Level of competition: mobile	*P*	C		
Level of competition: Internet service provider	*C*	C		
Government prioritization of ICT (scale 1–7)	..	4.3	4.1	3.8
ICT sector performance				
Access				
Telephone main lines (per 1,000 people)	377	331	*220*	*243*
International voice traffic (minutes per person)[a]	77	..	*39*	*28*
Mobile subscribers (per 1,000 people)	423	1,059	490	487
Population covered by mobile telephony (%)	99	99	*84*	*82*
Internet users (per 1,000 people)	97	*265*	*133*	*115*
Personal computers (per 1,000 people)	122	*196*	*99*	*73*
Households with television (%)	89	..	*92*	*92*
Quality				
Telephone faults (per 100 main lines per year)	17.0	*6.8*	*20.3*	*30.4*
Broadband subscribers (per 1,000 people)	0.2	16.5	*3.7*	*2.4*
International Internet bandwidth (bits per person)	602	*2,450*	176	148
Affordability				
Price basket for fixed line (US$ per month, residential)	12.1	*16.7*	*13.9*	*3.5*
Price basket for mobile (US$ per month)	..	15.1	11.1	10.3
Price basket for Internet (US$ per month)	..	*20.8*	*20.8*	*19.8*
Price of call to United States (US$ per 3 minutes)	0.97	1.06	1.03	1.06
Institutional efficiency and sustainability				
Total telecommunications revenue (% GDP)	4.6	3.7	3.4	3.5
Total telephone subscribers per employee	349	*512*	*402*	*150*
Total telecommunications investment (% revenue)	47.1	*24.8*	*18.6*	*19.0*
ICT applications				
ICT expenditure (% GDP)	7.6	*6.5*	*5.0*	*5.1*
E-government readiness index (scale 0–1)	..	0.55	0.49	0.39
Secure Internet servers (per 1 million people)	*26.7*	31.0	10.7	6.4
Schools connected to the Internet (%)	..	*90*	*60*	*65*

GNI per Capita, Atlas Method, 2000–4
Current US$

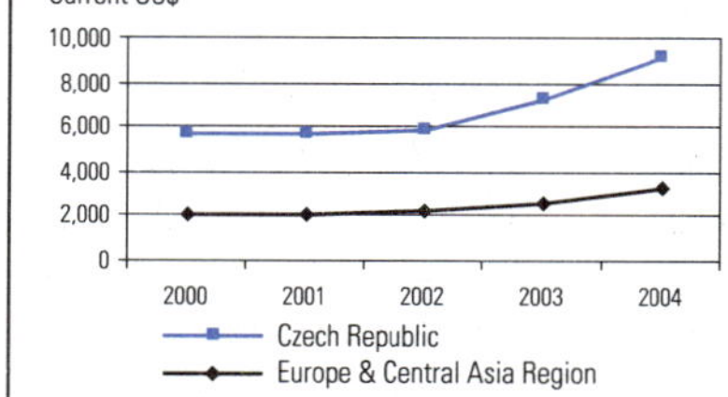

ICT MDG[b] Indicators, 2000–4
Number per 1,000 people

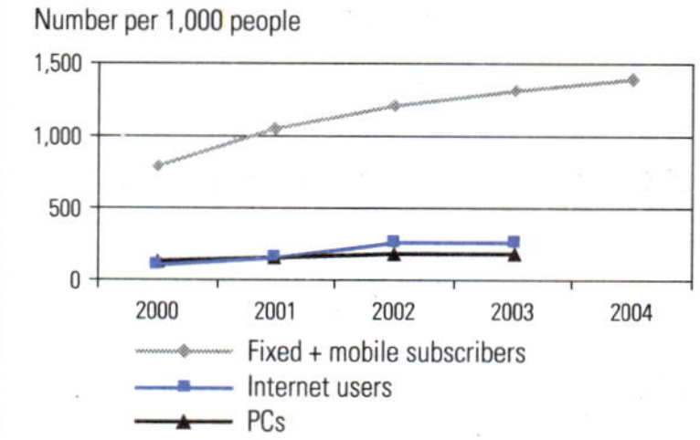

Price of Call to the United States, 2000–4
US$ per 3 minutes

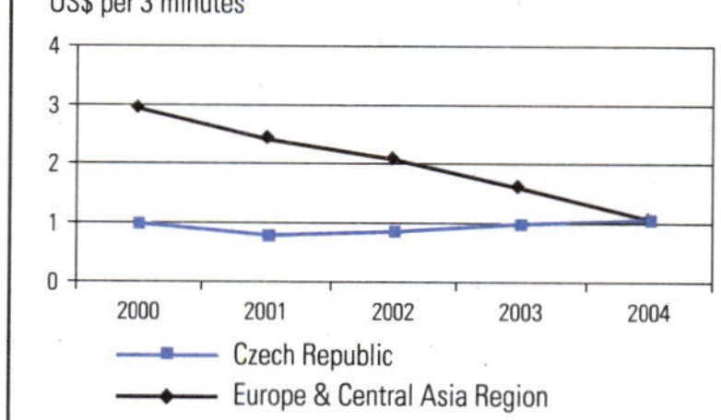

Total Telecommunications Revenue, 2000–4
Percentage of GDP

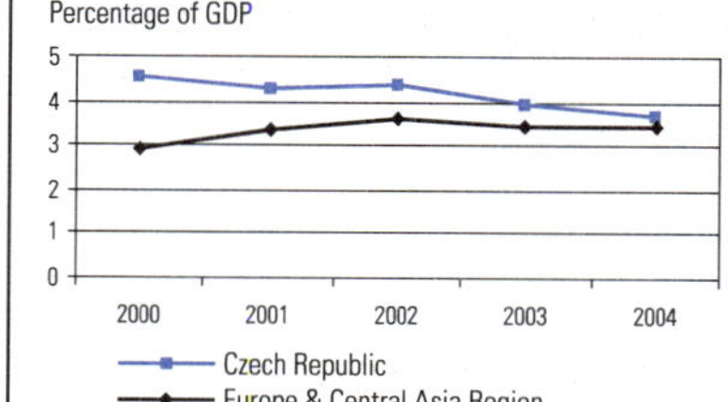

Notes: Figures in italics are for years other than those specified. .. indicates data are not available. C = competition; GDP = gross domestic product; GNI = gross national income; ICT = information and communication technology; M = monopoly; MDG = Millennium Development Goal; P = partial competition; and PCs = personal computers.

a. Outgoing and incoming.

b. Millennium Development Goal indicators 47, 48a, and 48b.

Sources: Economic and social context: UNDP and World Bank; ICT sector structure: ITU, WEF; ICT sector performance: Global Insight/WITSA, ITU, Netcraft, UNDESA, UNPAN, and World Bank. Produced by the Global Information and Communication Technologies Department and the Development Economics Data Group. For complete information, see Definitions and Data Sources.

Denmark

	Denmark 2000	Denmark 2004	High-income group 2004
Economic and social context			
Population, total (millions)	5	5	1,001
Urban population (% total population)	85	85	77
Poverty (% population below US$1 per day)	..	..	..
GNI per capita, Atlas method (current US$)	31,460	40,650	32,040
GDP growth, 1995–2000 and 2000–4 (%)	2.7	1.2	2.0
Adult literacy rate (% ages 15 and over)	..	..	..
Primary, secondary, tertiary school enrollment (% gross)	*96*	*102*	*93*
ICT sector structure			
Separate telecommunications regulator	Yes	Yes	
Status of main fixed-line operator	*Private*	Private	
Level of competition: international long distance	*C*	C	
Level of competition: mobile	*P*	P	
Level of competition: Internet service provider	*C*	C	
Government prioritization of ICT (scale 1–7)	..	5.8	4.9
ICT sector performance			
Access			
Telephone main lines (per 1,000 people)	713	644	558
International voice traffic (minutes per person)[a]	268	*288*	..
Mobile subscribers (per 1,000 people)	630	957	767
Population covered by mobile telephony (%)	96	99	98
Internet users (per 1,000 people)	391	556	*480*
Personal computers (per 1,000 people)	506	648	*504*
Households with television (%)	96	98	*98*
Quality			
Telephone faults (per 100 main lines per year)	*8.0*	*9.0*	..
Broadband subscribers (per 1,000 people)	10.6	168.8	126.2
International Internet bandwidth (bits per person)	1,408	34,917	4,718
Affordability			
Price basket for fixed line (US$ per month, residential)	20.8	25.7	25.8
Price basket for mobile (US$ per month)	..	19.9	17.8
Price basket for Internet (US$ per month)	..	*17.3*	*20.9*
Price of call to United States (US$ per 3 minutes)	1.30	0.89	0.77
Institutional efficiency and sustainability			
Total telecommunications revenue (% GDP)	2.6	2.6	2.9
Total telephone subscribers per employee	336	437	*485*
Total telecommunications investment (% revenue)	26.7	15.0	*12.3*
ICT applications			
ICT expenditure (% GDP)	6.2	*5.7*	*7.1*
E-government readiness index (scale 0–1)	..	0.93	0.70
Secure Internet servers (per 1 million people)	*73.9*	319.4	311.4
Schools connected to the Internet (%)	..	100	*99*

GNI per Capita, Atlas Method, 2000–4

Current US$

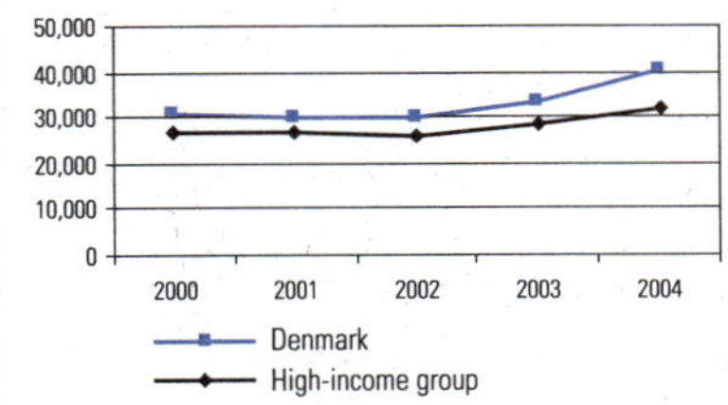

ICT MDG[b] Indicators, 2000–4

Number per 1,000 people

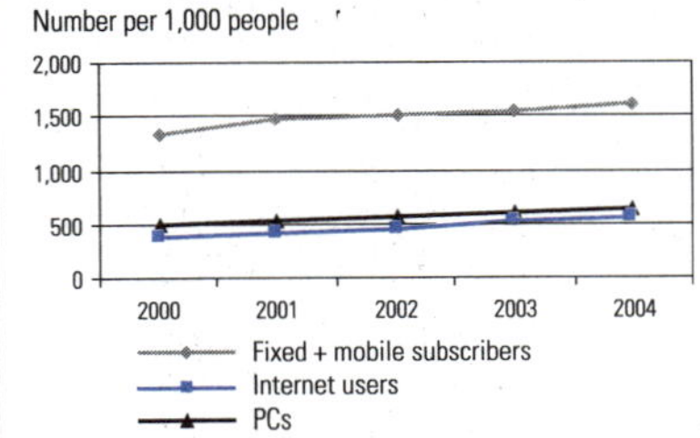

Price of Call to the United States, 2000–4

US$ per 3 minutes

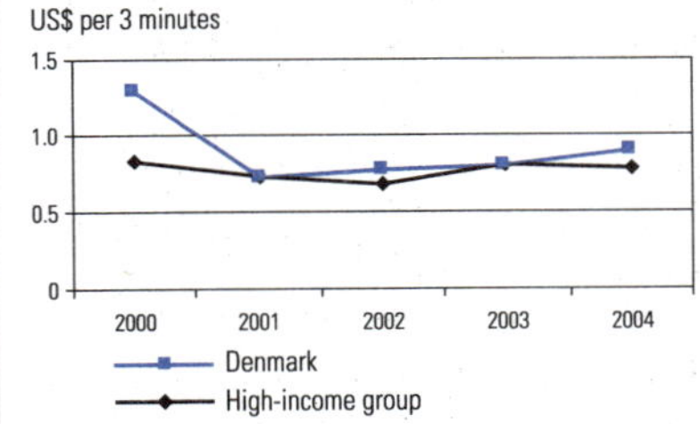

Total Telecommunications Revenue, 2000–4

Percentage of GDP

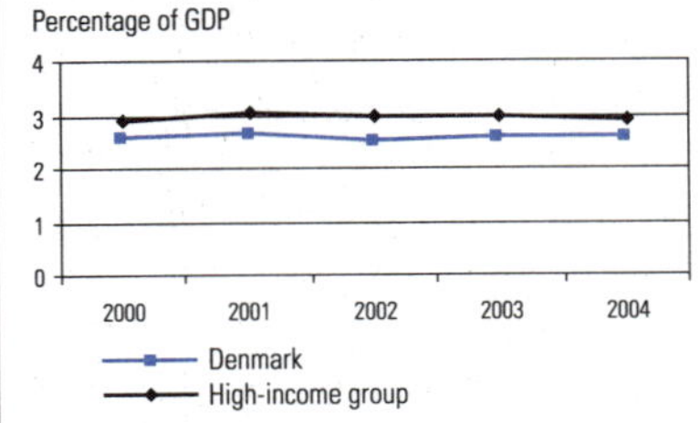

Notes: Figures in italics are for years other than those specified. .. indicates data are not available. C = competition; GDP = gross domestic product; GNI = gross national income; ICT = information and communication technology; M = monopoly; MDG = Millennium Development Goal; P = partial competition; and PCs = personal computers.

a. Outgoing and incoming.

b. Millennium Development Goal indicators 47, 48a, and 48b.

Sources: Economic and social context: UNDP and World Bank; ICT sector structure: ITU, WEF; ICT sector performance: Global Insight/WITSA, ITU, Netcraft, UNDESA, UNPAN, and World Bank. Produced by the Global Information and Communication Technologies Department and the Development Economics Data Group. For complete information, see Definitions and Data Sources.

Dominican Republic

	Dominican Republic 2000	Dominican Republic 2004	Lower-middle-income group 2004	Latin America & the Caribbean Region 2004
Economic and social context				
Population, total (millions)	8	9	2,430	541
Urban population (% total population)	58	60	49	77
Poverty (% population below US$1 per day)	*<2*	..	..	*9.5*
GNI per capita, Atlas method (current US$)	2,140	2,080	1,580	3,600
GDP growth, 1995–2000 and 2000–4 (%)	7.8	2.4	5.7	1.5
Adult literacy rate (% ages 15 and over)	88	..	90	89
Primary, secondary, tertiary school enrollment (% gross)	74	*76*	*72*	*82*
ICT sector structure				
Separate telecommunications regulator	Yes	Yes		
Status of main fixed-line operator	*Private*	Private		
Level of competition: international long distance	*C*	C		
Level of competition: mobile	*C*	C		
Level of competition: Internet service provider	*C*	C		
Government prioritization of ICT (scale 1–7)	..	3.4	3.8	3.5
ICT sector performance				
Access				
Telephone main lines (per 1,000 people)	107	106	192	181
International voice traffic (minutes per person)[a]	183	..	..	..
Mobile subscribers (per 1,000 people)	84	286	255	324
Population covered by mobile telephony (%)	..	*88*	*76*	*76*
Internet users (per 1,000 people)	31	60	70	*104*
Personal computers (per 1,000 people)	..	*46*	38	*75*
Households with television (%)	79	88	*89*	*88*
Quality				
Telephone faults (per 100 main lines per year)	..	..	..	..
Broadband subscribers (per 1,000 people)	0.0	3.8	12.6	*5.2*
International Internet bandwidth (bits per person)	*6*	19	58	165
Affordability				
Price basket for fixed line (US$ per month, residential)	18.1	*16.9*	*5.5*	*9.0*
Price basket for mobile (US$ per month)	..	7.0	8.9	9.1
Price basket for Internet (US$ per month)	..	*33.1*	*25.3*	*31.5*
Price of call to United States (US$ per 3 minutes)	..	0.22	1.45	0.90
Institutional efficiency and sustainability				
Total telecommunications revenue (% GDP)	5.4	*7.5*	3.6	3.3
Total telephone subscribers per employee	99	..	*195*	..
Total telecommunications investment (% revenue)	25.8	..	*25.5*	..
ICT applications				
ICT expenditure (% GDP)	..	..	*5.1*	*5.3*
E-government readiness index (scale 0–1)	..	0.36	0.26	0.39
Secure Internet servers (per 1 million people)	*0.9*	3.7	1.6	8.6
Schools connected to the Internet (%)	..	..	..	..

GNI per Capita, Atlas Method, 2000–4
Current US$

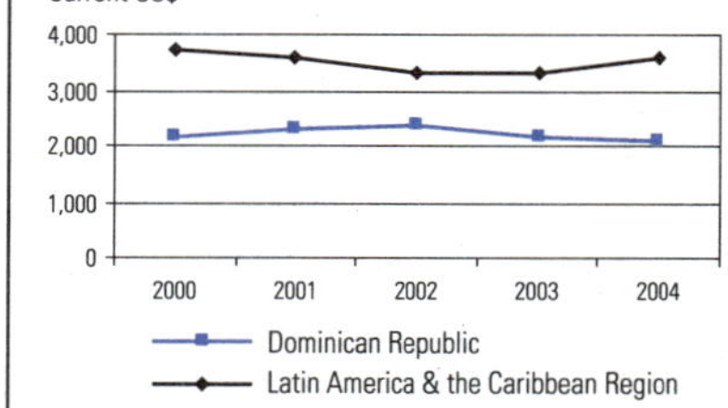

ICT MDG[b] Indicators, 2000–4
Number per 1,000 people

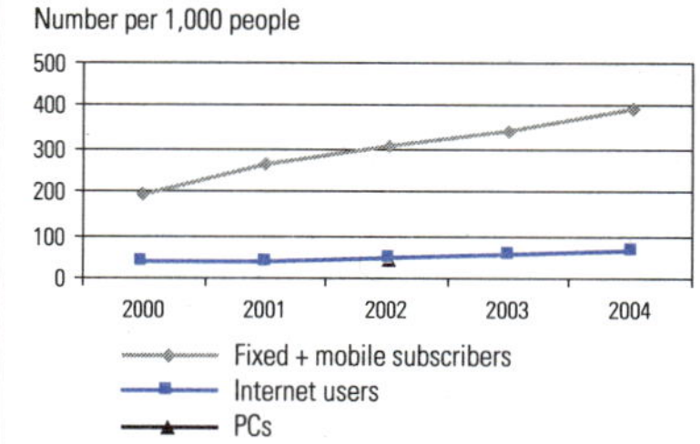

Price of Call to the United States, 2000–4
US$ per 3 minutes

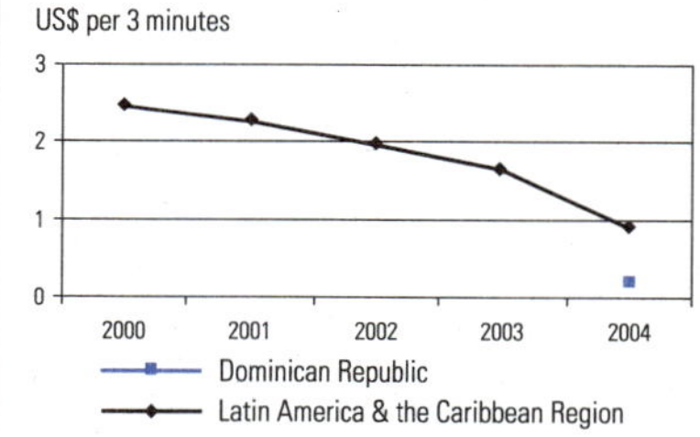

Total Telecommunications Revenue, 2000–4
Percentage of GDP

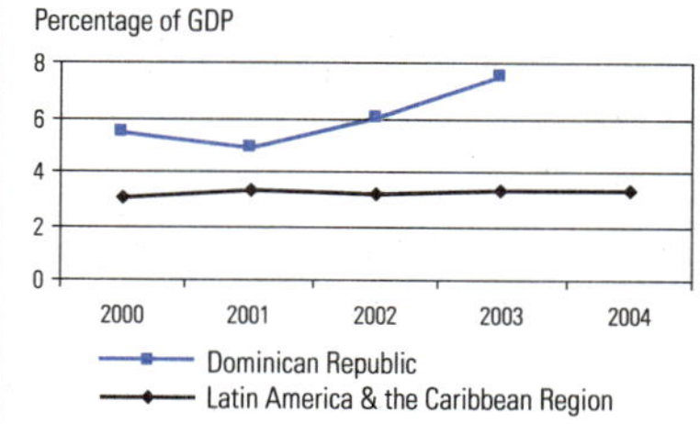

Notes: Figures in italics are for years other than those specified. .. indicates data are not available. C = competition; GDP = gross domestic product; GNI = gross national income; ICT = information and communication technology; M = monopoly; MDG = Millennium Development Goal; P = partial competition; and PCs = personal computers.

a. Outgoing and incoming.

b. Millennium Development Goal indicators 47, 48a, and 48b.

Sources: Economic and social context: UNDP and World Bank; ICT sector structure: ITU, WEF; ICT sector performance: Global Insight/WITSA, ITU, Netcraft, UNDESA, UNPAN, and World Bank. Produced by the Global Information and Communication Technologies Department and the Development Economics Data Group. For complete information, see Definitions and Data Sources.

Ecuador

	Ecuador 2000	Ecuador 2004	Lower-middle-income group 2004	Latin America & the Caribbean Region 2004
Economic and social context				
Population, total (millions)	12	13	2,430	541
Urban population (% total population)	60	62	49	77
Poverty (% population below US$1 per day)	*17.7*	..	..	*9.5*
GNI per capita, Atlas method (current US$)	1,330	2,180	1,580	3,600
GDP growth, 1995–2000 and 2000–4 (%)	0.7	4.2	5.7	1.5
Adult literacy rate (% ages 15 and over)	*91*	..	90	89
Primary, secondary, tertiary school enrollment (% gross)	..	..	*72*	*82*
ICT sector structure				
Separate telecommunications regulator	Yes	Yes		
Status of main fixed-line operator	*Public*	Public		
Level of competition: international long distance	*M*	M		
Level of competition: mobile	*P*	P		
Level of competition: Internet service provider	*P*	C		
Government prioritization of ICT (scale 1–7)	..	2.9	3.8	3.5
ICT sector performance				
Access				
Telephone main lines (per 1,000 people)	99	122	192	181
International voice traffic (minutes per person)[a]	42	..	..	..
Mobile subscribers (per 1,000 people)	39	261	255	324
Population covered by mobile telephony (%)	85	*88*	*76*	*76*
Internet users (per 1,000 people)	14	*44*	70	*104*
Personal computers (per 1,000 people)	22	*35*	38	*75*
Households with television (%)	76	*89*	*89*	*88*
Quality				
Telephone faults (per 100 main lines per year)	35.1	42.0	..	..
Broadband subscribers (per 1,000 people)	0.0	*0.0*	12.6	*5.2*
International Internet bandwidth (bits per person)	2	*38*	58	165
Affordability				
Price basket for fixed line (US$ per month, residential)	9.8	9.0	*5.5*	*9.0*
Price basket for mobile (US$ per month)	..	10.6	8.9	9.1
Price basket for Internet (US$ per month)	..	*31.8*	*25.3*	*31.5*
Price of call to United States (US$ per 3 minutes)	2.48	..	1.45	0.90
Institutional efficiency and sustainability				
Total telecommunications revenue (% GDP)	2.8	..	3.6	3.3
Total telephone subscribers per employee	244	..	*195*	..
Total telecommunications investment (% revenue)	..	..	*25.5*	..
ICT applications				
ICT expenditure (% GDP)	2.9	*3.7*	*5.1*	*5.3*
E-government readiness index (scale 0–1)	..	0.24	0.26	0.39
Secure Internet servers (per 1 million people)	*0.9*	2.9	1.6	8.6
Schools connected to the Internet (%)	..	..	..	..

GNI per Capita, Atlas Method, 2000–4
Current US$

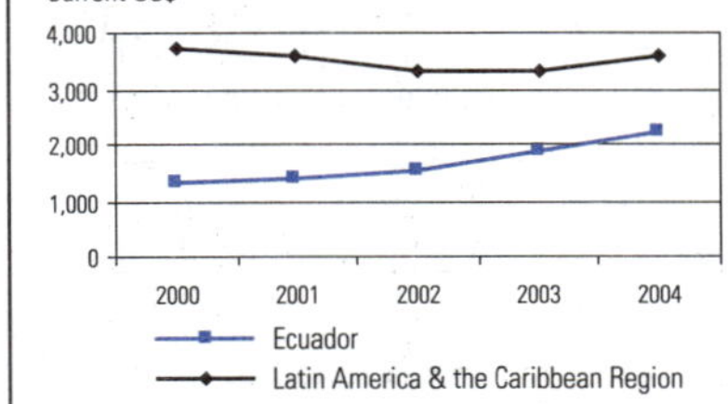

ICT MDG[b] Indicators, 2000–4
Number per 1,000 people

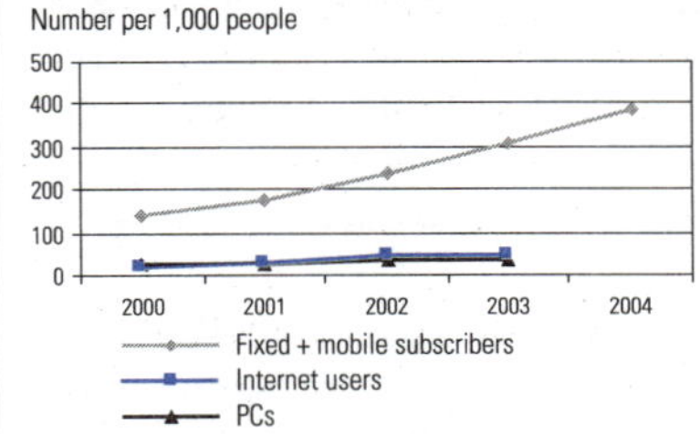

Price of Call to the United States, 2000–4
US$ per 3 minutes

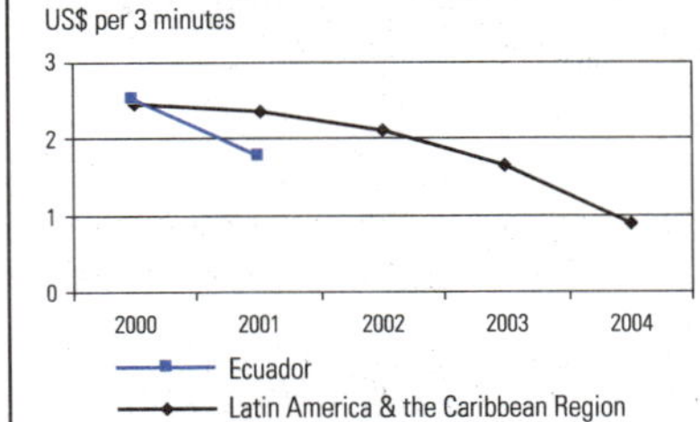

Total Telecommunications Revenue, 2000–4
Percentage of GDP

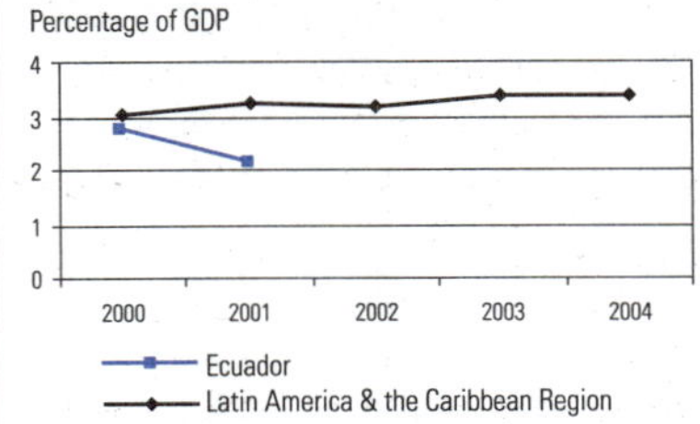

Notes: Figures in italics are for years other than those specified. .. indicates data are not available. C = competition; GDP = gross domestic product; GNI = gross national income; ICT = information and communication technology; M = monopoly; MDG = Millennium Development Goal; P = partial competition; and PCs = personal computers.

a. Outgoing and incoming.
b. Millennium Development Goal indicators 47, 48a, and 48b.

Sources: Economic and social context: UNDP and World Bank; ICT sector structure: ITU, WEF; ICT sector performance: Global Insight/WITSA, ITU, Netcraft, UNDESA, UNPAN, and World Bank. Produced by the Global Information and Communication Technologies Department and the Development Economics Data Group. For complete information, see Definitions and Data Sources.

Egypt, Arab Republic of

	Egypt, Arab Rep. of 2000	Egypt, Arab Rep. of 2004	Lower-middle-income group 2004	Middle East & North Africa Region 2004
Economic and social context				
Population, total (millions)	64	69	2,430	294
Urban population (% total population)	42	42	49	56
Poverty (% population below US$1 per day)	*3.1*	..	..	*2.4*
GNI per capita, Atlas method (current US$)	1,490	1,310	1,580	2,000
GDP growth, 1995–2000 and 2000–4 (%)	5.3	3.5	5.7	4.5
Adult literacy rate (% ages 15 and over)	..	..	90	..
Primary, secondary, tertiary school enrollment (% gross)	*80*	*74*	*72*	*68*
ICT sector structure				
Separate telecommunications regulator	Yes	Yes		
Status of main fixed-line operator	*Public*	Public		
Level of competition: international long distance	*M*	M		
Level of competition: mobile	*P*	P		
Level of competition: Internet service provider	*C*	C		
Government prioritization of ICT (scale 1–7)	..	5.5	3.8	..
ICT sector performance				
Access				
Telephone main lines (per 1,000 people)	86	138	192	*118*
International voice traffic (minutes per person)[a]	13	*21*	..	*20*
Mobile subscribers (per 1,000 people)	21	110	255	*88*
Population covered by mobile telephony (%)	96	91	*76*	..
Internet users (per 1,000 people)	10	57	70	*47*
Personal computers (per 1,000 people)	13	*22*	38	*30*
Households with television (%)	89	*95*	*89*	*88*
Quality				
Telephone faults (per 100 main lines per year)	2.0	*1.0*	..	..
Broadband subscribers (per 1,000 people)	..	0.4	12.6	*0.2*
International Internet bandwidth (bits per person)	0	23	58	*15*
Affordability				
Price basket for fixed line (US$ per month, residential)	4.3	3.8	*5.5*	*4.9*
Price basket for mobile (US$ per month)	..	4.1	8.9	8.1
Price basket for Internet (US$ per month)	..	*5.5*	*25.3*	*24.5*
Price of call to United States (US$ per 3 minutes)	3.33	1.45	1.45	1.64
Institutional efficiency and sustainability				
Total telecommunications revenue (% GDP)	2.8	3.5	3.6	2.8
Total telephone subscribers per employee	125	*274*	*195*	..
Total telecommunications investment (% revenue)	18.8	18.8	*25.5*	*27.8*
ICT applications				
ICT expenditure (% GDP)	1.1	*1.2*	*5.1*	..
E-government readiness index (scale 0–1)	..	0.10	0.26	0.16
Secure Internet servers (per 1 million people)	*0.2*	0.4	1.6	0.6
Schools connected to the Internet (%)	..	*66*	..	..

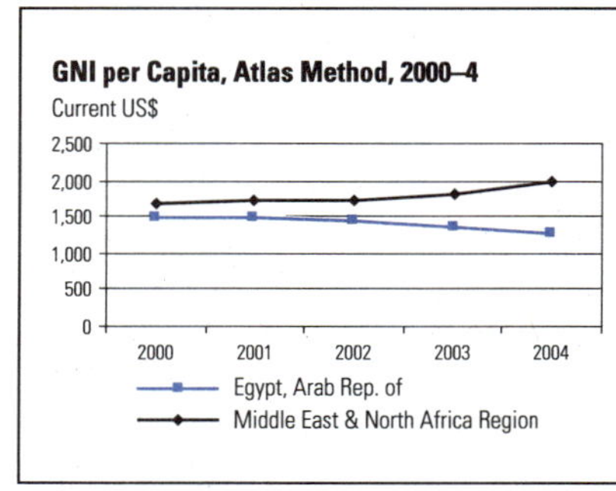

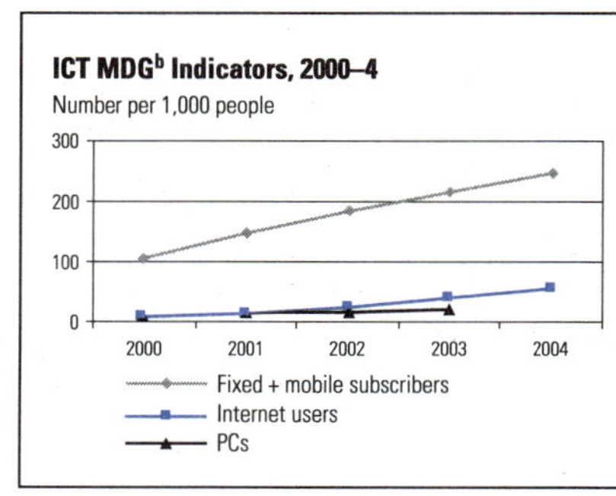

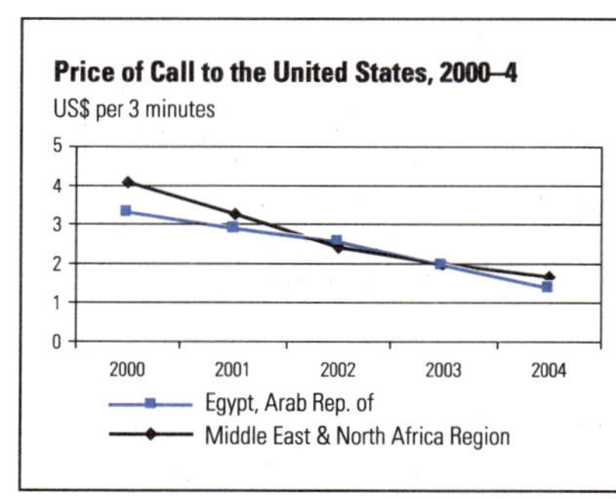

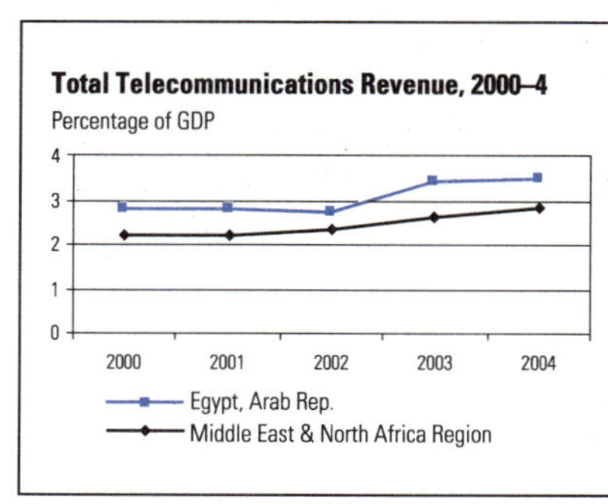

Notes: Figures in italics are for years other than those specified. .. indicates data are not available. C = competition; GDP = gross domestic product; GNI = gross national income; ICT = information and communication technology; M = monopoly; MDG = Millennium Development Goal; P = partial competition; and PCs = personal computers.

a. Outgoing and incoming.

b. Millennium Development Goal indicators 47, 48a, and 48b.

Sources: Economic and social context: UNDP and World Bank; ICT sector structure: ITU, WEF; ICT sector performance: Global Insight/WITSA, ITU, Netcraft, UNDESA, UNPAN, and World Bank. Produced by the Global Information and Communication Technologies Department and the Development Economics Data Group. For complete information, see Definitions and Data Sources.

El Salvador

	El Salvador 2000	El Salvador 2004	Lower-middle-income group 2004	Latin America & the Caribbean Region 2004
Economic and social context				
Population, total (millions)	6	7	2,430	541
Urban population (% total population)	58	60	49	77
Poverty (% population below US$1 per day)	31.1	..	..	*9.5*
GNI per capita, Atlas method (current US$)	2,020	2,350	1,580	3,600
GDP growth, 1995–2000 and 2000–4 (%)	3.3	1.9	5.7	1.5
Adult literacy rate (% ages 15 and over)	..	*80*	90	89
Primary, secondary, tertiary school enrollment (% gross)	65	*68*	*72*	*82*
ICT sector structure				
Separate telecommunications regulator	Yes	Yes		
Status of main fixed-line operator	*Mixed*	Mixed		
Level of competition: international long distance	*C*	C		
Level of competition: mobile	*C*	C		
Level of competition: Internet service provider	*C*	C		
Government prioritization of ICT (scale 1–7)	..	3.8	3.8	3.5
ICT sector performance				
Access				
Telephone main lines (per 1,000 people)	101	133	192	181
International voice traffic (minutes per person)[a]	137	327	..	..
Mobile subscribers (per 1,000 people)	120	216	255	324
Population covered by mobile telephony (%)	*85*	86	*76*	*76*
Internet users (per 1,000 people)	11	90	70	*104*
Personal computers (per 1,000 people)	19	*31*	38	*75*
Households with television (%)	88	..	*89*	*88*
Quality				
Telephone faults (per 100 main lines per year)	14.5	*35.2*	..	..
Broadband subscribers (per 1,000 people)	0.0	*2.9*	12.6	*5.2*
International Internet bandwidth (bits per person)	*7*	63	58	165
Affordability				
Price basket for fixed line (US$ per month, residential)	*16.4*	12.9	*5.5*	*9.0*
Price basket for mobile (US$ per month)	..	13.5	8.9	9.1
Price basket for Internet (US$ per month)	..	*48.1*	*25.3*	*31.5*
Price of call to United States (US$ per 3 minutes)	2.40	2.40	1.45	0.90
Institutional efficiency and sustainability				
Total telecommunications revenue (% GDP)	4.3	*4.1*	3.6	3.3
Total telephone subscribers per employee	323	..	*195*	..
Total telecommunications investment (% revenue)	155.8	*3.8*	*25.5*	..
ICT applications				
ICT expenditure (% GDP)	..	..	*5.1*	*5.3*
E-government readiness index (scale 0–1)	..	0.39	0.26	0.39
Secure Internet servers (per 1 million people)	*1.1*	5.3	1.6	8.6
Schools connected to the Internet (%)	..	..	..	..

GNI per Capita, Atlas Method, 2000–4

Current US$

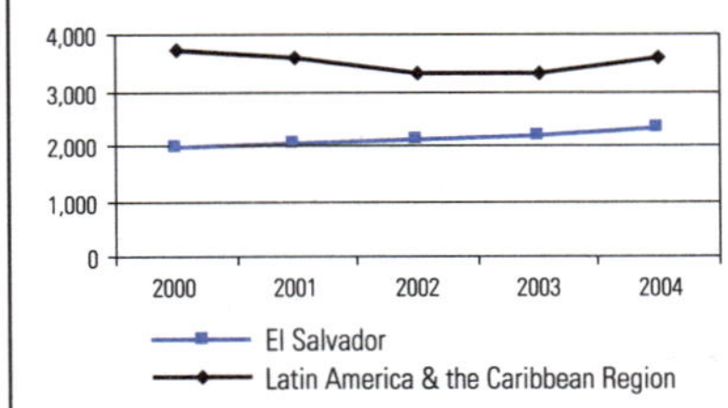

ICT MDG[b] Indicators, 2000–4

Number per 1,000 people

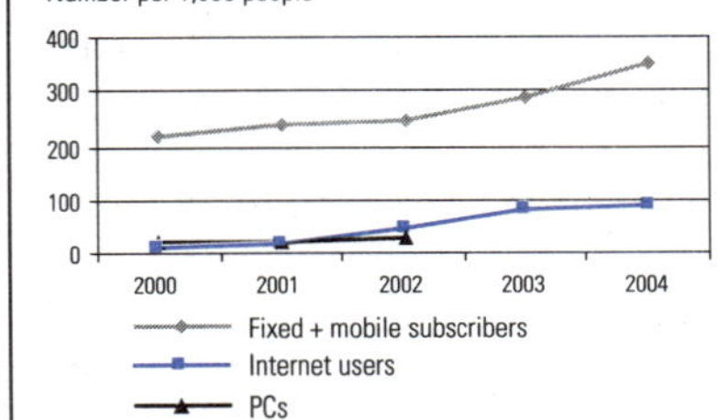

Price of Call to the United States, 2000–4

US$ per 3 minutes

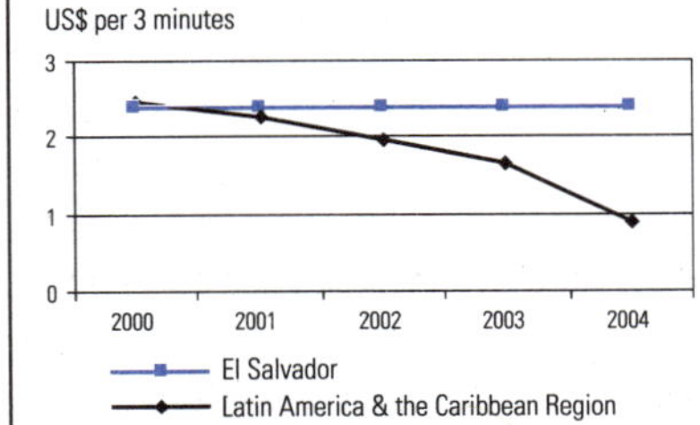

Total Telecommunications Revenue, 2000–4

Percentage of GDP

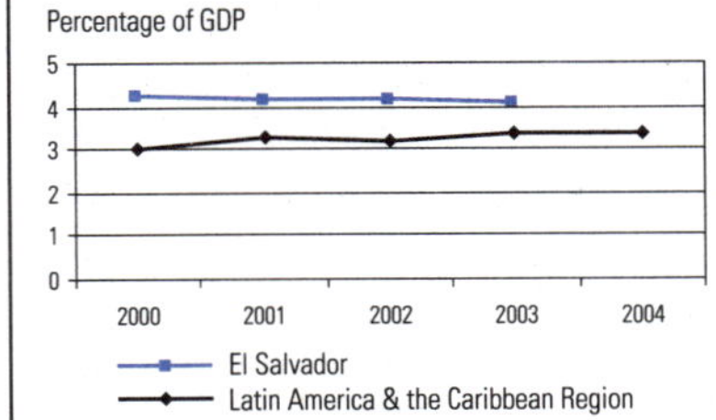

Notes: Figures in italics are for years other than those specified. .. indicates data are not available. C = competition; GDP = gross domestic product; GNI = gross national income; ICT = information and communication technology; M = monopoly; MDG = Millennium Development Goal; P = partial competition; and PCs = personal computers.

a. Outgoing and incoming.

b. Millennium Development Goal indicators 47, 48a, and 48b.

Sources: Economic and social context: UNDP and World Bank; ICT sector structure: ITU, WEF; ICT sector performance: Global Insight/WITSA, ITU, Netcraft, UNDESA, UNPAN, and World Bank. Produced by the Global Information and Communication Technologies Department and the Development Economics Data Group. For complete information, see Definitions and Data Sources.

Eritrea

	Eritrea 2000	Eritrea 2004	Low-income group 2004	Sub-Saharan Africa Region 2004
Economic and social context				
Population, total (millions)	4	4	2,338	719
Urban population (% total population)	19	20	31	37
Poverty (% population below US$1 per day)	..	..	..	*46.4*
GNI per capita, Atlas method (current US$)	160	180	510	600
GDP growth, 1995–2000 and 2000–4 (%)	1.5	3.3	5.4	3.9
Adult literacy rate (% ages 15 and over)	..	..	61	59
Primary, secondary, tertiary school enrollment (% gross)	32	*35*	*55*	*52*
ICT sector structure				
Separate telecommunications regulator	Yes	Yes		
Status of main fixed-line operator	*Public*	Public		
Level of competition: international long distance	*M*	M		
Level of competition: mobile	*C*	C		
Level of competition: Internet service provider	*C*	C		
Government prioritization of ICT (scale 1–7)	..	..	..	*4.4*
ICT sector performance				
Access				
Telephone main lines (per 1,000 people)	7	*9*	33	17
International voice traffic (minutes per person)[a]	5	*7*	*4*	..
Mobile subscribers (per 1,000 people)	0	*0*	48	86
Population covered by mobile telephony (%)	0	*0*	*43*	..
Internet users (per 1,000 people)	1	*2*	20	15
Personal computers (per 1,000 people)	2	*3*	*8*	*12*
Households with television (%)	11	*14*	*16*	*15*
Quality				
Telephone faults (per 100 main lines per year)	57.5	*51.1*	..	..
Broadband subscribers (per 1,000 people)	0.0	0.0	0.5	*0.1*
International Internet bandwidth (bits per person)	0	*0*	*3*	*4*
Affordability				
Price basket for fixed line (US$ per month, residential)	5.3	*4.9*	*6.6*	*8.5*
Price basket for mobile (US$ per month)	..	..	11.6	13.5
Price basket for Internet (US$ per month)	..	*26.8*	*45.5*	*54.8*
Price of call to United States (US$ per 3 minutes)	5.83	*3.55*	1.95	2.43
Institutional efficiency and sustainability				
Total telecommunications revenue (% GDP)	2.8	*2.8*	2.3	5.0
Total telephone subscribers per employee	67	*56*	*89*	*144*
Total telecommunications investment (% revenue)	127.5	*5.2*	*27.8*	*27.8*
ICT applications				
ICT expenditure (% GDP)	..	..	*4.1*	..
E-government readiness index (scale 0–1)	..	0.00	0.12	0.11
Secure Internet servers (per 1 million people)	..	..	0.3	1.9
Schools connected to the Internet (%)	..	..	..	..

GNI per Capita, Atlas Method, 2000–4
Current US$

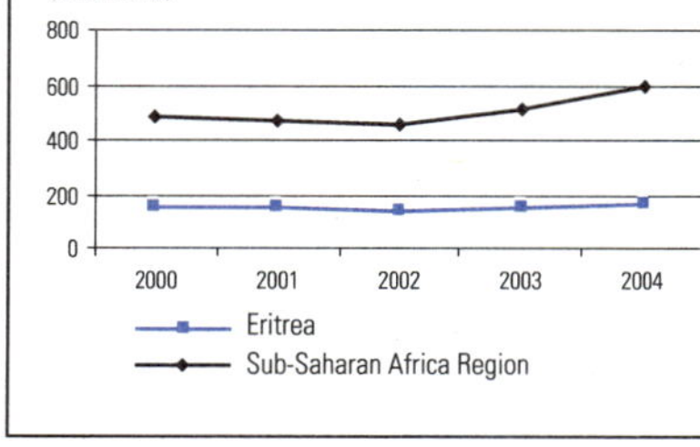

ICT MDG[b] Indicators, 2000–4
Number per 1,000 people

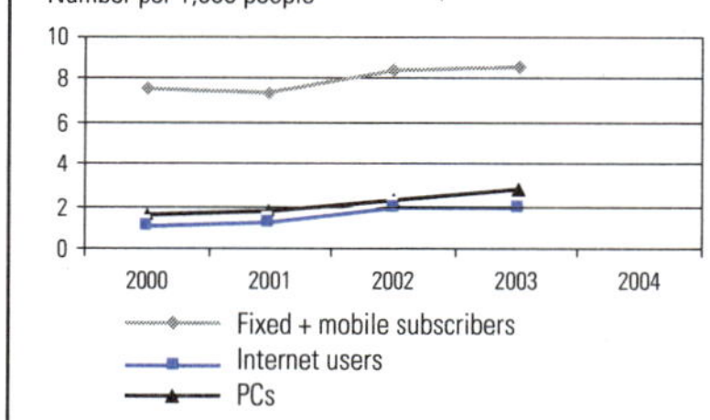

Price of Call to the United States, 2000–4
US$ per 3 minutes

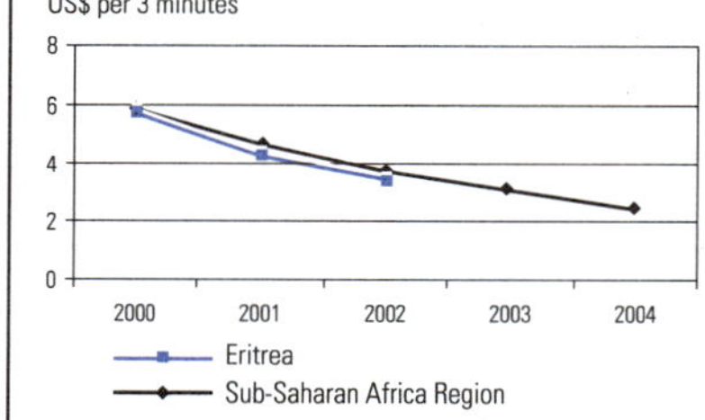

Total Telecommunications Revenue, 2000–4
Percentage of GDP

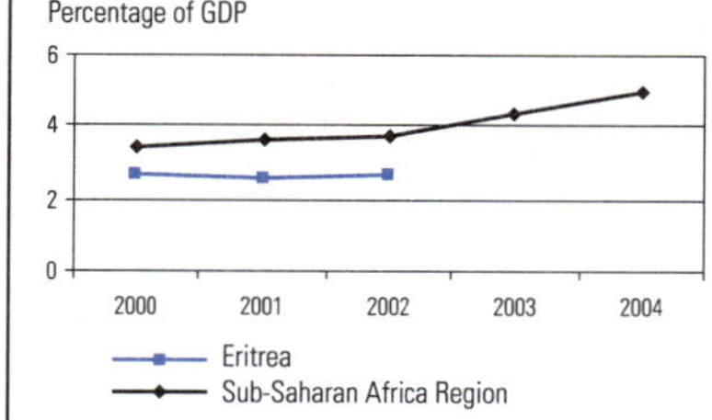

Notes: Figures in italics are for years other than those specified. .. indicates data are not available. C = competition; GDP = gross domestic product; GNI = gross national income; ICT = information and communication technology; M = monopoly; MDG = Millennium Development Goal; P = partial competition; and PCs = personal computers.

a. Outgoing and incoming.

b. Millennium Development Goal indicators 47, 48a, and 48b.

Sources: Economic and social context: UNDP and World Bank; ICT sector structure: ITU, WEF; ICT sector performance: Global Insight/WITSA, ITU, Netcraft, UNDESA, UNPAN, and World Bank. Produced by the Global Information and Communication Technologies Department and the Development Economics Data Group. For complete information, see Definitions and Data Sources.

Estonia

	Estonia 2000	Estonia 2004	Upper-middle-income group 2004	Europe & Central Asia Region 2004
Economic and social context				
Population, total (millions)	1	1	576	472
Urban population (% total population)	69	70	72	64
Poverty (% population below US$1 per day)	*<2*	..	..	*3.6*
GNI per capita, Atlas method (current US$)	4,070	7,010	4,770	3,290
GDP growth, 1995–2000 and 2000–4 (%)	5.4	6.2	2.7	5.0
Adult literacy rate (% ages 15 and over)	100	..	94	97
Primary, secondary, tertiary school enrollment (% gross)	91	*92*	*80*	*81*
ICT sector structure				
Separate telecommunications regulator	Yes	Yes		
Status of main fixed-line operator	*Mixed*	Mixed		
Level of competition: international long distance	*M*	C		
Level of competition: mobile	*C*	P		
Level of competition: Internet service provider	*C*	C		
Government prioritization of ICT (scale 1–7)	..	5.4	4.1	3.8
ICT sector performance				
Access				
Telephone main lines (per 1,000 people)	382	329	*220*	*243*
International voice traffic (minutes per person)[a]	128	..	*39*	*28*
Mobile subscribers (per 1,000 people)	407	967	490	487
Population covered by mobile telephony (%)	99	99	*84*	*82*
Internet users (per 1,000 people)	286	390	*133*	*115*
Personal computers (per 1,000 people)	161	*210*	*99*	*73*
Households with television (%)	91	*93*	*92*	*92*
Quality				
Telephone faults (per 100 main lines per year)	19.2	..	*20.3*	*30.4*
Broadband subscribers (per 1,000 people)	*12.7*	103.1	*3.7*	*2.4*
International Internet bandwidth (bits per person)	137	3,234	176	148
Affordability				
Price basket for fixed line (US$ per month, residential)	9.4	*14.2*	*13.9*	*3.5*
Price basket for mobile (US$ per month)	..	11.4	11.1	10.3
Price basket for Internet (US$ per month)	..	*13.6*	*20.8*	*19.8*
Price of call to United States (US$ per 3 minutes)	1.62	0.90	1.03	1.06
Institutional efficiency and sustainability				
Total telecommunications revenue (% GDP)	5.3	6.6	3.4	3.5
Total telephone subscribers per employee	354	..	*402*	*150*
Total telecommunications investment (% revenue)	17.6	..	*18.6*	*19.0*
ICT applications				
ICT expenditure (% GDP)	..	..	*5.0*	*5.1*
E-government readiness index (scale 0–1)	..	0.70	0.49	0.39
Secure Internet servers (per 1 million people)	*58.7*	84.0	10.7	6.4
Schools connected to the Internet (%)	..	*75*	*60*	*65*

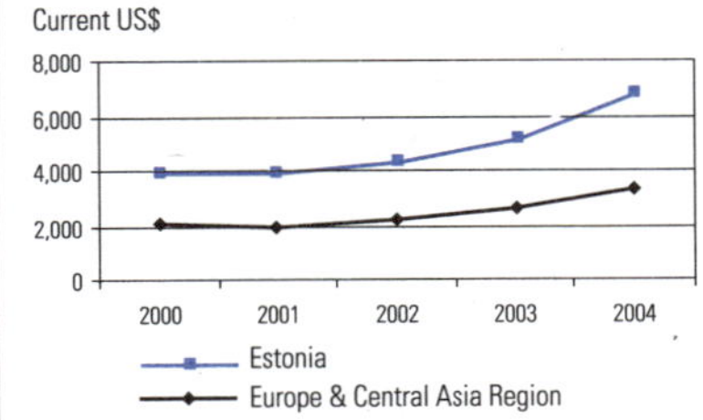

GNI per Capita, Atlas Method, 2000–4

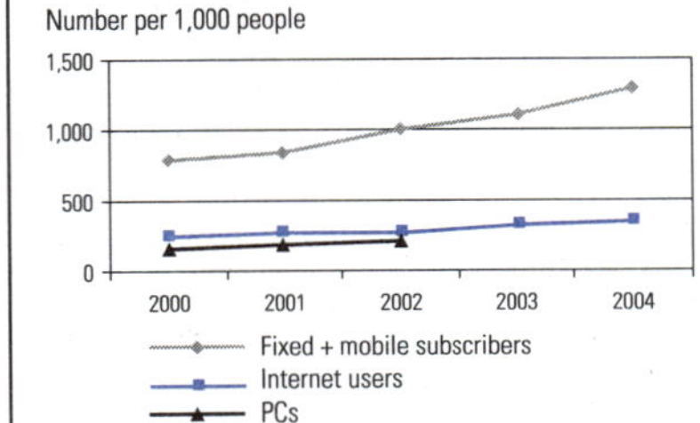

ICT MDG[b] Indicators, 2000–4

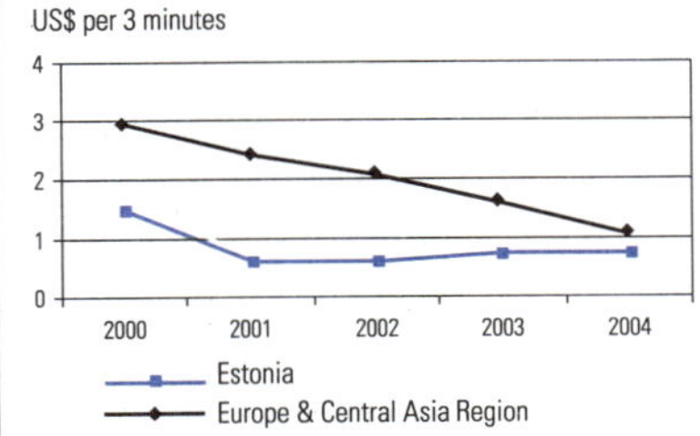

Price of Call to the United States, 2000–4

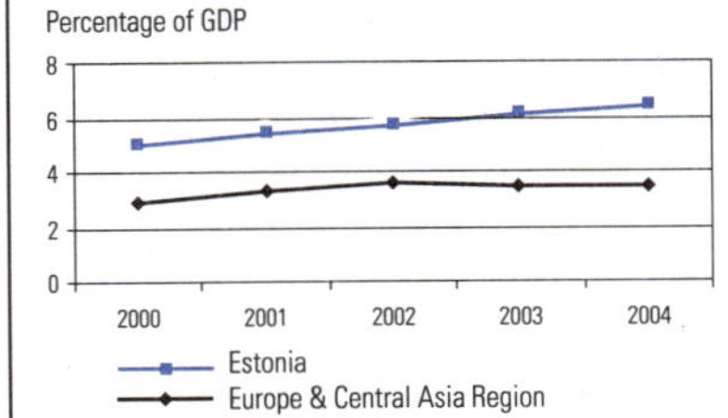

Total Telecommunications Revenue, 2000–4

Notes: Figures in italics are for years other than those specified. .. indicates data are not available. C = competition; GDP = gross domestic product; GNI = gross national income; ICT = information and communication technology; M = monopoly; MDG = Millennium Development Goal; P = partial competition; and PCs = personal computers.

a. Outgoing and incoming.
b. Millennium Development Goal indicators 47, 48a, and 48b.

Sources: Economic and social context: UNDP and World Bank; ICT sector structure: ITU, WEF; ICT sector performance: Global Insight/WITSA, ITU, Netcraft, UNDESA, UNPAN, and World Bank. Produced by the Global Information and Communication Technologies Department and the Development Economics Data Group. For complete information, see Definitions and Data Sources.

Ethiopia

	Ethiopia 2000	Ethiopia 2004	Low-income group 2004	Sub-Saharan Africa Region 2004
Economic and social context				
Population, total (millions)	64	70	2,338	719
Urban population (% total population)	15	16	31	37
Poverty (% population below US$1 per day)	23.0	..	..	*46.4*
GNI per capita, Atlas method (current US$)	110	110	510	600
GDP growth, 1995–2000 and 2000–4 (%)	4.3	3.7	5.4	3.9
Adult literacy rate (% ages 15 and over)	..	*42*	61	59
Primary, secondary, tertiary school enrollment (% gross)	32	*38*	*55*	*52*
ICT sector structure				
Separate telecommunications regulator	Yes	Yes		
Status of main fixed-line operator	*Public*	Public		
Level of competition: international long distance	*M*	M		
Level of competition: mobile	*M*	M		
Level of competition: Internet service provider	*M*	M		
Government prioritization of ICT (scale 1–7)	..	3.7	..	*4.4*
ICT sector performance				
Access				
Telephone main lines (per 1,000 people)	4	7	33	17
International voice traffic (minutes per person)[a]	1	..	*4*	..
Mobile subscribers (per 1,000 people)	0	3	48	86
Population covered by mobile telephony (%)	..	..	*43*	..
Internet users (per 1,000 people)	0	1	20	15
Personal computers (per 1,000 people)	1	*2*	*8*	*12*
Households with television (%)	2	*2*	*16*	*15*
Quality				
Telephone faults (per 100 main lines per year)	..	*100.0*	..	..
Broadband subscribers (per 1,000 people)	0.0	*0.0*	0.5	*0.1*
International Internet bandwidth (bits per person)	0	0	*3*	*4*
Affordability				
Price basket for fixed line (US$ per month, residential)	3.1	*2.9*	*6.6*	*8.5*
Price basket for mobile (US$ per month)	..	3.4	11.6	13.5
Price basket for Internet (US$ per month)	..	*27.4*	*45.5*	*54.8*
Price of call to United States (US$ per 3 minutes)	7.35	*7.05*	1.95	2.43
Institutional efficiency and sustainability				
Total telecommunications revenue (% GDP)	1.4	*1.7*	2.3	5.0
Total telephone subscribers per employee	35	*65*	*89*	*144*
Total telecommunications investment (% revenue)	42.6	*27.8*	*27.8*	*27.8*
ICT applications				
ICT expenditure (% GDP)	..	..	*4.1*	..
E-government readiness index (scale 0–1)	..	0.03	0.12	0.11
Secure Internet servers (per 1 million people)	*0.0*	0.0	0.3	1.9
Schools connected to the Internet (%)	..	*1*	..	..

GNI per Capita, Atlas Method, 2000–4
Current US$

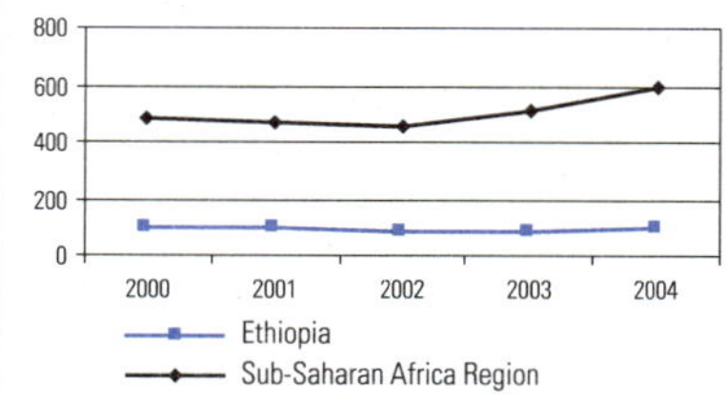

ICT MDG[b] Indicators, 2000–4
Number per 1,000 people

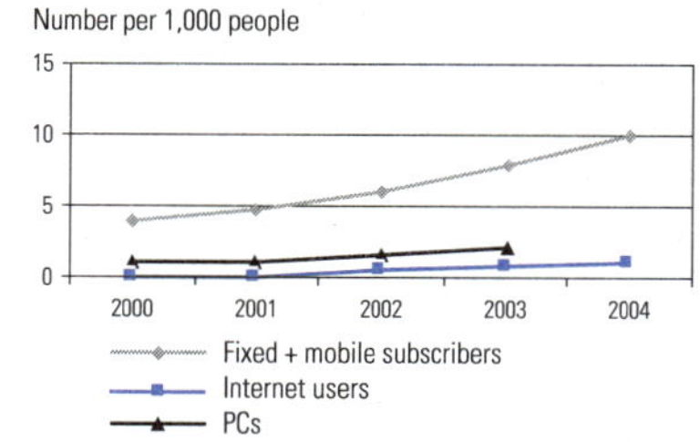

Price of Call to the United States, 2000–4
US$ per 3 minutes

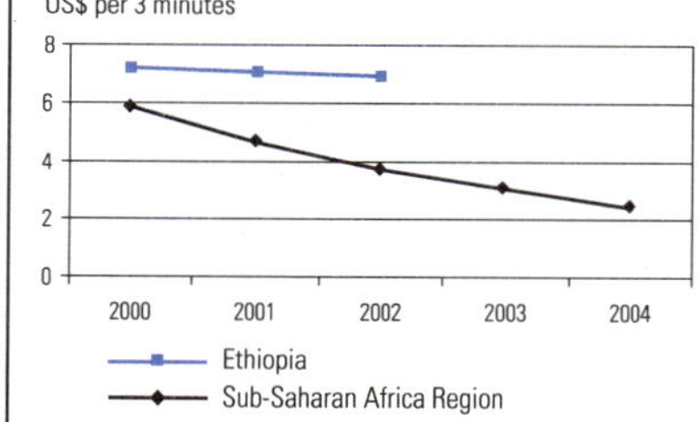

Total Telecommunications Revenue, 2000–4
Percentage of GDP

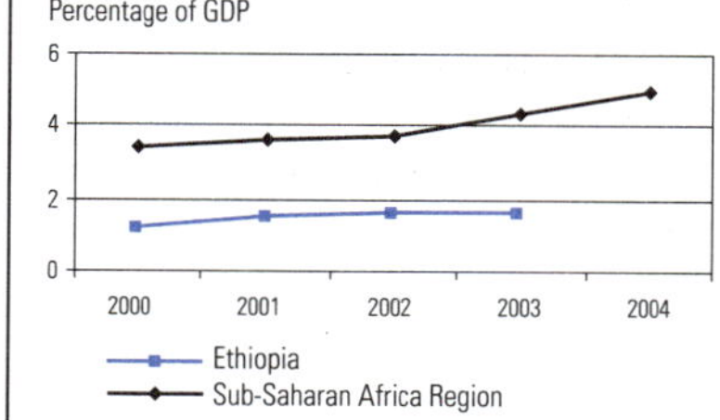

Notes: Figures in italics are for years other than those specified. .. indicates data are not available. C = competition; GDP = gross domestic product; GNI = gross national income; ICT = information and communication technology; M = monopoly; MDG = Millennium Development Goal; P = partial competition; and PCs = personal computers.

a. Outgoing and incoming.

b. Millennium Development Goal indicators 47, 48a, and 48b.

Sources: Economic and social context: UNDP and World Bank; ICT sector structure: ITU, WEF; ICT sector performance: Global Insight/WITSA, ITU, Netcraft, UNDESA, UNPAN, and World Bank. Produced by the Global Information and Communication Technologies Department and the Development Economics Data Group. For complete information, see Definitions and Data Sources.

Finland

	Finland 2000	Finland 2004	High-income group 2004
Economic and social context			
Population, total (millions)	5	5	1,001
Urban population (% total population)	61	61	77
Poverty (% population below US$1 per day)	..	..	..
GNI per capita, Atlas method (current US$)	24,940	32,790	32,040
GDP growth, 1995–2000 and 2000–4 (%)	4.8	2.2	2.0
Adult literacy rate (% ages 15 and over)	..	..	..
Primary, secondary, tertiary school enrollment (% gross)	106	*108*	*93*
ICT sector structure			
Separate telecommunications regulator	Yes	Yes	
Status of main fixed-line operator	*Mixed*	Mixed	
Level of competition: international long distance	*C*	C	
Level of competition: mobile	*C*	P	
Level of competition: Internet service provider	*C*	C	
Government prioritization of ICT (scale 1–7)	..	5.7	4.9
ICT sector performance			
Access			
Telephone main lines (per 1,000 people)	551	*489*	558
International voice traffic (minutes per person)[a]	178	..	..
Mobile subscribers (per 1,000 people)	721	863	767
Population covered by mobile telephony (%)	99	99	98
Internet users (per 1,000 people)	373	527	*480*
Personal computers (per 1,000 people)	396	*442*	*504*
Households with television (%)	92	*91*	*98*
Quality			
Telephone faults (per 100 main lines per year)	..	..	..
Broadband subscribers (per 1,000 people)	3.9	149.6	126.2
International Internet bandwidth (bits per person)	347	4,337	4,718
Affordability			
Price basket for fixed line (US$ per month, residential)	19.4	*24.7*	25.8
Price basket for mobile (US$ per month)	..	13.6	17.8
Price basket for Internet (US$ per month)	..	*22.5*	*20.9*
Price of call to United States (US$ per 3 minutes)	1.07	1.80	0.77
Institutional efficiency and sustainability			
Total telecommunications revenue (% GDP)	3.4	3.1	2.9
Total telephone subscribers per employee	272	*329*	*485*
Total telecommunications investment (% revenue)	20.3	*17.2*	*12.3*
ICT applications			
ICT expenditure (% GDP)	7.5	*6.9*	*7.1*
E-government readiness index (scale 0–1)	..	0.81	0.70
Secure Internet servers (per 1 million people)	*96.0*	246.0	311.4
Schools connected to the Internet (%)	..	*99*	*99*

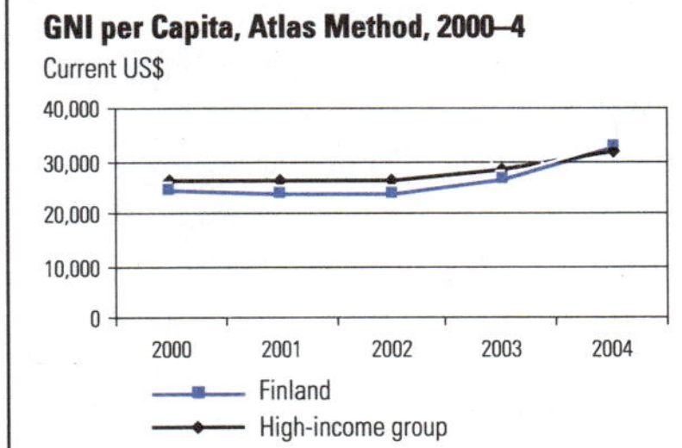

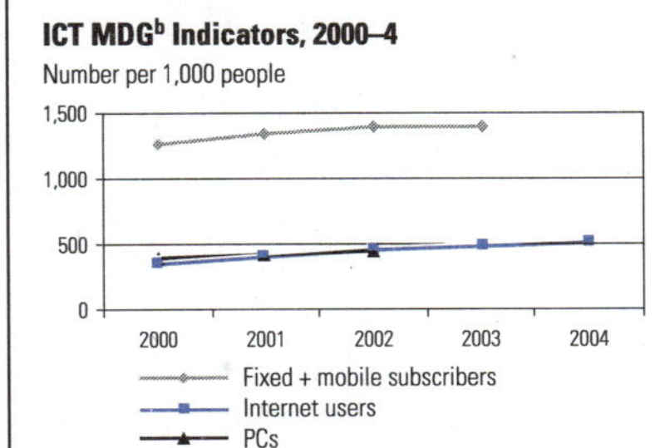

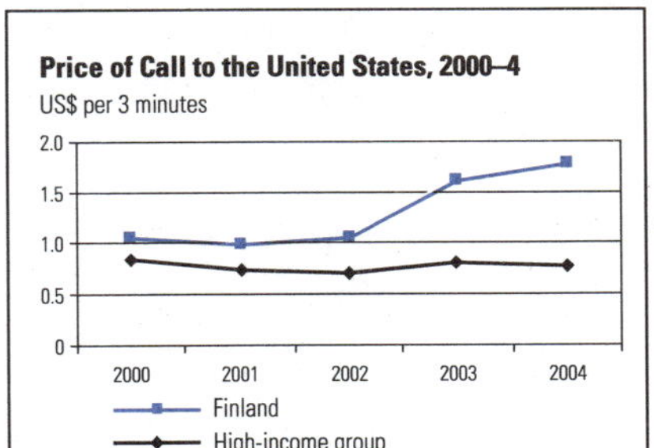

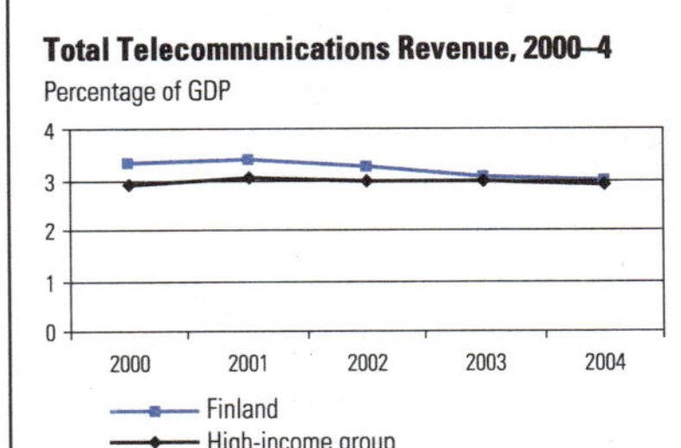

Notes: Figures in italics are for years other than those specified. .. indicates data are not available. C = competition; GDP = gross domestic product; GNI = gross national income; ICT = information and communication technology; M = monopoly; MDG = Millennium Development Goal; P = partial competition; and PCs = personal computers.

a. Outgoing and incoming.

b. Millennium Development Goal indicators 47, 48a, and 48b.

Sources: Economic and social context: UNDP and World Bank; ICT sector structure: ITU, WEF; ICT sector performance: Global Insight/WITSA, ITU, Netcraft, UNDESA, UNPAN, and World Bank. Produced by the Global Information and Communication Technologies Department and the Development Economics Data Group. For complete information, see Definitions and Data Sources.

France

	France 2000	France 2004	High-income group 2004
Economic and social context			
Population, total (millions)	59	60	1,001
Urban population (% total population)	76	76	77
Poverty (% population below US$1 per day)	..	..	..
GNI per capita, Atlas method (current US$)	23,990	30,090	32,040
GDP growth, 1995–2000 and 2000–4 (%)	2.7	1.4	2.0
Adult literacy rate (% ages 15 and over)	..	..	..
Primary, secondary, tertiary school enrollment (% gross)	91	*92*	*93*
ICT sector structure			
Separate telecommunications regulator	Yes	Yes	
Status of main fixed-line operator	*Mixed*	Mixed	
Level of competition: international long distance	*C*	C	
Level of competition: mobile	*P*	C	
Level of competition: Internet service provider	*C*	C	
Government prioritization of ICT (scale 1–7)	..	4.7	4.9
ICT sector performance			
Access			
Telephone main lines (per 1,000 people)	577	565	558
International voice traffic (minutes per person)[a]	173	212	..
Mobile subscribers (per 1,000 people)	493	719	767
Population covered by mobile telephony (%)	99	99	98
Internet users (per 1,000 people)	144	399	*480*
Personal computers (per 1,000 people)	304	414	*504*
Households with television (%)	94	*95*	*98*
Quality			
Telephone faults (per 100 main lines per year)	6.5	..	..
Broadband subscribers (per 1,000 people)	3.3	108.8	126.2
International Internet bandwidth (bits per person)	1,148	8,433	4,718
Affordability			
Price basket for fixed line (US$ per month, residential)	18.4	25.7	25.8
Price basket for mobile (US$ per month)	..	29.7	17.8
Price basket for Internet (US$ per month)	..	*14.2*	*20.9*
Price of call to United States (US$ per 3 minutes)	0.82	0.84	0.77
Institutional efficiency and sustainability			
Total telecommunications revenue (% GDP)	2.1	2.3	2.9
Total telephone subscribers per employee	405	542	*485*
Total telecommunications investment (% revenue)	26.5	10.9	*12.3*
ICT applications			
ICT expenditure (% GDP)	6.5	*5.9*	*7.1*
E-government readiness index (scale 0–1)	..	0.54	0.70
Secure Internet servers (per 1 million people)	*27.7*	64.3	311.4
Schools connected to the Internet (%)	..	*97*	*99*

GNI per Capita, Atlas Method, 2000–4

Current US$

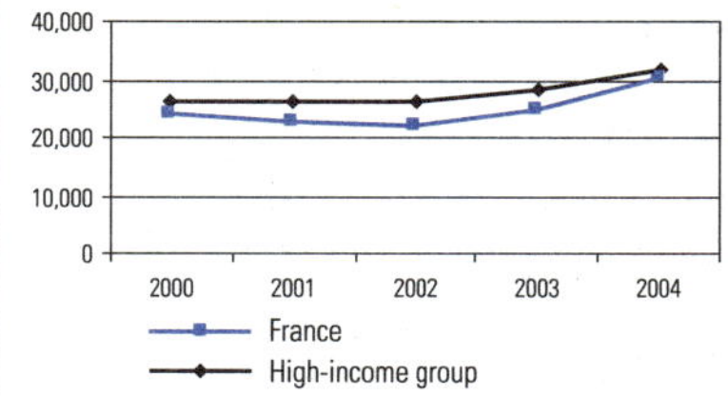

ICT MDG[b] Indicators, 2000–4

Number per 1,000 people

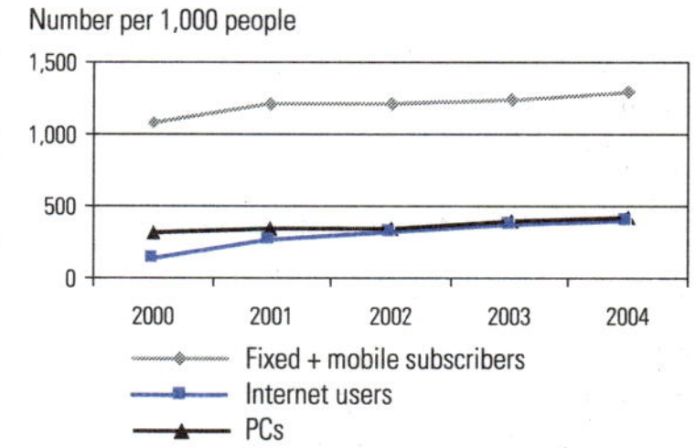

Price of Call to the United States, 2000–4

US$ per 3 minutes

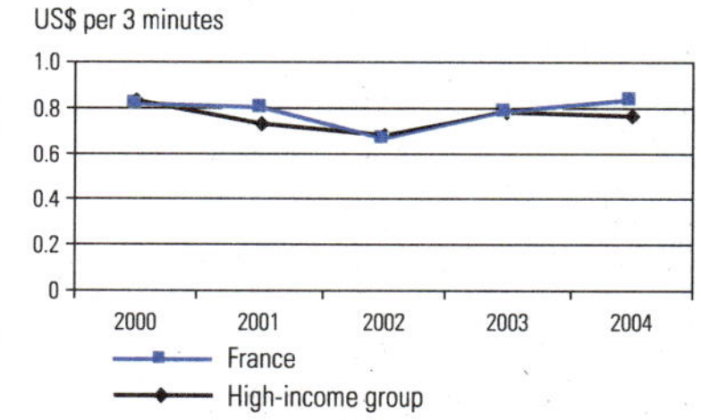

Total Telecommunications Revenue, 2000–4

Percentage of GDP

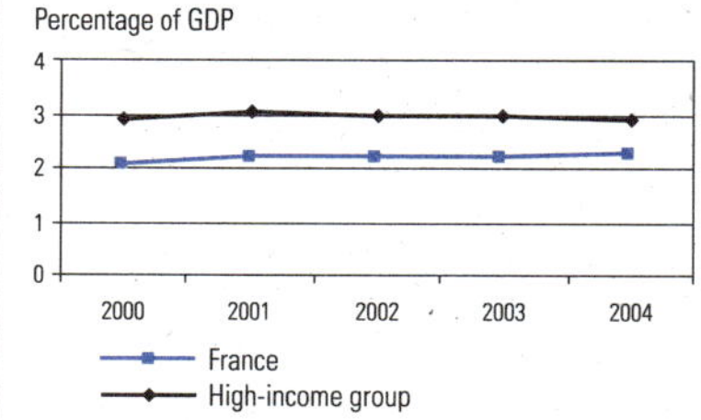

Notes: Figures in italics are for years other than those specified. .. indicates data are not available. C = competition; GDP = gross domestic product; GNI = gross national income; ICT = information and communication technology; M = monopoly; MDG = Millennium Development Goal; P = partial competition; and PCs = personal computers.

a. Outgoing and incoming.

b. Millennium Development Goal indicators 47, 48a, and 48b.

Sources: Economic and social context: UNDP and World Bank; ICT sector structure: ITU, WEF; ICT sector performance: Global Insight/WITSA, ITU, Netcraft, UNDESA, UNPAN, and World Bank. Produced by the Global Information and Communication Technologies Department and the Development Economics Data Group. For complete information, see Definitions and Data Sources.

Gabon

	Gabon 2000	Gabon 2004	Upper-middle-income group 2004	Sub-Saharan Africa Region 2004
Economic and social context				
Population, total (millions)	1	1	576	719
Urban population (% total population)	81	84	72	37
Poverty (% population below US$1 per day)	..	..	..	*46.4*
GNI per capita, Atlas method (current US$)	3,120	3,940	4,770	600
GDP growth, 1995–2000 and 2000–4 (%)	1.4	1.7	2.7	3.9
Adult literacy rate (% ages 15 and over)	..	..	94	59
Primary, secondary, tertiary school enrollment (% gross)	72	..	*80*	*52*
ICT sector structure				
Separate telecommunications regulator	No	Yes		
Status of main fixed-line operator	*Public*	Public		
Level of competition: international long distance	*M*	M		
Level of competition: mobile	*P*	C		
Level of competition: Internet service provider	*C*	C		
Government prioritization of ICT (scale 1–7)	..	..	4.1	*4.4*
ICT sector performance				
Access				
Telephone main lines (per 1,000 people)	31	*29*	*220*	17
International voice traffic (minutes per person)[a]	..	..	*39*	..
Mobile subscribers (per 1,000 people)	95	*223*	490	86
Population covered by mobile telephony (%)	*13*	*24*	*84*	..
Internet users (per 1,000 people)	12	*26*	*133*	15
Personal computers (per 1,000 people)	10	*22*	*99*	*12*
Households with television (%)	51	*54*	*92*	*15*
Quality				
Telephone faults (per 100 main lines per year)	*57.0*	*50.0*	*20.3*	..
Broadband subscribers (per 1,000 people)	*0.0*	*0.0*	*3.7*	*0.1*
International Internet bandwidth (bits per person)	0	*34*	176	*4*
Affordability				
Price basket for fixed line (US$ per month, residential)	27.3	*27.8*	*13.9*	*8.5*
Price basket for mobile (US$ per month)	..	16.8	11.1	13.5
Price basket for Internet (US$ per month)	..	*122.0*	*20.8*	*54.8*
Price of call to United States (US$ per 3 minutes)	*14.12*	5.68	1.03	2.43
Institutional efficiency and sustainability				
Total telecommunications revenue (% GDP)	2.2	*2.6*	3.4	5.0
Total telephone subscribers per employee	150	..	*402*	*144*
Total telecommunications investment (% revenue)	41.8	*8.8*	*18.6*	*27.8*
ICT applications				
ICT expenditure (% GDP)	..	..	*5.0*	..
E-government readiness index (scale 0–1)	..	0.08	0.49	0.11
Secure Internet servers (per 1 million people)	*0.8*	4.4	10.7	1.9
Schools connected to the Internet (%)	..	..	*60*	..

GNI per Capita, Atlas Method, 2000–4
Current US$

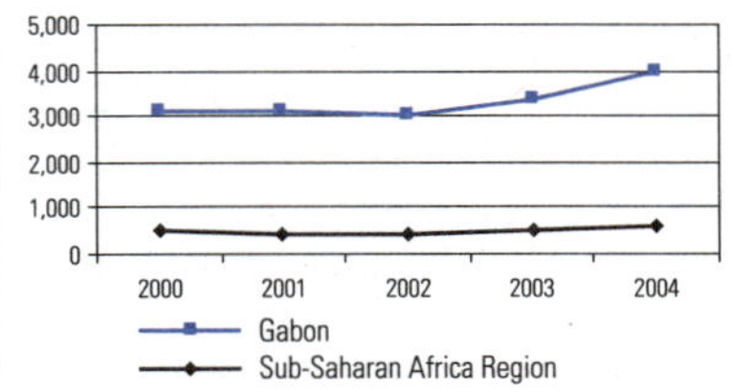

ICT MDG[b] Indicators, 2000–4
Number per 1,000 people

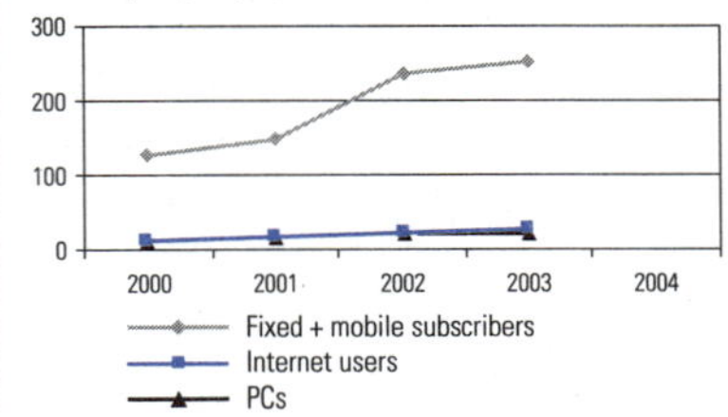

Price of Call to the United States, 2000–4
US$ per 3 minutes

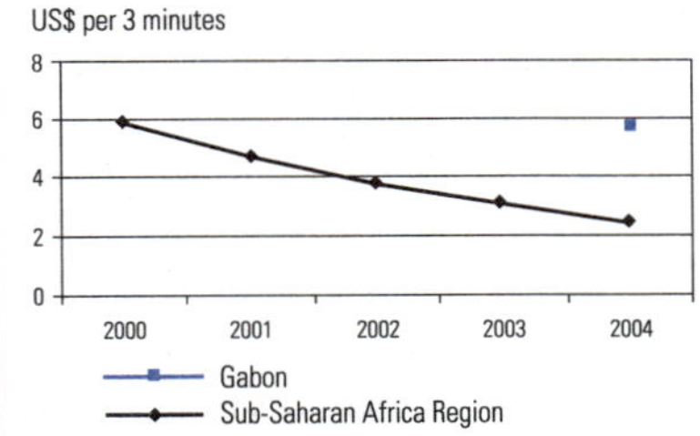

Total Telecommunications Revenue, 2000–4
Percentage of GDP

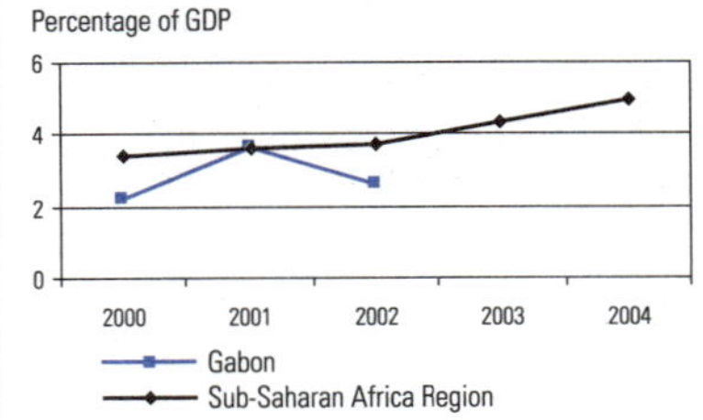

Notes: Figures in italics are for years other than those specified. .. indicates data are not available. C = competition; GDP = gross domestic product; GNI = gross national income; ICT = information and communication technology; M = monopoly; MDG = Millennium Development Goal; P = partial competition; and PCs = personal computers.

a. Outgoing and incoming.

b. Millennium Development Goal indicators 47, 48a, and 48b.

Sources: Economic and social context: UNDP and World Bank; ICT sector structure: ITU, WEF; ICT sector performance: Global Insight/WITSA, ITU, Netcraft, UNDESA, UNPAN, and World Bank. Produced by the Global Information and Communication Technologies Department and the Development Economics Data Group. For complete information, see Definitions and Data Sources.

The Gambia

	Gambia, The 2000	Gambia, The 2004	Low-income group 2004	Sub-Saharan Africa Region 2004
Economic and social context				
Population, total (millions)	1	1	2,338	719
Urban population (% total population)	26	26	31	37
Poverty (% population below US$1 per day)	..	..	..	*46.4*
GNI per capita, Atlas method (current US$)	320	290	510	600
GDP growth, 1995–2000 and 2000–4 (%)	4.6	3.8	5.4	3.9
Adult literacy rate (% ages 15 and over)	..	..	61	59
Primary, secondary, tertiary school enrollment (% gross)	45	*48*	*55*	*52*
ICT sector structure				
Separate telecommunications regulator	No	Yes		
Status of main fixed-line operator	*Public*	Public		
Level of competition: international long distance	*M*	M		
Level of competition: mobile	*M*	P		
Level of competition: Internet service provider	*C*	C		
Government prioritization of ICT (scale 1–7)	..	5.3	..	*4.4*
ICT sector performance				
Access				
Telephone main lines (per 1,000 people)	25	21	33	17
International voice traffic (minutes per person)[a]	..	..	*4*	..
Mobile subscribers (per 1,000 people)	4	131	48	86
Population covered by mobile telephony (%)	..	60	*43*	..
Internet users (per 1,000 people)	9	35	20	15
Personal computers (per 1,000 people)	11	*15*	*8*	*12*
Households with television (%)	12	*12*	*16*	*15*
Quality				
Telephone faults (per 100 main lines per year)	..	..	..	..
Broadband subscribers (per 1,000 people)	0.0	*0.0*	0.5	*0.1*
International Internet bandwidth (bits per person)	*0*	*3*	*3*	*4*
Affordability				
Price basket for fixed line (US$ per month, residential)	..	*3.9*	*6.6*	*8.5*
Price basket for mobile (US$ per month)	..	..	11.6	13.5
Price basket for Internet (US$ per month)	..	*27.1*	*45.5*	*54.8*
Price of call to United States (US$ per 3 minutes)	5.39	1.81	1.95	2.43
Institutional efficiency and sustainability				
Total telecommunications revenue (% GDP)	6.4	10.3	2.3	5.0
Total telephone subscribers per employee	41	*124*	*89*	*144*
Total telecommunications investment (% revenue)	23.5	*9.0*	*27.8*	*27.8*
ICT applications				
ICT expenditure (% GDP)	..	..	*4.1*	..
E-government readiness index (scale 0–1)	..	0.08	0.12	0.11
Secure Internet servers (per 1 million people)	..	..	0.3	1.9
Schools connected to the Internet (%)	..	..	..	..

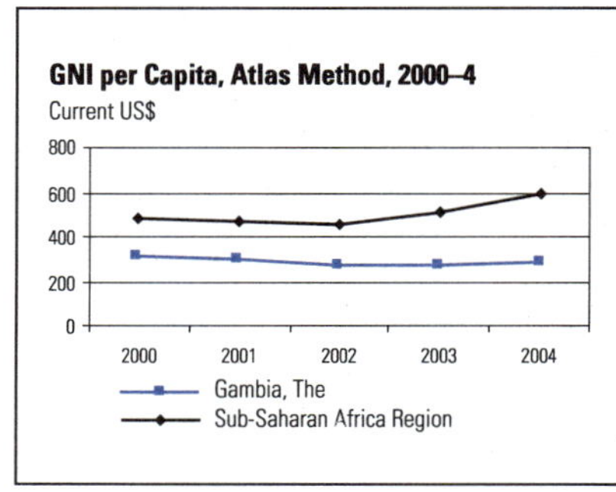

GNI per Capita, Atlas Method, 2000–4

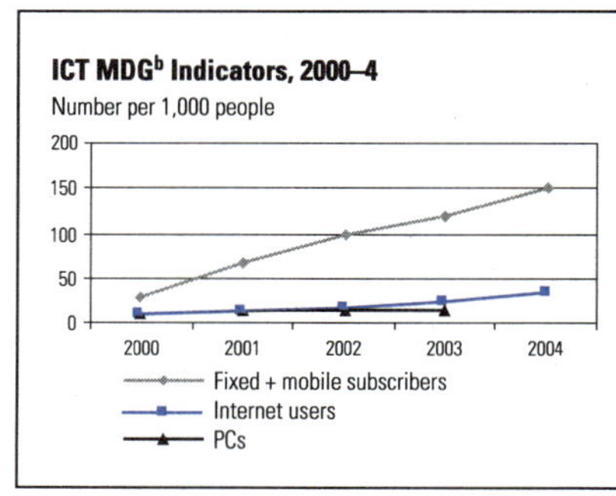

ICT MDG[b] Indicators, 2000–4

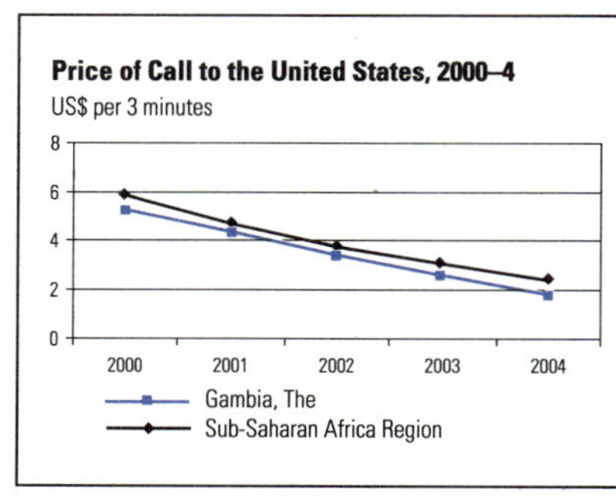

Price of Call to the United States, 2000–4

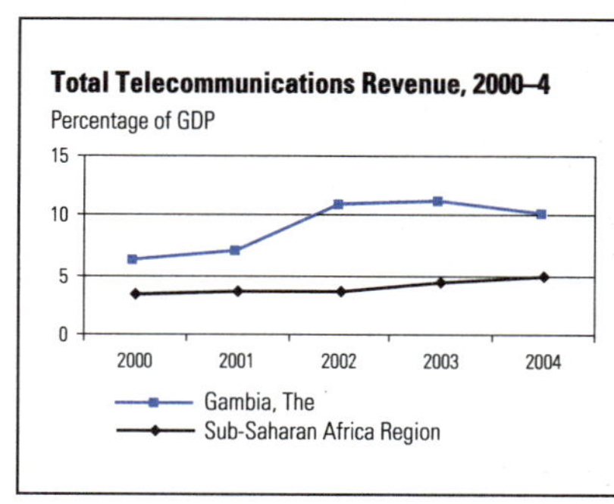

Total Telecommunications Revenue, 2000–4

Notes: Figures in italics are for years other than those specified. .. indicates data are not available. C = competition; GDP = gross domestic product; GNI = gross national income; ICT = information and communication technology; M = monopoly; MDG = Millennium Development Goal; P = partial competition; and PCs = personal computers.

a. Outgoing and incoming.

b. Millennium Development Goal indicators 47, 48a, and 48b.

Sources: Economic and social context: UNDP and World Bank; ICT sector structure: ITU, WEF; ICT sector performance: Global Insight/WITSA, ITU, Netcraft, UNDESA, UNPAN, and World Bank. Produced by the Global Information and Communication Technologies Department and the Development Economics Data Group. For complete information, see Definitions and Data Sources.

Georgia

	Georgia 2000	Georgia 2004	Lower-middle-income group 2004	Europe & Central Asia Region 2004
Economic and social context				
Population, total (millions)	5	5	2,430	472
Urban population (% total population)	53	52	49	64
Poverty (% population below US$1 per day)	2.8	*2.7*	..	*3.6*
GNI per capita, Atlas method (current US$)	660	1,040	1,580	3,290
GDP growth, 1995–2000 and 2000–4 (%)	5.7	7.6	5.7	5.0
Adult literacy rate (% ages 15 and over)	..	..	90	97
Primary, secondary, tertiary school enrollment (% gross)	69	*71*	*72*	*81*
ICT sector structure				
Separate telecommunications regulator	Yes	Yes		
Status of main fixed-line operator	*Mixed*	..		
Level of competition: international long distance	*P*	C		
Level of competition: mobile	*C*	C		
Level of competition: Internet service provider	..	C		
Government prioritization of ICT (scale 1–7)	..	3.3	3.8	3.8
ICT sector performance				
Access				
Telephone main lines (per 1,000 people)	108	*142*	192	*243*
International voice traffic (minutes per person)[a]	29	..	..	*28*
Mobile subscribers (per 1,000 people)	41	231	255	487
Population covered by mobile telephony (%)	*79*	..	*76*	*82*
Internet users (per 1,000 people)	5	*33*	70	*115*
Personal computers (per 1,000 people)	24	*34*	38	*73*
Households with television (%)	81	*76*	*89*	*92*
Quality				
Telephone faults (per 100 main lines per year)	26.3	*17.2*	..	*30.4*
Broadband subscribers (per 1,000 people)	..	*0.3*	12.6	*2.4*
International Internet bandwidth (bits per person)	..	*6*	58	148
Affordability				
Price basket for fixed line (US$ per month, residential)	4.2	*4.6*	*5.5*	*3.5*
Price basket for mobile (US$ per month)	..	6.5	8.9	10.3
Price basket for Internet (US$ per month)	..	*25.1*	*25.3*	*19.8*
Price of call to United States (US$ per 3 minutes)	2.88	*0.68*	1.45	1.06
Institutional efficiency and sustainability				
Total telecommunications revenue (% GDP)	3.7	*4.0*	3.6	3.5
Total telephone subscribers per employee	69	*162*	*195*	*150*
Total telecommunications investment (% revenue)	..	*3.5*	*25.5*	*19.0*
ICT applications				
ICT expenditure (% GDP)	..	..	*5.1*	*5.1*
E-government readiness index (scale 0–1)	..	0.15	0.26	0.39
Secure Internet servers (per 1 million people)	*2.1*	2.4	1.6	6.4
Schools connected to the Internet (%)	..	..	..	*65*

GNI per Capita, Atlas Method, 2000–4

Current US$

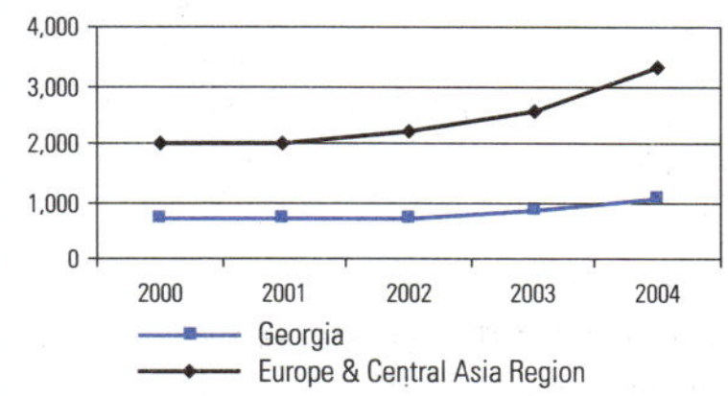

ICT MDG[b] Indicators, 2000–4

Number per 1,000 people

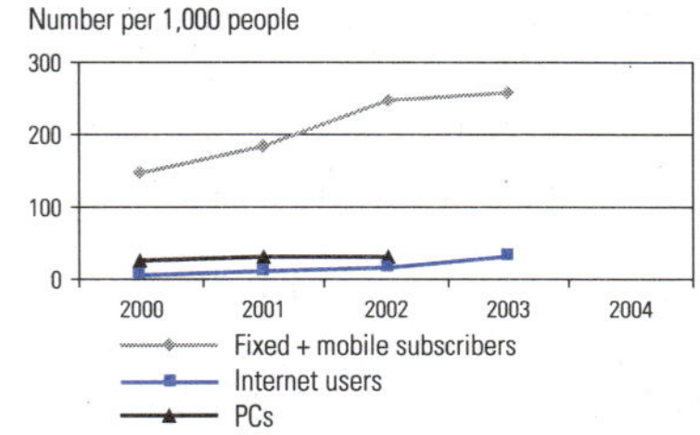

Price of Call to the United States, 2000–4

US$ per 3 minutes

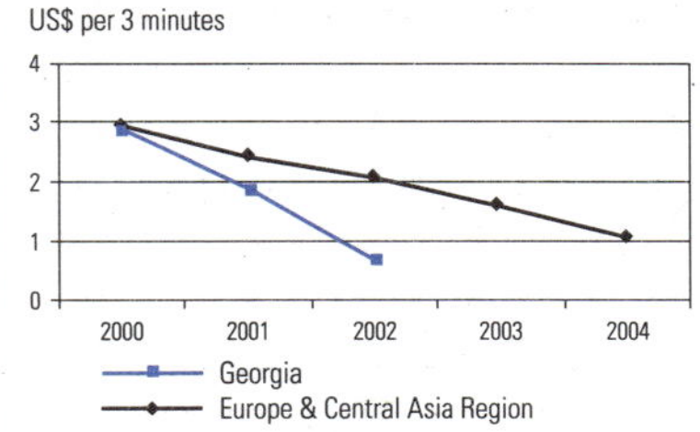

Total Telecommunications Revenue, 2000–4

Percentage of GDP

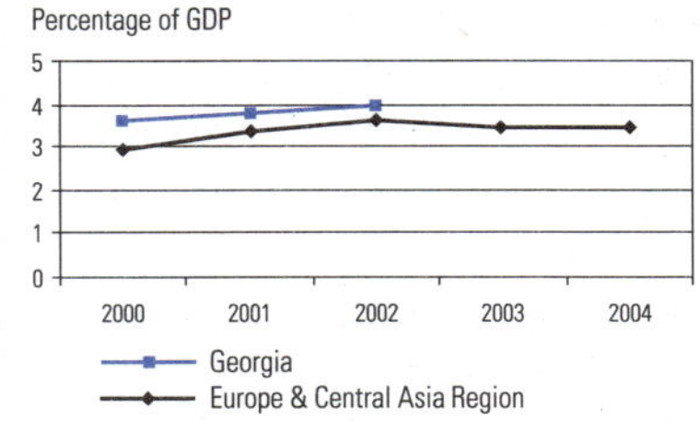

Notes: Figures in italics are for years other than those specified. .. indicates data are not available. C = competition; GDP = gross domestic product; GNI = gross national income; ICT = information and communication technology; M = monopoly; MDG = Millennium Development Goal; P = partial competition; and PCs = personal computers.

a. Outgoing and incoming.

b. Millennium Development Goal indicators 47, 48a, and 48b.

Sources: Economic and social context: UNDP and World Bank; ICT sector structure: ITU, WEF; ICT sector performance: Global Insight/WITSA, ITU, Netcraft, UNDESA, UNPAN, and World Bank. Produced by the Global Information and Communication Technologies Department and the Development Economics Data Group. For complete information, see Definitions and Data Sources.

Germany

	Germany 2000	Germany 2004	High-income group 2004
Economic and social context			
Population, total (millions)	82	83	1,001
Urban population (% total population)	88	88	77
Poverty (% population below US$1 per day)	..	..	..
GNI per capita, Atlas method (current US$)	25,140	30,120	32,040
GDP growth, 1995–2000 and 2000–4 (%)	1.8	0.5	2.0
Adult literacy rate (% ages 15 and over)	..	..	..
Primary, secondary, tertiary school enrollment (% gross)	89	*89*	*93*
ICT sector structure			
Separate telecommunications regulator	Yes	Yes	
Status of main fixed-line operator	*Mixed*	Mixed	
Level of competition: international long distance	*C*	C	
Level of competition: mobile	*P*	P	
Level of competition: Internet service provider	*C*	C	
Government prioritization of ICT (scale 1–7)	..	4.9	4.9
ICT sector performance			
Access			
Telephone main lines (per 1,000 people)	611	660	558
International voice traffic (minutes per person)[a]	191	..	..
Mobile subscribers (per 1,000 people)	586	858	767
Population covered by mobile telephony (%)	99	99	98
Internet users (per 1,000 people)	302	*472*	*480*
Personal computers (per 1,000 people)	336	440	*504*
Households with television (%)	94	*94*	*98*
Quality			
Telephone faults (per 100 main lines per year)	..	..	..
Broadband subscribers (per 1,000 people)	3.2	83.6	126.2
International Internet bandwidth (bits per person)	848	6,850	4,718
Affordability			
Price basket for fixed line (US$ per month, residential)	15.1	*17.5*	25.8
Price basket for mobile (US$ per month)	..	30.6	17.8
Price basket for Internet (US$ per month)	..	*14.1*	*20.9*
Price of call to United States (US$ per 3 minutes)	0.34	0.43	0.77
Institutional efficiency and sustainability			
Total telecommunications revenue (% GDP)	2.7	3.0	2.9
Total telephone subscribers per employee	409	*526*	*485*
Total telecommunications investment (% revenue)	17.4	*7.9*	*12.3*
ICT applications			
ICT expenditure (% GDP)	6.2	*5.7*	*7.1*
E-government readiness index (scale 0–1)	..	0.80	0.70
Secure Internet servers (per 1 million people)	*62.6*	167.6	311.4
Schools connected to the Internet (%)	..	*99*	*99*

GNI per Capita, Atlas Method, 2000–4
Current US$

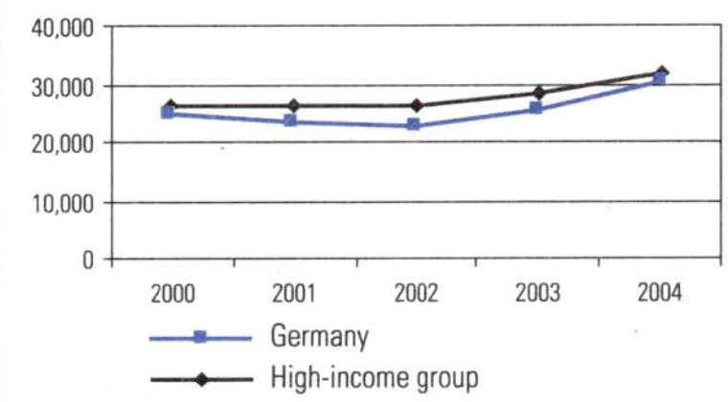

ICT MDG[b] Indicators, 2000–4
Number per 1,000 people

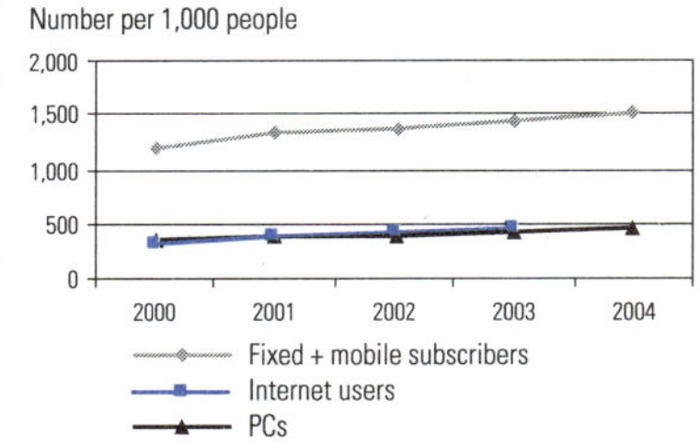

Price of Call to the United Stated, 2000–4
US$ per 3 minutes

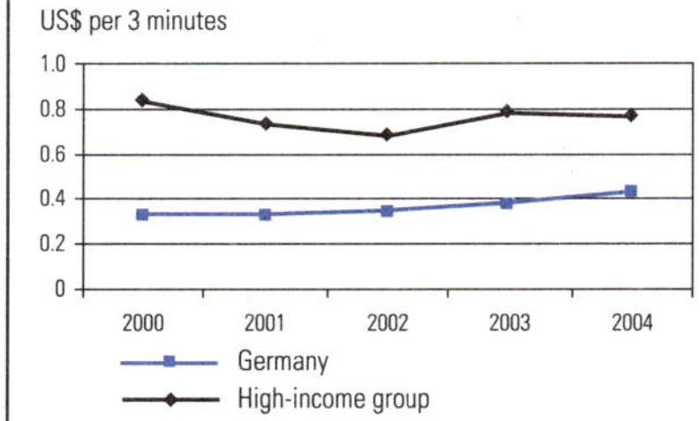

Total Telecommunications Revenue, 2000–4
Percentage of GDP

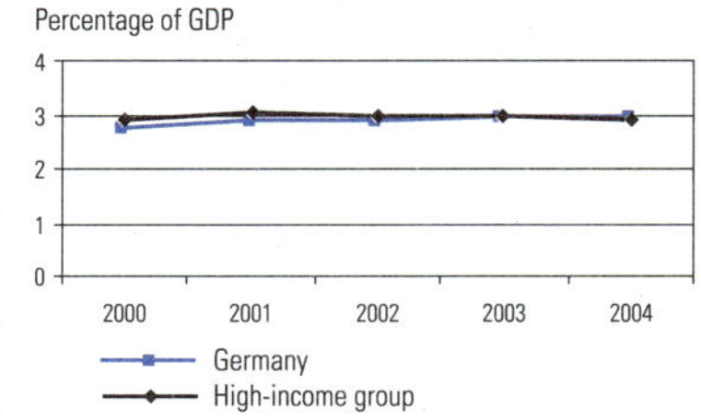

Notes: Figures in italics are for years other than those specified. .. indicates data are not available. C = competition; GDP = gross domestic product; GNI = gross national income; ICT = information and communication technology; M = monopoly; MDG = Millennium Development Goal; P = partial competition; and PCs = personal computers.

a. Outgoing and incoming.

b. Millennium Development Goal indicators 47, 48a, and 48b.

Sources: Economic and social context: UNDP and World Bank; ICT sector structure: ITU, WEF; ICT sector performance: Global Insight/WITSA, ITU, Netcraft, UNDESA, UNPAN, and World Bank. Produced by the Global Information and Communication Technologies Department and the Development Economics Data Group. For complete information, see Definitions and Data Sources.

Ghana

	Ghana 2000	Ghana 2004	Low-income group 2004	Sub-Saharan Africa Region 2004
Economic and social context				
Population, total (millions)	20	21	2,338	719
Urban population (% total population)	44	46	31	37
Poverty (% population below US$1 per day)	*44.8*	..	..	*46.4*
GNI per capita, Atlas method (current US$)	330	380	510	600
GDP growth, 1995–2000 and 2000–4 (%)	4.4	4.8	5.4	3.9
Adult literacy rate (% ages 15 and over)	54	..	61	59
Primary, secondary, tertiary school enrollment (% gross)	45	*48*	*55*	*52*
ICT sector structure				
Separate telecommunications regulator	Yes	Yes		
Status of main fixed-line operator	*Mixed*	Mixed		
Level of competition: international long distance	*P*	P		
Level of competition: mobile	*C*	P		
Level of competition: Internet service provider	*C*	C		
Government prioritization of ICT (scale 1–7)	..	5.7	..	*4.4*
ICT sector performance				
Access				
Telephone main lines (per 1,000 people)	12	*15*	33	17
International voice traffic (minutes per person)[a]	11	*10*	*4*	..
Mobile subscribers (per 1,000 people)	7	71	48	86
Population covered by mobile telephony (%)	..	*28*	*43*	..
Internet users (per 1,000 people)	2	24	20	15
Personal computers (per 1,000 people)	3	*4*	*8*	*12*
Households with television (%)	22	*21*	*16*	*15*
Quality				
Telephone faults (per 100 main lines per year)	*67.4*	..	..	..
Broadband subscribers (per 1,000 people)	0.0	*0.0*	0.5	*0.1*
International Internet bandwidth (bits per person)	*0*	*1*	*3*	*4*
Affordability				
Price basket for fixed line (US$ per month, residential)	3.9	*3.6*	*6.6*	*8.5*
Price basket for mobile (US$ per month)	..	11.1	11.6	13.5
Price basket for Internet (US$ per month)	..	*40.0*	*45.5*	*54.8*
Price of call to United States (US$ per 3 minutes)	1.65	0.39	1.95	2.43
Institutional efficiency and sustainability				
Total telecommunications revenue (% GDP)	1.8	5.3	2.3	5.0
Total telephone subscribers per employee	97	*150*	*89*	*144*
Total telecommunications investment (% revenue)	*29.5*	*46.4*	*27.8*	*27.8*
ICT applications				
ICT expenditure (% GDP)	..	..	*4.1*	..
E-government readiness index (scale 0–1)	..	0.05	0.12	0.11
Secure Internet servers (per 1 million people)	*0.1*	0.0	0.3	1.9
Schools connected to the Internet (%)	..	1	..	..

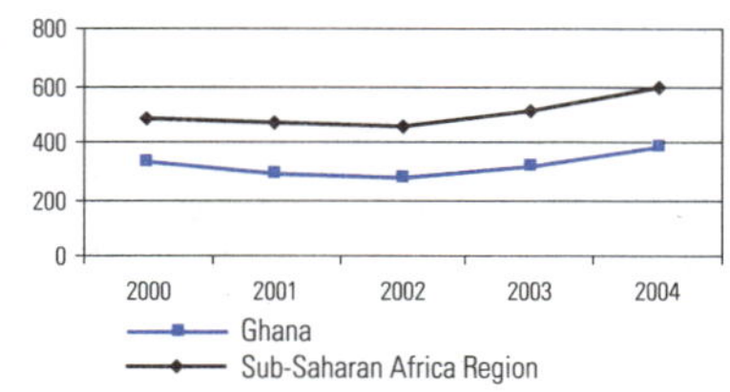

GNI per Capita, Atlas Method, 2000–4

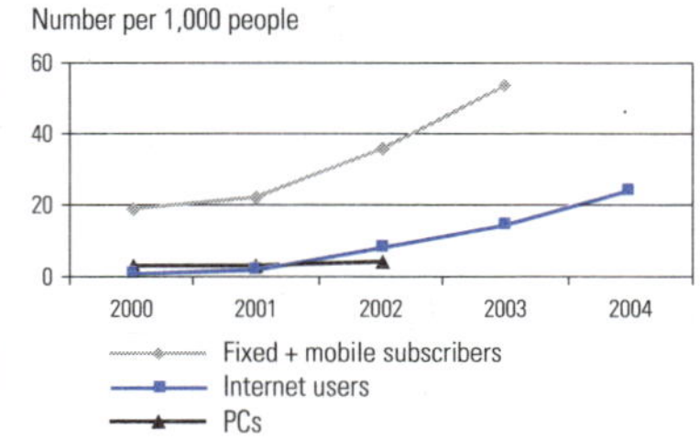

ICT MDG[b] Indicators, 2000–4

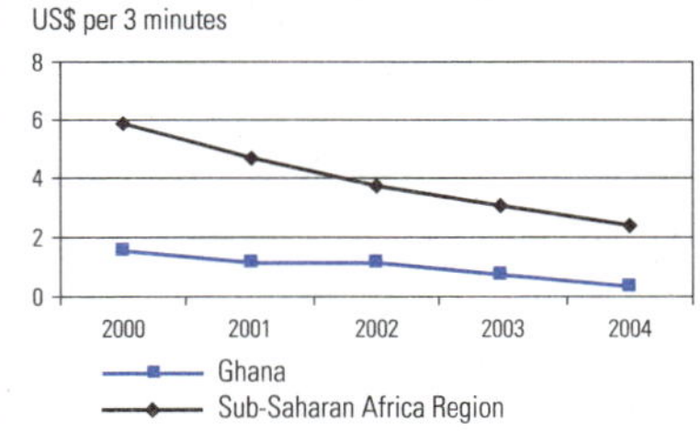

Price of Call to the United States, 2000–4

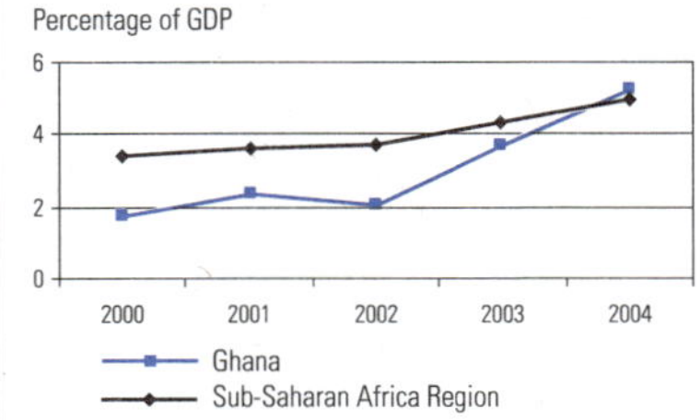

Total Telecommunications Revenue, 2000–4

Notes: Figures in italics are for years other than those specified. .. indicates data are not available. C = competition; GDP = gross domestic product; GNI = gross national income; ICT = information and communication technology; M = monopoly; MDG = Millennium Development Goal; P = partial competition; and PCs = personal computers.

a. Outgoing and incoming.

b. Millennium Development Goal indicators 47, 48a, and 48b.

Sources: Economic and social context: UNDP and World Bank; ICT sector structure: ITU, WEF; ICT sector performance: Global Insight/WITSA, ITU, Netcraft, UNDESA, UNPAN, and World Bank. Produced by the Global Information and Communication Technologies Department and the Development Economics Data Group. For complete information, see Definitions and Data Sources.

Greece

	Greece 2000	Greece 2004	High-income group 2004
Economic and social context			
Population, total (millions)	11	11	1,001
Urban population (% total population)	60	61	77
Poverty (% population below US$1 per day)	..	..	..
GNI per capita, Atlas method (current US$)	11,290	16,610	32,040
GDP growth, 1995–2000 and 2000–4 (%)	3.4	4.1	2.0
Adult literacy rate (% ages 15 and over)	..	*91*	..
Primary, secondary, tertiary school enrollment (% gross)	86	*92*	*93*
ICT sector structure			
Separate telecommunications regulator	Yes	Yes	
Status of main fixed-line operator	*Mixed*	Mixed	
Level of competition: international long distance	*M*	C	
Level of competition: mobile	*P*	P	
Level of competition: Internet service provider	*C*	C	
Government prioritization of ICT (scale 1–7)	..	4.3	4.9
ICT sector performance			
Access			
Telephone main lines (per 1,000 people)	518	573	558
International voice traffic (minutes per person)[a]	154	*167*	..
Mobile subscribers (per 1,000 people)	543	988	767
Population covered by mobile telephony (%)	99	99	98
Internet users (per 1,000 people)	92	*156*	*480*
Personal computers (per 1,000 people)	69	*82*	*504*
Households with television (%)	97	98	*98*
Quality			
Telephone faults (per 100 main lines per year)	11.3	*13.6*	..
Broadband subscribers (per 1,000 people)	0.0	4.6	126.2
International Internet bandwidth (bits per person)	51	588	4,718
Affordability			
Price basket for fixed line (US$ per month, residential)	10.2	*14.3*	25.8
Price basket for mobile (US$ per month)	..	19.3	17.8
Price basket for Internet (US$ per month)	..	*23.0*	*20.9*
Price of call to United States (US$ per 3 minutes)	0.69	1.09	0.77
Institutional efficiency and sustainability			
Total telecommunications revenue (% GDP)	4.1	*4.0*	2.9
Total telephone subscribers per employee	451	..	*485*
Total telecommunications investment (% revenue)	42.3	*22.7*	*12.3*
ICT applications			
ICT expenditure (% GDP)	4.5	*4.3*	*7.1*
E-government readiness index (scale 0–1)	..	0.41	0.70
Secure Internet servers (per 1 million people)	*10.6*	26.2	311.4
Schools connected to the Internet (%)	..	*59*	*99*

GNI per Capita, Atlas Method, 2000–4

Current US$

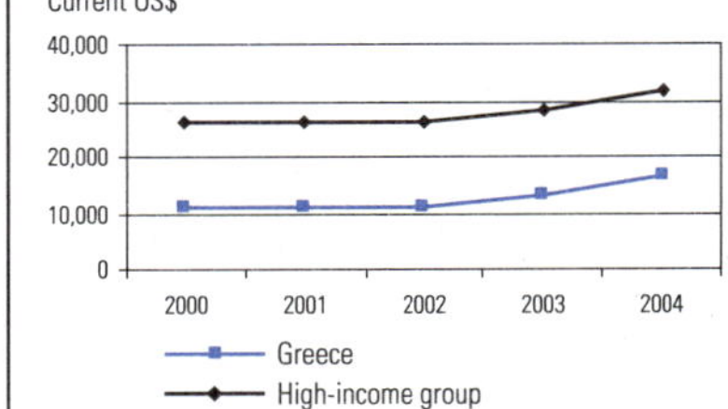

ICT MDG[b] Indicators, 2000–4

Number per 1,000 people

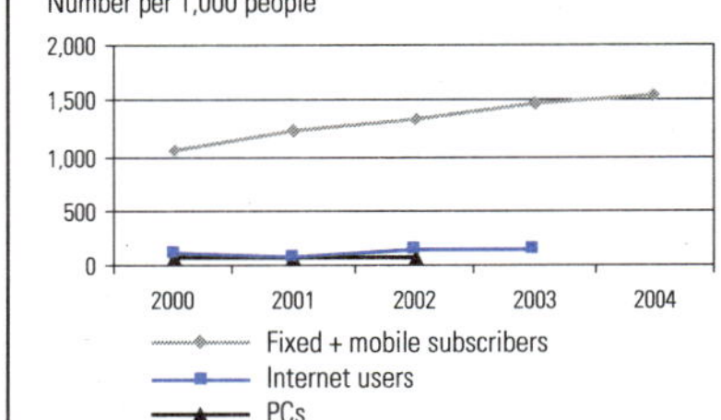

Price of Call to the United States, 2000–4

US$ per 3 minutes

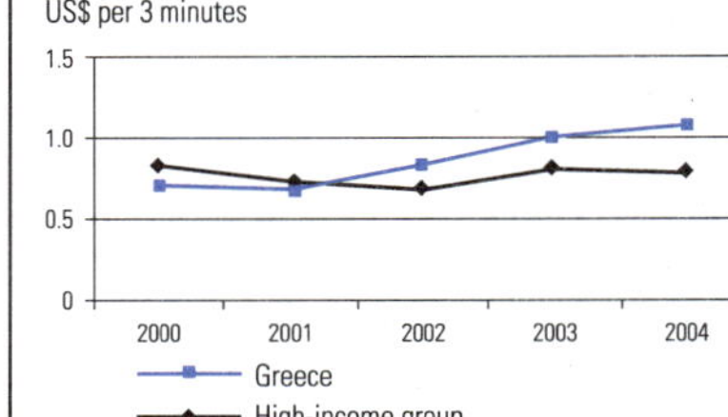

Total Telecommunications Revenue, 2000–4

Percentage of GDP

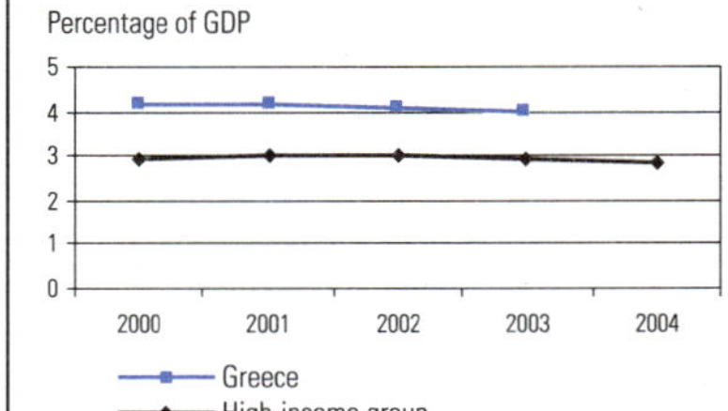

Notes: Figures in italics are for years other than those specified. .. indicates data are not available. C = competition; GDP = gross domestic product; GNI = gross national income; ICT = information and communication technology; M = monopoly; MDG = Millennium Development Goal; P = partial competition; and PCs = personal computers.

a. Outgoing and incoming.

b. Millennium Development Goal indicators 47, 48a, and 48b.

Sources: Economic and social context: UNDP and World Bank; ICT sector structure: ITU, WEF; ICT sector performance: Global Insight/WITSA, ITU, Netcraft, UNDESA, UNPAN, and World Bank. Produced by the Global Information and Communication Technologies Department and the Development Economics Data Group. For complete information, see Definitions and Data Sources.

Guatemala

	Guatemala 2000	Guatemala 2004	Lower-middle-income group 2004	Latin America & the Caribbean Region 2004
Economic and social context				
Population, total (millions)	11	13	2,430	541
Urban population (% total population)	45	47	49	*77*
Poverty (% population below US$1 per day)	16.0	..	..	*9.5*
GNI per capita, Atlas method (current US$)	1,700	2,130	1,580	3,600
GDP growth, 1995–2000 and 2000–4 (%)	4.1	2.3	5.7	1.5
Adult literacy rate (% ages 15 and over)	..	*69*	90	89
Primary, secondary, tertiary school enrollment (% gross)	57	*61*	*72*	*82*
ICT sector structure				
Separate telecommunications regulator	Yes	Yes		
Status of main fixed-line operator	*Private*	Private		
Level of competition: international long distance	*C*	C		
Level of competition: mobile	*C*	C		
Level of competition: Internet service provider	*C*	C		
Government prioritization of ICT (scale 1–7)	..	3.3	3.8	3.5
ICT sector performance				
Access				
Telephone main lines (per 1,000 people)	59	87	192	181
International voice traffic (minutes per person)[a]	37	*80*	..	..
Mobile subscribers (per 1,000 people)	75	216	255	324
Population covered by mobile telephony (%)	54	*78*	*76*	*76*
Internet users (per 1,000 people)	7	48	70	*104*
Personal computers (per 1,000 people)	19	*35*	38	*75*
Households with television (%)	39	..	*89*	*88*
Quality				
Telephone faults (per 100 main lines per year)	..	..	..	..
Broadband subscribers (per 1,000 people)	0.0	*0.0*	12.6	*5.2*
International Internet bandwidth (bits per person)	1	56	58	165
Affordability				
Price basket for fixed line (US$ per month, residential)	11.0	*10.5*	*5.5*	*9.0*
Price basket for mobile (US$ per month)	..	4.3	8.9	9.1
Price basket for Internet (US$ per month)	..	*31.2*	*25.3*	*31.5*
Price of call to United States (US$ per 3 minutes)	*0.76*	*1.21*	1.45	0.90
Institutional efficiency and sustainability				
Total telecommunications revenue (% GDP)	2.2	*2.8*	3.6	3.3
Total telephone subscribers per employee	307	..	*195*	..
Total telecommunications investment (% revenue)	..	..	*25.5*	..
ICT applications				
ICT expenditure (% GDP)	..	..	*5.1*	*5.3*
E-government readiness index (scale 0–1)	..	0.32	0.26	0.39
Secure Internet servers (per 1 million people)	*1.0*	4.0	1.6	8.6
Schools connected to the Internet (%)	..	..	..	..

GNI per Capita, Atlas Method, 2000–4
Current US$

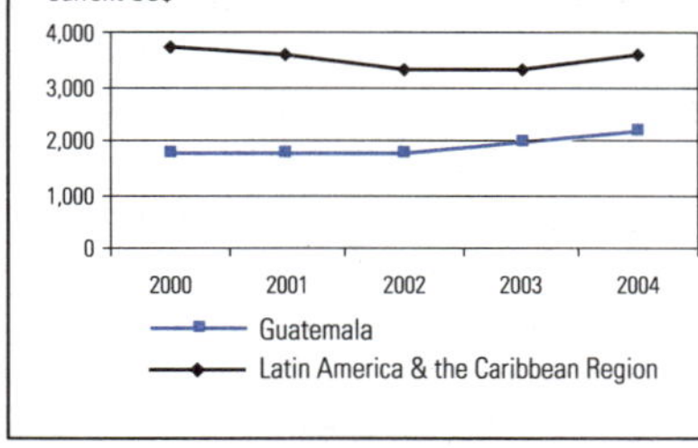

ICT MDG[b] Indicators, 2000–4
Number per 1,000 people

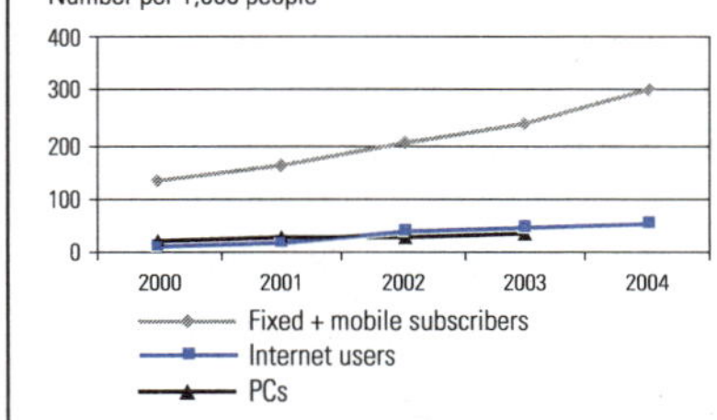

Price of Call to the United States, 2000–4
US$ per 3 minutes

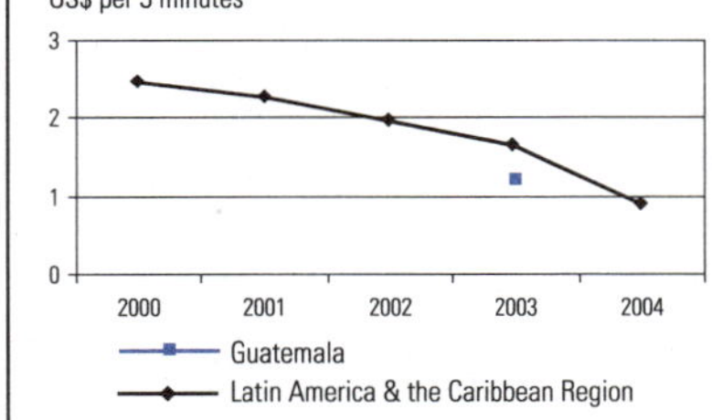

Total Telecommunications Revenue, 2000–4
Percentage of GDP

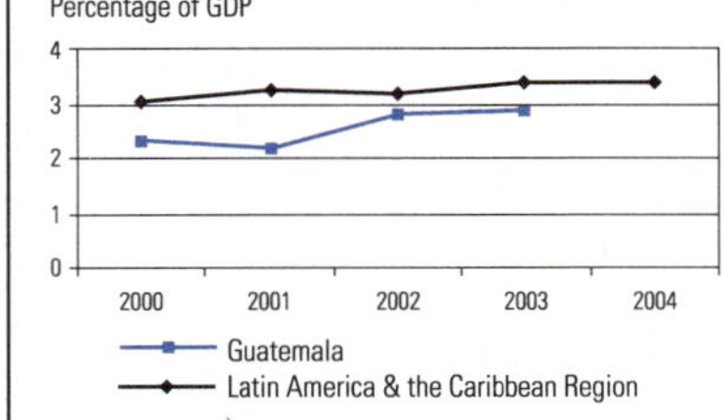

Notes: Figures in italics are for years other than those specified. .. indicates data are not available. C = competition; GDP = gross domestic product; GNI = gross national income; ICT = information and communication technology; M = monopoly; MDG = Millennium Development Goal; P = partial competition; and PCs = personal computers.

a. Outgoing and incoming.

b. Millennium Development Goal indicators 47, 48a, and 48b.

Sources: Economic and social context: UNDP and World Bank; ICT sector structure: ITU, WEF; ICT sector performance: Global Insight/WITSA, ITU, Netcraft, UNDESA, UNPAN, and World Bank. Produced by the Global Information and Communication Technologies Department and the Development Economics Data Group. For complete information, see Definitions and Data Sources.

Guinea

	Guinea 2000	Guinea 2004	Low-income group 2004	Sub-Saharan Africa Region 2004
Economic and social context				
Population, total (millions)	7	8	2,338	719
Urban population (% total population)	33	36	31	37
Poverty (% population below US$1 per day)	..	..	..	*46.4*
GNI per capita, Atlas method (current US$)	450	460	510	600
GDP growth, 1995–2000 and 2000–4 (%)	4.4	2.9	5.4	3.9
Adult literacy rate (% ages 15 and over)	..	..	61	59
Primary, secondary, tertiary school enrollment (% gross)	34	*41*	*55*	*52*
ICT sector structure				
Separate telecommunications regulator	Yes	Yes		
Status of main fixed-line operator	*Mixed*	Mixed		
Level of competition: international long distance	*M*	P		
Level of competition: mobile	*C*	P		
Level of competition: Internet service provider	*C*	C		
Government prioritization of ICT (scale 1–7)	..	..	..	*4.4*
ICT sector performance				
Access				
Telephone main lines (per 1,000 people)	3	*3*	33	17
International voice traffic (minutes per person)[a]	5	*7*	*4*	..
Mobile subscribers (per 1,000 people)	6	*14*	48	86
Population covered by mobile telephony (%)	..	..	*43*	..
Internet users (per 1,000 people)	1	*5*	20	15
Personal computers (per 1,000 people)	4	*5*	*8*	*12*
Households with television (%)	9	*9*	*16*	*15*
Quality				
Telephone faults (per 100 main lines per year)	*62.6*	*1.6*	..	..
Broadband subscribers (per 1,000 people)	0.0	*0.0*	0.5	*0.1*
International Internet bandwidth (bits per person)	0	*0*	*3*	*4*
Affordability				
Price basket for fixed line (US$ per month, residential)	10.6	*9.4*	*6.6*	*8.5*
Price basket for mobile (US$ per month)	..	..	11.6	13.5
Price basket for Internet (US$ per month)	..	*63.3*	*45.5*	*54.8*
Price of call to United States (US$ per 3 minutes)	5.15	..	1.95	2.43
Institutional efficiency and sustainability				
Total telecommunications revenue (% GDP)	0.9	*1.1*	2.3	5.0
Total telephone subscribers per employee	82	*150*	*89*	*144*
Total telecommunications investment (% revenue)	17.8	..	*27.8*	*27.8*
ICT applications				
ICT expenditure (% GDP)	..	..	*4.1*	..
E-government readiness index (scale 0–1)	..	0.03	0.12	0.11
Secure Internet servers (per 1 million people)	..	..	0.3	1.9
Schools connected to the Internet (%)	..	..	..	..

GNI per Capita, Atlas Method, 2000–4
Current US$

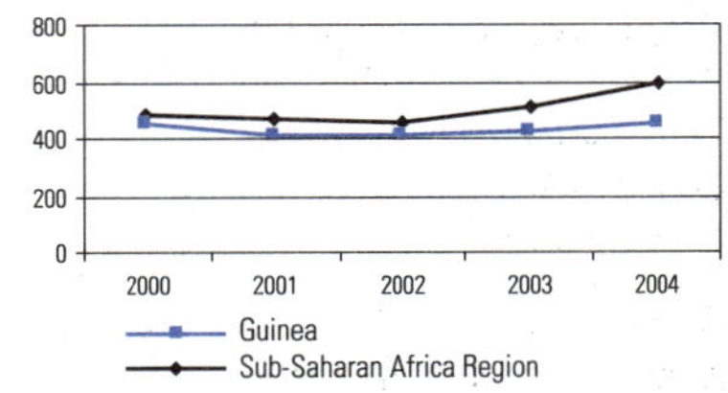

ICT MDG[b] Indicators, 2000–4
Number per 1,000 people

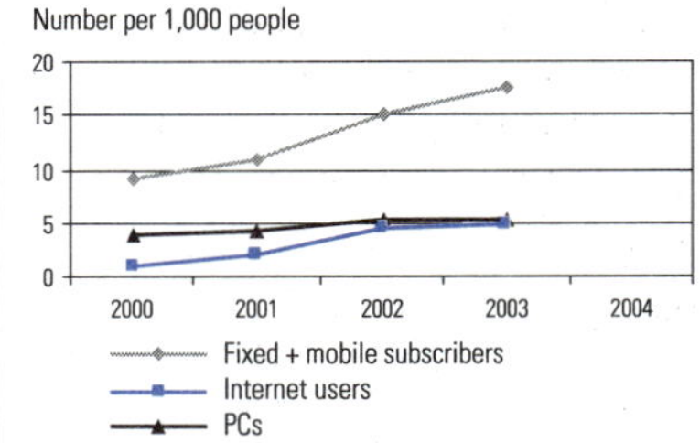

Price of Call to the United States, 2000–4
US$ per 3 minutes

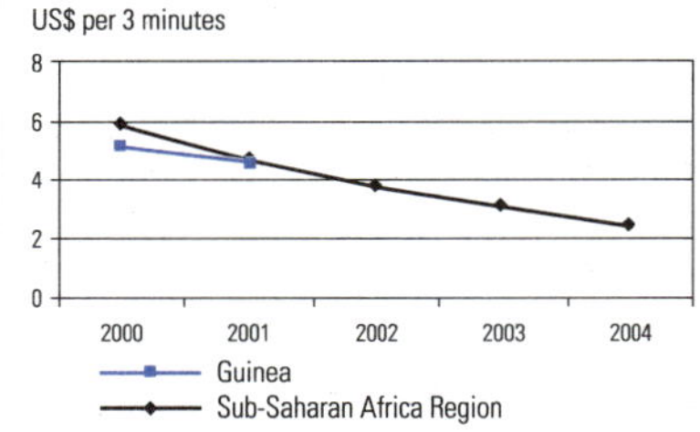

Total Telecommunications Revenue, 2000–4
Percentage of GDP

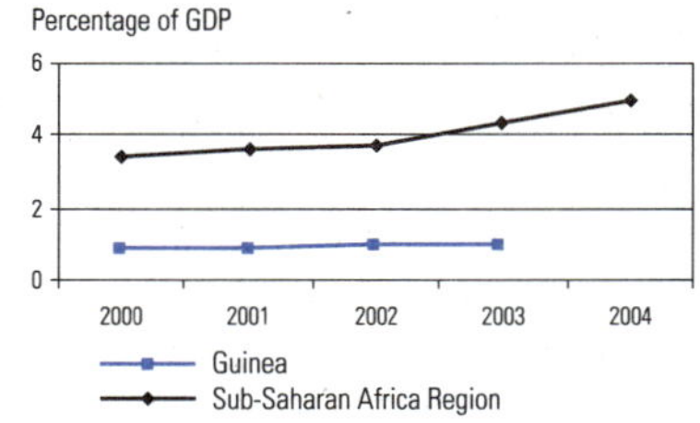

Notes: Figures in italics are for years other than those specified. .. indicates data are not available. C = competition; GDP = gross domestic product; GNI = gross national income; ICT = information and communication technology; M = monopoly; MDG = Millennium Development Goal; P = partial competition; and PCs = personal computers.

a. Outgoing and incoming.

b. Millennium Development Goal indicators 47, 48a, and 48b.

Sources: Economic and social context: UNDP and World Bank; ICT sector structure: ITU, WEF; ICT sector performance: Global Insight/WITSA, ITU, Netcraft, UNDESA, UNPAN, and World Bank. Produced by the Global Information and Communication Technologies Department and the Development Economics Data Group. For complete information, see Definitions and Data Sources.

Guinea-Bissau

	Guinea-Bissau 2000	Guinea-Bissau 2004	Low-income group 2004	Sub-Saharan Africa Region 2004
Economic and social context				
Population, total (millions)	1	2	2,338	719
Urban population (% total population)	32	35	31	37
Poverty (% population below US$1 per day)	..	..	..	*46.4*
GNI per capita, Atlas method (current US$)	160	160	510	600
GDP growth, 1995–2000 and 2000–4 (%)	–2.7	–1.2	5.4	3.9
Adult literacy rate (% ages 15 and over)	..	..	61	59
Primary, secondary, tertiary school enrollment (% gross)	37	..	*55*	*52*
ICT sector structure				
Separate telecommunications regulator	Yes	Yes		
Status of main fixed-line operator	*Mixed*	Mixed		
Level of competition: international long distance	*M*	M		
Level of competition: mobile	..	P		
Level of competition: Internet service provider	..	C		
Government prioritization of ICT (scale 1–7)	..	..	..	*4.4*
ICT sector performance				
Access				
Telephone main lines (per 1,000 people)	8	*7*	33	17
International voice traffic (minutes per person)[a]	9	..	*4*	..
Mobile subscribers (per 1,000 people)	0	*1*	48	86
Population covered by mobile telephony (%)	..	..	*43*	..
Internet users (per 1,000 people)	2	*13*	20	15
Personal computers (per 1,000 people)	..	..	*8*	*12*
Households with television (%)	20	*26*	*16*	*15*
Quality				
Telephone faults (per 100 main lines per year)	70.5	..	..	..
Broadband subscribers (per 1,000 people)	*0.0*	*0.0*	0.5	*0.1*
International Internet bandwidth (bits per person)	0	..	*3*	*4*
Affordability				
Price basket for fixed line (US$ per month, residential)	..	..	*6.6*	*8.5*
Price basket for mobile (US$ per month)	..	..	11.6	13.5
Price basket for Internet (US$ per month)	..	*105.0*	*45.5*	*54.8*
Price of call to United States (US$ per 3 minutes)	..	..	1.95	2.43
Institutional efficiency and sustainability				
Total telecommunications revenue (% GDP)	..	..	2.3	5.0
Total telephone subscribers per employee	46	..	*89*	*144*
Total telecommunications investment (% revenue)	..	..	*27.8*	*27.8*
ICT applications				
ICT expenditure (% GDP)	..	..	*4.1*	..
E-government readiness index (scale 0–1)	..	0.00	0.12	0.11
Secure Internet servers (per 1 million people)	..	..	0.3	1.9
Schools connected to the Internet (%)	..	..	..	..

GNI per Capita, Atlas Method, 2000–4

Current US$

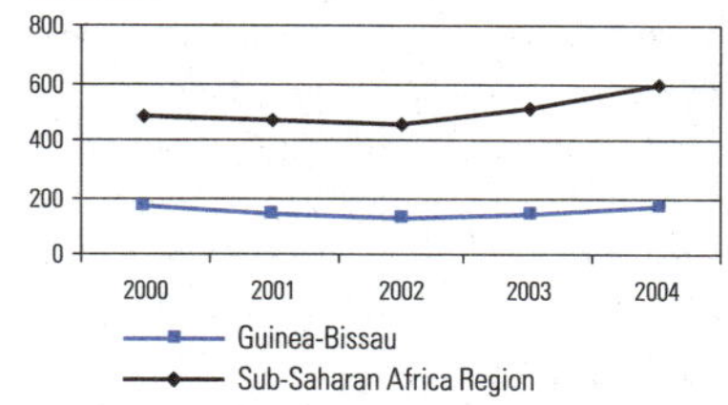

ICT MDG[b] Indicators, 2000–4

Number per 1,000 people

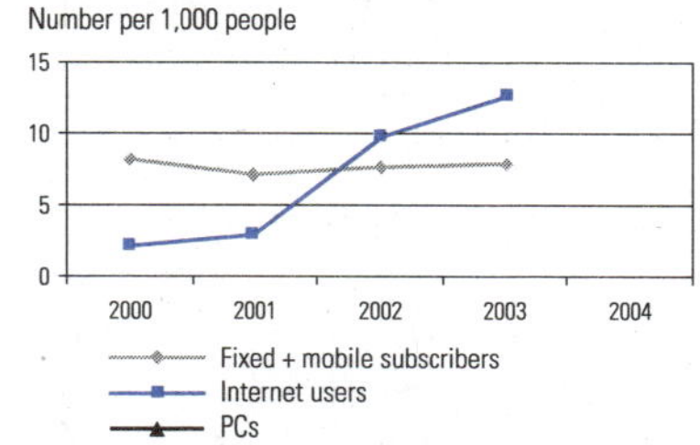

Price of Call to the United States, 2000–4

US$ per 3 minutes

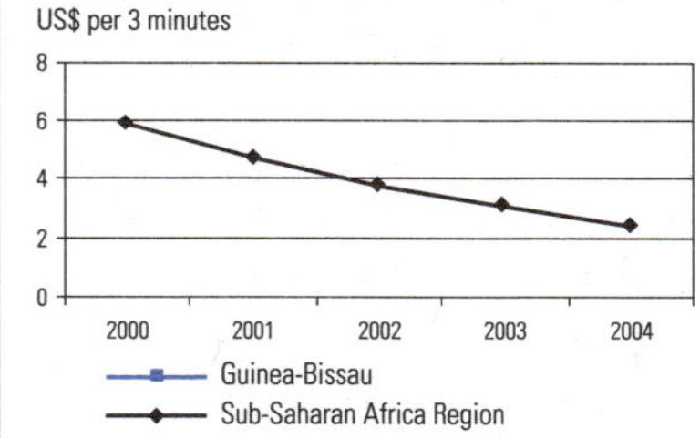

Total Telecommunications Revenue, 2000–4

Percentage of GDP

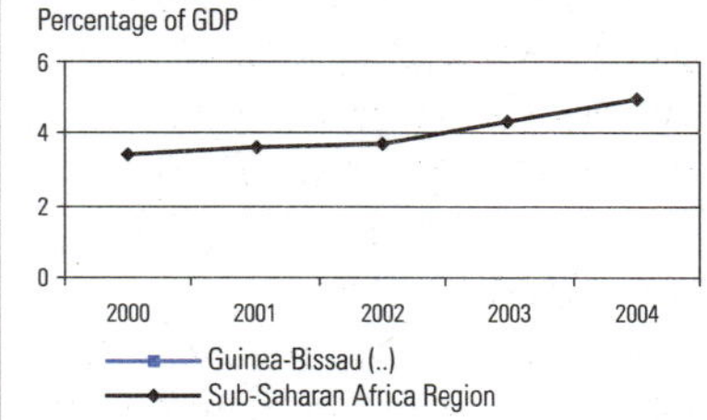

Notes: Figures in italics are for years other than those specified. .. indicates data are not available. C = competition; GDP = gross domestic product; GNI = gross national income; ICT = information and communication technology; M = monopoly; MDG = Millennium Development Goal; P = partial competition; and PCs = personal computers.

a. Outgoing and incoming.

b. Millennium Development Goal indicators 47, 48a, and 48b.

Sources: Economic and social context: UNDP and World Bank; ICT sector structure: ITU, WEF; ICT sector performance: Global Insight/WITSA, ITU, Netcraft, UNDESA, UNPAN, and World Bank. Produced by the Global Information and Communication Technologies Department and the Development Economics Data Group. For complete information, see Definitions and Data Sources.

Honduras

	Honduras 2000	Honduras 2004	Lower-middle-income group 2004	Latin America & the Caribbean Region 2004
Economic and social context				
Population, total (millions)	6	7	2,430	541
Urban population (% total population)	44	46	49	77
Poverty (% population below US$1 per day)	*20.7*	..	..	*9.5*
GNI per capita, Atlas method (current US$)	860	1,030	1,580	3,600
GDP growth, 1995–2000 and 2000–4 (%)	2.8	3.3	5.7	1.5
Adult literacy rate (% ages 15 and over)	*80*	..	90	89
Primary, secondary, tertiary school enrollment (% gross)	62	..	*72*	*82*
ICT sector structure				
Separate telecommunications regulator	Yes	Yes		
Status of main fixed-line operator	*Public*	Public		
Level of competition: international long distance	*C*	M		
Level of competition: mobile	*P*	M		
Level of competition: Internet service provider	*C*	..		
Government prioritization of ICT (scale 1–7)	..	2.9	3.8	3.5
ICT sector performance				
Access				
Telephone main lines (per 1,000 people)	46	55	192	181
International voice traffic (minutes per person)[a]	45	81	..	..
Mobile subscribers (per 1,000 people)	24	99	255	324
Population covered by mobile telephony (%)	..	*49*	*76*	*76*
Internet users (per 1,000 people)	9	31	70	*104*
Personal computers (per 1,000 people)	11	*14*	38	*75*
Households with television (%)	46	58	*89*	*88*
Quality				
Telephone faults (per 100 main lines per year)	*24.0*	*3.6*	..	..
Broadband subscribers (per 1,000 people)	0.0	*0.0*	12.6	*5.2*
International Internet bandwidth (bits per person)	2	*3*	58	165
Affordability				
Price basket for fixed line (US$ per month, residential)	6.2	5.9	*5.5*	*9.0*
Price basket for mobile (US$ per month)	..	6.9	8.9	9.1
Price basket for Internet (US$ per month)	..	*40.6*	*25.3*	*31.5*
Price of call to United States (US$ per 3 minutes)	3.97	2.52	1.45	0.90
Institutional efficiency and sustainability				
Total telecommunications revenue (% GDP)	5.1	5.8	3.6	3.3
Total telephone subscribers per employee	*158*	186	*195*	..
Total telecommunications investment (% revenue)	16.2	25.5	*25.5*	..
ICT applications				
ICT expenditure (% GDP)	4.2	*4.6*	*5.1*	*5.3*
E-government readiness index (scale 0–1)	..	0.24	0.26	0.39
Secure Internet servers (per 1 million people)	*0.6*	4.3	1.6	8.6
Schools connected to the Internet (%)	..	..	..	..

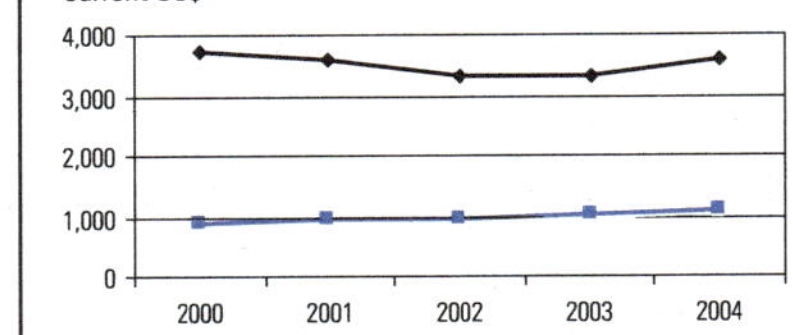

GNI per Capita, Atlas Method, 2000–4

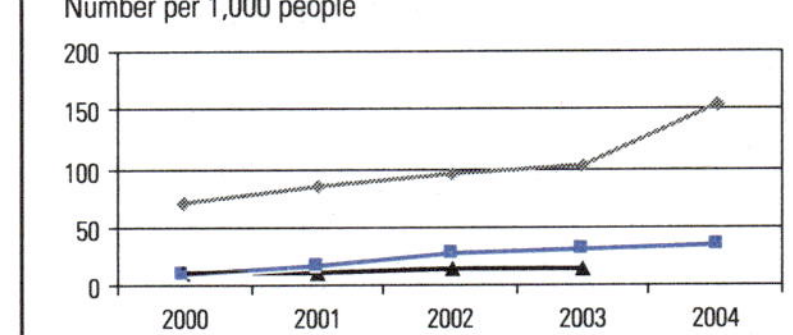

ICT MDG[b] Indicators, 2000–4

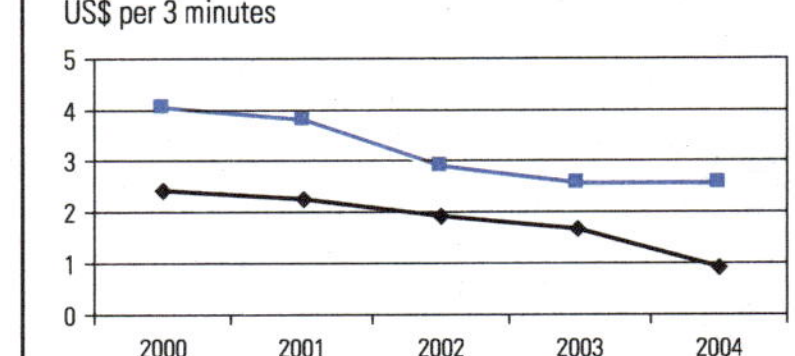

Price of Call to the United States, 2000–4

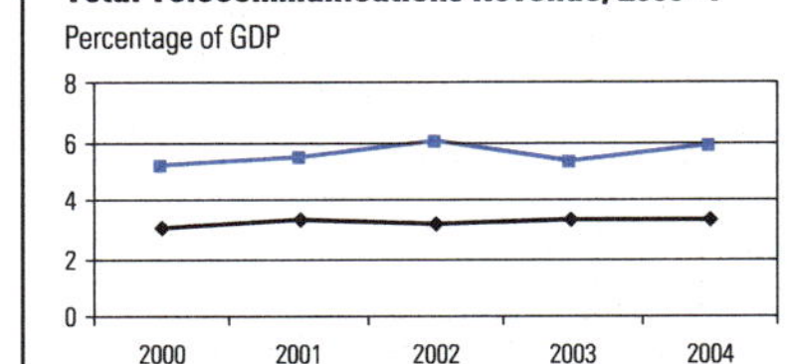

Total Telecommunications Revenue, 2000–4

Notes: Figures in italics are for years other than those specified. .. indicates data are not available. C = competition; GDP = gross domestic product; GNI = gross national income; ICT = information and communication technology; M = monopoly; MDG = Millennium Development Goal; P = partial competition; and PCs = personal computers.

a. Outgoing and incoming.

b. Millennium Development Goal indicators 47, 48a, and 48b.

Sources: Economic and social context: UNDP and World Bank; ICT sector structure: ITU, WEF; ICT sector performance: Global Insight/WITSA, ITU, Netcraft, UNDESA, UNPAN, and World Bank. Produced by the Global Information and Communication Technologies Department and the Development Economics Data Group. For complete information, see Definitions and Data Sources.

Hong Kong, China

	Hong Kong, China 2000	Hong Kong, China 2004	High-income group 2004
Economic and social context			
Population, total (millions)	7	7	1,001
Urban population (% total population)	100	100	77
Poverty (% population below US$1 per day)	..	..	..
GNI per capita, Atlas method (current US$)	26,820	26,810	32,040
GDP growth, 1995–2000 and 2000–4 (%)	2.6	3.2	2.0
Adult literacy rate (% ages 15 and over)	..	..	..
Primary, secondary, tertiary school enrollment (% gross)	71	*74*	*93*
ICT sector structure			
Separate telecommunications regulator	..	..	
Status of main fixed-line operator	..	..	
Level of competition: international long distance	..	..	
Level of competition: mobile	..	..	
Level of competition: Internet service provider	..	..	
Government prioritization of ICT (scale 1–7)	..	5.3	4.9
ICT sector performance			
Access			
Telephone main lines (per 1,000 people)	589	552	558
International voice traffic (minutes per person)[a]	754	*894*	..
Mobile subscribers (per 1,000 people)	817	1,192	767
Population covered by mobile telephony (%)	100	100	98
Internet users (per 1,000 people)	278	508	*480*
Personal computers (per 1,000 people)	354	453	*504*
Households with television (%)	98	99	*98*
Quality			
Telephone faults (per 100 main lines per year)	..	*1.3*	..
Broadband subscribers (per 1,000 people)	66.7	216.9	126.2
International Internet bandwidth (bits per person)	627	4,819	4,718
Affordability			
Price basket for fixed line (US$ per month, residential)	12.6	15.1	25.8
Price basket for mobile (US$ per month)	..	3.4	17.8
Price basket for Internet (US$ per month)	..	*3.9*	*20.9*
Price of call to United States (US$ per 3 minutes)	2.62	*2.62*	0.77
Institutional efficiency and sustainability			
Total telecommunications revenue (% GDP)	4.5	*3.9*	2.9
Total telephone subscribers per employee	243	*532*	*485*
Total telecommunications investment (% revenue)	12.0	*16.9*	*12.3*
ICT applications			
ICT expenditure (% GDP)	7.2	*8.4*	*7.1*
E-government readiness index (scale 0–1)	..	..	0.70
Secure Internet servers (per 1 million people)	*80.0*	141.0	311.4
Schools connected to the Internet (%)	..	100	*99*

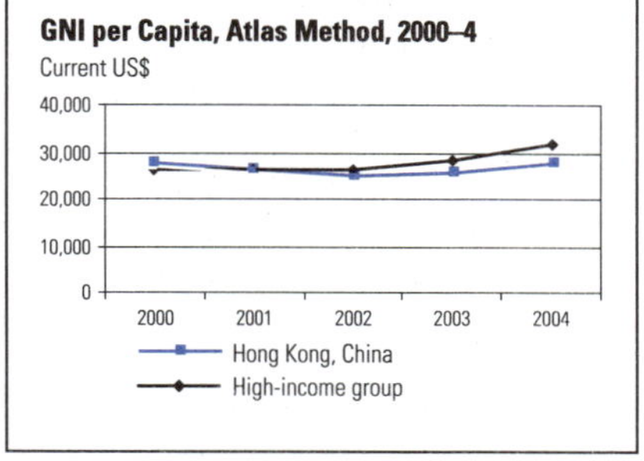

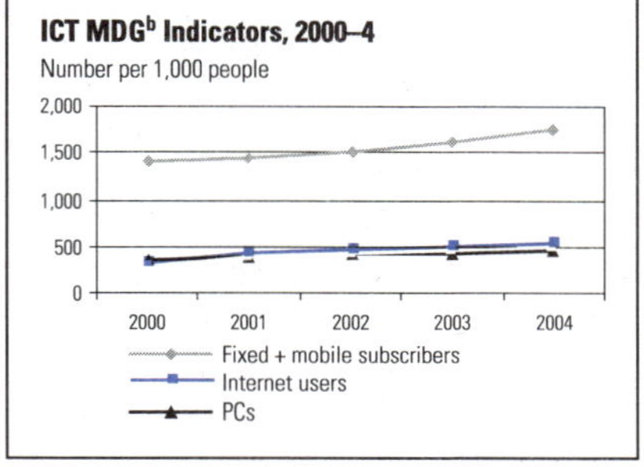

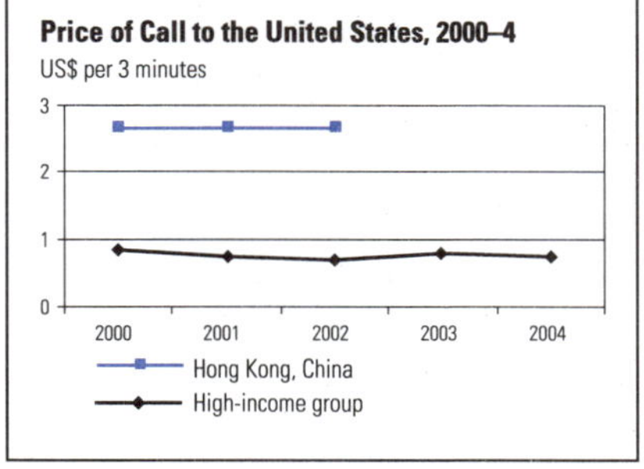

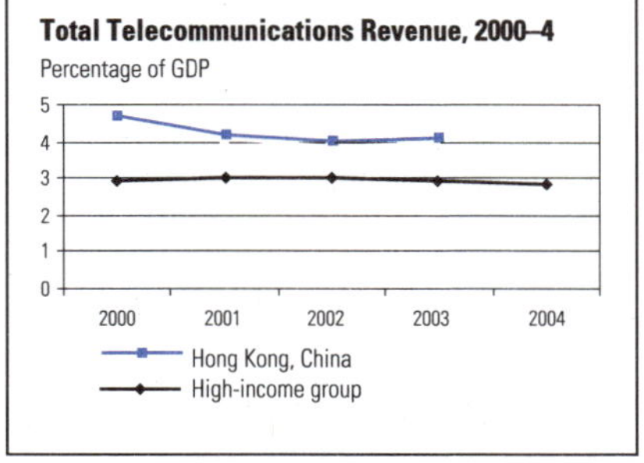

Notes: Figures in italics are for years other than those specified. .. indicates data are not available. C = competition; GDP = gross domestic product; GNI = gross national income; ICT = information and communication technology; M = monopoly; MDG = Millennium Development Goal; P = partial competition; and PCs = personal computers.

a. Outgoing and incoming.

b. Millennium Development Goal indicators 47, 48a, and 48b.

Sources: Economic and social context: UNDP and World Bank; ICT sector structure: ITU, WEF; ICT sector performance: Global Insight/WITSA, ITU, Netcraft, UNDESA, UNPAN, and World Bank. Produced by the Global Information and Communication Technologies Department and the Development Economics Data Group. For complete information, see Definitions and Data Sources.

Hungary

	Hungary 2000	Hungary 2004	Upper-middle-income group 2004	Europe & Central Asia Region 2004
Economic and social context				
Population, total (millions)	10	10	576	472
Urban population (% total population)	64	66	72	64
Poverty (% population below US$1 per day)	*<2*	*<2*	..	*3.6*
GNI per capita, Atlas method (current US$)	4,650	8,270	4,770	3,290
GDP growth, 1995–2000 and 2000–4 (%)	4.2	3.5	2.7	5.0
Adult literacy rate (% ages 15 and over)	*99*	..	94	97
Primary, secondary, tertiary school enrollment (% gross)	83	*89*	*80*	*81*
ICT sector structure				
Separate telecommunications regulator	Yes	Yes		
Status of main fixed-line operator	*Private*	Private		
Level of competition: international long distance	*M*	C		
Level of competition: mobile	*P*	P		
Level of competition: Internet service provider	*C*	C		
Government prioritization of ICT (scale 1–7)	..	4.4	4.1	3.8
ICT sector performance				
Access				
Telephone main lines (per 1,000 people)	379	*356*	*220*	*243*
International voice traffic (minutes per person)[a]	68	..	*39*	*28*
Mobile subscribers (per 1,000 people)	307	832	490	487
Population covered by mobile telephony (%)	95	*99*	*84*	*82*
Internet users (per 1,000 people)	71	*237*	*133*	*115*
Personal computers (per 1,000 people)	87	*119*	*99*	*73*
Households with television (%)	91	92	*92*	*92*
Quality				
Telephone faults (per 100 main lines per year)	..	..	*20.3*	*30.4*
Broadband subscribers (per 1,000 people)	0.3	36.3	*3.7*	*2.4*
International Internet bandwidth (bits per person)	102	1,060	176	148
Affordability				
Price basket for fixed line (US$ per month, residential)	14.4	*20.3*	*13.9*	*3.5*
Price basket for mobile (US$ per month)	..	13.3	11.1	10.3
Price basket for Internet (US$ per month)	..	*18.2*	*20.8*	*19.8*
Price of call to United States (US$ per 3 minutes)	1.28	1.01	1.03	1.06
Institutional efficiency and sustainability				
Total telecommunications revenue (% GDP)	6.9	*5.7*	3.4	3.5
Total telephone subscribers per employee	330	*507*	*402*	*150*
Total telecommunications investment (% revenue)	16.9	*18.4*	*18.6*	*19.0*
ICT applications				
ICT expenditure (% GDP)	7.3	*6.1*	*5.0*	*5.1*
E-government readiness index (scale 0–1)	..	0.54	0.49	0.39
Secure Internet servers (per 1 million people)	*12.5*	20.8	10.7	6.4
Schools connected to the Internet (%)	..	*85*	*60*	*65*

GNI per Capita, Atlas Method, 2000–4
Current US$

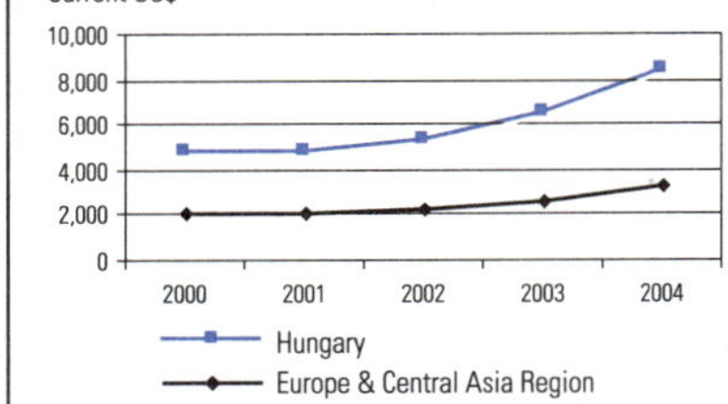

ICT MDG[b] Indicators, 2000–4
Number per 1,000 people

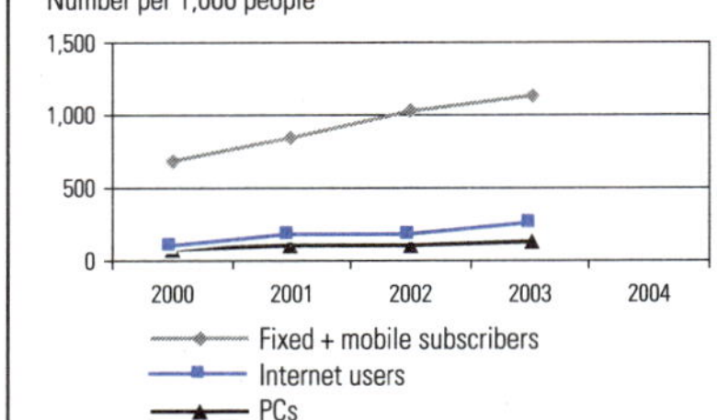

Price of Call to the United States, 2000–4
US$ per 3 minutes

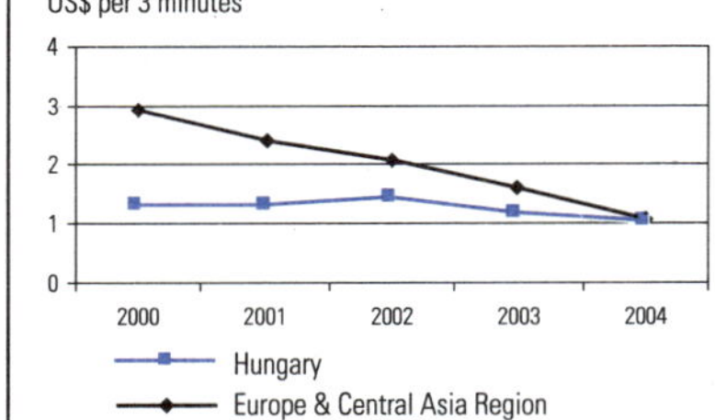

Total Telecommunications Revenue, 2000–4
Percentage of GDP

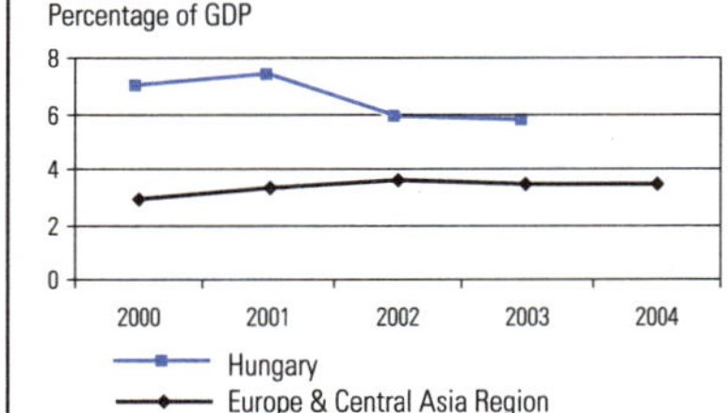

Notes: Figures in italics are for years other than those specified. .. indicates data are not available. C = competition; GDP = gross domestic product; GNI = gross national income; ICT = information and communication technology; M = monopoly; MDG = Millennium Development Goal; P = partial competition; and PCs = personal computers.

a. Outgoing and incoming.

b. Millennium Development Goal indicators 47, 48a, and 48b.

Sources: Economic and social context: UNDP and World Bank; ICT sector structure: ITU, WEF; ICT sector performance: Global Insight/WITSA, ITU, Netcraft, UNDESA, UNPAN, and World Bank. Produced by the Global Information and Communication Technologies Department and the Development Economics Data Group. For complete information, see Definitions and Data Sources.

India

	India 2000	India 2004	Low-income group 2004	South Asia Region 2004
Economic and social context				
Population, total (millions)	1,016	1,080	2,338	1,448
Urban population (% total population)	28	29	31	28
Poverty (% population below US$1 per day)	*35.3*	..	..	*31.3*
GNI per capita, Atlas method (current US$)	450	620	510	590
GDP growth, 1995–2000 and 2000–4 (%)	5.8	6.2	5.4	5.8
Adult literacy rate (% ages 15 and over)	*61*	..	61	58
Primary, secondary, tertiary school enrollment (% gross)	55	*60*	*55*	*56*
ICT sector structure				
Separate telecommunications regulator	Yes	Yes		
Status of main fixed-line operator	*Mixed*	Mixed		
Level of competition: international long distance	*M*	M		
Level of competition: mobile	*P*	P		
Level of competition: Internet service provider	*C*	C		
Government prioritization of ICT (scale 1–7)	..	5.7	..	5.3
ICT sector performance				
Access				
Telephone main lines (per 1,000 people)	32	43	33	41
International voice traffic (minutes per person)[a]	2	*3*	*4*	*4*
Mobile subscribers (per 1,000 people)	4	48	48	47
Population covered by mobile telephony (%)	..	*41*	*43*	*43*
Internet users (per 1,000 people)	5	23	20	21
Personal computers (per 1,000 people)	5	11	*8*	11
Households with television (%)	30	37	*16*	*32*
Quality				
Telephone faults (per 100 main lines per year)	165.5	*113.0*	..	*88.1*
Broadband subscribers (per 1,000 people)	0.0	0.6	0.5	0.6
International Internet bandwidth (bits per person)	1	4	*3*	4
Affordability				
Price basket for fixed line (US$ per month, residential)	6.0	3.2	*6.6*	3.2
Price basket for mobile (US$ per month)	..	3.2	11.6	3.2
Price basket for Internet (US$ per month)	..	*8.7*	*45.5*	*15.1*
Price of call to United States (US$ per 3 minutes)	3.36	1.19	1.95	1.21
Institutional efficiency and sustainability				
Total telecommunications revenue (% GDP)	1.5	1.9	2.3	1.9
Total telephone subscribers per employee	85	..	*89*	*89*
Total telecommunications investment (% revenue)	49.3	..	*27.8*	*15.3*
ICT applications				
ICT expenditure (% GDP)	3.6	*3.7*	*4.1*	*4.1*
E-government readiness index (scale 0–1)	..	0.57	0.12	0.34
Secure Internet servers (per 1 million people)	*0.1*	0.4	0.3	0.4
Schools connected to the Internet (%)	..	..	..	..

GNI per Capita, Atlas Method, 2000–4
Current US$

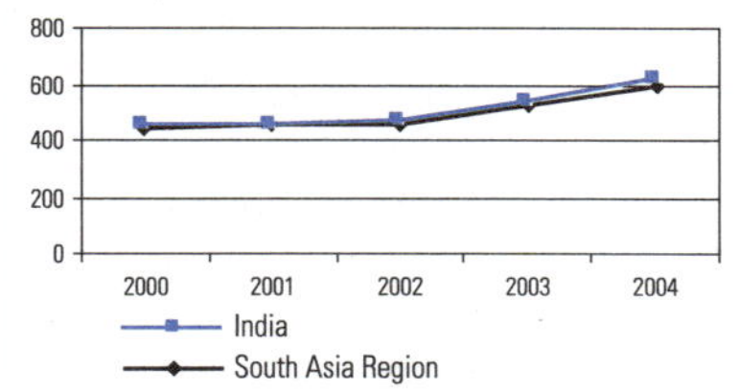

ICT MDG[b] Indicators, 2000–4
Number per 1,000 people

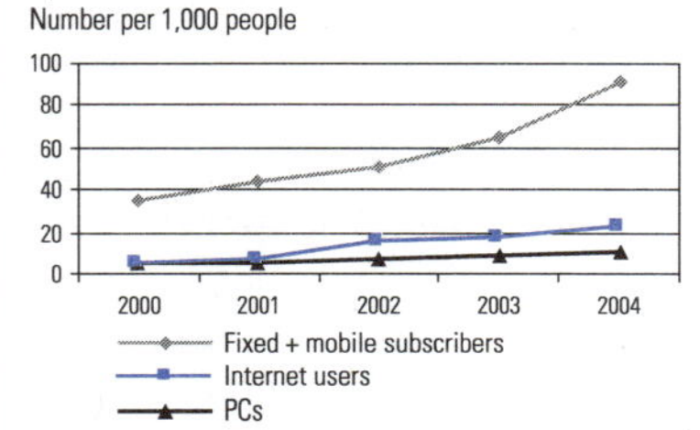

Price of Call to the United States, 2000–4
US$ per 3 minutes

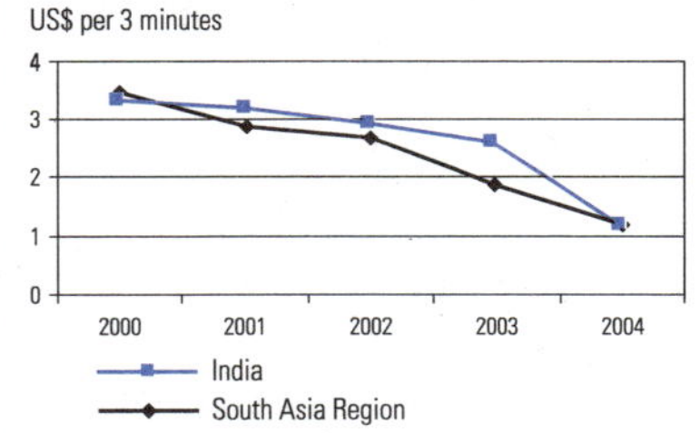

Total Telecommunications Revenue, 2000–4
Percentage of GDP

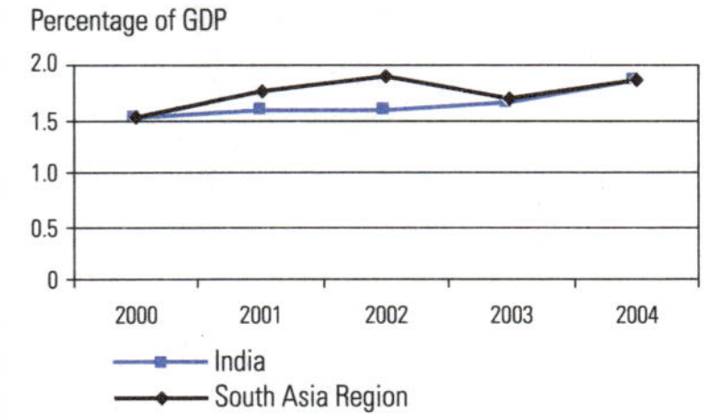

Notes: Figures in italics are for years other than those specified. .. indicates data are not available. C = competition; GDP = gross domestic product; GNI = gross national income; ICT = information and communication technology; M = monopoly; MDG = Millennium Development Goal; P = partial competition; and PCs = personal computers.

a. Outgoing and incoming.

b. Millennium Development Goal indicators 47, 48a, and 48b.

Sources: Economic and social context: UNDP and World Bank; ICT sector structure: ITU, WEF; ICT sector performance: Global Insight/WITSA, ITU, Netcraft, UNDESA, UNPAN, and World Bank. Produced by the Global Information and Communication Technologies Department and the Development Economics Data Group. For complete information, see Definitions and Data Sources.

Indonesia

	Indonesia 2000	Indonesia 2004	Lower-middle-income group 2004	East Asia & Pacific Region 2004
Economic and social context				
Population, total (millions)	206	218	2,430	1,870
Urban population (% total population)	42	47	49	41
Poverty (% population below US$1 per day)	7.2	*7.5*	..	*14.9*
GNI per capita, Atlas method (current US$)	590	1,140	1,580	1,280
GDP growth, 1995–2000 and 2000–4 (%)	–0.6	4.6	5.7	7.5
Adult literacy rate (% ages 15 and over)	..	*88*	90	90
Primary, secondary, tertiary school enrollment (% gross)	64	*66*	*72*	*68*
ICT sector structure				
Separate telecommunications regulator	No	Yes		
Status of main fixed-line operator	*Mixed*	*Mixed*		
Level of competition: international long distance	*P*	P		
Level of competition: mobile	*C*	C		
Level of competition: Internet service provider	*C*	C		
Government prioritization of ICT (scale 1–7)	..	4.2	3.8	4.8
ICT sector performance				
Access				
Telephone main lines (per 1,000 people)	32	46	192	194
International voice traffic (minutes per person)[a]	4	5	..	..
Mobile subscribers (per 1,000 people)	18	141	255	248
Population covered by mobile telephony (%)	..	*85*	*76*	*73*
Internet users (per 1,000 people)	9	52	70	75
Personal computers (per 1,000 people)	11	19	38	37
Households with television (%)	56	66	*89*	*80*
Quality				
Telephone faults (per 100 main lines per year)	49.0	*49.3*	..	..
Broadband subscribers (per 1,000 people)	0.0	0.3	12.6	13.4
International Internet bandwidth (bits per person)	1	18	58	52
Affordability				
Price basket for fixed line (US$ per month, residential)	4.5	6.2	*5.5*	*4.5*
Price basket for mobile (US$ per month)	..	4.6	8.9	5.1
Price basket for Internet (US$ per month)	..	*22.3*	*25.3*	*19.9*
Price of call to United States (US$ per 3 minutes)	3.90	2.79	1.45	1.20
Institutional efficiency and sustainability				
Total telecommunications revenue (% GDP)	1.4	2.3	3.6	3.6
Total telephone subscribers per employee	259	*665*	*195*	..
Total telecommunications investment (% revenue)	11.4	42.4	*25.5*	*31.0*
ICT applications				
ICT expenditure (% GDP)	2.5	*3.0*	*5.1*	*5.0*
E-government readiness index (scale 0–1)	..	0.32	0.26	0.21
Secure Internet servers (per 1 million people)	*0.3*	0.4	1.6	0.6
Schools connected to the Internet (%)	..	..	..	..

GNI per Capita, Atlas Method, 2000–4

Current US$

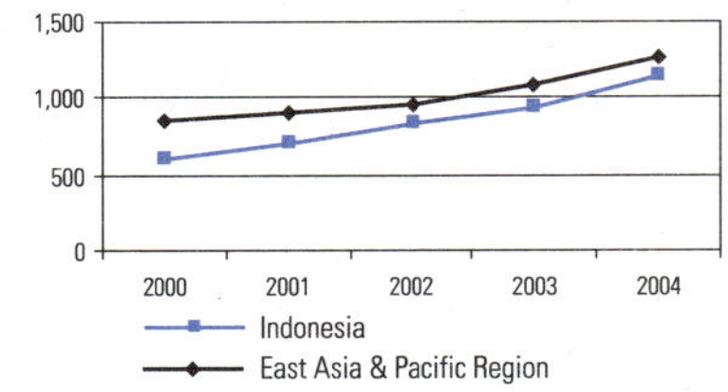

ICT MDG[b] Indicators, 2000–4

Number per 1,000 people

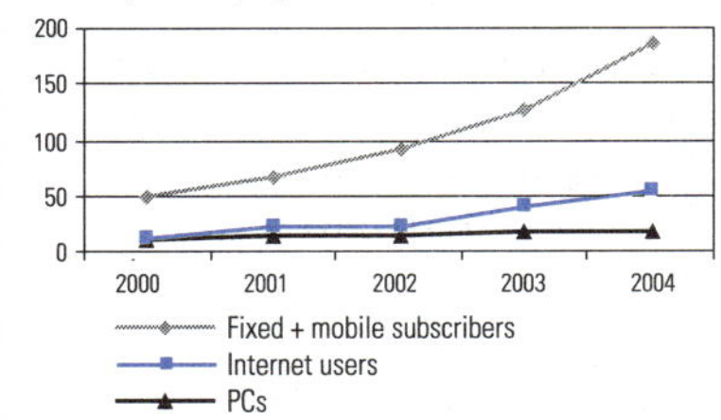

Price of Call to the United States, 2000–4

US$ per 3 minutes

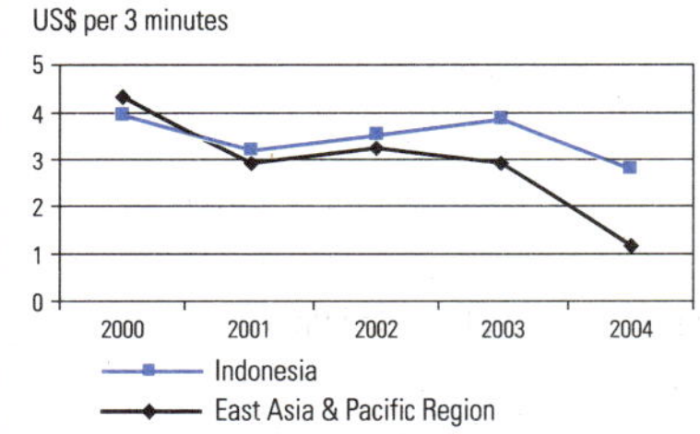

Total Telecommunications Revenue, 2000–4

Percentage of GDP

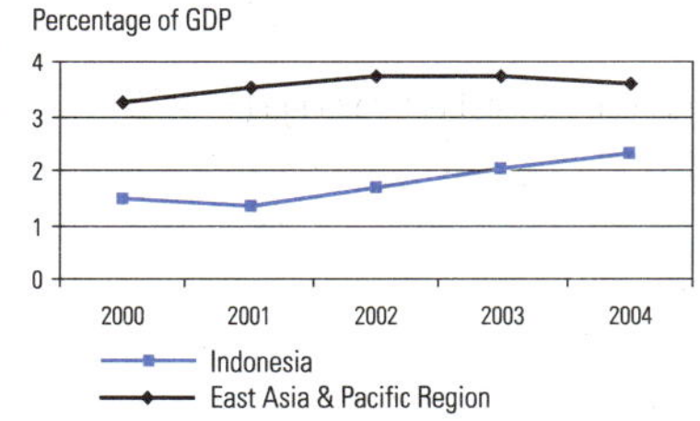

Notes: Figures in italics are for years other than those specified. .. indicates data are not available. C = competition; GDP = gross domestic product; GNI = gross national income; ICT = information and communication technology; M = monopoly; MDG = Millennium Development Goal; P = partial competition; and PCs = personal computers.

a. Outgoing and incoming.

b. Millennium Development Goal indicators 47, 48a, and 48b.

Sources: Economic and social context: UNDP and World Bank; ICT sector structure: ITU, WEF; ICT sector performance: Global Insight/WITSA, ITU, Netcraft, UNDESA, UNPAN, and World Bank. Produced by the Global Information and Communication Technologies Department and the Development Economics Data Group. For complete information, see Definitions and Data Sources.

Iran, Islamic Republic of

	Iran, Islamic Rep. of 2000	Iran, Islamic Rep. of 2004	Lower-middle-income group 2004	Middle East & North Africa Region 2004
Economic and social context				
Population, total (millions)	64	67	2,430	294
Urban population (% total population)	64	67	49	56
Poverty (% population below US$1 per day)	*<2*	..	..	*2.4*
GNI per capita, Atlas method (current US$)	1,650	2,300	1,580	2,000
GDP growth, 1995–2000 and 2000–4 (%)	3.2	6.2	5.7	4.5
Adult literacy rate (% ages 15 and over)	..	*77*	90	..
Primary, secondary, tertiary school enrollment (% gross)	70	*69*	*72*	*68*
ICT sector structure				
Separate telecommunications regulator	..	..		
Status of main fixed-line operator	*Mixed*	Mixed		
Level of competition: international long distance	*M*	M		
Level of competition: mobile	*M*	M		
Level of competition: Internet service provider	..	M		
Government prioritization of ICT (scale 1–7)	..	..	3.8	..
ICT sector performance				
Access				
Telephone main lines (per 1,000 people)	149	*219*	192	*118*
International voice traffic (minutes per person)[a]	7	*7*	..	*20*
Mobile subscribers (per 1,000 people)	15	*51*	255	*88*
Population covered by mobile telephony (%)	..	..	*76*	..
Internet users (per 1,000 people)	10	*65*	70	*47*
Personal computers (per 1,000 people)	63	*75*	38	*30*
Households with television (%)	68	*77*	*89*	*88*
Quality				
Telephone faults (per 100 main lines per year)	..	..	..	..
Broadband subscribers (per 1,000 people)	0.0	*0.2*	12.6	*0.2*
International Internet bandwidth (bits per person)	1	*15*	58	*15*
Affordability				
Price basket for fixed line (US$ per month, residential)	10.8	*2.8*	*5.5*	*4.9*
Price basket for mobile (US$ per month)	..	2.9	8.9	8.1
Price basket for Internet (US$ per month)	..	*5.9*	*25.3*	*24.5*
Price of call to United States (US$ per 3 minutes)	7.65	0.55	1.45	1.64
Institutional efficiency and sustainability				
Total telecommunications revenue (% GDP)	1.2	*1.1*	3.6	2.8
Total telephone subscribers per employee	221	*304*	*195*	..
Total telecommunications investment (% revenue)	5.8	*122.8*	*25.5*	*27.8*
ICT applications				
ICT expenditure (% GDP)	1.6	*2.2*	*5.1*	..
E-government readiness index (scale 0–1)	..	0.16	0.26	0.16
Secure Internet servers (per 1 million people)	*0.0*	0.2	1.6	0.6
Schools connected to the Internet (%)	..	..	..	..

GNI per Capita, Atlas Method, 2000–4
Current US$

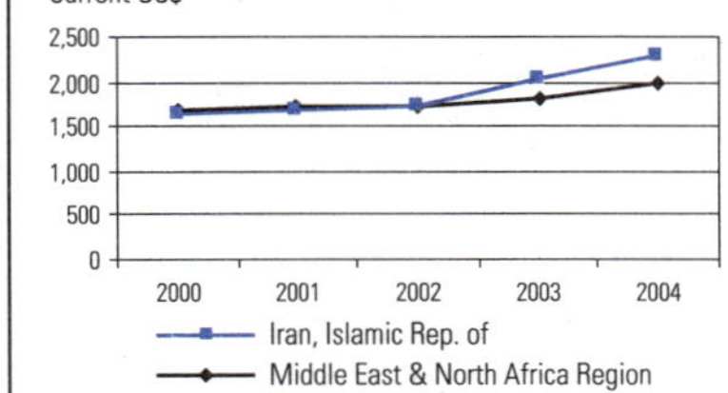

ICT MDG[b] Indicators, 2000–4
Number per 1,000 people

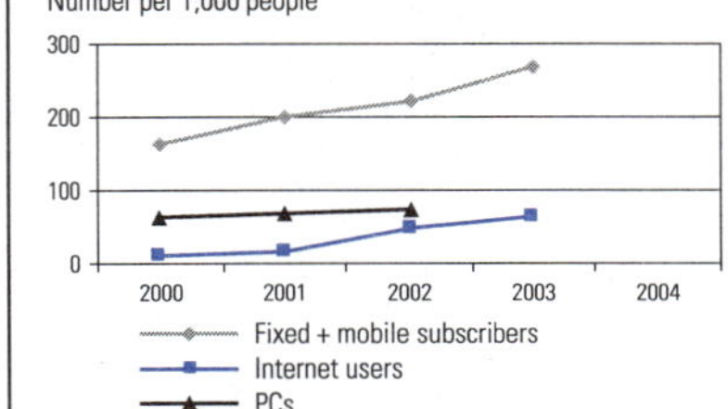

Price of Call to the United States, 2000–4
US$ per 3 minutes

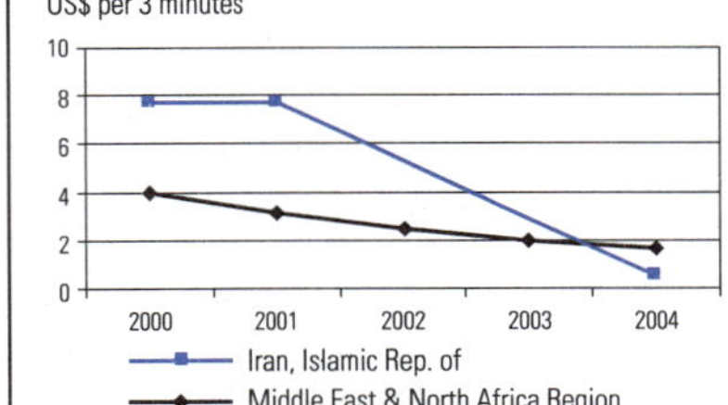

Total Telecommunications Revenue, 2000–4
Percentage of GDP

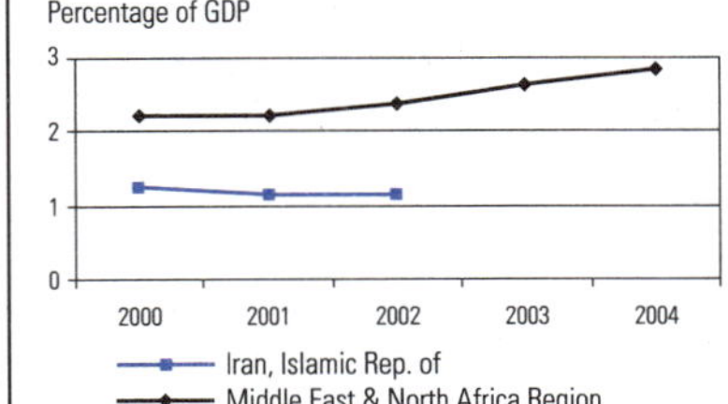

Notes: Figures in italics are for years other than those specified. .. indicates data are not available. C = competition; GDP = gross domestic product; GNI = gross national income; ICT = information and communication technology; M = monopoly; MDG = Millennium Development Goal; P = partial competition; and PCs = personal computers.

a. Outgoing and incoming.
b. Millennium Development Goal indicators 47, 48a, and 48b.

Sources: Economic and social context: UNDP and World Bank; ICT sector structure: ITU, WEF; ICT sector performance: Global Insight/WITSA, ITU, Netcraft, UNDESA, UNPAN, and World Bank. Produced by the Global Information and Communication Technologies Department and the Development Economics Data Group. For complete information, see Definitions and Data Sources.

Ireland

	Ireland 2000	Ireland 2004	High-income group 2004
Economic and social context			
Population, total (millions)	4	4	1,001
Urban population (% total population)	59	60	77
Poverty (% population below US$1 per day)	..	..	..
GNI per capita, Atlas method (current US$)	23,030	34,280	32,040
GDP growth, 1995–2000 and 2000–4 (%)	9.9	5.4	2.0
Adult literacy rate (% ages 15 and over)	..	..	..
Primary, secondary, tertiary school enrollment (% gross)	90	*93*	*93*
ICT sector structure			
Separate telecommunications regulator	No	Yes	
Status of main fixed-line operator	*Private*	Private	
Level of competition: international long distance	*C*	C	
Level of competition: mobile	*P*	C	
Level of competition: Internet service provider	*C*	C	
Government prioritization of ICT (scale 1–7)	..	5.1	4.9
ICT sector performance			
Access			
Telephone main lines (per 1,000 people)	480	502	558
International voice traffic (minutes per person)[a]	708	995	..
Mobile subscribers (per 1,000 people)	645	921	767
Population covered by mobile telephony (%)	98	99	98
Internet users (per 1,000 people)	178	298	*480*
Personal computers (per 1,000 people)	357	510	*504*
Households with television (%)	95	95	*98*
Quality			
Telephone faults (per 100 main lines per year)	*7.1*	*6.0*	..
Broadband subscribers (per 1,000 people)	0.0	34.4	126.2
International Internet bandwidth (bits per person)	240	6,118	4,718
Affordability			
Price basket for fixed line (US$ per month, residential)	22.6	39.7	25.8
Price basket for mobile (US$ per month)	..	19.1	17.8
Price basket for Internet (US$ per month)	..	*28.3*	*20.9*
Price of call to United States (US$ per 3 minutes)	0.80	0.71	0.77
Institutional efficiency and sustainability			
Total telecommunications revenue (% GDP)	2.8	2.7	2.9
Total telephone subscribers per employee	246	395	*485*
Total telecommunications investment (% revenue)	14.3	*10.8*	*12.3*
ICT applications			
ICT expenditure (% GDP)	5.7	*3.9*	*7.1*
E-government readiness index (scale 0–1)	..	0.66	0.70
Secure Internet servers (per 1 million people)	*90.6*	309.8	311.4
Schools connected to the Internet (%)	..	*99*	*99*

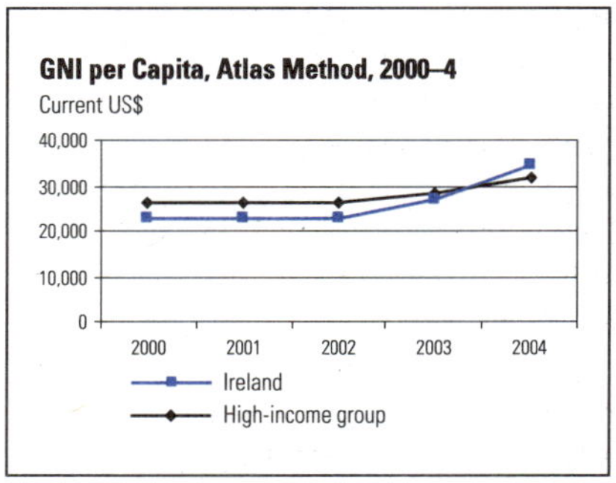

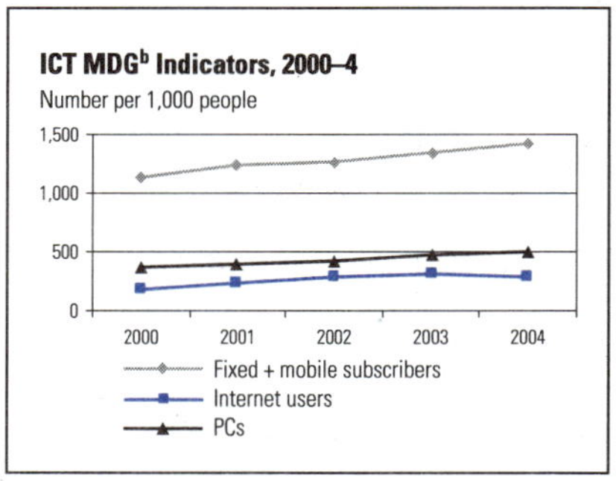

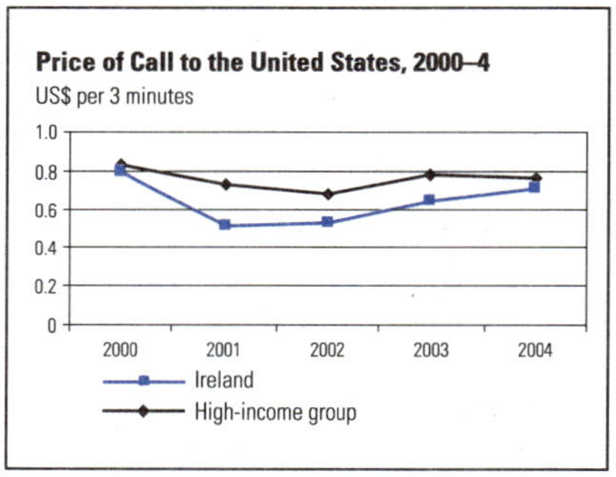

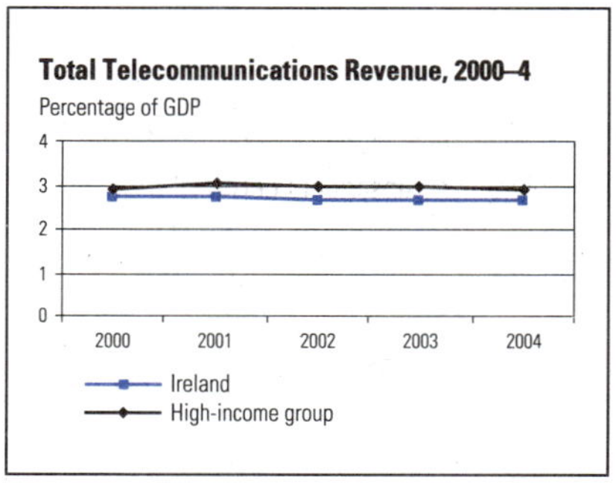

Notes: Figures in italics are for years other than those specified. .. indicates data are not available. C = competition; GDP = gross domestic product; GNI = gross national income; ICT = information and communication technology; M = monopoly; MDG = Millennium Development Goal; P = partial competition; and PCs = personal computers.

a. Outgoing and incoming.

b. Millennium Development Goal indicators 47, 48a, and 48b.

Sources: Economic and social context: UNDP and World Bank; ICT sector structure: ITU, WEF; ICT sector performance: Global Insight/WITSA, ITU, Netcraft, UNDESA, UNPAN, and World Bank. Produced by the Global Information and Communication Technologies Department and the Development Economics Data Group. For complete information, see Definitions and Data Sources.

Israel

	Israel 2000	Israel 2004	High-income group 2004
Economic and social context			
Population, total (millions)	6	7	1,001
Urban population (% total population)	91	92	77
Poverty (% population below US$1 per day)	..	..	..
GNI per capita, Atlas method (current US$)	17,060	17,380	32,040
GDP growth, 1995–2000 and 2000–4 (%)	3.9	0.9	2.0
Adult literacy rate (% ages 15 and over)	..	97	..
Primary, secondary, tertiary school enrollment (% gross)	90	*91*	*93*
ICT sector structure			
Separate telecommunications regulator	..	..	
Status of main fixed-line operator	*Mixed*	Mixed	
Level of competition: international long distance	*C*	C	
Level of competition: mobile	*C*	C	
Level of competition: Internet service provider	*C*	*C*	
Government prioritization of ICT (scale 1–7)	..	5.1	4.9
ICT sector performance			
Access			
Telephone main lines (per 1,000 people)	473	426	558
International voice traffic (minutes per person)[a]	268	*306*	..
Mobile subscribers (per 1,000 people)	700	1,076	767
Population covered by mobile telephony (%)	..	*97*	98
Internet users (per 1,000 people)	202	397	*480*
Personal computers (per 1,000 people)	253	*327*	*504*
Households with television (%)	92	*93*	*98*
Quality			
Telephone faults (per 100 main lines per year)	..	..	..
Broadband subscribers (per 1,000 people)	1.0	135.3	126.2
International Internet bandwidth (bits per person)	53	471	4,718
Affordability			
Price basket for fixed line (US$ per month, residential)	11.6	14.9	25.8
Price basket for mobile (US$ per month)	..	9.3	17.8
Price basket for Internet (US$ per month)	..	*29.8*	*20.9*
Price of call to United States (US$ per 3 minutes)	..	0.59	0.77
Institutional efficiency and sustainability			
Total telecommunications revenue (% GDP)	3.2	*4.5*	2.9
Total telephone subscribers per employee	618	*825*	*485*
Total telecommunications investment (% revenue)	16.1	*39.1*	*12.3*
ICT applications			
ICT expenditure (% GDP)	8.2	*7.9*	*7.1*
E-government readiness index (scale 0–1)	..	0.69	0.70
Secure Internet servers (per 1 million people)	*46.7*	127.8	311.4
Schools connected to the Internet (%)	..	*95*	*99*

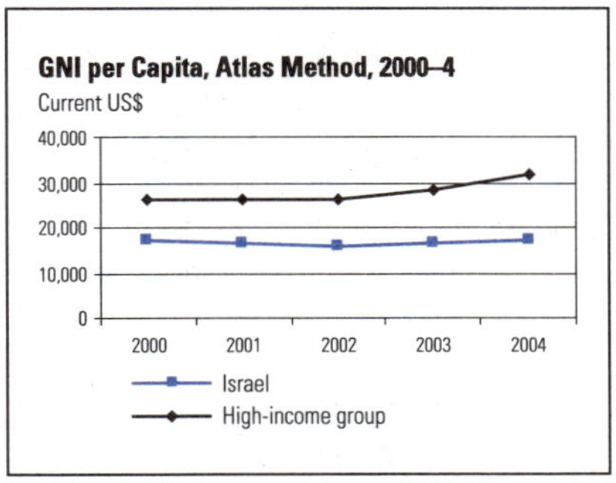

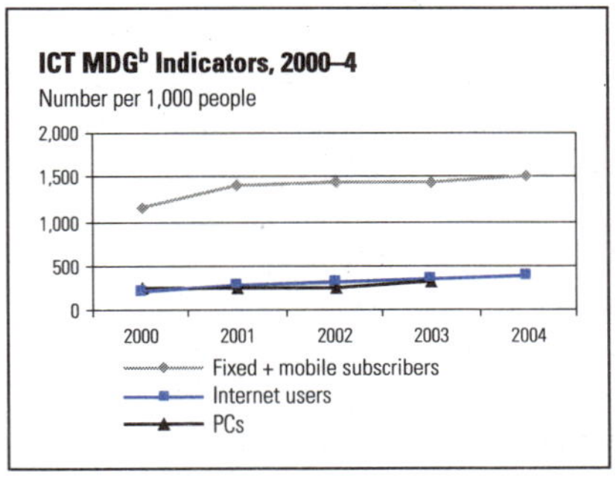

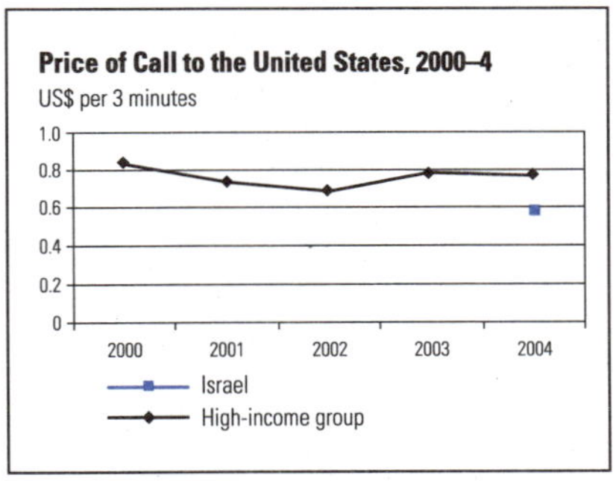

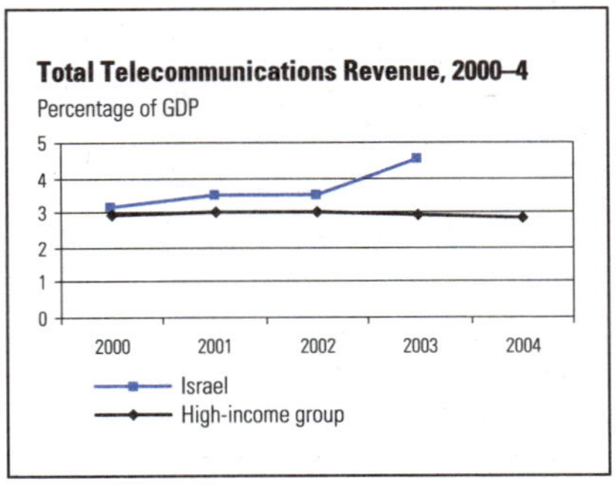

Notes: Figures in italics are for years other than those specified. .. indicates data are not available. C = competition; GDP = gross domestic product; GNI = gross national income; ICT = information and communication technology; M = monopoly; MDG = Millennium Development Goal; P = partial competition; and PCs = personal computers.

a. Outgoing and incoming.

b. Millennium Development Goal indicators 47, 48a, and 48b.

Sources: Economic and social context: UNDP and World Bank; ICT sector structure: ITU, WEF; ICT sector performance: Global Insight/WITSA, ITU, Netcraft, UNDESA, UNPAN, and World Bank. Produced by the Global Information and Communication Technologies Department and the Development Economics Data Group. For complete information, see Definitions and Data Sources.

Italy

	Italy 2000	Italy 2004	High-income group 2004
Economic and social context			
Population, total (millions)	58	58	1,001
Urban population (% total population)	67	67	77
Poverty (% population below US$1 per day)	..	..	..
GNI per capita, Atlas method (current US$)	20,160	26,120	32,040
GDP growth, 1995–2000 and 2000–4 (%)	1.9	0.8	2.0
Adult literacy rate (% ages 15 and over)	..	..	..
Primary, secondary, tertiary school enrollment (% gross)	82	*87*	*93*
ICT sector structure			
Separate telecommunications regulator	Yes	Yes	
Status of main fixed-line operator	*Mixed*	Mixed	
Level of competition: international long distance	*C*	C	
Level of competition: mobile	*C*	C	
Level of competition: Internet service provider	*C*	C	
Government prioritization of ICT (scale 1–7)	..	3.9	4.9
ICT sector performance			
Access			
Telephone main lines (per 1,000 people)	471	*461*	558
International voice traffic (minutes per person)[a]	156	..	..
Mobile subscribers (per 1,000 people)	732	1,091	767
Population covered by mobile telephony (%)	100	*100*	98
Internet users (per 1,000 people)	229	*321*	*480*
Personal computers (per 1,000 people)	179	*232*	*504*
Households with television (%)	98	..	*98*
Quality			
Telephone faults (per 100 main lines per year)	17.4	..	..
Broadband subscribers (per 1,000 people)	2.0	81.7	126.2
International Internet bandwidth (bits per person)	165	1,572	4,718
Affordability			
Price basket for fixed line (US$ per month, residential)	18.9	*23.8*	25.8
Price basket for mobile (US$ per month)	..	14.0	17.8
Price basket for Internet (US$ per month)	..	*16.5*	*20.9*
Price of call to United States (US$ per 3 minutes)	0.81	0.79	0.77
Institutional efficiency and sustainability			
Total telecommunications revenue (% GDP)	2.3	*2.6*	2.9
Total telephone subscribers per employee	915	..	*485*
Total telecommunications investment (% revenue)	35.3	..	*12.3*
ICT applications			
ICT expenditure (% GDP)	4.8	*4.1*	*7.1*
E-government readiness index (scale 0–1)	..	0.55	0.70
Secure Internet servers (per 1 million people)	*18.0*	34.6	311.4
Schools connected to the Internet (%)	..	*88*	*99*

GNI per Capita, Atlas Method, 2000–4
Current US$

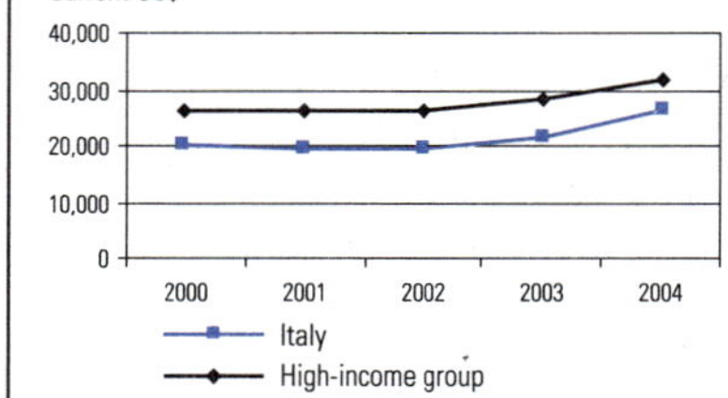

ICT MDG[b] Indicators, 2000–4
Number per 1,000 people

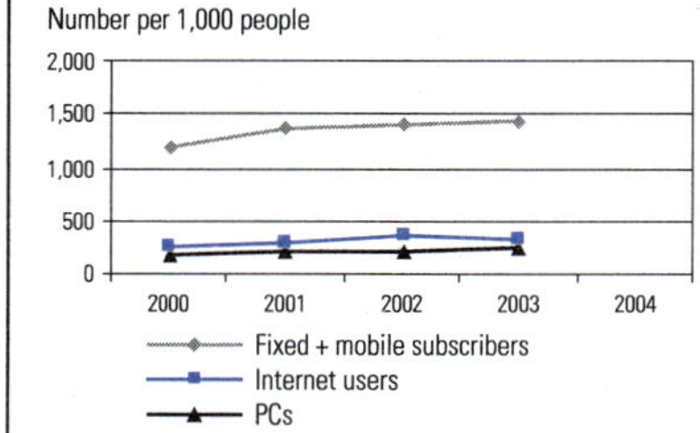

Price of Call to the United States, 2000–4
US$ per 3 minutes

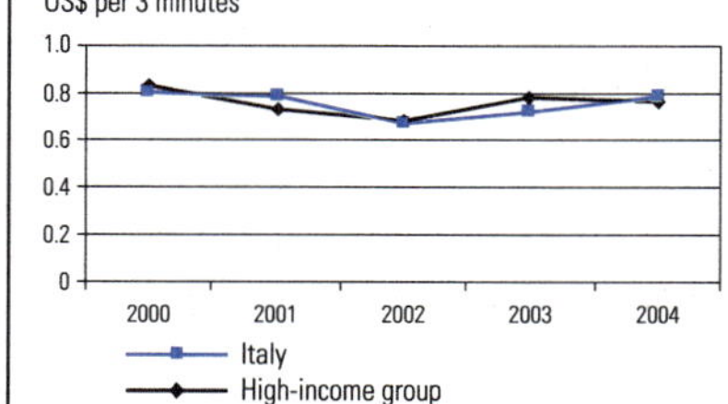

Total Telecommunications Revenue, 2000–4
Percentage of GDP

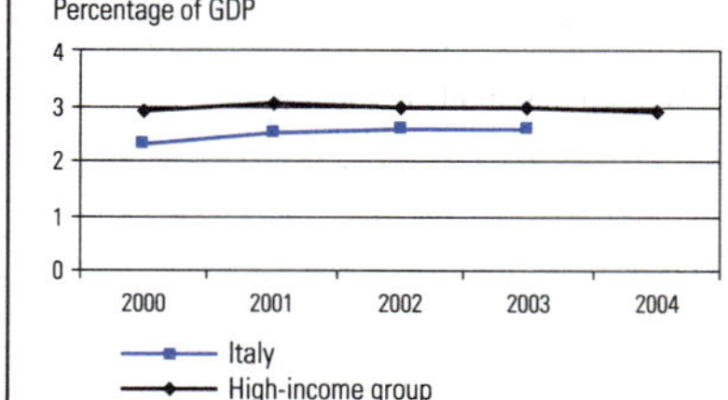

Notes: Figures in italics are for years other than those specified. .. indicates data are not available. C = competition; GDP = gross domestic product; GNI = gross national income; ICT = information and communication technology; M = monopoly; MDG = Millennium Development Goal; P = partial competition; and PCs = personal computers.

a. Outgoing and incoming.

b. Millennium Development Goal indicators 47, 48a, and 48b.

Sources: Economic and social context: UNDP and World Bank; ICT sector structure: ITU, WEF; ICT sector performance: Global Insight/WITSA, ITU, Netcraft, UNDESA, UNPAN, and World Bank. Produced by the Global Information and Communication Technologies Department and the Development Economics Data Group. For complete information, see Definitions and Data Sources.

Jamaica

	Jamaica 2000	Jamaica 2004	Lower-middle-income group 2004	Latin America & the Caribbean Region 2004
Economic and social context				
Population, total (millions)	3	3	2,430	541
Urban population (% total population)	52	52	49	77
Poverty (% population below US$1 per day)	<2	..	..	*9.5*
GNI per capita, Atlas method (current US$)	2,710	2,900	1,580	3,600
GDP growth, 1995–2000 and 2000–4 (%)	–1.9	1.7	5.7	1.5
Adult literacy rate (% ages 15 and over)	..	*88*	90	89
Primary, secondary, tertiary school enrollment (% gross)	74	*74*	*72*	*82*
ICT sector structure				
Separate telecommunications regulator	Yes	Yes		
Status of main fixed-line operator	*Private*	Mixed		
Level of competition: international long distance	*M*	C		
Level of competition: mobile	*C*	*C*		
Level of competition: Internet service provider	*C*	C		
Government prioritization of ICT (scale 1–7)	..	4.7	3.8	3.5
ICT sector performance				
Access				
Telephone main lines (per 1,000 people)	198	*174*	192	181
International voice traffic (minutes per person)[a]	156	*230*	..	..
Mobile subscribers (per 1,000 people)	142	*615*	255	324
Population covered by mobile telephony (%)	*80*	*95*	*76*	*76*
Internet users (per 1,000 people)	31	*264*	70	*104*
Personal computers (per 1,000 people)	47	*54*	38	*75*
Households with television (%)	69	*70*	*89*	*88*
Quality				
Telephone faults (per 100 main lines per year)	48.0	*39.7*	..	..
Broadband subscribers (per 1,000 people)	..	*9.5*	12.6	*5.2*
International Internet bandwidth (bits per person)	*28*	43	58	165
Affordability				
Price basket for fixed line (US$ per month, residential)	5.0	*10.3*	*5.5*	*9.0*
Price basket for mobile (US$ per month)	..	8.1	8.9	9.1
Price basket for Internet (US$ per month)	..	*39.8*	*25.3*	*31.5*
Price of call to United States (US$ per 3 minutes)	..	0.87	1.45	0.90
Institutional efficiency and sustainability				
Total telecommunications revenue (% GDP)	6.4	6.0	3.6	3.3
Total telephone subscribers per employee	274	*345*	*195*	..
Total telecommunications investment (% revenue)	32.5	*24.9*	*25.5*	..
ICT applications				
ICT expenditure (% GDP)	11.2	*12.4*	*5.1*	*5.3*
E-government readiness index (scale 0–1)	..	0.41	0.26	0.39
Secure Internet servers (per 1 million people)	*1.9*	9.0	1.6	8.6
Schools connected to the Internet (%)	..	*10*	..	..

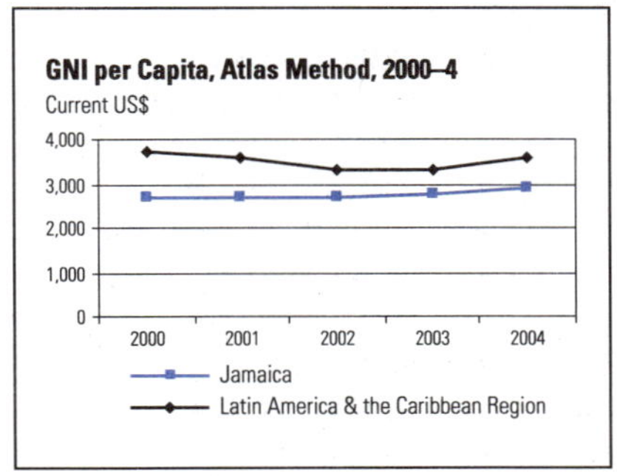

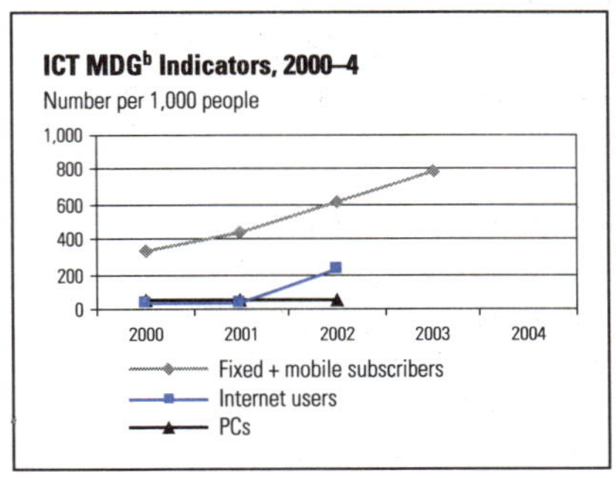

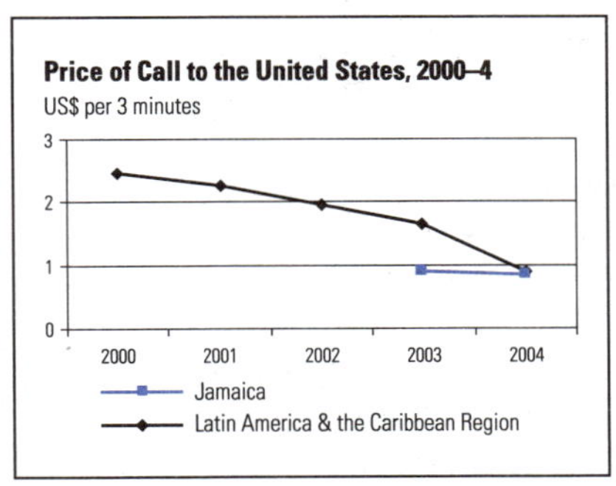

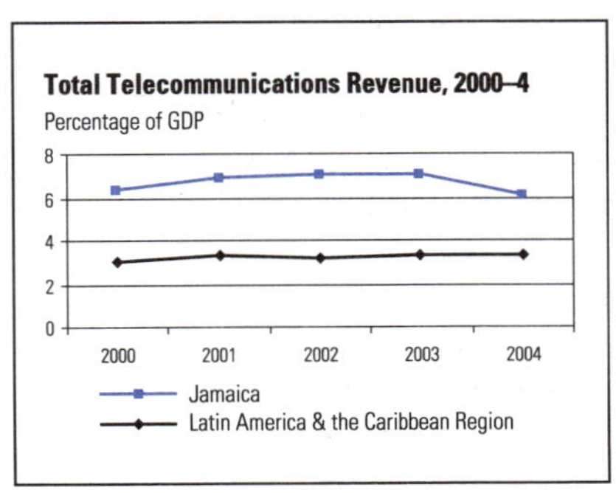

Notes: Figures in italics are for years other than those specified. .. indicates data are not available. C = competition; GDP = gross domestic product; GNI = gross national income; ICT = information and communication technology; M = monopoly; MDG = Millennium Development Goal; P = partial competition; and PCs = personal computers.

a. Outgoing and incoming.
b. Millennium Development Goal indicators 47, 48a, and 48b.

Sources: Economic and social context: UNDP and World Bank; ICT sector structure: ITU, WEF; ICT sector performance: Global Insight/WITSA, ITU, Netcraft, UNDESA, UNPAN, and World Bank. Produced by the Global Information and Communication Technologies Department and the Development Economics Data Group. For complete information, see Definitions and Data Sources.

Japan

	Japan 2000	Japan 2004	High-income group 2004
Economic and social context			
Population, total (millions)	127	128	1,001
Urban population (% total population)	65	66	77
Poverty (% population below US$1 per day)	..	..	..
GNI per capita, Atlas method (current US$)	35,280	37,180	32,040
GDP growth, 1995–2000 and 2000–4 (%)	1.0	1.3	2.0
Adult literacy rate (% ages 15 and over)	..	..	..
Primary, secondary, tertiary school enrollment (% gross)	83	*84*	*93*
ICT sector structure			
Separate telecommunications regulator	..	..	
Status of main fixed-line operator	*Mixed*	Private	
Level of competition: international long distance	*C*	C	
Level of competition: mobile	*C*	C	
Level of competition: Internet service provider	*C*	C	
Government prioritization of ICT (scale 1–7)	..	6.1	4.9
ICT sector performance			
Access			
Telephone main lines (per 1,000 people)	586	531	558
International voice traffic (minutes per person)[a]	30	*37*	..
Mobile subscribers (per 1,000 people)	526	669	767
Population covered by mobile telephony (%)	99	99	98
Internet users (per 1,000 people)	300	*606*	*480*
Personal computers (per 1,000 people)	359	425	*504*
Households with television (%)	99	*99*	*98*
Quality			
Telephone faults (per 100 main lines per year)	..	..	..
Broadband subscribers (per 1,000 people)	6.7	145.8	126.2
International Internet bandwidth (bits per person)	57	1,038	4,718
Affordability			
Price basket for fixed line (US$ per month, residential)	33.1	26.0	25.8
Price basket for mobile (US$ per month)	..	29.1	17.8
Price basket for Internet (US$ per month)	..	*21.1*	*20.9*
Price of call to United States (US$ per 3 minutes)	1.67	1.66	0.77
Institutional efficiency and sustainability			
Total telecommunications revenue (% GDP)	2.6	3.0	2.9
Total telephone subscribers per employee	910	..	*485*
Total telecommunications investment (% revenue)	26.8	*14.4*	*12.3*
ICT applications			
ICT expenditure (% GDP)	8.4	*7.4*	*7.1*
E-government readiness index (scale 0–1)	..	0.63	0.70
Secure Internet servers (per 1 million people)	*40.5*	160.2	311.4
Schools connected to the Internet (%)	..	*99*	*99*

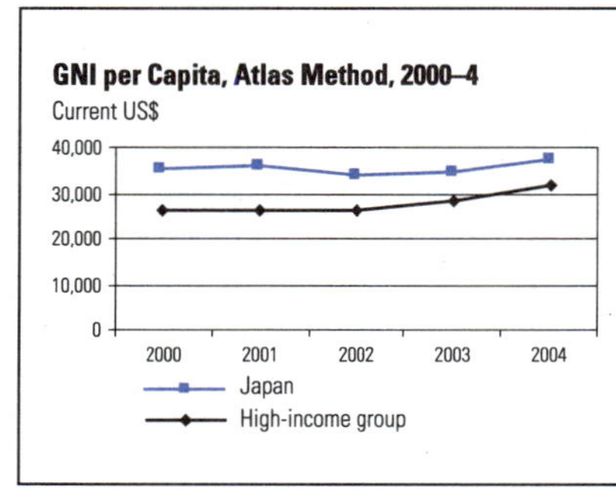

GNI per Capita, Atlas Method, 2000–4

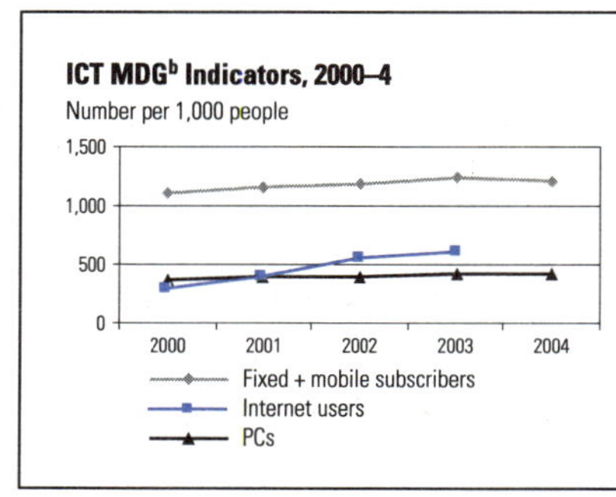

ICT MDG[b] Indicators, 2000–4

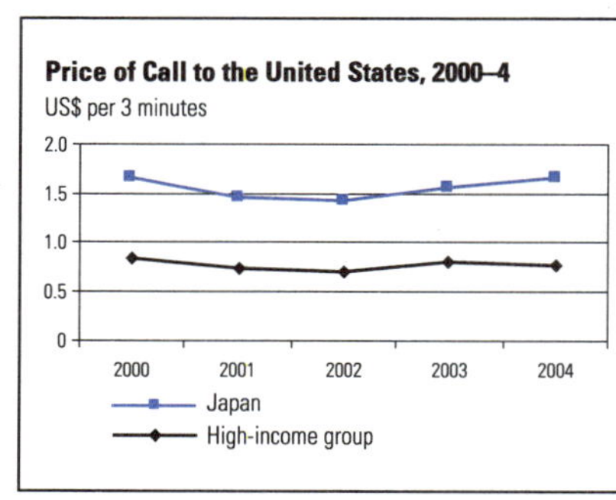

Price of Call to the United States, 2000–4

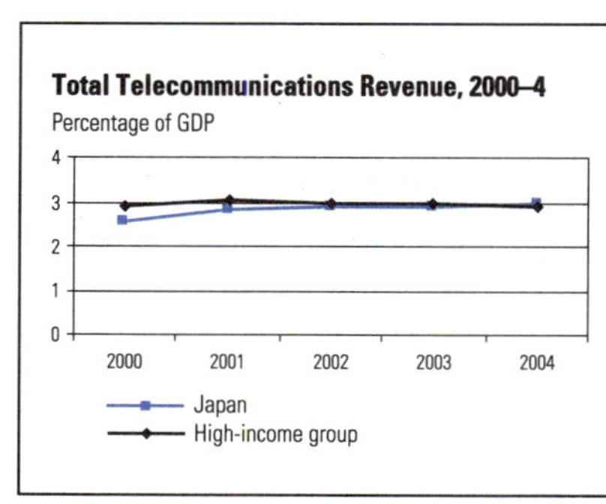

Total Telecommunications Revenue, 2000–4

Notes: Figures in italics are for years other than those specified. .. indicates data are not available. C = competition; GDP = gross domestic product; GNI = gross national income; ICT = information and communication technology; M = monopoly; MDG = Millennium Development Goal; P = partial competition; and PCs = personal computers.

a. Outgoing and incoming.

b. Millennium Development Goal indicators 47, 48a, and 48b.

Sources: Economic and social context: UNDP and World Bank; ICT sector structure: ITU, WEF; ICT sector performance: Global Insight/WITSA, ITU, Netcraft, UNDESA, UNPAN, and World Bank. Produced by the Global Information and Communication Technologies Department and the Development Economics Data Group. For complete information, see Definitions and Data Sources.

Jordan

	Jordan 2000	Jordan 2004	Lower-middle-income group 2004	Middle East & North Africa Region 2004
Economic and social context				
Population, total (millions)	5	5	2,430	294
Urban population (% total population)	79	79	49	56
Poverty (% population below US$1 per day)	*<2*	..	..	*2.4*
GNI per capita, Atlas method (current US$)	1,750	2,140	1,580	2,000
GDP growth, 1995–2000 and 2000–4 (%)	3.1	5.1	5.7	4.5
Adult literacy rate (% ages 15 and over)	..	*90*	90	..
Primary, secondary, tertiary school enrollment (% gross)	*75*	*78*	*72*	*68*
ICT sector structure				
Separate telecommunications regulator	Yes	Yes		
Status of main fixed-line operator	*Mixed*	Mixed		
Level of competition: international long distance	*M*	M		
Level of competition: mobile	*P*	P		
Level of competition: Internet service provider	*C*	C		
Government prioritization of ICT (scale 1–7)	..	5.3	3.8	..
ICT sector performance				
Access				
Telephone main lines (per 1,000 people)	127	117	192	*118*
International voice traffic (minutes per person)[a]	79	*84*	..	*20*
Mobile subscribers (per 1,000 people)	80	272	255	*88*
Population covered by mobile telephony (%)	99	*99*	*76*	..
Internet users (per 1,000 people)	26	*86*	70	*47*
Personal computers (per 1,000 people)	31	*55*	38	*30*
Households with television (%)	97	*97*	*89*	*88*
Quality				
Telephone faults (per 100 main lines per year)	18.2	*12.6*	..	..
Broadband subscribers (per 1,000 people)	*0.1*	*0.9*	12.6	*0.2*
International Internet bandwidth (bits per person)	..	*29*	58	*15*
Affordability				
Price basket for fixed line (US$ per month, residential)	8.8	10.0	*5.5*	*4.9*
Price basket for mobile (US$ per month)	..	9.4	8.9	8.1
Price basket for Internet (US$ per month)	..	*26.3*	*25.3*	*24.5*
Price of call to United States (US$ per 3 minutes)	2.86	1.44	1.45	1.64
Institutional efficiency and sustainability				
Total telecommunications revenue (% GDP)	6.7	8.4	3.6	2.8
Total telephone subscribers per employee	157	*303*	*195*	..
Total telecommunications investment (% revenue)	38.5	*25.4*	*25.5*	*27.8*
ICT applications				
ICT expenditure (% GDP)	8.8	*8.7*	*5.1*	..
E-government readiness index (scale 0–1)	..	0.35	0.26	0.16
Secure Internet servers (per 1 million people)	*0.4*	3.9	1.6	0.6
Schools connected to the Internet (%)	..	*18*	..	..

GNI per Capita, Atlas Method, 2000–4

Current US$

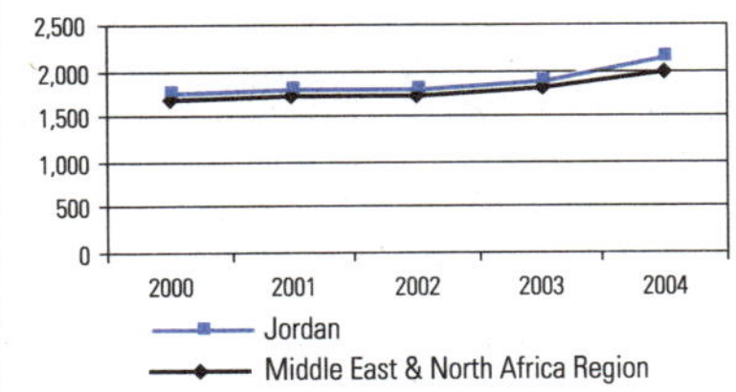

ICT MDG[b] Indicators, 2000–4

Number per 1,000 people

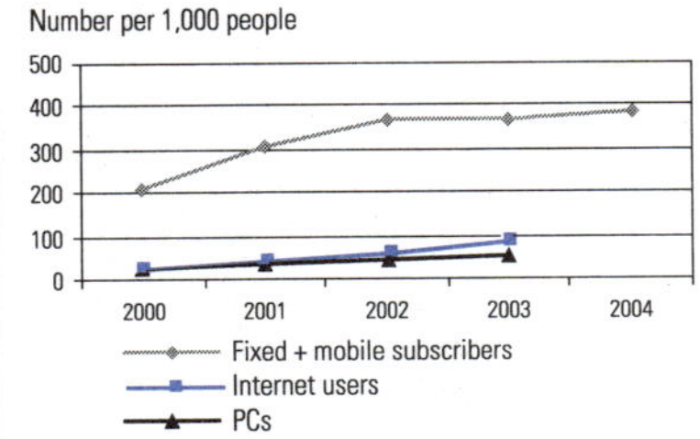

Price of Call to the United States, 2000–4

US$ per 3 minutes

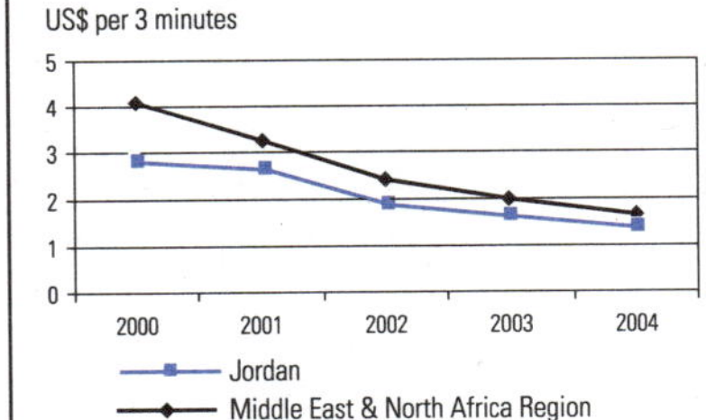

Total Telecommunications Revenue, 2000–4

Percentage of GDP

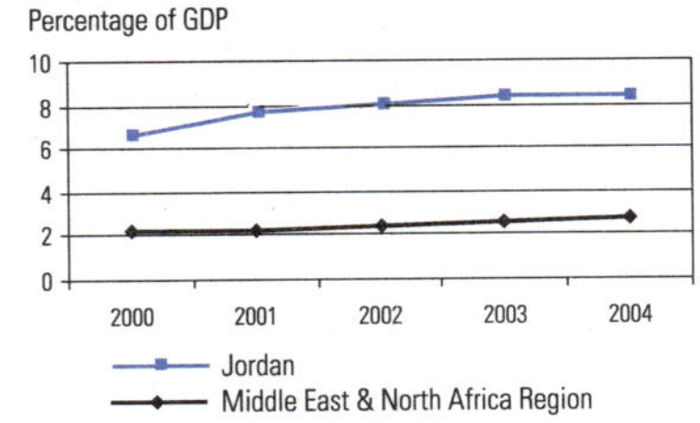

Notes: Figures in italics are for years other than those specified. .. indicates data are not available. C = competition; GDP = gross domestic product; GNI = gross national income; ICT = information and communication technology; M = monopoly; MDG = Millennium Development Goal; P = partial competition; and PCs = personal computers.

a. Outgoing and incoming.

b. Millennium Development Goal indicators 47, 48a, and 48b.

Sources: Economic and social context: UNDP and World Bank; ICT sector structure: ITU, WEF; ICT sector performance: Global Insight/WITSA, ITU, Netcraft, UNDESA, UNPAN, and World Bank. Produced by the Global Information and Communication Technologies Department and the Development Economics Data Group. For complete information, see Definitions and Data Sources.

Kazakhstan

	Kazakhstan 2000	Kazakhstan 2004	Lower-middle-incomegroup 2004	Europe & Central Asia Region 2004
Economic and social context				
Population, total (millions)	15	15	2,430	472
Urban population (% total population)	56	56	49	64
Poverty (% population below US$1 per day)	*<2*	*<2*	..	*3.6*
GNI per capita, Atlas method (current US$)	1,250	2,260	1,580	3,290
GDP growth, 1995–2000 and 2000–4 (%)	1.9	10.3	5.7	5.0
Adult literacy rate (% ages 15 and over)	*100*	..	90	97
Primary, secondary, tertiary school enrollment (% gross)	79	85	72	81
ICT sector structure				
Separate telecommunications regulator	..	..		
Status of main fixed-line operator	*Mixed*	Mixed		
Level of competition: international long distance	*C*	C		
Level of competition: mobile	*P*	P		
Level of competition: Internet service provider	..	..		
Government prioritization of ICT (scale 1–7)	..	..	3.8	3.8
ICT sector performance				
Access				
Telephone main lines (per 1,000 people)	122	*150*	192	*243*
International voice traffic (minutes per person)[a]	19	*26*	..	*28*
Mobile subscribers (per 1,000 people)	13	174	255	487
Population covered by mobile telephony (%)	*94*	..	*76*	*82*
Internet users (per 1,000 people)	7	*20.2*	70	*115*
Personal computers (per 1,000 people)	23	..	38	*73*
Households with television (%)	92	95	*89*	*92*
Quality				
Telephone faults (per 100 main lines per year)	..	*88.0*	..	*30.4*
Broadband subscribers (per 1,000 people)	0.0	*0.0*	12.6	*2.4*
International Internet bandwidth (bits per person)	*1*	*3*	58	148
Affordability				
Price basket for fixed line (US$ per month, residential)	4.0	*3.8*	*5.5*	*3.5*
Price basket for mobile (US$ per month)	..	4.7	8.9	10.3
Price basket for Internet (US$ per month)	..	*34.5*	*25.3*	*19.8*
Price of call to United States (US$ per 3 minutes)	*2.76*	..	1.45	1.06
Institutional efficiency and sustainability				
Total telecommunications revenue (% GDP)	1.5	3.0	3.6	3.5
Total telephone subscribers per employee	66	*110*	*195*	*150*
Total telecommunications investment (% revenue)	26.7	*19.6*	*25.5*	*19.0*
ICT applications				
ICT expenditure (% GDP)	..	..	*5.1*	*5.1*
E-government readiness index (scale 0–1)	..	0.32	0.26	0.39
Secure Internet servers (per 1 million people)	*0.5*	0.4	1.6	6.4
Schools connected to the Internet (%)	..	..	..	*65*

GNI per Capita, Atlas Method, 2000–4
Current US$

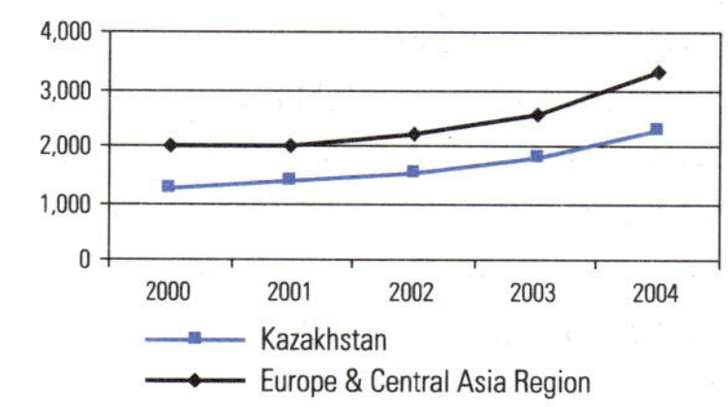

ICT MDG[b] Indicators, 2000–4
Number per 1,000 people

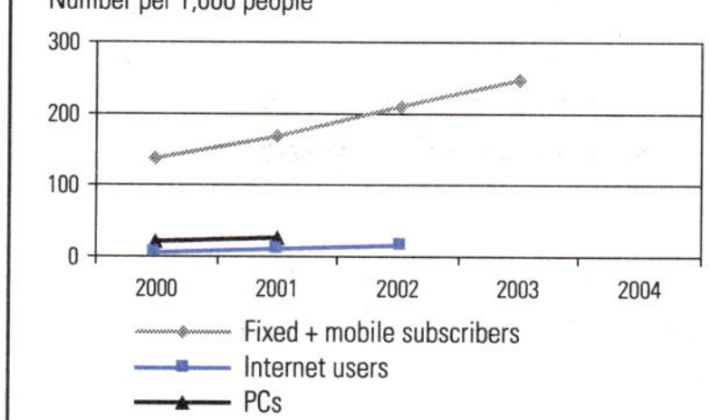

Price of Call to the United States, 2000–4
US$ per 3 minutes

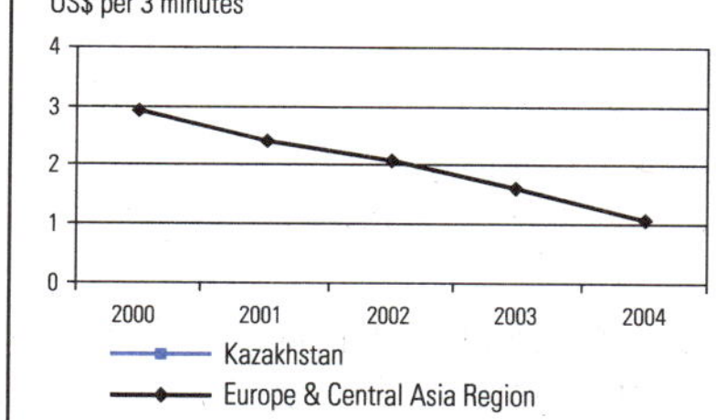

Total Telecommunications Revenue, 2000–4
Percentage of GDP

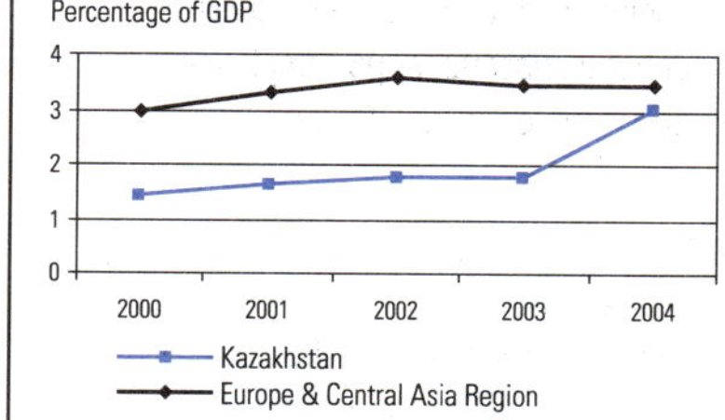

Notes: Figures in italics are for years other than those specified. .. indicates data are not available. C = competition; GDP = gross domestic product; GNI = gross national income; ICT = information and communication technology; M = monopoly; MDG = Millennium Development Goal; P = partial competition; and PCs = personal computers.

a. Outgoing and incoming.

b. Millennium Development Goal indicators 47, 48a, and 48b.

Sources: Economic and social context: UNDP and World Bank; ICT sector structure: ITU, WEF; ICT sector performance: Global Insight/WITSA, ITU, Netcraft, UNDESA, UNPAN, and World Bank. Produced by the Global Information and Communication Technologies Department and the Development Economics Data Group. For complete information, see Definitions and Data Sources.

Kenya

	Kenya 2000	Kenya 2004	Low-income group 2004	Sub-Saharan Africa Region 2004
Economic and social context				
Population, total (millions)	30	32	2,338	719
Urban population (% total population)	36	40	31	37
Poverty (% population below US$1 per day)	*22.8*	..	..	*46.4*
GNI per capita, Atlas method (current US$)	360	460	510	600
GDP growth, 1995–2000 and 2000–4 (%)	1.7	1.5	5.4	3.9
Adult literacy rate (% ages 15 and over)	74	..	61	59
Primary, secondary, tertiary school enrollment (% gross)	53	*52*	*55*	*52*
ICT sector structure				
Separate telecommunications regulator	Yes	Yes		
Status of main fixed-line operator	*Public*	Public		
Level of competition: international long distance	*M*	P		
Level of competition: mobile	*P*	P		
Level of competition: Internet service provider	*C*	C		
Government prioritization of ICT (scale 1–7)	..	4.0	..	*4.4*
ICT sector performance				
Access				
Telephone main lines (per 1,000 people)	10	10	33	17
International voice traffic (minutes per person)[a]	3	*2*	*4*	..
Mobile subscribers (per 1,000 people)	4	108	48	86
Population covered by mobile telephony (%)	..	..	*43*	..
Internet users (per 1,000 people)	3	15	20	15
Personal computers (per 1,000 people)	5	*8*	*8*	*12*
Households with television (%)	15	*19*	*16*	*15*
Quality				
Telephone faults (per 100 main lines per year)	220.9	*149.0*	..	..
Broadband subscribers (per 1,000 people)	0.0	*0.0*	0.5	*0.1*
International Internet bandwidth (bits per person)	0	2	*3*	*4*
Affordability				
Price basket for fixed line (US$ per month, residential)	6.6	*12.5*	*6.6*	*8.5*
Price basket for mobile (US$ per month)	..	14.0	11.6	13.5
Price basket for Internet (US$ per month)	..	*45.7*	*45.5*	*54.8*
Price of call to United States (US$ per 3 minutes)	7.35	3.00	1.95	2.43
Institutional efficiency and sustainability				
Total telecommunications revenue (% GDP)	2.9	4.9	2.3	5.0
Total telephone subscribers per employee	21	*80*	*89*	*144*
Total telecommunications investment (% revenue)	17.0	*7.8*	*27.8*	*27.8*
ICT applications				
ICT expenditure (% GDP)	3.5	*3.1*	*4.1*	..
E-government readiness index (scale 0–1)	..	0.14	0.12	0.11
Secure Internet servers (per 1 million people)	*0.0*	0.2	0.3	1.9
Schools connected to the Internet (%)	..	..	..	..

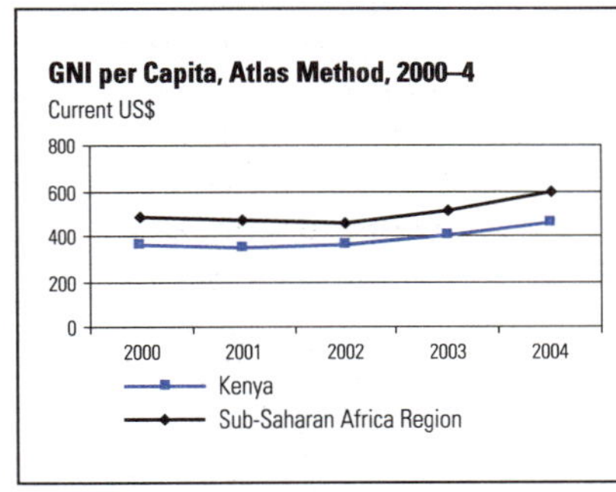

GNI per Capita, Atlas Method, 2000–4

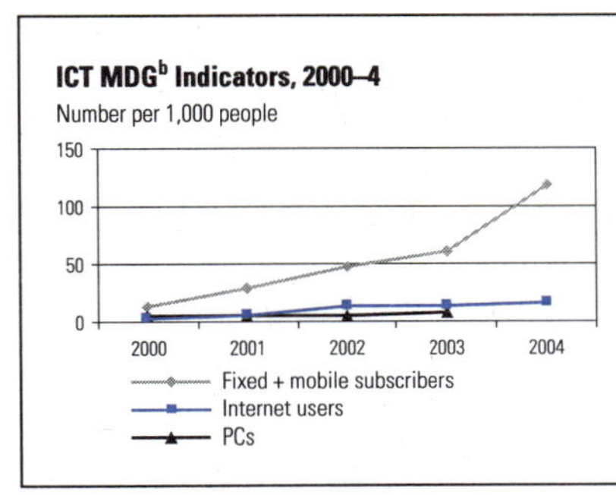

ICT MDG[b] Indicators, 2000–4

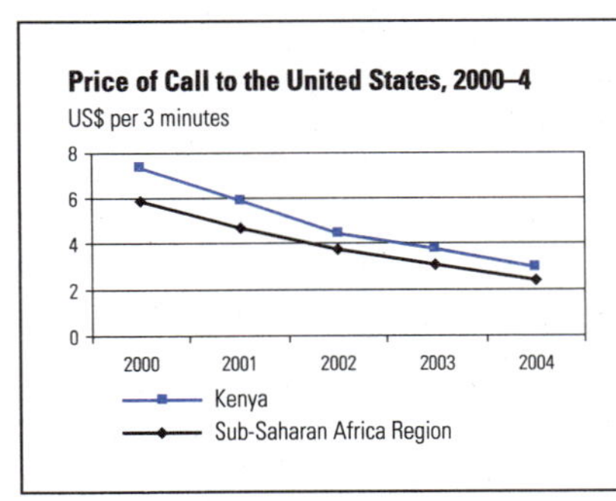

Price of Call to the United States, 2000–4

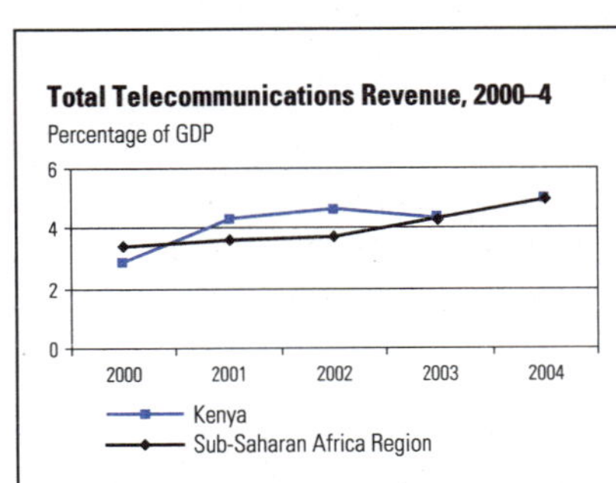

Total Telecommunications Revenue, 2000–4

Notes: Figures in italics are for years other than those specified. .. indicates data are not available. C = competition; GDP = gross domestic product; GNI = gross national income; ICT = information and communication technology; M = monopoly; MDG = Millennium Development Goal; P = partial competition; and PCs = personal computers.

a. Outgoing and incoming.

b. Millennium Development Goal indicators 47, 48a, and 48b.

Sources: Economic and social context: UNDP and World Bank; ICT sector structure: ITU, WEF; ICT sector performance: Global Insight/WITSA, ITU, Netcraft, UNDESA, UNPAN, and World Bank. Produced by the Global Information and Communication Technologies Department and the Development Economics Data Group. For complete information, see Definitions and Data Sources.

Korea, Republic of

	Korea, Rep. of 2000	Korea, Rep. of 2004	High-income group 2004
Economic and social context			
Population, total (millions)	47	48	1,001
Urban population (% total population)	80	81	77
Poverty (% population below US$1 per day)	*<2*	..	..
GNI per capita, Atlas method (current US$)	9,790	13,980	32,040
GDP growth, 1995–2000 and 2000–4 (%)	3.5	4.7	2.0
Adult literacy rate (% ages 15 and over)	..	..	..
Primary, secondary, tertiary school enrollment (% gross)	91	*94*	*93*
ICT sector structure			
Separate telecommunications regulator	Yes	Yes	
Status of main fixed-line operator	*Mixed*	Private	
Level of competition: international long distance	*C*	C	
Level of competition: mobile	*C*	P	
Level of competition: Internet service provider	*C*	C	
Government prioritization of ICT (scale 1–7)	..	5.4	4.9
ICT sector performance			
Access			
Telephone main lines (per 1,000 people)	467	467	558
International voice traffic (minutes per person)[a]	38	*41*	..
Mobile subscribers (per 1,000 people)	570	760	767
Population covered by mobile telephony (%)	99	99	98
Internet users (per 1,000 people)	405	656	*480*
Personal computers (per 1,000 people)	396	*558*	*504*
Households with television (%)	94	93	*98*
Quality			
Telephone faults (per 100 main lines per year)	1.8	*1.0*	..
Broadband subscribers (per 1,000 people)	82.3	247.6	126.2
International Internet bandwidth (bits per person)	48	993	4,718
Affordability			
Price basket for fixed line (US$ per month, residential)	7.1	7.3	25.8
Price basket for mobile (US$ per month)	..	2.1	17.8
Price basket for Internet (US$ per month)	..	*9.7*	*20.9*
Price of call to United States (US$ per 3 minutes)	1.93	0.76	0.77
Institutional efficiency and sustainability			
Total telecommunications revenue (% GDP)	3.5	4.2	2.9
Total telephone subscribers per employee	702	..	*485*
Total telecommunications investment (% revenue)	43.9	0.0	*12.3*
ICT applications			
ICT expenditure (% GDP)	6.8	*6.6*	*7.1*
E-government readiness index (scale 0–1)	..	0.95	0.70
Secure Internet servers (per 1 million people)	*7.3*	18.6	311.4
Schools connected to the Internet (%)	..	100	*99*

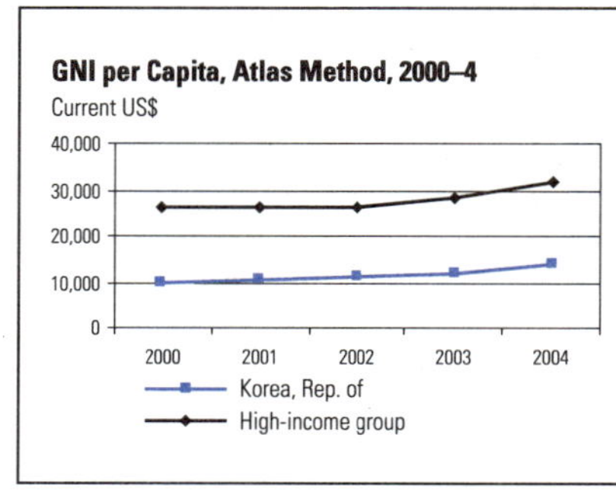

GNI per Capita, Atlas Method, 2000–4

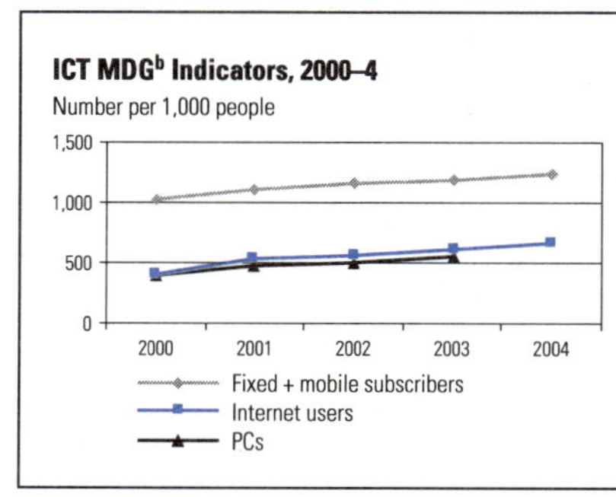

ICT MDG[b] Indicators, 2000–4

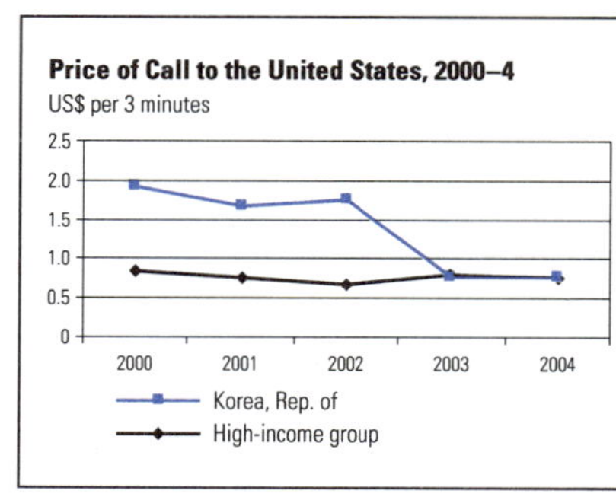

Price of Call to the United States, 2000–4

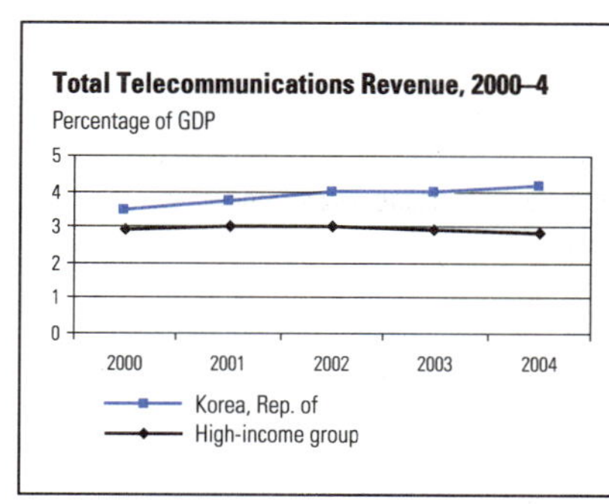

Total Telecommunications Revenue, 2000–4

Notes: Figures in italics are for years other than those specified. .. indicates data are not available. C = competition; GDP = gross domestic product; GNI = gross national income; ICT = information and communication technology; M = monopoly; MDG = Millennium Development Goal; P = partial competition; and PCs = personal computers.

a. Outgoing and incoming.

b. Millennium Development Goal indicators 47, 48a, and 48b.

Sources: Economic and social context: UNDP and World Bank; ICT sector structure: ITU, WEF; ICT sector performance: Global Insight/WITSA, ITU, Netcraft, UNDESA, UNPAN, and World Bank. Produced by the Global Information and Communication Technologies Department and the Development Economics Data Group. For complete information, see Definitions and Data Sources.

Kuwait

	Kuwait 2000	Kuwait 2004	High-income group 2004
Economic and social context			
Population, total (millions)	2	2	1,001
Urban population (% total population)	96	96	77
Poverty (% population below US$1 per day)	..	..	..
GNI per capita, Atlas method (current US$)	16,290	*17,970*	32,040
GDP growth, 1995–2000 and 2000–4 (%)	0.9	*2.4*	2.0
Adult literacy rate (% ages 15 and over)	..	*83*	..
Primary, secondary, tertiary school enrollment (% gross)	74	*74*	*93*
ICT sector structure			
Separate telecommunications regulator	..	..	
Status of main fixed-line operator	*Public*	Public	
Level of competition: international long distance	*C*	M	
Level of competition: mobile	*P*	P	
Level of competition: Internet service provider	*P*	P	
Government prioritization of ICT (scale 1–7)	..	..	4.9
ICT sector performance			
Access			
Telephone main lines (per 1,000 people)	213	202	558
International voice traffic (minutes per person)[a]	178	*184*	..
Mobile subscribers (per 1,000 people)	217	855	767
Population covered by mobile telephony (%)	99	99	98
Internet users (per 1,000 people)	68	*237*	*480*
Personal computers (per 1,000 people)	114	*122*	*504*
Households with television (%)	..	*95*	*98*
Quality			
Telephone faults (per 100 main lines per year)	*30.0*	*27.0*	..
Broadband subscribers (per 1,000 people)	*2.2*	5.4	126.2
International Internet bandwidth (bits per person)	*26*	*120*	4,718
Affordability			
Price basket for fixed line (US$ per month, residential)	9.9	*10.3*	25.8
Price basket for mobile (US$ per month)	..	7.4	17.8
Price basket for Internet (US$ per month)	..	*24.7*	*20.9*
Price of call to United States (US$ per 3 minutes)	1.94	*1.50*	0.77
Institutional efficiency and sustainability			
Total telecommunications revenue (% GDP)	1.7	*2.4*	2.9
Total telephone subscribers per employee	123	*234*	*485*
Total telecommunications investment (% revenue)	..	..	*12.3*
ICT applications			
ICT expenditure (% GDP)	1.7	*1.7*	*7.1*
E-government readiness index (scale 0–1)	..	0.14	0.70
Secure Internet servers (per 1 million people)	*1.8*	21.1	311.4
Schools connected to the Internet (%)	..	..	*99*

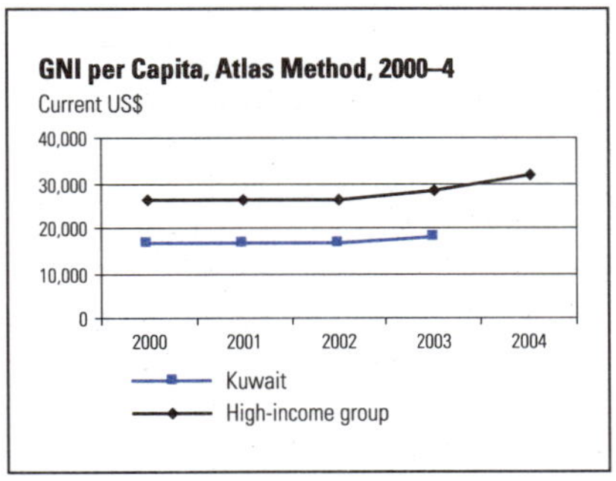

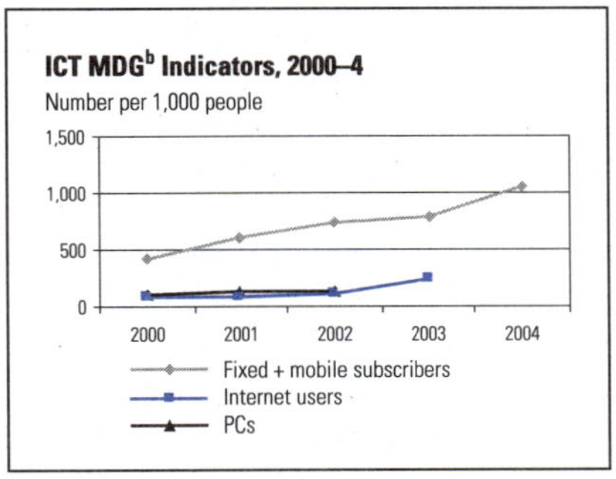

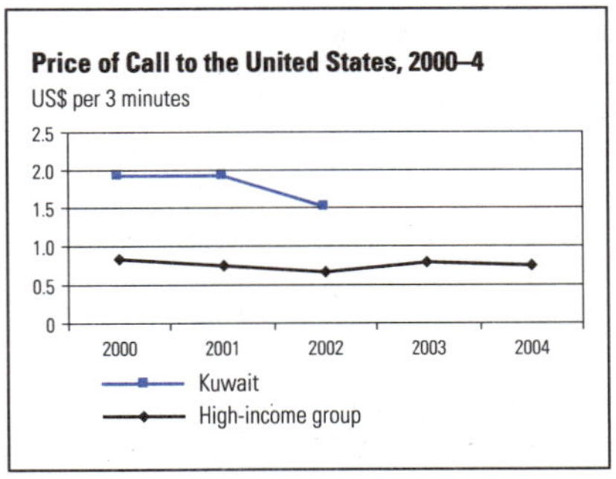

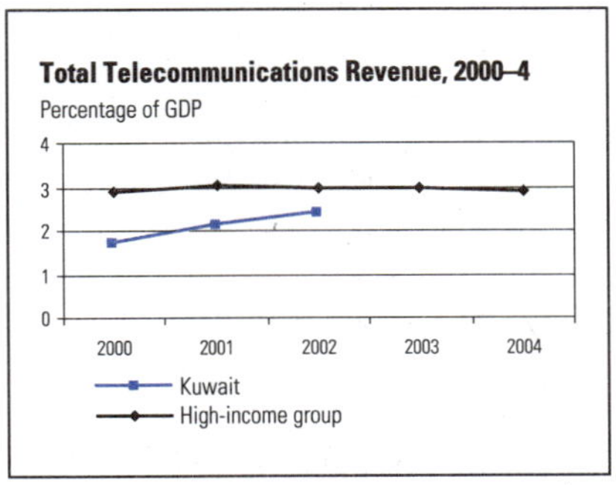

Notes: Figures in italics are for years other than those specified. .. indicates data are not available. C = competition; GDP = gross domestic product; GNI = gross national income; ICT = information and communication technology; M = monopoly; MDG = Millennium Development Goal; P = partial competition; and PCs = personal computers.

a. Outgoing and incoming.

b. Millennium Development Goal indicators 47, 48a, and 48b.

Sources: Economic and social context: UNDP and World Bank; ICT sector structure: ITU, WEF; ICT sector performance: Global Insight/WITSA, ITU, Netcraft, UNDESA, UNPAN, and World Bank. Produced by the Global Information and Communication Technologies Department and the Development Economics Data Group. For complete information, see Definitions and Data Sources.

Kyrgyz Republic

	Kyrgyz Republic 2000	Kyrgyz Republic 2004	Low-income group 2004	Europe & Central Asia Region 2004
Economic and social context				
Population, total (millions)	5	5	2,338	472
Urban population (% total population)	34	34	31	64
Poverty (% population below US$1 per day)	*<2*	*<2*	..	*3.6*
GNI per capita, Atlas method (current US$)	280	400	510	3,290
GDP growth, 1995–2000 and 2000–4 (%)	5.4	4.5	5.4	5.0
Adult literacy rate (% ages 15 and over)	*99*	..	61	97
Primary, secondary, tertiary school enrollment (% gross)	80	*82*	*55*	*81*
ICT sector structure				
Separate telecommunications regulator	Yes	Yes		
Status of main fixed-line operator	*Public*	*Public*		
Level of competition: international long distance	*M*	C		
Level of competition: mobile	*P*	C		
Level of competition: Internet service provider	*P*	C		
Government prioritization of ICT (scale 1–7)	..	..	..	3.8
ICT sector performance				
Access				
Telephone main lines (per 1,000 people)	77	*78*	33	*243*
International voice traffic (minutes per person)[a]	12	*13*	*4*	*28*
Mobile subscribers (per 1,000 people)	2	*18*	48	487
Population covered by mobile telephony (%)	..	..	*43*	*82*
Internet users (per 1,000 people)	10	*30*	20	*115*
Personal computers (per 1,000 people)	10	*18*	*8*	*73*
Households with television (%)	85	..	*16*	*92*
Quality				
Telephone faults (per 100 main lines per year)	..	*82.0*	..	*30.4*
Broadband subscribers (per 1,000 people)	0.0	*0.0*	0.5	*2.4*
International Internet bandwidth (bits per person)	..	*3*	*3*	148
Affordability				
Price basket for fixed line (US$ per month, residential)	0.7	1.4	*6.6*	*3.5*
Price basket for mobile (US$ per month)	..	10.8	11.6	10.3
Price basket for Internet (US$ per month)	..	*13.0*	*45.5*	*19.8*
Price of call to United States (US$ per 3 minutes)	9.84	..	1.95	1.06
Institutional efficiency and sustainability				
Total telecommunications revenue (% GDP)	2.0	3.8	2.3	3.5
Total telephone subscribers per employee	53	*65*	*89*	*150*
Total telecommunications investment (% revenue)	8.0	*20.2*	*27.8*	*19.0*
ICT applications				
ICT expenditure (% GDP)	..	..	*4.1*	*5.1*
E-government readiness index (scale 0–1)	..	0.39	0.12	0.39
Secure Internet servers (per 1 million people)	*0.4*	0.4	0.3	6.4
Schools connected to the Internet (%)	1	..	..	*65*

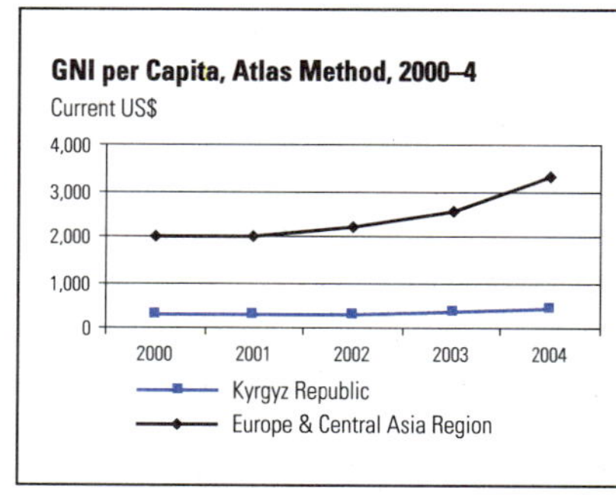

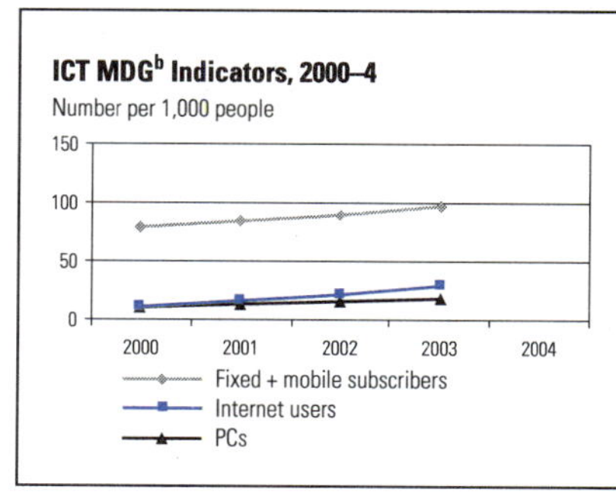

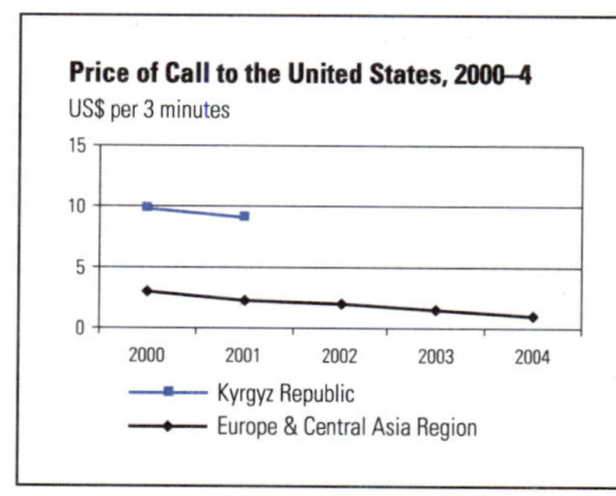

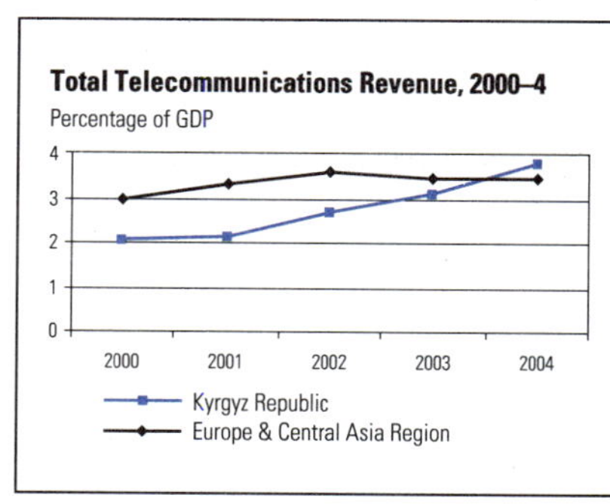

Notes: Figures in italics are for years other than those specified. .. indicates data are not available. C = competition; GDP = gross domestic product; GNI = gross national income; ICT = information and communication technology; M = monopoly; MDG = Millennium Development Goal; P = partial competition; and PCs = personal computers.

a. Outgoing and incoming.

b. Millennium Development Goal indicators 47, 48a, and 48b.

Sources: Economic and social context: UNDP and World Bank; ICT sector structure: ITU, WEF; ICT sector performance: Global Insight/WITSA, ITU, Netcraft, UNDESA, UNPAN, and World Bank. Produced by the Global Information and Communication Technologies Department and the Development Economics Data Group. For complete information, see Definitions and Data Sources.

Lao People's Democratic Republic

	Lao PDR 2000	Lao PDR 2004	Low-income group 2004	East Asia & Pacific Region 2004
Economic and social context				
Population, total (millions)	5	6	2,338	1,870
Urban population (% total population)	19	21	31	41
Poverty (% population below US$1 per day)	*26.3*	..	..	*14.9*
GNI per capita, Atlas method (current US$)	280	390	510	1,280
GDP growth, 1995–2000 and 2000–4 (%)	6.1	5.7	5.4	7.5
Adult literacy rate (% ages 15 and over)	*69*	..	61	90
Primary, secondary, tertiary school enrollment (% gross)	57	*61*	*55*	*68*
ICT sector structure				
Separate telecommunications regulator	..	..		
Status of main fixed-line operator	*Mixed*	Mixed		
Level of competition: international long distance	*M*	M		
Level of competition: mobile	*M*	C		
Level of competition: Internet service provider	..	C		
Government prioritization of ICT (scale 1–7)	..	..	..	4.8
ICT sector performance				
Access				
Telephone main lines (per 1,000 people)	8	12	33	194
International voice traffic (minutes per person)[a]	5	..	*4*	..
Mobile subscribers (per 1,000 people)	2	35	48	248
Population covered by mobile telephony (%)	..	*7*	*43*	*73*
Internet users (per 1,000 people)	1	*3*	20	75
Personal computers (per 1,000 people)	3	*3*	*8*	37
Households with television (%)	30	..	*16*	*80*
Quality				
Telephone faults (per 100 main lines per year)	..	..	..	..
Broadband subscribers (per 1,000 people)	*0.0*	*0.0*	0.5	13.4
International Internet bandwidth (bits per person)	0	..	*3*	52
Affordability				
Price basket for fixed line (US$ per month, residential)	2.9	*2.3*	*6.6*	*4.5*
Price basket for mobile (US$ per month)	..	2.2	11.6	5.1
Price basket for Internet (US$ per month)	..	*31.9*	*45.5*	*19.9*
Price of call to United States (US$ per 3 minutes)	9.20	1.11	1.95	1.20
Institutional efficiency and sustainability				
Total telecommunications revenue (% GDP)	1.5	1.6	2.3	3.6
Total telephone subscribers per employee	47	*85*	*89*	..
Total telecommunications investment (% revenue)	31.3	*40.5*	*27.8*	*31.0*
ICT applications				
ICT expenditure (% GDP)	..	..	*4.1*	*5.0*
E-government readiness index (scale 0–1)	..	0.06	0.12	0.21
Secure Internet servers (per 1 million people)	..	..	0.3	0.6
Schools connected to the Internet (%)	..	..	..	..

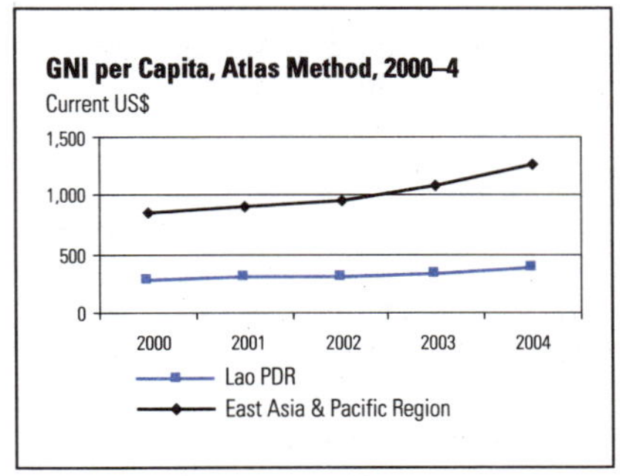

GNI per Capita, Atlas Method, 2000–4

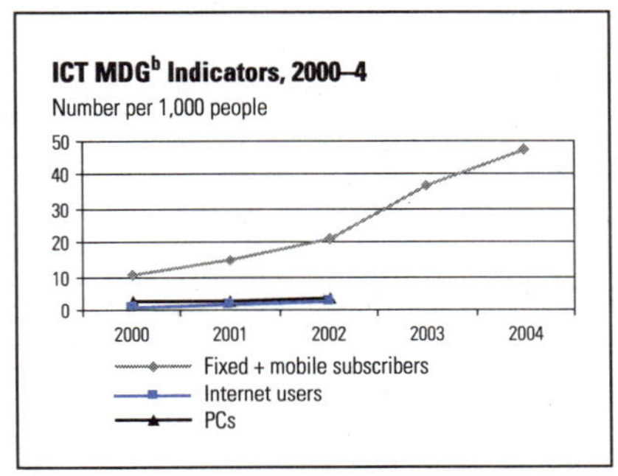

ICT MDG[b] Indicators, 2000–4

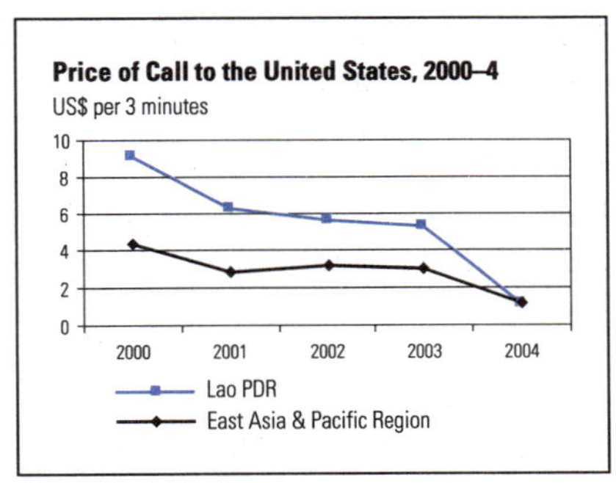

Price of Call to the United States, 2000–4

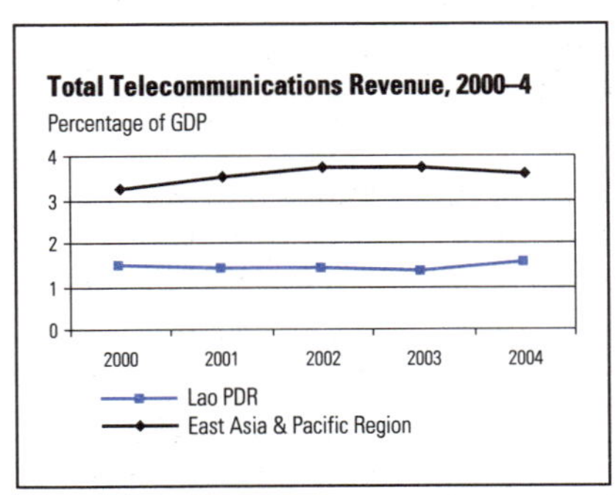

Total Telecommunications Revenue, 2000–4

Notes: Figures in italics are for years other than those specified. .. indicates data are not available. C = competition; GDP = gross domestic product; GNI = gross national income; ICT = information and communication technology; M = monopoly; MDG = Millennium Development Goal; P = partial competition; and PCs = personal computers.

a. Outgoing and incoming.

b. Millennium Development Goal indicators 47, 48a, and 48b.

Sources: Economic and social context: UNDP and World Bank; ICT sector structure: ITU, WEF; ICT sector performance: Global Insight/WITSA, ITU, Netcraft, UNDESA, UNPAN, and World Bank. Produced by the Global Information and Communication Technologies Department and the Development Economics Data Group. For complete information, see Definitions and Data Sources.

Latvia

	Latvia		Upper-middle-income group	Europe & Central Asia Region
	2000	2004	2004	2004
Economic and social context				
Population, total (millions)	2	2	576	472
Urban population (% total population)	67	66	72	64
Poverty (% population below US$1 per day)	*<2*	..	..	*3.6*
GNI per capita, Atlas method (current US$)	3,190	5,460	4,770	3,290
GDP growth, 1995–2000 and 2000–4 (%)	5.4	7.5	2.7	5.0
Adult literacy rate (% ages 15 and over)	100	..	94	97
Primary, secondary, tertiary school enrollment (% gross)	87	*90*	*80*	*81*
ICT sector structure				
Separate telecommunications regulator	No	Yes		
Status of main fixed-line operator	*Mixed*	*Public*		
Level of competition: international long distance	*M*	C		
Level of competition: mobile	*C*	C		
Level of competition: Internet service provider	*C*	C		
Government prioritization of ICT (scale 1–7)	..	3.8	4.1	3.8
ICT sector performance				
Access				
Telephone main lines (per 1,000 people)	310	274	*220*	*243*
International voice traffic (minutes per person)[a]	62	*69*	*39*	*28*
Mobile subscribers (per 1,000 people)	169	673	490	487
Population covered by mobile telephony (%)	89	97	*84*	*82*
Internet users (per 1,000 people)	63	*403*	*133*	*115*
Personal computers (per 1,000 people)	143	*171*	*99*	*73*
Households with television (%)	76	*85*	*92*	*92*
Quality				
Telephone faults (per 100 main lines per year)	28.7	*20.3*	*20.3*	*30.4*
Broadband subscribers (per 1,000 people)	0.1	16.9	*3.7*	*2.4*
International Internet bandwidth (bits per person)	65	729	176	148
Affordability				
Price basket for fixed line (US$ per month, residential)	12.5	15.6	*13.9*	*3.5*
Price basket for mobile (US$ per month)	..	14.9	11.1	10.3
Price basket for Internet (US$ per month)	..	*58.1*	*20.8*	*19.8*
Price of call to United States (US$ per 3 minutes)	2.05	1.63	1.03	1.06
Institutional efficiency and sustainability				
Total telecommunications revenue (% GDP)	5.6	6.4	3.4	3.5
Total telephone subscribers per employee	263	*415*	*402*	*150*
Total telecommunications investment (% revenue)	16.9	*15.8*	*18.6*	*19.0*
ICT applications				
ICT expenditure (% GDP)	..	..	*5.0*	*5.1*
E-government readiness index (scale 0–1)	..	0.39	0.49	0.39
Secure Internet servers (per 1 million people)	*18.2*	34.7	10.7	6.4
Schools connected to the Internet (%)	..	*97*	*60*	*65*

GNI per Capita, Atlas Method, 2000–4
Current US$

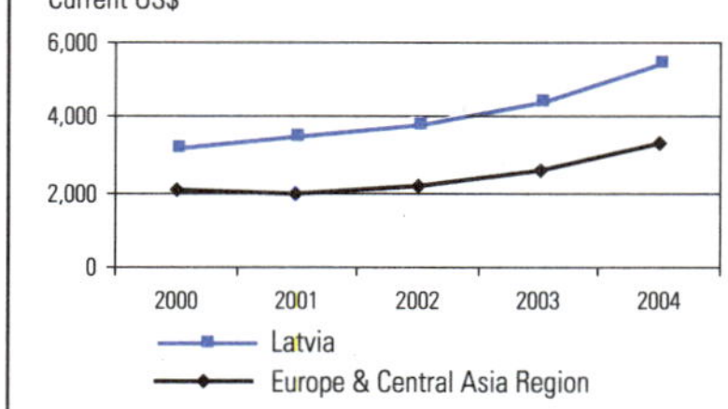

ICT MDG[b] Indicators, 2000–4
Number per 1,000 people

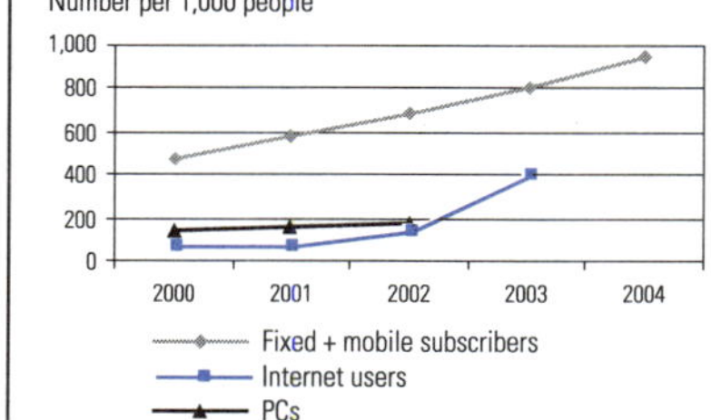

Price of Call to the United States, 2000–4
US$ per 3 minutes

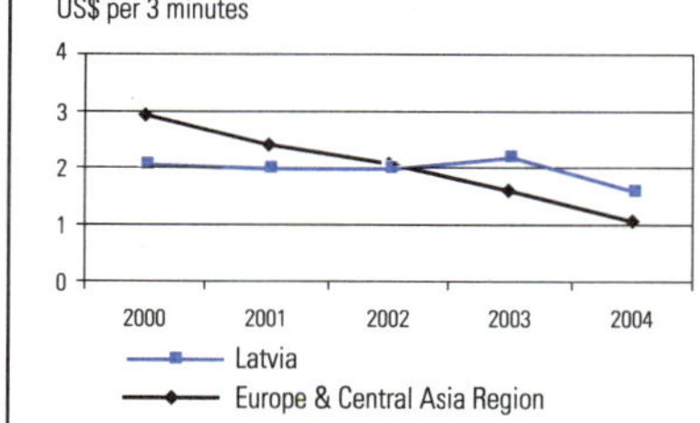

Total Telecommunications Revenue, 2000–4
Percentage of GDP

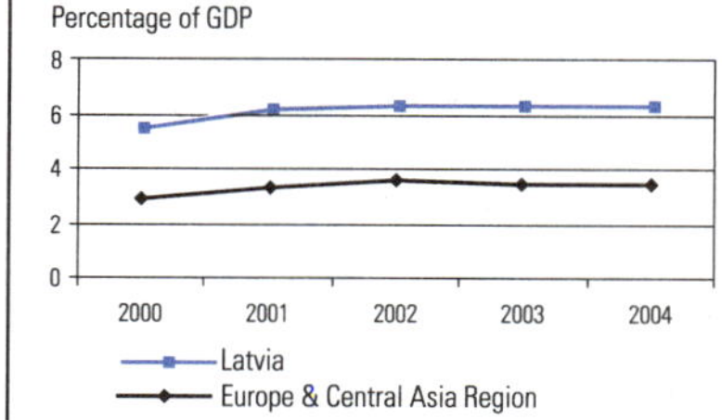

Notes: Figures in italics are for years other than those specified. .. indicates data are not available. C = competition; GDP = gross domestic product; GNI = gross national income; ICT = information and communication technology; M = monopoly; MDG = Millennium Development Goal; P = partial competition; and PCs = personal computers.

a. Outgoing and incoming.
b. Millennium Development Goal indicators 47, 48a, and 48b.

Sources: Economic and social context: UNDP and World Bank; ICT sector structure: ITU, WEF; ICT sector performance: Global Insight/WITSA, ITU, Netcraft, UNDESA, UNPAN, and World Bank. Produced by the Global Information and Communication Technologies Department and the Development Economics Data Group. For complete information, see Definitions and Data Sources.

Lebanon

	Lebanon 2000	Lebanon 2004	Upper-middle-income group 2004	Middle East & North Africa Region 2004
Economic and social context				
Population, total (millions)	4	5	576	294
Urban population (% total population)	87	88	72	56
Poverty (% population below US$1 per day)	..	..	..	*2.4*
GNI per capita, Atlas method (current US$)	4,080	4,980	4,770	2,000
GDP growth, 1995–2000 and 2000–4 (%)	2.0	4.4	2.7	4.5
Adult literacy rate (% ages 15 and over)	..	..	94	..
Primary, secondary, tertiary school enrollment (% gross)	77	*79*	*80*	*68*
ICT sector structure				
Separate telecommunications regulator	..	..		
Status of main fixed-line operator	*Public*	Public		
Level of competition: international long distance	*M*	M		
Level of competition: mobile	..	..		
Level of competition: Internet service provider	*C*	..		
Government prioritization of ICT (scale 1–7)	..	..	4.1	..
ICT sector performance				
Access				
Telephone main lines (per 1,000 people)	133	*156*	*220*	*118*
International voice traffic (minutes per person)[a]	..	..	*39*	*20*
Mobile subscribers (per 1,000 people)	172	*178*	490	*88*
Population covered by mobile telephony (%)	..	..	*84*	..
Internet users (per 1,000 people)	69	*122*	*133*	*47*
Personal computers (per 1,000 people)	40	*62*	*99*	*30*
Households with television (%)	92	*93*	*92*	*88*
Quality				
Telephone faults (per 100 main lines per year)	..	*31.0*	*20.3*	..
Broadband subscribers (per 1,000 people)	..	*7.9*	*3.7*	*0.2*
International Internet bandwidth (bits per person)	..	*14*	176	*15*
Affordability				
Price basket for fixed line (US$ per month, residential)	21.3	20.4	*13.9*	*4.9*
Price basket for mobile (US$ per month)	..	20.1	11.1	8.1
Price basket for Internet (US$ per month)	..	*36.9*	*20.8*	*24.5*
Price of call to United States (US$ per 3 minutes)	*4.48*	2.19	1.03	1.64
Institutional efficiency and sustainability				
Total telecommunications revenue (% GDP)	6.4	4.9	3.4	2.8
Total telephone subscribers per employee	..	*300*	*402*	..
Total telecommunications investment (% revenue)	..	..	*18.6*	*27.8*
ICT applications				
ICT expenditure (% GDP)	..	..	*5.0*	..
E-government readiness index (scale 0–1)	..	0.24	0.49	0.16
Secure Internet servers (per 1 million people)	*4.3*	6.4	10.7	0.6
Schools connected to the Internet (%)	..	*20*	*60*	..

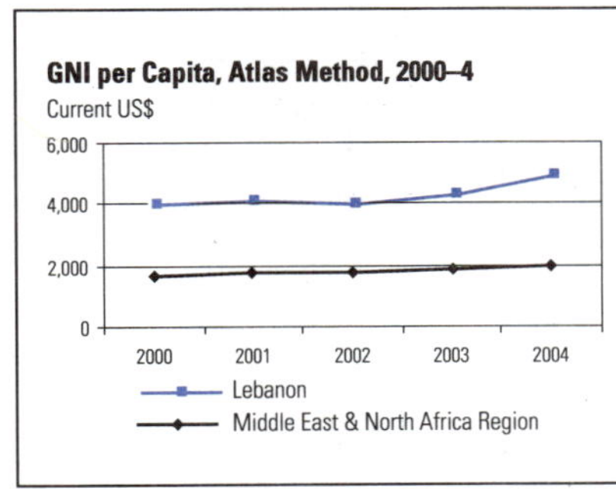

GNI per Capita, Atlas Method, 2000–4

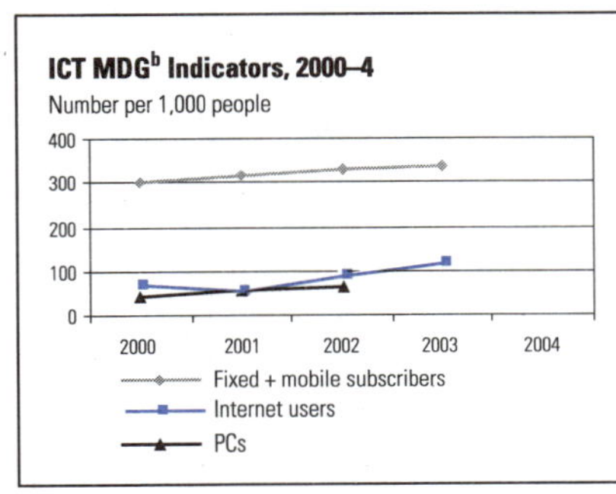

ICT MDG[b] Indicators, 2000–4

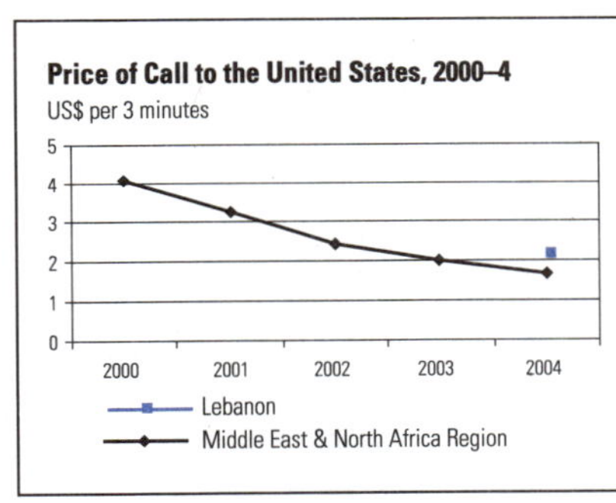

Price of Call to the United States, 2000–4

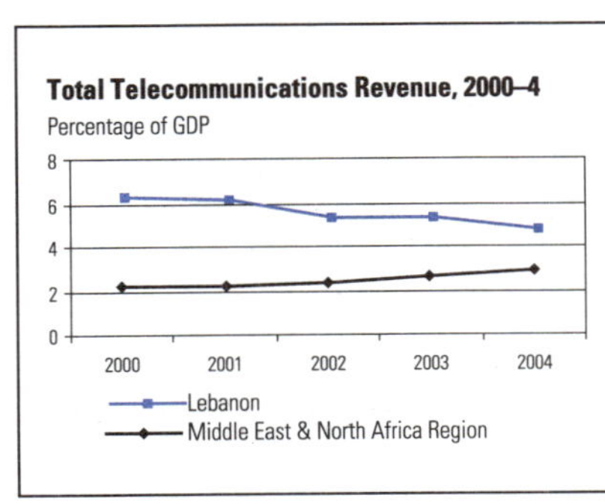

Total Telecommunications Revenue, 2000–4

Notes: Figures in italics are for years other than those specified. .. indicates data are not available. C = competition; GDP = gross domestic product; GNI = gross national income; ICT = information and communication technology; M = monopoly; MDG = Millennium Development Goal; P = partial competition; and PCs = personal computers.

a. Outgoing and incoming.

b. Millennium Development Goal indicators 47, 48a, and 48b.

Sources: Economic and social context: UNDP and World Bank; ICT sector structure: ITU, WEF; ICT sector performance: Global Insight/WITSA, ITU, Netcraft, UNDESA, UNPAN, and World Bank. Produced by the Global Information and Communication Technologies Department and the Development Economics Data Group. For complete information, see Definitions and Data Sources.

Lesotho

	Lesotho 2000	Lesotho 2004	Low-income group 2004	Sub-Saharan Africa Region 2004
Economic and social context				
Population, total (millions)	2	2	2,338	719
Urban population (% total population)	18	18	31	37
Poverty (% population below US$1 per day)	*36.4*	..	..	*46.4*
GNI per capita, Atlas method (current US$)	640	740	510	600
GDP growth, 1995–2000 and 2000–4 (%)	2.2	3.3	5.4	3.9
Adult literacy rate (% ages 15 and over)	*81*	..	61	59
Primary, secondary, tertiary school enrollment (% gross)	64	*66*	*55*	*52*
ICT sector structure				
Separate telecommunications regulator	Yes	Yes		
Status of main fixed-line operator	*Public*	Mixed		
Level of competition: international long distance	*M*	P		
Level of competition: mobile	*M*	C		
Level of competition: Internet service provider	..	C		
Government prioritization of ICT (scale 1–7)	..	..	..	*4.4*
ICT sector performance				
Access				
Telephone main lines (per 1,000 people)	13	*20*	33	17
International voice traffic (minutes per person)[a]	..	..	*4*	..
Mobile subscribers (per 1,000 people)	12	94	48	86
Population covered by mobile telephony (%)	..	*80*	*43*	..
Internet users (per 1,000 people)	2	17	20	15
Personal computers (per 1,000 people)	..	*8*	*8*	*12*
Households with television (%)	*16*	*17*	*16*	*15*
Quality				
Telephone faults (per 100 main lines per year)	*69.2*	*72.8*	..	..
Broadband subscribers (per 1,000 people)	0.0	*0.0*	0.5	*0.1*
International Internet bandwidth (bits per person)	0	*1*	*3*	*4*
Affordability				
Price basket for fixed line (US$ per month, residential)	*9.7*	18.6	*6.6*	*8.5*
Price basket for mobile (US$ per month)	..	14.3	11.6	13.5
Price basket for Internet (US$ per month)	..	*43.4*	*45.5*	*54.8*
Price of call to United States (US$ per 3 minutes)	*2.31*	3.28	1.95	2.43
Institutional efficiency and sustainability				
Total telecommunications revenue (% GDP)	1.3	*2.5*	2.3	5.0
Total telephone subscribers per employee	126	..	*89*	*144*
Total telecommunications investment (% revenue)	9.8	*40.8*	*27.8*	*27.8*
ICT applications				
ICT expenditure (% GDP)	..	..	*4.1*	..
E-government readiness index (scale 0–1)	..	0.19	0.12	0.11
Secure Internet servers (per 1 million people)	..	..	0.3	1.9
Schools connected to the Internet (%)	..	..	..	..

GNI per Capita, Atlas Method, 2000–4
Current US$

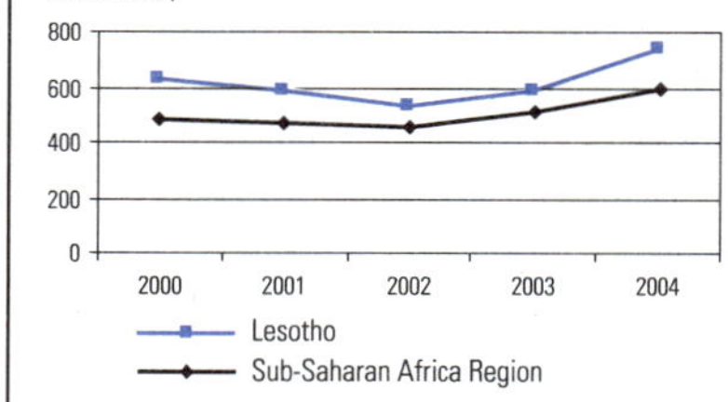

ICT MDG[b] Indicators, 2000–4
Number per 1,000 people

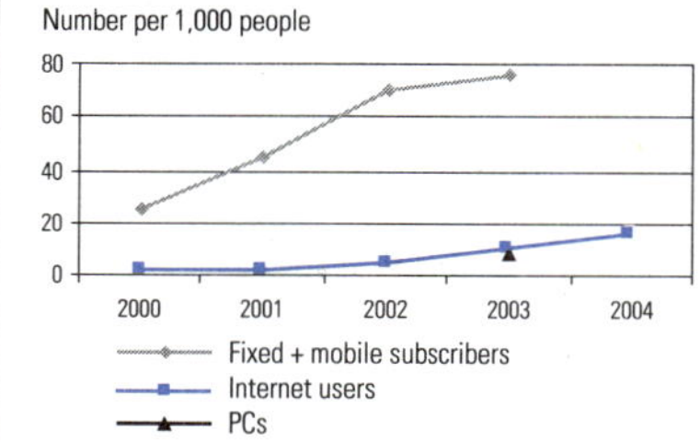

Price of Call to the United States, 2000–4
US$ per 3 minutes

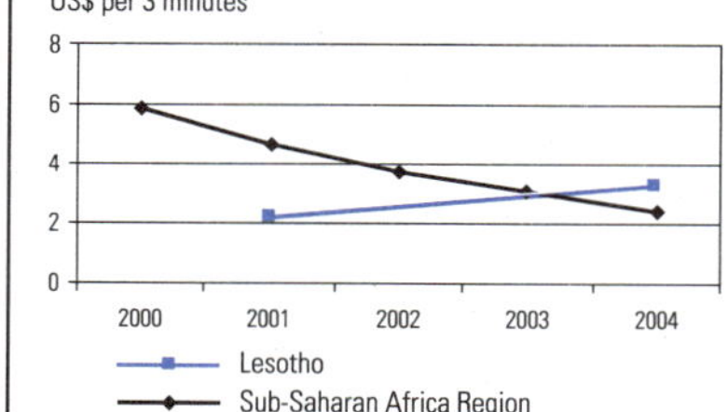

Total Telecommunications Revenue, 2000–4
Percentage of GDP

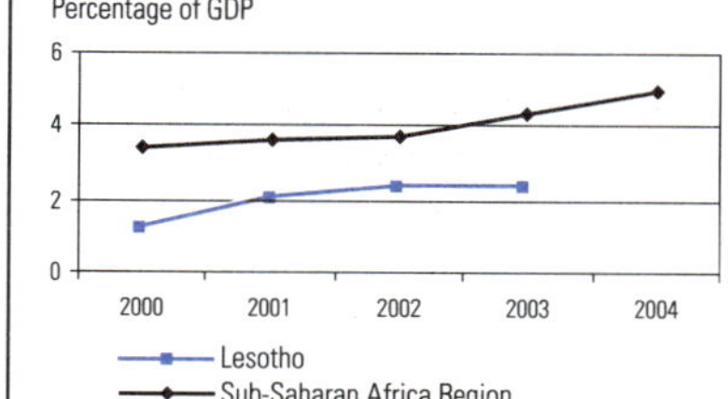

Notes: Figures in italics are for years other than those specified. .. indicates data are not available. C = competition; GDP = gross domestic product; GNI = gross national income; ICT = information and communication technology; M = monopoly; MDG = Millennium Development Goal; P = partial competition; and PCs = personal computers.

a. Outgoing and incoming.

b. Millennium Development Goal indicators 47, 48a, and 48b.

Sources: Economic and social context: UNDP and World Bank; ICT sector structure: ITU, WEF; ICT sector performance: Global Insight/WITSA, ITU, Netcraft, UNDESA, UNPAN, and World Bank. Produced by the Global Information and Communication Technologies Department and the Development Economics Data Group. For complete information, see Definitions and Data Sources.

Lithuania

	Lithuania 2000	Lithuania 2004	Upper-middle-income group 2004	Europe & Central Asia Region 2004
Economic and social context				
Population, total (millions)	4	3	576	472
Urban population (% total population)	67	67	72	64
Poverty (% population below US$1 per day)	<2	..	..	*3.6*
GNI per capita, Atlas method (current US$)	3,170	5,740	4,770	3,290
GDP growth, 1995–2000 and 2000–4 (%)	4.3	7.5	2.7	5.0
Adult literacy rate (% ages 15 and over)	*100*	..	94	97
Primary, secondary, tertiary school enrollment (% gross)	*88*	*94*	*80*	*81*
ICT sector structure				
Separate telecommunications regulator	No	Yes		
Status of main fixed-line operator	*Mixed*	Mixed		
Level of competition: international long distance	*M*	C		
Level of competition: mobile	*C*	C		
Level of competition: Internet service provider	*C*	C		
Government prioritization of ICT (scale 1–7)	..	4.7	4.1	3.8
ICT sector performance				
Access				
Telephone main lines (per 1,000 people)	339	*239*	*220*	*243*
International voice traffic (minutes per person)[a]	43	..	*39*	*28*
Mobile subscribers (per 1,000 people)	150	828	490	487
Population covered by mobile telephony (%)	100	*100*	*84*	*82*
Internet users (per 1,000 people)	64	*201*	*133*	*115*
Personal computers (per 1,000 people)	68	*110*	*99*	*73*
Households with television (%)	95	*97*	*92*	*92*
Quality				
Telephone faults (per 100 main lines per year)	19.8	*16.3*	*20.3*	*30.4*
Broadband subscribers (per 1,000 people)	0.0	37.5	*3.7*	*2.4*
International Internet bandwidth (bits per person)	14	194	176	148
Affordability				
Price basket for fixed line (US$ per month, residential)	9.5	*14.6*	*13.9*	*3.5*
Price basket for mobile (US$ per month)	..	6.9	11.1	10.3
Price basket for Internet (US$ per month)	..	*34.1*	*20.8*	*19.8*
Price of call to United States (US$ per 3 minutes)	3.10	*2.31*	1.03	1.06
Institutional efficiency and sustainability				
Total telecommunications revenue (% GDP)	2.3	3.5	3.4	3.5
Total telephone subscribers per employee	283	..	*402*	*150*
Total telecommunications investment (% revenue)	49.8	..	*18.6*	*19.0*
ICT applications				
ICT expenditure (% GDP)	..	..	*5.0*	*5.1*
E-government readiness index (scale 0–1)	..	0.43	0.49	0.39
Secure Internet servers (per 1 million people)	*12.3*	13.7	10.7	6.4
Schools connected to the Internet (%)	..	*56*	*60*	*65*

GNI per Capita, Atlas Method, 2000–4
Current US$

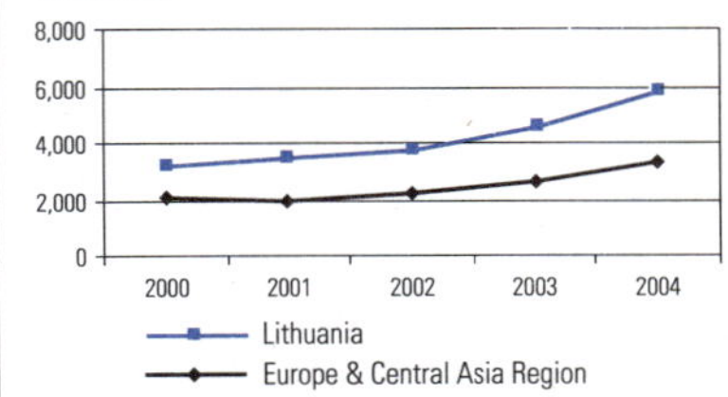

ICT MDG[b] Indicators, 2000–4
Number per 1,000 people

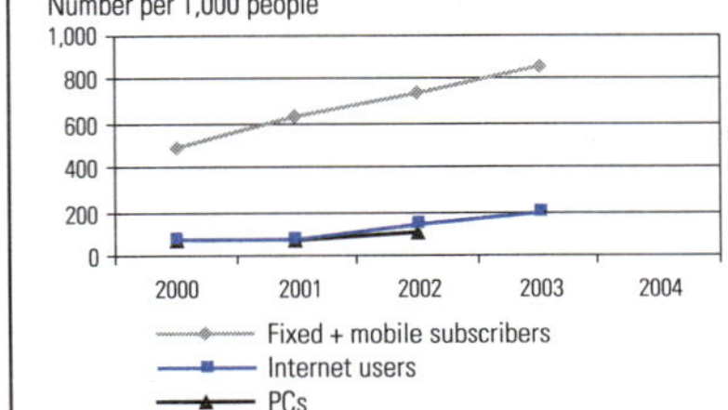

Price of Call to the United States, 2000–4
US$ per 3 minutes

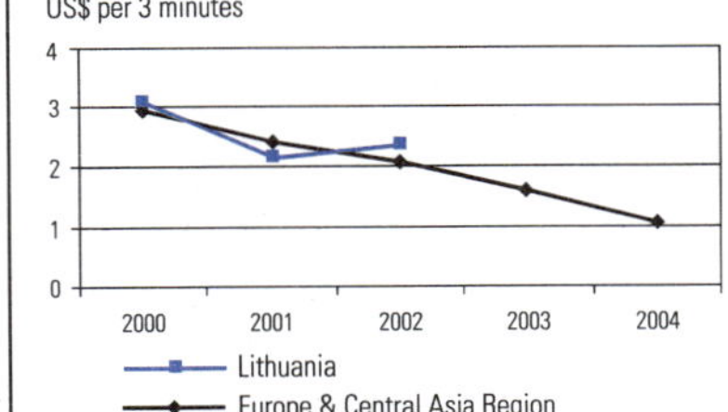

Total Telecommunications Revenue, 2000–4
Percentage of GDP

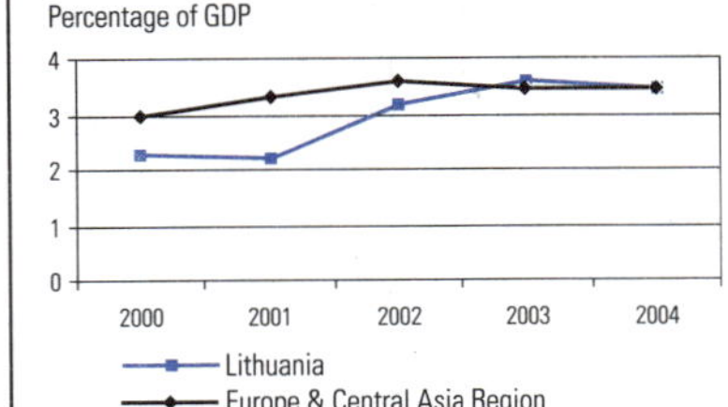

Notes: Figures in italics are for years other than those specified. .. indicates data are not available. C = competition; GDP = gross domestic product; GNI = gross national income; ICT = information and communication technology; M = monopoly; MDG = Millennium Development Goal; P = partial competition; and PCs = personal computers.

a. Outgoing and incoming.

b. Millennium Development Goal indicators 47, 48a, and 48b.

Sources: Economic and social context: UNDP and World Bank; ICT sector structure: ITU, WEF; ICT sector performance: Global Insight/WITSA, ITU, Netcraft, UNDESA, UNPAN, and World Bank. Produced by the Global Information and Communication Technologies Department and the Development Economics Data Group. For complete information, see Definitions and Data Sources.

Macedonia, Former Yugoslav Republic of

	Macedonia, FYR 2000	Macedonia, FYR 2004	Lower-middle-income group 2004	Europe & Central Asia Region 2004
Economic and social context				
Population, total (millions)	2	2	2,430	472
Urban population (% total population)	59	60	49	64
Poverty (% population below US$1 per day)	*<2*	..	..	*3.6*
GNI per capita, Atlas method (current US$)	1,830	2,350	1,580	3,290
GDP growth, 1995–2000 and 2000–4 (%)	3.0	0.8	5.7	5.0
Adult literacy rate (% ages 15 and over)	..	*96*	90	97
Primary, secondary, tertiary school enrollment (% gross)	70	*70*	*72*	*81*
ICT sector structure				
Separate telecommunications regulator	..	..		
Status of main fixed-line operator	*Public*	Mixed		
Level of competition: international long distance	*M*	M		
Level of competition: mobile	*M*	C		
Level of competition: Internet service provider	*C*	C		
Government prioritization of ICT (scale 1–7)	..	3.6	3.8	3.8
ICT sector performance				
Access				
Telephone main lines (per 1,000 people)	250	304	192	*243*
International voice traffic (minutes per person)[a]	118	*127*	..	*28*
Mobile subscribers (per 1,000 people)	57	478	255	487
Population covered by mobile telephony (%)	90	99	*76*	*82*
Internet users (per 1,000 people)	25	97	70	*115*
Personal computers (per 1,000 people)	*35*	68	38	*73*
Households with television (%)	82	..	*89*	*92*
Quality				
Telephone faults (per 100 main lines per year)	..	17.2	..	*30.4*
Broadband subscribers (per 1,000 people)	0.0	1.5	12.6	*2.4*
International Internet bandwidth (bits per person)	..	*25*	58	148
Affordability				
Price basket for fixed line (US$ per month, residential)	5.3	..	*5.5*	*3.5*
Price basket for mobile (US$ per month)	..	..	8.9	10.3
Price basket for Internet (US$ per month)	..	*18.9*	*25.3*	*19.8*
Price of call to United States (US$ per 3 minutes)	*3.95*	..	1.45	1.06
Institutional efficiency and sustainability				
Total telecommunications revenue (% GDP)	5.2	*6.7*	3.6	3.5
Total telephone subscribers per employee	168	..	*195*	*150*
Total telecommunications investment (% revenue)	*23.0*	*12.6*	*25.5*	*19.0*
ICT applications				
ICT expenditure (% GDP)	..	..	*5.1*	*5.1*
E-government readiness index (scale 0–1)	..	0.12	0.26	0.39
Secure Internet servers (per 1 million people)	..	0.5	1.6	6.4
Schools connected to the Internet (%)	..	..	..	*65*

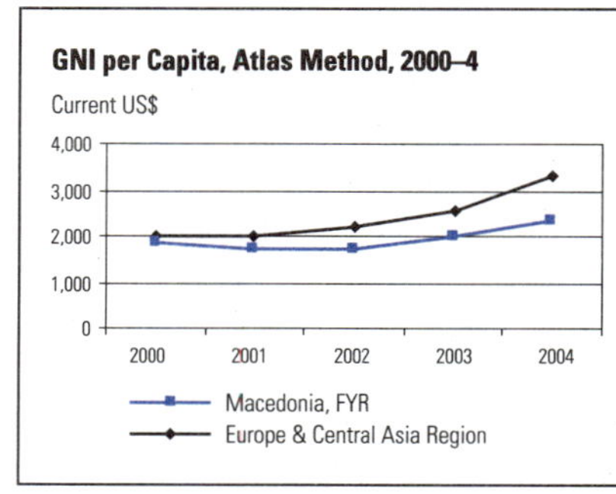

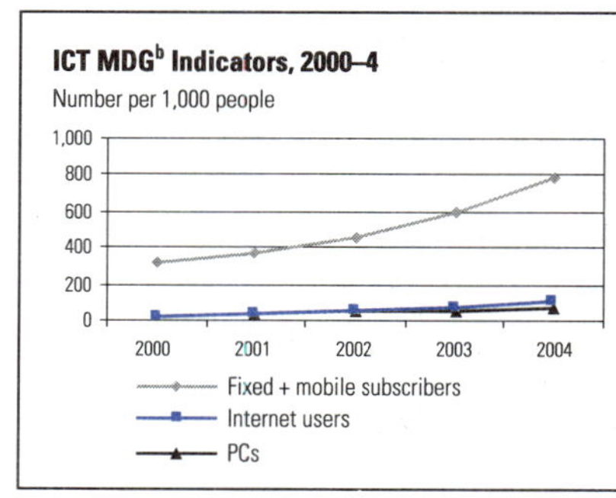

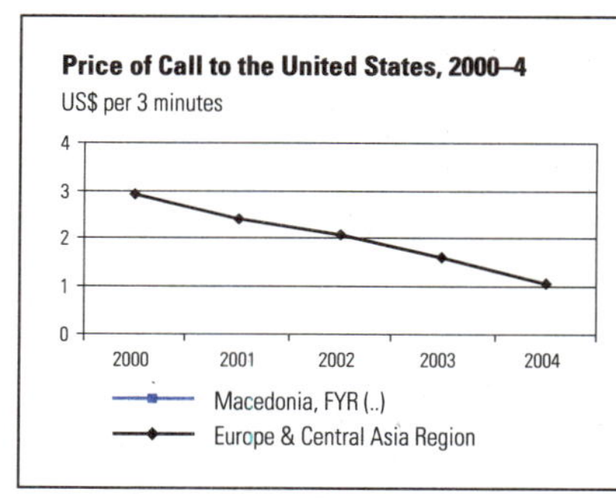

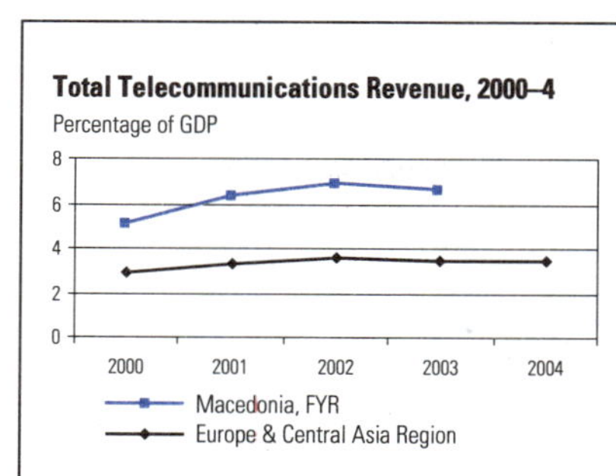

Notes: Figures in italics are for years other than those specified. .. indicates data are not available. C = competition; GDP = gross domestic product; GNI = gross national income; ICT = information and communication technology; M = monopoly; MDG = Millennium Development Goal; P = partial competition; and PCs = personal computers.

a. Outgoing and incoming.

b. Millennium Development Goal indicators 47, 48a, and 48b.

Sources: Economic and social context: UNDP and World Bank; ICT sector structure: ITU, WEF; ICT sector performance: Global Insight/WITSA, ITU, Netcraft, UNDESA, UNPAN, and World Bank. Produced by the Global Information and Communication Technologies Department and the Development Economics Data Group. For complete information, see Definitions and Data Sources.

Madagascar

	Madagascar 2000	Madagascar 2004	Low-income group 2004	Sub-Saharan Africa Region 2004
Economic and social context				
Population, total (millions)	16	17	2,338	719
Urban population (% total population)	26	27	31	37
Poverty (% population below US$1 per day)	*49.1*	*61.0*	..	*46.4*
GNI per capita, Atlas method (current US$)	250	300	510	600
GDP growth, 1995–2000 and 2000–4 (%)	3.9	0.9	5.4	3.9
Adult literacy rate (% ages 15 and over)	71	..	61	59
Primary, secondary, tertiary school enrollment (% gross)	44	*51*	*55*	*52*
ICT sector structure				
Separate telecommunications regulator	Yes	Yes		
Status of main fixed-line operator	*Mixed*	Mixed		
Level of competition: international long distance	*C*	*C*		
Level of competition: mobile	*C*	C		
Level of competition: Internet service provider	*C*	C		
Government prioritization of ICT (scale 1–7)	..	4.7	..	*4.4*
ICT sector performance				
Access				
Telephone main lines (per 1,000 people)	4	3	33	17
International voice traffic (minutes per person)[a]	2	1	*4*	..
Mobile subscribers (per 1,000 people)	4	19	48	86
Population covered by mobile telephony (%)	..	*30*	*43*	..
Internet users (per 1,000 people)	2	*4*	20	15
Personal computers (per 1,000 people)	2	*5*	*8*	*12*
Households with television (%)	7	*8*	*16*	*15*
Quality				
Telephone faults (per 100 main lines per year)	79.0	*42.5*	..	..
Broadband subscribers (per 1,000 people)	0.0	*0.0*	0.5	*0.1*
International Internet bandwidth (bits per person)	0	*0*	*3*	*4*
Affordability				
Price basket for fixed line (US$ per month, residential)	11.3	*7.5*	*6.6*	*8.5*
Price basket for mobile (US$ per month)	..	4.0	11.6	13.5
Price basket for Internet (US$ per month)	..	*67.3*	*45.5*	*54.8*
Price of call to United States (US$ per 3 minutes)	8.98	0.59	1.95	2.43
Institutional efficiency and sustainability				
Total telecommunications revenue (% GDP)	10.3	*12.5*	2.3	5.0
Total telephone subscribers per employee	44	*93*	*89*	*144*
Total telecommunications investment (% revenue)	14.0	..	*27.8*	*27.8*
ICT applications				
ICT expenditure (% GDP)	..	..	*4.1*	..
E-government readiness index (scale 0–1)	..	0.08	0.12	0.11
Secure Internet servers (per 1 million people)	..	0.1	0.3	1.9
Schools connected to the Internet (%)	..	..	..	..

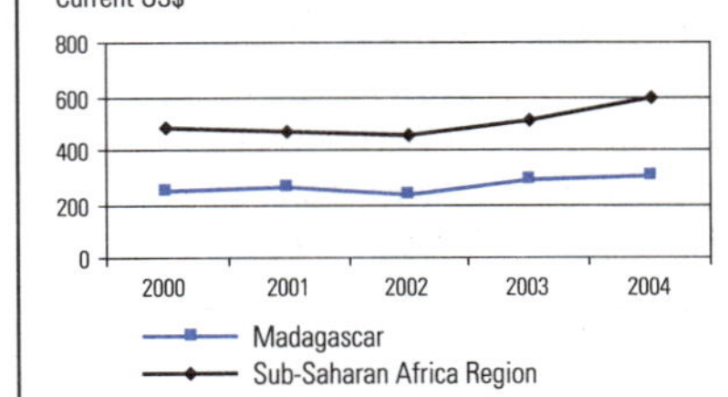

GNI per Capita, Atlas Method, 2000–4

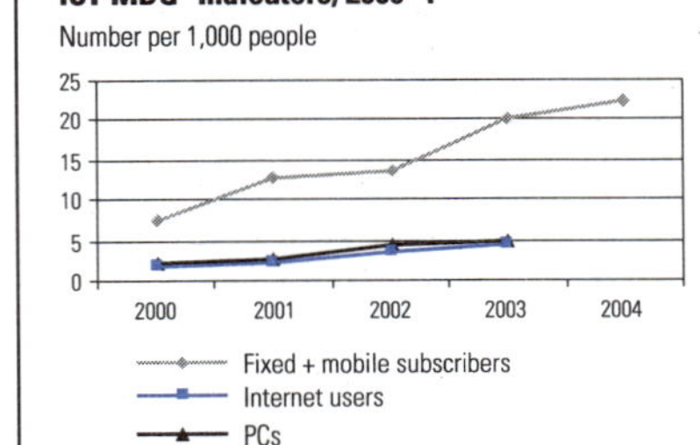

ICT MDG[b] Indicators, 2000–4

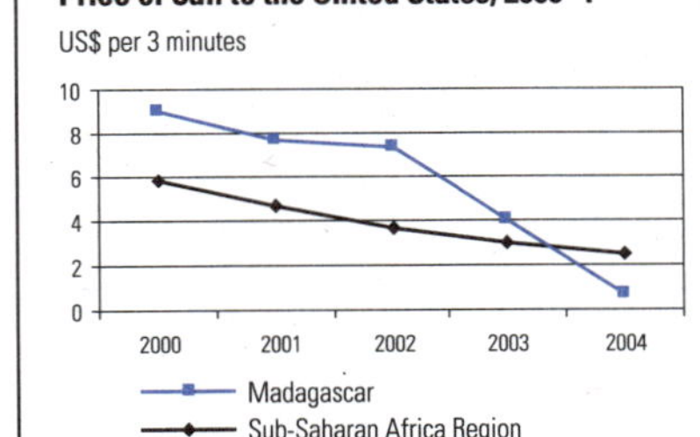

Price of Call to the United States, 2000–4

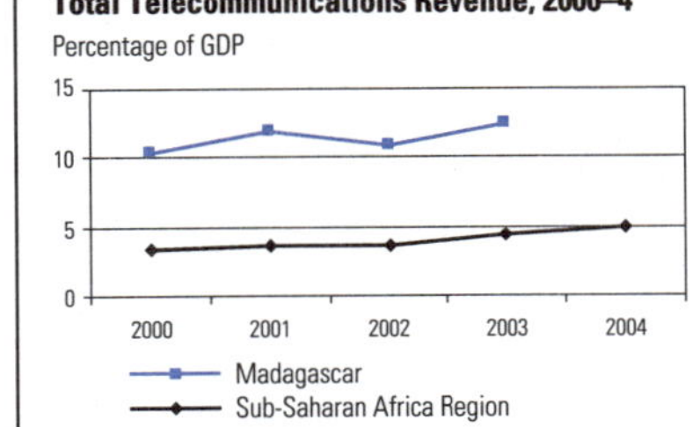

Total Telecommunications Revenue, 2000–4

Notes: Figures in italics are for years other than those specified. .. indicates data are not available. C = competition; GDP = gross domestic product; GNI = gross national income; ICT = information and communication technology; M = monopoly; MDG = Millennium Development Goal; P = partial competition; and PCs = personal computers.

a. Outgoing and incoming.

b. Millennium Development Goal indicators 47, 48a, and 48b.

Sources: Economic and social context: UNDP and World Bank; ICT sector structure: ITU, WEF; ICT sector performance: Global Insight/WITSA, ITU, Netcraft, UNDESA, UNPAN, and World Bank. Produced by the Global Information and Communication Technologies Department and the Development Economics Data Group. For complete information, see Definitions and Data Sources.

Malawi

	Malawi 2000	Malawi 2004	Low-income group 2004	Sub-Saharan Africa Region 2004
Economic and social context				
Population, total (millions)	10	11	2,338	719
Urban population (% total population)	15	17	31	37
Poverty (% population below US$1 per day)	*41.7*	..	..	*46.4*
GNI per capita, Atlas method (current US$)	170	170	510	600
GDP growth, 1995–2000 and 2000–4 (%)	3.8	1.8	5.4	3.9
Adult literacy rate (% ages 15 and over)	*64*	..	61	59
Primary, secondary, tertiary school enrollment (% gross)	72	*72*	*55*	*52*
ICT sector structure				
Separate telecommunications regulator	Yes	Yes		
Status of main fixed-line operator	*Public*	Public		
Level of competition: international long distance	*M*	P		
Level of competition: mobile	*M*	P		
Level of competition: Internet service provider	..	P		
Government prioritization of ICT (scale 1–7)	..	3.5	..	*4.4*
ICT sector performance				
Access				
Telephone main lines (per 1,000 people)	5	*8*	33	17
International voice traffic (minutes per person)[a]	2	..	*4*	..
Mobile subscribers (per 1,000 people)	5	*12*	48	86
Population covered by mobile telephony (%)	..	*70*	*43*	..
Internet users (per 1,000 people)	1	*3*	20	15
Personal computers (per 1,000 people)	1	*1*	*8*	*12*
Households with television (%)	1	*2*	*16*	*15*
Quality				
Telephone faults (per 100 main lines per year)	..	..	..	..
Broadband subscribers (per 1,000 people)	*0.0*	*0.0*	0.5	*0.1*
International Internet bandwidth (bits per person)	0	..	*3*	*4*
Affordability				
Price basket for fixed line (US$ per month, residential)	5.8	*4.5*	*6.6*	*8.5*
Price basket for mobile (US$ per month)	..	20.0	11.6	13.5
Price basket for Internet (US$ per month)	..	*62.0*	*45.5*	*54.8*
Price of call to United States (US$ per 3 minutes)	4.32	..	1.95	2.43
Institutional efficiency and sustainability				
Total telecommunications revenue (% GDP)	1.7	2.0	2.3	5.0
Total telephone subscribers per employee	33	*49*	*89*	*144*
Total telecommunications investment (% revenue)	..	..	*27.8*	*27.8*
ICT applications				
ICT expenditure (% GDP)	..	..	*4.1*	..
E-government readiness index (scale 0–1)	..	0.15	0.12	0.11
Secure Internet servers (per 1 million people)	..	0.1	0.3	1.9
Schools connected to the Internet (%)	..	*1*	..	..

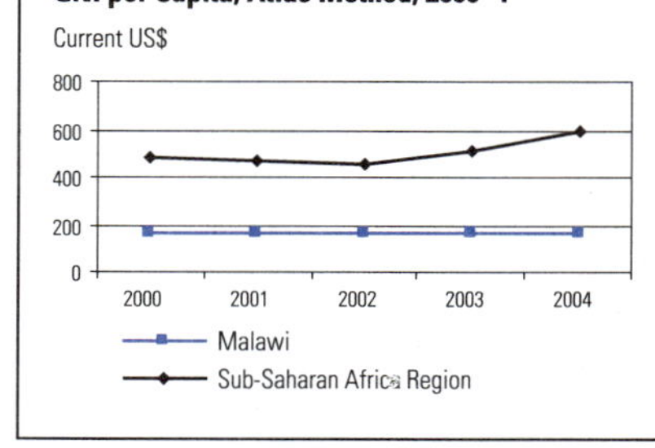

GNI per Capita, Atlas Method, 2000–4

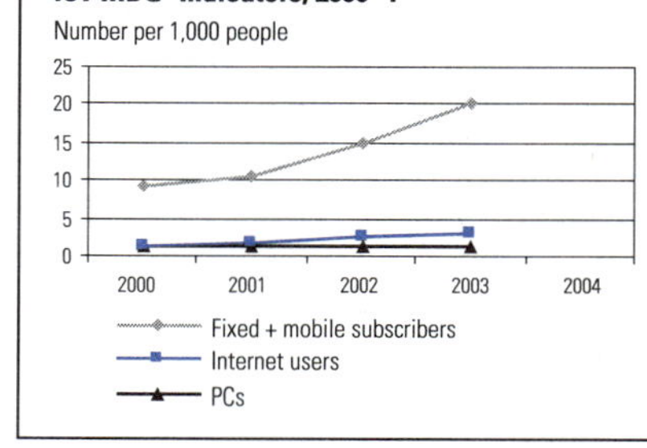

ICT MDG[b] Indicators, 2000–4

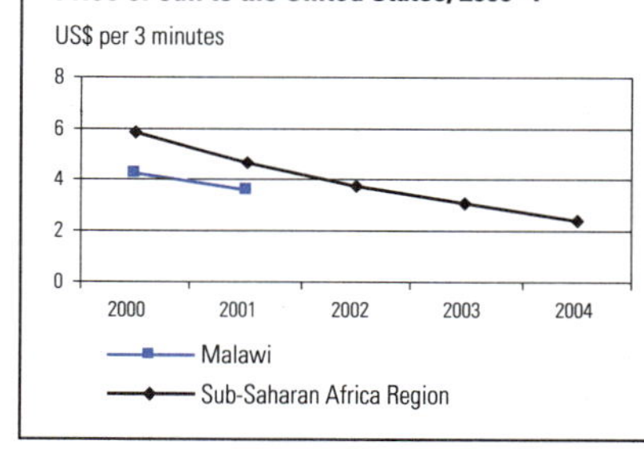

Price of Call to the United States, 2000–4

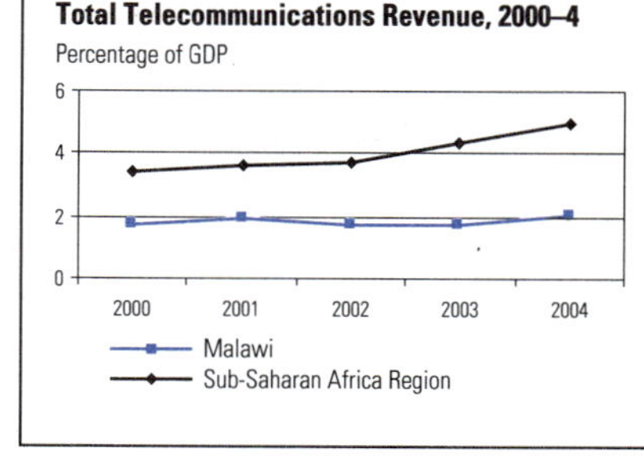

Total Telecommunications Revenue, 2000–4

Notes: Figures in italics are for years other than those specified. .. indicates data are not available. C = competition; GDP = gross domestic product; GNI = gross national income; ICT = information and communication technology; M = monopoly; MDG = Millennium Development Goal; P = partial competition; and PCs = personal computers.

a. Outgoing and incoming.

b. Millennium Development Goal indicators 47, 48a, and 48b.

Sources: Economic and social context: UNDP and World Bank; ICT sector structure: ITU, WEF; ICT sector performance: Global Insight/WITSA, ITU, Netcraft, UNDESA, UNPAN, and World Bank. Produced by the Global Information and Communication Technologies Department and the Development Economics Data Group. For complete information, see Definitions and Data Sources.

Malaysia

	Malaysia 2000	Malaysia 2004	Upper-middle-income group 2004	East Asia & Pacific Region 2004
Economic and social context				
Population, total (millions)	23	25	576	1,870
Urban population (% total population)	62	64	72	41
Poverty (% population below US$1 per day)	*<2*	..	..	*14.9*
GNI per capita, Atlas method (current US$)	3,390	4,650	4,770	1,280
GDP growth, 1995–2000 and 2000–4 (%)	3.7	4.3	2.7	7.5
Adult literacy rate (% ages 15 and over)	89	..	94	90
Primary, secondary, tertiary school enrollment (% gross)	70	*71*	*80*	*68*
ICT sector structure				
Separate telecommunications regulator	Yes	Yes		
Status of main fixed-line operator	*Mixed*	Mixed		
Level of competition: international long distance	*P*	C		
Level of competition: mobile	*P*	C		
Level of competition: Internet service provider	*C*	C		
Government prioritization of ICT (scale 1–7)	..	5.2	4.1	4.8
ICT sector performance				
Access				
Telephone main lines (per 1,000 people)	199	176	*220*	194
International voice traffic (minutes per person)[a]	63	..	*39*	..
Mobile subscribers (per 1,000 people)	220	573	490	248
Population covered by mobile telephony (%)	*95*	96	*84*	*73*
Internet users (per 1,000 people)	214	392	*133*	75
Personal computers (per 1,000 people)	95	*170*	*99*	37
Households with television (%)	84	98	*92*	*80*
Quality				
Telephone faults (per 100 main lines per year)	40.0	28.0	*20.3*	..
Broadband subscribers (per 1,000 people)	0.0	10.0	*3.7*	13.4
International Internet bandwidth (bits per person)	23	127	176	52
Affordability				
Price basket for fixed line (US$ per month, residential)	6.9	8.7	*13.9*	*4.5*
Price basket for mobile (US$ per month)	..	5.6	11.1	5.1
Price basket for Internet (US$ per month)	..	*8.4*	*20.8*	*19.9*
Price of call to United States (US$ per 3 minutes)	2.37	0.71	1.03	1.20
Institutional efficiency and sustainability				
Total telecommunications revenue (% GDP)	4.3	4.5	3.4	3.6
Total telephone subscribers per employee	394	722	*402*	..
Total telecommunications investment (% revenue)	27.7	18.9	*18.6*	*31.0*
ICT applications				
ICT expenditure (% GDP)	7.5	*6.9*	*5.0*	*5.0*
E-government readiness index (scale 0–1)	..	0.49	0.49	0.21
Secure Internet servers (per 1 million people)	*6.1*	11.3	10.7	0.6
Schools connected to the Internet (%)	15	..	*60*	..

GNI per Capita, Atlas Method, 2000–4

Current US$

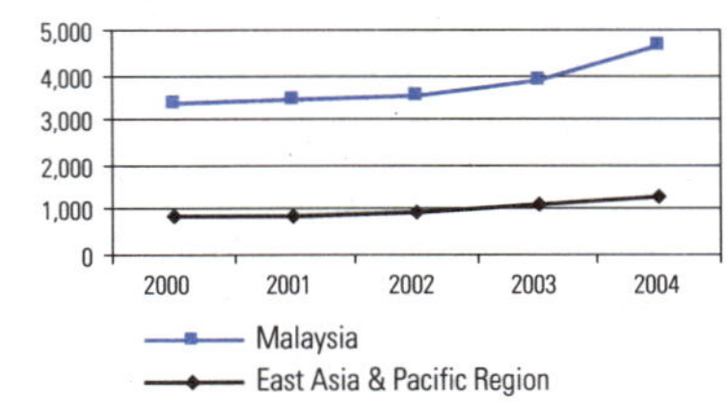

ICT MDG[b] Indicators, 2000–4

Number per 1,000 people

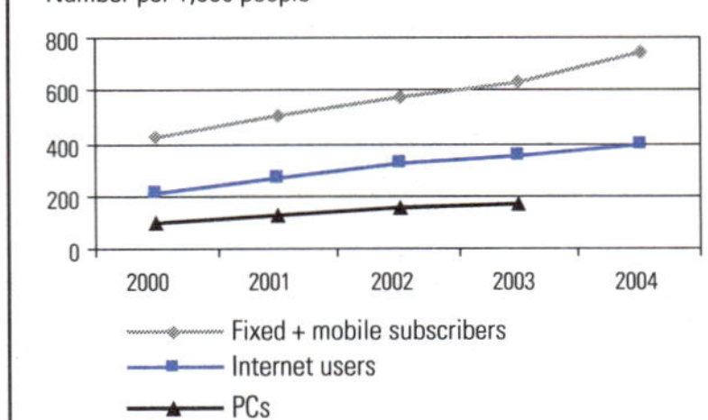

Price of Call to the United States, 2000–4

US$ per 3 minutes

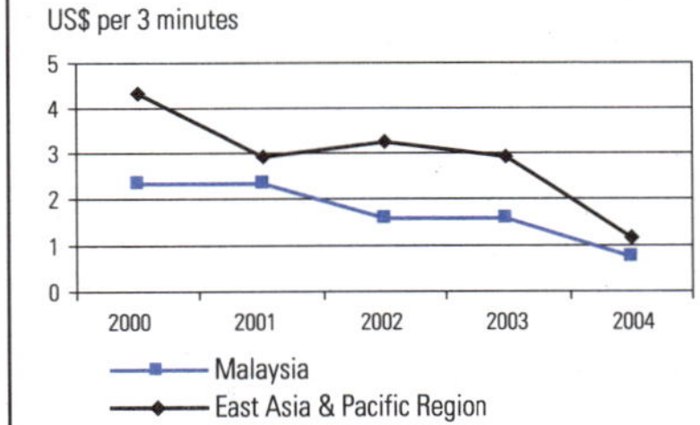

Total Telecommunications Revenue, 2000–4

Percentage of GDP

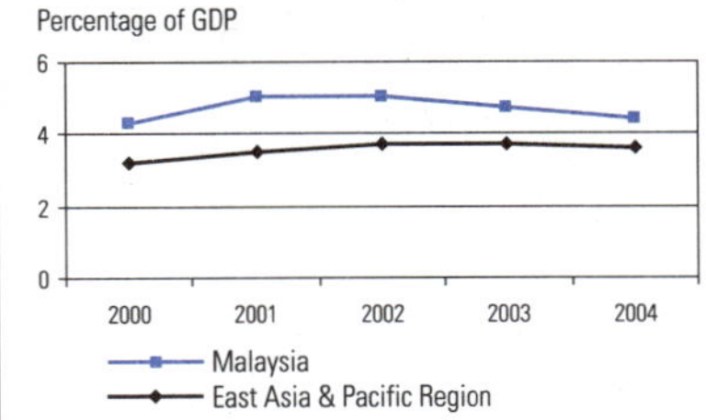

Notes: Figures in italics are for years other than those specified. .. indicates data are not available. C = competition; GDP = gross domestic product; GNI = gross national income; ICT = information and communication technology; M = monopoly; MDG = Millennium Development Goal; P = partial competition; and PCs = personal computers.

a. Outgoing and incoming.

b. Millennium Development Goal indicators 47, 48a, and 48b.

Sources: Economic and social context: UNDP and World Bank; ICT sector structure: ITU, WEF; ICT sector performance: Global Insight/WITSA, ITU, Netcraft, UNDESA, UNPAN, and World Bank. Produced by the Global Information and Communication Technologies Department and the Development Economics Data Group. For complete information, see Definitions and Data Sources.

Mali

	Mali 2000	Mali 2004	Low-income group 2004	Sub-Saharan Africa Region 2004
Economic and social context				
Population, total (millions)	11	12	2,338	719
Urban population (% total population)	30	33	31	37
Poverty (% population below US$1 per day)	..	..	..	*46.4*
GNI per capita, Atlas method (current US$)	240	360	510	600
GDP growth, 1995–2000 and 2000–4 (%)	5.5	6.3	5.4	3.9
Adult literacy rate (% ages 15 and over)	*19*	..	61	59
Primary, secondary, tertiary school enrollment (% gross)	*27*	*32*	*55*	*52*
ICT sector structure				
Separate telecommunications regulator	Yes	Yes		
Status of main fixed-line operator	*Public*	Public		
Level of competition: international long distance	*P*	P		
Level of competition: mobile	*P*	P		
Level of competition: Internet service provider	*C*	C		
Government prioritization of ICT (scale 1–7)	..	5.0	..	*4.4*
ICT sector performance				
Access				
Telephone main lines (per 1,000 people)	4	*5*	33	17
International voice traffic (minutes per person)[a]	7	..	*4*	..
Mobile subscribers (per 1,000 people)	1	*21*	48	86
Population covered by mobile telephony (%)	15	..	*43*	..
Internet users (per 1,000 people)	2	5	20	15
Personal computers (per 1,000 people)	1	*1*	*8*	*12*
Households with television (%)	13	*15*	*16*	*15*
Quality				
Telephone faults (per 100 main lines per year)	177.6	..	..	..
Broadband subscribers (per 1,000 people)	0.0	*0.0*	0.5	*0.1*
International Internet bandwidth (bits per person)	0	*1*	*3*	*4*
Affordability				
Price basket for fixed line (US$ per month, residential)	8.3	*8.5*	*6.6*	*8.5*
Price basket for mobile (US$ per month)	..	13.5	11.6	13.5
Price basket for Internet (US$ per month)	..	*58.0*	*45.5*	*54.8*
Price of call to United States (US$ per 3 minutes)	12.64	..	1.95	2.43
Institutional efficiency and sustainability				
Total telecommunications revenue (% GDP)	2.5	*3.0*	2.3	5.0
Total telephone subscribers per employee	37	*71*	*89*	*144*
Total telecommunications investment (% revenue)	28.1	..	*27.8*	*27.8*
ICT applications				
ICT expenditure (% GDP)	..	..	*4.1*	..
E-government readiness index (scale 0–1)	..	0.01	0.12	0.11
Secure Internet servers (per 1 million people)	*0.1*	..	0.3	1.9
Schools connected to the Internet (%)	..	..	..	..

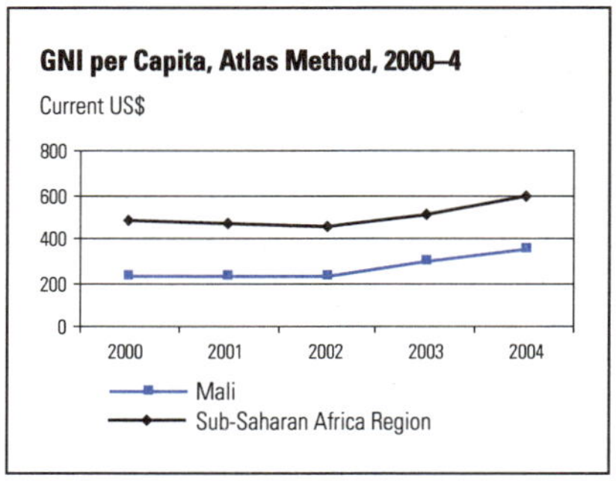

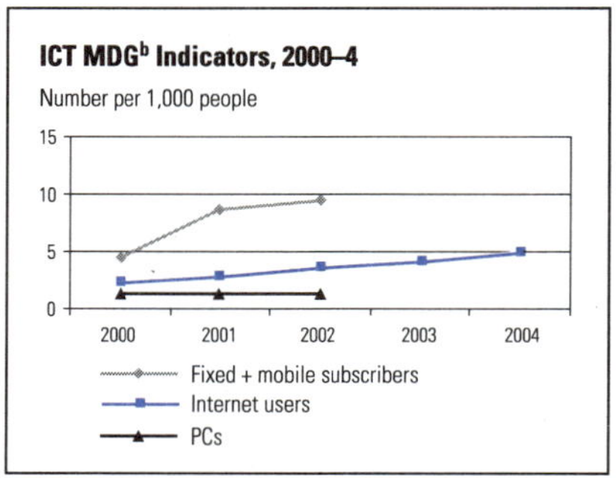

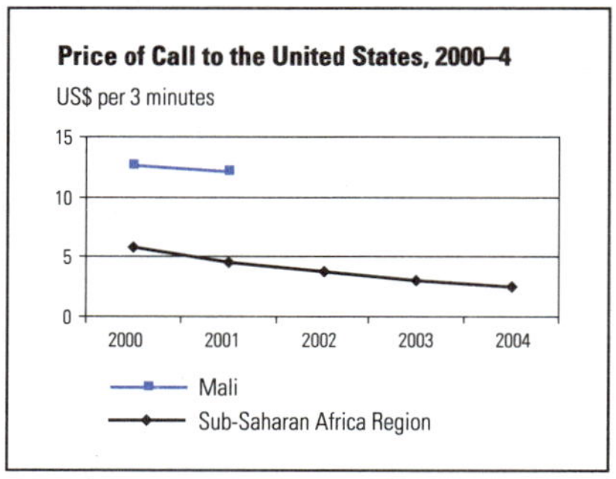

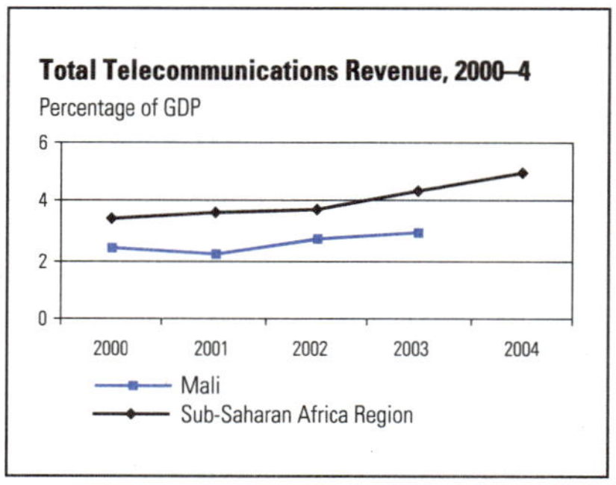

Notes: Figures in italics are for years other than those specified. .. indicates data are not available. C = competition; GDP = gross domestic product; GNI = gross national income; ICT = information and communication technology; M = monopoly; MDG = Millennium Development Goal; P = partial competition; and PCs = personal computers.

a. Outgoing and incoming.

b. Millennium Development Goal indicators 47, 48a, and 48b.

Sources: Economic and social context: UNDP and World Bank; ICT sector structure: ITU, WEF; ICT sector performance: Global Insight/WITSA, ITU, Netcraft, UNDESA, UNPAN, and World Bank. Produced by the Global Information and Communication Technologies Department and the Development Economics Data Group. For complete information, see Definitions and Data Sources.

Mauritania

	Mauritania 2000	Mauritania 2004	Low-income group 2004	Sub-Saharan Africa Region 2004
Economic and social context				
Population, total (millions)	3	3	2,338	719
Urban population (% total population)	58	63	31	37
Poverty (% population below US$1 per day)	25.9	..	..	*46.4*
GNI per capita, Atlas method (current US$)	390	420	510	600
GDP growth, 1995–2000 and 2000–4 (%)	4.0	5.3	5.4	3.9
Adult literacy rate (% ages 15 and over)	51	..	61	59
Primary, secondary, tertiary school enrollment (% gross)	44	*45*	*55*	*52*
ICT sector structure				
Separate telecommunications regulator	Yes	Yes		
Status of main fixed-line operator	*Public*	Mixed		
Level of competition: international long distance	*M*	C		
Level of competition: mobile	*M*	C		
Level of competition: Internet service provider	*C*	C		
Government prioritization of ICT (scale 1–7)	..	..	..	*4.4*
ICT sector performance				
Access				
Telephone main lines (per 1,000 people)	7	*13*	33	17
International voice traffic (minutes per person)[a]	..	..	*4*	..
Mobile subscribers (per 1,000 people)	6	*123*	48	86
Population covered by mobile telephony (%)	..	..	*43*	..
Internet users (per 1,000 people)	2	*5*	20	15
Personal computers (per 1,000 people)	9	*10*	*8*	*12*
Households with television (%)	19	*21*	*16*	*15*
Quality				
Telephone faults (per 100 main lines per year)	*115.0*	*128.0*	..	..
Broadband subscribers (per 1,000 people)	*0.0*	*0.0*	0.5	*0.1*
International Internet bandwidth (bits per person)	0	*3*	*3*	*4*
Affordability				
Price basket for fixed line (US$ per month, residential)	14.1	*12.3*	*6.6*	*8.5*
Price basket for mobile (US$ per month)	..	..	11.6	13.5
Price basket for Internet (US$ per month)	..	*38.6*	*45.5*	*54.8*
Price of call to United States (US$ per 3 minutes)	*4.36*	..	1.95	2.43
Institutional efficiency and sustainability				
Total telecommunications revenue (% GDP)	2.7	8.4	2.3	5.0
Total telephone subscribers per employee	48	..	*89*	*144*
Total telecommunications investment (% revenue)	..	..	*27.8*	*27.8*
ICT applications				
ICT expenditure (% GDP)	..	..	*4.1*	..
E-government readiness index (scale 0–1)	..	0.07	0.12	0.11
Secure Internet servers (per 1 million people)	*0.4*	..	0.3	1.9
Schools connected to the Internet (%)	..	..	..	..

GNI per Capita, Atlas Method, 2000–4

Current US$

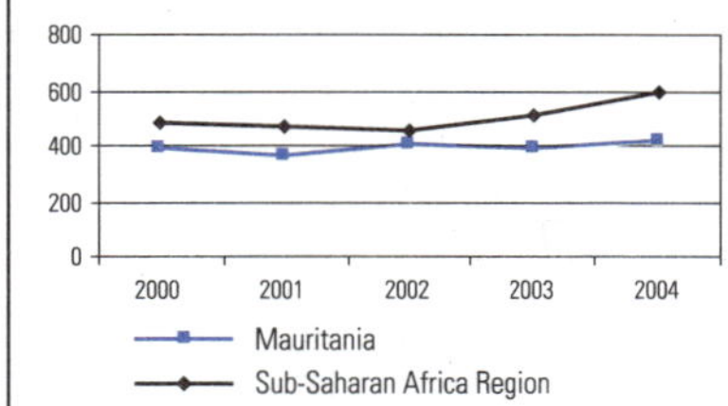

ICT MDG[b] Indicators, 2000–4

Number per 1,000 people

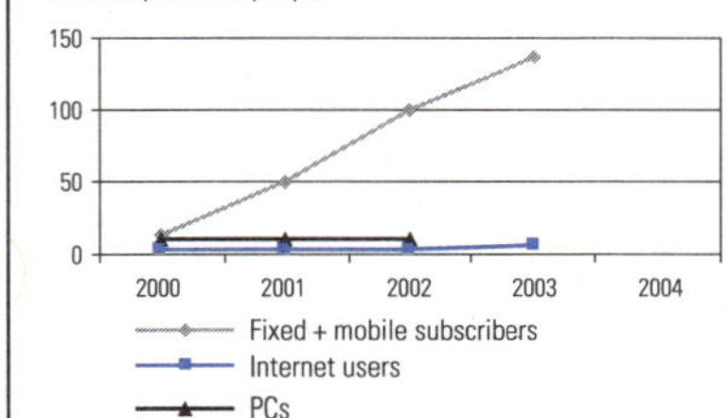

Price of Call to the United States, 2000–4

US$ per 3 minutes

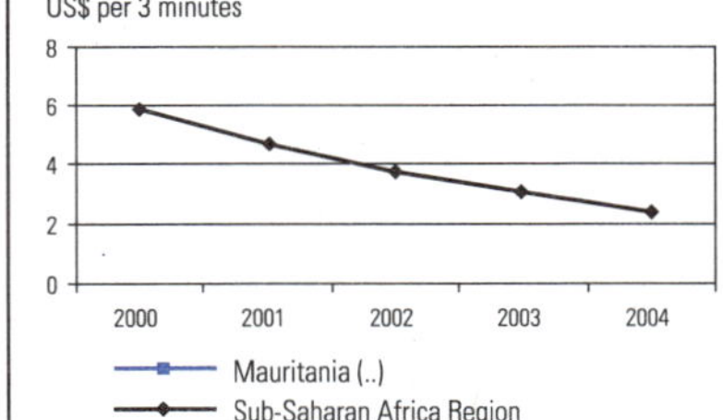

Total Telecommunications Revenue, 2000–4

Percentage of GDP

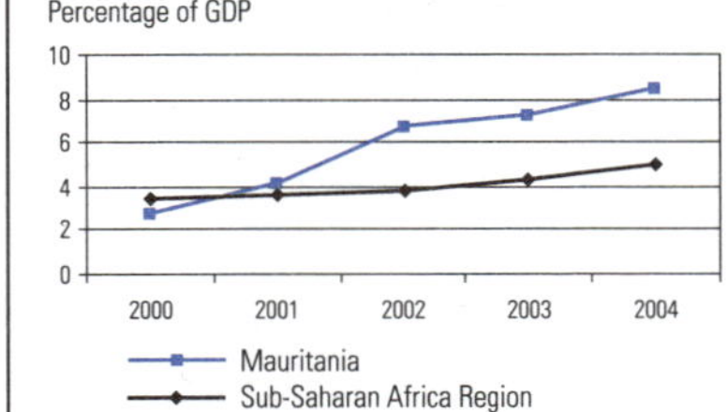

Notes: Figures in italics are for years other than those specified. .. indicates data are not available. C = competition; GDP = gross domestic product; GNI = gross national income; ICT = information and communication technology; M = monopoly; MDG = Millennium Development Goal; P = partial competition; and PCs = personal computers.

a. Outgoing and incoming.

b. Millennium Development Goal indicators 47, 48a, and 48b.

Sources: Economic and social context: UNDP and World Bank; ICT sector structure: ITU, WEF; ICT sector performance: Global Insight/WITSA, ITU, Netcraft, UNDESA, UNPAN, and World Bank. Produced by the Global Information and Communication Technologies Department and the Development Economics Data Group. For complete information, see Definitions and Data Sources.

Mauritius

	Mauritius 2000	Mauritius 2004	Upper-middle-income group 2004	Sub-Saharan Africa Region 2004
Economic and social context				
Population, total (millions)	1	1	576	719
Urban population (% total population)	43	44	72	37
Poverty (% population below US$1 per day)	..	..	..	*46.4*
GNI per capita, Atlas method (current US$)	3,690	4,640	4,770	600
GDP growth, 1995–2000 and 2000–4 (%)	5.4	4.4	2.7	3.9
Adult literacy rate (% ages 15 and over)	84	..	94	59
Primary, secondary, tertiary school enrollment (% gross)	68	*71*	*80*	*52*
ICT sector structure				
Separate telecommunications regulator	Yes	Yes		
Status of main fixed-line operator	*Public*	Mixed		
Level of competition: international long distance	*M*	C		
Level of competition: mobile	*P*	C		
Level of competition: Internet service provider	*M*	C		
Government prioritization of ICT (scale 1–7)	..	5.8	4.1	*4.4*
ICT sector performance				
Access				
Telephone main lines (per 1,000 people)	237	287	*220*	17
International voice traffic (minutes per person)[a]	72	*83*	*39*	..
Mobile subscribers (per 1,000 people)	152	505	490	86
Population covered by mobile telephony (%)	100	99	*84*	..
Internet users (per 1,000 people)	73	146	*133*	15
Personal computers (per 1,000 people)	101	*176*	*99*	*12*
Households with television (%)	91	94	*92*	*15*
Quality				
Telephone faults (per 100 main lines per year)	56.4	*41.5*	*20.3*	..
Broadband subscribers (per 1,000 people)	0.0	2.0	*3.7*	*0.1*
International Internet bandwidth (bits per person)	5	146	176	*4*
Affordability				
Price basket for fixed line (US$ per month, residential)	5.2	*7.4*	*13.9*	*8.5*
Price basket for mobile (US$ per month)	..	4.8	11.1	13.5
Price basket for Internet (US$ per month)	..	*15.0*	*20.8*	*54.8*
Price of call to United States (US$ per 3 minutes)	4.00	1.67	1.03	2.43
Institutional efficiency and sustainability				
Total telecommunications revenue (% GDP)	3.4	3.7	3.4	5.0
Total telephone subscribers per employee	251	*373*	*402*	*144*
Total telecommunications investment (% revenue)	37.6	*36.0*	*18.6*	*27.8*
ICT applications				
ICT expenditure (% GDP)	..	..	*5.0*	..
E-government readiness index (scale 0–1)	..	0.54	0.49	0.11
Secure Internet servers (per 1 million people)	*10.0*	15.4	10.7	1.9
Schools connected to the Internet (%)	..	*19*	*60*	..

GNI per Capita, Atlas Method, 2000–4

Current US$

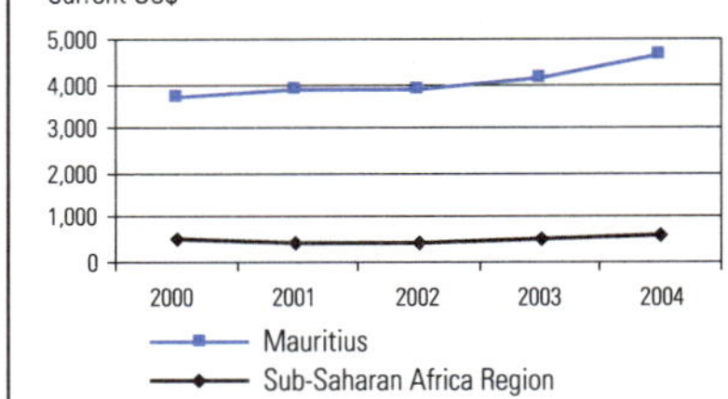

ICT MDG[b] Indicators, 2000–4

Number per 1,000 people

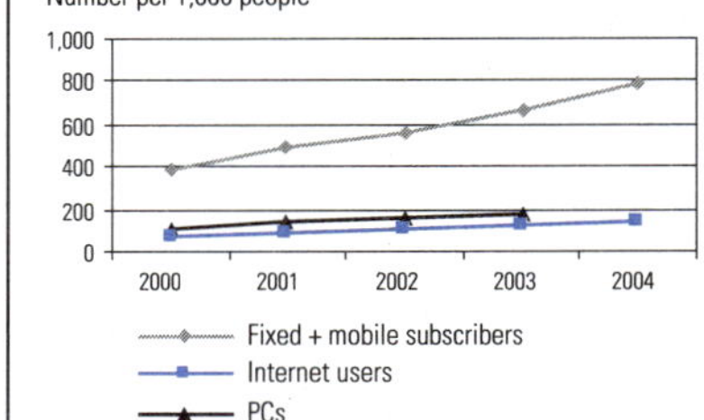

Price of Call to the United States, 2000–4

US$ per 3 minutes

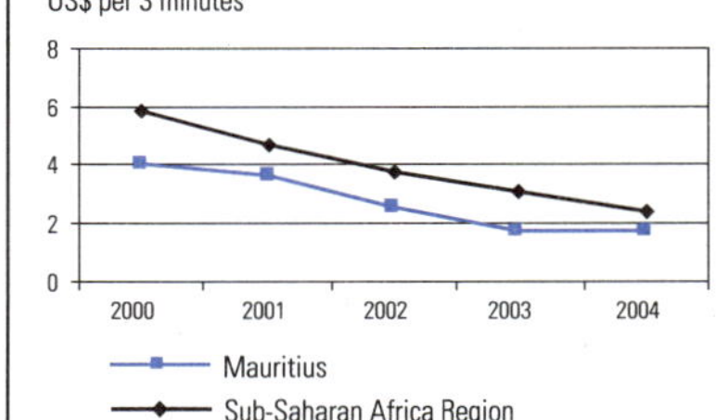

Total Telecommunications Revenue, 2000–4

Percentage of GDP

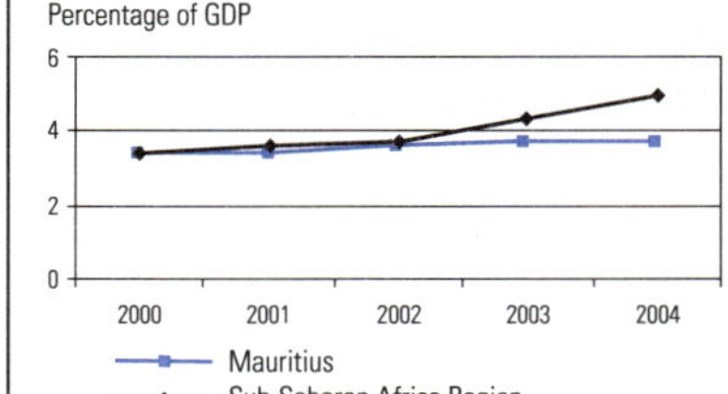

Notes: Figures in italics are for years other than those specified. .. indicates data are not available. C = competition; GDP = gross domestic product; GNI = gross national income; ICT = information and communication technology; M = monopoly; MDG = Millennium Development Goal; P = partial competition; and PCs = personal computers.

a. Outgoing and incoming.

b. Millennium Development Goal indicators 47, 48a, and 48b.

Sources: Economic and social context: UNDP and World Bank; ICT sector structure: ITU, WEF; ICT sector performance: Global Insight/WITSA, ITU, Netcraft, UNDESA, UNPAN, and World Bank. Produced by the Global Information and Communication Technologies Department and the Development Economics Data Group. For complete information, see Definitions and Data Sources.

Mexico

	Mexico 2000	Mexico 2004	Upper-middle-income group 2004	Latin America & the Caribbean Region 2004
Economic and social context				
Population, total (millions)	98	104	576	541
Urban population (% total population)	75	76	72	77
Poverty (% population below US$1 per day)	9.9	..	..	*9.5*
GNI per capita, Atlas method (current US$)	5,110	6,770	4,770	3,600
GDP growth, 1995–2000 and 2000–4 (%)	5.4	1.5	2.7	1.5
Adult literacy rate (% ages 15 and over)	..	*90*	94	89
Primary, secondary, tertiary school enrollment (% gross)	72	*75*	*80*	*82*
ICT sector structure				
Separate telecommunications regulator	Yes	Yes		
Status of main fixed-line operator	*Private*	Private		
Level of competition: international long distance	*C*	C		
Level of competition: mobile	*P*	C		
Level of competition: Internet service provider	*C*	C		
Government prioritization of ICT (scale 1–7)	..	4.2	4.1	3.5
ICT sector performance				
Access				
Telephone main lines (per 1,000 people)	126	174	*220*	181
International voice traffic (minutes per person)[a]	79	*82*	*39*	..
Mobile subscribers (per 1,000 people)	144	370	490	324
Population covered by mobile telephony (%)	86	86	*84*	*76*
Internet users (per 1,000 people)	52	135	*133*	*104*
Personal computers (per 1,000 people)	58	*84*	*99*	*75*
Households with television (%)	90	92	*92*	*88*
Quality				
Telephone faults (per 100 main lines per year)	1.9	..	*20.3*	..
Broadband subscribers (per 1,000 people)	0.2	*3.1*	*3.7*	*5.2*
International Internet bandwidth (bits per person)	9	108	176	165
Affordability				
Price basket for fixed line (US$ per month, residential)	17.5	15.5	*13.9*	*9.0*
Price basket for mobile (US$ per month)	..	11.4	11.1	9.1
Price basket for Internet (US$ per month)	..	*22.6*	*20.8*	*31.5*
Price of call to United States (US$ per 3 minutes)	3.01	..	1.03	0.90
Institutional efficiency and sustainability				
Total telecommunications revenue (% GDP)	2.1	2.7	3.4	3.3
Total telephone subscribers per employee	272	*504*	*402*	..
Total telecommunications investment (% revenue)	41.5	19.5	*18.6*	..
ICT applications				
ICT expenditure (% GDP)	3.1	*3.1*	*5.0*	*5.3*
E-government readiness index (scale 0–1)	..	0.78	0.49	0.39
Secure Internet servers (per 1 million people)	*2.6*	6.1	10.7	8.6
Schools connected to the Internet (%)	..	*60*	*60*	..

GNI per Capita, Atlas Method, 2000–4

Current US$

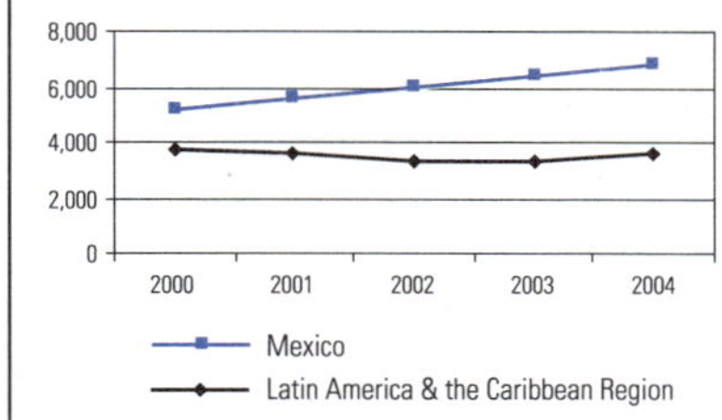

ICT MDG[b] Indicators, 2000–4

Number per 1,000 people

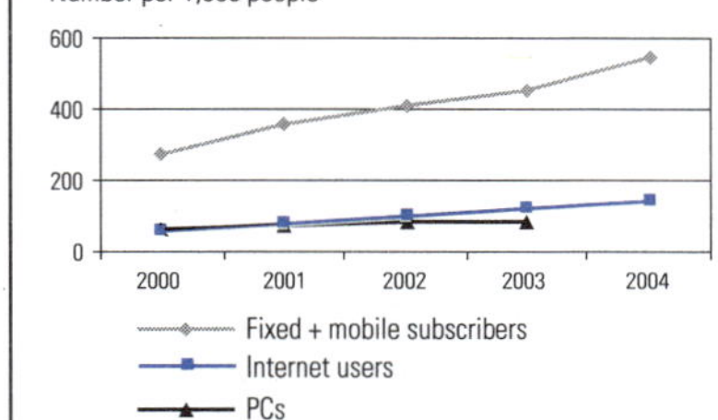

Price of Call to the United States, 2000–4

US$ per 3 minutes

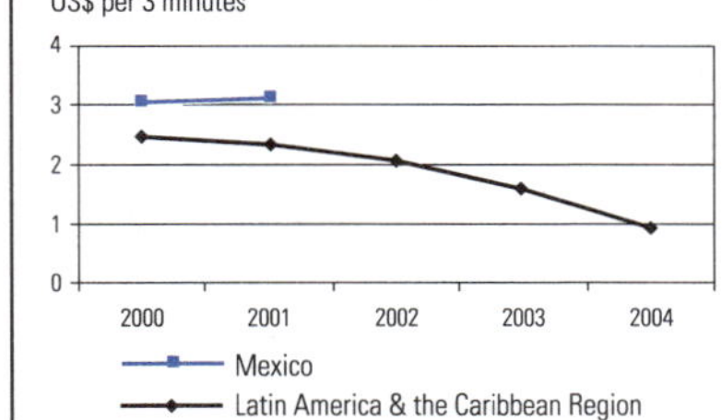

Total Telecommunications Revenue, 2000–4

Percentage of GDP

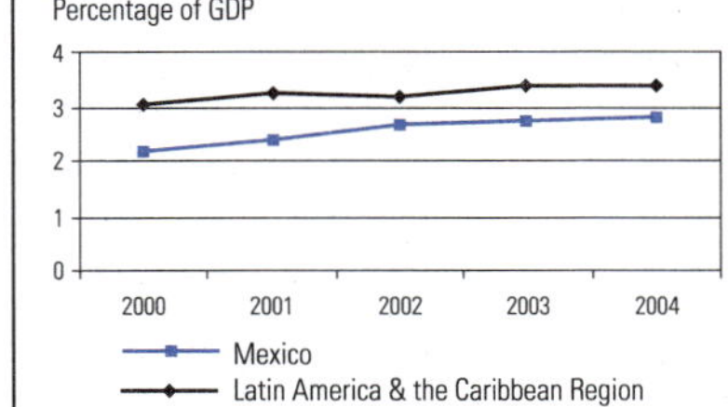

Notes: Figures in italics are for years other than those specified. .. indicates data are not available. C = competition; GDP = gross domestic product; GNI = gross national income; ICT = information and communication technology; M = monopoly; MDG = Millennium Development Goal; P = partial competition; and PCs = personal computers.

a. Outgoing and incoming.

b. Millennium Development Goal indicators 47, 48a, and 48b.

Sources: Economic and social context: UNDP and World Bank; ICT sector structure: ITU, WEF; ICT sector performance: Global Insight/WITSA, ITU, Netcraft, UNDESA, UNPAN, and World Bank. Produced by the Global Information and Communication Technologies Department and the Development Economics Data Group. For complete information, see Definitions and Data Sources.

Moldova

	Moldova 2000	Moldova 2004	Low-income group 2004	Europe & Central Asia Region 2004
Economic and social context				
Population, total (millions)	4	4	2,338	472
Urban population (% total population)	46	46	31	64
Poverty (% population below US$1 per day)	*32.2*	*21.8*	..	*3.6*
GNI per capita, Atlas method (current US$)	390	710	510	3,290
GDP growth, 1995–2000 and 2000–4 (%)	–2.7	6.9	5.4	5.0
Adult literacy rate (% ages 15 and over)	96	..	61	97
Primary, secondary, tertiary school enrollment (% gross)	61	*62*	*55*	*81*
ICT sector structure				
Separate telecommunications regulator	Yes	Yes		
Status of main fixed-line operator	*Public*	Public		
Level of competition: international long distance	*M*	M		
Level of competition: mobile	*C*	C		
Level of competition: Internet service provider	*C*	C		
Government prioritization of ICT (scale 1–7)	..	..	..	3.8
ICT sector performance				
Access				
Telephone main lines (per 1,000 people)	136	205	33	*243*
International voice traffic (minutes per person)[a]	38	*43*	*4*	*28*
Mobile subscribers (per 1,000 people)	32	142	48	487
Population covered by mobile telephony (%)	70	92	*43*	*82*
Internet users (per 1,000 people)	12	96	20	*115*
Personal computers (per 1,000 people)	15	23	*8*	*73*
Households with television (%)	..	*75*	*16*	*92*
Quality				
Telephone faults (per 100 main lines per year)	*55.3*	*35.0*	..	*30.4*
Broadband subscribers (per 1,000 people)	..	0.7	0.5	*2.4*
International Internet bandwidth (bits per person)	2	*8*	*3*	148
Affordability				
Price basket for fixed line (US$ per month, residential)	2.2	*1.8*	*6.6*	*3.5*
Price basket for mobile (US$ per month)	..	27.6	11.6	10.3
Price basket for Internet (US$ per month)	..	*19.0*	*45.5*	*19.8*
Price of call to United States (US$ per 3 minutes)	4.10	*2.21*	1.95	1.06
Institutional efficiency and sustainability				
Total telecommunications revenue (% GDP)	3.9	7.5	2.3	3.5
Total telephone subscribers per employee	97	*158*	*89*	*150*
Total telecommunications investment (% revenue)	67.9	36.9	*27.8*	*19.0*
ICT applications				
ICT expenditure (% GDP)	..	..	*4.1*	*5.1*
E-government readiness index (scale 0–1)	..	0.09	0.12	0.39
Secure Internet servers (per 1 million people)	*0.7*	2.1	0.3	6.4
Schools connected to the Internet (%)	..	*50*	..	*65*

GNI per Capita, Atlas Method, 2000–4

Current US$

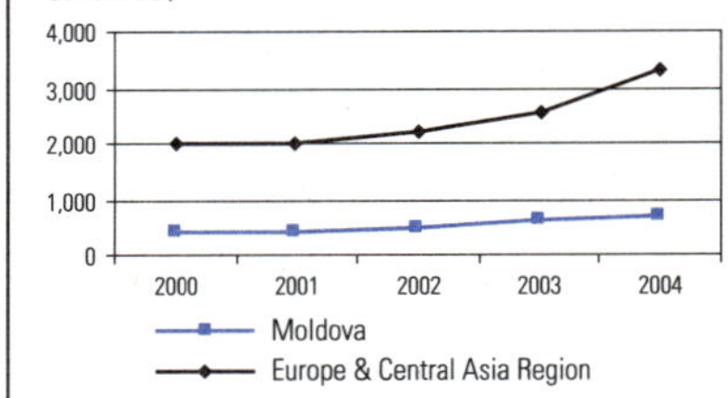

ICT MDG[b] Indicators, 2000–4

Number per 1,000 people

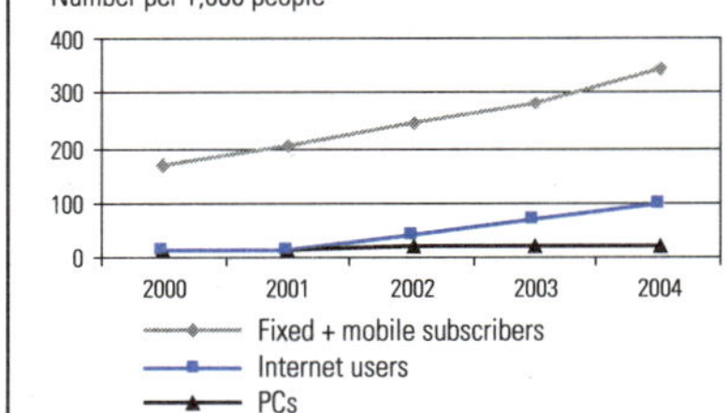

Price of Call to the United States, 2000–4

US$ per 3 minutes

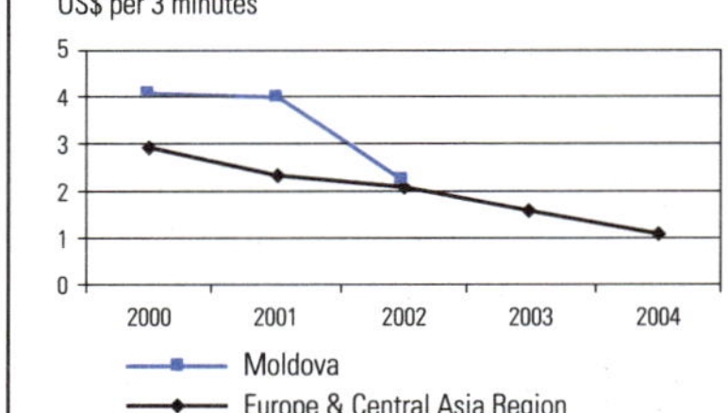

Total Telecommunications Revenue, 2000–4

Percentage of GDP

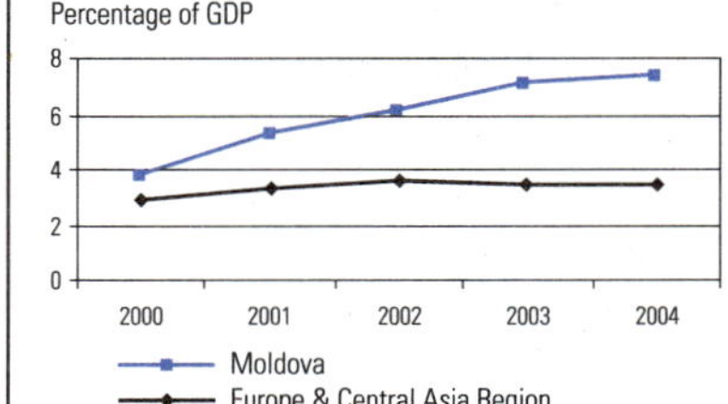

Notes: Figures in italics are for years other than those specified. .. indicates data are not available. C = competition; GDP = gross domestic product; GNI = gross national income; ICT = information and communication technology; M = monopoly; MDG = Millennium Development Goal; P = partial competition; and PCs = personal computers.

a. Outgoing and incoming.

b. Millennium Development Goal indicators 47, 48a, and 48b.

Sources: Economic and social context: UNDP and World Bank; ICT sector structure: ITU, WEF; ICT sector performance: Global Insight/WITSA, ITU, Netcraft, UNDESA, UNPAN, and World Bank. Produced by the Global Information and Communication Technologies Department and the Development Economics Data Group. For complete information, see Definitions and Data Sources.

Mongolia

	Mongolia 2000	Mongolia 2004	Low-income group 2004	East Asia & Pacific Region 2004
Economic and social context				
Population, total (millions)	2	3	2,338	1,870
Urban population (% total population)	57	57	31	41
Poverty (% population below US$1 per day)	*27.0*	..	..	*14.9*
GNI per capita, Atlas method (current US$)	390	590	510	1,280
GDP growth, 1995–2000 and 2000–4 (%)	3.0	5.2	5.4	7.5
Adult literacy rate (% ages 15 and over)	98	..	61	90
Primary, secondary, tertiary school enrollment (% gross)	68	*74*	*55*	*68*
ICT sector structure				
Separate telecommunications regulator	No	Yes		
Status of main fixed-line operator	*Mixed*	Mixed		
Level of competition: international long distance	*M*	C		
Level of competition: mobile	*P*	C		
Level of competition: Internet service provider	*C*	C		
Government prioritization of ICT (scale 1–7)	..	..	..	4.8
ICT sector performance				
Access				
Telephone main lines (per 1,000 people)	49	*55*	33	194
International voice traffic (minutes per person)[a]	10	..	*4*	..
Mobile subscribers (per 1,000 people)	64	*129*	48	248
Population covered by mobile telephony (%)	58	*64*	*43*	*73*
Internet users (per 1,000 people)	13	*58*	20	75
Personal computers (per 1,000 people)	13	*28*	*8*	37
Households with television (%)	28	*29*	*16*	*80*
Quality				
Telephone faults (per 100 main lines per year)	51.6	*20.4*	..	..
Broadband subscribers (per 1,000 people)	*0.0*	*0.2*	0.5	13.4
International Internet bandwidth (bits per person)	3	*14*	*3*	52
Affordability				
Price basket for fixed line (US$ per month, residential)	2.6	*2.5*	*6.6*	*4.5*
Price basket for mobile (US$ per month)	..	9.6	11.6	5.1
Price basket for Internet (US$ per month)	..	*17.8*	*45.5*	*19.9*
Price of call to United States (US$ per 3 minutes)	4.92	..	1.95	1.20
Institutional efficiency and sustainability				
Total telecommunications revenue (% GDP)	5.0	3.7	2.3	3.6
Total telephone subscribers per employee	60	*82*	*89*	..
Total telecommunications investment (% revenue)	4.7	*9.1*	*27.8*	*31.0*
ICT applications				
ICT expenditure (% GDP)	..	..	*4.1*	*5.0*
E-government readiness index (scale 0–1)	..	0.19	0.12	0.21
Secure Internet servers (per 1 million people)	*0.4*	2.0	0.3	0.6
Schools connected to the Internet (%)	..	*19*	..	..

GNI per Capita, Atlas Method, 2000–4

Current US$

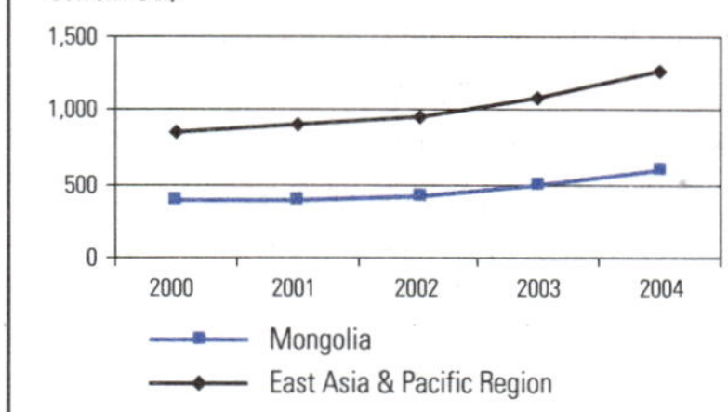

ICT MDG[b] Indicators, 2000–4

Number per 1,000 people

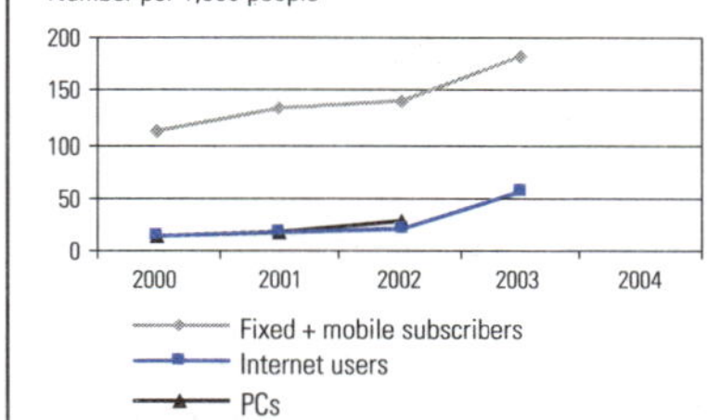

Price of Call to the United States, 2000–4

US$ per 3 minutes

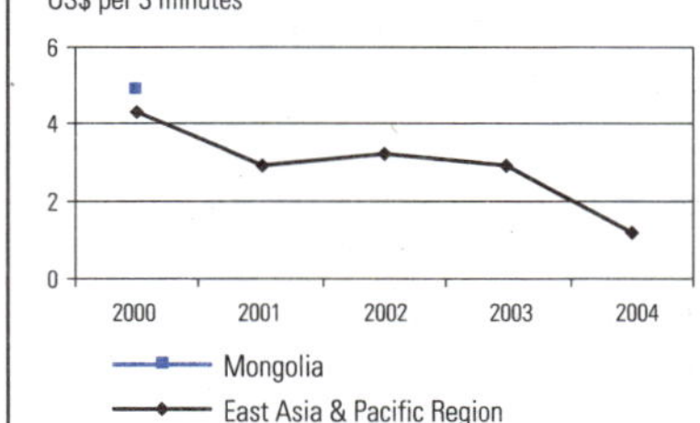

Total Telecommunications Revenue, 2000–4

Percentage of GDP

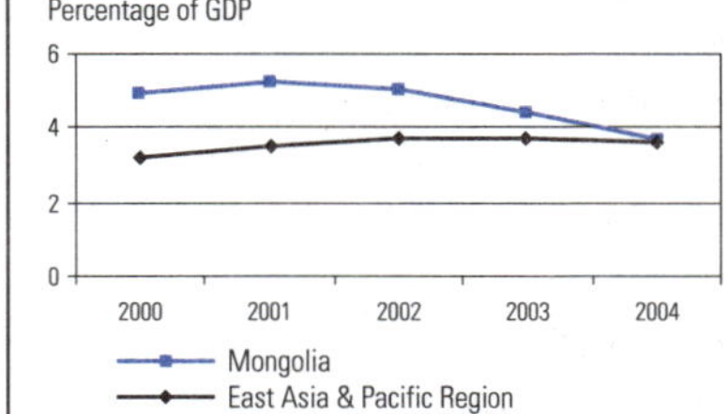

Notes: Figures in italics are for years other than those specified. .. indicates data are not available. C = competition; GDP = gross domestic product; GNI = gross national income; ICT = information and communication technology; M = monopoly; MDG = Millennium Development Goal; P = partial competition; and PCs = personal computers.

a. Outgoing and incoming.

b. Millennium Development Goal indicators 47, 48a, and 48b.

Sources: Economic and social context: UNDP and World Bank; ICT sector structure: ITU, WEF; ICT sector performance: Global Insight/WITSA, ITU, Netcraft, UNDESA, UNPAN, and World Bank. Produced by the Global Information and Communication Technologies Department and the Development Economics Data Group. For complete information, see Definitions and Data Sources.

Morocco

	Morocco 2000	Morocco 2004	Lower-middle-income group 2004	Middle East & North Africa Region 2004
Economic and social context				
Population, total (millions)	29	31	2,430	294
Urban population (% total population)	55	58	49	56
Poverty (% population below US$1 per day)	*<2*	..	..	*2.4*
GNI per capita, Atlas method (current US$)	1,180	1,520	1,580	2,000
GDP growth, 1995–2000 and 2000–4 (%)	3.2	4.5	5.7	4.5
Adult literacy rate (% ages 15 and over)	..	*51*	90	..
Primary, secondary, tertiary school enrollment (% gross)	54	*58*	*72*	*68*
ICT sector structure				
Separate telecommunications regulator	Yes	Yes		
Status of main fixed-line operator	*Mixed*	Mixed		
Level of competition: international long distance	*M*	M		
Level of competition: mobile	*C*	C		
Level of competition: Internet service provider	*C*	C		
Government prioritization of ICT (scale 1–7)	..	4.3	3.8	..
ICT sector performance				
Access				
Telephone main lines (per 1,000 people)	50	43	192	*118*
International voice traffic (minutes per person)[a]	26	*28*	..	*20*
Mobile subscribers (per 1,000 people)	82	305	255	*88*
Population covered by mobile telephony (%)	95	95	*76*	..
Internet users (per 1,000 people)	7	82	70	*47*
Personal computers (per 1,000 people)	12	*20*	38	*30*
Households with television (%)	72	76	*89*	*88*
Quality				
Telephone faults (per 100 main lines per year)	24.8	..	..	..
Broadband subscribers (per 1,000 people)	0.0	2.1	12.6	*0.2*
International Internet bandwidth (bits per person)	4	25	58	*15*
Affordability				
Price basket for fixed line (US$ per month, residential)	10.8	18.4	*5.5*	*4.9*
Price basket for mobile (US$ per month)	..	16.0	8.9	8.1
Price basket for Internet (US$ per month)	..	*25.3*	*25.3*	*24.5*
Price of call to United States (US$ per 3 minutes)	2.03	1.41	1.45	1.64
Institutional efficiency and sustainability				
Total telecommunications revenue (% GDP)	3.5	4.9	3.6	2.8
Total telephone subscribers per employee	251	..	*195*	..
Total telecommunications investment (% revenue)	50.7	*13.6*	*25.5*	*27.8*
ICT applications				
ICT expenditure (% GDP)	5.4	*5.6*	*5.1*	..
E-government readiness index (scale 0–1)	..	0.23	0.26	0.16
Secure Internet servers (per 1 million people)	*0.2*	0.6	1.6	0.6
Schools connected to the Internet (%)	..	..	..	..

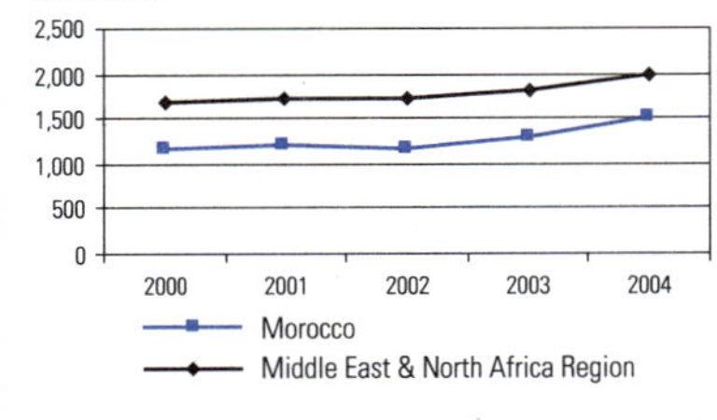

GNI per Capita, Atlas Method, 2000–4

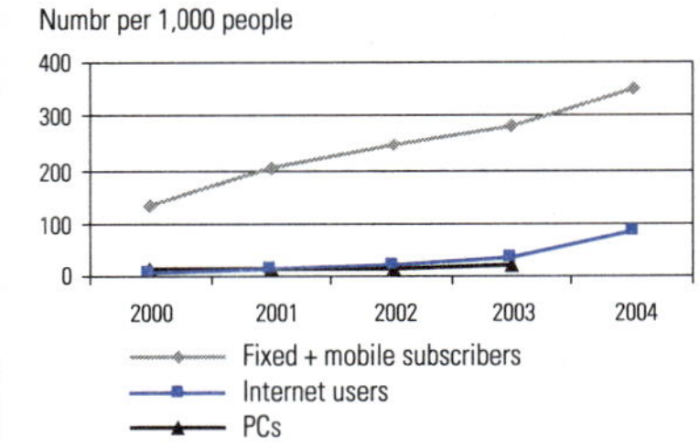

ICT MDG[b] Indicators, 2000–4

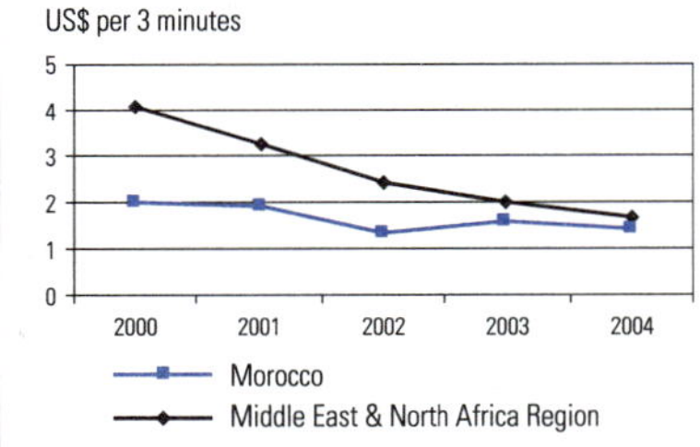

Price of Call to the United States, 2000–4

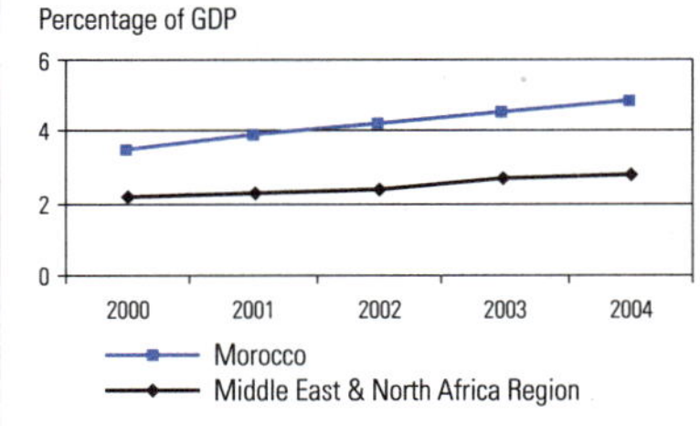

Total Telecommunications Revenue, 2000–4

Notes: Figures in italics are for years other than those specified. .. indicates data are not available. C = competition; GDP = gross domestic product; GNI = gross national income; ICT = information and communication technology; M = monopoly; MDG = Millennium Development Goal; P = partial competition; and PCs = personal computers.

a. Outgoing and incoming.
b. Millennium Development Goal indicators 47, 48a, and 48b.

Sources: Economic and social context: UNDP and World Bank; ICT sector structure: ITU, WEF; ICT sector performance: Global Insight/WITSA, ITU, Netcraft, UNDESA, UNPAN, and World Bank. Produced by the Global Information and Communication Technologies Department and the Development Economics Data Group. For complete information, see Definitions and Data Sources.

Mozambique

	Mozambique 2000	Mozambique 2004	Low-income group 2004	Sub-Saharan Africa Region 2004
Economic and social context				
Population, total (millions)	18	19	2,338	719
Urban population (% total population)	32	37	31	37
Poverty (% population below US$1 per day)	*37.9*	..	..	*46.4*
GNI per capita, Atlas method (current US$)	210	250	510	600
GDP growth, 1995–2000 and 2000–4 (%)	8.7	8.5	5.4	3.9
Adult literacy rate (% ages 15 and over)	..	*46*	61	59
Primary, secondary, tertiary school enrollment (% gross)	38	*43*	*55*	*52*
ICT sector structure				
Separate telecommunications regulator	Yes	Yes		
Status of main fixed-line operator	*Public*	Mixed		
Level of competition: international long distance	*M*	M		
Level of competition: mobile	*C*	C		
Level of competition: Internet service provider	*C*	C		
Government prioritization of ICT (scale 1–7)	..	4.5	..	*4.4*
ICT sector performance				
Access				
Telephone main lines (per 1,000 people)	5	*4*	33	17
International voice traffic (minutes per person)[a]	3	..	*4*	..
Mobile subscribers (per 1,000 people)	3	47	48	86
Population covered by mobile telephony (%)	..	..	*43*	..
Internet users (per 1,000 people)	1	5	20	15
Personal computers (per 1,000 people)	3	*4*	*8*	*12*
Households with television (%)	4	*6*	*16*	*15*
Quality				
Telephone faults (per 100 main lines per year)	80.0	*65.0*	..	..
Broadband subscribers (per 1,000 people)	0.0	*0.0*	0.5	*0.1*
International Internet bandwidth (bits per person)	0	*1*	*3*	*4*
Affordability				
Price basket for fixed line (US$ per month, residential)	14.6	16.5	*6.6*	*8.5*
Price basket for mobile (US$ per month)	..	10.9	11.6	13.5
Price basket for Internet (US$ per month)	..	*40.8*	*45.5*	*54.8*
Price of call to United States (US$ per 3 minutes)	*5.22*	1.17	1.95	2.43
Institutional efficiency and sustainability				
Total telecommunications revenue (% GDP)	3.0	3.0	2.3	5.0
Total telephone subscribers per employee	60	*158*	*89*	*144*
Total telecommunications investment (% revenue)	49.6	*46.1*	*27.8*	*27.8*
ICT applications				
ICT expenditure (% GDP)	..	..	*4.1*	..
E-government readiness index (scale 0–1)	..	0.17	0.12	0.11
Secure Internet servers (per 1 million people)	..	0.1	0.3	1.9
Schools connected to the Internet (%)	..	*0*	..	..

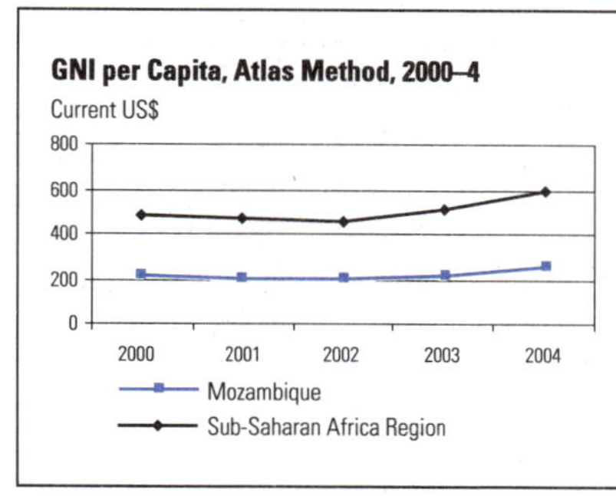

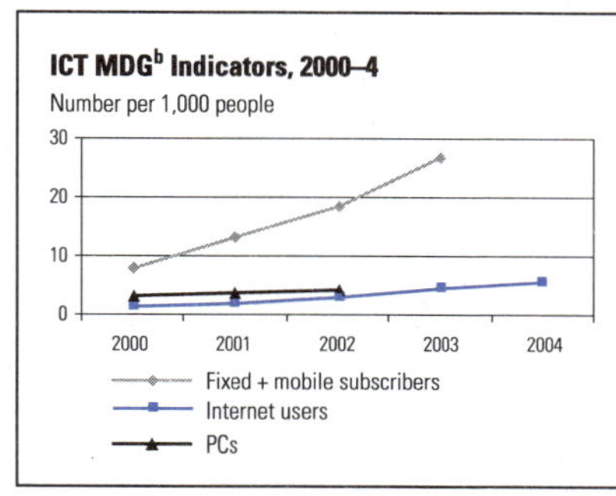

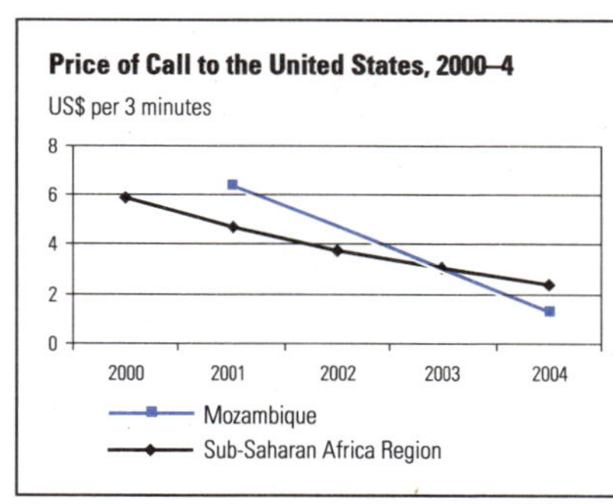

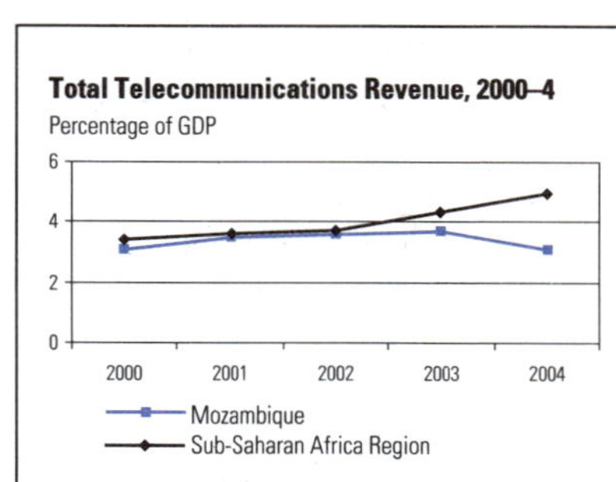

Notes: Figures in italics are for years other than those specified. .. indicates data are not available. C = competition; GDP = gross domestic product; GNI = gross national income; ICT = information and communication technology; M = monopoly; MDG = Millennium Development Goal; P = partial competition; and PCs = personal computers.

a. Outgoing and incoming.

b. Millennium Development Goal indicators 47, 48a, and 48b.

Sources: Economic and social context: UNDP and World Bank; ICT sector structure: ITU, WEF; ICT sector performance: Global Insight/WITSA, ITU, Netcraft, UNDESA, UNPAN, and World Bank. Produced by the Global Information and Communication Technologies Department and the Development Economics Data Group. For complete information, see Definitions and Data Sources.

Myanmar

	Myanmar 2000	Myanmar 2004	Low-income group 2004	East Asia & Pacific Region 2004
Economic and social context				
Population, total (millions)	48	50	2,338	1,870
Urban population (% total population)	28	30	31	41
Poverty (% population below US$1 per day)	..	..	..	*14.9*
GNI per capita, Atlas method (current US$)	..	..	510	1,280
GDP growth, 1995–2000 and 2000–4 (%)	8.1	..	5.4	7.5
Adult literacy rate (% ages 15 and over)	90	..	61	90
Primary, secondary, tertiary school enrollment (% gross)	47	*48*	*55*	*68*
ICT sector structure				
Separate telecommunications regulator	..	..		
Status of main fixed-line operator	*Public*	Public		
Level of competition: international long distance	*M*	M		
Level of competition: mobile	*M*	M		
Level of competition: Internet service provider	*M*	P		
Government prioritization of ICT (scale 1–7)	..	..	..	4.8
ICT sector performance				
Access				
Telephone main lines (per 1,000 people)	6	*7*	33	194
International voice traffic (minutes per person)[a]	1	*2*	*4*	..
Mobile subscribers (per 1,000 people)	0	*1*	48	248
Population covered by mobile telephony (%)	..	..	*43*	*73*
Internet users (per 1,000 people)	0	*1*	20	75
Personal computers (per 1,000 people)	2	*5*	*8*	37
Households with television (%)	3	*3*	*16*	*80*
Quality				
Telephone faults (per 100 main lines per year)	172.0	*155*	..	..
Broadband subscribers (per 1,000 people)	*0.0*	*0.0*	0.5	13.4
International Internet bandwidth (bits per person)	0	*0*	*3*	52
Affordability				
Price basket for fixed line (US$ per month, residential)	0.6	*2.9*	*6.6*	*4.5*
Price basket for mobile (US$ per month)	..	..	11.6	5.1
Price basket for Internet (US$ per month)	..	*42.5*	*45.5*	*19.9*
Price of call to United States (US$ per 3 minutes)	0.44	..	1.95	1.20
Institutional efficiency and sustainability				
Total telecommunications revenue (% GDP)	0.2	..	2.3	3.6
Total telephone subscribers per employee	37	*49*	*89*	..
Total telecommunications investment (% revenue)	25.3	*21.7*	*27.8*	*31.0*
ICT applications				
ICT expenditure (% GDP)	..	..	*4.1*	*5.0*
E-government readiness index (scale 0–1)	..	0.19	0.12	0.21
Secure Internet servers (per 1 million people)	..	0.0	0.3	0.6
Schools connected to the Internet (%)	..	..	..	..

GNI per Capita, Atlas Method, 2000–4

Current US$

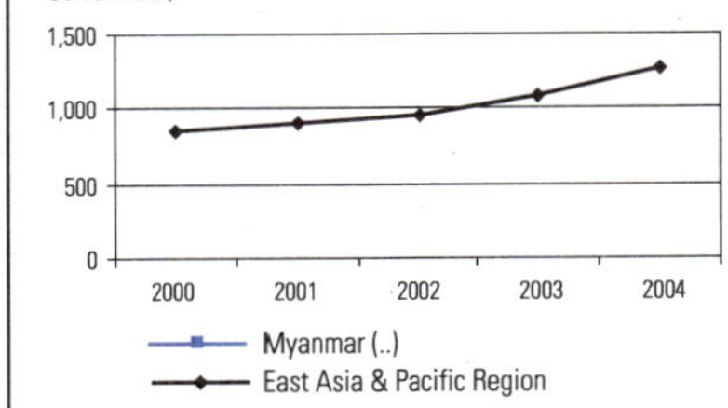

ICT MDG[b] Indicators, 2000–4

Number per 1,000 people

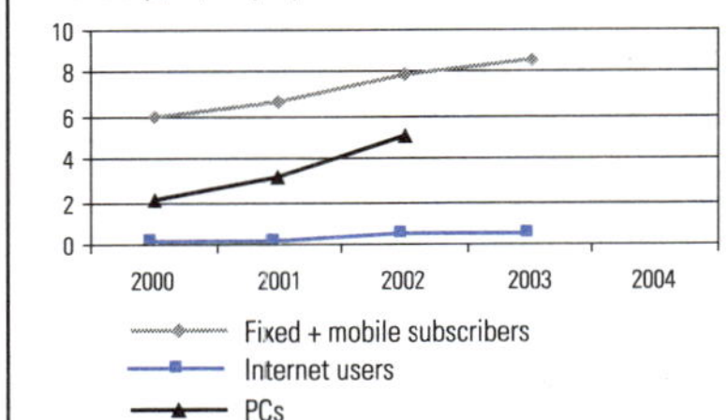

Price of Call to the United States, 2000–4

US$ per 3 minutes

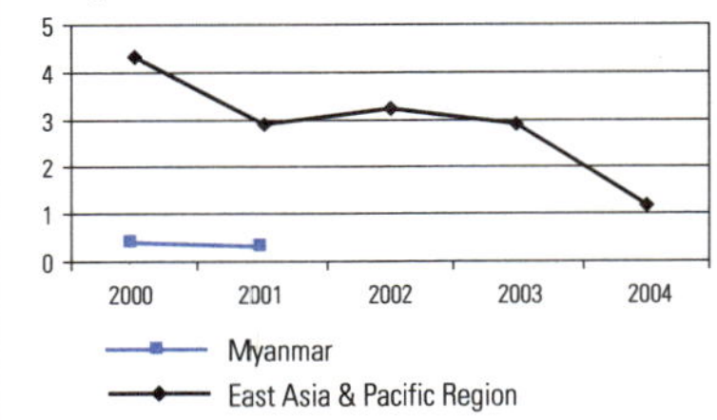

Total Telecommunications Revenue, 2000–4

Percentage of GDP

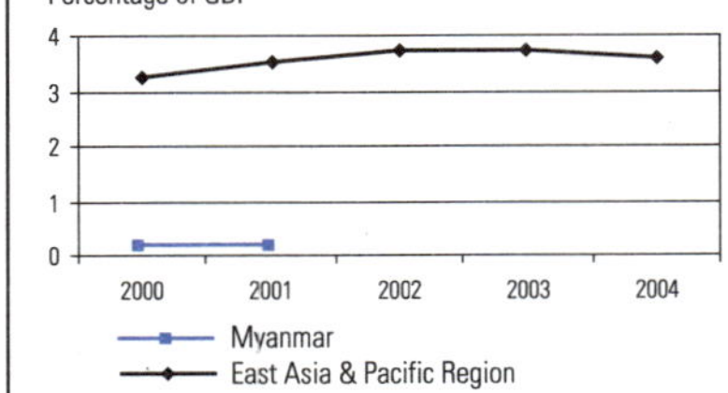

Notes: Figures in italics are for years other than those specified. .. indicates data are not available. C = competition; GDP = gross domestic product; GNI = gross national income; ICT = information and communication technology; M = monopoly; MDG = Millennium Development Goal; P = partial competition; and PCs = personal computers.

a. Outgoing and incoming.

b. Millennium Development Goal indicators 47, 48a, and 48b.

Sources: Economic and social context: UNDP and World Bank; ICT sector structure: ITU, WEF; ICT sector performance: Global Insight/WITSA, ITU, Netcraft, UNDESA, UNPAN, and World Bank. Produced by the Global Information and Communication Technologies Department and the Development Economics Data Group. For complete information, see Definitions and Data Sources.

Namibia

	Namibia 2000	Namibia 2004	Lower-middle-income group 2004	Sub-Saharan Africa Region 2004
Economic and social context				
Population, total (millions)	2	2	2,430	719
Urban population (% total population)	31	33	49	37
Poverty (% population below US$1 per day)	..	..	..	*46.4*
GNI per capita, Atlas method (current US$)	1,870	2,370	1,580	600
GDP growth, 1995–2000 and 2000–4 (%)	3.5	3.2	5.7	3.9
Adult literacy rate (% ages 15 and over)	*85*	..	90	59
Primary, secondary, tertiary school enrollment (% gross)	71	*71*	*72*	*52*
ICT sector structure				
Separate telecommunications regulator	Yes	Yes		
Status of main fixed-line operator	*Public*	Public		
Level of competition: international long distance	*M*	M		
Level of competition: mobile	*M*	M		
Level of competition: Internet service provider	*C*	C		
Government prioritization of ICT (scale 1–7)	..	4.4	3.8	*4.4*
ICT sector performance				
Access				
Telephone main lines (per 1,000 people)	58	63	192	17
International voice traffic (minutes per person)[a]	59	*57*	..	..
Mobile subscribers (per 1,000 people)	43	*111*	255	86
Population covered by mobile telephony (%)	84	88	*76*	..
Internet users (per 1,000 people)	16	39	70	15
Personal computers (per 1,000 people)	40	*95*	38	*12*
Households with television (%)	37	*39*	*89*	*15*
Quality				
Telephone faults (per 100 main lines per year)	*51.5*	32.6	..	..
Broadband subscribers (per 1,000 people)	0.0	0.0	12.6	*0.1*
International Internet bandwidth (bits per person)	2	*4*	58	*4*
Affordability				
Price basket for fixed line (US$ per month, residential)	9.4	15.8	*5.5*	*8.5*
Price basket for mobile (US$ per month)	..	14.7	8.9	13.5
Price basket for Internet (US$ per month)	..	*33.4*	*25.3*	*54.8*
Price of call to United States (US$ per 3 minutes)	4.28	..	1.45	2.43
Institutional efficiency and sustainability				
Total telecommunications revenue (% GDP)	3.6	*4.6*	3.6	5.0
Total telephone subscribers per employee	115	*181*	*195*	*144*
Total telecommunications investment (% revenue)	29.6	*7.5*	*25.5*	*27.8*
ICT applications				
ICT expenditure (% GDP)	..	..	*5.1*	..
E-government readiness index (scale 0–1)	..	0.12	0.26	0.11
Secure Internet servers (per 1 million people)	*1.5*	4.4	1.6	1.9
Schools connected to the Internet (%)	..	*4*	..	..

GNI per Capita, Atlas Method, 2000–4

Current US$

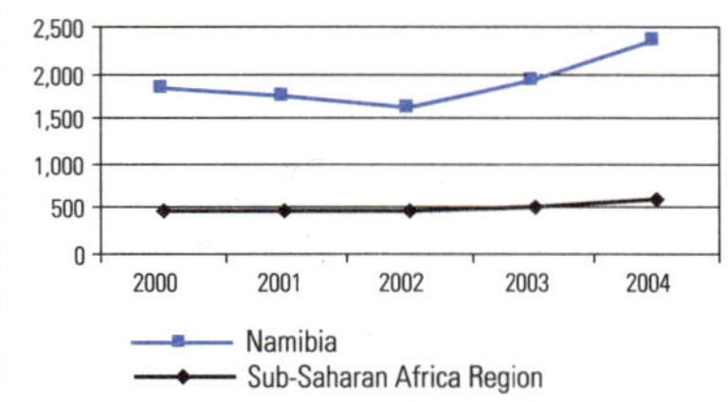

ICT MDG[b] Indicators, 2000–4

Number per 1,000 people

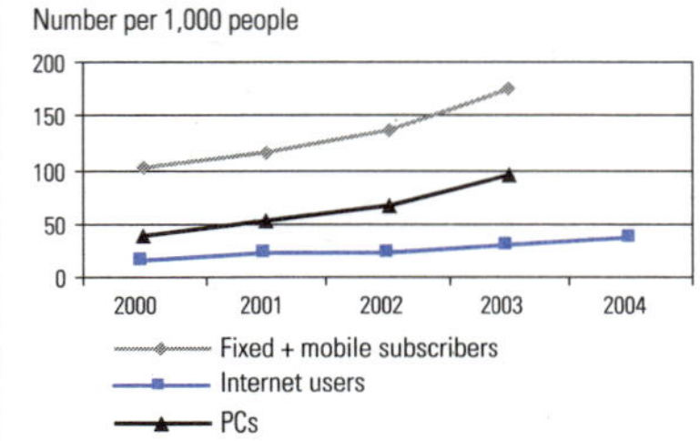

Price of Call to the United States, 2000–4

US$ per 3 minutes

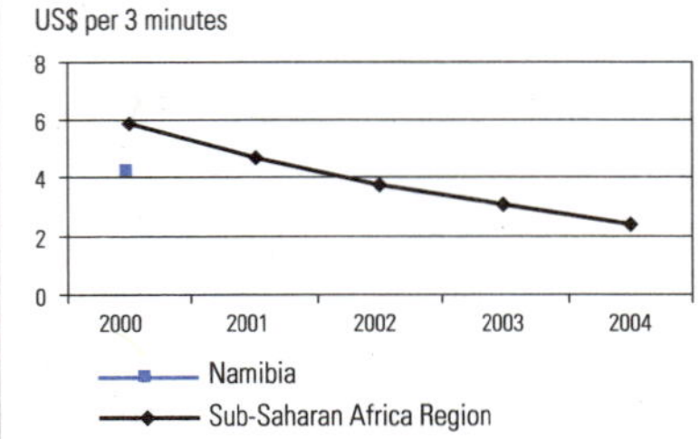

Total Telecommunications Revenue, 2000–4

Percentage of GDP

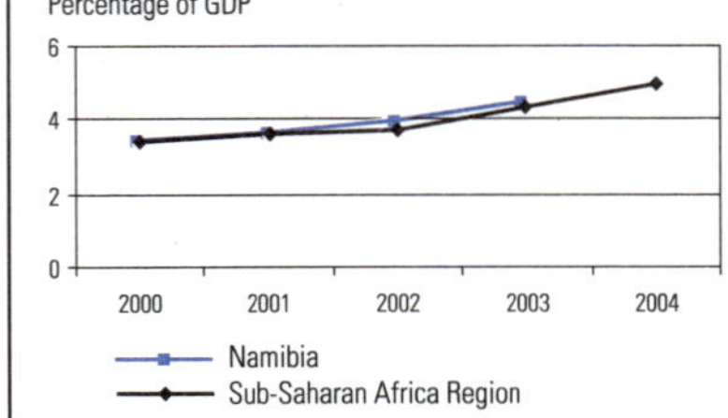

Notes: Figures in italics are for years other than those specified. .. indicates data are not available. C = competition; GDP = gross domestic product; GNI = gross national income; ICT = information and communication technology; M = monopoly; MDG = Millennium Development Goal; P = partial competition; and PCs = personal computers.

a. Outgoing and incoming.

b. Millennium Development Goal indicators 47, 48a, and 48b.

Sources: Economic and social context: UNDP and World Bank; ICT sector structure: ITU, WEF; ICT sector performance: Global Insight/WITSA, ITU, Netcraft, UNDESA, UNPAN, and World Bank. Produced by the Global Information and Communication Technologies Department and the Development Economics Data Group. For complete information, see Definitions and Data Sources.

Nepal

	Nepal 2000	Nepal 2004	Low-income group 2004	South Asia Region 2004
Economic and social context				
Population, total (millions)	23	25	2,338	1,448
Urban population (% total population)	14	15	31	28
Poverty (% population below US$1 per day)	*39.1*	..	..	*31.3*
GNI per capita, Atlas method (current US$)	230	260	510	590
GDP growth, 1995–2000 and 2000–4 (%)	4.6	2.6	5.4	5.8
Adult literacy rate (% ages 15 and over)	*49*	..	61	58
Primary, secondary, tertiary school enrollment (% gross)	58	*61*	*55*	*56*
ICT sector structure				
Separate telecommunications regulator	Yes	Yes		
Status of main fixed-line operator	*Public*	Public		
Level of competition: international long distance	*M*	C		
Level of competition: mobile	*M*	C		
Level of competition: Internet service provider	*C*	*C*		
Government prioritization of ICT (scale 1–7)	..	..	..	5.3
ICT sector performance				
Access				
Telephone main lines (per 1,000 people)	12	17	33	41
International voice traffic (minutes per person)[a]	2	*4*	*4*	*4*
Mobile subscribers (per 1,000 people)	0	10	48	47
Population covered by mobile telephony (%)	..	..	*43*	*43*
Internet users (per 1,000 people)	2	9	20	21
Personal computers (per 1,000 people)	3	*4*	*8*	11
Households with television (%)	3	..	*16*	*32*
Quality				
Telephone faults (per 100 main lines per year)	100.2	*88.1*	..	*88.1*
Broadband subscribers (per 1,000 people)	0.0	0.0	0.5	0.6
International Internet bandwidth (bits per person)	*0*	*1*	*3*	4
Affordability				
Price basket for fixed line (US$ per month, residential)	2.6	3.1	*6.6*	3.2
Price basket for mobile (US$ per month)	..	2.8	11.6	3.2
Price basket for Internet (US$ per month)	..	*13.5*	*45.5*	*15.1*
Price of call to United States (US$ per 3 minutes)	*5.28*	2.04	1.95	1.21
Institutional efficiency and sustainability				
Total telecommunications revenue (% GDP)	1.3	1.7	2.3	1.9
Total telephone subscribers per employee	60	*88*	*89*	*89*
Total telecommunications investment (% revenue)	26.7	..	*27.8*	*15.3*
ICT applications				
ICT expenditure (% GDP)	..	..	*4.1*	*4.1*
E-government readiness index (scale 0–1)	..	0.34	0.12	0.34
Secure Internet servers (per 1 million people)	..	0.3	0.3	0.4
Schools connected to the Internet (%)	..	..	..	..

GNI per Capita, Atlas Method, 2000–4

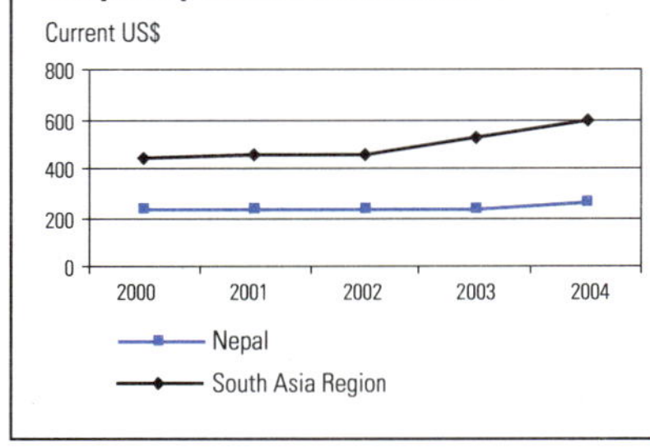

ICT MDG[b] Indicators, 2000–4

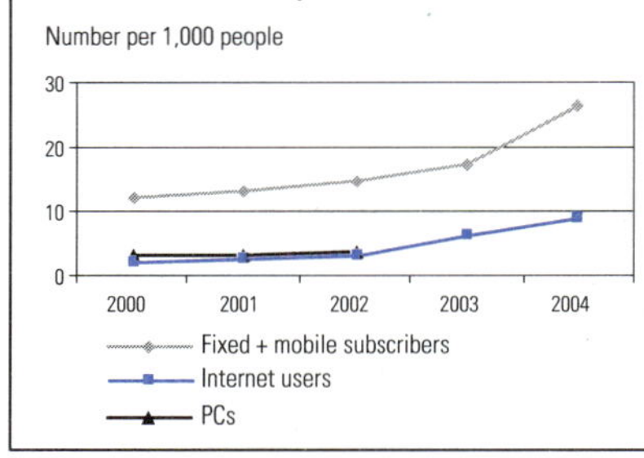

Price of Call to the United States, 2000–4

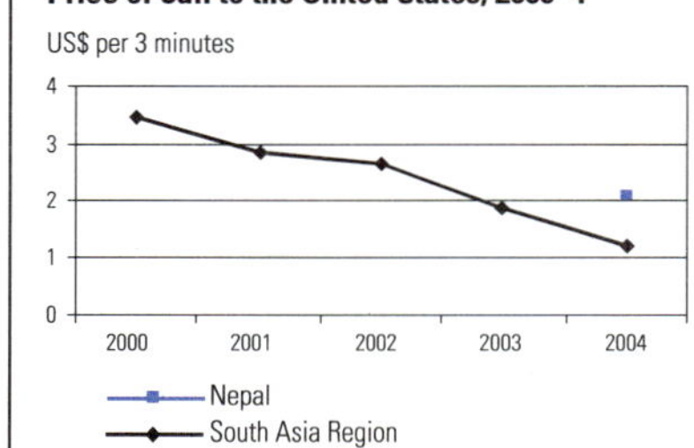

Total Telecommunications Revenue, 2000–4

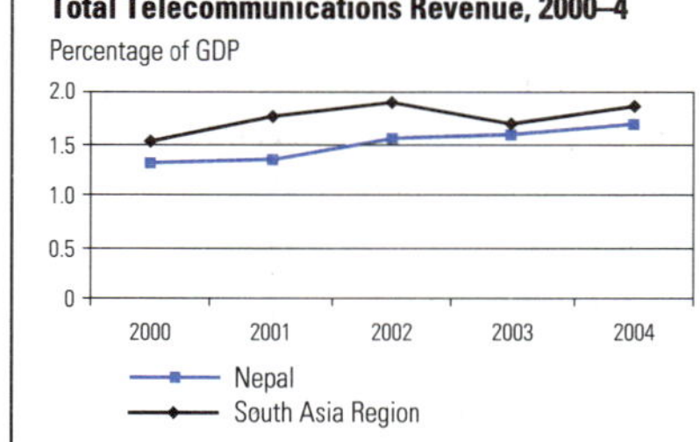

Notes: Figures in italics are for years other than those specified. .. indicates data are not available. C = competition; GDP = gross domestic product; GNI = gross national income; ICT = information and communication technology; M = monopoly; MDG = Millennium Development Goal; P = partial competition; and PCs = personal computers.

a. Outgoing and incoming.

b. Millennium Development Goal indicators 47, 48a, and 48b.

Sources: Economic and social context: UNDP and World Bank; ICT sector structure: ITU, WEF; ICT sector performance: Global Insight/WITSA, ITU, Netcraft, UNDESA, UNPAN, and World Bank. Produced by the Global Information and Communication Technologies Department and the Development Economics Data Group. For complete information, see Definitions and Data Sources.

Netherlands

	Netherlands 2000	Netherlands 2004	High-income group 2004
Economic and social context			
Population, total (millions)	16	16	1,001
Urban population (% total population)	64	66	77
Poverty (% population below US$1 per day)	..	..	..
GNI per capita, Atlas method (current US$)	25,210	31,700	32,040
GDP growth, 1995–2000 and 2000–4 (%)	3.8	0.3	2.0
Adult literacy rate (% ages 15 and over)	..	..	..
Primary, secondary, tertiary school enrollment (% gross)	99	*99*	*93*
ICT sector structure			
Separate telecommunications regulator	Yes	Yes	
Status of main fixed-line operator	*Mixed*	Mixed	
Level of competition: international long distance	*C*	C	
Level of competition: mobile	*C*	*P*	
Level of competition: Internet service provider	*C*	P	
Government prioritization of ICT (scale 1–7)	..	4.9	4.9
ICT sector performance			
Access			
Telephone main lines (per 1,000 people)	621	588	558
International voice traffic (minutes per person)[a]	311	..	..
Mobile subscribers (per 1,000 people)	676	992	767
Population covered by mobile telephony (%)	100	100	98
Internet users (per 1,000 people)	440	*524*	*480*
Personal computers (per 1,000 people)	396	*524*	*504*
Households with television (%)	95	*99*	*98*
Quality			
Telephone faults (per 100 main lines per year)	2.7	..	..
Broadband subscribers (per 1,000 people)	16.3	189.8	126.2
International Internet bandwidth (bits per person)	4,277	20,589	4,718
Affordability			
Price basket for fixed line (US$ per month, residential)	20.9	31.7	25.8
Price basket for mobile (US$ per month)	..	24.5	17.8
Price basket for Internet (US$ per month)	..	*24.1*	*20.9*
Price of call to United States (US$ per 3 minutes)	0.56	0.32	0.77
Institutional efficiency and sustainability			
Total telecommunications revenue (% GDP)	3.9	3.3	2.9
Total telephone subscribers per employee	353	*657*	*485*
Total telecommunications investment (% revenue)	21.6	..	*12.3*
ICT applications			
ICT expenditure (% GDP)	6.9	*6.4*	*7.1*
E-government readiness index (scale 0–1)	..	0.72	0.70
Secure Internet servers (per 1 million people)	*49.8*	232.6	311.4
Schools connected to the Internet (%)	..	*92*	*99*

GNI per Capita, Atlas Method, 2000–4
Current US$

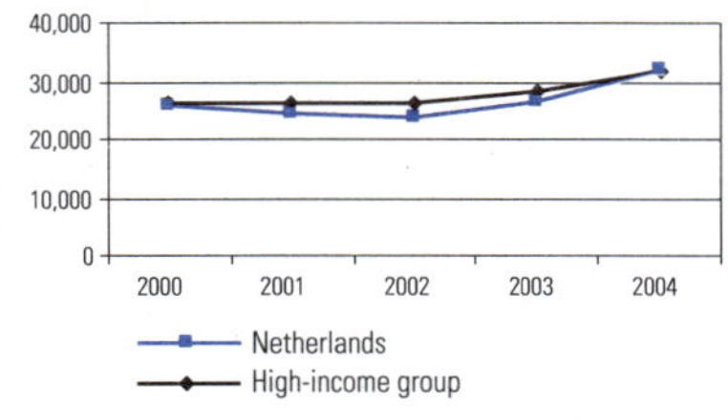

ICT MDG[b] Indicators, 2000–4
Number per 1,000 people

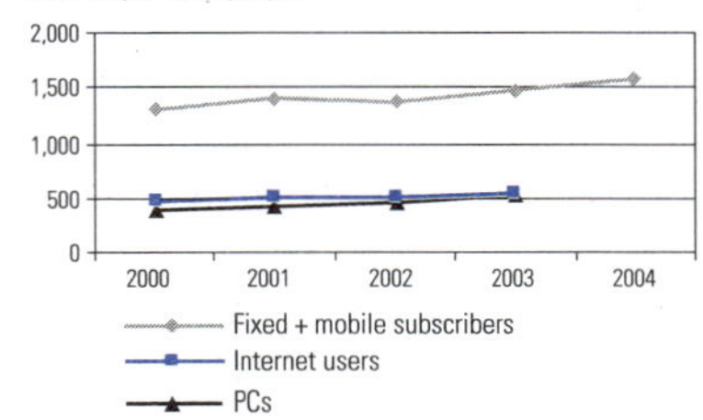

Price of Call to the United States, 2000–4
US$ per 3 minutes

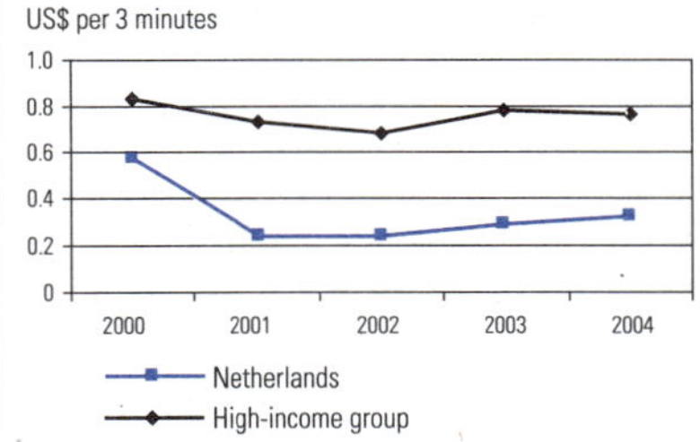

Total Telecommunications Revenue, 2000–4
Percentage of GDP

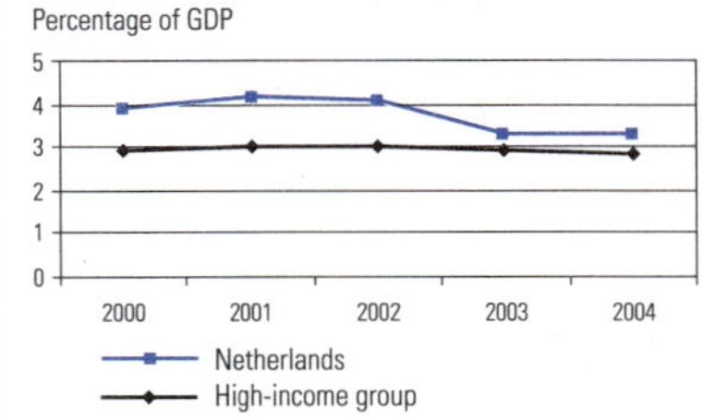

Notes: Figures in italics are for years other than those specified. .. indicates data are not available. C = competition; GDP = gross domestic product; GNI = gross national income; ICT = information and communication technology; M = monopoly; MDG = Millennium Development Goal; P = partial competition; and PCs = personal computers.

a. Outgoing and incoming.

b. Millennium Development Goal indicators 47, 48a, and 48b.

Sources: Economic and social context: UNDP and World Bank; ICT sector structure: ITU, WEF; ICT sector performance: Global Insight/WITSA, ITU, Netcraft, UNDESA, UNPAN, and World Bank. Produced by the Global Information and Communication Technologies Department and the Development Economics Data Group. For complete information, see Definitions and Data Sources.

New Zealand

	New Zealand 2000	New Zealand 2004	High-income group 2004
Economic and social context			
Population, total (millions)	4	4	1,001
Urban population (% total population)	86	86	77
Poverty (% population below US$1 per day)	..	..	..
GNI per capita, Atlas method (current US$)	13,700	20,310	32,040
GDP growth, 1995–2000 and 2000–4 (%)	2.4	3.9	2.0
Adult literacy rate (% ages 15 and over)	..	..	..
Primary, secondary, tertiary school enrollment (% gross)	99	*106*	*93*
ICT sector structure			
Separate telecommunications regulator	..	..	
Status of main fixed-line operator	*Private*	Private	
Level of competition: international long distance	*C*	C	
Level of competition: mobile	*C*	C	
Level of competition: Internet service provider	*C*	C	
Government prioritization of ICT (scale 1–7)	..	4.4	4.9
ICT sector performance			
Access			
Telephone main lines (per 1,000 people)	475	443	558
International voice traffic (minutes per person)[a]	289	*468*	..
Mobile subscribers (per 1,000 people)	400	811	767
Population covered by mobile telephony (%)	97	97	98
Internet users (per 1,000 people)	393	*526*	*480*
Personal computers (per 1,000 people)	358	*417*	*504*
Households with television (%)	97	98	*98*
Quality			
Telephone faults (per 100 main lines per year)	19.3	..	..
Broadband subscribers (per 1,000 people)	1.2	*18.0*	126.2
International Internet bandwidth (bits per person)	65	1,127	4,718
Affordability			
Price basket for fixed line (US$ per month, residential)	17.0	*18.1*	25.8
Price basket for mobile (US$ per month)	..	19.8	17.8
Price basket for Internet (US$ per month)	..	*12.9*	*20.9*
Price of call to United States (US$ per 3 minutes)	0.80	1.30	0.77
Institutional efficiency and sustainability			
Total telecommunications revenue (% GDP)	3.5	3.3	2.9
Total telephone subscribers per employee	630	..	*485*
Total telecommunications investment (% revenue)	16.2	*13.2*	*12.3*
ICT applications			
ICT expenditure (% GDP)	11.3	*10.0*	*7.1*
E-government readiness index (scale 0–1)	..	0.74	0.70
Secure Internet servers (per 1 million people)	*156.9*	426.7	311.4
Schools connected to the Internet (%)	..	*99*	*99*

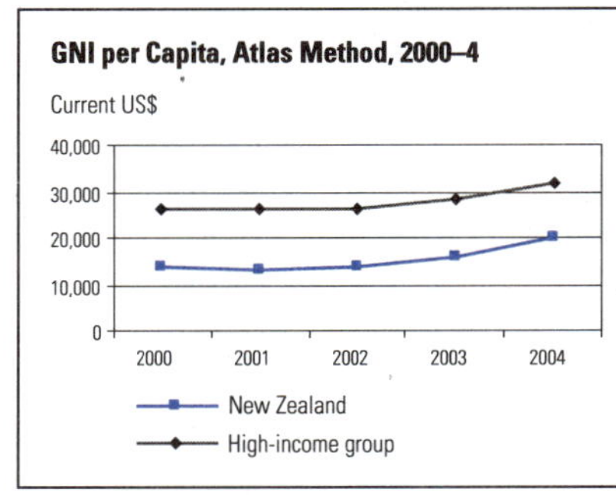

GNI per Capita, Atlas Method, 2000–4

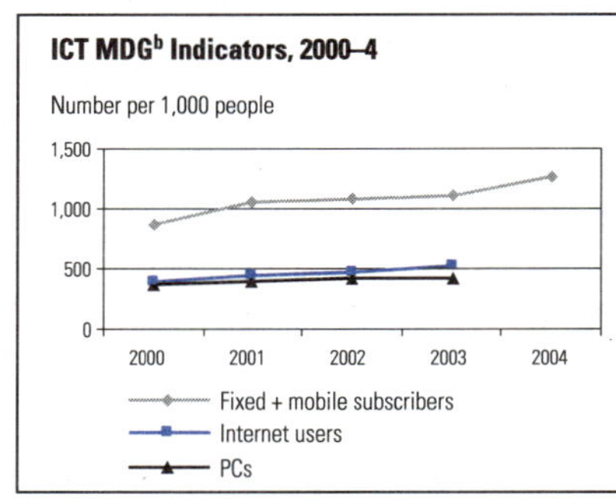

ICT MDG[b] Indicators, 2000–4

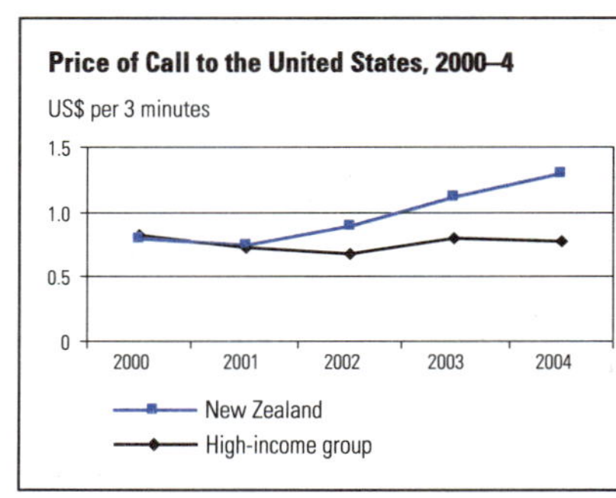

Price of Call to the United States, 2000–4

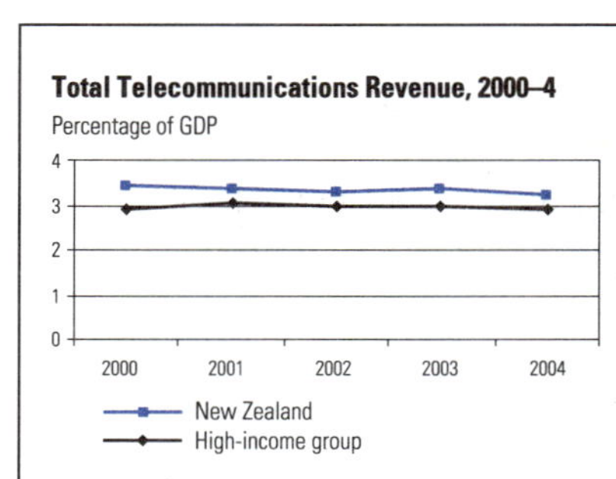

Total Telecommunications Revenue, 2000–4

Notes: Figures in italics are for years other than those specified. .. indicates data are not available. C = competition; GDP = gross domestic product; GNI = gross national income; ICT = information and communication technology; M = monopoly; MDG = Millennium Development Goal; P = partial competition; and PCs = personal computers.

a. Outgoing and incoming.

b. Millennium Development Goal indicators 47, 48a, and 48b.

Sources: Economic and social context: UNDP and World Bank; ICT sector structure: ITU, WEF; ICT sector performance: Global Insight/WITSA, ITU, Netcraft, UNDESA, UNPAN, and World Bank. Produced by the Global Information and Communication Technologies Department and the Development Economics Data Group. For complete information, see Definitions and Data Sources.

Nicaragua

	Nicaragua 2000	Nicaragua 2004	Low-income group 2004	Latin America & the Caribbean Region 2004
Economic and social context				
Population, total (millions)	5	6	2,338	541
Urban population (% total population)	56	58	31	77
Poverty (% population below US$1 per day)	*44.7*	*45.1*	..	*9.5*
GNI per capita, Atlas method (current US$)	740	790	510	3,600
GDP growth, 1995–2000 and 2000–4 (%)	5.0	2.3	5.4	1.5
Adult literacy rate (% ages 15 and over)	*77*	..	61	89
Primary, secondary, tertiary school enrollment (% gross)	69	*69*	*55*	*82*
ICT sector structure				
Separate telecommunications regulator	Yes	Yes		
Status of main fixed-line operator	*Public*	Private		
Level of competition: international long distance	*M*	M		
Level of competition: mobile	*C*	C		
Level of competition: Internet service provider	*C*	*C*		
Government prioritization of ICT (scale 1–7)	..	3.6	..	3.5
ICT sector performance				
Access				
Telephone main lines (per 1,000 people)	32	38	33	181
International voice traffic (minutes per person)[a]	35	*19*	*4*	..
Mobile subscribers (per 1,000 people)	18	132	48	324
Population covered by mobile telephony (%)	..	48	*43*	*76*
Internet users (per 1,000 people)	10	19	20	*104*
Personal computers (per 1,000 people)	24	*31.9*	*8*	*75*
Households with television (%)	59	..	*16*	*88*
Quality				
Telephone faults (per 100 main lines per year)	*79.3*	*4.6*	..	..
Broadband subscribers (per 1,000 people)	0.1	*0.4*	0.5	*5.2*
International Internet bandwidth (bits per person)	..	*6*	*3*	165
Affordability				
Price basket for fixed line (US$ per month, residential)	17.7	*14.3*	*6.6*	*9.0*
Price basket for mobile (US$ per month)	..	16.0	11.6	9.1
Price basket for Internet (US$ per month)	..	*51.1*	*45.5*	*31.5*
Price of call to United States (US$ per 3 minutes)	3.20	*3.20*	1.95	0.90
Institutional efficiency and sustainability				
Total telecommunications revenue (% GDP)	2.6	*2.5*	2.3	3.3
Total telephone subscribers per employee	114	*179*	*89*	..
Total telecommunications investment (% revenue)	5.9	*40.1*	*27.8*	..
ICT applications				
ICT expenditure (% GDP)	..	..	*4.1*	*5.3*
E-government readiness index (scale 0–1)	..	0.27	0.12	0.39
Secure Internet servers (per 1 million people)	*1.2*	2.5	0.3	8.6
Schools connected to the Internet (%)	..	..	..	..

GNI per Capita, Atlas Method, 2000–4

Current US$

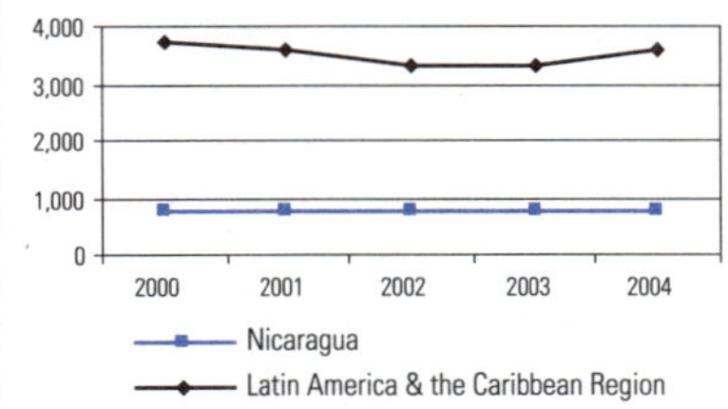

ICT MDG[b] Indicators, 2000–4

Number per 1,000 people

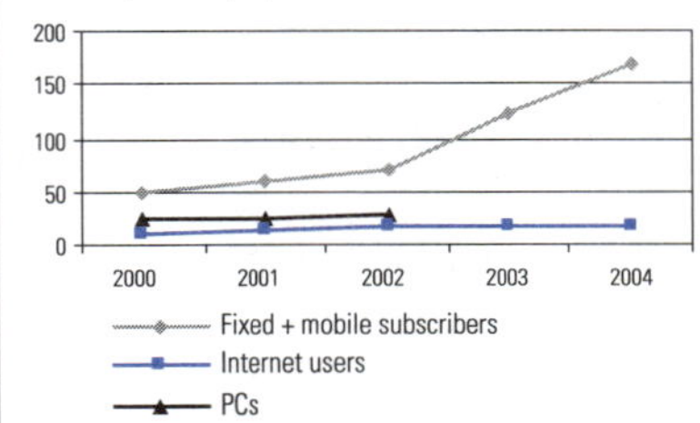

Price of Call to the United States, 2000–4

US$ per 3 minutes

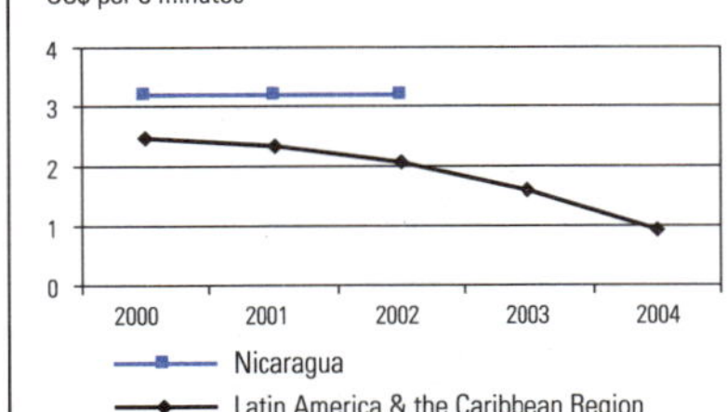

Total Telecommunications Revenue, 2000–4

Percentage of GDP

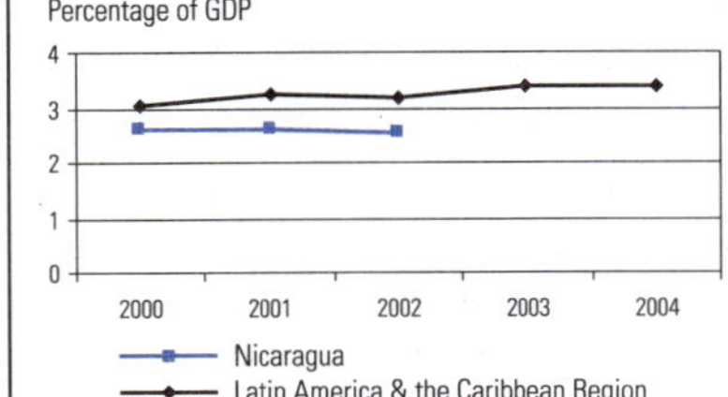

Notes: Figures in italics are for years other than those specified. .. indicates data are not available. C = competition; GDP = gross domestic product; GNI = gross national income; ICT = information and communication technology; M = monopoly; MDG = Millennium Development Goal; P = partial competition; and PCs = personal computers.

a. Outgoing and incoming.

b. Millennium Development Goal indicators 47, 48a, and 48b.

Sources: Economic and social context: UNDP and World Bank; ICT sector structure: ITU, WEF; ICT sector performance: Global Insight/WITSA, ITU, Netcraft, UNDESA, UNPAN, and World Bank. Produced by the Global Information and Communication Technologies Department and the Development Economics Data Group. For complete information, see Definitions and Data Sources.

Niger

	Niger 2000	Niger 2004	Low-income group 2004	Sub-Saharan Africa Region 2004
Economic and social context				
Population, total (millions)	11	12	2,338	719
Urban population (% total population)	21	23	31	37
Poverty (% population below US$1 per day)	*60.6*	..	..	*46.4*
GNI per capita, Atlas method (current US$)	180	230	510	600
GDP growth, 1995–2000 and 2000–4 (%)	3.4	4.1	5.4	3.9
Adult literacy rate (% ages 15 and over)	*14*	..	61	59
Primary, secondary, tertiary school enrollment (% gross)	18	*21*	*55*	*52*
ICT sector structure				
Separate telecommunications regulator	..	..		
Status of main fixed-line operator	*Public*	Mixed		
Level of competition: international long distance	*M*	M		
Level of competition: mobile	*C*	C		
Level of competition: Internet service provider	*C*	M		
Government prioritization of ICT (scale 1–7)	..	..	..	*4.4*
ICT sector performance				
Access				
Telephone main lines (per 1,000 people)	2	2	33	17
International voice traffic (minutes per person)[a]	*1*	..	*4*	..
Mobile subscribers (per 1,000 people)	0	*2*	48	86
Population covered by mobile telephony (%)	..	*13*	*43*	..
Internet users (per 1,000 people)	0	*1*	20	15
Personal computers (per 1,000 people)	0	*1*	*8*	*12*
Households with television (%)	5	*5*	*16*	*15*
Quality				
Telephone faults (per 100 main lines per year)	94.8	..	..	..
Broadband subscribers (per 1,000 people)	*0.0*	*0.0*	0.5	*0.1*
International Internet bandwidth (bits per person)	0	*0*	*3*	*4*
Affordability				
Price basket for fixed line (US$ per month, residential)	9.4	..	*6.6*	*8.5*
Price basket for mobile (US$ per month)	..	19.3	11.6	13.5
Price basket for Internet (US$ per month)	..	*96.9*	*45.5*	*54.8*
Price of call to United States (US$ per 3 minutes)	9.03	..	1.95	2.43
Institutional efficiency and sustainability				
Total telecommunications revenue (% GDP)	0.9	..	2.3	5.0
Total telephone subscribers per employee	16	..	*89*	*144*
Total telecommunications investment (% revenue)	..	..	*27.8*	*27.8*
ICT applications				
ICT expenditure (% GDP)	..	..	*4.1*	..
E-government readiness index (scale 0–1)	..	0.01	0.12	0.11
Secure Internet servers (per 1 million people)	..	..	0.3	1.9
Schools connected to the Internet (%)	..	..	..	..

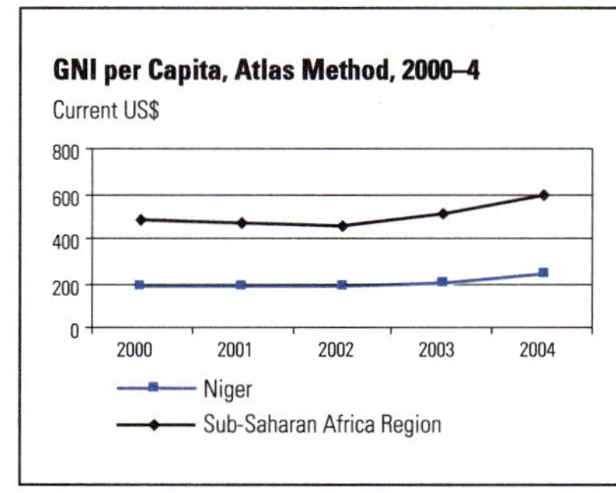

GNI per Capita, Atlas Method, 2000–4

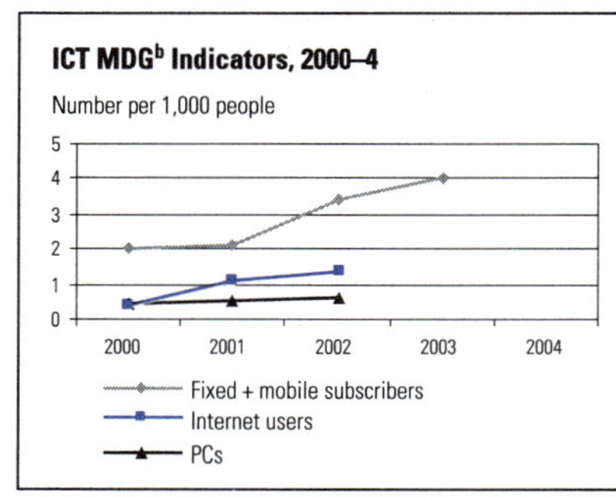

ICT MDG[b] Indicators, 2000–4

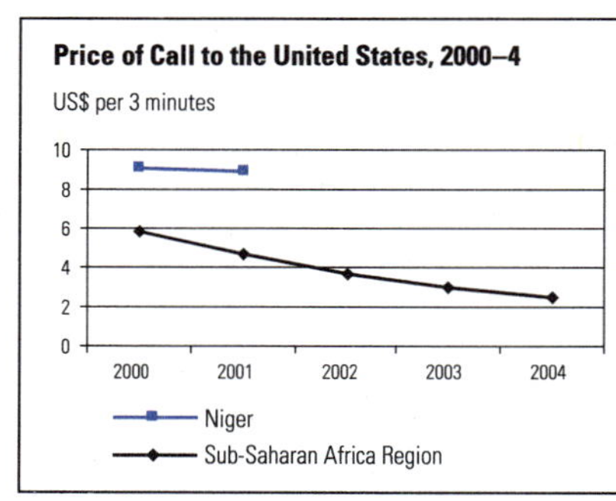

Price of Call to the United States, 2000–4

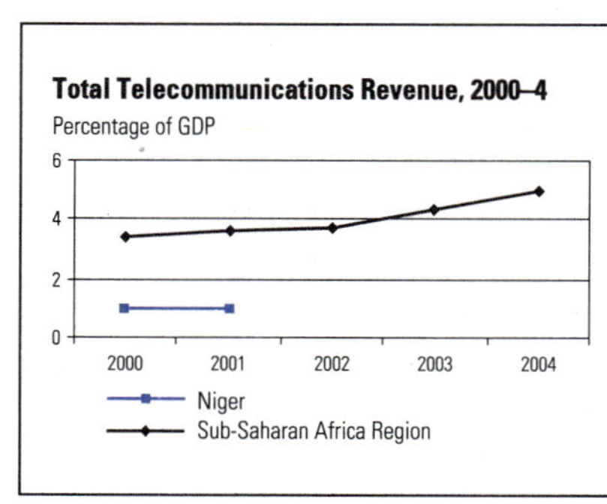

Total Telecommunications Revenue, 2000–4

Notes: Figures in italics are for years other than those specified. .. indicates data are not available. C = competition; GDP = gross domestic product; GNI = gross national income; ICT = information and communication technology; M = monopoly; MDG = Millennium Development Goal; P = partial competition; and PCs = personal computers.

a. Outgoing and incoming.

b. Millennium Development Goal indicators 47, 48a, and 48b.

Sources: Economic and social context: UNDP and World Bank; ICT sector structure: ITU, WEF; ICT sector performance: Global Insight/WITSA, ITU, Netcraft, UNDESA, UNPAN, and World Bank. Produced by the Global Information and Communication Technologies Department and the Development Economics Data Group. For complete information, see Definitions and Data Sources.

Nigeria

	Nigeria 2000	Nigeria 2004	Low-income group 2004	Sub-Saharan Africa Region 2004
Economic and social context				
Population, total (millions)	127	140	2,338	719
Urban population (% total population)	44	47	31	37
Poverty (% population below US$1 per day)	*70.2*	..	..	*46.4*
GNI per capita, Atlas method (current US$)	260	390	510	600
GDP growth, 1995–2000 and 2000–4 (%)	2.6	4.9	5.4	3.9
Adult literacy rate (% ages 15 and over)	..	*67*	61	59
Primary, secondary, tertiary school enrollment (% gross)	..	*64*	*55*	*52*
ICT sector structure				
Separate telecommunications regulator	Yes	Yes		
Status of main fixed-line operator	*Public*	Public		
Level of competition: international long distance	*M*	P		
Level of competition: mobile	..	P		
Level of competition: Internet service provider	*C*	C		
Government prioritization of ICT (scale 1–7)	..	3.9	..	*4.4*
ICT sector performance				
Access				
Telephone main lines (per 1,000 people)	4	8	33	17
International voice traffic (minutes per person)[a]	2	..	*4*	..
Mobile subscribers (per 1,000 people)	0	66	48	86
Population covered by mobile telephony (%)	..	58	*43*	..
Internet users (per 1,000 people)	1	7	20	15
Personal computers (per 1,000 people)	6	*6*	*8*	*12*
Households with television (%)	26	*26*	*16*	*15*
Quality				
Telephone faults (per 100 main lines per year)	..	..	..	..
Broadband subscribers (per 1,000 people)	0.0	*0.0*	0.5	*0.1*
International Internet bandwidth (bits per person)	0	*1*	*3*	*4*
Affordability				
Price basket for fixed line (US$ per month, residential)	*12.8*	13.7	*6.6*	*8.5*
Price basket for mobile (US$ per month)	..	11.2	11.6	13.5
Price basket for Internet (US$ per month)	..	*85.5*	*45.5*	*54.8*
Price of call to United States (US$ per 3 minutes)	*7.15*	1.49	1.95	2.43
Institutional efficiency and sustainability				
Total telecommunications revenue (% GDP)	0.8	4.4	2.3	5.0
Total telephone subscribers per employee	50	*192*	*89*	*144*
Total telecommunications investment (% revenue)	37.2	*164.2*	*27.8*	*27.8*
ICT applications				
ICT expenditure (% GDP)	..	..	*4.1*	..
E-government readiness index (scale 0–1)	..	0.14	0.12	0.11
Secure Internet servers (per 1 million people)	*0.0*	0.1	0.3	1.9
Schools connected to the Internet (%)	..	..	..	..

GNI per Capita, Atlas Method, 2000–4

Current US$

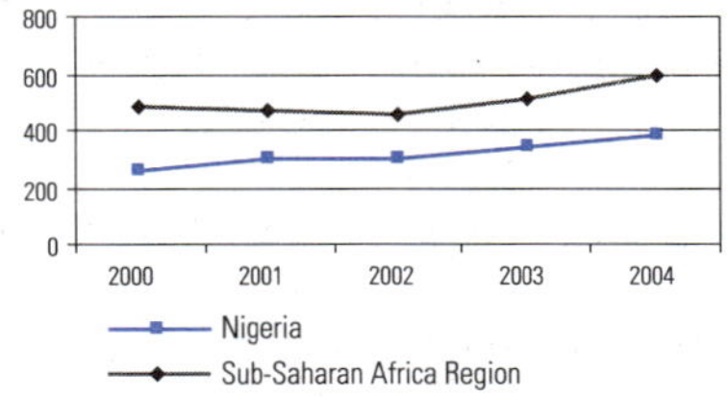

ICT MDG[b] Indicators, 2000–4

Number per 1,000 people

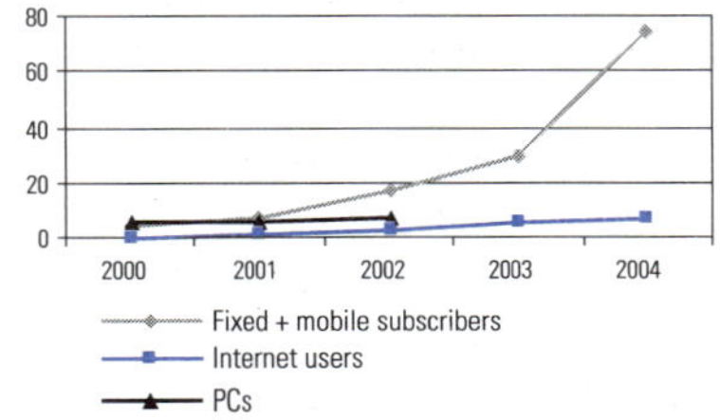

Price of Call to the United States, 2000–4

US$ per 3 minutes

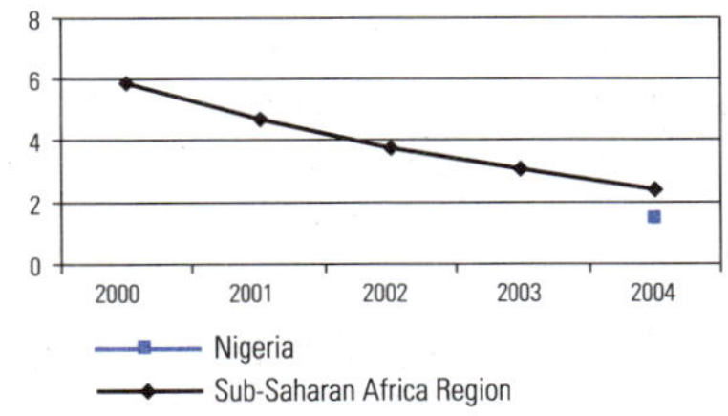

Total Telecommunications Revenue, 2000–4

Percentage of GDP

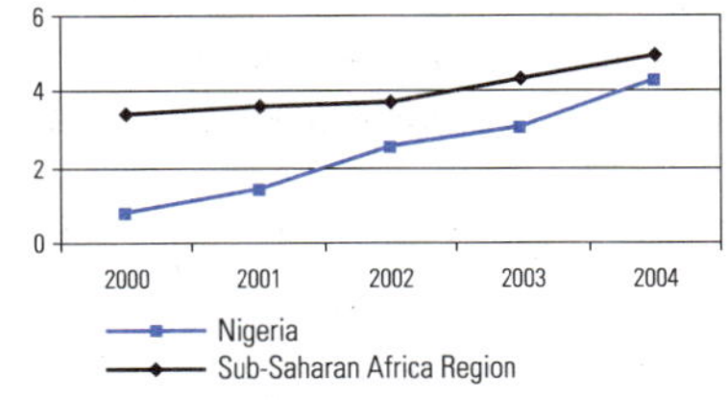

Notes: Figures in italics are for years other than those specified. .. indicates data are not available. C = competition; GDP = gross domestic product; GNI = gross national income; ICT = information and communication technology; M = monopoly; MDG = Millennium Development Goal; P = partial competition; and PCs = personal computers.

a. Outgoing and incoming.

b. Millennium Development Goal indicators 47, 48a, and 48b.

Sources: Economic and social context: UNDP and World Bank; ICT sector structure: ITU, WEF; ICT sector performance: Global Insight/WITSA, ITU, Netcraft, UNDESA, UNPAN, and World Bank. Produced by the Global Information and Communication Technologies Department and the Development Economics Data Group. For complete information, see Definitions and Data Sources.

Norway

	Norway 2000	Norway 2004	High-income group 2004
Economic and social context			
Population, total (millions)	4	5	1,001
Urban population (% total population)	76	80	77
Poverty (% population below US$1 per day)	..	..	..
GNI per capita, Atlas method (current US$)	35,660	52,030	32,040
GDP growth, 1995–2000 and 2000–4 (%)	3.5	1.7	2.0
Adult literacy rate (% ages 15 and over)	..	..	..
Primary, secondary, tertiary school enrollment (% gross)	98	*101*	*93*
ICT sector structure			
Separate telecommunications regulator	Yes	Yes	
Status of main fixed-line operator	*Public*	*Mixed*	
Level of competition: international long distance	*C*	C	
Level of competition: mobile	*P*	P	
Level of competition: Internet service provider	*C*	C	
Government prioritization of ICT (scale 1–7)	..	5.2	4.9
ICT sector performance			
Access			
Telephone main lines (per 1,000 people)	735	670	558
International voice traffic (minutes per person)[a]	213	..	..
Mobile subscribers (per 1,000 people)	750	863	767
Population covered by mobile telephony (%)	96	*99*	98
Internet users (per 1,000 people)	434	546	*480*
Personal computers (per 1,000 people)	490	567	*504*
Households with television (%)	99	*100*	*98*
Quality			
Telephone faults (per 100 main lines per year)	..	..	..
Broadband subscribers (per 1,000 people)	5.2	*87.4*	126.2
International Internet bandwidth (bits per person)	875	9,389	4,718
Affordability			
Price basket for fixed line (US$ per month, residential)	25.4	*29.9*	25.8
Price basket for mobile (US$ per month)	..	6.4	17.8
Price basket for Internet (US$ per month)	..	*26.3*	*20.9*
Price of call to United States (US$ per 3 minutes)	0.40	*0.31*	0.77
Institutional efficiency and sustainability			
Total telecommunications revenue (% GDP)	1.5	1.8	2.9
Total telephone subscribers per employee	298	*478*	*485*
Total telecommunications investment (% revenue)	82.4	*9.9*	*12.3*
ICT applications			
ICT expenditure (% GDP)	5.7	*5.1*	*7.1*
E-government readiness index (scale 0–1)	..	0.69	0.70
Secure Internet servers (per 1 million people)	*81.8*	246.6	311.4
Schools connected to the Internet (%)	..	*99*	*99*

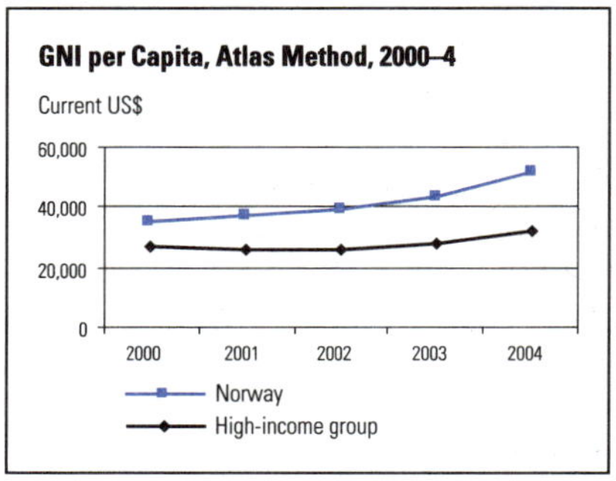

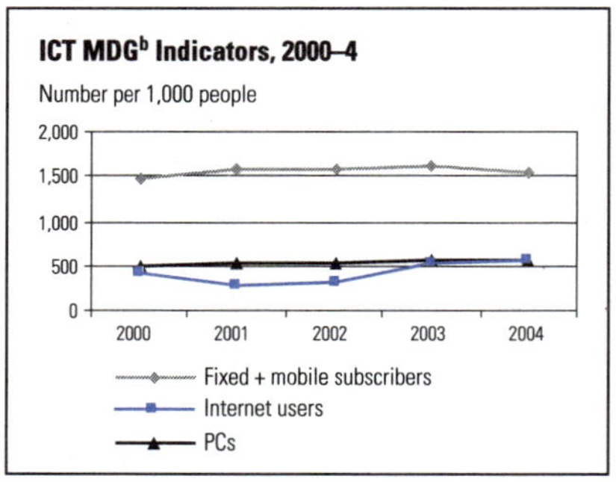

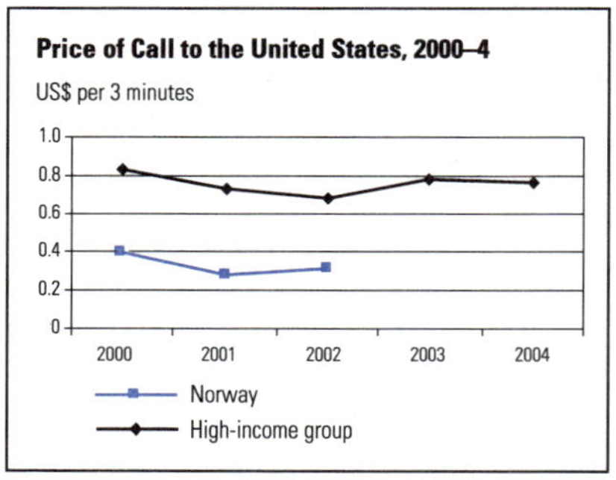

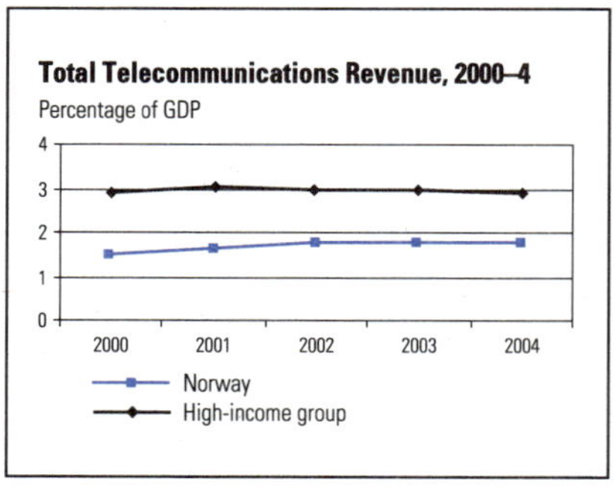

Notes: Figures in italics are for years other than those specified. .. indicates data are not available. C = competition; GDP = gross domestic product; GNI = gross national income; ICT = information and communication technology; M = monopoly; MDG = Millennium Development Goal; P = partial competition; and PCs = personal computers.

a. Outgoing and incoming.

b. Millennium Development Goal indicators 47, 48a, and 48b.

Sources: Economic and social context: UNDP and World Bank; ICT sector structure: ITU, WEF; ICT sector performance: Global Insight/WITSA, ITU, Netcraft, UNDESA, UNPAN, and World Bank. Produced by the Global Information and Communication Technologies Department and the Development Economics Data Group. For complete information, see Definitions and Data Sources.

Oman

	Oman 2000	Oman 2004	Upper-middle-Income group 2004	Middle East & North Africa Region 2004
Economic and social context				
Population, total (millions)	2	3	576	294
Urban population (% total population)	76	78	72	56
Poverty (% population below US$1 per day)	..	..	..	*2.4*
GNI per capita, Atlas method (current US$)	6,710	*7,890*	4,770	2,000
GDP growth, 1995–2000 and 2000–4 (%)	3.2	*3.5*	2.7	4.5
Adult literacy rate (% ages 15 and over)	..	*74*	94	..
Primary, secondary, tertiary school enrollment (% gross)	63	*63*	*80*	*68*
ICT sector structure				
Separate telecommunications regulator	No	Yes		
Status of main fixed-line operator	*Public*	Public		
Level of competition: international long distance	*M*	M		
Level of competition: mobile	*M*	M		
Level of competition: Internet service provider	..	M		
Government prioritization of ICT (scale 1–7)	..	..	4.1	..
ICT sector performance				
Access				
Telephone main lines (per 1,000 people)	92	*91*	*220*	*118*
International voice traffic (minutes per person)[a]	*108*	..	*39*	*20*
Mobile subscribers (per 1,000 people)	68	*228*	490	*88*
Population covered by mobile telephony (%)	..	..	*84*	..
Internet users (per 1,000 people)	37	*71*	*133*	*47*
Personal computers (per 1,000 people)	33	*37*	*99*	*30*
Households with television (%)	79	*79*	*92*	*88*
Quality				
Telephone faults (per 100 main lines per year)	..	*19.3*	20.3	..
Broadband subscribers (per 1,000 people)	*0.0*	*0.0*	*3.7*	*0.2*
International Internet bandwidth (bits per person)	..	*15*	176	*15*
Affordability				
Price basket for fixed line (US$ per month, residential)	13.1	*12.9*	*13.9*	*4.9*
Price basket for mobile (US$ per month)	..	5.1	11.1	8.1
Price basket for Internet (US$ per month)	..	*23.6*	*20.8*	*24.5*
Price of call to United States (US$ per 3 minutes)	7.89	1.87	1.03	1.64
Institutional efficiency and sustainability				
Total telecommunications revenue (% GDP)	1.9	*2.5*	3.4	2.8
Total telephone subscribers per employee	187	*322*	*402*	..
Total telecommunications investment (% revenue)	19.0	*25.0*	*18.6*	*27.8*
ICT applications				
ICT expenditure (% GDP)	..	..	*5.0*	..
E-government readiness index (scale 0–1)	..	0.05	0.49	0.16
Secure Internet servers (per 1 million people)	*0.8*	2.3	10.7	0.6
Schools connected to the Internet (%)	..	..	*60*	..

GNI per Capita, Atlas Method, 2000–4

Current US$

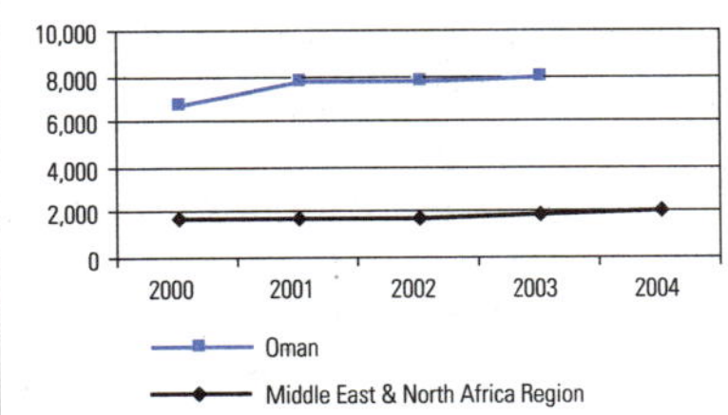

ICT MDG[b] Indicators, 2000–4

Number per 1,000 people

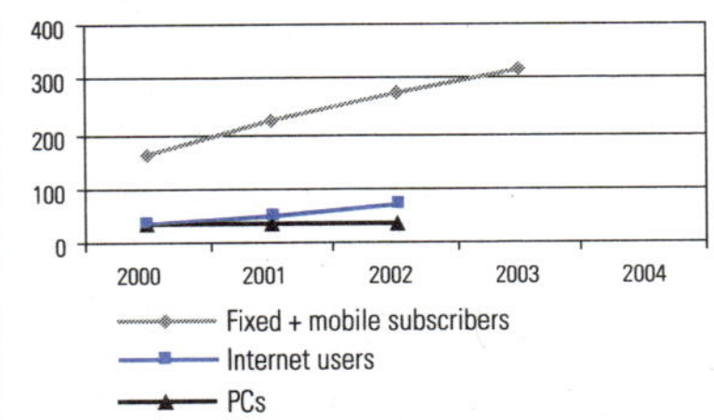

Price of Call to the United States, 2000–4

US$ per 3 minutes

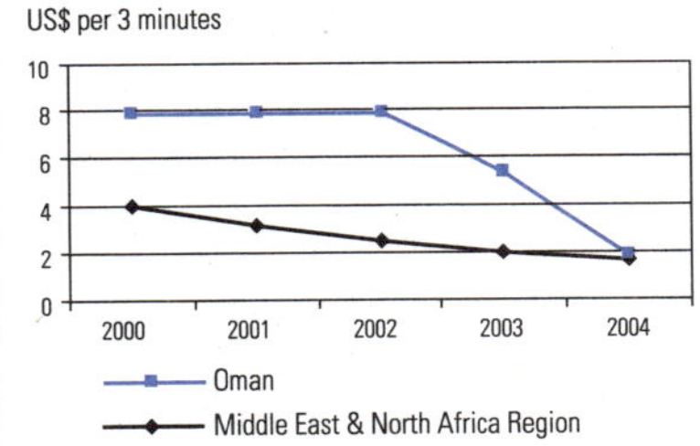

Total Telecommunications Revenue, 2000–4

Percentage of GDP

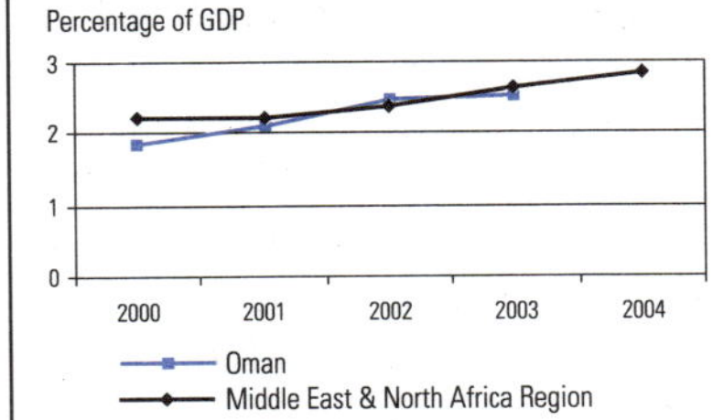

Notes: Figures in italics are for years other than those specified. .. indicates data are not available. C = competition; GDP = gross domestic product; GNI = gross national income; ICT = information and communication technology; M = monopoly; MDG = Millennium Development Goal; P = partial competition; and PCs = personal computers.

a. Outgoing and incoming.

b. Millennium Development Goal indicators 47, 48a, and 48b.

Sources: Economic and social context: UNDP and World Bank; ICT sector structure: ITU, WEF; ICT sector performance: Global Insight/WITSA, ITU, Netcraft, UNDESA, UNPAN, and World Bank. Produced by the Global Information and Communication Technologies Department and the Development Economics Data Group. For complete information, see Definitions and Data Sources.

Pakistan

	Pakistan 2000	Pakistan 2004	Low-income group 2004	South Asia Region 2004
Economic and social context				
Population, total (millions)	138	152	2,338	1,448
Urban population (% total population)	33	34	31	28
Poverty (% population below US$1 per day)	*13.4*	..	..	*31.3*
GNI per capita, Atlas method (current US$)	480	600	510	590
GDP growth, 1995–2000 and 2000–4 (%)	3.0	4.1	5.4	5.8
Adult literacy rate (% ages 15 and over)	..	49	61	58
Primary, secondary, tertiary school enrollment (% gross)	37	*35*	*55*	*56*
ICT sector structure				
Separate telecommunications regulator	Yes	Yes		
Status of main fixed-line operator	*Mixed*	Mixed		
Level of competition: international long distance	*M*	C		
Level of competition: mobile	*P*	C		
Level of competition: Internet service provider	*C*	C		
Government prioritization of ICT (scale 1–7)	..	5.7	..	5.3
ICT sector performance				
Access				
Telephone main lines (per 1,000 people)	22	32	33	41
International voice traffic (minutes per person)[a]	7	*11*	*4*	*4*
Mobile subscribers (per 1,000 people)	3	52	48	47
Population covered by mobile telephony (%)	..	*45*	*43*	*43*
Internet users (per 1,000 people)	2	13	20	21
Personal computers (per 1,000 people)	4	*5*	*8*	11
Households with television (%)	37	*39*	*16*	*32*
Quality				
Telephone faults (per 100 main lines per year)	..	..	..	*88.1*
Broadband subscribers (per 1,000 people)	0.0	0.0	0.5	0.6
International Internet bandwidth (bits per person)	0	*4*	*3*	4
Affordability				
Price basket for fixed line (US$ per month, residential)	..	*6.1*	*6.6*	3.2
Price basket for mobile (US$ per month)	..	2.9	11.6	3.2
Price basket for Internet (US$ per month)	..	*15.6*	*45.5*	*15.1*
Price of call to United States (US$ per 3 minutes)	3.60	1.03	1.95	1.21
Institutional efficiency and sustainability				
Total telecommunications revenue (% GDP)	1.8	2.1	2.3	1.9
Total telephone subscribers per employee	61	*89*	*89*	*89*
Total telecommunications investment (% revenue)	19.8	11.7	*27.8*	*15.3*
ICT applications				
ICT expenditure (% GDP)	6.6	*7.3*	*4.1*	*4.1*
E-government readiness index (scale 0–1)	..	0.48	0.12	0.34
Secure Internet servers (per 1 million people)	*0.0*	0.2	0.3	0.4
Schools connected to the Internet (%)	..	..	..	..

GNI per Capita, Atlas Method, 2000–4

Current US$

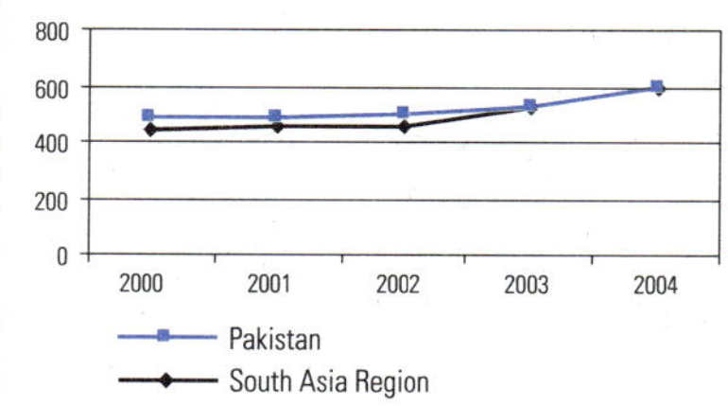

ICT MDG[b] Indicators, 2000–4

Number per 1,000 people

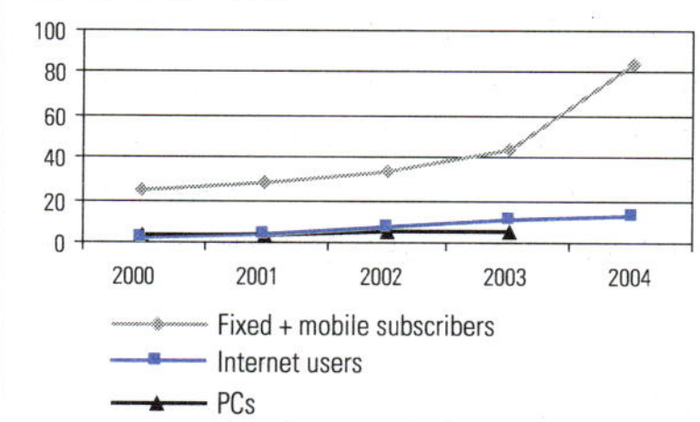

Price of Call to the United States, 2000–4

US$ per 3 minutes

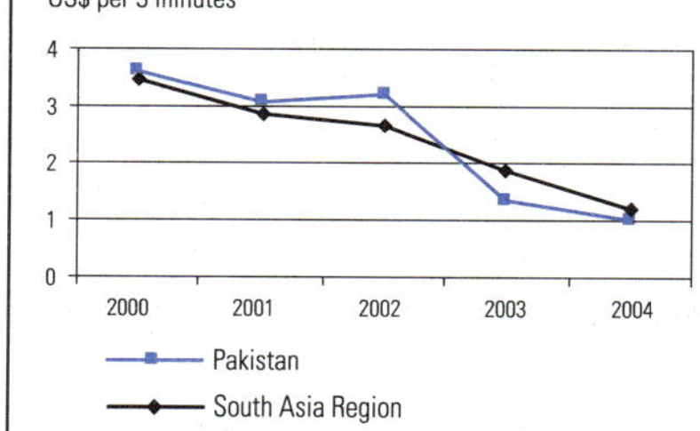

Total Telecommunications Revenue, 2000–4

Percentage of GDP

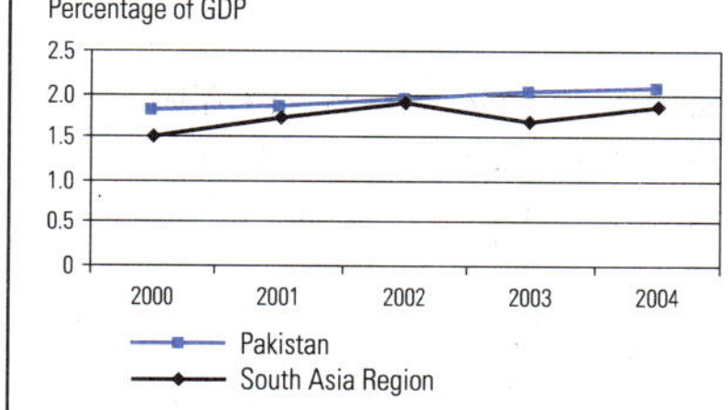

Notes: Figures in italics are for years other than those specified. .. indicates data are not available. C = competition; GDP = gross domestic product; GNI = gross national income; ICT = information and communication technology; M = monopoly; MDG = Millennium Development Goal; P = partial competition; and PCs = personal computers.

a. Outgoing and incoming.

b. Millennium Development Goal indicators 47, 48a, and 48b.

Sources: Economic and social context: UNDP and World Bank; ICT sector structure: ITU, WEF; ICT sector performance: Global Insight/WITSA, ITU, Netcraft, UNDESA, UNPAN, and World Bank. Produced by the Global Information and Communication Technologies Department and the Development Economics Data Group. For complete information, see Definitions and Data Sources.

Panama

	Panama 2000	Panama 2004	Upper-middle-income group 2004	Latin America & the Caribbean Region 2004
Economic and social context				
Population, total (millions)	3	3	576	541
Urban population (% total population)	56	57	72	77
Poverty (% population below US$1 per day)	7.2	..	..	*9.5*
GNI per capita, Atlas method (current US$)	3,870	4,450	4,770	3,600
GDP growth, 1995–2000 and 2000–4 (%)	5.0	3.3	2.7	1.5
Adult literacy rate (% ages 15 and over)	92	..	94	89
Primary, secondary, tertiary school enrollment (% gross)	77	*79*	*80*	*82*
ICT sector structure				
Separate telecommunications regulator	Yes	Yes		
Status of main fixed-line operator	*Mixed*	Mixed		
Level of competition: international long distance	*M*	C		
Level of competition: mobile	*P*	P		
Level of competition: Internet service provider	*C*	C		
Government prioritization of ICT (scale 1–7)	..	3.1	4.1	3.5
ICT sector performance				
Access				
Telephone main lines (per 1,000 people)	150	124	*220*	181
International voice traffic (minutes per person)[a]	57	..	*39*	..
Mobile subscribers (per 1,000 people)	144	283	490	324
Population covered by mobile telephony (%)	74	87	*84*	*76*
Internet users (per 1,000 people)	32	78	*133*	*104*
Personal computers (per 1,000 people)	37	*40*	*99*	*75*
Households with television (%)	77	*77*	*92*	*88*
Quality				
Telephone faults (per 100 main lines per year)	48.0	13.9	*20.3*	..
Broadband subscribers (per 1,000 people)	0.0	6.0	*3.7*	*5.2*
International Internet bandwidth (bits per person)	5	*5*	176	165
Affordability				
Price basket for fixed line (US$ per month, residential)	10.9	..	*13.9*	*9.0*
Price basket for mobile (US$ per month)	..	18.1	11.1	9.1
Price basket for Internet (US$ per month)	..	*36.0*	*20.8*	*31.5*
Price of call to United States (US$ per 3 minutes)	4.36	*3.64*	1.03	0.90
Institutional efficiency and sustainability				
Total telecommunications revenue (% GDP)	3.8	4.7	3.4	3.3
Total telephone subscribers per employee	153	188	*402*	..
Total telecommunications investment (% revenue)	..	..	*18.6*	..
ICT applications				
ICT expenditure (% GDP)	9.0	*9.2*	*5.0*	*5.3*
E-government readiness index (scale 0–1)	..	0.52	0.49	0.39
Secure Internet servers (per 1 million people)	*10.0*	49.2	10.7	8.6
Schools connected to the Internet (%)	..	..	*60*	..

GNI per Capita, Atlas Method, 2000–4

Current US$

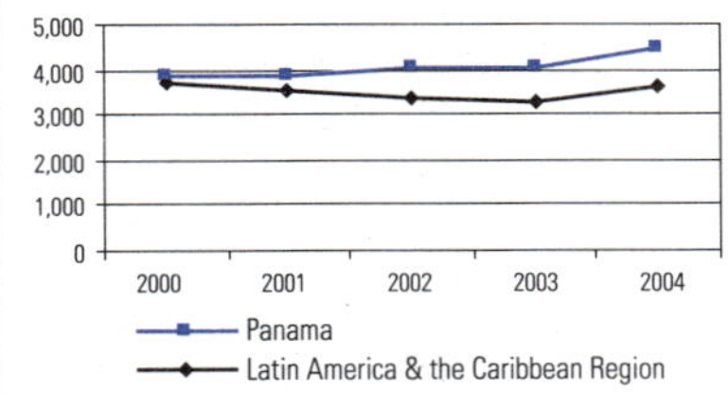

ICT MDG[b] Indicators, 2000–4

Number per 1,000 people

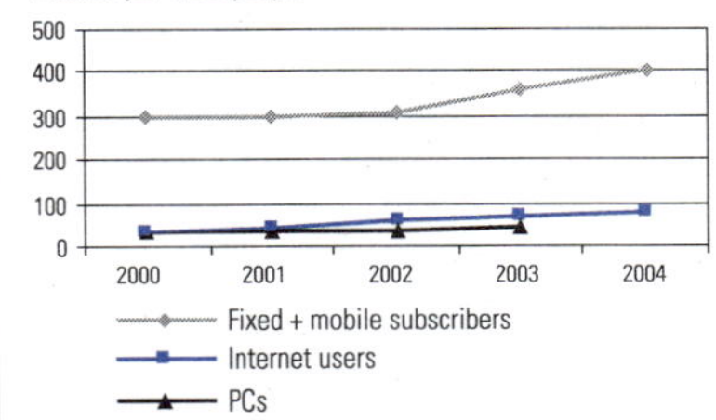

Price of Call to the United States, 2000–4

US$ per 3 minutes

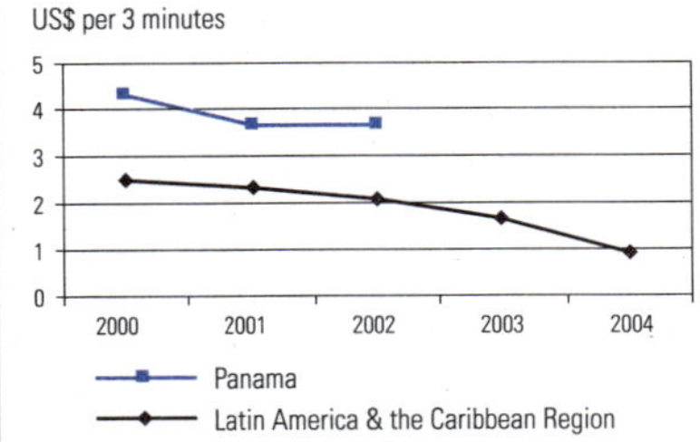

Total Telecommunications Revenue, 2000–4

Percentage of GDP

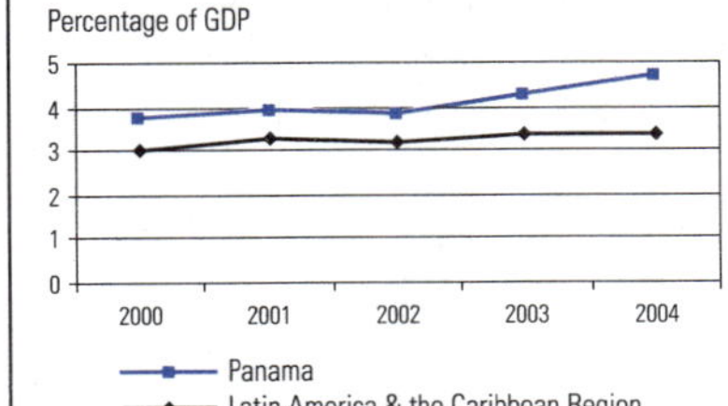

Notes: Figures in italics are for years other than those specified. .. indicates data are not available. C = competition; GDP = gross domestic product; GNI = gross national income; ICT = information and communication technology; M = monopoly; MDG = Millennium Development Goal; P = partial competition; and PCs = personal computers.

a. Outgoing and incoming.

b. Millennium Development Goal indicators 47, 48a, and 48b.

Sources: Economic and social context: UNDP and World Bank; ICT sector structure: ITU, WEF; ICT sector performance: Global Insight/WITSA, ITU, Netcraft, UNDESA, UNPAN, and World Bank. Produced by the Global Information and Communication Technologies Department and the Development Economics Data Group. For complete information, see Definitions and Data Sources.

Papua New Guinea

	Papua New Guinea 2000	Papua New Guinea 2004	Low-income group 2004	East Asia & Pacific Region 2004
Economic and social context				
Population, total (millions)	5	6	2,338	1,870
Urban population (% total population)	13	13	31	41
Poverty (% population below US$1 per day)	..	..	..	*14.9*
GNI per capita, Atlas method (current US$)	670	580	510	1,280
GDP growth, 1995–2000 and 2000–4 (%)	0.7	0.6	5.4	7.5
Adult literacy rate (% ages 15 and over)	57	..	61	90
Primary, secondary, tertiary school enrollment (% gross)	41	*41*	*55*	*68*
ICT sector structure				
Separate telecommunications regulator	Yes	Yes		
Status of main fixed-line operator	*Public*	Public		
Level of competition: international long distance	*M*	M		
Level of competition: mobile	*M*	M		
Level of competition: Internet service provider	..	P		
Government prioritization of ICT (scale 1–7)	..	..	..	4.8
ICT sector performance				
Access				
Telephone main lines (per 1,000 people)	13	12	33	194
International voice traffic (minutes per person)[a]	8	..	*4*	..
Mobile subscribers (per 1,000 people)	2	7	48	248
Population covered by mobile telephony (%)	..	..	*43*	*73*
Internet users (per 1,000 people)	9	*18.2*	20	75
Personal computers (per 1,000 people)	55	*60*	*8*	37
Households with television (%)	8	..	*16*	*80*
Quality				
Telephone faults (per 100 main lines per year)	..	..	..	..
Broadband subscribers (per 1,000 people)	*0.0*	*0.0*	0.5	13.4
International Internet bandwidth (bits per person)	1	*1*	*3*	52
Affordability				
Price basket for fixed line (US$ per month, residential)	20.1	*6.6*	*6.6*	*4.5*
Price basket for mobile (US$ per month)	..	8.4	11.6	5.1
Price basket for Internet (US$ per month)	..	*20.0*	*45.5*	*19.9*
Price of call to United States (US$ per 3 minutes)	4.32	..	1.95	1.20
Institutional efficiency and sustainability				
Total telecommunications revenue (% GDP)	2.3	..	2.3	3.6
Total telephone subscribers per employee	41	..	*89*	..
Total telecommunications investment (% revenue)	82.6	..	*27.8*	*31.0*
ICT applications				
ICT expenditure (% GDP)	..	..	*4.1*	*5.0*
E-government readiness index (scale 0–1)	..	0.12	0.12	0.21
Secure Internet servers (per 1 million people)	..	0.2	0.3	0.6
Schools connected to the Internet (%)	..	..	..	..

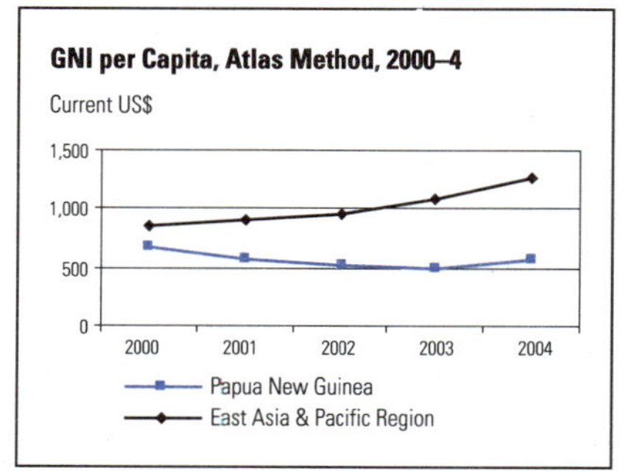

GNI per Capita, Atlas Method, 2000–4
Current US$

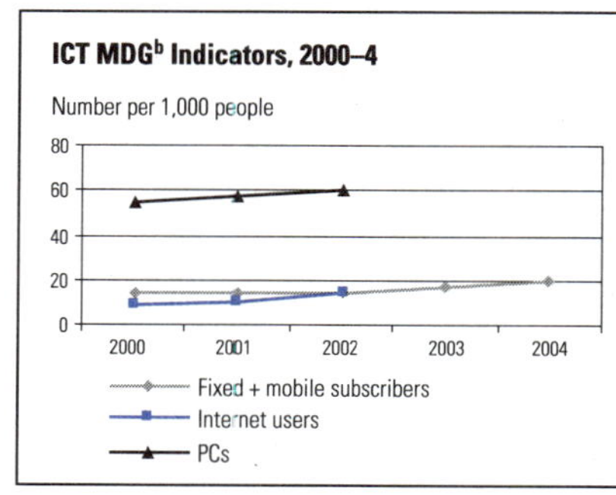

ICT MDG[b] Indicators, 2000–4
Number per 1,000 people

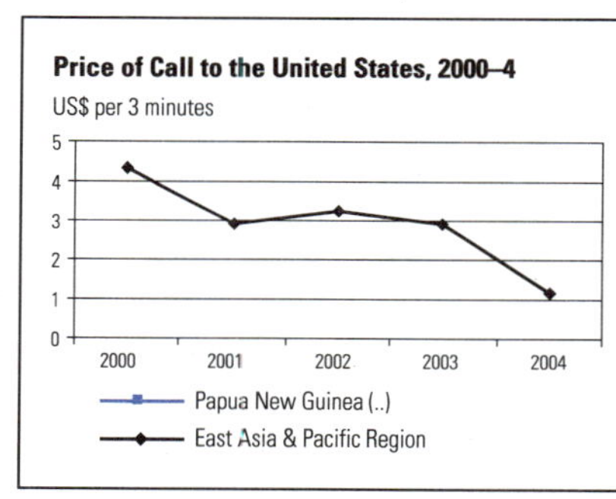

Price of Call to the United States, 2000–4
US$ per 3 minutes

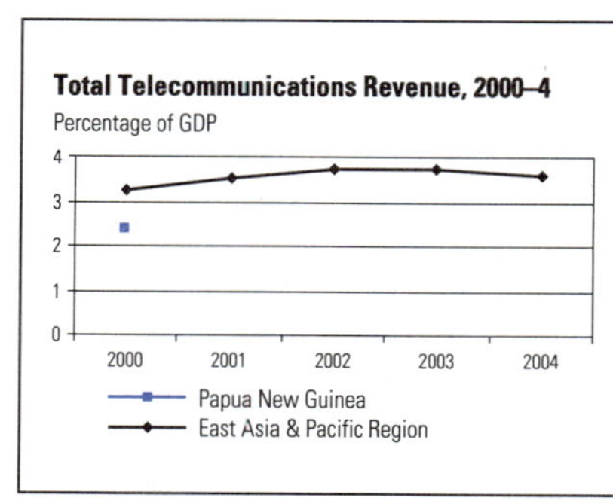

Total Telecommunications Revenue, 2000–4
Percentage of GDP

Notes: Figures in italics are for years other than those specified. .. indicates data are not available. C = competition; GDP = gross domestic product; GNI = gross national income; ICT = information and communication technology; M = monopoly; MDG = Millennium Development Goal; P = partial competition; and PCs = personal computers.

a. Outgoing and incoming.

b. Millennium Development Goal indicators 47, 48a, and 48b.

Sources: Economic and social context: UNDP and World Bank; ICT sector structure: ITU, WEF; ICT sector performance: Global Insight/WITSA, ITU, Netcraft, UNDESA, UNPAN, and World Bank. Produced by the Global Information and Communication Technologies Department and the Development Economics Data Group. For complete information, see Definitions and Data Sources.

Paraguay

	Paraguay 2000	Paraguay 2004	Lower-middle-income group 2004	Latin America & the Caribbean Region 2004
Economic and social context				
Population, total (millions)	5	6	2,430	541
Urban population (% total population)	55	58	49	77
Poverty (% population below US$1 per day)	*14.9*	*16.4*	..	*9.5*
GNI per capita, Atlas method (current US$)	1,510	1,170	1,580	3,600
GDP growth, 1995–2000 and 2000–4 (%)	0.7	1.2	5.7	1.5
Adult literacy rate (% ages 15 and over)	*92*	..	90	89
Primary, secondary, tertiary school enrollment (% gross)	71	*73*	*72*	*82*
ICT sector structure				
Separate telecommunications regulator	Yes	Yes		
Status of main fixed-line operator	*Public*	Public		
Level of competition: international long distance	*M*	M		
Level of competition: mobile	*C*	C		
Level of competition: Internet service provider	*C*	C		
Government prioritization of ICT (scale 1–7)	..	2.6	3.8	3.5
ICT sector performance				
Access				
Telephone main lines (per 1,000 people)	54	52	192	181
International voice traffic (minutes per person)[a]	20	..	..	..
Mobile subscribers (per 1,000 people)	156	208	255	324
Population covered by mobile telephony (%)	..	60	*76*	*76*
Internet users (per 1,000 people)	8	24	70	*104*
Personal computers (per 1,000 people)	13	*36*	38	*75*
Households with television (%)	75	..	*89*	*88*
Quality				
Telephone faults (per 100 main lines per year)	4.1	..	..	..
Broadband subscribers (per 1,000 people)	0.0	*0.1*	12.6	*5.2*
International Internet bandwidth (bits per person)	2	*18*	58	165
Affordability				
Price basket for fixed line (US$ per month, residential)	18.0	..	*5.5*	*9.0*
Price basket for mobile (US$ per month)	..	7.3	8.9	9.1
Price basket for Internet (US$ per month)	..	*36.3*	*25.3*	*31.5*
Price of call to United States (US$ per 3 minutes)	0.97	0.90	1.45	0.90
Institutional efficiency and sustainability				
Total telecommunications revenue (% GDP)	4.2	4.1	3.6	3.3
Total telephone subscribers per employee	*123*	..	*195*	..
Total telecommunications investment (% revenue)	26.3	..	*25.5*	..
ICT applications				
ICT expenditure (% GDP)	..	..	*5.1*	*5.3*
E-government readiness index (scale 0–1)	..	0.11	0.26	0.39
Secure Internet servers (per 1 million people)	*0.7*	1.6	1.6	8.6
Schools connected to the Internet (%)	..	..	..	..

GNI per Capita, Atlas Method, 2000–4

Current US$

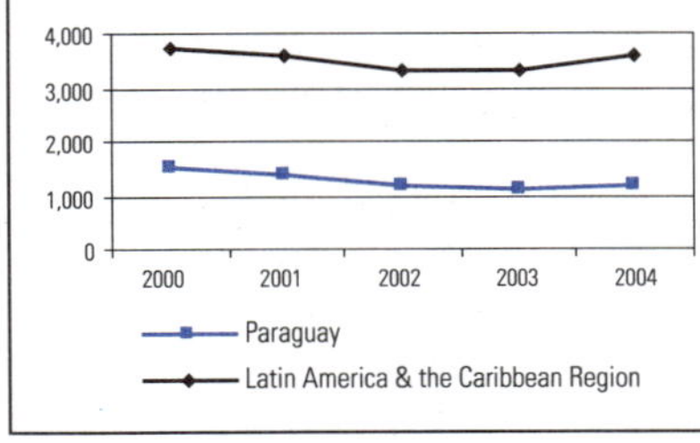

ICT MDG[b] Indicators, 2000–4

Number per 1,000 people

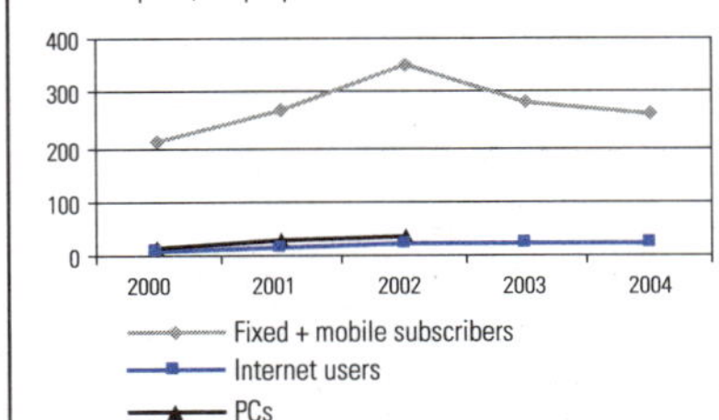

Price of Call to the United States, 2000–4

US$ per 3 minutes

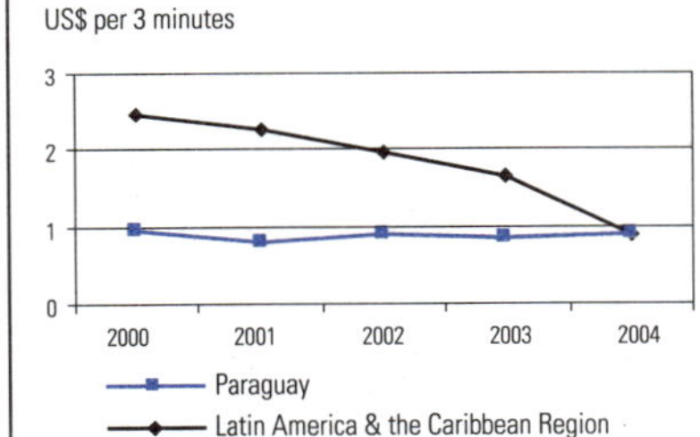

Total Telecommunications Revenue, 2000–4

Percentage of GDP

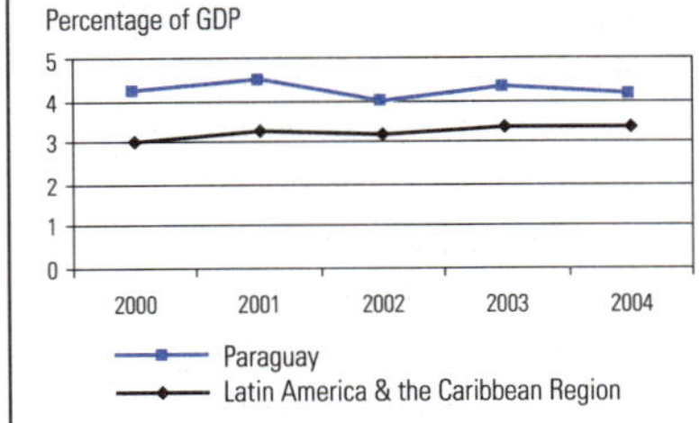

Notes: Figures in italics are for years other than those specified. .. indicates data are not available. C = competition; GDP = gross domestic product; GNI = gross national income; ICT = information and communication technology; M = monopoly; MDG = Millennium Development Goal; P = partial competition; and PCs = personal computers.

a. Outgoing and incoming.

b. Millennium Development Goal indicators 47, 48a, and 48b.

Sources: Economic and social context: UNDP and World Bank; ICT sector structure: ITU, WEF; ICT sector performance: Global Insight/WITSA, ITU, Netcraft, UNDESA, UNPAN, and World Bank. Produced by the Global Information and Communication Technologies Department and the Development Economics Data Group. For complete information, see Definitions and Data Sources.

Peru

	Peru 2000	Peru 2004	Lower-middle-income group 2004	Latin America & the Caribbean Region 2004
Economic and social context				
Population, total (millions)	26	28	2,430	541
Urban population (% total population)	73	74	49	77
Poverty (% population below US$1 per day)	18.1	..	..	*9.5*
GNI per capita, Atlas method (current US$)	2,050	2,360	1,580	3,600
GDP growth, 1995–2000 and 2000–4 (%)	2.3	3.6	5.7	1.5
Adult literacy rate (% ages 15 and over)	..	88	90	89
Primary, secondary, tertiary school enrollment (% gross)	*88*	*87*	*72*	*82*
ICT sector structure				
Separate telecommunications regulator	Yes	Yes		
Status of main fixed-line operator	*Private*	Private		
Level of competition: international long distance	*C*	C		
Level of competition: mobile	*C*	C		
Level of competition: Internet service provider	*C*	C		
Government prioritization of ICT (scale 1–7)	..	2.8	3.8	3.5
ICT sector performance				
Access				
Telephone main lines (per 1,000 people)	66	74	192	181
International voice traffic (minutes per person)[a]	17	*46*	..	..
Mobile subscribers (per 1,000 people)	49	149	255	324
Population covered by mobile telephony (%)	..	*75*	*76*	*76*
Internet users (per 1,000 people)	31	*105*	70	*104*
Personal computers (per 1,000 people)	40	*52*	38	*75*
Households with television (%)	67	..	*89*	*88*
Quality				
Telephone faults (per 100 main lines per year)	*17.1*	..	..	..
Broadband subscribers (per 1,000 people)	0.0	7.6	12.6	*5.2*
International Internet bandwidth (bits per person)	4	*55*	58	165
Affordability				
Price basket for fixed line (US$ per month, residential)	18.9	19.4	*5.5*	*9.0*
Price basket for mobile (US$ per month)	..	21.9	8.9	9.1
Price basket for Internet (US$ per month)	..	*32.8*	*25.3*	*31.5*
Price of call to United States (US$ per 3 minutes)	2.08	1.80	1.45	0.90
Institutional efficiency and sustainability				
Total telecommunications revenue (% GDP)	2.7	3.0	3.6	3.3
Total telephone subscribers per employee	317	..	*195*	..
Total telecommunications investment (% revenue)	22.2	*19.2*	*25.5*	..
ICT applications				
ICT expenditure (% GDP)	6.9	*6.9*	*5.1*	*5.3*
E-government readiness index (scale 0–1)	..	0.52	0.26	0.39
Secure Internet servers (per 1 million people)	*1.3*	4.7	1.6	8.6
Schools connected to the Internet (%)	*3*	..	..	..

GNI per Capita, Atlas Method, 2000–4

Current US$

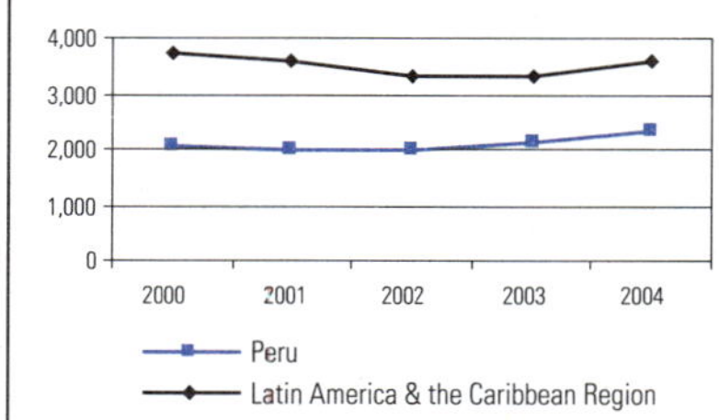

ICT MDG[b] Indicators, 2000–4

Number per 1,000 people

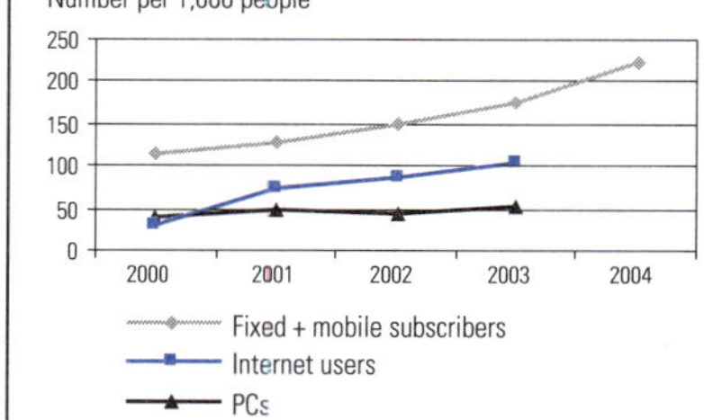

Price of Call to the United States, 2000–4

US$ per 3 minutes

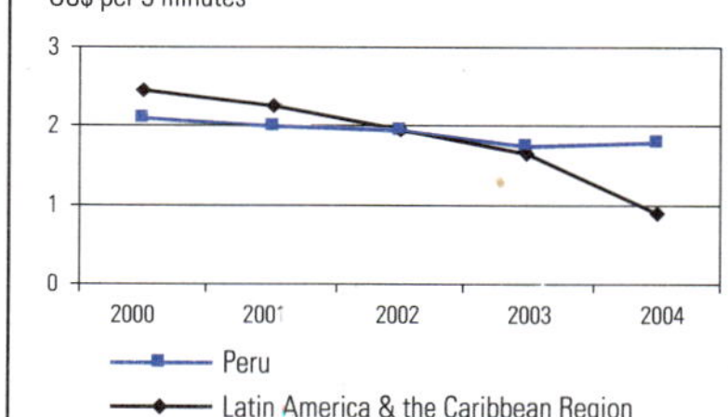

Total Telecommunications Revenue, 2000–4

Percentage of GDP

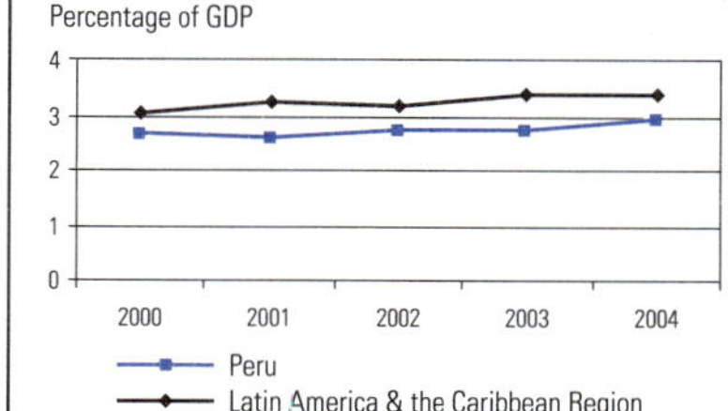

Notes: Figures in italics are for years other than those specified. .. indicates data are not available. C = competition; GDP = gross domestic product; GNI = gross national income; ICT = information and communication technology; M = monopoly; MDG = Millennium Development Goal; P = partial competition; and PCs = personal computers.

a. Outgoing and incoming.

b. Millennium Development Goal indicators 47, 48a, and 48b.

Sources: Economic and social context: UNDP and World Bank; ICT sector structure: ITU, WEF; ICT sector performance: Global Insight/WITSA, ITU, Netcraft, UNDESA, UNPAN, and World Bank. Produced by the Global Information and Communication Technologies Department and the Development Economics Data Group. For complete information, see Definitions and Data Sources.

Philippines

	Philippines 2000	Philippines 2004	Lower-middle-income group 2004	East Asia & Pacific Region 2004
Economic and social context				
Population, total (millions)	77	83	2,430	1,870
Urban population (% total population)	59	62	49	41
Poverty (% population below US$1 per day)	15.5	..	..	*14.9*
GNI per capita, Atlas method (current US$)	1,030	1,170	1,580	1,280
GDP growth, 1995–2000 and 2000–4 (%)	3.5	4.2	5.7	7.5
Adult literacy rate (% ages 15 and over)	93	..	90	90
Primary, secondary, tertiary school enrollment (% gross)	80	*82*	*72*	*68*
ICT sector structure				
Separate telecommunications regulator	Yes	Yes		
Status of main fixed-line operator	*Private*	Private		
Level of competition: international long distance	*C*	C		
Level of competition: mobile	*C*	C		
Level of competition: Internet service provider	*C*	C		
Government prioritization of ICT (scale 1–7)	..	4.5	3.8	4.8
ICT sector performance				
Access				
Telephone main lines (per 1,000 people)	40	41	192	194
International voice traffic (minutes per person)[a]	28	*35*	..	..
Mobile subscribers (per 1,000 people)	84	387	255	248
Population covered by mobile telephony (%)	70	*80*	*76*	*73*
Internet users (per 1,000 people)	20	58	70	75
Personal computers (per 1,000 people)	19	29	38	37
Households with television (%)	53	*76*	*89*	*80*
Quality				
Telephone faults (per 100 main lines per year)	..	..	..	..
Broadband subscribers (per 1,000 people)	0.0	*0.3*	12.6	13.4
International Internet bandwidth (bits per person)	2	*12*	58	52
Affordability				
Price basket for fixed line (US$ per month, residential)	14.0	12.2	*5.5*	*4.5*
Price basket for mobile (US$ per month)	..	4.0	8.9	5.1
Price basket for Internet (US$ per month)	..	*17.1*	*25.3*	*19.9*
Price of call to United States (US$ per 3 minutes)	2.07	1.20	1.45	1.20
Institutional efficiency and sustainability				
Total telecommunications revenue (% GDP)	2.9	3.7	3.6	3.6
Total telephone subscribers per employee	476	..	*195*	..
Total telecommunications investment (% revenue)	47.4	*25.5*	*25.5*	*31.0*
ICT applications				
ICT expenditure (% GDP)	4.5	*5.9*	*5.1*	*5.0*
E-government readiness index (scale 0–1)	..	0.59	0.26	0.21
Secure Internet servers (per 1 million people)	*0.9*	1.9	1.6	0.6
Schools connected to the Internet (%)	*2*	..	..	..

GNI per Capita, Atlas Method, 2000–4

Current US$

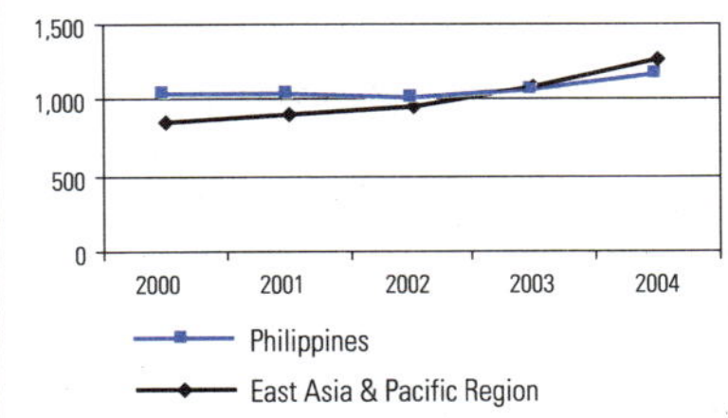

ICT MDG[b] Indicators, 2000–4

Number per 1,000 people

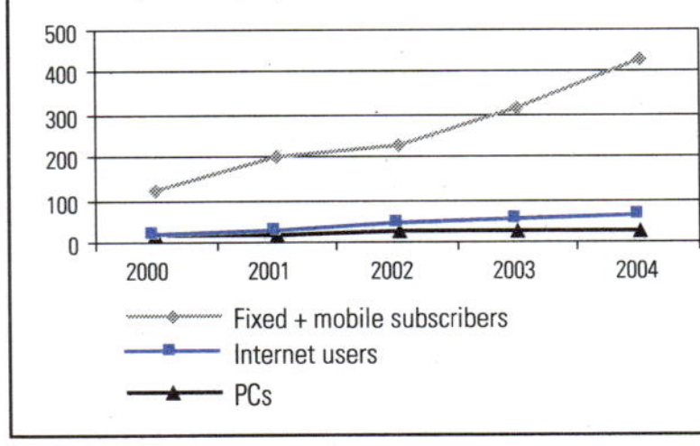

Price of Call to the United States, 2000–4

US$ per 3 minutes

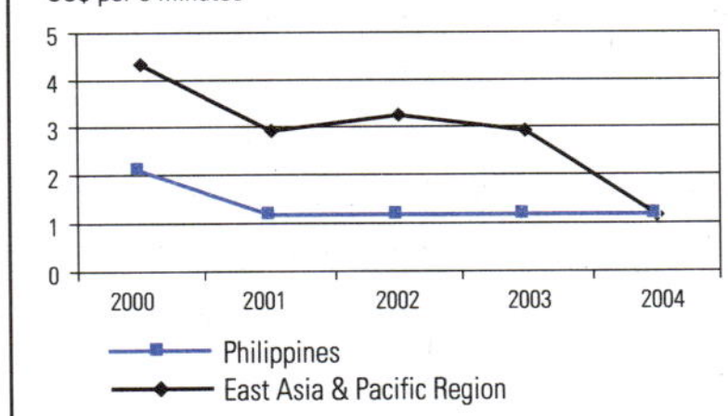

Total Telecommunications Revenue, 2000–4

Percentage of GDP

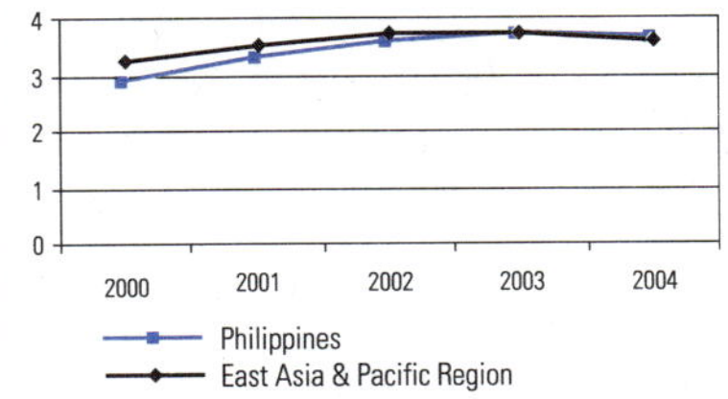

Notes: Figures in italics are for years other than those specified. .. indicates data are not available. C = competition; GDP = gross domestic product; GNI = gross national income; ICT = information and communication technology; M = monopoly; MDG = Millennium Development Goal; P = partial competition; and PCs = personal computers.

a. Outgoing and incoming.

b. Millennium Development Goal indicators 47, 48a, and 48b.

Sources: Economic and social context: UNDP and World Bank; ICT sector structure: ITU, WEF; ICT sector performance: Global Insight/WITSA, ITU, Netcraft, UNDESA, UNPAN, and World Bank. Produced by the Global Information and Communication Technologies Department and the Development Economics Data Group. For complete information, see Definitions and Data Sources.

Poland

	Poland 2000	Poland 2004	Upper-middle-income group 2004	Europe & Central Asia Region 2004
Economic and social context				
Population, total (millions)	39	38	576	472
Urban population (% total population)	62	62	72	64
Poverty (% population below US$1 per day)	*<2*	*<2*	..	*3.6*
GNI per capita, Atlas method (current US$)	4,430	6,090	4,770	3,290
GDP growth, 1995–2000 and 2000–4 (%)	5.1	2.8	2.7	5.0
Adult literacy rate (% ages 15 and over)	..	..	94	97
Primary, secondary, tertiary school enrollment (% gross)	88	*90*	*80*	*81*
ICT sector structure				
Separate telecommunications regulator	Yes	Yes		
Status of main fixed-line operator	*Mixed*	Mixed		
Level of competition: international long distance	*M*	C		
Level of competition: mobile	*P*	C		
Level of competition: Internet service provider	*C*	C		
Government prioritization of ICT (scale 1–7)	..	2.9	4.1	3.8
ICT sector performance				
Access				
Telephone main lines (per 1,000 people)	283	*321*	*220*	*243*
International voice traffic (minutes per person)[a]	50	..	*39*	*28*
Mobile subscribers (per 1,000 people)	175	603	490	487
Population covered by mobile telephony (%)	95	98	*84*	*82*
Internet users (per 1,000 people)	72	*235*	*133*	*115*
Personal computers (per 1,000 people)	69	*127*	*99*	*73*
Households with television (%)	92	*92*	*92*	*92*
Quality				
Telephone faults (per 100 main lines per year)	*17.2*	..	*20.3*	*30.4*
Broadband subscribers (per 1,000 people)	0.0	32.7	*3.7*	*2.4*
International Internet bandwidth (bits per person)	20	*340*	176	148
Affordability				
Price basket for fixed line (US$ per month, residential)	14.0	*17.3*	*13.9*	*3.5*
Price basket for mobile (US$ per month)	..	7.7	11.1	10.3
Price basket for Internet (US$ per month)	..	*15.7*	*20.8*	*19.8*
Price of call to United States (US$ per 3 minutes)	2.92	0.99	1.03	1.06
Institutional efficiency and sustainability				
Total telecommunications revenue (% GDP)	3.2	3.5	3.4	3.5
Total telephone subscribers per employee	256	*602*	*402*	*150*
Total telecommunications investment (% revenue)	25.8	*18.8*	*18.6*	*19.0*
ICT applications				
ICT expenditure (% GDP)	4.2	*4.5*	*5.0*	*5.1*
E-government readiness index (scale 0–1)	..	0.58	0.49	0.39
Secure Internet servers (per 1 million people)	*8.5*	14.8	10.7	6.4
Schools connected to the Internet (%)	..	*90*	*60*	*65*

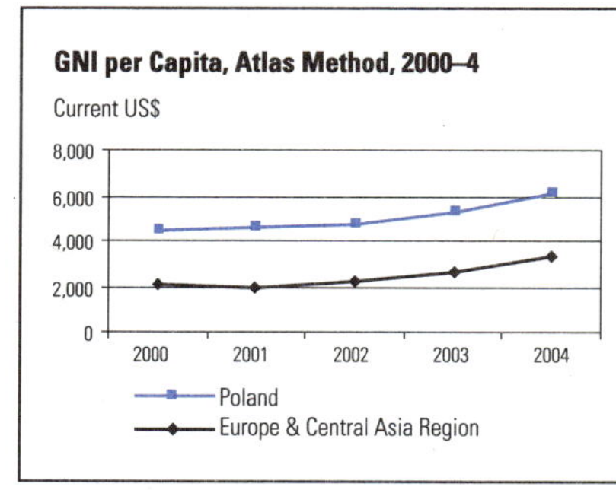
GNI per Capita, Atlas Method, 2000–4

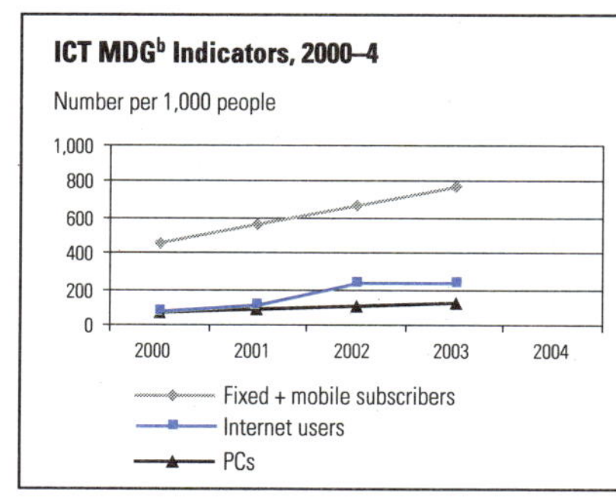
ICT MDG[b] Indicators, 2000–4

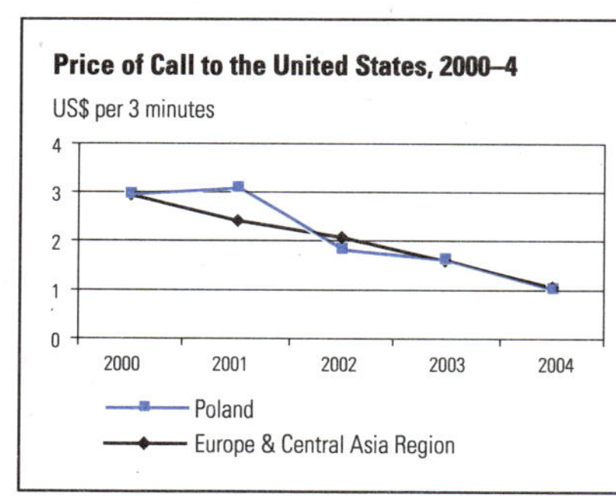
Price of Call to the United States, 2000–4

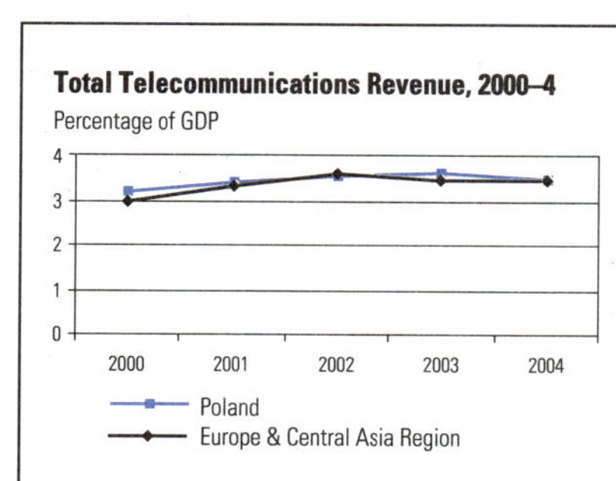
Total Telecommunications Revenue, 2000–4

Notes: Figures in italics are for years other than those specified. .. indicates data are not available. C = competition; GDP = gross domestic product; GNI = gross national income; ICT = information and communication technology; M = monopoly; MDG = Millennium Development Goal; P = partial competition; and PCs = personal computers.

a. Outgoing and incoming.

b. Millennium Development Goal indicators 47, 48a, and 48b.

Sources: Economic and social context: UNDP and World Bank; ICT sector structure: ITU, WEF; ICT sector performance: Global Insight/WITSA, ITU, Netcraft, UNDESA, UNPAN, and World Bank. Produced by the Global Information and Communication Technologies Department and the Development Economics Data Group. For complete information, see Definitions and Data Sources.

Portugal

	Portugal 2000	Portugal 2004	High-income group 2004
Economic and social context			
Population, total (millions)	10	10	1,001
Urban population (% total population)	53	55	77
Poverty (% population below US$1 per day)	..	..	..
GNI per capita, Atlas method (current US$)	10,930	14,350	32,040
GDP growth, 1995–2000 and 2000–4 (%)	3.9	0.3	2.0
Adult literacy rate (% ages 15 and over)	..	..	..
Primary, secondary, tertiary school enrollment (% gross)	93	*94*	*93*
ICT sector structure			
Separate telecommunications regulator	Yes	Yes	
Status of main fixed-line operator	*Mixed*	Mixed	
Level of competition: international long distance	*C*	C	
Level of competition: mobile	*P*	C	
Level of competition: Internet service provider	*C*	C	
Government prioritization of ICT (scale 1–7)	..	4.9	4.9
ICT sector performance			
Access			
Telephone main lines (per 1,000 people)	422	406	558
International voice traffic (minutes per person)[a]	138	195	..
Mobile subscribers (per 1,000 people)	651	954	767
Population covered by mobile telephony (%)	99	99	98
Internet users (per 1,000 people)	244	*195*	*480*
Personal computers (per 1,000 people)	103	165	*504*
Households with television (%)	100	99	*98*
Quality			
Telephone faults (per 100 main lines per year)	10.5	*10.1*	..
Broadband subscribers (per 1,000 people)	2.5	82.3	126.2
International Internet bandwidth (bits per person)	49	838	4,718
Affordability			
Price basket for fixed line (US$ per month, residential)	17.4	25.8	25.8
Price basket for mobile (US$ per month)	..	31.7	17.8
Price basket for Internet (US$ per month)	..	*20.6*	*20.9*
Price of call to United States (US$ per 3 minutes)	0.83	1.04	0.77
Institutional efficiency and sustainability			
Total telecommunications revenue (% GDP)	4.8	5.4	2.9
Total telephone subscribers per employee	594	*883*	*485*
Total telecommunications investment (% revenue)	39.3	*11.3*	*12.3*
ICT applications			
ICT expenditure (% GDP)	4.4	*4.2*	*7.1*
E-government readiness index (scale 0–1)	..	0.39	0.70
Secure Internet servers (per 1 million people)	*13.4*	43.9	311.4
Schools connected to the Internet (%)	..	*92*	*99*

GNI per Capita, Atlas Method, 2000–4

Current US$

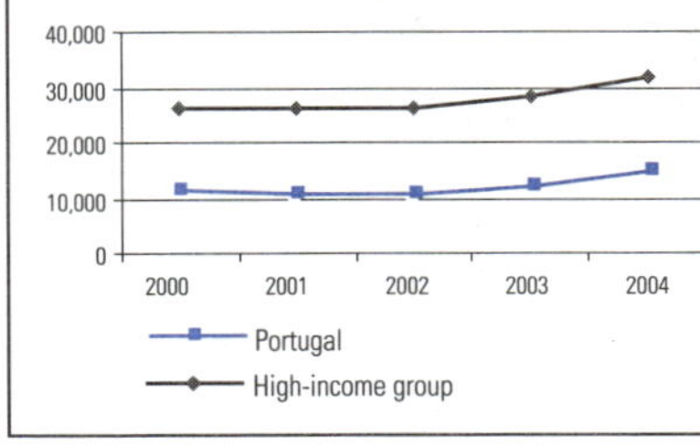

ICT MDG[b] Indicators, 2000–4

Number per 1,000 people

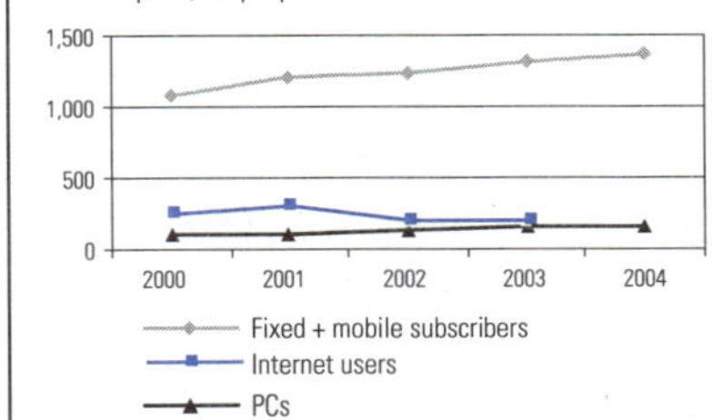

Price of Call to the United States, 2000–4

US$ per 3 minutes

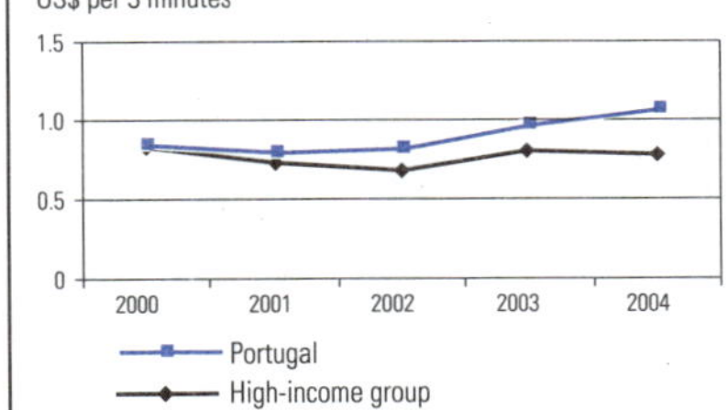

Total Telecommunications Revenue, 2000–4

Percentage of GDP

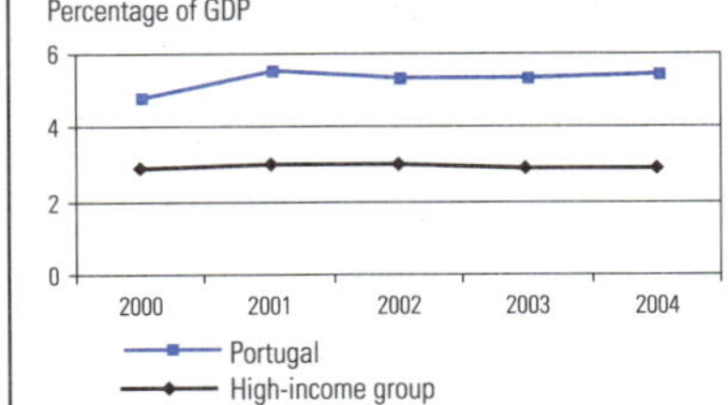

Notes: Figures in italics are for years other than those specified. .. indicates data are not available. C = competition; GDP = gross domestic product; GNI = gross national income; ICT = information and communication technology; M = monopoly; MDG = Millennium Development Goal; P = partial competition; and PCs = personal computers.

a. Outgoing and incoming.

b. Millennium Development Goal indicators 47, 48a, and 48b.

Sources: Economic and social context: UNDP and World Bank; ICT sector structure: ITU, WEF; ICT sector performance: Global Insight/WITSA, ITU, Netcraft, UNDESA, UNPAN, and World Bank. Produced by the Global Information and Communication Technologies Department and the Development Economics Data Group. For complete information, see Definitions and Data Sources.

Romania

	Romania 2000	Romania 2004	Lower-middle-income group 2004	Europe & Central Asia Region 2004
Economic and social context				
Population, total (millions)	22	22	2,430	472
Urban population (% total population)	55	55	49	64
Poverty (% population below US$1 per day)	2.1	*<2*	..	*3.6*
GNI per capita, Atlas method (current US$)	1,680	2,920	1,580	3,290
GDP growth, 1995–2000 and 2000–4 (%)	–2.3	5.5	5.7	5.0
Adult literacy rate (% ages 15 and over)	..	*97*	90	97
Primary, secondary, tertiary school enrollment (% gross)	68	*72*	*72*	*81*
ICT sector structure				
Separate telecommunications regulator	No	Yes		
Status of main fixed-line operator	*Mixed*	Mixed		
Level of competition: international long distance	*M*	C		
Level of competition: mobile	*C*	C		
Level of competition: Internet service provider	*C*	C		
Government prioritization of ICT (scale 1–7)	..	4.3	3.8	3.8
ICT sector performance				
Access				
Telephone main lines (per 1,000 people)	174	198	192	*243*
International voice traffic (minutes per person)[a]	34	*49*	..	*28*
Mobile subscribers (per 1,000 people)	111	464	255	487
Population covered by mobile telephony (%)	97	97	*76*	*82*
Internet users (per 1,000 people)	36	*184*	70	*115*
Personal computers (per 1,000 people)	32	*83*	38	*73*
Households with television (%)	96	..	*89*	*92*
Quality				
Telephone faults (per 100 main lines per year)	35.7	*8.9*	..	*30.4*
Broadband subscribers (per 1,000 people)	*0.3*	*0.7*	12.6	*2.4*
International Internet bandwidth (bits per person)	4	*107*	58	148
Affordability				
Price basket for fixed line (US$ per month, residential)	10.6	*9.6*	*5.5*	*3.5*
Price basket for mobile (US$ per month)	..	8.8	8.9	10.3
Price basket for Internet (US$ per month)	..	*25.3*	*25.3*	*19.8*
Price of call to United States (US$ per 3 minutes)	2.49	0.82	1.45	1.06
Institutional efficiency and sustainability				
Total telecommunications revenue (% GDP)	2.3	*3.8*	3.6	3.5
Total telephone subscribers per employee	151	*252*	*195*	*150*
Total telecommunications investment (% revenue)	67.8	*17.5*	*25.5*	*19.0*
ICT applications				
ICT expenditure (% GDP)	3.1	*2.8*	*5.1*	*5.1*
E-government readiness index (scale 0–1)	..	0.61	0.26	0.39
Secure Internet servers (per 1 million people)	*2.4*	3.0	1.6	6.4
Schools connected to the Internet (%)	..	*57*	..	*65*

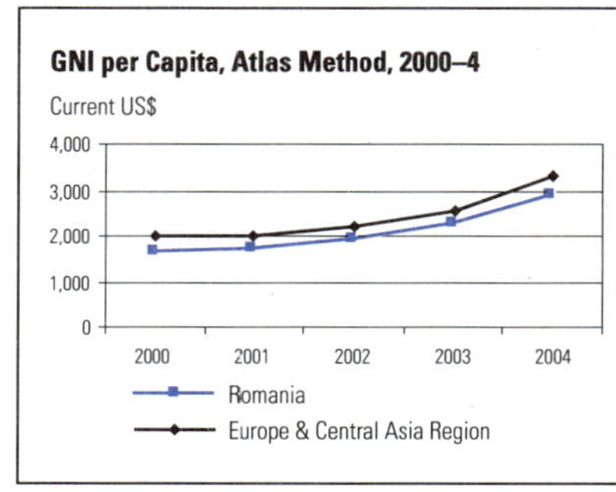

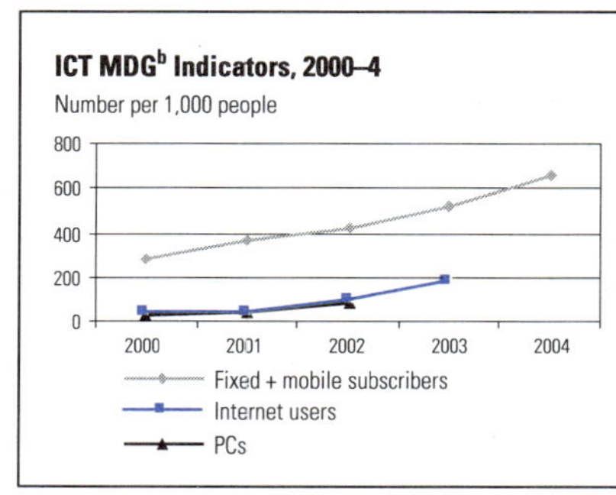

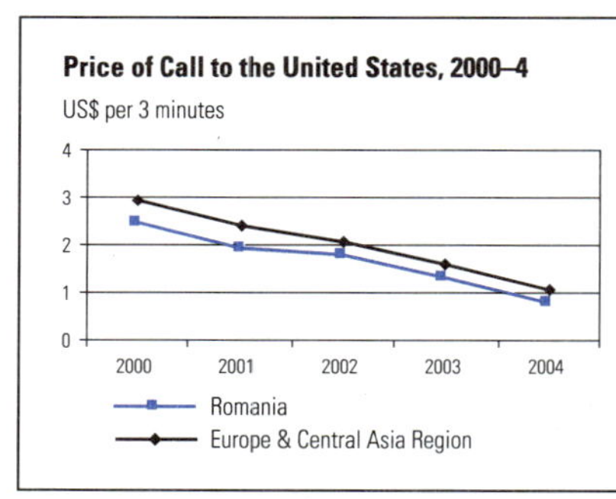

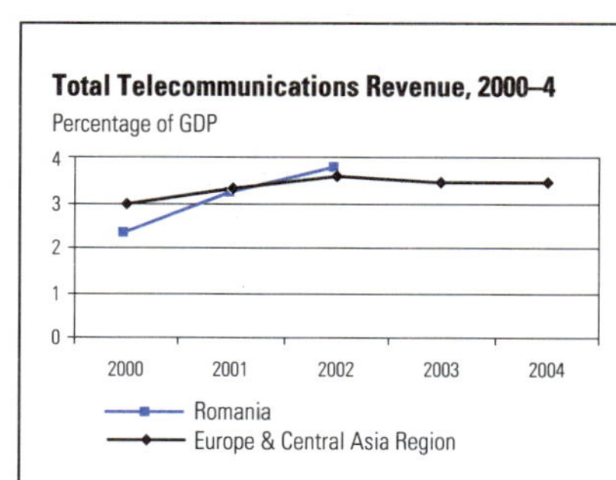

Notes: Figures in italics are for years other than those specified. .. indicates data are not available. C = competition; GDP = gross domestic product; GNI = gross national income; ICT = information and communication technology; M = monopoly; MDG = Millennium Development Goal; P = partial competition; and PCs = personal computers.

a. Outgoing and incoming.

b. Millennium Development Goal indicators 47, 48a, and 48b.

Sources: Economic and social context: UNDP and World Bank; ICT sector structure: ITU, WEF; ICT sector performance: Global Insight/WITSA, ITU, Netcraft, UNDESA, UNPAN, and World Bank. Produced by the Global Information and Communication Technologies Department and the Development Economics Data Group. For complete information, see Definitions and Data Sources.

Russian Federation

	Russian Federation 2000	Russian Federation 2004	Upper-middle-income group 2004	Europe & Central Asia Region 2004
Economic and social context				
Population, total (millions)	146	143	576	472
Urban population (% total population)	73	73	72	64
Poverty (% population below US$1 per day)	6.1	*<2*	..	*3.6*
GNI per capita, Atlas method (current US$)	1,720	3,410	4,770	3,290
GDP growth, 1995–2000 and 2000–4 (%)	1.2	6.1	2.7	5.0
Adult literacy rate (% ages 15 and over)	..	*99*	94	97
Primary, secondary, tertiary school enrollment (% gross)	90	*90*	*80*	*81*
ICT sector structure				
Separate telecommunications regulator	..	..		
Status of main fixed-line operator	*Mixed*	Mixed		
Level of competition: international long distance	*P*	P		
Level of competition: mobile	*C*	C		
Level of competition: Internet service provider	..	..		
Government prioritization of ICT (scale 1–7)	..	4.0	4.1	3.8
ICT sector performance				
Access				
Telephone main lines (per 1,000 people)	220	*261*	*220*	*243*
International voice traffic (minutes per person)[a]	*13*	*15*	*39*	*28*
Mobile subscribers (per 1,000 people)	24	517	490	487
Population covered by mobile telephony (%)	..	*78*	*84*	*82*
Internet users (per 1,000 people)	20	*91*	*133*	*115*
Personal computers (per 1,000 people)	64	*113*	*99*	*73*
Households with television (%)	89	*98*	*92*	*92*
Quality				
Telephone faults (per 100 main lines per year)	*35.2*	*29.0*	*20.3*	*30.4*
Broadband subscribers (per 1,000 people)	0.0	*0.9*	*3.7*	*2.4*
International Internet bandwidth (bits per person)	21	101	176	148
Affordability				
Price basket for fixed line (US$ per month, residential)	..	*7.8*	*13.9*	*3.5*
Price basket for mobile (US$ per month)	..	6.3	11.1	10.3
Price basket for Internet (US$ per month)	..	*10.0*	*20.8*	*19.8*
Price of call to United States (US$ per 3 minutes)	2.56	2.03	1.03	1.06
Institutional efficiency and sustainability				
Total telecommunications revenue (% GDP)	2.0	3.2	3.4	3.5
Total telephone subscribers per employee	83	*194*	*402*	*150*
Total telecommunications investment (% revenue)	11.5	*25.5*	*18.6*	*19.0*
ICT applications				
ICT expenditure (% GDP)	3.5	*3.7*	*5.0*	*5.1*
E-government readiness index (scale 0–1)	..	0.39	0.49	0.39
Secure Internet servers (per 1 million people)	*2.0*	2.1	10.7	6.4
Schools connected to the Internet (%)	..	*65*	*60*	*65*

GNI per Capita, Atlas Method, 2000–4
Current US$

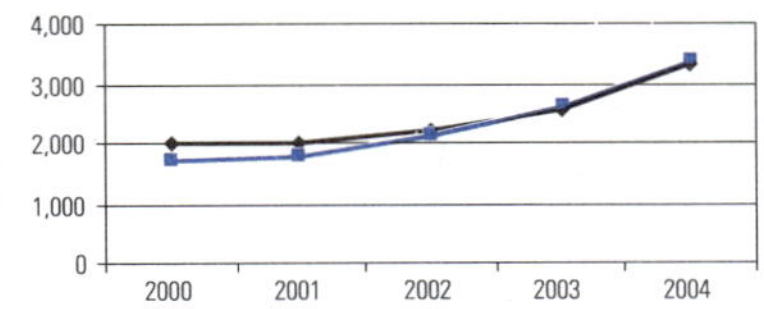

ICT MDG[b] Indicators, 2000–4
Number per 1,000 people

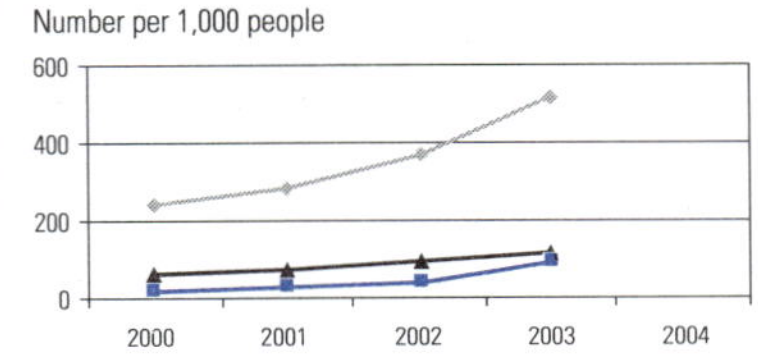

Price of Call to the United States, 2000–4
US$ per 3 minutes

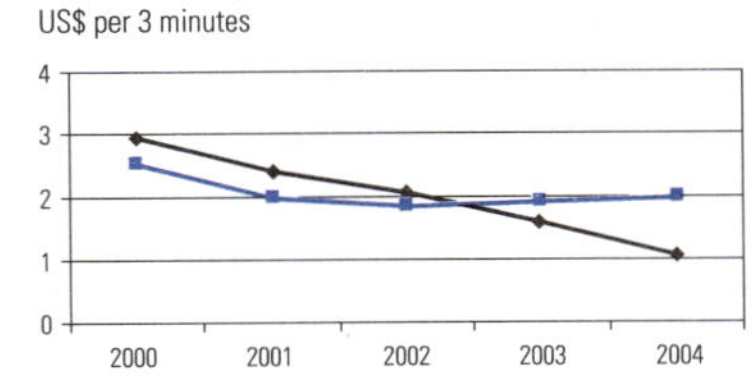

Total Telecommunications Revenue, 2000–4
Percentage of GDP

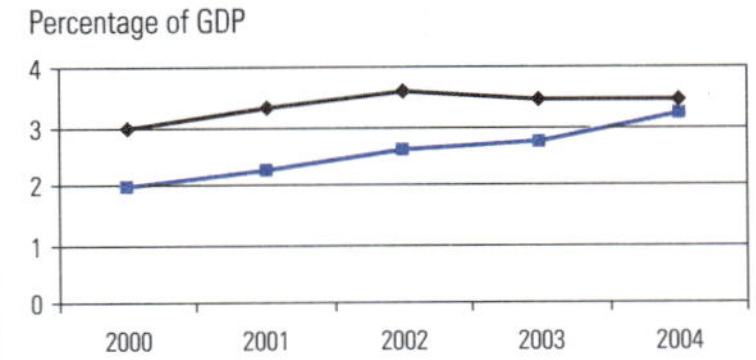

Notes: Figures in italics are for years other than those specified. .. indicates data are not available. C = competition; GDP = gross domestic product; GNI = gross national income; ICT = information and communication technology; M = monopoly; MDG = Millennium Development Goal; P = partial competition; and PCs = personal computers.

a. Outgoing and incoming.
b. Millennium Development Goal indicators 47, 48a, and 48b.

Sources: Economic and social context: UNDP and World Bank; ICT sector structure: ITU, WEF; ICT sector performance: Global Insight/WITSA, ITU, Netcraft, UNDESA, UNPAN, and World Bank. Produced by the Global Information and Communication Technologies Department and the Development Economics Data Group. For complete information, see Definitions and Data Sources.

Rwanda

	Rwanda 2000	Rwanda 2004	Low-income group 2004	Sub-Saharan Africa Region 2004
Economic and social context				
Population, total (millions)	8	8	2,338	719
Urban population (% total population)	14	20	31	37
Poverty (% population below US$1 per day)	51.7	..	..	*46.4*
GNI per capita, Atlas method (current US$)	260	220	510	600
GDP growth, 1995–2000 and 2000–4 (%)	9.8	5.1	5.4	3.9
Adult literacy rate (% ages 15 and over)	64	..	61	59
Primary, secondary, tertiary school enrollment (% gross)	53	*55*	*55*	*52*
ICT sector structure				
Separate telecommunications regulator	No	Yes		
Status of main fixed-line operator	*Mixed*	Mixed		
Level of competition: international long distance	*M*	*M*		
Level of competition: mobile	*M*	C		
Level of competition: Internet service provider	..	C		
Government prioritization of ICT (scale 1–7)	..	..	..	*4.4*
ICT sector performance				
Access				
Telephone main lines (per 1,000 people)	2	3	33	17
International voice traffic (minutes per person)[a]	1	*1*	*4*	..
Mobile subscribers (per 1,000 people)	5	16	48	86
Population covered by mobile telephony (%)	..	65	*43*	..
Internet users (per 1,000 people)	1	4	20	15
Personal computers (per 1,000 people)	..	..	*8*	*12*
Households with television (%)	2	*2*	*16*	*15*
Quality				
Telephone faults (per 100 main lines per year)	*16.0*	..	..	..
Broadband subscribers (per 1,000 people)	0.0	0.0	0.5	*0.1*
International Internet bandwidth (bits per person)	*0*	*1*	*3*	*4*
Affordability				
Price basket for fixed line (US$ per month, residential)	8.4	*7.9*	*6.6*	*8.5*
Price basket for mobile (US$ per month)	..	24.8	11.6	13.5
Price basket for Internet (US$ per month)	..	*66.8*	*45.5*	*54.8*
Price of call to United States (US$ per 3 minutes)	*11.23*	2.43	1.95	2.43
Institutional efficiency and sustainability				
Total telecommunications revenue (% GDP)	2.0	*4.2*	2.3	5.0
Total telephone subscribers per employee	189	..	*89*	*144*
Total telecommunications investment (% revenue)	*15.9*	*11.3*	*27.8*	*27.8*
ICT applications				
ICT expenditure (% GDP)	..	..	*4.1*	..
E-government readiness index (scale 0–1)	..	0.12	0.12	0.11
Secure Internet servers (per 1 million people)	*0.1*	..	0.3	1.9
Schools connected to the Internet (%)	..	..	..	..

GNI per Capita, Atlas Method, 2000–4
Current US$

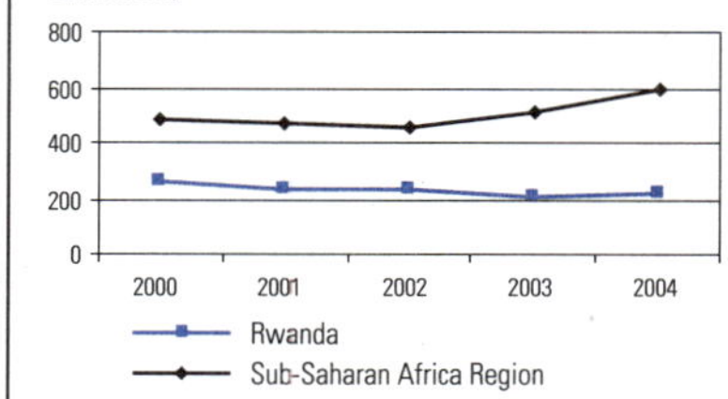

ICT MDG[b] Indicators, 2000–4
Number per 1,000 people

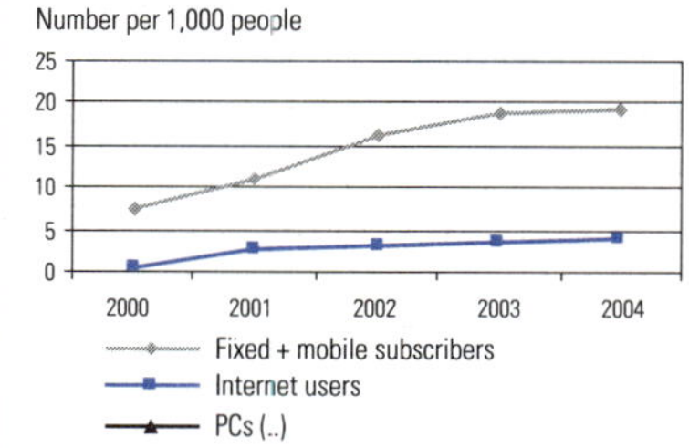

Price of Call to the United States, 2000–4
US$ per 3 minutes

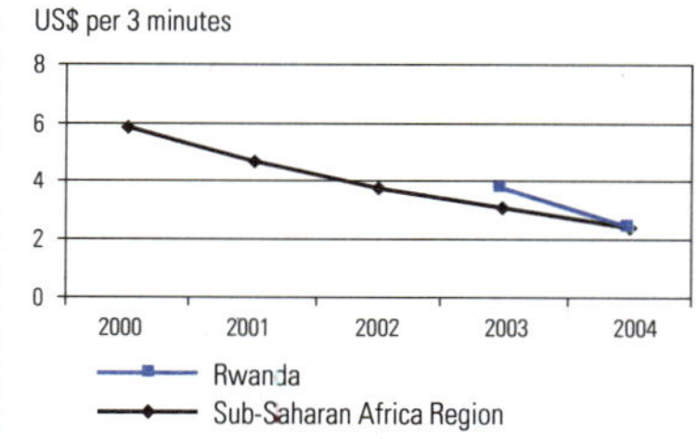

Total Telecommunications Revenue, 2000–4
Percentage of GDP

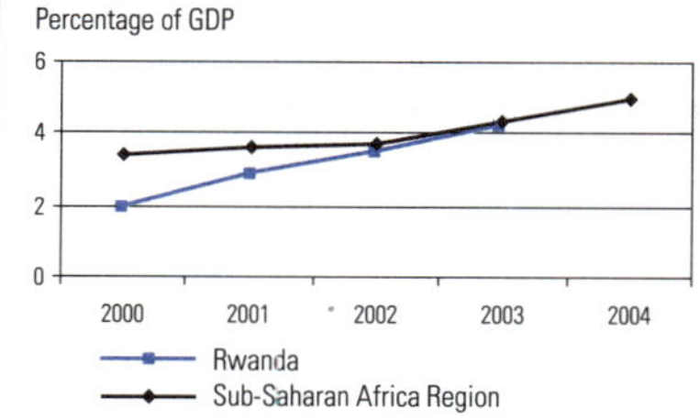

Notes: Figures in italics are for years other than those specified. .. indicates data are not available. C = competition; GDP = gross domestic product; GNI = gross national income; ICT = information and communication technology; M = monopoly; MDG = Millennium Development Goal; P = partial competition; and PCs = personal computers.

a. Outgoing and incoming.

b. Millennium Development Goal indicators 47, 48a, and 48b.

Sources: Economic and social context: UNDP and World Bank; ICT sector structure: ITU, WEF; ICT sector performance: Global Insight/WITSA, ITU, Netcraft, UNDESA, UNPAN, and World Bank. Produced by the Global Information and Communication Technologies Department and the Development Economics Data Group. For complete information, see Definitions and Data Sources.

Saudi Arabia

	Saudi Arabia 2000	Saudi Arabia 2004	High-income group 2004
Economic and social context			
Population, total (millions)	21	23	1,001
Urban population (% total population)	86	88	77
Poverty (% population below US$1 per day)	..	..	..
GNI per capita, Atlas method (current US$)	8,110	10,430	32,040
GDP growth, 1995–2000 and 2000–4 (%)	2.3	3.4	2.0
Adult literacy rate (% ages 15 and over)	79	..	..
Primary, secondary, tertiary school enrollment (% gross)	58	*57*	*93*
ICT sector structure			
Separate telecommunications regulator	No	Yes	
Status of main fixed-line operator	*Public*	Mixed	
Level of competition: international long distance	*M*	M	
Level of competition: mobile	*M*	M	
Level of competition: Internet service provider	..	C	
Government prioritization of ICT (scale 1–7)	..	..	4.9
ICT sector performance			
Access			
Telephone main lines (per 1,000 people)	143	159	558
International voice traffic (minutes per person)[a]	87	*125*	..
Mobile subscribers (per 1,000 people)	66	395	767
Population covered by mobile telephony (%)	*92*	..	98
Internet users (per 1,000 people)	22	95	*480*
Personal computers (per 1,000 people)	63	*137*	*504*
Households with television (%)	93	*99*	*98*
Quality			
Telephone faults (per 100 main lines per year)	2.0	*1.7*	..
Broadband subscribers (per 1,000 people)	0.0	*0.4*	126.2
International Internet bandwidth (bits per person)	..	56	4,718
Affordability			
Price basket for fixed line (US$ per month, residential)	11.7	*11.7*	25.8
Price basket for mobile (US$ per month)	..	9.6	17.8
Price basket for Internet (US$ per month)	..	*34.7*	*20.9*
Price of call to United States (US$ per 3 minutes)	5.20	*2.40*	0.77
Institutional efficiency and sustainability			
Total telecommunications revenue (% GDP)	2.4	3.2	2.9
Total telephone subscribers per employee	189	..	*485*
Total telecommunications investment (% revenue)	33.9	*24.5*	*12.3*
ICT applications			
ICT expenditure (% GDP)	2.4	*2.5*	*7.1*
E-government readiness index (scale 0–1)	..	0.31	0.70
Secure Internet servers (per 1 million people)	*0.5*	2.5	311.4
Schools connected to the Internet (%)	..	..	*99*

GNI per Capita, Atlas Method, 2000–4

Current US$

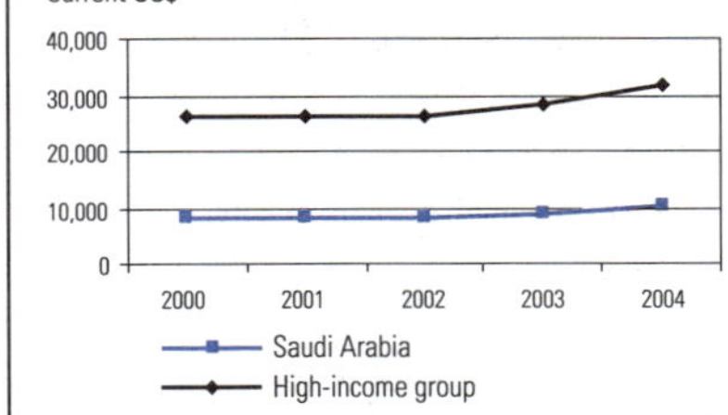

ICT MDG[b] Indicators, 2000–4

Number per 1,000 people

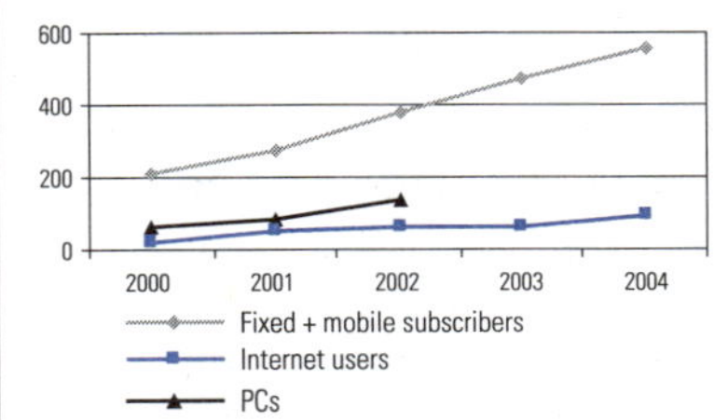

Price of Call to the United States, 2000–4

US$ per 3 minutes

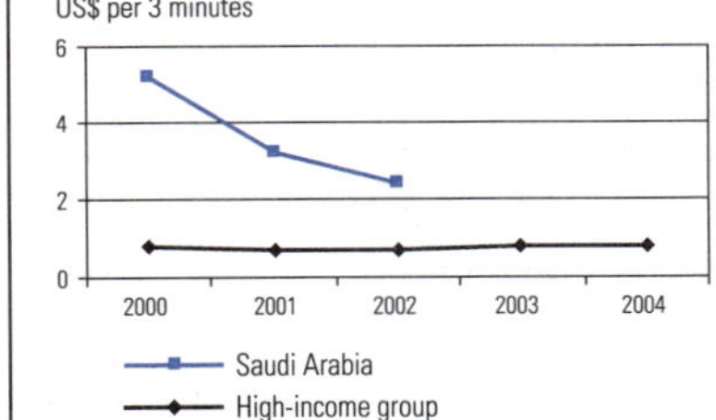

Total Telecommunications Revenue, 2000–4

Percentage of GDP

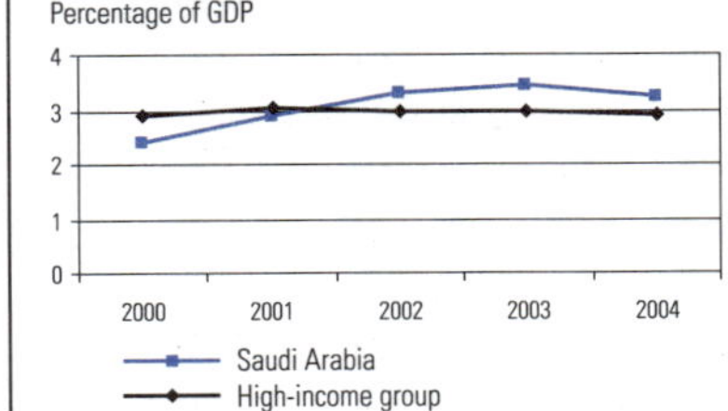

Notes: Figures in italics are for years other than those specified. .. indicates data are not available. C = competition; GDP = gross domestic product; GNI = gross national income; ICT = information and communication technology; M = monopoly; MDG = Millennium Development Goal; P = partial competition; and PCs = personal computers.

a. Outgoing and incoming.

b. Millennium Development Goal indicators 47, 48a, and 48b.

Sources: Economic and social context: UNDP and World Bank; ICT sector structure: ITU, WEF; ICT sector performance: Global Insight/WITSA, ITU, Netcraft, UNDESA, UNPAN, and World Bank. Produced by the Global Information and Communication Technologies Department and the Development Economics Data Group. For complete information, see Definitions and Data Sources.

Senegal

	Senegal 2000	Senegal 2004	Low-income group 2004	Sub-Saharan Africa Region 2004
Economic and social context				
Population, total (millions)	10	10	2,338	719
Urban population (% total population)	47	50	31	37
Poverty (% population below US$1 per day)	*22.3*	..	..	*46.4*
GNI per capita, Atlas method (current US$)	490	670	510	600
GDP growth, 1995–2000 and 2000–4 (%)	5.3	4.6	5.4	3.9
Adult literacy rate (% ages 15 and over)	..	*39*	61	59
Primary, secondary, tertiary school enrollment (% gross)	37	*40*	*55*	*52*
ICT sector structure				
Separate telecommunications regulator	No	Yes		
Status of main fixed-line operator	*Mixed*	Mixed		
Level of competition: international long distance	*M*	C		
Level of competition: mobile	*P*	C		
Level of competition: Internet service provider	*C*	C		
Government prioritization of ICT (scale 1–7)	..	*5.8*	..	*4.4*
ICT sector performance				
Access				
Telephone main lines (per 1,000 people)	22	23	33	17
International voice traffic (minutes per person)[a]	18	..	*4*	..
Mobile subscribers (per 1,000 people)	26	107	48	86
Population covered by mobile telephony (%)	..	85	*43*	..
Internet users (per 1,000 people)	4	19	20	15
Personal computers (per 1,000 people)	17	*21*	*8*	*12*
Households with television (%)	26	*29*	*16*	*15*
Quality				
Telephone faults (per 100 main lines per year)	17.0	..	..	..
Broadband subscribers (per 1,000 people)	0.0	*0.2*	0.5	*0.1*
International Internet bandwidth (bits per person)	4	30	*3*	*4*
Affordability				
Price basket for fixed line (US$ per month, residential)	8.7	*14.5*	*6.6*	*8.5*
Price basket for mobile (US$ per month)	..	13.5	11.6	13.5
Price basket for Internet (US$ per month)	..	*40.6*	*45.5*	*54.8*
Price of call to United States (US$ per 3 minutes)	2.23	1.02	1.95	2.43
Institutional efficiency and sustainability				
Total telecommunications revenue (% GDP)	4.2	7.1	2.3	5.0
Total telephone subscribers per employee	324	..	*89*	*144*
Total telecommunications investment (% revenue)	42.5	*42.8*	*27.8*	*27.8*
ICT applications				
ICT expenditure (% GDP)	7.2	*7.5*	*4.1*	..
E-government readiness index (scale 0–1)	..	0.29	0.12	0.11
Secure Internet servers (per 1 million people)	*0.1*	0.3	0.3	1.9
Schools connected to the Internet (%)	..	..	..	..

GNI per Capita, Atlas Method, 2000–4

Current US$

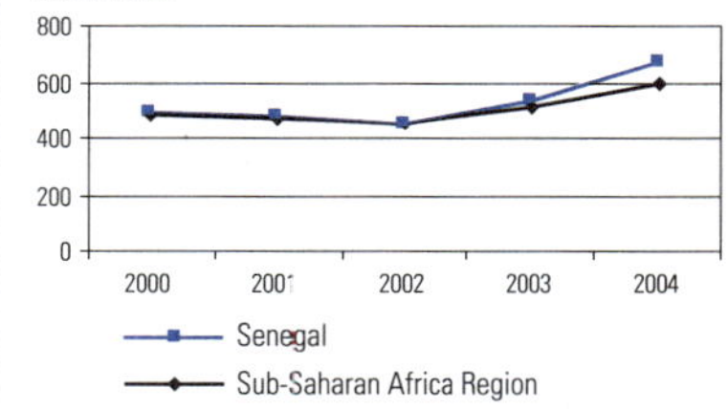

ICT MDG[b] Indicators, 2000–4

Number per 1,000 people

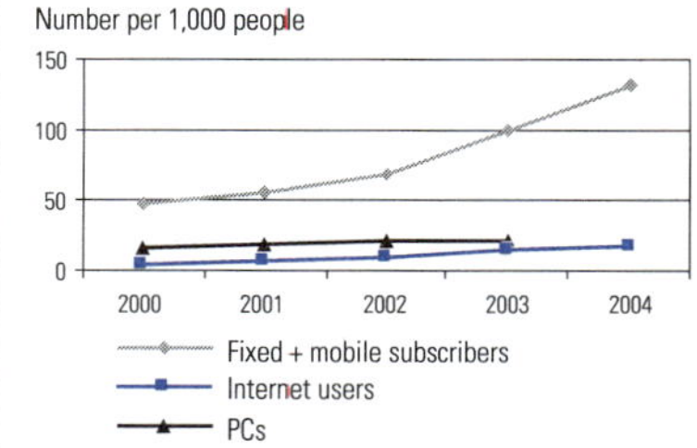

Price of Call to the United States, 2000–4

US$ per 3 minutes

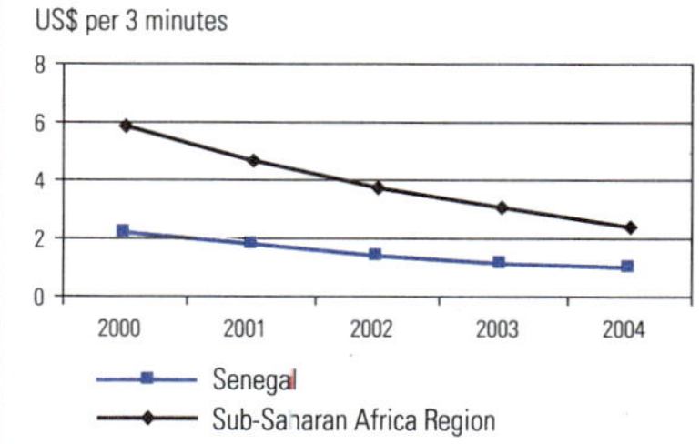

Total Telecommunications Revenue, 2000–4

Percentage of GDP

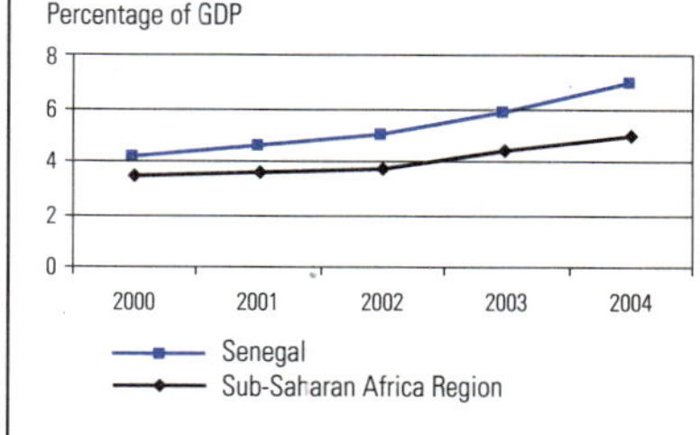

Notes: Figures in italics are for years other than those specified. .. indicates data are not available. C = competition; GDP = gross domestic product; GNI = gross national income; ICT = information and communication technology; M = monopoly; MDG = Millennium Development Goal; P = partial competition; and PCs = personal computers.

a. Outgoing and incoming.

b. Millennium Development Goal indicators 47, 48a, and 48b.

Sources: Economic and social context: UNDP and World Bank; ICT sector structure: ITU, WEF; ICT sector performance: Global Insight/WITSA, ITU, Netcraft, UNDESA, UNPAN, and World Bank. Produced by the Global Information and Communication Technologies Department and the Development Economics Data Group. For complete information, see Definitions and Data Sources.

Serbia and Montenegro

	Serbia and Montenegro 2000	Serbia and Montenegro 2004	Lower-middle-income group 2004	Europe & Central Asia Region 2004
Economic and social context				
Population, total (millions)	11	8	2,430	472
Urban population (% total population)	52	52	49	64
Poverty (% population below US$1 per day)	..	..	..	*3.6*
GNI per capita, Atlas method (current US$)	1,220	2,620	1,580	3,290
GDP growth, 1995–2000 and 2000–4 (%)	–0.8	4.5	5.7	5.0
Adult literacy rate (% ages 15 and over)	..	*96*	90	97
Primary, secondary, tertiary school enrollment (% gross)	74	..	*72*	*81*
ICT sector structure				
Separate telecommunications regulator	..	..		
Status of main fixed-line operator	..	Mixed		
Level of competition: international long distance	..	..		
Level of competition: mobile	..	C		
Level of competition: Internet service provider	..	..		
Government prioritization of ICT (scale 1–7)	..	3.5	3.8	3.8
ICT sector performance				
Access				
Telephone main lines (per 1,000 people)	226	322	192	*243*
International voice traffic (minutes per person)[a]	96	*118*	..	*28*
Mobile subscribers (per 1,000 people)	123	528	255	487
Population covered by mobile telephony (%)	77	95	*76*	*82*
Internet users (per 1,000 people)	38	*104*	70	*115*
Personal computers (per 1,000 people)	23	*36*	38	*73*
Households with television (%)	..	*92*	*89*	*92*
Quality				
Telephone faults (per 100 main lines per year)	..	..	..	*30.4*
Broadband subscribers (per 1,000 people)	..	*0.0*	12.6	*2.4*
International Internet bandwidth (bits per person)	*1*	23	58	148
Affordability				
Price basket for fixed line (US$ per month, residential)	*2.8*	*2.3*	*5.5*	*3.5*
Price basket for mobile (US$ per month)	..	6.4	8.9	10.3
Price basket for Internet (US$ per month)	..	*13.2*	*25.3*	*19.8*
Price of call to United States (US$ per 3 minutes)	1.52	*2.08*	1.45	1.06
Institutional efficiency and sustainability				
Total telecommunications revenue (% GDP)	3.3	..	3.6	3.5
Total telephone subscribers per employee	247	*374*	*195*	*150*
Total telecommunications investment (% revenue)	28.0	..	*25.5*	*19.0*
ICT applications				
ICT expenditure (% GDP)	..	..	*5.1*	*5.1*
E-government readiness index (scale 0–1)	..	0.34	0.26	0.39
Secure Internet servers (per 1 million people)	*0.7*	1.1	1.6	6.4
Schools connected to the Internet (%)	..	*70*	..	*65*

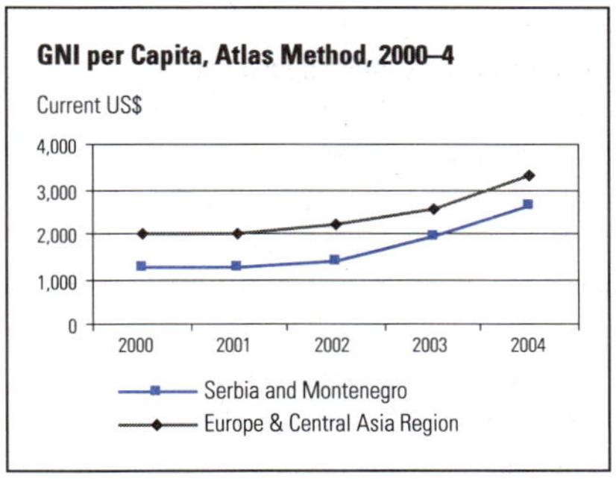

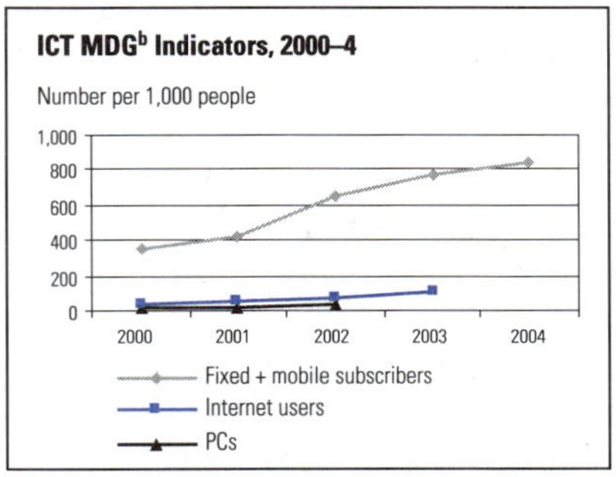

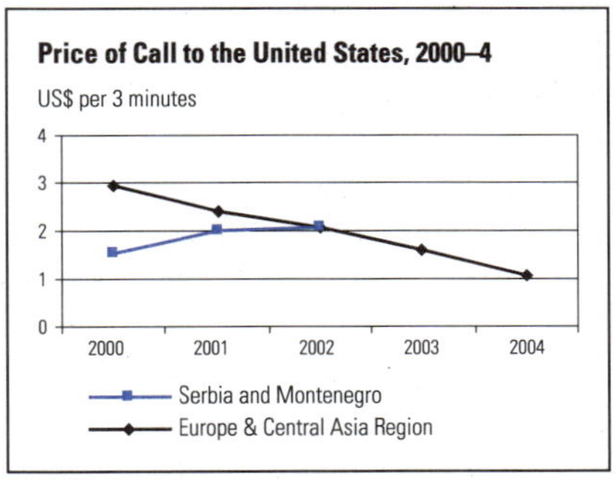

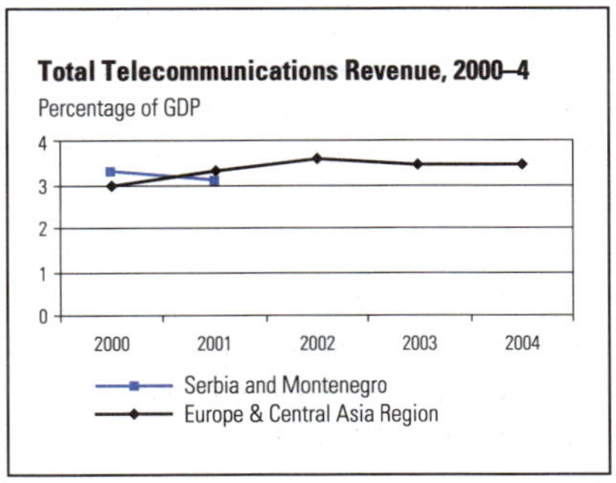

Notes: Figures in italics are for years other than those specified. .. indicates data are not available. C = competition; GDP = gross domestic product; GNI = gross national income; ICT = information and communication technology; M = monopoly; MDG = Millennium Development Goal; P = partial competition; and PCs = personal computers.

a. Outgoing and incoming.

b. Millennium Development Goal indicators 47, 48a, and 48b.

Sources: Economic and social context: UNDP and World Bank; ICT sector structure: ITU, WEF; ICT sector performance: Global Insight/WITSA, ITU, Netcraft, UNDESA, UNPAN, and World Bank. Produced by the Global Information and Communication Technologies Department and the Development Economics Data Group. For complete information, see Definitions and Data Sources.

Sierra Leone

	Sierra Leone 2000	Sierra Leone 2004	Low-income group 2004	Sub-Saharan Africa Region 2004
Economic and social context				
Population, total (millions)	5	5	2,338	719
Urban population (% total population)	37	40	31	37
Poverty (% population below US$1 per day)	..	..	..	*46.4*
GNI per capita, Atlas method (current US$)	130	200	510	600
GDP growth, 1995–2000 and 2000–4 (%)	–5.2	15.8	5.4	3.9
Adult literacy rate (% ages 15 and over)	30	..	61	59
Primary, secondary, tertiary school enrollment (% gross)	45	..	*55*	*52*
ICT sector structure				
Separate telecommunications regulator	..	..		
Status of main fixed-line operator	*Public*	Public		
Level of competition: international long distance	*M*	P		
Level of competition: mobile	*C*	C		
Level of competition: Internet service provider	*C*	P		
Government prioritization of ICT (scale 1–7)	..	..	..	4.4
ICT sector performance				
Access				
Telephone main lines (per 1,000 people)	4	5	33	17
International voice traffic (minutes per person)[a]	..	..	*4*	..
Mobile subscribers (per 1,000 people)	2	22	48	86
Population covered by mobile telephony (%)	..	35	*43*	..
Internet users (per 1,000 people)	1	2	20	15
Personal computers (per 1,000 people)	..	..	*8*	*12*
Households with television (%)	4	*7*	*16*	*15*
Quality				
Telephone faults (per 100 main lines per year)	..	..	..	..
Broadband subscribers (per 1,000 people)	0.0	0.0	0.5	*0.1*
International Internet bandwidth (bits per person)	0	*0*	*3*	*4*
Affordability				
Price basket for fixed line (US$ per month, residential)	3.0	*3.0*	*6.6*	*8.5*
Price basket for mobile (US$ per month)	..	13.6	11.6	13.5
Price basket for Internet (US$ per month)	..	*100.0*	*45.5*	*54.8*
Price of call to United States (US$ per 3 minutes)	*2.74*	..	1.95	2.43
Institutional efficiency and sustainability				
Total telecommunications revenue (% GDP)	..	*2.8*	2.3	5.0
Total telephone subscribers per employee	31	..	*89*	*144*
Total telecommunications investment (% revenue)	..	..	*27.8*	*27.8*
ICT applications				
ICT expenditure (% GDP)	..	..	*4.1*	..
E-government readiness index (scale 0–1)	..	0.10	0.12	0.11
Secure Internet servers (per 1 million people)	*0.2*	..	0.3	1.9
Schools connected to the Internet (%)	..	..	..	..

GNI per Capita, Atlas Method, 2000–4

Current US$

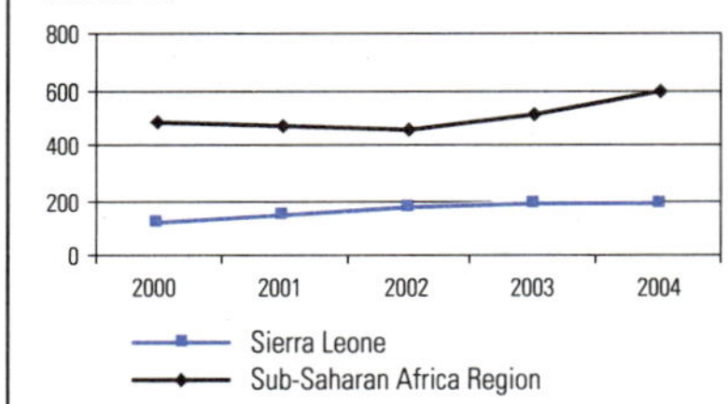

ICT MDG[b] Indicators, 2000–4

Number per 1,000 people

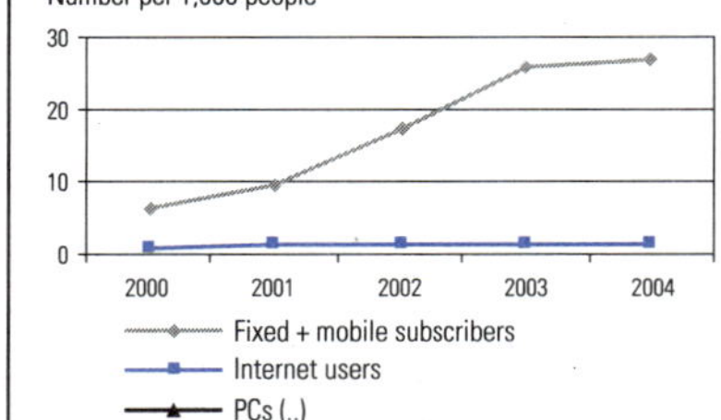

Price of Call to the United States, 2000–4

US$ per 3 minutes

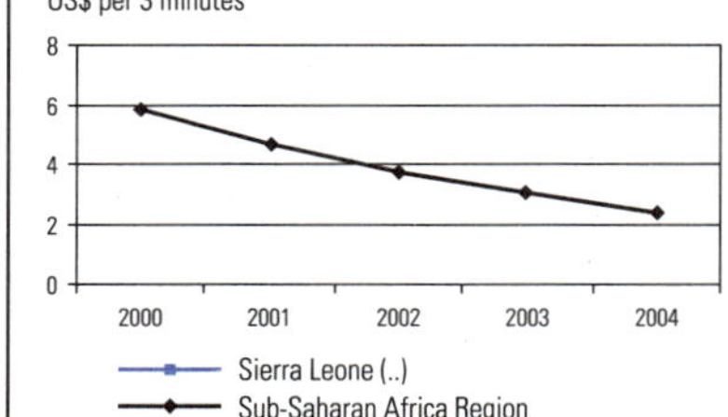

Total Telecommunications Revenue, 2000–4

Percentage of GDP

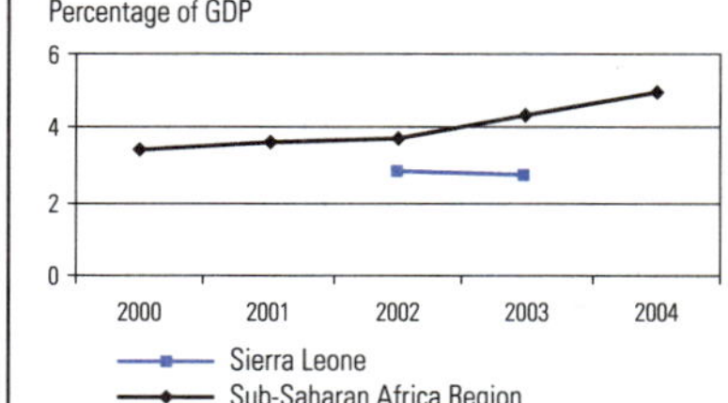

Notes: Figures in italics are for years other than those specified. .. indicates data are not available. C = competition; GDP = gross domestic product; GNI = gross national income; ICT = information and communication technology; M = monopoly; MDG = Millennium Development Goal; P = partial competition; and PCs = personal computers.

a. Outgoing and incoming.

b. Millennium Development Goal indicators 47, 48a, and 48b.

Sources: Economic and social context: UNDP and World Bank; ICT sector structure: ITU, WEF; ICT sector performance: Global Insight/WITSA, ITU, Netcraft, UNDESA, UNPAN, and World Bank. Produced by the Global Information and Communication Technologies Department and the Development Economics Data Group. For complete information, see Definitions and Data Sources.

Singapore

	Singapore 2000	Singapore 2004	High-income group 2004
Economic and social context			
Population, total (millions)	4	4	1,001
Urban population (% total population)	100	100	77
Poverty (% population below US$1 per day)	..	..	..
GNI per capita, Atlas method (current US$)	22,890	24,220	32,040
GDP growth, 1995–2000 and 2000–4 (%)	5.6	2.8	2.0
Adult literacy rate (% ages 15 and over)	93	..	..
Primary, secondary, tertiary school enrollment (% gross)	..	..	*93*
ICT sector structure			
Separate telecommunications regulator	Yes	Yes	
Status of main fixed-line operator	*Mixed*	Mixed	
Level of competition: international long distance	*C*	C	
Level of competition: mobile	*C*	C	
Level of competition: Internet service provider	*C*	C	
Government prioritization of ICT (scale 1–7)	..	6.1	4.9
ICT sector performance			
Access			
Telephone main lines (per 1,000 people)	484	430	558
International voice traffic (minutes per person)[a]	646	..	..
Mobile subscribers (per 1,000 people)	684	891	767
Population covered by mobile telephony (%)	100	100	98
Internet users (per 1,000 people)	324	559	*480*
Personal computers (per 1,000 people)	483	*565*	*504*
Households with television (%)	98	98	*98*
Quality			
Telephone faults (per 100 main lines per year)	2.4	..	..
Broadband subscribers (per 1,000 people)	18.9	118.2	126.2
International Internet bandwidth (bits per person)	560	5,699	4,718
Affordability			
Price basket for fixed line (US$ per month, residential)	6.2	6.7	25.8
Price basket for mobile (US$ per month)	..	5.7	17.8
Price basket for Internet (US$ per month)	..	*11.0*	*20.9*
Price of call to United States (US$ per 3 minutes)	0.68	0.69	0.77
Institutional efficiency and sustainability			
Total telecommunications revenue (% GDP)	3.4	3.4	2.9
Total telephone subscribers per employee	310	403	*485*
Total telecommunications investment (% revenue)	14.9	11.2	*12.3*
ICT applications			
ICT expenditure (% GDP)	9.9	*10.4*	*7.1*
E-government readiness index (scale 0–1)	..	0.97	0.70
Secure Internet servers (per 1 million people)	*127.1*	226.3	311.4
Schools connected to the Internet (%)	..	100	*99*

GNI per Capita, Atlas Method, 2000–4

Current US$

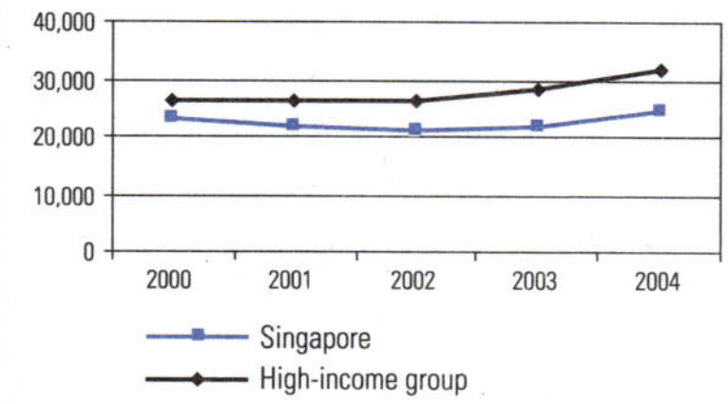

ICT MDG[b] Indicators, 2000–4

Number per 1,000 people

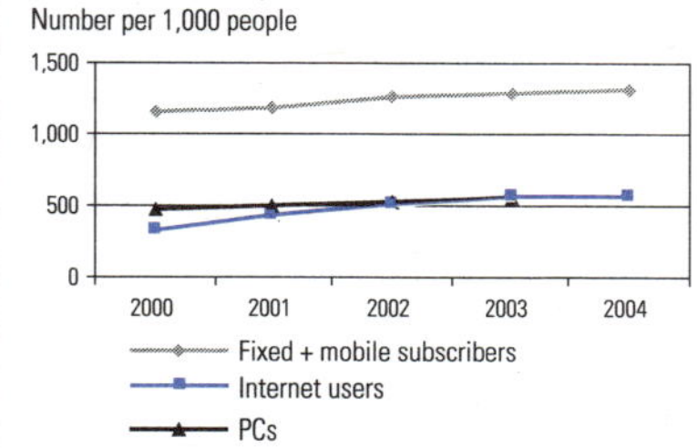

Price of Call to the United States, 2000–4

US$ per 3 minutes

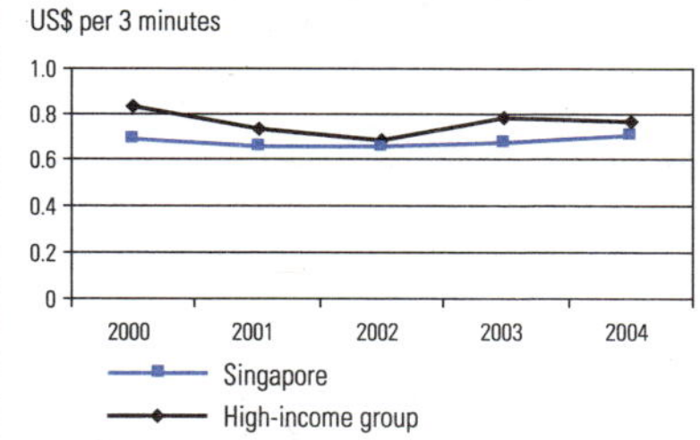

Total Telecommunications Revenue, 2000–4

Percentage of GDP

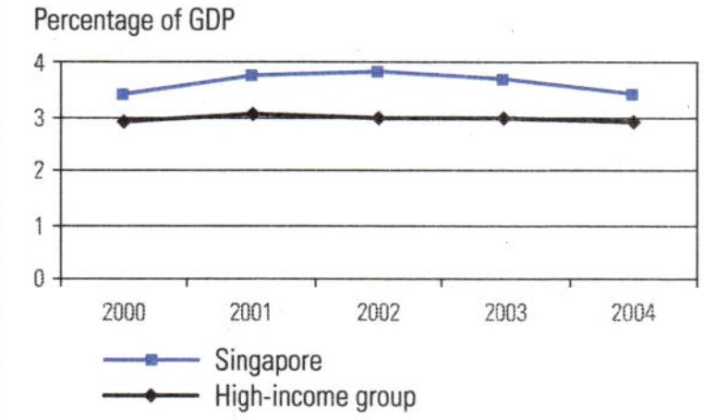

Notes: Figures in italics are for years other than those specified. .. indicates data are not available. C = competition; GDP = gross domestic product; GNI = gross national income; ICT = information and communication technology; M = monopoly; MDG = Millennium Development Goal; P = partial competition; and PCs = personal computers.

a. Outgoing and incoming.

b. Millennium Development Goal indicators 47, 48a, and 48b.

Sources: Economic and social context: UNDP and World Bank; ICT sector structure: ITU, WEF; ICT sector performance: Global Insight/WITSA, ITU, Netcraft, UNDESA, UNPAN, and World Bank. Produced by the Global Information and Communication Technologies Department and the Development Economics Data Group. For complete information, see Definitions and Data Sources.

Slovak Republic

	Slovak Republic 2000	Slovak Republic 2004	Upper-middle-income group 2004	Europe & Central Asia Region 2004
Economic and social context				
Population, total (millions)	5	5	576	472
Urban population (% total population)	57	58	72	64
Poverty (% population below US$1 per day)	*<2*	..	..	*3.6*
GNI per capita, Atlas method (current US$)	3,870	6,480	4,770	3,290
GDP growth, 1995–2000 and 2000–4 (%)	3.6	4.6	2.7	5.0
Adult literacy rate (% ages 15 and over)	*100*	..	94	97
Primary, secondary, tertiary school enrollment (% gross)	73	*75*	*80*	*81*
ICT sector structure				
Separate telecommunications regulator	Yes	Yes		
Status of main fixed-line operator	*Mixed*	Mixed		
Level of competition: international long distance	*M*	C		
Level of competition: mobile	*C*	C		
Level of competition: Internet service provider	*C*	C		
Government prioritization of ICT (scale 1–7)	..	4.1	4.1	3.8
ICT sector performance				
Access				
Telephone main lines (per 1,000 people)	315	232	*220*	*243*
International voice traffic (minutes per person)[a]	59	88	*39*	*28*
Mobile subscribers (per 1,000 people)	206	793	490	487
Population covered by mobile telephony (%)	98	99	*84*	*82*
Internet users (per 1,000 people)	94	422	*133*	*115*
Personal computers (per 1,000 people)	137	241	*99*	*73*
Households with television (%)	90	100	*92*	*92*
Quality				
Telephone faults (per 100 main lines per year)	27.0	9.5	*20.3*	*30.4*
Broadband subscribers (per 1,000 people)	0.0	11.6	*3.7*	*2.4*
International Internet bandwidth (bits per person)	41	1,855	176	148
Affordability				
Price basket for fixed line (US$ per month, residential)	9.2	..	*13.9*	*3.5*
Price basket for mobile (US$ per month)	..	10.3	11.1	10.3
Price basket for Internet (US$ per month)	..	*20.7*	*20.8*	*19.8*
Price of call to United States (US$ per 3 minutes)	1.13	1.06	1.03	1.06
Institutional efficiency and sustainability				
Total telecommunications revenue (% GDP)	3.9	4.0	3.4	3.5
Total telephone subscribers per employee	186	508	*402*	*150*
Total telecommunications investment (% revenue)	22.6	15.5	*18.6*	*19.0*
ICT applications				
ICT expenditure (% GDP)	5.9	*5.3*	*5.0*	*5.1*
E-government readiness index (scale 0–1)	..	0.49	0.49	0.39
Secure Internet servers (per 1 million people)	*14.7*	11.7	10.7	6.4
Schools connected to the Internet (%)	..	*65*	*60*	*65*

GNI per Capita, Atlas Method, 2000–4

Current US$

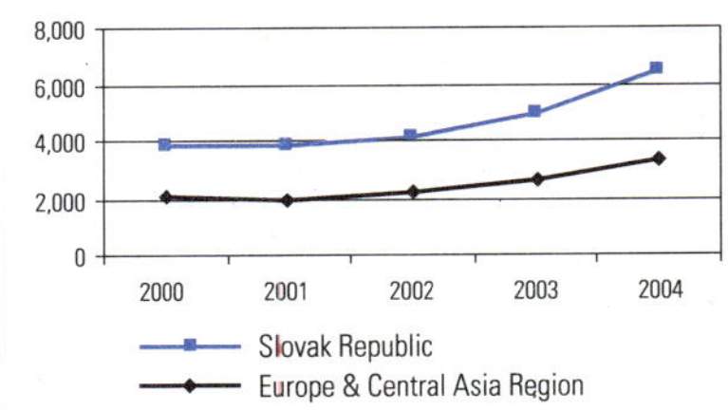

ICT MDG[b] Indicators, 2000–4

Number per 1,000 people

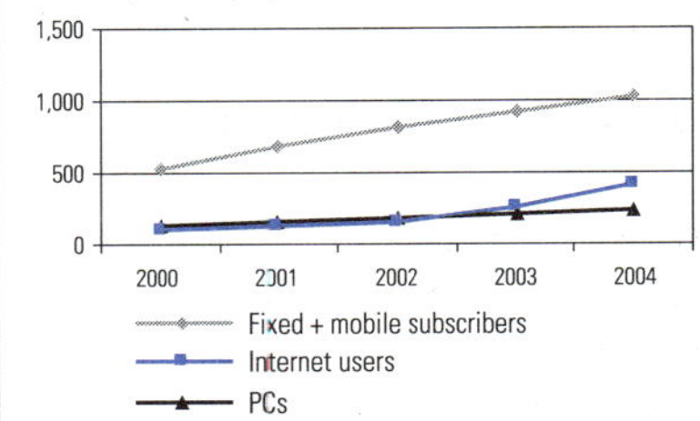

Price of Call to the United States, 2000–4

US$ per 3 minutes

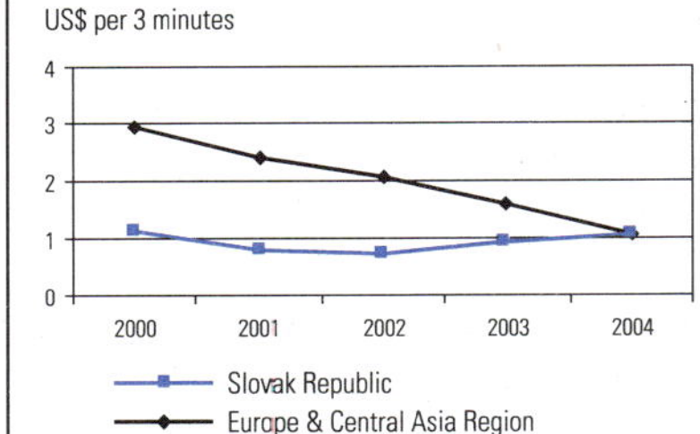

Total Telecommunications Revenue, 2000–4

Percentage of GDP

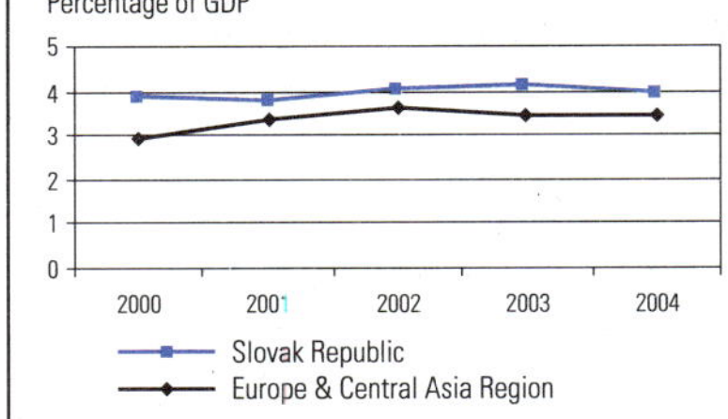

Notes: Figures in italics are for years other than those specified. .. indicates data are not available. C = competition; GDP = gross domestic product; GNI = gross national income; ICT = information and communication technology; M = monopoly; MDG = Millennium Development Goal; P = partial competition; and PCs = personal computers.

a. Outgoing and incoming.

b. Millennium Development Goal indicators 47, 48a, and 48b.

Sources: Economic and social context: UNDP and World Bank; ICT sector structure: ITU, WEF; ICT sector performance: Global Insight/WITSA, ITU, Netcraft, UNDESA, UNPAN, and World Bank. Produced by the Global Information and Communication Technologies Department and the Development Economics Data Group. For complete information, see Definitions and Data Sources.

Slovenia

	Slovenia 2000	Slovenia 2004	High-income group 2004
Economic and social context			
Population, total (millions)	2	2	1,001
Urban population (% total population)	51	51	77
Poverty (% population below US$1 per day)	*<2*	..	..
GNI per capita, Atlas method (current US$)	10,630	14,810	32,040
GDP growth, 1995–2000 and 2000–4 (%)	4.3	3.2	2.0
Adult literacy rate (% ages 15 and over)	..	*100*	..
Primary, secondary, tertiary school enrollment (% gross)	90	*95*	*93*
ICT sector structure			
Separate telecommunications regulator	No	Yes	
Status of main fixed-line operator	*Mixed*	Mixed	
Level of competition: international long distance	*M*	M	
Level of competition: mobile	*P*	P	
Level of competition: Internet service provider	*C*	C	
Government prioritization of ICT (scale 1–7)	..	4.3	4.9
ICT sector performance			
Access			
Telephone main lines (per 1,000 people)	395	*407*	558
International voice traffic (minutes per person)[a]	..	..	..
Mobile subscribers (per 1,000 people)	611	952	767
Population covered by mobile telephony (%)	98	99	98
Internet users (per 1,000 people)	158	319	*480*
Personal computers (per 1,000 people)	276	*301*	*504*
Households with television (%)	97	98	*98*
Quality			
Telephone faults (per 100 main lines per year)	20.5	*22.5*	..
Broadband subscribers (per 1,000 people)	*2.8*	59.2	126.2
International Internet bandwidth (bits per person)	95	1,086	4,718
Affordability			
Price basket for fixed line (US$ per month, residential)	10.1	*12.6*	25.8
Price basket for mobile (US$ per month)	..	11.7	17.8
Price basket for Internet (US$ per month)	..	*25.4*	*20.9*
Price of call to United States (US$ per 3 minutes)	0.81	0.65	0.77
Institutional efficiency and sustainability			
Total telecommunications revenue (% GDP)	1.8	2.8	2.9
Total telephone subscribers per employee	541	*556*	*485*
Total telecommunications investment (% revenue)	101.1	*22.1*	*12.3*
ICT applications			
ICT expenditure (% GDP)	..	..	*7.1*
E-government readiness index (scale 0–1)	..	0.51	0.70
Secure Internet servers (per 1 million people)	*51.2*	65.2	311.4
Schools connected to the Internet (%)	..	*99*	*99*

GNI per Capita, Atlas Method, 2000–4
Current US$

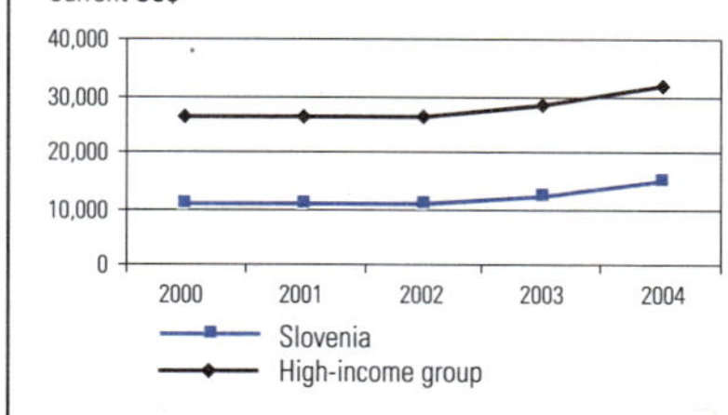

ICT MDG[b] Indicators, 2000–4
Number per 1,000 people

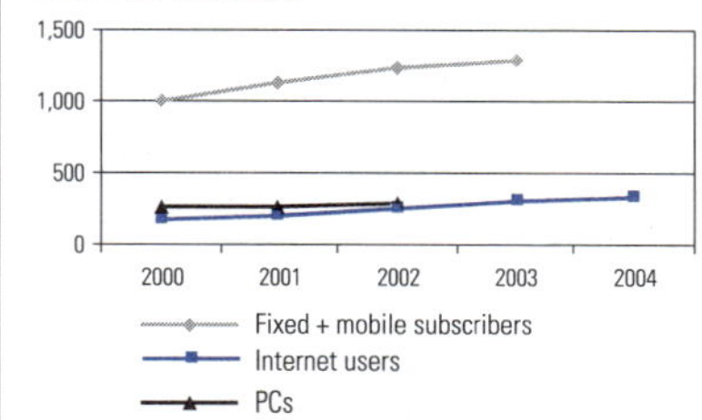

Price of Call to the United States, 2000–4
US$ per 3 minutes

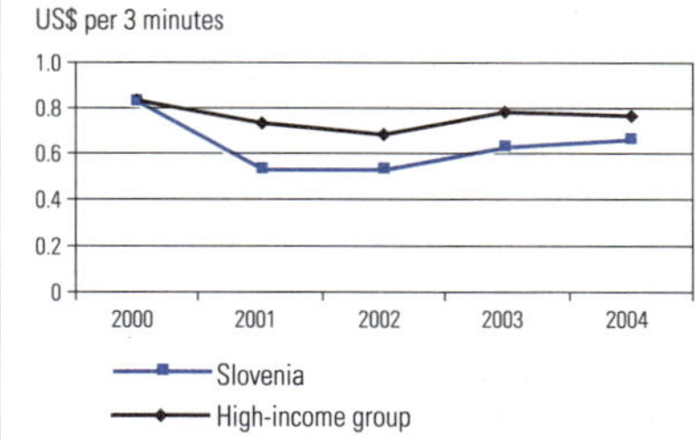

Total Telecommunications Revenue, 2000–4
Percentage of GDP

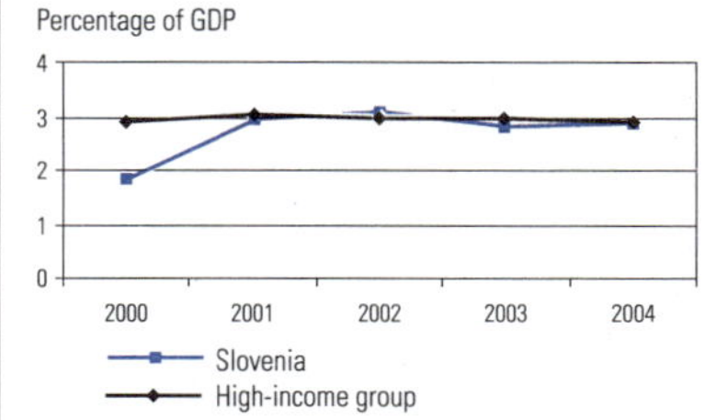

Notes: Figures in italics are for years other than those specified. .. indicates data are not available. C = competition; GDP = gross domestic product; GNI = gross national income; ICT = information and communication technology; M = monopoly; MDG = Millennium Development Goal; P = partial competition; and PCs = personal computers.

a. Outgoing and incoming.

b. Millennium Development Goal indicators 47, 48a, and 48b.

Sources: Economic and social context: UNDP and World Bank; ICT sector structure: ITU, WEF; ICT sector performance: Global Insight/WITSA, ITU, Netcraft, UNDESA, UNPAN, and World Bank. Produced by the Global Information and Communication Technologies Department and the Development Economics Data Group. For complete information, see Definitions and Data Sources.

South Africa

	South Africa 2000	South Africa 2004	Upper-middle-income group 2004	Sub-Saharan Africa Region 2004
Economic and social context				
Population, total (millions)	44	46	576	719
Urban population (% total population)	55	57	72	37
Poverty (% population below US$1 per day)	10.7	..	..	*46.4*
GNI per capita, Atlas method (current US$)	3,050	3,630	4,770	600
GDP growth, 1995–2000 and 2000–4 (%)	2.5	3.2	2.7	3.9
Adult literacy rate (% ages 15 and over)	..	..	94	59
Primary, secondary, tertiary school enrollment (% gross)	76	*78*	*80*	*52*
ICT sector structure				
Separate telecommunications regulator	Yes	Yes		
Status of main fixed-line operator	*Mixed*	Mixed		
Level of competition: international long distance	*M*	C		
Level of competition: mobile	*C*	P		
Level of competition: Internet service provider	..	C		
Government prioritization of ICT (scale 1–7)	..	4.8	4.1	*4.4*
ICT sector performance				
Access				
Telephone main lines (per 1,000 people)	113	104	*220*	17
International voice traffic (minutes per person)[a]	22	40	*39*	..
Mobile subscribers (per 1,000 people)	190	471	490	86
Population covered by mobile telephony (%)	92	96	*84*	..
Internet users (per 1,000 people)	55	81	*133*	15
Personal computers (per 1,000 people)	66	88	*99*	*12*
Households with television (%)	55	*54*	*92*	*15*
Quality				
Telephone faults (per 100 main lines per year)	51.8	43.3	20.3	..
Broadband subscribers (per 1,000 people)	0.0	1.3	*3.7*	*0.1*
International Internet bandwidth (bits per person)	8	29	176	*4*
Affordability				
Price basket for fixed line (US$ per month, residential)	13.3	21.6	*13.9*	*8.5*
Price basket for mobile (US$ per month)	..	13.5	11.1	13.5
Price basket for Internet (US$ per month)	..	*33.3*	*20.8*	*54.8*
Price of call to United States (US$ per 3 minutes)	1.98	0.79	1.03	2.43
Institutional efficiency and sustainability				
Total telecommunications revenue (% GDP)	5.1	5.8	3.4	5.0
Total telephone subscribers per employee	264	672	*402*	*144*
Total telecommunications investment (% revenue)	25.5	11.3	*18.6*	*27.8*
ICT applications				
ICT expenditure (% GDP)	7.9	*7.8*	*5.0*	..
E-government readiness index (scale 0–1)	..	0.52	0.49	0.11
Secure Internet servers (per 1 million people)	*11.6*	19.9	10.7	1.9
Schools connected to the Internet (%)	..	*27*	*60*	..

GNI per Capita, Atlas Method, 2000–4

Current US$

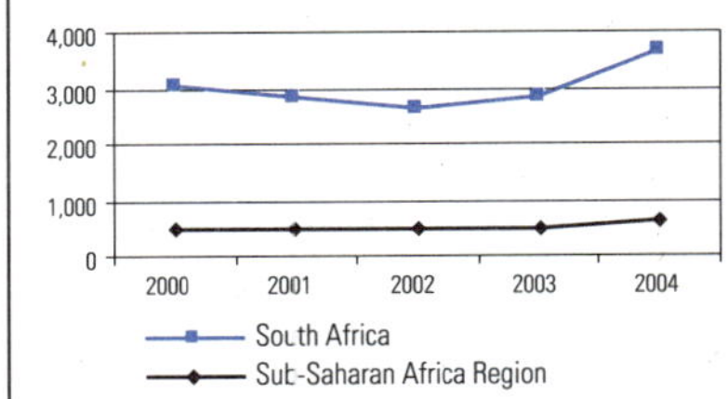

ICT MDG[b] Indicators, 2000–4

Number per 1,000 people

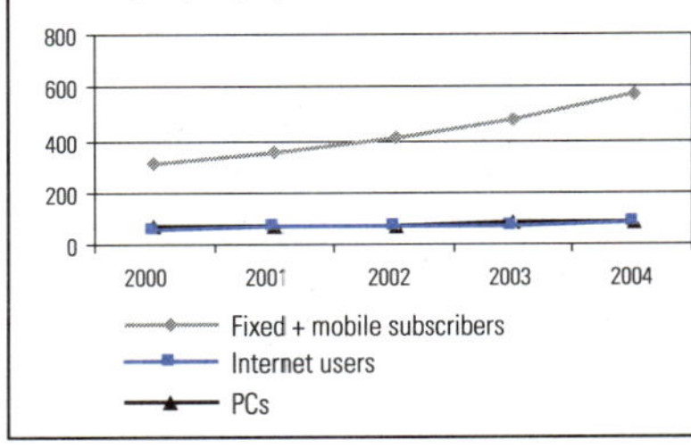

Price of Call to the United States, 2000–4

US$ per 3 minutes

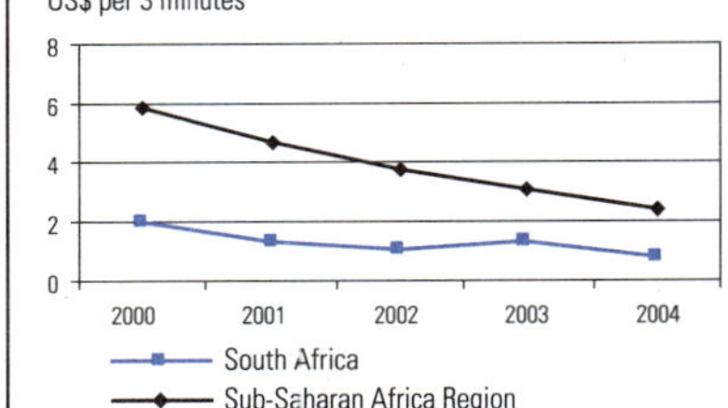

Total Telecommunications Revenue, 2000–4

Percentage of GDP

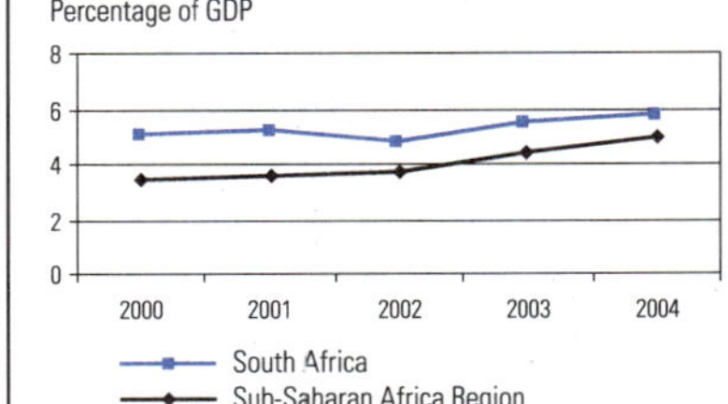

Notes: Figures in italics are for years other than those specified. .. indicates data are not available. C = competition; GDP = gross domestic product; GNI = gross national income; ICT = information and communication technology; M = monopoly; MDG = Millennium Development Goal; P = partial competition; and PCs = personal computers.

a. Outgoing and incoming.

b. Millennium Development Goal indicators 47, 48a, and 48b.

Sources: Economic and social context: UNDP and World Bank; ICT sector structure: ITU, WEF; ICT sector performance: Global Insight/WITSA, ITU, Netcraft, UNDESA, UNPAN, and World Bank. Produced by the Global Information and Communication Technologies Department and the Development Economics Data Group. For complete information, see Definitions and Data Sources.

Spain

	Spain 2000	Spain 2004	High-income group 2004
Economic and social context			
Population, total (millions)	40	41	1,001
Urban population (% total population)	76	77	77
Poverty (% population below US$1 per day)	..	..	..
GNI per capita, Atlas method (current US$)	14,790	21,210	32,040
GDP growth, 1995–2000 and 2000–4 (%)	4.0	2.5	2.0
Adult literacy rate (% ages 15 and over)	..	..	..
Primary, secondary, tertiary school enrollment (% gross)	91	*94*	*93*
ICT sector structure			
Separate telecommunications regulator	Yes	Yes	
Status of main fixed-line operator	*Private*	Private	
Level of competition: international long distance	*C*	C	
Level of competition: mobile	*C*	C	
Level of competition: Internet service provider	*C*	C	
Government prioritization of ICT (scale 1–7)	..	4.5	4.9
ICT sector performance			
Access			
Telephone main lines (per 1,000 people)	422	434	558
International voice traffic (minutes per person)[a]	139	..	..
Mobile subscribers (per 1,000 people)	599	935	767
Population covered by mobile telephony (%)	99	99	98
Internet users (per 1,000 people)	135	317	*480*
Personal computers (per 1,000 people)	143	*199*	*504*
Households with television (%)	99	*99*	*98*
Quality			
Telephone faults (per 100 main lines per year)	*1.5*	*4.0*	..
Broadband subscribers (per 1,000 people)	1.9	83.6	126.2
International Internet bandwidth (bits per person)	295	2,918	4,718
Affordability			
Price basket for fixed line (US$ per month, residential)	14.6	*20.3*	25.8
Price basket for mobile (US$ per month)	..	21.5	17.8
Price basket for Internet (US$ per month)	..	*20.7*	*20.9*
Price of call to United States (US$ per 3 minutes)	1.08	0.60	0.77
Institutional efficiency and sustainability			
Total telecommunications revenue (% GDP)	4.1	4.7	2.9
Total telephone subscribers per employee	638	871	*485*
Total telecommunications investment (% revenue)	29.6	*13.2*	*12.3*
ICT applications			
ICT expenditure (% GDP)	4.1	*3.8*	*7.1*
E-government readiness index (scale 0–1)	..	0.39	0.70
Secure Internet servers (per 1 million people)	*23.0*	68.7	311.4
Schools connected to the Internet (%)	..	*94*	*99*

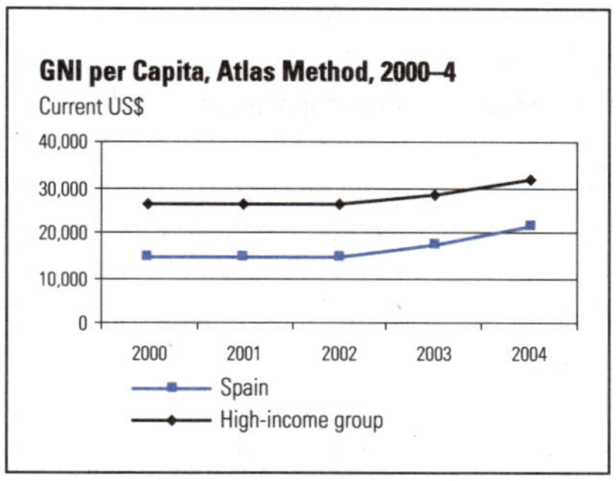

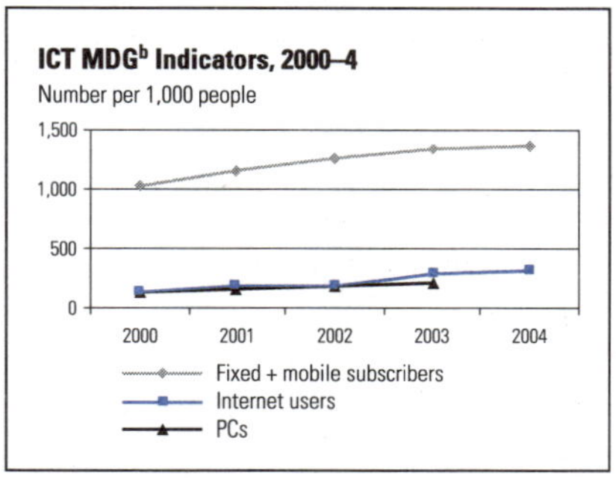

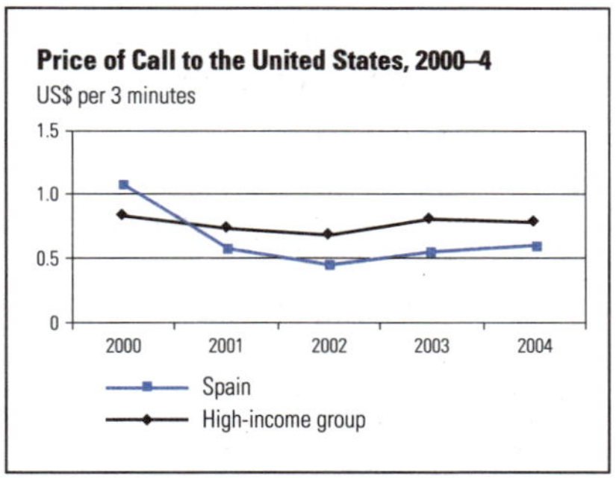

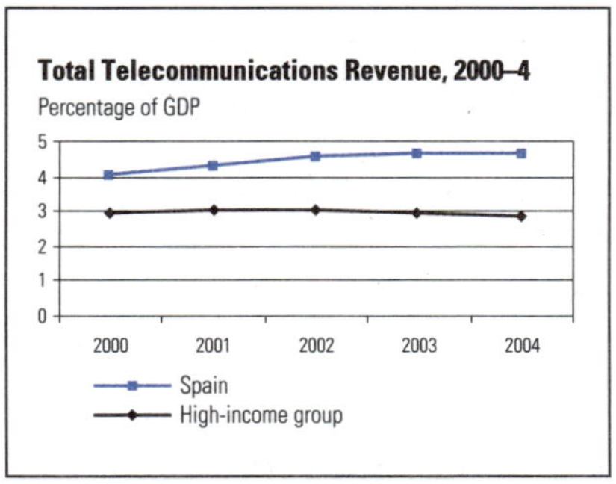

Notes: Figures in italics are for years other than those specified. .. indicates data are not available. C = competition; GDP = gross domestic product; GNI = gross national income; ICT = information and communication technology; M = monopoly; MDG = Millennium Development Goal; P = partial competition; and PCs = personal computers.

a. Outgoing and incoming.

b. Millennium Development Goal indicators 47, 48a, and 48b.

Sources: Economic and social context: UNDP and World Bank; ICT sector structure: ITU, WEF; ICT sector performance: Global Insight/WITSA, ITU, Netcraft, UNDESA, UNPAN, and World Bank. Produced by the Global Information and Communication Technologies Department and the Development Economics Data Group. For complete information, see Definitions and Data Sources.

Sri Lanka

	Sri Lanka 2000	Sri Lanka 2004	Lower-middle-income group 2004	South Asia Region 2004
Economic and social context				
Population, total (millions)	18	19	2,430	1,448
Urban population (% total population)	21	21	49	28
Poverty (% population below US$1 per day)	7.6	..	..	*31.3*
GNI per capita, Atlas method (current US$)	850	1,010	1,580	590
GDP growth, 1995–2000 and 2000–4 (%)	5.0	3.8	5.7	5.8
Adult literacy rate (% ages 15 and over)	*90*	..	90	58
Primary, secondary, tertiary school enrollment (% gross)	..	*68*	*72*	*56*
ICT sector structure				
Separate telecommunications regulator	Yes	Yes		
Status of main fixed-line operator	*Mixed*	Mixed		
Level of competition: international long distance	*M*	P		
Level of competition: mobile	*P*	C		
Level of competition: Internet service provider	*C*	C		
Government prioritization of ICT (scale 1–7)	..	4.8	3.8	5.3
ICT sector performance				
Access				
Telephone main lines (per 1,000 people)	42	51	192	41
International voice traffic (minutes per person)[a]	10	*20*	..	*4*
Mobile subscribers (per 1,000 people)	23	114	255	47
Population covered by mobile telephony (%)	..	40	*76*	*43*
Internet users (per 1,000 people)	7	14	70	21
Personal computers (per 1,000 people)	7	*13*	38	11
Households with television (%)	22	*32*	*89*	*32*
Quality				
Telephone faults (per 100 main lines per year)	132.0	*18.5*	..	*88.1*
Broadband subscribers (per 1,000 people)	*0.0*	*0.1*	12.6	0.6
International Internet bandwidth (bits per person)	1	17	58	4
Affordability				
Price basket for fixed line (US$ per month, residential)	6.9	7.3	*5.5*	3.2
Price basket for mobile (US$ per month)	..	3.7	8.9	3.2
Price basket for Internet (US$ per month)	..	*15.1*	*25.3*	*15.1*
Price of call to United States (US$ per 3 minutes)	3.29	2.11	1.45	1.21
Institutional efficiency and sustainability				
Total telecommunications revenue (% GDP)	1.9	2.2	3.6	1.9
Total telephone subscribers per employee	101	*166*	*195*	*89*
Total telecommunications investment (% revenue)	35.8	*24.6*	*25.5*	*15.3*
ICT applications				
ICT expenditure (% GDP)	5.0	*5.7*	*5.1*	*4.1*
E-government readiness index (scale 0–1)	..	0.27	0.26	0.34
Secure Internet servers (per 1 million people)	*0.3*	1.5	1.6	0.4
Schools connected to the Internet (%)	..	..	..	..

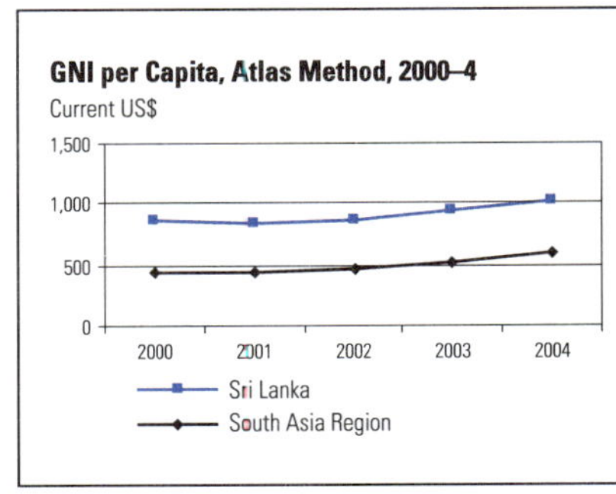

GNI per Capita, Atlas Method, 2000–4

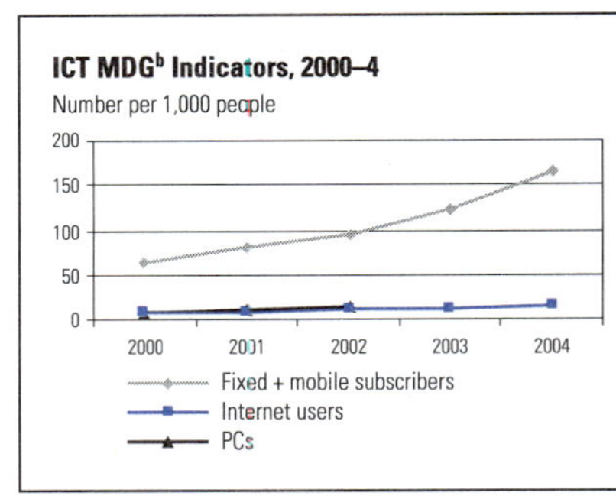

ICT MDG[b] Indicators, 2000–4

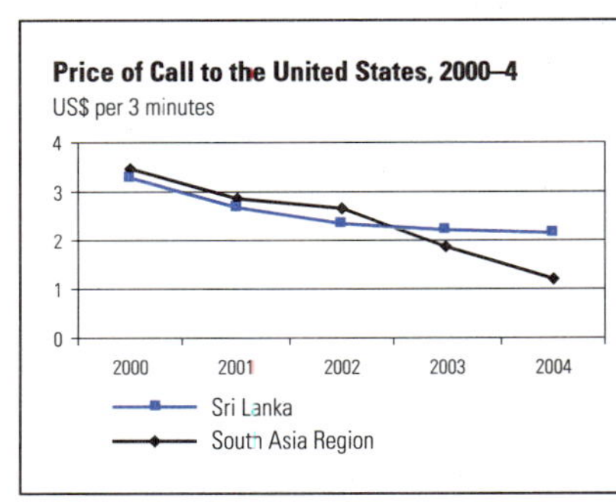

Price of Call to the United States, 2000–4

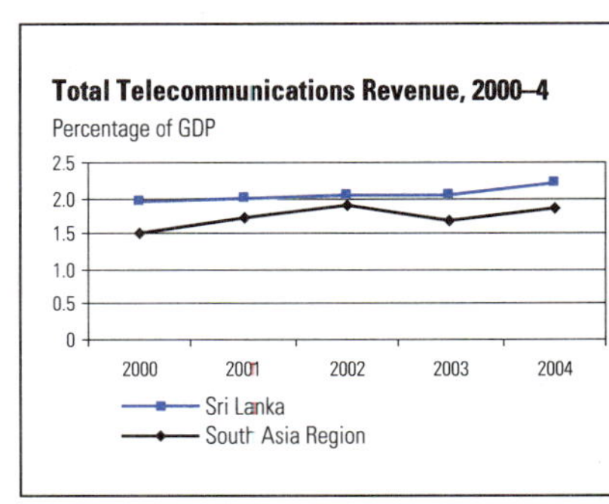

Total Telecommunications Revenue, 2000–4

Notes: Figures in italics are for years other than those specified. .. indicates data are not available. C = competition; GDP = gross domestic product; GNI = gross national income; ICT = information and communication technology; M = monopoly; MDG = Millennium Development Goal; P = partial competition; and PCs = personal computers.

a. Outgoing and incoming.

b. Millennium Development Goal indicators 47, 48a, and 48b.

Sources: Economic and social context: UNDP and World Bank; ICT sector structure: ITU, WEF; ICT sector performance: Global Insight/WITSA, ITU, Netcraft, UNDESA, UNPAN, and World Bank. Produced by the Global Information and Communication Technologies Department and the Development Economics Data Group. For complete information, see Definitions and Data Sources.

Sudan

	Sudan 2000	Sudan 2004	Low-income group 2004	Sub-Saharan Africa Region 2004
Economic and social context				
Population, total (millions)	31	34	2,338	719
Urban population (% total population)	36	40	31	37
Poverty (% population below US$1 per day)	..	..	..	*46.4*
GNI per capita, Atlas method (current US$)	330	530	510	600
GDP growth, 1995–2000 and 2000–4 (%)	6.3	6.0	5.4	3.9
Adult literacy rate (% ages 15 and over)	59	..	61	59
Primary, secondary, tertiary school enrollment (% gross)	36	*38*	55	52
ICT sector structure				
Separate telecommunications regulator	Yes	Yes		
Status of main fixed-line operator	*Mixed*	Mixed		
Level of competition: international long distance	*M*	M		
Level of competition: mobile	*M*	P		
Level of competition: Internet service provider	*C*	C		
Government prioritization of ICT (scale 1–7)	..	..	..	*4.4*
ICT sector performance				
Access				
Telephone main lines (per 1,000 people)	12	31	33	17
International voice traffic (minutes per person)[a]	6	*9*	*4*	..
Mobile subscribers (per 1,000 people)	1	22	48	86
Population covered by mobile telephony (%)	60	*60*	*43*	..
Internet users (per 1,000 people)	1	9	20	15
Personal computers (per 1,000 people)	3	*6*	*8*	*12*
Households with television (%)	45	*49*	*16*	*15*
Quality				
Telephone faults (per 100 main lines per year)	*5.0*	17.0	..	..
Broadband subscribers (per 1,000 people)	0.0	0.1	0.5	*0.1*
International Internet bandwidth (bits per person)	0	6	*3*	*4*
Affordability				
Price basket for fixed line (US$ per month, residential)	3.8	*4.4*	*6.6*	*8.5*
Price basket for mobile (US$ per month)	..	3.2	11.6	13.5
Price basket for Internet (US$ per month)	..	*160.7*	*45.5*	*54.8*
Price of call to United States (US$ per 3 minutes)	42.02	..	1.95	2.43
Institutional efficiency and sustainability				
Total telecommunications revenue (% GDP)	1.1	2.4	2.3	5.0
Total telephone subscribers per employee	146	*225*	*89*	*144*
Total telecommunications investment (% revenue)	71.2	*70.7*	*27.8*	*27.8*
ICT applications				
ICT expenditure (% GDP)	..	..	*4.1*	..
E-government readiness index (scale 0–1)	..	0.14	0.12	0.11
Secure Internet servers (per 1 million people)	..	..	0.3	1.9
Schools connected to the Internet (%)	..	..	..	..

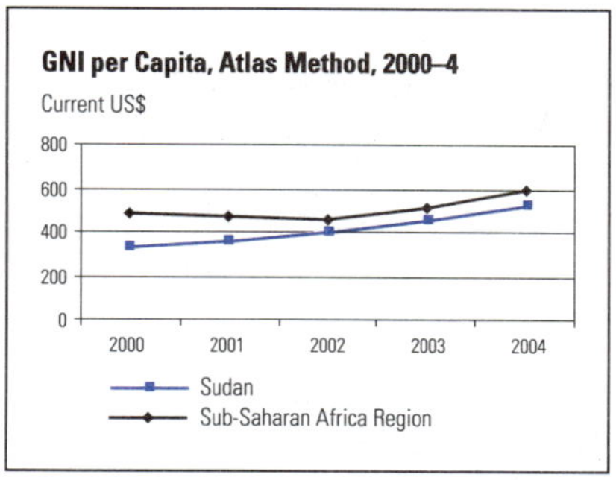
GNI per Capita, Atlas Method, 2000–4

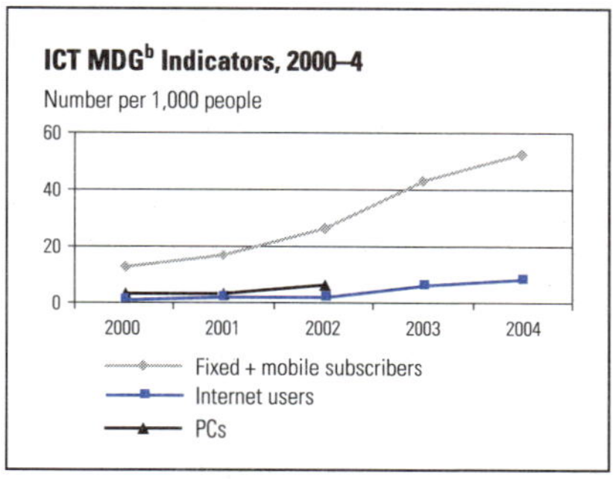
ICT MDG[b] Indicators, 2000–4

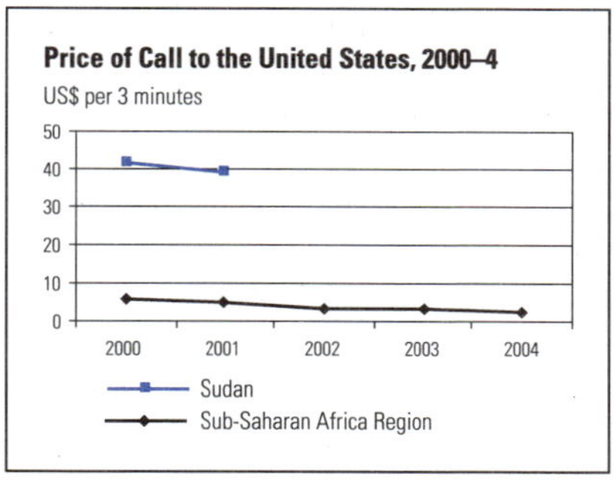
Price of Call to the United States, 2000–4

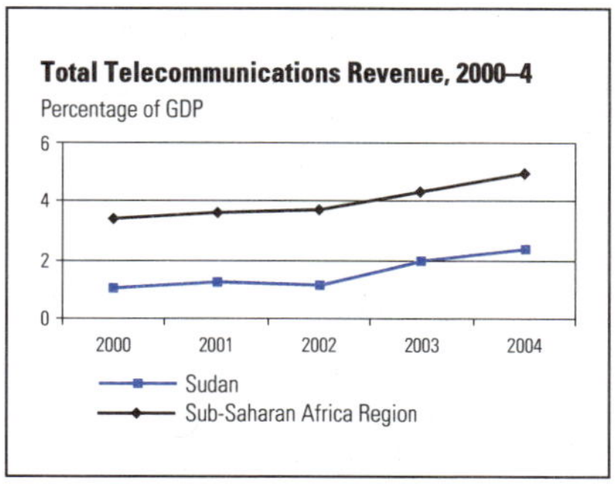
Total Telecommunications Revenue, 2000–4

Notes: Figures in italics are for years other than those specified. .. indicates data are not available. C = competition; GDP = gross domestic product; GNI = gross national income; ICT = information and communication technology; M = monopoly; MDG = Millennium Development Goal; P = partial competition; and PCs = personal computers.

a. Outgoing and incoming.

b. Millennium Development Goal indicators 47, 48a, and 48b.

Sources: Economic and social context: UNDP and World Bank; ICT sector structure: ITU, WEF; ICT sector performance: Global Insight/WITSA, ITU, Netcraft, UNDESA, UNPAN, and World Bank. Produced by the Global Information and Communication Technologies Department and the Development Economics Data Group. For complete information, see Definitions and Data Sources.

Swaziland

	Swaziland 2000	Swaziland 2004	Lower-middle-income group 2004	Sub-Saharan Africa Region 2004
Economic and social context				
Population, total (millions)	1	1	2,430	719
Urban population (% total population)	23	24	49	37
Poverty (% population below US$1 per day)	..	..	..	*46.4*
GNI per capita, Atlas method (current US$)	1,370	1,660	1,580	600
GDP growth, 1995–2000 and 2000–4 (%)	3.4	2.3	5.7	3.9
Adult literacy rate (% ages 15 and over)	79	..	90	59
Primary, secondary, tertiary school enrollment (% gross)	62	*60*	*72*	*52*
ICT sector structure				
Separate telecommunications regulator	..	..		
Status of main fixed-line operator	*Public*	Public		
Level of competition: international long distance	*M*	M		
Level of competition: mobile	*M*	M		
Level of competition: Internet service provider	*C*	C		
Government prioritization of ICT (scale 1–7)	..	..	3.8	*4.4*
ICT sector performance				
Access				
Telephone main lines (per 1,000 people)	30	*42*	192	17
International voice traffic (minutes per person)[a]	47	*38*	..	..
Mobile subscribers (per 1,000 people)	32	129	255	86
Population covered by mobile telephony (%)	70	90	*76*	..
Internet users (per 1,000 people)	10	*24*	70	15
Personal computers (per 1,000 people)	11	*27*	38	*12*
Households with television (%)	18	*18*	*89*	*15*
Quality				
Telephone faults (per 100 main lines per year)	160.0	*100.0*	..	..
Broadband subscribers (per 1,000 people)	0.0	0.0	12.6	*0.1*
International Internet bandwidth (bits per person)	1	*1*	58	*4*
Affordability				
Price basket for fixed line (US$ per month, residential)	4.1	8.3	*5.5*	*8.5*
Price basket for mobile (US$ per month)	..	16.6	8.9	13.5
Price basket for Internet (US$ per month)	..	*20.6*	*25.3*	*54.8*
Price of call to United States (US$ per 3 minutes)	3.68	2.97	1.45	2.43
Institutional efficiency and sustainability				
Total telecommunications revenue (% GDP)	2.7	3.7	3.6	5.0
Total telephone subscribers per employee	137	*209*	*195*	*144*
Total telecommunications investment (% revenue)	9.9	*56.1*	*25.5*	*27.8*
ICT applications				
ICT expenditure (% GDP)	..	..	*5.1*	..
E-government readiness index (scale 0–1)	..	0.26	0.26	0.11
Secure Internet servers (per 1 million people)	*0.9*	1.8	1.6	1.9
Schools connected to the Internet (%)	..	..	..	..

GNI per Capita, Atlas Method, 2000–4

Current US$

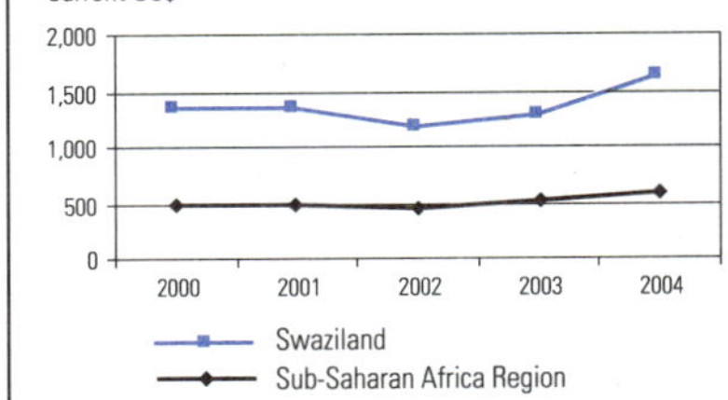

ICT MDG[b] Indicators, 2000–4

Number per 1,000 people

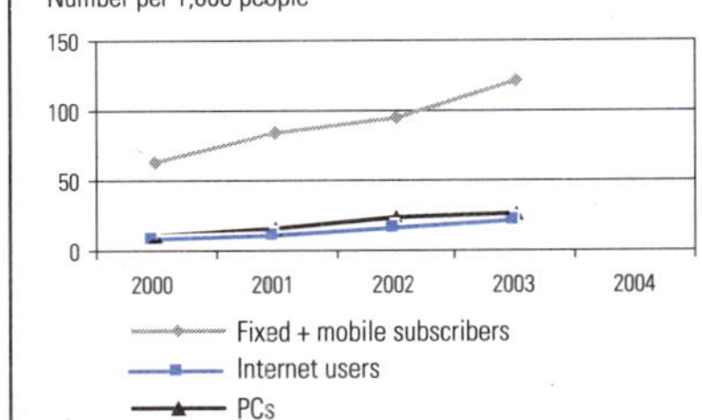

Price of Call to the United States, 2000–4

US$ per 3 minutes

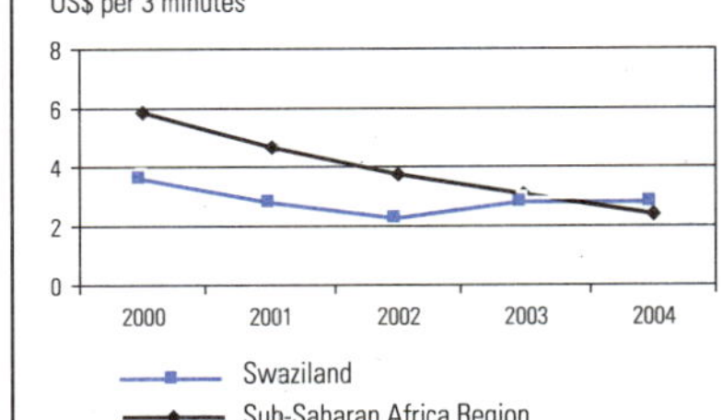

Total Telecommunications Revenue, 2000–4

Percentage of GDP

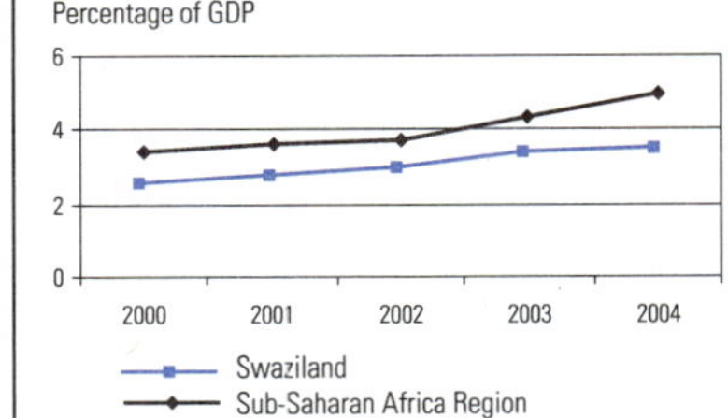

Notes: Figures in italics are for years other than those specified. .. indicates data are not available. C = competition; GDP = gross domestic product; GNI = gross national income; ICT = information and communication technology; M = monopoly; MDG = Millennium Development Goal; P = partial competition; and PCs = personal computers.

a. Outgoing and incoming.

b. Millennium Development Goal indicators 47, 48a, and 48b.

Sources: Economic and social context: UNDP and World Bank; ICT sector structure: ITU, WEF; ICT sector performance: Global Insight/WITSA, ITU, Netcraft, UNDESA, UNPAN, and World Bank. Produced by the Global Information and Communication Technologies Department and the Development Economics Data Group. For complete information, see Definitions and Data Sources.

Sweden

	Sweden 2000	Sweden 2004	High-income group 2004
Economic and social context			
Population, total (millions)	9	9	1,001
Urban population (% total population)	83	83	77
Poverty (% population below US$1 per day)	..	..	..
GNI per capita, Atlas method (current US$)	28,650	35,770	32,040
GDP growth, 1995–2000 and 2000–4 (%)	3.3	2.0	2.0
Adult literacy rate (% ages 15 and over)	..	..	..
Primary, secondary, tertiary school enrollment (% gross)	113	*114*	*93*
ICT sector structure			
Separate telecommunications regulator	Yes	Yes	
Status of main fixed-line operator	*Mixed*	Mixed	
Level of competition: international long distance	*C*	C	
Level of competition: mobile	*P*	P	
Level of competition: Internet service provider	*C*	*C*	
Government prioritization of ICT (scale 1–7)	..	5.1	4.9
ICT sector performance			
Access			
Telephone main lines (per 1,000 people)	759	709	558
International voice traffic (minutes per person)[a]	245	*264*	..
Mobile subscribers (per 1,000 people)	718	1,026	767
Population covered by mobile telephony (%)	99	99	98
Internet users (per 1,000 people)	456	592	*480*
Personal computers (per 1,000 people)	507	*640*	*504*
Households with television (%)	94	*94*	*98*
Quality			
Telephone faults (per 100 main lines per year)	..	..	..
Broadband subscribers (per 1,000 people)	9.3	152.7	126.2
International Internet bandwidth (bits per person)	2,098	17,544	4,718
Affordability			
Price basket for fixed line (US$ per month, residential)	19.6	27.1	25.8
Price basket for mobile (US$ per month)	..	15.8	17.8
Price basket for Internet (US$ per month)	..	*22.4*	*20.9*
Price of call to United States (US$ per 3 minutes)	0.36	0.41	0.77
Institutional efficiency and sustainability			
Total telecommunications revenue (% GDP)	3.2	1.9	2.9
Total telephone subscribers per employee	456	*772*	*485*
Total telecommunications investment (% revenue)	31.9	*10.9*	*12.3*
ICT applications			
ICT expenditure (% GDP)	7.5	*7.0*	*7.1*
E-government readiness index (scale 0–1)	..	0.77	0.70
Secure Internet servers (per 1 million people)	*116.1*	262.0	311.4
Schools connected to the Internet (%)	..	*99*	*99*

GNI per Capita, Atlas Method, 2000–4

Current US$

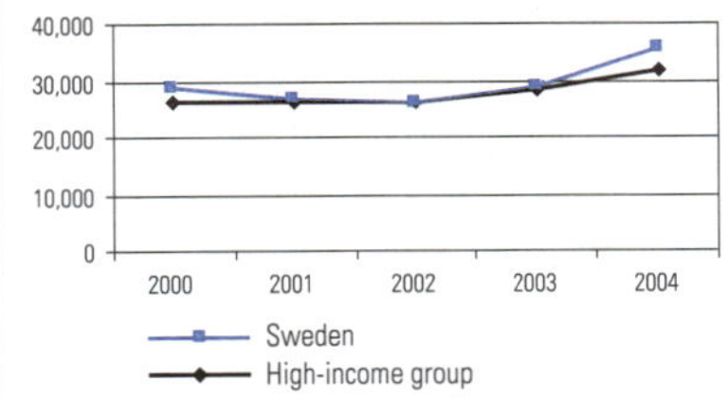

ICT MDG[b] Indicators, 2000–4

Number per 1,000 people

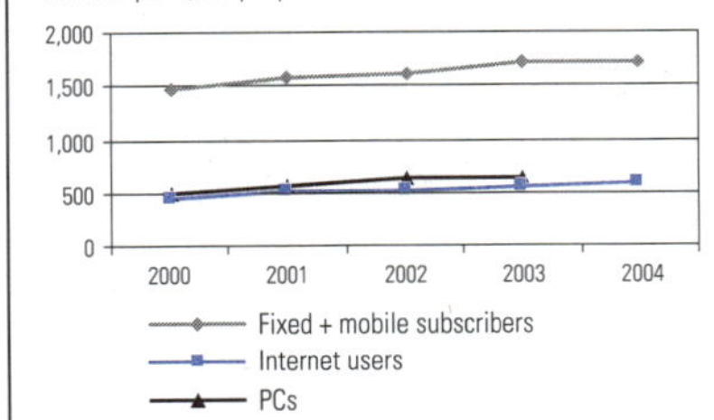

Price of Call to the United States, 2000–4

US$ per 3 minutes

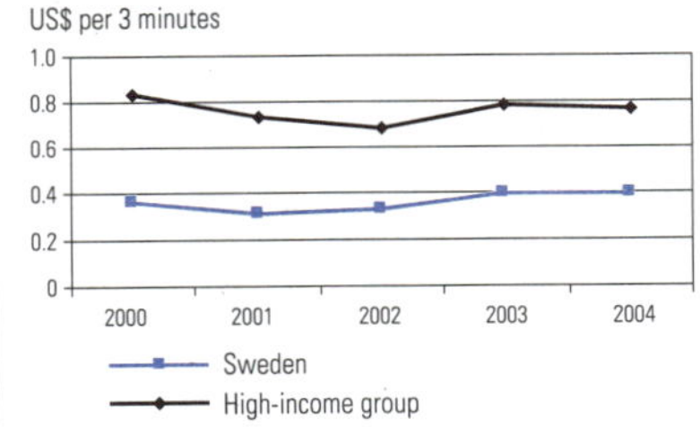

Total Telecommunications Revenue, 2000–4

Percentage of GDP

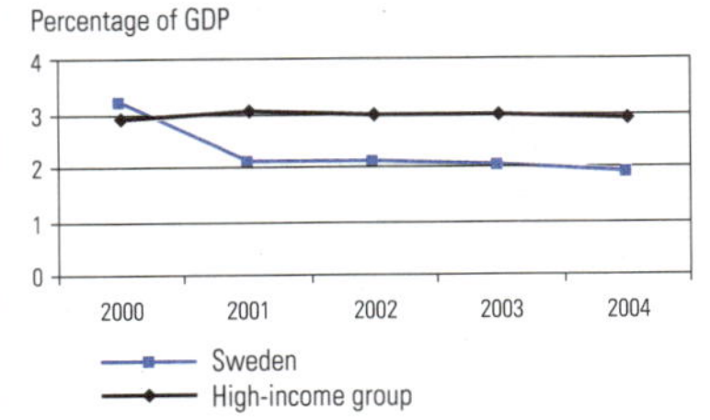

Notes: Figures in italics are for years other than those specified. .. indicates data are not available. C = competition; GDP = gross domestic product; GNI = gross national income; ICT = information and communication technology; M = monopoly; MDG = Millennium Development Goal; P = partial competition; and PCs = personal computers.

a. Outgoing and incoming.

b. Millennium Development Goal indicators 47, 48a, and 48b.

Sources: Economic and social context: UNDP and World Bank; ICT sector structure: ITU, WEF; ICT sector performance: Global Insight/WITSA, ITU, Netcraft, UNDESA, UNPAN, and World Bank. Produced by the Global Information and Communication Technologies Department and the Development Economics Data Group. For complete information, see Definitions and Data Sources.

Switzerland

	Switzerland 2000	Switzerland 2004	High-income group 2004
Economic and social context			
Population, total (millions)	7	7	1,001
Urban population (% total population)	68	68	77
Poverty (% population below US$1 per day)	..	..	..
GNI per capita, Atlas method (current US$)	40,160	48,230	32,040
GDP growth, 1995–2000 and 2000–4 (%)	2.0	0.5	2.0
Adult literacy rate (% ages 15 and over)	..	..	..
Primary, secondary, tertiary school enrollment (% gross)	88	*90*	*93*
ICT sector structure			
Separate telecommunications regulator	Yes	Yes	
Status of main fixed-line operator	*Mixed*	Mixed	
Level of competition: international long distance	*C*	C	
Level of competition: mobile	*C*	P	
Level of competition: Internet service provider	*C*	C	
Government prioritization of ICT (scale 1–7)	..	4.7	4.9
ICT sector performance			
Access			
Telephone main lines (per 1,000 people)	729	701	558
International voice traffic (minutes per person)[a]	634	*704*	..
Mobile subscribers (per 1,000 people)	646	852	767
Population covered by mobile telephony (%)	98	99	98
Internet users (per 1,000 people)	292	421	*480*
Personal computers (per 1,000 people)	655	827	*504*
Households with television (%)	99	99	*98*
Quality			
Telephone faults (per 100 main lines per year)	*18.5*	*8.0*	..
Broadband subscribers (per 1,000 people)	6.0	173.7	126.2
International Internet bandwidth (bits per person)	2,942	9,681	4,718
Affordability			
Price basket for fixed line (US$ per month, residential)	20.3	29.6	25.8
Price basket for mobile (US$ per month)	..	33.0	17.8
Price basket for Internet (US$ per month)	..	*22.4*	*20.9*
Price of call to United States (US$ per 3 minutes)	0.21	0.29	0.77
Institutional efficiency and sustainability			
Total telecommunications revenue (% GDP)	3.4	3.6	2.9
Total telephone subscribers per employee	409	*513*	*485*
Total telecommunications investment (% revenue)	27.2	*14.1*	*12.3*
ICT applications			
ICT expenditure (% GDP)	7.8	*7.2*	*7.1*
E-government readiness index (scale 0–1)	..	0.59	0.70
Secure Internet servers (per 1 million people)	*149.2*	382.1	311.4
Schools connected to the Internet (%)	*66*	..	*99*

GNI per Capita, Atlas Method, 2000–4

Current US$

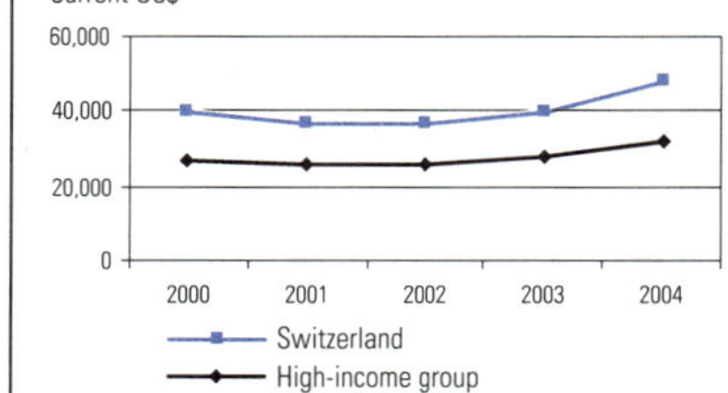

ICT MDG[b] Indicators, 2000–4

Number per 1,000 people

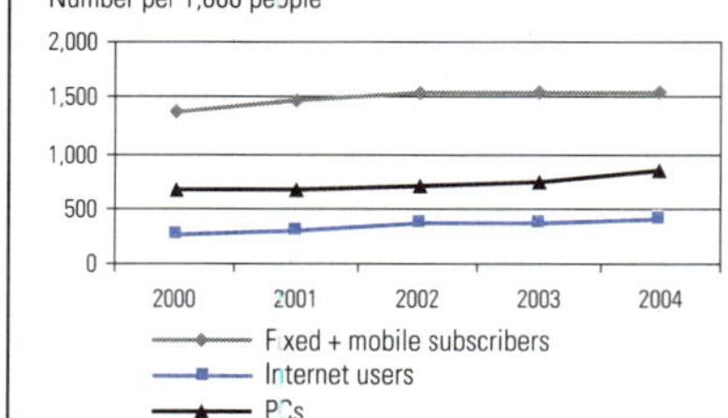

Price of Call to the United States, 2000–4

US$ per 3 minutes

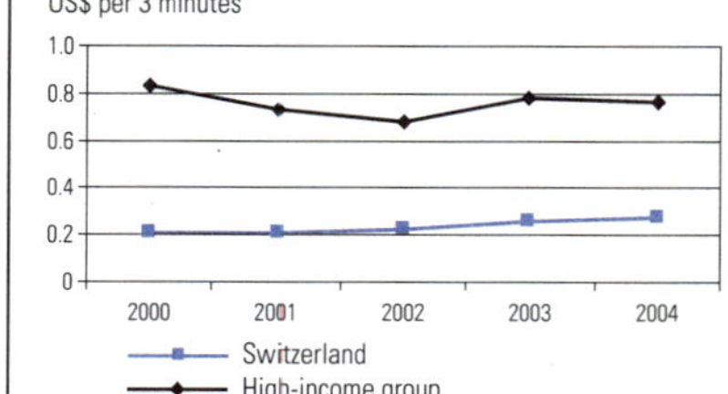

Total Telecommunications Revenue, 2000–4

Percentage of GDP

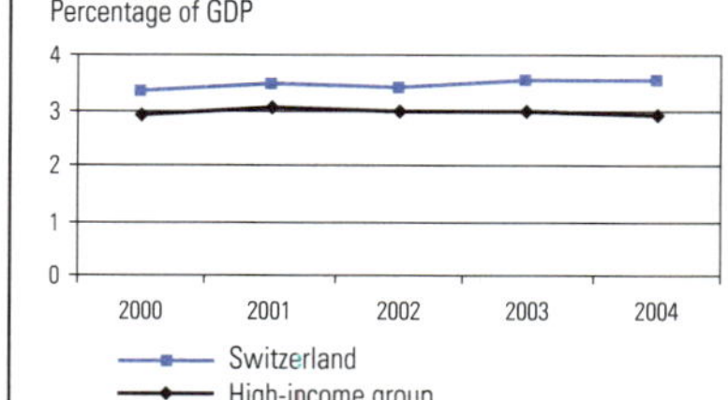

Notes: Figures in italics are for years other than those specified. .. indicates data are not available. C = competition; GDP = gross domestic product; GNI = gross national income; ICT = information and communication technology; M = monopoly; MDG = Millennium Development Goal; P = partial competition; and PCs = personal computers.

a. Outgoing and incoming.

b. Millennium Development Goal indicators 47, 48a, and 48b.

Sources: Economic and social context: UNDP and World Bank; ICT sector structure: ITU, WEF; ICT sector performance: Global Insight/WITSA, ITU, Netcraft, UNDESA, UNPAN, and World Bank. Produced by the Global Information and Communication Technologies Department and the Development Economics Data Group. For complete information, see Definitions and Data Sources.

Syrian Arab Republic

	Syrian Arab Republic 2000	Syrian Arab Republic 2004	Lower-middle-income group 2004	Middle East & North Africa Region 2004
Economic and social context				
Population, total (millions)	16	18	2,430	294
Urban population (% total population)	50	50	49	56
Poverty (% population below US$1 per day)	..	..	..	*2.4*
GNI per capita, Atlas method (current US$)	950	1,190	1,580	2,000
GDP growth, 1995–2000 and 2000–4 (%)	2.5	3.1	5.7	4.5
Adult literacy rate (% ages 15 and over)	..	*83*	90	..
Primary, secondary, tertiary school enrollment (% gross)	59	*62*	*72*	*68*
ICT sector structure				
Separate telecommunications regulator	..	..		
Status of main fixed-line operator	*Public*	Public		
Level of competition: international long distance	*M*	M		
Level of competition: mobile	..	P		
Level of competition: Internet service provider	..	P		
Government prioritization of ICT (scale 1–7)	..	..	3.8	..
ICT sector performance				
Access				
Telephone main lines (per 1,000 people)	103	132	192	*118*
International voice traffic (minutes per person)[a]	24	..	..	*20*
Mobile subscribers (per 1,000 people)	2	141	255	*88*
Population covered by mobile telephony (%)	50	99	*76*	..
Internet users (per 1,000 people)	2	45	70	*47*
Personal computers (per 1,000 people)	15	*19*	38	*30*
Households with television (%)	72	*80*	*89*	*88*
Quality				
Telephone faults (per 100 main lines per year)	50.0	..	..	..
Broadband subscribers (per 1,000 people)	0.0	*0.0*	12.6	*0.2*
International Internet bandwidth (bits per person)	..	*1*	58	*15*
Affordability				
Price basket for fixed line (US$ per month, residential)	3.3	*3.0*	*5.5*	*4.9*
Price basket for mobile (US$ per month)	..	48.2	8.9	8.1
Price basket for Internet (US$ per month)	..	*55.2*	*25.3*	*24.5*
Price of call to United States (US$ per 3 minutes)	4.81	..	1.45	1.64
Institutional efficiency and sustainability				
Total telecommunications revenue (% GDP)	2.0	..	3.6	2.8
Total telephone subscribers per employee	80	..	*195*	..
Total telecommunications investment (% revenue)	58.2	..	*25.5*	*27.8*
ICT applications				
ICT expenditure (% GDP)	..	..	*5.1*	..
E-government readiness index (scale 0–1)	..	0.05	0.26	0.16
Secure Internet servers (per 1 million people)	*0.1*	..	1.6	0.6
Schools connected to the Internet (%)	..	..	..	..

GNI per Capita, Atlas Method, 2000–4
Current US$

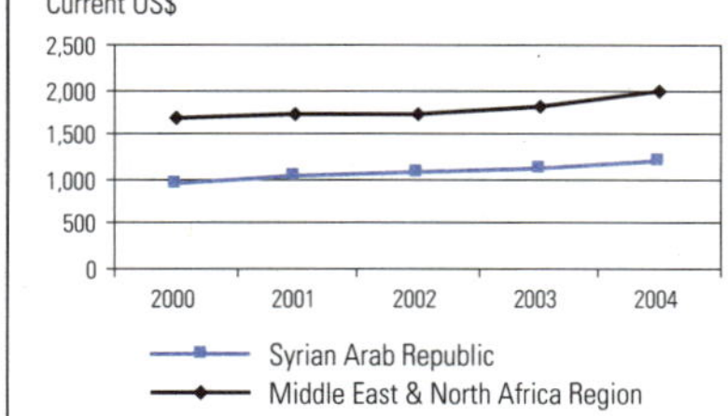

ICT MDG[b] Indicators, 2000–4
Number per 1,000 people

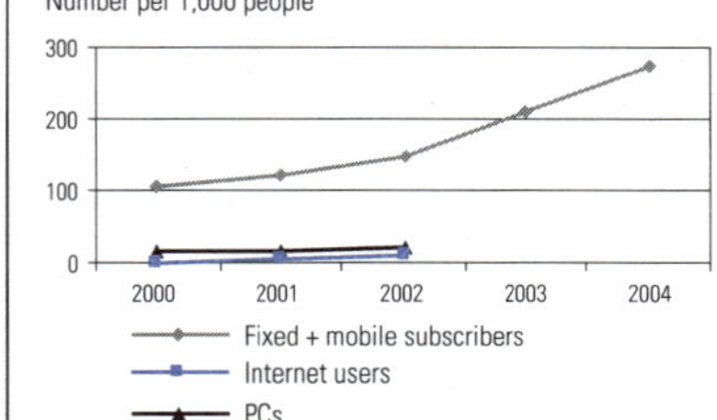

Price of Call to the United States, 2000–4
US$ per 3 minutes

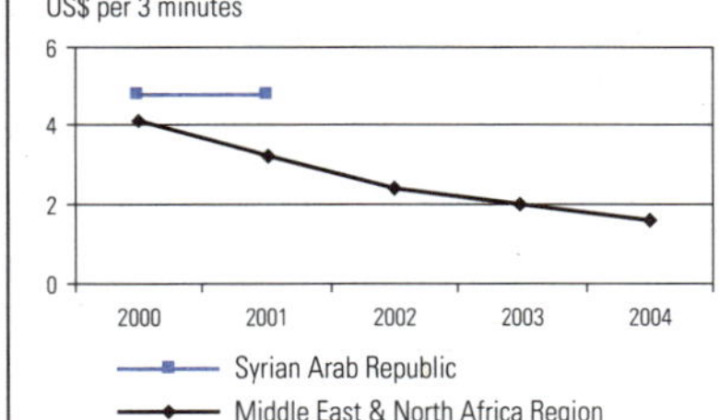

Total Telecommunications Revenue, 2000–4
Percentage of GDP

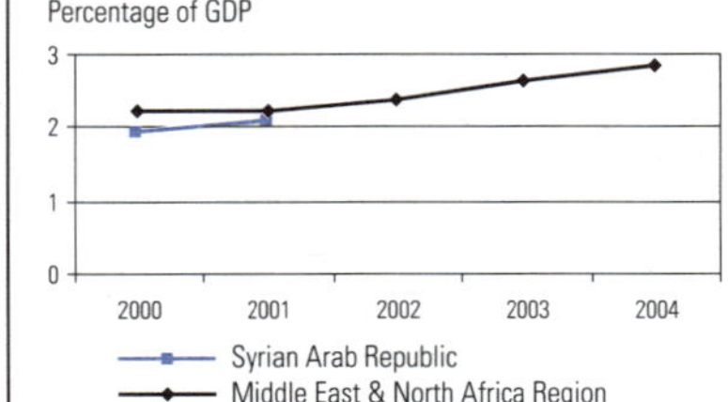

Notes: Figures in italics are for years other than those specified. .. indicates data are not available. C = competition; GDP = gross domestic product; GNI = gross national income; ICT = information and communication technology; M = monopoly; MDG = Millennium Development Goal; P = partial competition; and PCs = personal computers.

a. Outgoing and incoming.
b. Millennium Development Goal indicators 47, 48a, and 48b.

Sources: Economic and social context: UNDP and World Bank; ICT sector structure: ITU, WEF; ICT sector performance: Global Insight/WITSA, ITU, Netcraft, UNDESA, UNPAN, and World Bank. Produced by the Global Information and Communication Technologies Department and the Development Economics Data Group. For complete information, see Definitions and Data Sources.

Tajikistan

	Tajikistan 2000	Tajikistan 2004	Low-income group 2004	Europe & Central Asia Region 2004
Economic and social context				
Population, total (millions)	6	6	2,338	472
Urban population (% total population)	26	25	31	64
Poverty (% population below US$1 per day)	*13.9*	*7.4*	..	*3.6*
GNI per capita, Atlas method (current US$)	180	280	510	3,290
GDP growth, 1995–2000 and 2000–4 (%)	1.1	9.9	5.4	5.0
Adult literacy rate (% ages 15 and over)	99	..	61	97
Primary, secondary, tertiary school enrollment (% gross)	71	*76*	*55*	*81*
ICT sector structure				
Separate telecommunications regulator	..	..		
Status of main fixed-line operator	*Mixed*	Mixed		
Level of competition: international long distance	*M*	M		
Level of competition: mobile	..	..		
Level of competition: Internet service provider	..	..		
Government prioritization of ICT (scale 1–7)	..	..	..	3.8
ICT sector performance				
Access				
Telephone main lines (per 1,000 people)	35	*38*	33	*243*
International voice traffic (minutes per person)[a]	4	*8*	*4*	*28*
Mobile subscribers (per 1,000 people)	0	*7*	48	487
Population covered by mobile telephony (%)	..	..	*43*	*82*
Internet users (per 1,000 people)	0	*1*	20	*115*
Personal computers (per 1,000 people)	..	..	*8*	*73*
Households with television (%)	80	..	*16*	*92*
Quality				
Telephone faults (per 100 main lines per year)	124.9	*144.0*	..	*30.4*
Broadband subscribers (per 1,000 people)	*0.0*	*0.0*	0.5	*2.4*
International Internet bandwidth (bits per person)	..	*0*	*3*	148
Affordability				
Price basket for fixed line (US$ per month, residential)	*1.3*	*1.0*	*6.6*	*3.5*
Price basket for mobile (US$ per month)	..	12.3	11.6	10.3
Price basket for Internet (US$ per month)	..	*54.4*	*45.5*	*19.8*
Price of call to United States (US$ per 3 minutes)	*9.57*	*6.96*	1.95	1.06
Institutional efficiency and sustainability				
Total telecommunications revenue (% GDP)	*0.6*	2.1	2.3	3.5
Total telephone subscribers per employee	45	*57*	*89*	*150*
Total telecommunications investment (% revenue)	1.0	*2.2*	*27.8*	*19.0*
ICT applications				
ICT expenditure (% GDP)	..	..	*4.1*	*5.1*
E-government readiness index (scale 0–1)	..	0.00	0.12	0.39
Secure Internet servers (per 1 million people)	..	..	0.3	6.4
Schools connected to the Internet (%)	..	..	..	*65*

GNI per Capita, Atlas Method, 2000–4

Current US$

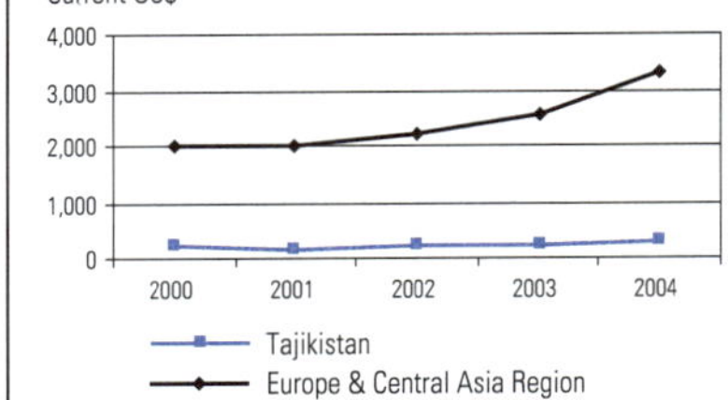

ICT MDG[b] Indicators, 2000–4

Number per 1,000 people

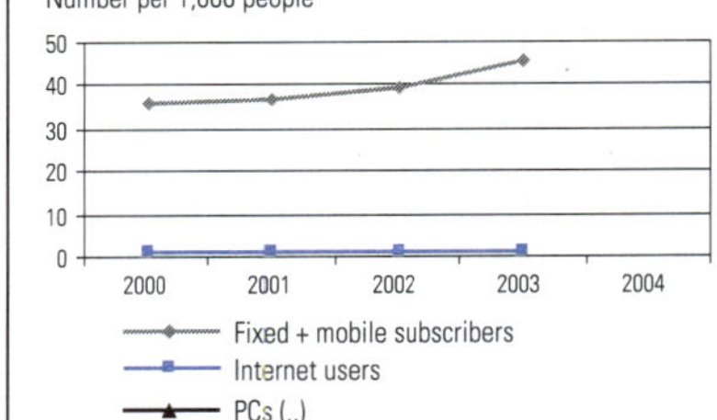

Price of Call to the United States, 2000–4

US$ per 3 minutes

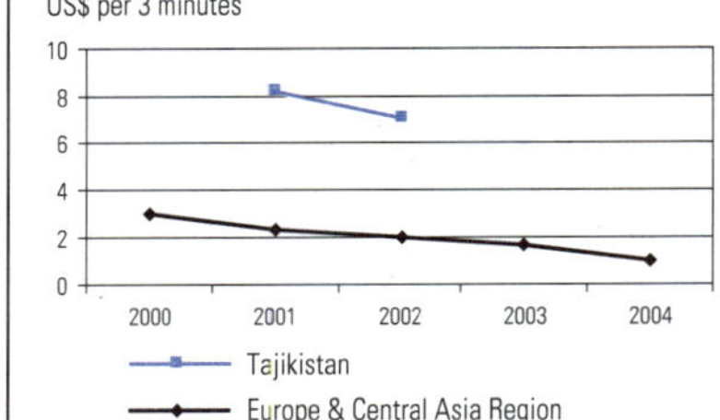

Total Telecommunications Revenue, 2000–4

Percentage of GDP

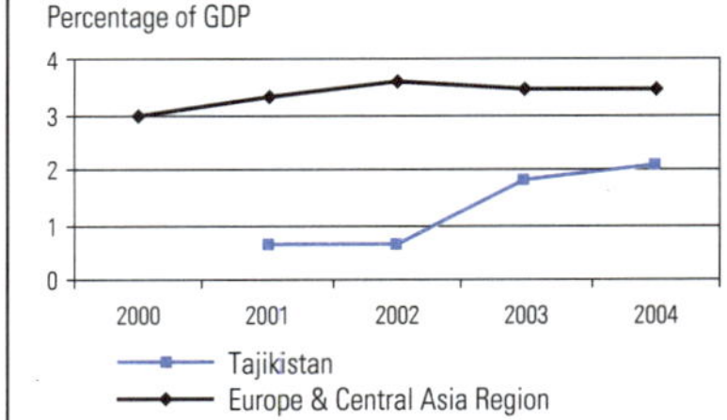

Notes: Figures in italics are for years other than those specified. .. indicates data are not available. C = competition; GDP = gross domestic product; GNI = gross national income; ICT = information and communication technology; M = monopoly; MDG = Millennium Development Goal; P = partial competition; and PCs = personal computers.

a. Outgoing and incoming.

b. Millennium Development Goal indicators 47, 48a, and 48b.

Sources: Economic and social context: UNDP and World Bank; ICT sector structure: ITU, WEF; ICT sector performance: Global Insight/WITSA, ITU, Netcraft, UNDESA, UNPAN, and World Bank. Produced by the Global Information and Communication Technologies Department and the Development Economics Data Group. For complete information, see Definitions and Data Sources.

Tanzania

	Tanzania 2000	Tanzania 2004	Low-income group 2004	Sub-Saharan Africa Region 2004
Economic and social context				
Population, total (millions)	34	37	2,338	719
Urban population (% total population)	32	36	31	37
Poverty (% population below US$1 per day)	..	..	..	*46.4*
GNI per capita, Atlas method (current US$)	280	330	510	600
GDP growth, 1995–2000 and 2000–4 (%)	3.9	6.8	5.4	3.9
Adult literacy rate (% ages 15 and over)	..	*69*	61	59
Primary, secondary, tertiary school enrollment (% gross)	31	*41*	*55*	*52*
ICT sector structure				
Separate telecommunications regulator	Yes	Yes		
Status of main fixed-line operator	*Mixed*	Mixed		
Level of competition: international long distance	*P*	M		
Level of competition: mobile	*C*	C		
Level of competition: Internet service provider	..	C		
Government prioritization of ICT (scale 1–7)	..	4.5	..	*4.4*
ICT sector performance				
Access				
Telephone main lines (per 1,000 people)	5	4	33	17
International voice traffic (minutes per person)[a]	1	*1*	*4*	..
Mobile subscribers (per 1,000 people)	5	51	48	86
Population covered by mobile telephony (%)	..	*25*	*43*	..
Internet users (per 1,000 people)	1	*7*	20	15
Personal computers (per 1,000 people)	3	*6*	*8*	*12*
Households with television (%)	9	*14*	*16*	*15*
Quality				
Telephone faults (per 100 main lines per year)	21.0	*24.0*	..	..
Broadband subscribers (per 1,000 people)	*0.0*	0.0	0.5	*0.1*
International Internet bandwidth (bits per person)	0	*0*	*3*	*4*
Affordability				
Price basket for fixed line (US$ per month, residential)	9.3	*11.6*	*6.6*	*8.5*
Price basket for mobile (US$ per month)	..	11.1	11.6	13.5
Price basket for Internet (US$ per month)	..	*117.0*	*45.5*	*54.8*
Price of call to United States (US$ per 3 minutes)	10.70	3.17	1.95	2.43
Institutional efficiency and sustainability				
Total telecommunications revenue (% GDP)	1.9	*2.4*	2.3	5.0
Total telephone subscribers per employee	97	*262*	*89*	*144*
Total telecommunications investment (% revenue)	12.6	..	*27.8*	*27.8*
ICT applications				
ICT expenditure (% GDP)	..	..	*4.1*	..
E-government readiness index (scale 0–1)	..	0.23	0.12	0.11
Secure Internet servers (per 1 million people)	..	0.0	0.3	1.9
Schools connected to the Internet (%)	..	..	..	..

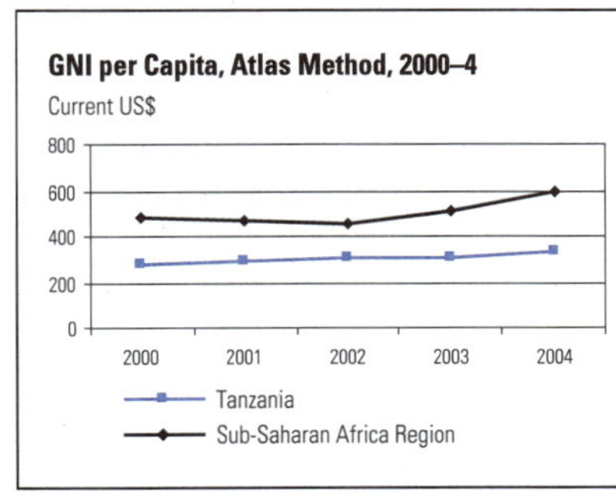

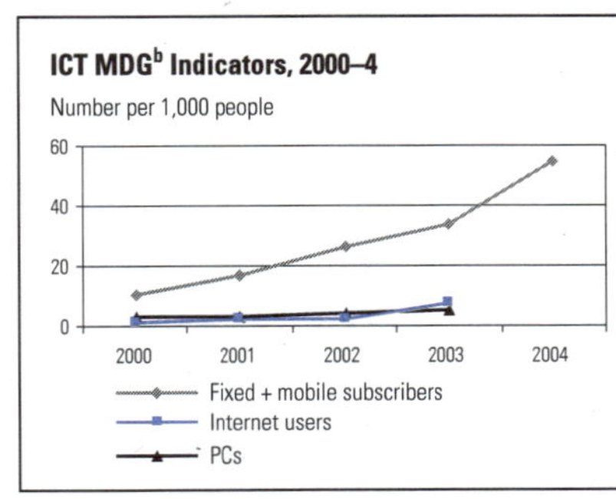

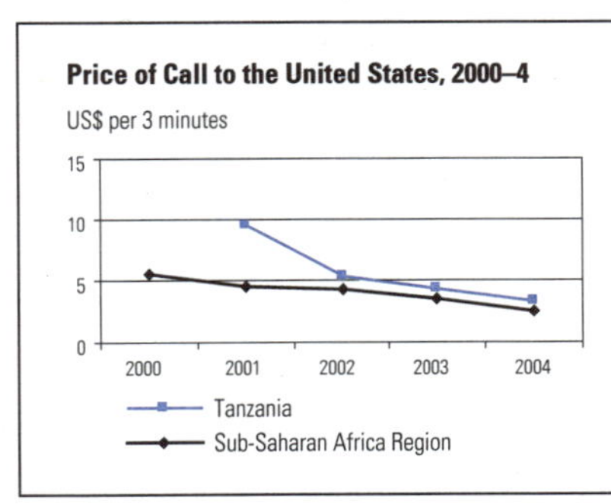

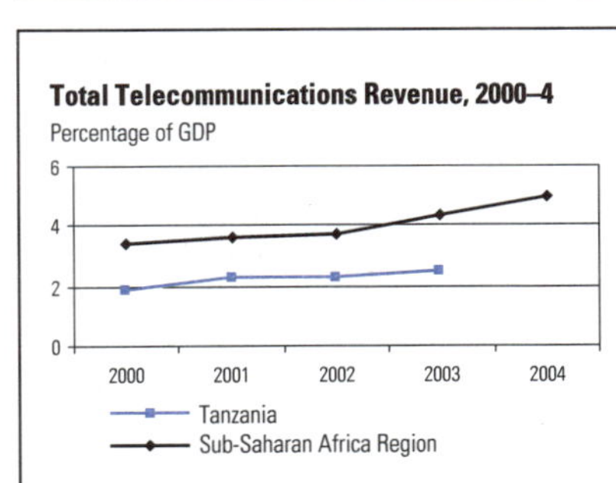

Notes: Figures in italics are for years other than those specified. .. indicates data are not available. C = competition; GDP = gross domestic product; GNI = gross national income; ICT = information and communication technology; M = monopoly; MDG = Millennium Development Goal; P = partial competition; and PCs = personal computers.

a. Outgoing and incoming.

b. Millennium Development Goal indicators 47, 48a, and 48b.

Sources: Economic and social context: UNDP and World Bank; ICT sector structure: ITU, WEF; ICT sector performance: Global Insight/WITSA, ITU, Netcraft, UNDESA, UNPAN, and World Bank. Produced by the Global Information and Communication Technologies Department and the Development Economics Data Group. For complete information, see Definitions and Data Sources.

Thailand

	Thailand 2000	Thailand 2004	Lower-middle-income group 2004	East Asia & Pacific Region 2004
Economic and social context				
Population, total (millions)	61	62	2,430	1,870
Urban population (% total population)	31	32	49	41
Poverty (% population below US$1 per day)	<2	..	..	*14.9*
GNI per capita, Atlas method (current US$)	2,010	2,540	1,580	1,280
GDP growth, 1995–2000 and 2000–4 (%)	–0.7	5.3	5.7	7.5
Adult literacy rate (% ages 15 and over)	93	..	90	90
Primary, secondary, tertiary school enrollment (% gross)	73	*72*	*72*	*68*
ICT sector structure				
Separate telecommunications regulator	No	Yes		
Status of main fixed-line operator	*Public*	Public		
Level of competition: international long distance	*M*	M		
Level of competition: mobile	*C*	P		
Level of competition: Internet service provider	*C*	C		
Government prioritization of ICT (scale 1–7)	..	5.3	3.8	4.8
ICT sector performance				
Access				
Telephone main lines (per 1,000 people)	92	*106*	192	194
International voice traffic (minutes per person)[a]	10	*11*	..	..
Mobile subscribers (per 1,000 people)	50	420	255	248
Population covered by mobile telephony (%)	..	*92*	*76*	*73*
Internet users (per 1,000 people)	38	112	70	75
Personal computers (per 1,000 people)	28	74	38	37
Households with television (%)	91	92	*89*	*80*
Quality				
Telephone faults (per 100 main lines per year)	19.6	..	..	..
Broadband subscribers (per 1,000 people)	0.0	*0.2*	12.6	13.4
International Internet bandwidth (bits per person)	4	48	58	52
Affordability				
Price basket for fixed line (US$ per month, residential)	8.4	8.3	*5.5*	*4.5*
Price basket for mobile (US$ per month)	..	6.8	8.9	5.1
Price basket for Internet (US$ per month)	..	*7.0*	*25.3*	*19.9*
Price of call to United States (US$ per 3 minutes)	2.19	0.67	1.45	1.20
Institutional efficiency and sustainability				
Total telecommunications revenue (% GDP)	2.6	3.6	3.6	3.6
Total telephone subscribers per employee	265	..	*195*	..
Total telecommunications investment (% revenue)	27.0	*36.5*	*25.5*	*31.0*
ICT applications				
ICT expenditure (% GDP)	3.5	*3.5*	*5.1*	*5.0*
E-government readiness index (scale 0–1)	..	0.53	0.26	0.21
Secure Internet servers (per 1 million people)	*1.9*	4.1	1.6	0.6
Schools connected to the Internet (%)	..	*37*	..	..

GNI per Capita, Atlas Method, 2000–4
Current US$

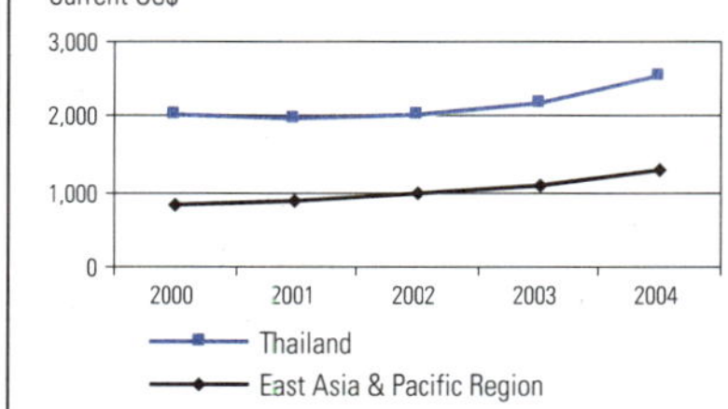

ICT MDG[b] Indicators, 2000–4
Number per 1,000 people

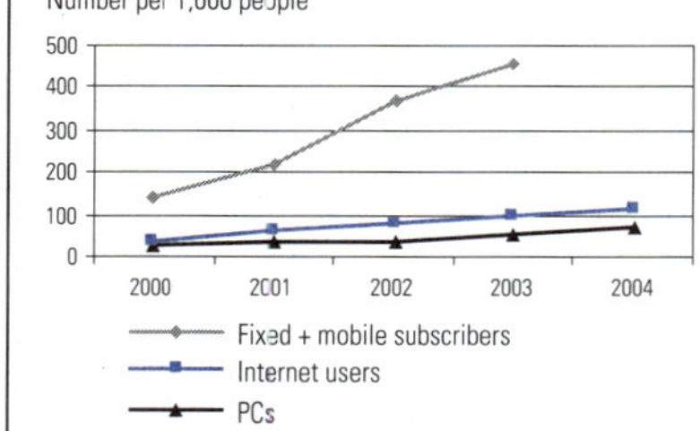

Price of Call to the United States, 2000–4
US$ per 3 minutes

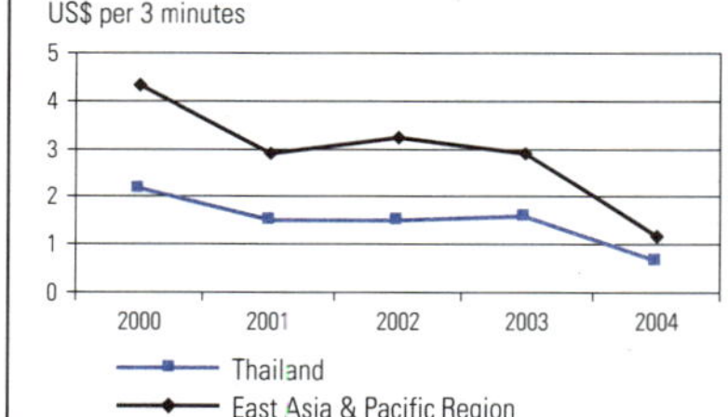

Total Telecommunications Revenue, 2000–4
Percentage of GDP

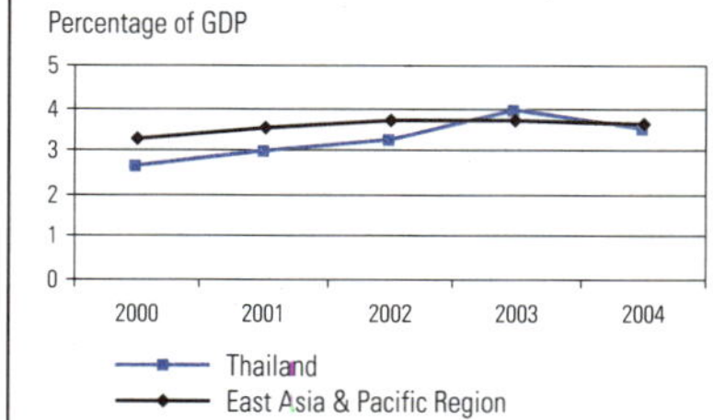

Notes: Figures in italics are for years other than those specified. .. indicates data are not available. C = competition; GDP = gross domestic product; GNI = gross national income; ICT = information and communication technology; M = monopoly; MDG = Millennium Development Goal; P = partial competition; and PCs = personal computers.

a. Outgoing and incoming.

b. Millennium Development Goal indicators 47, 48a, and 48b.

Sources: Economic and social context: UNDP and World Bank; ICT sector structure: ITU, WEF; ICT sector performance: Global Insight/WITSA, ITU, Netcraft, UNDESA, UNPAN, and World Bank. Produced by the Global Information and Communication Technologies Department and the Development Economics Data Group. For complete information, see Definitions and Data Sources.

Togo

	Togo 2000	Togo 2004	Low-income group 2004	Sub-Saharan Africa Region 2004
Economic and social context				
Population, total (millions)	5	5	2,338	719
Urban population (% total population)	33	36	31	37
Poverty (% population below US$1 per day)	..	..	..	*46.4*
GNI per capita, Atlas method (current US$)	320	380	510	600
GDP growth, 1995–2000 and 2000–4 (%)	4.2	2.6	5.4	3.9
Adult literacy rate (% ages 15 and over)	53	..	61	59
Primary, secondary, tertiary school enrollment (% gross)	65	*66*	*55*	*52*
ICT sector structure				
Separate telecommunications regulator	Yes	Yes		
Status of main fixed-line operator	*Public*	Public		
Level of competition: international long distance	*P*	P		
Level of competition: mobile	*P*	P		
Level of competition: Internet service provider	*C*	C		
Government prioritization of ICT (scale 1–7)	..	..	..	*4.4*
ICT sector performance				
Access				
Telephone main lines (per 1,000 people)	9	*12*	33	17
International voice traffic (minutes per person)[a]	9	*16*	*4*	..
Mobile subscribers (per 1,000 people)	11	*45*	48	86
Population covered by mobile telephony (%)	80	80	*43*	..
Internet users (per 1,000 people)	22	70	20	15
Personal computers (per 1,000 people)	22	*33*	*8*	*12*
Households with television (%)	17	*51*	*16*	*15*
Quality				
Telephone faults (per 100 main lines per year)	6.0	*6.2*	..	..
Broadband subscribers (per 1,000 people)	0.0	0.0	0.5	*0.1*
International Internet bandwidth (bits per person)	0	*3*	*3*	*4*
Affordability				
Price basket for fixed line (US$ per month, residential)	10.2	*10.4*	*6.6*	*8.5*
Price basket for mobile (US$ per month)	..	13.4	11.6	13.5
Price basket for Internet (US$ per month)	..	*30.4*	*45.5*	*54.8*
Price of call to United States (US$ per 3 minutes)	7.90	3.98	1.95	2.43
Institutional efficiency and sustainability				
Total telecommunications revenue (% GDP)	2.9	*3.4*	2.3	5.0
Total telephone subscribers per employee	106	*248*	*89*	*144*
Total telecommunications investment (% revenue)	45.3	*71.2*	*27.8*	*27.8*
ICT applications				
ICT expenditure (% GDP)	..	..	*4.1*	..
E-government readiness index (scale 0–1)	..	0.05	0.12	0.11
Secure Internet servers (per 1 million people)	..	0.2	0.3	1.9
Schools connected to the Internet (%)	..	..	..	..

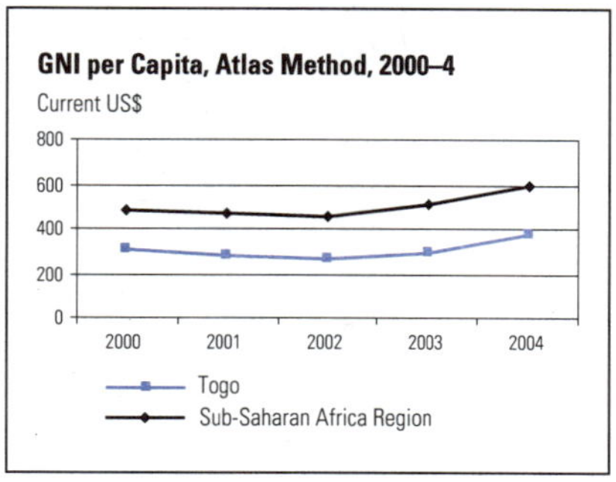

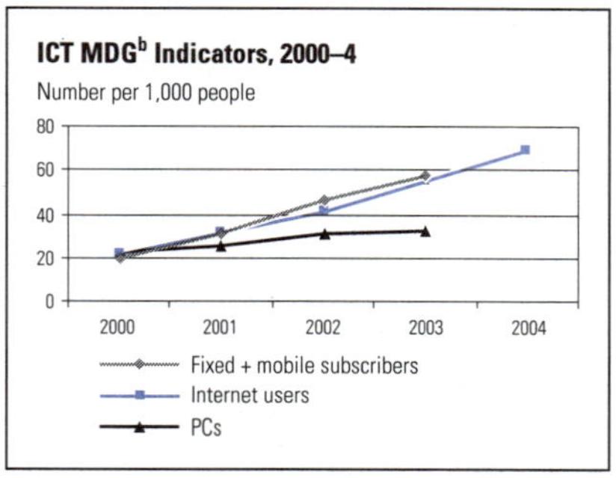

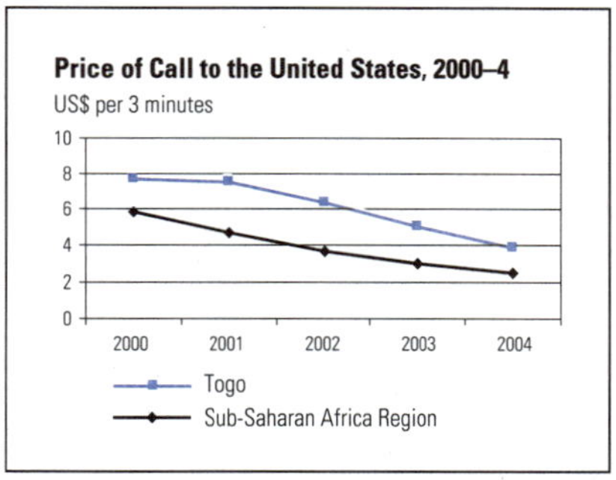

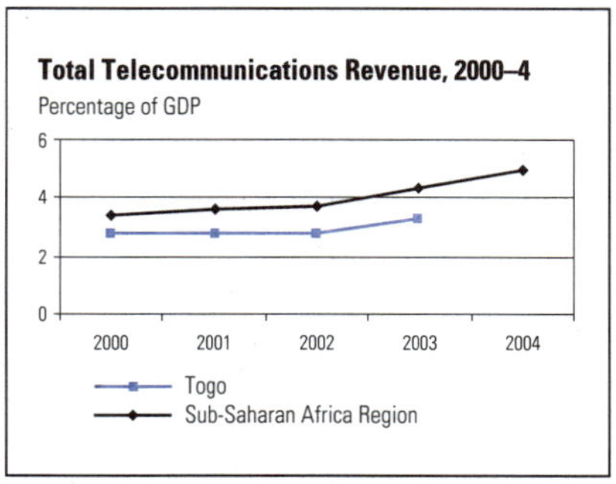

Notes: Figures in italics are for years other than those specified. .. indicates data are not available. C = competition; GDP = gross domestic product; GNI = gross national income; ICT = information and communication technology; M = monopoly; MDG = Millennium Development Goal; P = partial competition; and PCs = personal computers.

a. Outgoing and incoming.

b. Millennium Development Goal indicators 47, 48a, and 48b.

Sources: Economic and social context: UNDP and World Bank; ICT sector structure: ITU, WEF; ICT sector performance: Global Insight/WITSA, ITU, Netcraft, UNDESA, UNPAN, and World Bank. Produced by the Global Information and Communication Technologies Department and the Development Economics Data Group. For complete information, see Definitions and Data Sources.

Trinidad and Tobago

	Trinidad and Tobago 2000	Trinidad and Tobago 2004	Upper-middle-income group 2004	Latin America & the Caribbean Region 2004
Economic and social context				
Population, total (millions)	1	1	576	541
Urban population (% total population)	74	76	72	77
Poverty (% population below US$1 per day)	..	..	..	*9.5*
GNI per capita, Atlas method (current US$)	5,220	8,580	4,770	3,600
GDP growth, 1995–2000 and 2000–4 (%)	5.0	7.2	2.7	1.5
Adult literacy rate (% ages 15 and over)	..	*98*	94	89
Primary, secondary, tertiary school enrollment (% gross)	67	*66*	*80*	*82*
ICT sector structure				
Separate telecommunications regulator	No	Yes		
Status of main fixed-line operator	*Mixed*	Public		
Level of competition: international long distance	*M*	M		
Level of competition: mobile	*C*	M		
Level of competition: Internet service provider	*C*	C		
Government prioritization of ICT (scale 1–7)	..	4.0	4.1	3.5
ICT sector performance				
Access				
Telephone main lines (per 1,000 people)	246	*249*	*220*	181
International voice traffic (minutes per person)[a]	183	..	*39*	..
Mobile subscribers (per 1,000 people)	126	*396*	490	324
Population covered by mobile telephony (%)	..	..	*84*	*76*
Internet users (per 1,000 people)	78	*114*	*133*	*104*
Personal computers (per 1,000 people)	62	*79*	*99*	*75*
Households with television (%)	86	*88*	*92*	*88*
Quality				
Telephone faults (per 100 main lines per year)	..	..	*20.3*	..
Broadband subscribers (per 1,000 people)	0.0	*0.1*	*3.7*	*5.2*
International Internet bandwidth (bits per person)	47	67	176	165
Affordability				
Price basket for fixed line (US$ per month, residential)	7.0	*7.0*	*13.9*	*9.0*
Price basket for mobile (US$ per month)	..	7.8	11.1	9.1
Price basket for Internet (US$ per month)	..	*13.4*	*20.8*	*31.5*
Price of call to United States (US$ per 3 minutes)	2.47	0.95	1.03	0.90
Institutional efficiency and sustainability				
Total telecommunications revenue (% GDP)	3.0	*3.7*	3.4	3.3
Total telephone subscribers per employee	157	..	*402*	..
Total telecommunications investment (% revenue)	39.2	..	*18.6*	..
ICT applications				
ICT expenditure (% GDP)	..	..	*5.0*	*5.3*
E-government readiness index (scale 0–1)	..	0.33	0.49	0.39
Secure Internet servers (per 1 million people)	*9.3*	11.3	10.7	8.6
Schools connected to the Internet (%)	..	*15*	*60*	..

GNI per Capita, Atlas Method, 2000–4
Current US$

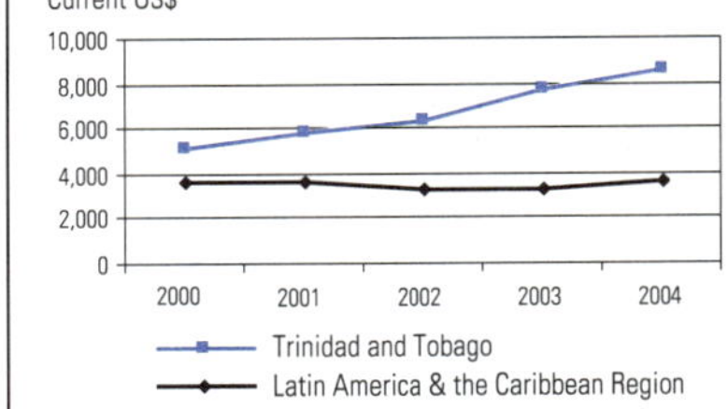

ICT MDG[b] Indicators, 2000–4
Number per 1,000 people

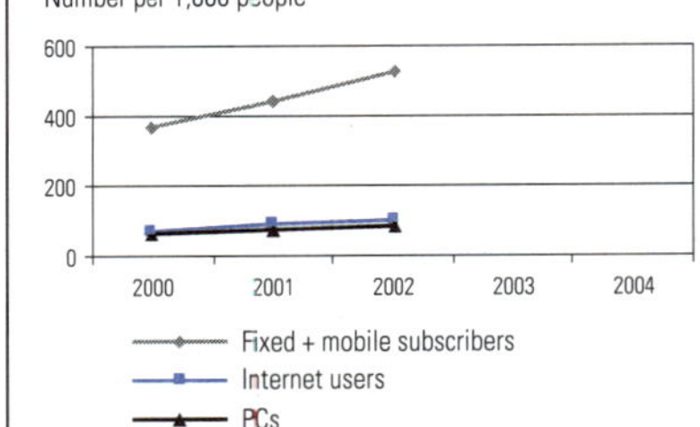

Price of Call to the United States, 2000–4
US$ per 3 minutes

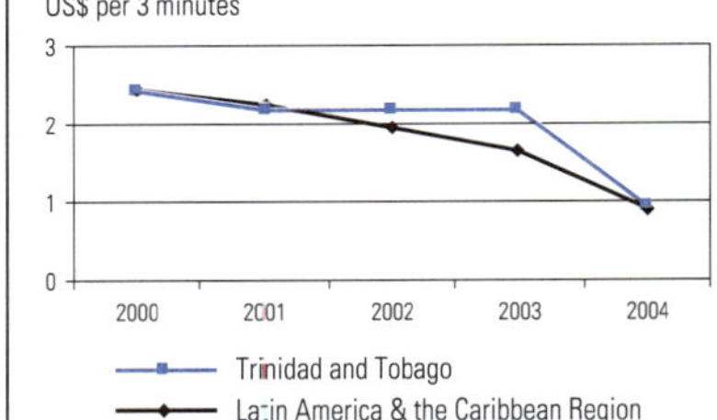

Total Telecommunications Revenue, 2000–4
Percentage of GDP

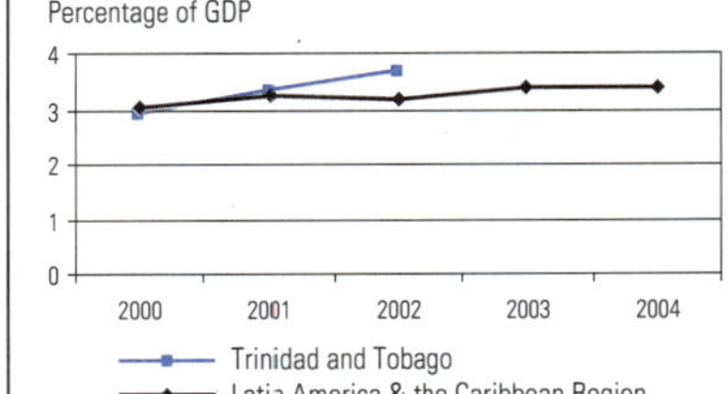

Notes: Figures in italics are for years other than those specified. .. indicates data are not available. C = competition; GDP = gross domestic product; GNI = gross national income; ICT = information and communication technology; M = monopoly; MDG = Millennium Development Goal; P = partial competition; and PCs = personal computers.

a. Outgoing and incoming.

b. Millennium Development Goal indicators 47, 48a, and 48b.

Sources: Economic and social context: UNDP and World Bank; ICT sector structure: ITU, WEF; ICT sector performance: Global Insight/WITSA, ITU, Netcraft, UNDESA, UNPAN, and World Bank. Produced by the Global Information and Communication Technologies Department and the Development Economics Data Group. For complete information, see Definitions and Data Sources.

Tunisia

	Tunisia 2000	Tunisia 2004	Lower-middle-income group 2004	Middle East & North Africa Region 2004
Economic and social context				
Population, total (millions)	10	10	2,430	294
Urban population (% total population)	63	64	49	56
Poverty (% population below US$1 per day)	<2	..	..	*2.4*
GNI per capita, Atlas method (current US$)	2,080	2,630	1,580	2,000
GDP growth, 1995–2000 and 2000–4 (%)	5.5	4.3	5.7	4.5
Adult literacy rate (% ages 15 and over)	..	74	90	..
Primary, secondary, tertiary school enrollment (% gross)	75	*74*	*72*	*68*
ICT sector structure				
Separate telecommunications regulator	No	Yes		
Status of main fixed-line operator	*Public*	Public		
Level of competition: international long distance	*M*	M		
Level of competition: mobile	*M*	C		
Level of competition: Internet service provider	*C*	C		
Government prioritization of ICT (scale 1–7)	..	5.5	3.8	..
ICT sector performance				
Access				
Telephone main lines (per 1,000 people)	100	120	192	*118*
International voice traffic (minutes per person)[a]	45	*61*	..	*20*
Mobile subscribers (per 1,000 people)	12	373	255	*88*
Population covered by mobile telephony (%)	..	95	*76*	..
Internet users (per 1,000 people)	27	83	70	*47*
Personal computers (per 1,000 people)	22	47	38	*30*
Households with television (%)	87	90	*89*	*88*
Quality				
Telephone faults (per 100 main lines per year)	34.0	*28.0*	..	..
Broadband subscribers (per 1,000 people)	0.0	0.7	12.6	*0.2*
International Internet bandwidth (bits per person)	5	22	58	*15*
Affordability				
Price basket for fixed line (US$ per month, residential)	4.2	4.7	*5.5*	*4.9*
Price basket for mobile (US$ per month)	..	6.8	8.9	8.1
Price basket for Internet (US$ per month)	..	*17.3*	*25.3*	*24.5*
Price of call to United States (US$ per 3 minutes)	*2.25*	*2.28*	1.45	1.64
Institutional efficiency and sustainability				
Total telecommunications revenue (% GDP)	2.1	4.0	3.6	2.8
Total telephone subscribers per employee	153	*224*	*195*	..
Total telecommunications investment (% revenue)	39.8	*52.8*	*25.5*	*27.8*
ICT applications				
ICT expenditure (% GDP)	4.8	*5.2*	*5.1*	..
E-government readiness index (scale 0–1)	..	0.15	0.26	0.16
Secure Internet servers (per 1 million people)	*0.4*	1.9	1.6	0.6
Schools connected to the Internet (%)	..	*25*	..	..

GNI per Capita, Atlas Method, 2000–4
Current US$

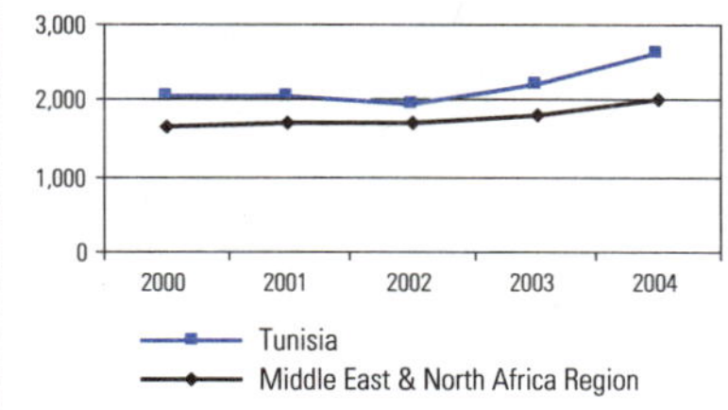

ICT MDG[b] Indicators, 2000–4
Number per 1,000 people

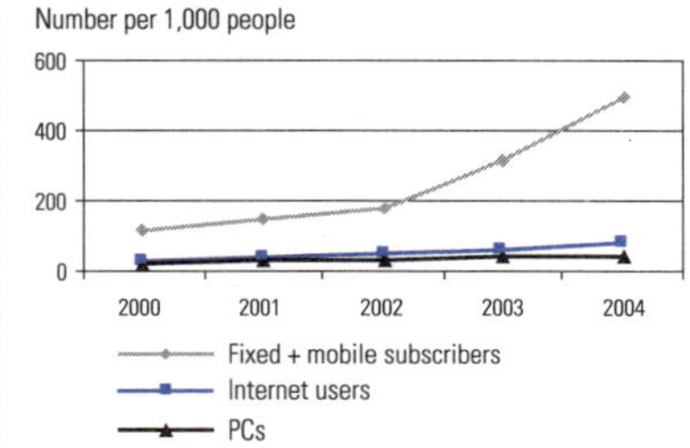

Price of Call to the United States, 2000–4
US$ per 3 minutes

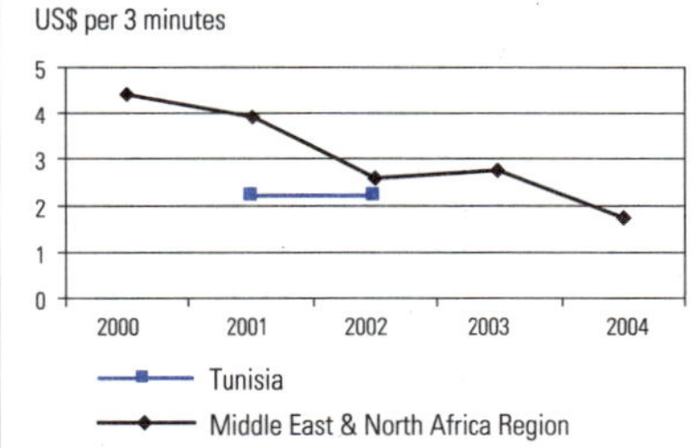

Total Telecommunications Revenue, 2000–4
Percentage of GDP

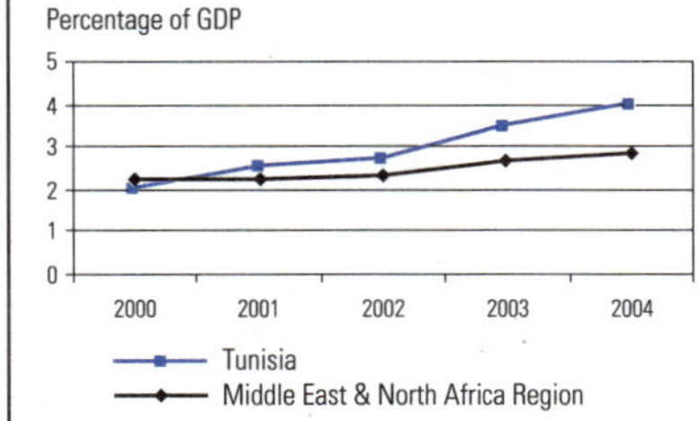

Notes: Figures in italics are for years other than those specified. .. indicates data are not available. C = competition; GDP = gross domestic product; GNI = gross national income; ICT = information and communication technology; M = monopoly; MDG = Millennium Development Goal; P = partial competition; and PCs = personal computers.

a. Outgoing and incoming.
b. Millennium Development Goal indicators 47, 48a, and 48b.

Sources: Economic and social context: UNDP and World Bank; ICT sector structure: ITU, WEF; ICT sector performance: Global Insight/WITSA, ITU, Netcraft, UNDESA, UNPAN, and World Bank. Produced by the Global Information and Communication Technologies Department and the Development Economics Data Group. For complete information, see Definitions and Data Sources.

Turkey

	Turkey 2000	Turkey 2004	Upper-middle-income group 2004	Europe & Central Asia Region 2004
Economic and social context				
Population, total (millions)	67	72	576	472
Urban population (% total population)	65	67	72	64
Poverty (% population below US$1 per day)	..	*<2*	..	*3.6*
GNI per capita, Atlas method (current US$)	2,980	3,750	4,770	3,290
GDP growth, 1995–2000 and 2000–4 (%)	3.4	4.2	2.7	5.0
Adult literacy rate (% ages 15 and over)	..	*88*	94	97
Primary, secondary, tertiary school enrollment (% gross)	66	*68*	*80*	*81*
ICT sector structure				
Separate telecommunications regulator	Yes	Yes		
Status of main fixed-line operator	*Public*	Public		
Level of competition: international long distance	*M*	C		
Level of competition: mobile	*C*	P		
Level of competition: Internet service provider	*C*	C		
Government prioritization of ICT (scale 1–7)	..	3.8	4.1	3.8
ICT sector performance				
Access				
Telephone main lines (per 1,000 people)	273	267	*220*	*243*
International voice traffic (minutes per person)[a]	29	*26*	*39*	*28*
Mobile subscribers (per 1,000 people)	239	494	490	487
Population covered by mobile telephony (%)	*88*	*68*	*84*	*82*
Internet users (per 1,000 people)	30	*78*	*133*	*115*
Personal computers (per 1,000 people)	37	*45*	*99*	*73*
Households with television (%)	96	..	*92*	*92*
Quality				
Telephone faults (per 100 main lines per year)	55.4	*30.4*	*20.3*	*30.4*
Broadband subscribers (per 1,000 people)	0.0	*0.8*	*3.7*	*2.4*
International Internet bandwidth (bits per person)	9	40	176	148
Affordability				
Price basket for fixed line (US$ per month, residential)	10.4	*10.3*	*13.9*	*3.5*
Price basket for mobile (US$ per month)	..	6.4	11.1	10.3
Price basket for Internet (US$ per month)	..	*19.8*	*20.8*	*19.8*
Price of call to United States (US$ per 3 minutes)	3.30	*2.09*	1.03	1.06
Institutional efficiency and sustainability				
Total telecommunications revenue (% GDP)	2.5	3.0	3.4	3.5
Total telephone subscribers per employee	477	*665*	*402*	*150*
Total telecommunications investment (% revenue)	12.6	*2.7*	*18.6*	*19.0*
ICT applications				
ICT expenditure (% GDP)	7.9	*7.3*	*5.0*	*5.1*
E-government readiness index (scale 0–1)	..	0.53	0.49	0.39
Secure Internet servers (per 1 million people)	*3.2*	12.3	10.7	6.4
Schools connected to the Internet (%)	..	*40*	*60*	*65*

GNI per Capita, Atlas Method, 2000–4
Current US$

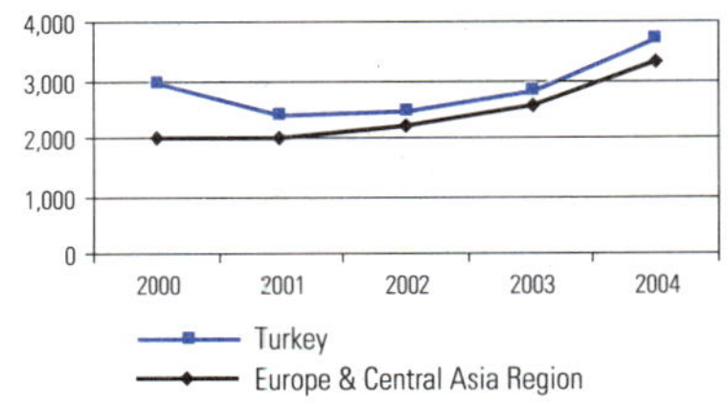

ICT MDG[b] Indicators, 2000–4
Number per 1,000 people

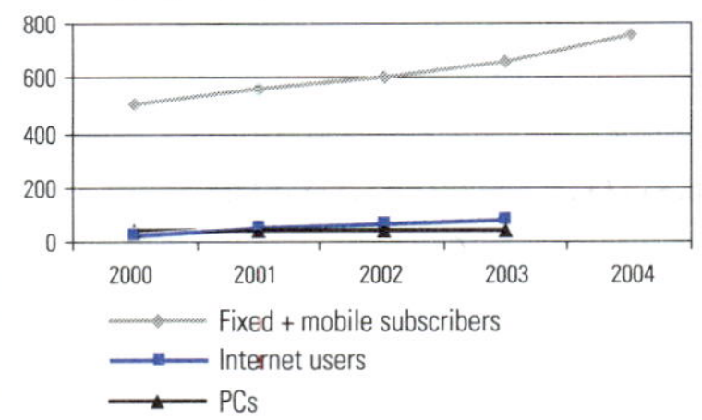

Price of Call to the United States, 2000–4
US$ per 3 minutes

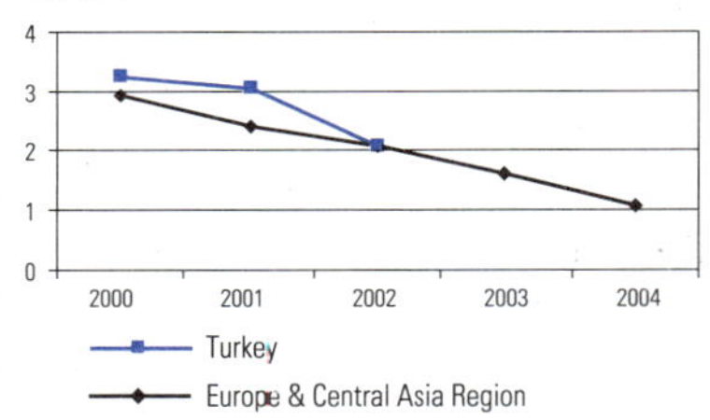

Total Telecommunications Revenue, 2000–4
Percentage of GDP

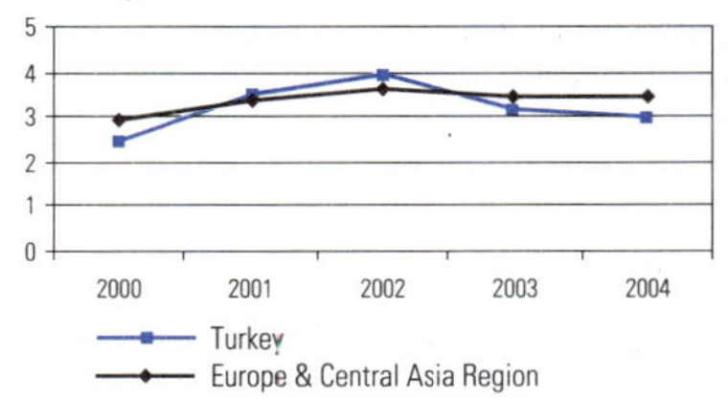

Notes: Figures in italics are for years other than those specified. .. indicates data are not available. C = competition; GDP = gross domestic product; GNI = gross national income; ICT = information and communication technology; M = monopoly; MDG = Millennium Development Goal; P = partial competition; and PCs = personal computers.

a. Outgoing and incoming.
b. Millennium Development Goal indicators 47, 48a, and 48b.

Sources: Economic and social context: UNDP and World Bank; ICT sector structure: ITU, WEF; ICT sector performance: Global Insight/WITSA, ITU, Netcraft, UNDESA, UNPAN, and World Bank. Produced by the Global Information and Communication Technologies Department and the Development Economics Data Group. For complete information, see Definitions and Data Sources.

Turkmenistan

	Turkmenistan 2000	Turkmenistan 2004	Lower-middle-income group 2004	Europe & Central Asia Region 2004
Economic and social context				
Population, total (millions)	5	5	2,430	472
Urban population (% total population)	45	46	49	64
Poverty (% population below US$1 per day)	*12.1*	..	..	*3.6*
GNI per capita, Atlas method (current US$)	620	1,340	1,580	3,290
GDP growth, 1995–2000 and 2000–4 (%)	3.9	18.5	5.7	5.0
Adult literacy rate (% ages 15 and over)	..	..	90	97
Primary, secondary, tertiary school enrollment (% gross)	..	..	*72*	*81*
ICT sector structure				
Separate telecommunications regulator	..	..		
Status of main fixed-line operator	*Public*	Public		
Level of competition: international long distance	*M*	M		
Level of competition: mobile	*C*	C		
Level of competition: Internet service provider	..	..		
Government prioritization of ICT (scale 1–7)	..	..	3.8	3.8
ICT sector performance				
Access				
Telephone main lines (per 1,000 people)	78	*77*	192	*243*
International voice traffic (minutes per person)[a]	6	..	..	*28*
Mobile subscribers (per 1,000 people)	2	*2*	255	487
Population covered by mobile telephony (%)	..	..	*76*	*82*
Internet users (per 1,000 people)	1	*2*	70	*115*
Personal computers (per 1,000 people)	..	..	38	*73*
Households with television (%)	94	94	*89*	*92*
Quality				
Telephone faults (per 100 main lines per year)	93.9	*96.0*	..	*30.4*
Broadband subscribers (per 1,000 people)	0.0	0.0	12.6	*2.4*
International Internet bandwidth (bits per person)	0	0	58	148
Affordability				
Price basket for fixed line (US$ per month, residential)	1.5	*1.5*	*5.5*	*3.5*
Price basket for mobile (US$ per month)	..	..	8.9	10.3
Price basket for Internet (US$ per month)	..	*20.0*	*25.3*	*19.8*
Price of call to United States (US$ per 3 minutes)	..	..	1.45	1.06
Institutional efficiency and sustainability				
Total telecommunications revenue (% GDP)	0.9	0.8	3.6	3.5
Total telephone subscribers per employee	50	*59*	*195*	*150*
Total telecommunications investment (% revenue)	14.9	*14.1*	*25.5*	*19.0*
ICT applications				
ICT expenditure (% GDP)	..	..	*5.1*	*5.1*
E-government readiness index (scale 0–1)	..	0.07	0.26	0.39
Secure Internet servers (per 1 million people)	..	..	1.6	6.4
Schools connected to the Internet (%)	..	..	..	*65*

GNI per Capita, Atlas Method, 2000–4

Current US$

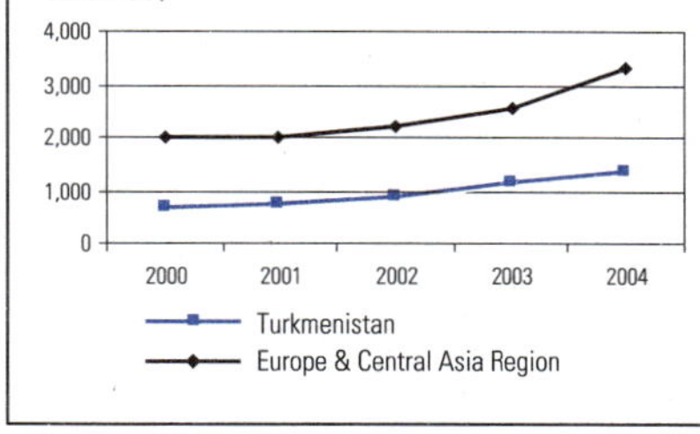

ICT MDG[b] Indicators, 2000–4

Number per 1,000 people

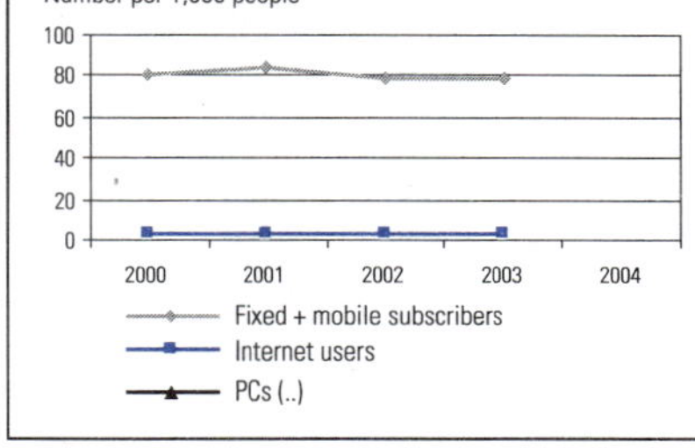

Price of Call to the United States, 2000–4

US$ per 3 minutes

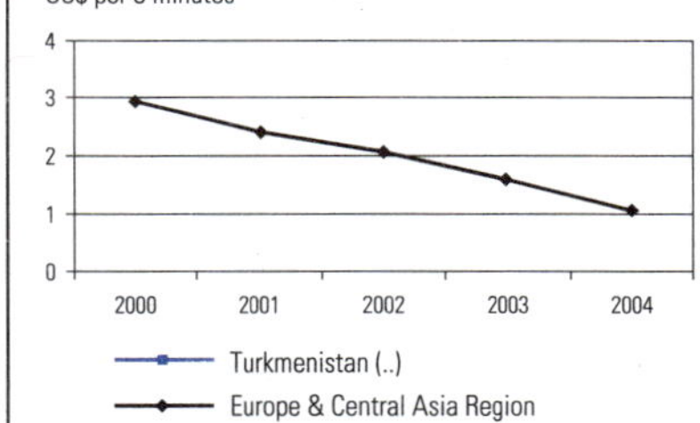

Total Telecommunications Revenue, 2000–4

Percentage of GDP

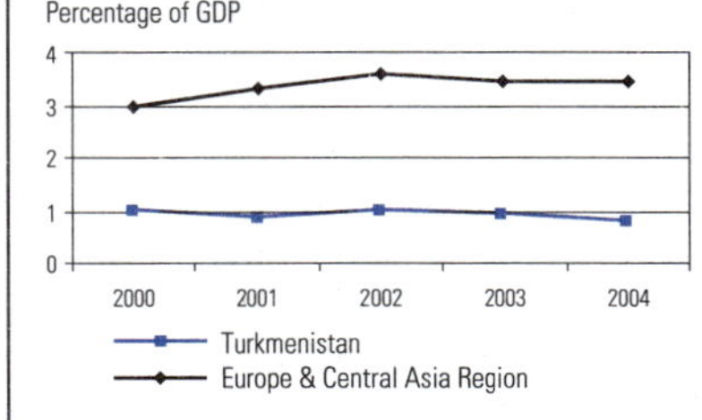

Notes: Figures in italics are for years other than those specified. .. indicates data are not available. C = competition; GDP = gross domestic product; GNI = gross national income; ICT = information and communication technology; M = monopoly; MDG = Millennium Development Goal; P = partial competition; and PCs = personal computers.

a. Outgoing and incoming.

b. Millennium Development Goal indicators 47, 48a, and 48b.

Sources: Economic and social context: UNDP and World Bank; ICT sector structure: ITU, WEF; ICT sector performance: Global Insight/WITSA, ITU, Netcraft, UNDESA, UNPAN, and World Bank. Produced by the Global Information and Communication Technologies Department and the Development Economics Data Group. For complete information, see Definitions and Data Sources.

Uganda

	Uganda 2000	Uganda 2004	Low-income group 2004	Sub-Saharan Africa Region 2004
Economic and social context				
Population, total (millions)	23	26	2,338	719
Urban population (% total population)	12	12	31	37
Poverty (% population below US$1 per day)	..	..	..	*46.4*
GNI per capita, Atlas method (current US$)	270	270	510	600
GDP growth, 1995–2000 and 2000–4 (%)	6.3	5.8	5.4	3.9
Adult literacy rate (% ages 15 and over)	..	*69*	61	59
Primary, secondary, tertiary school enrollment (% gross)	70	*74*	*55*	*52*
ICT sector structure				
Separate telecommunications regulator	Yes	Yes		
Status of main fixed-line operator	*Mixed*	*Mixed*		
Level of competition: international long distance	*C*	P		
Level of competition: mobile	*C*	P		
Level of competition: Internet service provider	*C*	C		
Government prioritization of ICT (scale 1–7)	..	4.9	..	*4.4*
ICT sector performance				
Access				
Telephone main lines (per 1,000 people)	3	3	33	17
International voice traffic (minutes per person)[a]	1	..	*4*	..
Mobile subscribers (per 1,000 people)	5	45	48	86
Population covered by mobile telephony (%)	16	*70*	*43*	..
Internet users (per 1,000 people)	2	6	20	15
Personal computers (per 1,000 people)	3	5	*8*	*12*
Households with television (%)	5	*6*	*16*	*15*
Quality				
Telephone faults (per 100 main lines per year)	..	..	..	..
Broadband subscribers (per 1,000 people)	0.0	0.0	0.5	*0.1*
International Internet bandwidth (bits per person)	0	3	*3*	*4*
Affordability				
Price basket for fixed line (US$ per month, residential)	14.4	*16.6*	*6.6*	*8.5*
Price basket for mobile (US$ per month)	..	7.9	11.6	13.5
Price basket for Internet (US$ per month)	..	*96.8*	*45.5*	*54.8*
Price of call to United States (US$ per 3 minutes)	*3.63*	*3.51*	1.95	2.43
Institutional efficiency and sustainability				
Total telecommunications revenue (% GDP)	1.6	*4.6*	2.3	5.0
Total telephone subscribers per employee	79	..	*89*	*144*
Total telecommunications investment (% revenue)	..	*7.7*	*27.8*	*27.8*
ICT applications				
ICT expenditure (% GDP)	..	..	*4.1*	..
E-government readiness index (scale 0–1)	..	0.29	0.12	0.11
Secure Internet servers (per 1 million people)	..	0.1	0.3	1.9
Schools connected to the Internet (%)	..	*1*	..	..

GNI per Capita, Atlas Method, 2000–4

Current US$

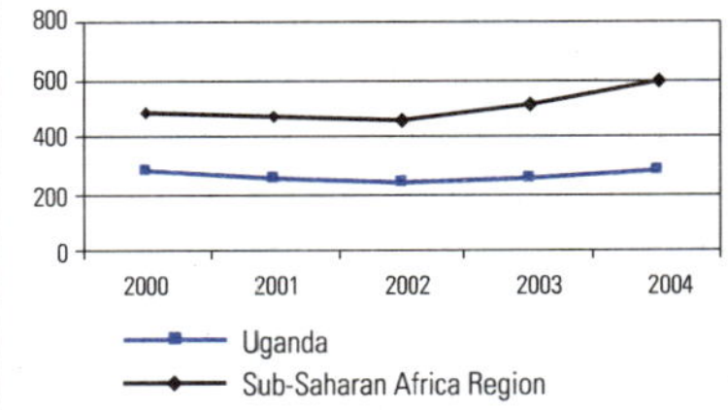

ICT MDG[b] Indicators, 2000–4

Number per 1,000 people

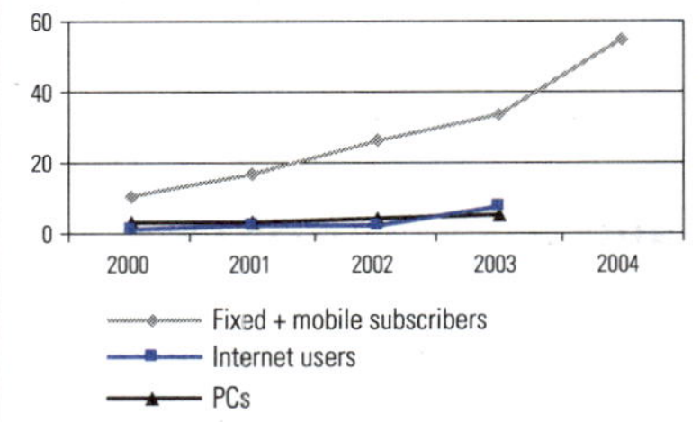

Price of Call to the United States, 2000–4

US$ per 3 minutes

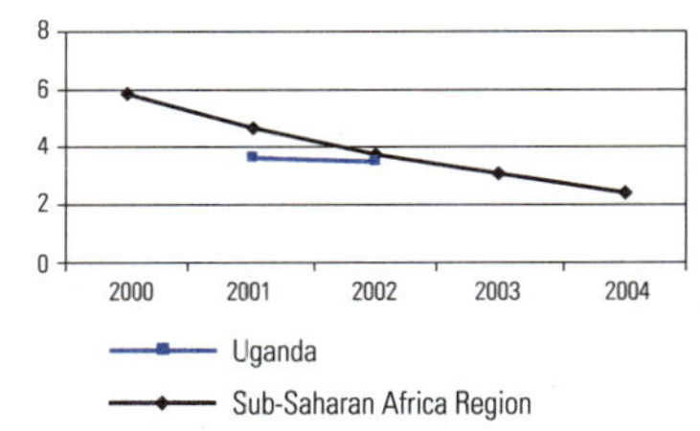

Total Telecommunications Revenue, 2000–4

Percentage of GDP

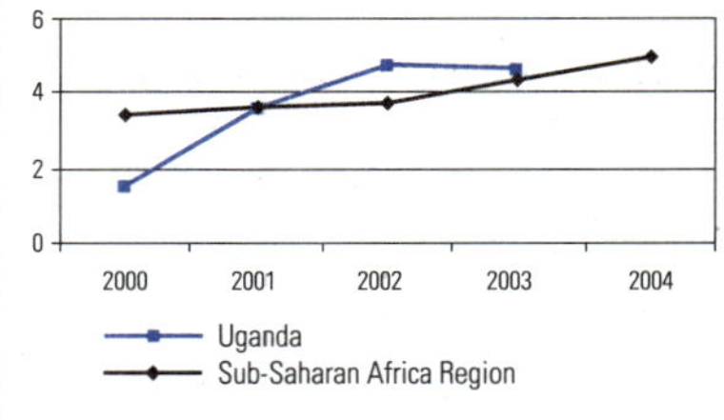

Notes: Figures in italics are for years other than those specified. .. indicates data are not available. C = competition; GDP = gross domestic product; GNI = gross national income; ICT = information and communication technology; M = monopoly; MDG = Millennium Development Goal; P = partial competition; and PCs = personal computers.

a. Outgoing and incoming.

b. Millennium Development Goal indicators 47, 48a, and 48b.

Sources: Economic and social context: UNDP and World Bank; ICT sector structure: ITU, WEF; ICT sector performance: Global Insight/WITSA, ITU, Netcraft, UNDESA, UNPAN, and World Bank. Produced by the Global Information and Communication Technologies Department and the Development Economics Data Group. For complete information, see Definitions and Data Sources.

Ukraine

	Ukraine 2000	Ukraine 2004	Lower-middle-income group 2004	Europe & Central Asia Region 2004
Economic and social context				
Population, total (millions)	50	48	2,430	472
Urban population (% total population)	67	67	49	64
Poverty (% population below US$1 per day)	*2.9*	..	..	*3.6*
GNI per capita, Atlas method (current US$)	690	1,260	1,580	3,290
GDP growth, 1995–2000 and 2000–4 (%)	–1.9	8.6	5.7	5.0
Adult literacy rate (% ages 15 and over)	*99*	..	90	97
Primary, secondary, tertiary school enrollment (% gross)	86	*86*	*72*	*81*
ICT sector structure				
Separate telecommunications regulator	..	..		
Status of main fixed-line operator	*Public*	Public		
Level of competition: international long distance	*C*	C		
Level of competition: mobile	*C*	C		
Level of competition: Internet service provider	..	..		
Government prioritization of ICT (scale 1–7)	..	3.7	3.8	3.8
ICT sector performance				
Access				
Telephone main lines (per 1,000 people)	210	*230*	192	*243*
International voice traffic (minutes per person)[a]	13	*15*	..	*28*
Mobile subscribers (per 1,000 people)	17	285	255	487
Population covered by mobile telephony (%)	..	*75*	*76*	*82*
Internet users (per 1,000 people)	7	*62*	70	*115*
Personal computers (per 1,000 people)	18	*20*	38	*73*
Households with television (%)	..	*97*	*89*	*92*
Quality				
Telephone faults (per 100 main lines per year)	*34.5*	*43.0*	..	*30.4*
Broadband subscribers (per 1,000 people)	0.0	*0.0*	12.6	*2.4*
International Internet bandwidth (bits per person)	1	17	58	148
Affordability				
Price basket for fixed line (US$ per month, residential)	2.4	*2.5*	*5.5*	*3.5*
Price basket for mobile (US$ per month)	..	10.3	8.9	10.3
Price basket for Internet (US$ per month)	..	*16.7*	*25.3*	*19.8*
Price of call to United States (US$ per 3 minutes)	..	*1.65*	1.45	1.06
Institutional efficiency and sustainability				
Total telecommunications revenue (% GDP)	3.8	6.1	3.6	3.5
Total telephone subscribers per employee	88	*142*	*195*	*150*
Total telecommunications investment (% revenue)	22.3	*33.0*	*25.5*	*19.0*
ICT applications				
ICT expenditure (% GDP)	7.9	*7.0*	*5.1*	*5.1*
E-government readiness index (scale 0–1)	..	0.56	0.26	0.39
Secure Internet servers (per 1 million people)	*0.9*	1.1	1.6	6.4
Schools connected to the Internet (%)	..	..	..	*65*

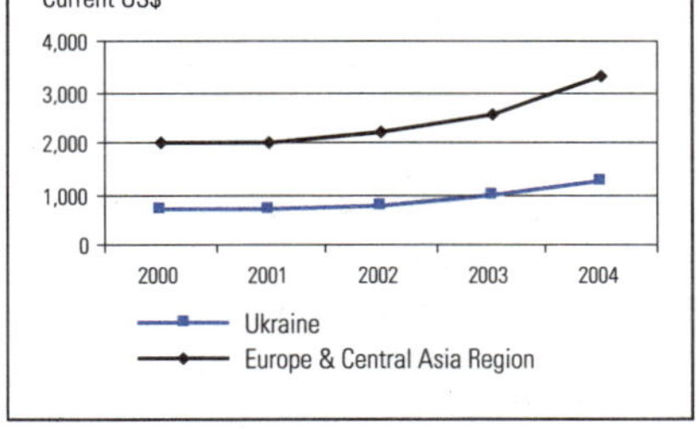

GNI per Capita, Atlas Method, 2000–4

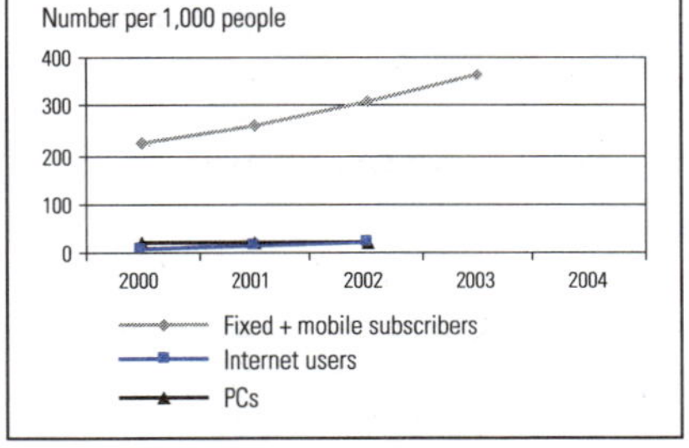

ICT MDG[b] Indicators, 2000–4

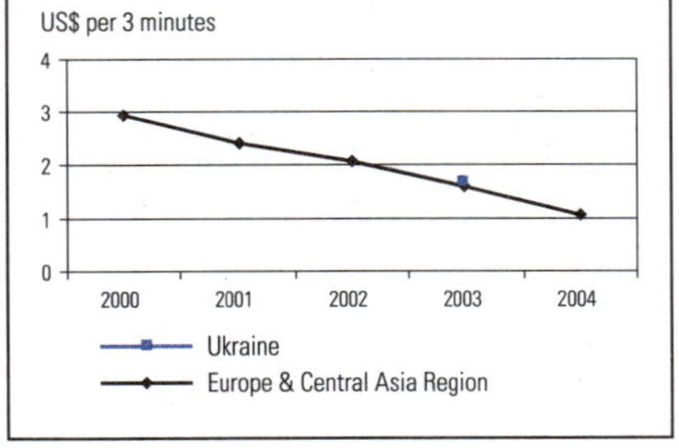

Price of Call to the United States, 2000–4

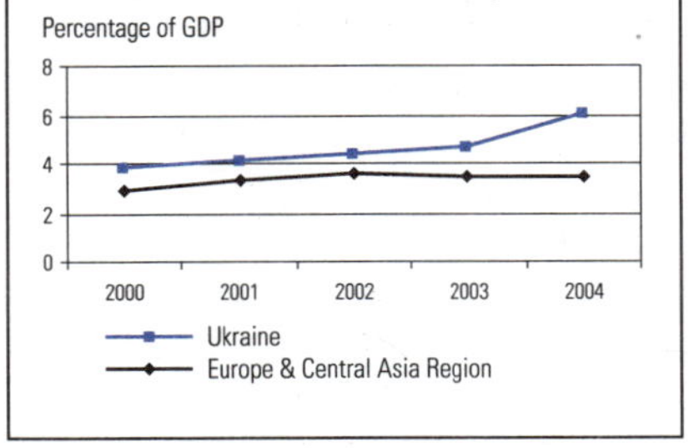

Total Telecommunications Revenue, 2000–4

Notes: Figures in italics are for years other than those specified. .. indicates data are not available. C = competition; GDP = gross domestic product; GNI = gross national income; ICT = information and communication technology; M = monopoly; MDG = Millennium Development Goal; P = partial competition; and PCs = personal computers.

a. Outgoing and incoming.

b. Millennium Development Goal indicators 47, 48a, and 48b.

Sources: Economic and social context: UNDP and World Bank; ICT sector structure: ITU, WEF; ICT sector performance: Global Insight/WITSA, ITU, Netcraft, UNDESA, UNPAN, and World Bank. Produced by the Global Information and Communication Technologies Department and the Development Economics Data Group. For complete information, see Definitions and Data Sources.

United Arab Emirates

	United Arab Emirates 2000	United Arab Emirates 2004	High-income group 2004
Economic and social context			
Population, total (millions)	3	4	1,001
Urban population (% total population)	85	85	77
Poverty (% population below US$1 per day)	..	..	..
GNI per capita, Atlas method (current US$)	*17,790*	..	32,040
GDP growth, 1995–2000 and 2000–4 (%)	5.0	*2.6*	2.0
Adult literacy rate (% ages 15 and over)	..	*77*	..
Primary, secondary, tertiary school enrollment (% gross)	72	*74*	*93*
ICT sector structure			
Separate telecommunications regulator	..	..	
Status of main fixed-line operator	*Mixed*	Mixed	
Level of competition: international long distance	*M*	M	
Level of competition: mobile	*M*	M	
Level of competition: Internet service provider	*M*	M	
Government prioritization of ICT (scale 1–7)	..	5.9	4.9
ICT sector performance			
Access			
Telephone main lines (per 1,000 people)	314	277	558
International voice traffic (minutes per person)[a]	485	*757*	..
Mobile subscribers (per 1,000 people)	440	860	767
Population covered by mobile telephony (%)	99	99	98
Internet users (per 1,000 people)	236	397	*480*
Personal computers (per 1,000 people)	123	117	*504*
Households with television (%)	86	*86*	*98*
Quality			
Telephone faults (per 100 main lines per year)	0.2	*0.3*	..
Broadband subscribers (per 1,000 people)	0.4	13.1	126.2
International Internet bandwidth (bits per person)	5	543	4,718
Affordability			
Price basket for fixed line (US$ per month, residential)	5.0	5.0	25.8
Price basket for mobile (US$ per month)	..	3.5	17.8
Price basket for Internet (US$ per month)	..	*13.1*	*20.9*
Price of call to United States (US$ per 3 minutes)	3.51	1.73	0.77
Institutional efficiency and sustainability			
Total telecommunications revenue (% GDP)	2.7	*3.1*	2.9
Total telephone subscribers per employee	294	500	*485*
Total telecommunications investment (% revenue)	28.9	13.0	*12.3*
ICT applications			
ICT expenditure (% GDP)	..	..	*7.1*
E-government readiness index (scale 0–1)	..	0.31	0.70
Secure Internet servers (per 1 million people)	*8.9*	40.4	311.4
Schools connected to the Internet (%)	..	..	*99*

GNI per Capita, Atlas Method, 2000–4

Current US$

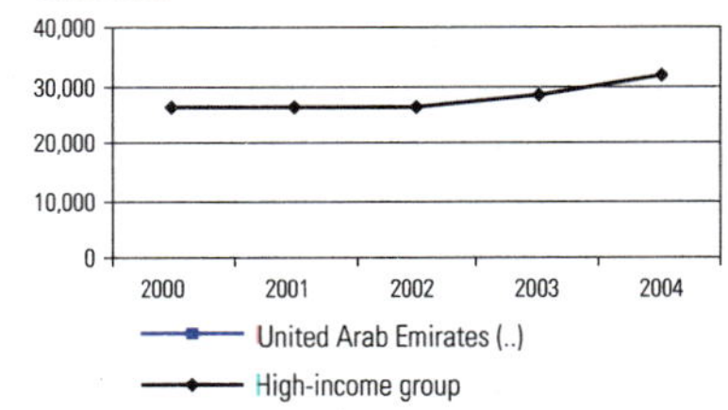

ICT MDG[b] Indicators, 2000–4

Number per 1,000 people

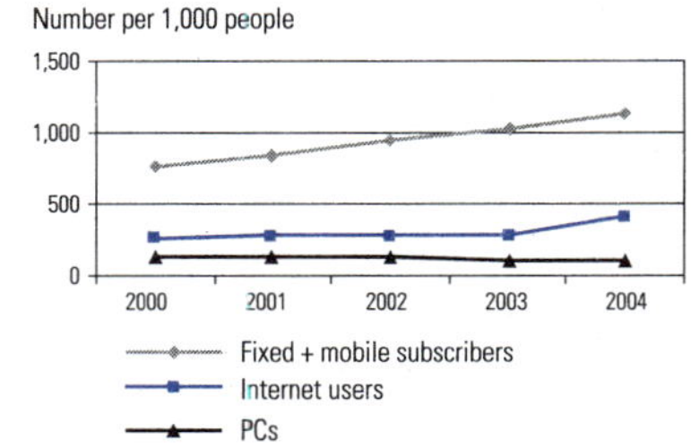

Price of Call to the United States, 2000–4

US$ per 3 minutes

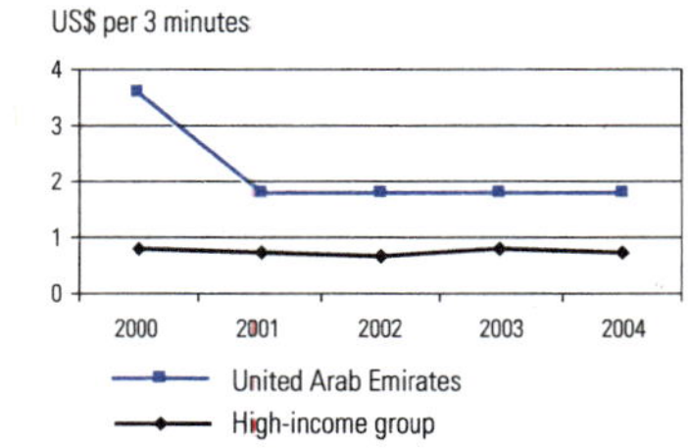

Total Telecommunications Revenue, 2000–4

Percentage of GDP

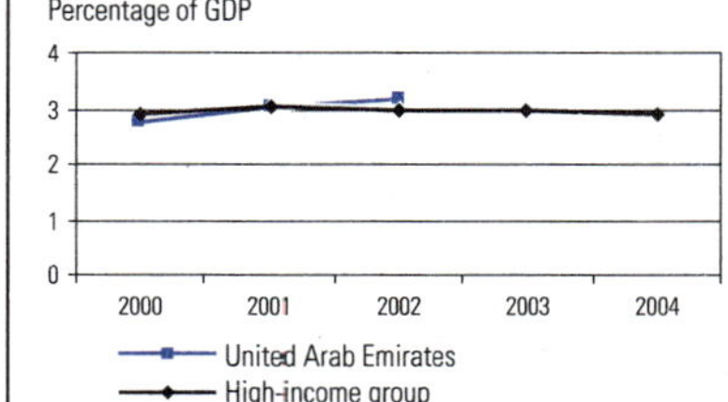

Notes: Figures in italics are for years other than those specified. .. indicates data are not available. C = competition; GDP = gross domestic product; GNI = gross national income; ICT = information and communication technology; M = monopoly; MDG = Millennium Development Goal; P = partial competition; and PCs = personal computers.

a. Outgoing and incoming.

b. Millennium Development Goal indicators 47, 48a, and 48b.

Sources: Economic and social context: UNDP and World Bank; ICT sector structure: ITU, WEF; ICT sector performance: Global Insight/WITSA, ITU, Netcraft, UNDESA, UNPAN, and World Bank. Produced by the Global Information and Communication Technologies Department and the Development Economics Data Group. For complete information, see Definitions and Data Sources.

United Kingdom

	United Kingdom 2000	United Kingdom 2004	High-income group 2004
Economic and social context			
Population, total (millions)	59	59	1,001
Urban population (% total population)	89	89	77
Poverty (% population below US$1 per day)	..	..	..
GNI per capita, Atlas method (current US$)	25,410	33,940	32,040
GDP growth, 1995–2000 and 2000–4 (%)	3.1	2.2	2.0
Adult literacy rate (% ages 15 and over)	..	..	..
Primary, secondary, tertiary school enrollment (% gross)	113	*123*	*93*
ICT sector structure			
Separate telecommunications regulator	Yes	Yes	
Status of main fixed-line operator	*Private*	Private	
Level of competition: international long distance	*C*	C	
Level of competition: mobile	*C*	P	
Level of competition: Internet service provider	*C*	C	
Government prioritization of ICT (scale 1–7)	..	5.0	4.9
ICT sector performance			
Access			
Telephone main lines (per 1,000 people)	598	567	558
International voice traffic (minutes per person)[a]	263	..	..
Mobile subscribers (per 1,000 people)	738	1,042	767
Population covered by mobile telephony (%)	99	99	98
Internet users (per 1,000 people)	268	533	*480*
Personal computers (per 1,000 people)	343	*496*	*504*
Households with television (%)	99	99	*98*
Quality			
Telephone faults (per 100 main lines per year)	4.5	*11.0*	..
Broadband subscribers (per 1,000 people)	1.8	103.3	126.2
International Internet bandwidth (bits per person)	1,469	13,156	4,718
Affordability			
Price basket for fixed line (US$ per month, residential)	25.3	29.5	25.8
Price basket for mobile (US$ per month)	..	19.1	17.8
Price basket for Internet (US$ per month)	..	*23.9*	*20.9*
Price of call to United States (US$ per 3 minutes)	1.07	0.77	0.77
Institutional efficiency and sustainability			
Total telecommunications revenue (% GDP)	3.8	3.8	2.9
Total telephone subscribers per employee	380	*358*	*485*
Total telecommunications investment (% revenue)	30.5	*22.1*	*12.3*
ICT applications			
ICT expenditure (% GDP)	8.1	*7.3*	*7.1*
E-government readiness index (scale 0–1)	..	0.97	0.70
Secure Internet servers (per 1 million people)	*109.5*	354.1	311.4
Schools connected to the Internet (%)	..	*99*	*99*

GNI per Capita, Atlas Method, 2000–4
Current US$

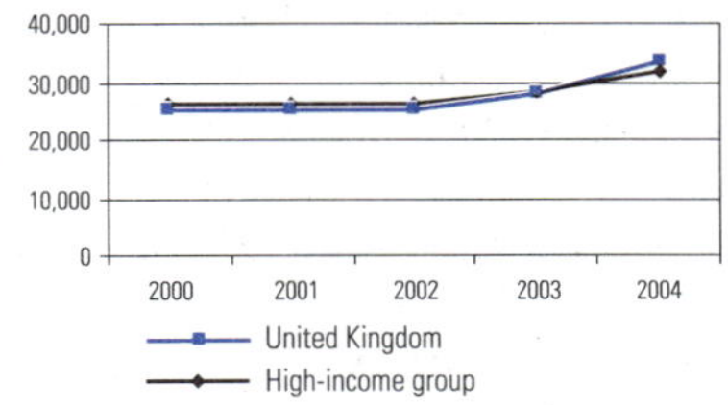

ICT MDG[b] Indicators, 2000–4
Number per 1,000 people

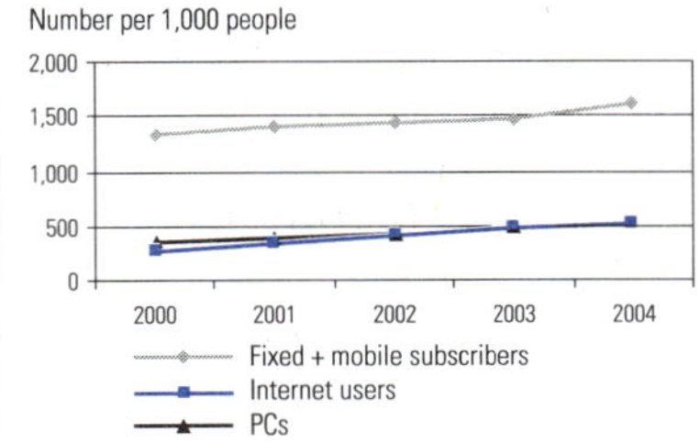

Price of Call to the United States, 2000–4
US$ per 3 minutes

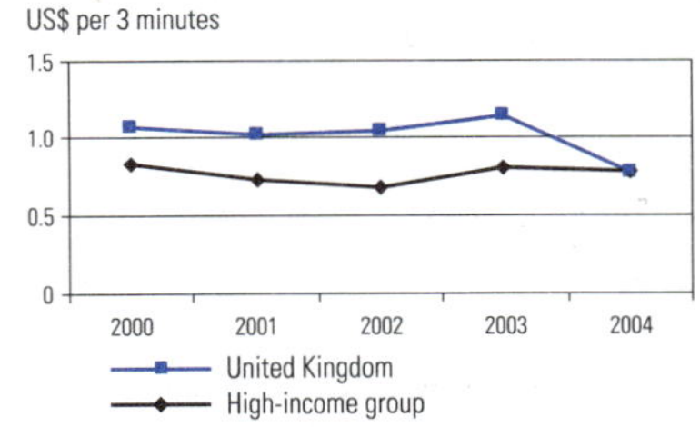

Total Telecommunications Revenue, 2000–4
Percentage of GDP

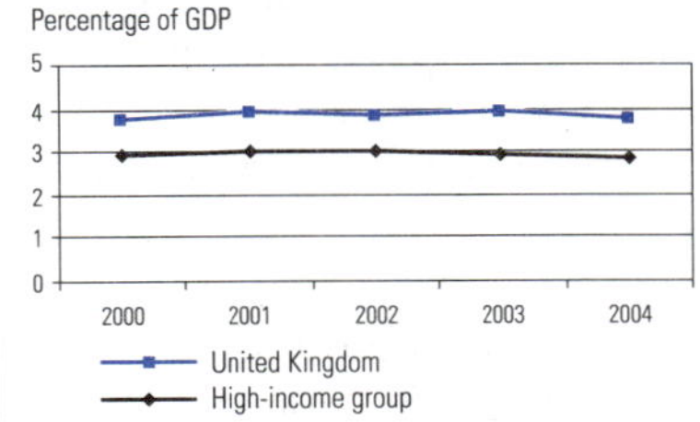

Notes: Figures in italics are for years other than those specified. .. indicates data are not available. C = competition; GDP = gross domestic product; GNI = gross national income; ICT = information and communication technology; M = monopoly; MDG = Millennium Development Goal; P = partial competition; and PCs = personal computers.

a. Outgoing and incoming.

b. Millennium Development Goal indicators 47, 48a, and 48b.

Sources: Economic and social context: UNDP and World Bank; ICT sector structure: ITU, WEF; ICT sector performance: Global Insight/WITSA, ITU, Netcraft, UNDESA, UNPAN, and World Bank. Produced by the Global Information and Communication Technologies Department and the Development Economics Data Group. For complete information, see Definitions and Data Sources.

United States

	United States 2000	United States 2004	High-income group 2004
Economic and social context			
Population, total (millions)	282	294	1,001
Urban population (% total population)	79	80	77
Poverty (% population below US$1 per day)	..	..	..
GNI per capita, Atlas method (current US$)	34,400	41,400	32,040
GDP growth, 1995–2000 and 2000–4 (%)	4.2	2.6	2.0
Adult literacy rate (% ages 15 and over)	..	..	..
Primary, secondary, tertiary school enrollment (% gross)	92	*93*	*93*
ICT sector structure			
Separate telecommunications regulator	Yes	Yes	
Status of main fixed-line operator	*Private*	Private	
Level of competition: international long distance	*C*	C	
Level of competition: mobile	*C*	C	
Level of competition: Internet service provider	*C*	C	
Government prioritization of ICT (scale 1–7)	..	5.2	4.9
ICT sector performance			
Access			
Telephone main lines (per 1,000 people)	663	606	558
International voice traffic (minutes per person)[a]	149	*199*	..
Mobile subscribers (per 1,000 people)	388	615	767
Population covered by mobile telephony (%)	..	95	98
Internet users (per 1,000 people)	439	569	*480*
Personal computers (per 1,000 people)	570	760	*504*
Households with television (%)	97	97	*98*
Quality			
Telephone faults (per 100 main lines per year)	14.2	*11.7*	..
Broadband subscribers (per 1,000 people)	25.1	129.1	126.2
International Internet bandwidth (bits per person)	394	3,308	4,718
Affordability			
Price basket for fixed line (US$ per month, residential)	21.5	25.0	25.8
Price basket for mobile (US$ per month)	..	10.8	17.8
Price basket for Internet (US$ per month)	..	*15.0*	*20.9*
Price of call to United States (US$ per 3 minutes)	..	..	0.77
Institutional efficiency and sustainability			
Total telecommunications revenue (% GDP)	3.0	2.5	2.9
Total telephone subscribers per employee	261	344	*485*
Total telecommunications investment (% revenue)	38.7	18.0	*12.3*
ICT applications			
ICT expenditure (% GDP)	9.5	*8.8*	*7.1*
E-government readiness index (scale 0–1)	..	1.00	0.70
Secure Internet servers (per 1 million people)	*273.8*	674.9	311.4
Schools connected to the Internet (%)	..	*99*	*99*

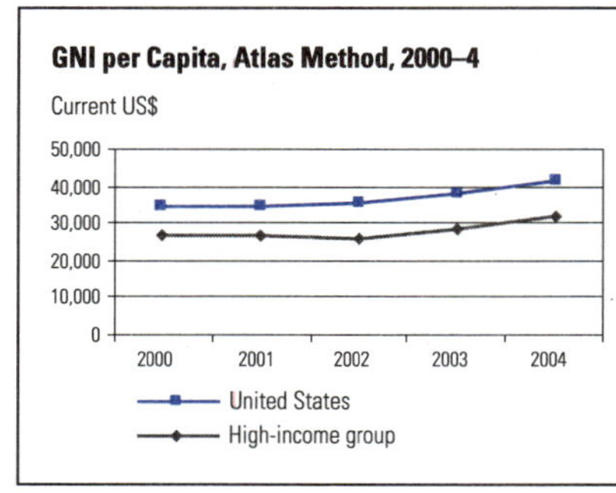

GNI per Capita, Atlas Method, 2000–4

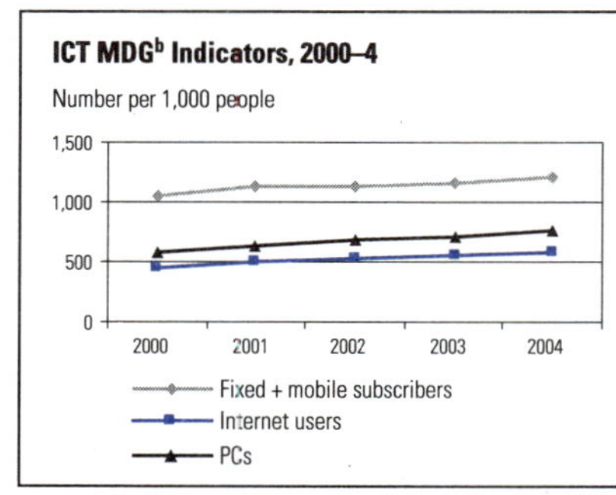

ICT MDG[b] Indicators, 2000–4

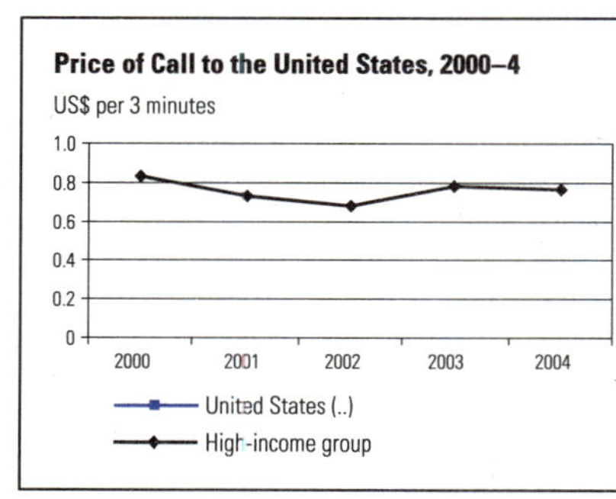

Price of Call to the United States, 2000–4

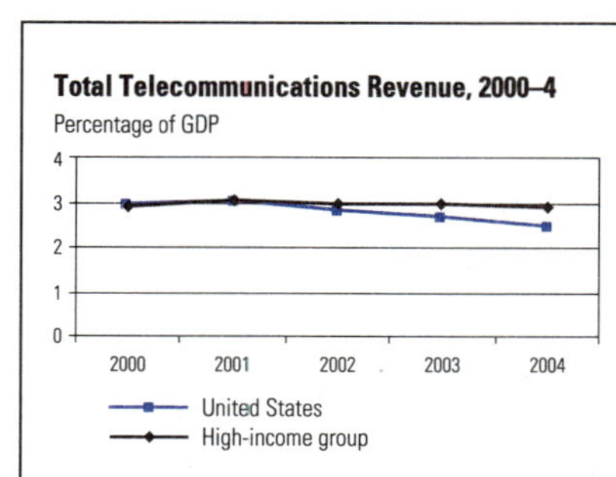

Total Telecommunications Revenue, 2000–4

Notes: Figures in italics are for years other than those specified. .. indicates data are not available. C = competition; GDP = gross domestic product; GNI = gross national income; ICT = information and communication technology; M = monopoly; MDG = Millennium Development Goal; P = partial competition; and PCs = personal computers.

a. Outgoing and incoming.

b. Millennium Development Goal indicators 47, 48a, and 48b.

Sources: Economic and social context: UNDP and World Bank; ICT sector structure: ITU, WEF; ICT sector performance: Global Insight/WITSA, ITU, Netcraft, UNDESA, UNPAN, and World Bank. Produced by the Global Information and Communication Technologies Department and the Development Economics Data Group. For complete information, see Definitions and Data Sources.

Uruguay

	Uruguay 2000	Uruguay 2004	Upper-middle-income group 2004	Latin America & the Caribbean Region 2004
Economic and social context				
Population, total (millions)	3	3	576	541
Urban population (% total population)	92	93	72	77
Poverty (% population below US$1 per day)	<2	..	..	*9.5*
GNI per capita, Atlas method (current US$)	6,120	3,950	4,770	3,600
GDP growth, 1995–2000 and 2000–4 (%)	2.2	−1.2	2.7	1.5
Adult literacy rate (% ages 15 and over)	..	*98*	94	89
Primary, secondary, tertiary school enrollment (% gross)	84	*88*	*80*	*82*
ICT sector structure				
Separate telecommunications regulator	No	Yes		
Status of main fixed-line operator	*Public*	Public		
Level of competition: international long distance	*M*	P		
Level of competition: mobile	*C*	C		
Level of competition: Internet service provider	*P*	C		
Government prioritization of ICT (scale 1–7)	..	4.0	4.1	3.5
ICT sector performance				
Access				
Telephone main lines (per 1,000 people)	280	*278*	*220*	181
International voice traffic (minutes per person)[a]	61	..	*39*	..
Mobile subscribers (per 1,000 people)	124	165	490	324
Population covered by mobile telephony (%)	99	99	*84*	*76*
Internet users (per 1,000 people)	111	*170*	*133*	*104*
Personal computers (per 1,000 people)	105	*121*	*99*	*75*
Households with television (%)	99	..	*92*	*88*
Quality				
Telephone faults (per 100 main lines per year)	*5.6*	..	*20.3*	..
Broadband subscribers (per 1,000 people)	*0.4*	*3.3*	*3.7*	*5.2*
International Internet bandwidth (bits per person)	8	177	176	165
Affordability				
Price basket for fixed line (US$ per month, residential)	15.6	9.0	*13.9*	*9.0*
Price basket for mobile (US$ per month)	..	7.4	11.1	9.1
Price basket for Internet (US$ per month)	..	*26.5*	*20.8*	*31.5*
Price of call to United States (US$ per 3 minutes)	4.88	0.52	1.03	0.90
Institutional efficiency and sustainability				
Total telecommunications revenue (% GDP)	3.9	2.9	3.4	3.3
Total telephone subscribers per employee	243	..	*402*	..
Total telecommunications investment (% revenue)	13.5	*7.0*	*18.6*	..
ICT applications				
ICT expenditure (% GDP)	6.2	*7.1*	*5.0*	*5.3*
E-government readiness index (scale 0–1)	..	0.48	0.49	0.39
Secure Internet servers (per 1 million people)	*11.1*	23.2	10.7	8.6
Schools connected to the Internet (%)	..	*50*	*60*	..

GNI per Capita, Atlas Method, 2000–4

Current US$

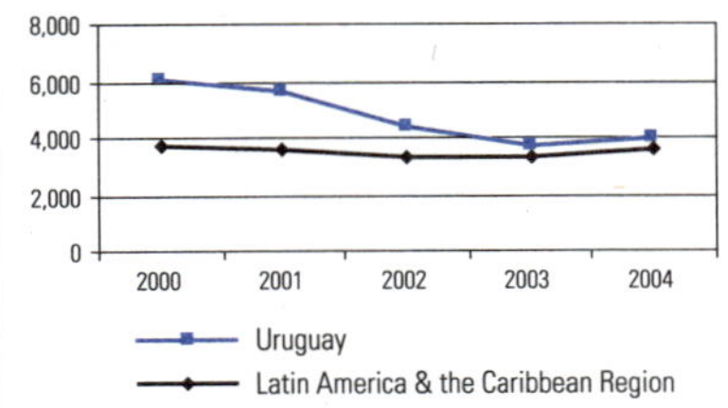

ICT MDG[b] Indicators, 2000–4

Number per 1,000 people

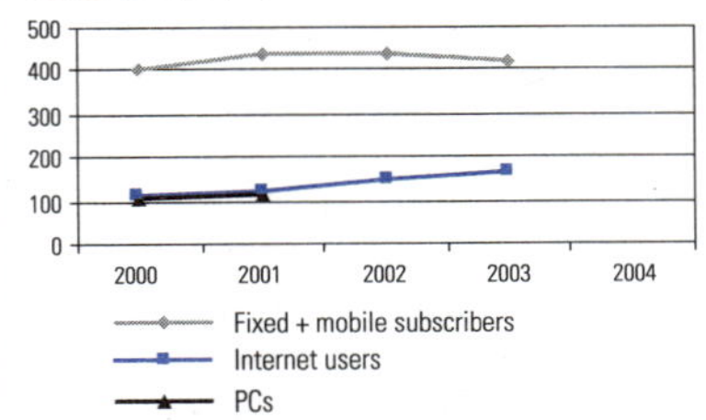

Price of Call to the United States, 2000–4

US$ per 3 minutes

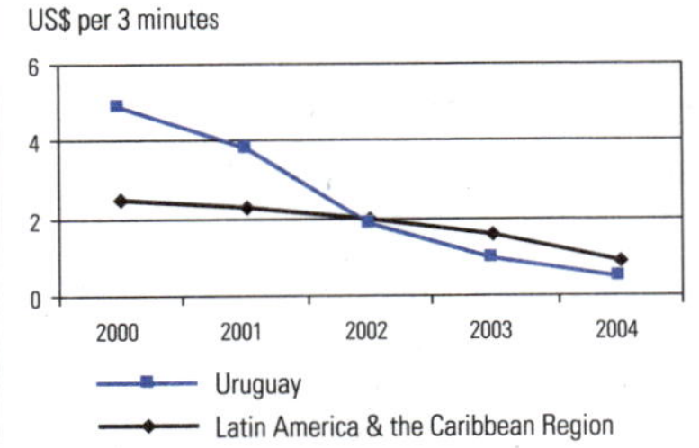

Total Telecommunications Revenue, 2000–4

Percentage of GDP

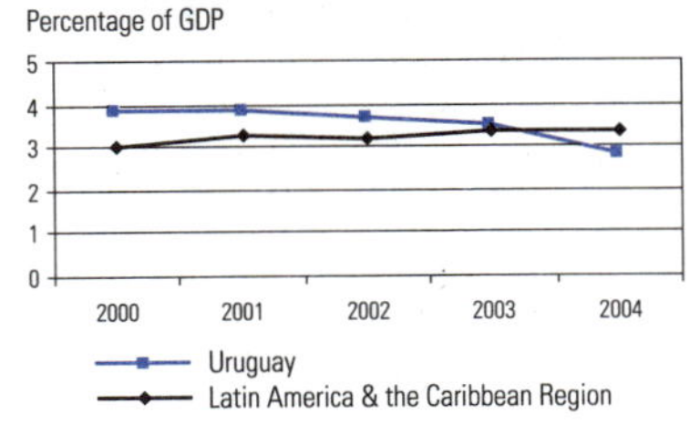

Notes: Figures in italics are for years other than those specified. .. indicates data are not available. C = competition; GDP = gross domestic product; GNI = gross national income; ICT = information and communication technology; M = monopoly; MDG = Millennium Development Goal; P = partial competition; and PCs = personal computers.

a. Outgoing and incoming.

b. Millennium Development Goal indicators 47, 48a, and 48b.

Sources: Economic and social context: UNDP and World Bank; ICT sector structure: ITU, WEF; ICT sector performance: Global Insight/WITSA, ITU, Netcraft, UNDESA, UNPAN, and World Bank. Produced by the Global Information and Communication Technologies Department and the Development Economics Data Group. For complete information, see Definitions and Data Sources.

Uzbekistan

	Uzbekistan 2000	Uzbekistan 2004	Low-income group 2004	Europe & Central Asia Region 2004
Economic and social context				
Population, total (millions)	25	26	2,338	472
Urban population (% total population)	37	37	31	64
Poverty (% population below US$1 per day)	17.3	..	..	*3.6*
GNI per capita, Atlas method (current US$)	630	460	510	3,290
GDP growth, 1995–2000 and 2000–4 (%)	4.1	4.8	5.4	5.0
Adult literacy rate (% ages 15 and over)	..	*99*	61	97
Primary, secondary, tertiary school enrollment (% gross)	*76*	*76*	*55*	*81*
ICT sector structure				
Separate telecommunications regulator	..	..		
Status of main fixed-line operator	*Public*	Public		
Level of competition: international long distance	*P*	P		
Level of competition: mobile	*C*	C		
Level of competition: Internet service provider	..	..		
Government prioritization of ICT (scale 1–7)	..	..	..	3.8
ICT sector performance				
Access				
Telephone main lines (per 1,000 people)	67	*67*	33	*243*
International voice traffic (minutes per person)[a]	6	..	*4*	*28*
Mobile subscribers (per 1,000 people)	2	*13*	48	487
Population covered by mobile telephony (%)	..	*75*	*43*	*82*
Internet users (per 1,000 people)	5	*19*	20	*115*
Personal computers (per 1,000 people)	7	..	*8*	*73*
Households with television (%)	*93*	..	*16*	*92*
Quality				
Telephone faults (per 100 main lines per year)	92.6	*97.0*	..	*30.4*
Broadband subscribers (per 1,000 people)	0.0	*0.1*	0.5	*2.4*
International Internet bandwidth (bits per person)	0	*1*	*3*	148
Affordability				
Price basket for fixed line (US$ per month, residential)	2.6	..	*6.6*	*3.5*
Price basket for mobile (US$ per month)	..	4.6	11.6	10.3
Price basket for Internet (US$ per month)	..	*20.2*	*45.5*	*19.8*
Price of call to United States (US$ per 3 minutes)	13.95	..	1.95	1.06
Institutional efficiency and sustainability				
Total telecommunications revenue (% GDP)	1.6	2.1	2.3	3.5
Total telephone subscribers per employee	66	*87*	*89*	*150*
Total telecommunications investment (% revenue)	25.3	..	*27.8*	*19.0*
ICT applications				
ICT expenditure (% GDP)	..	..	*4.1*	*5.1*
E-government readiness index (scale 0–1)	..	0.23	0.12	0.39
Secure Internet servers (per 1 million people)	..	0.0	0.3	6.4
Schools connected to the Internet (%)	..	..	..	*65*

GNI per Capita, Atlas Method, 2000–4

Current US$

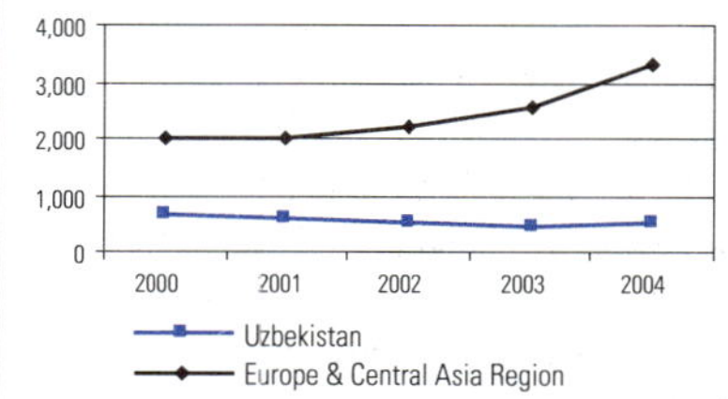

ICT MDG[b] Indicators, 2000–4

Number per 1,000 people

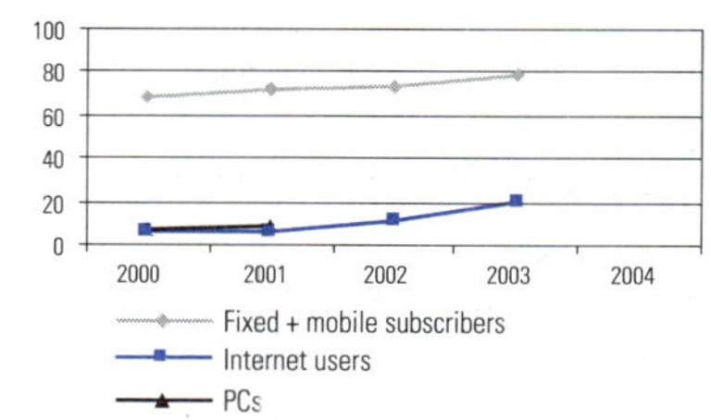

Price of Call to the United States, 2000–4

US$ per 3 minutes

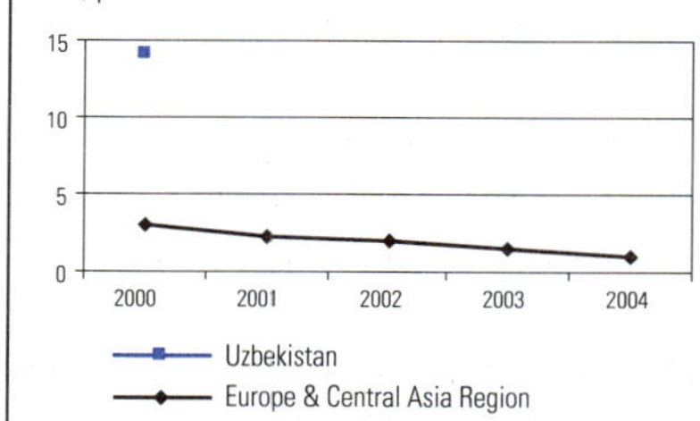

Total Telecommunications Revenue, 2000–4

Percentage of GDP

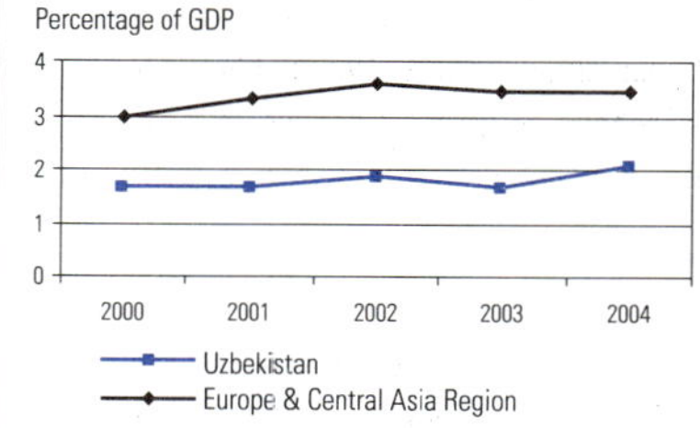

Notes: Figures in italics are for years other than those specified. .. indicates data are not available. C = competition; GDP = gross domestic product; GNI = gross national income; ICT = information and communication technology; M = monopoly; MDG = Millennium Development Goal; P = partial competition; and PCs = personal computers.

a. Outgoing and incoming.

b. Millennium Development Goal indicators 47, 48a, and 48b.

Sources: Economic and social context: UNDP and World Bank; ICT sector structure: ITU, WEF; ICT sector performance: Global Insight/WITSA, ITU, Netcraft, UNDESA, UNPAN, and World Bank. Produced by the Global Information and Communication Technologies Department and the Development Economics Data Group. For complete information, see Definitions and Data Sources.

Venezuela, República Bolivariana de

	Venezuela, R.B. de 2000	Venezuela, R.B. de 2004	Upper-middle-income group 2004	Latin America & the Caribbean Region 2004
Economic and social context				
Population, total (millions)	24	26	576	541
Urban population (% total population)	87	88	72	77
Poverty (% population below US$1 per day)	*14.3*	..	..	*9.5*
GNI per capita, Atlas method (current US$)	4,100	4,020	4,770	3,600
GDP growth, 1995–2000 and 2000–4 (%)	0.6	−1.3	2.7	1.5
Adult literacy rate (% ages 15 and over)	*93*	..	94	89
Primary, secondary, tertiary school enrollment (% gross)	72	*75*	*80*	*82*
ICT sector structure				
Separate telecommunications regulator	Yes	Yes		
Status of main fixed-line operator	*Mixed*	Private		
Level of competition: international long distance	*M*	C		
Level of competition: mobile	*C*	C		
Level of competition: Internet service provider	*C*	C		
Government prioritization of ICT (scale 1–7)	..	3.4	4.1	3.5
ICT sector performance				
Access				
Telephone main lines (per 1,000 people)	104	128	*220*	181
International voice traffic (minutes per person)[a]	23	*35*	*39*	..
Mobile subscribers (per 1,000 people)	224	322	490	324
Population covered by mobile telephony (%)	..	90	*84*	*76*
Internet users (per 1,000 people)	34	89	*133*	*104*
Personal computers (per 1,000 people)	45	*66*	*99*	*75*
Households with television (%)	79	*90*	*92*	*88*
Quality				
Telephone faults (per 100 main lines per year)	2.0	..	*20.3*	..
Broadband subscribers (per 1,000 people)	0.3	8.0	*3.7*	*5.2*
International Internet bandwidth (bits per person)	6	38	176	165
Affordability				
Price basket for fixed line (US$ per month, residential)	..	*16.2*	*13.9*	*9.0*
Price basket for mobile (US$ per month)	..	14.5	11.1	9.1
Price basket for Internet (US$ per month)	..	*19.5*	*20.8*	*31.5*
Price of call to United States (US$ per 3 minutes)	..	*0.84*	1.03	0.90
Institutional efficiency and sustainability				
Total telecommunications revenue (% GDP)	3.3	3.0	3.4	3.3
Total telephone subscribers per employee	386	..	*402*	..
Total telecommunications investment (% revenue)	26.3	*23.0*	*18.6*	..
ICT applications				
ICT expenditure (% GDP)	3.7	*5.3*	*5.0*	*5.3*
E-government readiness index (scale 0–1)	..	0.52	0.49	0.39
Secure Internet servers (per 1 million people)	*3.7*	4.4	10.7	8.6
Schools connected to the Internet (%)	..	..	*60*	..

GNI per Capita, Atlas Method, 2000–4

Current US$

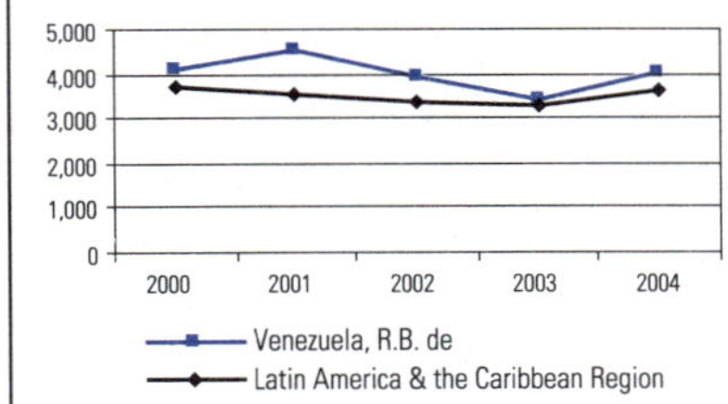

ICT MDG[b] Indicators, 2000–4

Number per 1,000 people

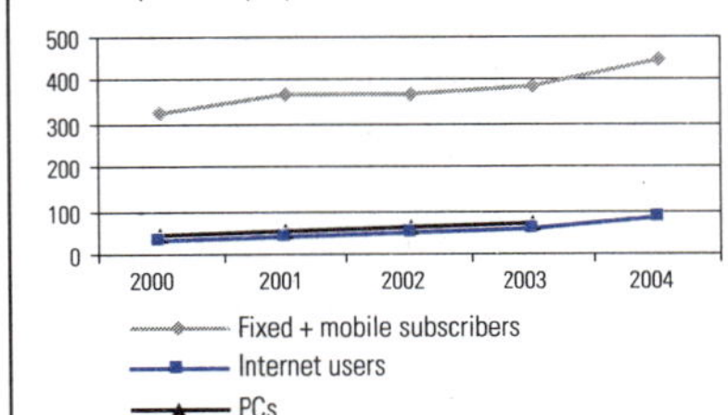

Price of Call to the United States, 2000–4

US$ per 3 minutes

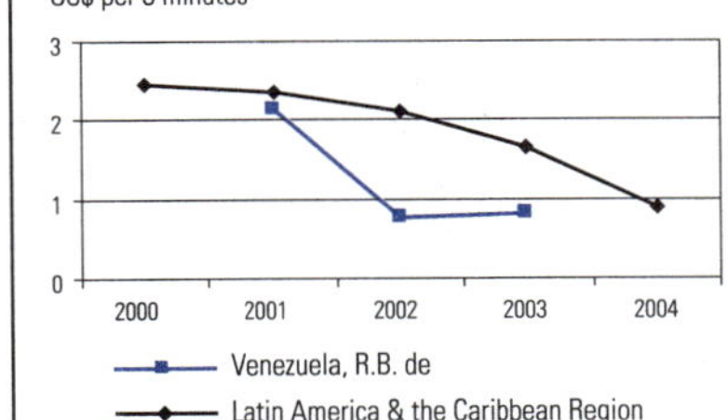

Total Telecommunications Revenue, 2000–4

Percentage of GDP

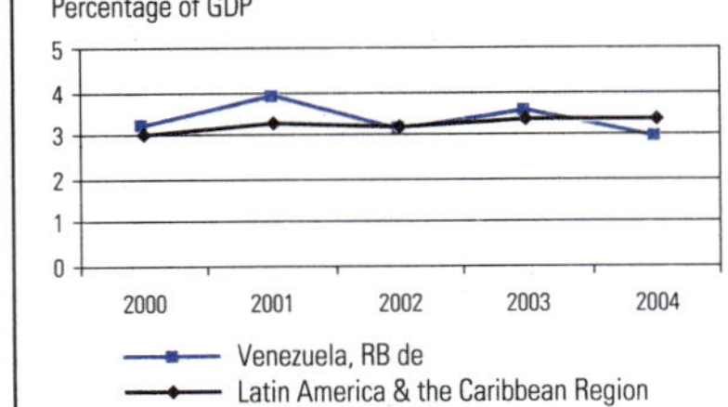

Notes: Figures in italics are for years other than those specified. .. indicates data are not available. C = competition; GDP = gross domestic product; GNI = gross national income; ICT = information and communication technology; M = monopoly; MDG = Millennium Development Goal; P = partial competition; and PCs = personal computers.

a. Outgoing and incoming.

b. Millennium Development Goal indicators 47, 48a, and 48b.

Sources: Economic and social context: UNDP and World Bank; ICT sector structure: ITU, WEF; ICT sector performance: Global Insight/WITSA, ITU, Netcraft, UNDESA, UNPAN, and World Bank. Produced by the Global Information and Communication Technologies Department and the Development Economics Data Group. For complete information, see Definitions and Data Sources.

Vietnam

	Vietnam 2000	Vietnam 2004	Low-income group 2004	East Asia & Pacific Region 2004
Economic and social context				
Population, total (millions)	79	82	2,338	1,870
Urban population (% total population)	24	26	31	41
Poverty (% population below US$1 per day)	*3.8*	*<2*	..	*14.9*
GNI per capita, Atlas method (current US$)	380	550	510	1,280
GDP growth, 1995–2000 and 2000–4 (%)	6.7	7.2	5.4	7.5
Adult literacy rate (% ages 15 and over)	*90*	..	61	90
Primary, secondary, tertiary school enrollment (% gross)	64	*64*	*55*	*68*
ICT sector structure				
Separate telecommunications regulator	..	..		
Status of main fixed-line operator	*Public*	Public		
Level of competition: international long distance	*M*	M		
Level of competition: mobile	*M*	M		
Level of competition: Internet service provider	*P*	P		
Government prioritization of ICT (scale 1–7)	..	4.9	..	4.8
ICT sector performance				
Access				
Telephone main lines (per 1,000 people)	32	70	33	194
International voice traffic (minutes per person)[a]	7	*8*	*4*	..
Mobile subscribers (per 1,000 people)	10	53	48	248
Population covered by mobile telephony (%)	..	67	*43*	*73*
Internet users (per 1,000 people)	3	65	20	75
Personal computers (per 1,000 people)	7	11	*8*	37
Households with television (%)	78	*83*	*16*	*80*
Quality				
Telephone faults (per 100 main lines per year)	..	..	..	..
Broadband subscribers (per 1,000 people)	0.0	0.6	0.5	13.4
International Internet bandwidth (bits per person)	0	27	*3*	52
Affordability				
Price basket for fixed line (US$ per month, residential)	5.4	*4.3*	*6.6*	*4.5*
Price basket for mobile (US$ per month)	..	6.9	11.6	5.1
Price basket for Internet (US$ per month)	..	*19.9*	*45.5*	*19.9*
Price of call to United States (US$ per 3 minutes)	9.29	1.95	1.95	1.20
Institutional efficiency and sustainability				
Total telecommunications revenue (% GDP)	2.5	3.5	2.3	3.6
Total telephone subscribers per employee	43	*73*	*89*	..
Total telecommunications investment (% revenue)	..	..	*27.8*	*31.0*
ICT applications				
ICT expenditure (% GDP)	..	..	*4.1*	*5.0*
E-government readiness index (scale 0–1)	..	0.14	0.12	0.21
Secure Internet servers (per 1 million people)	*0.1*	0.1	0.3	0.6
Schools connected to the Internet (%)	20	..	..	..

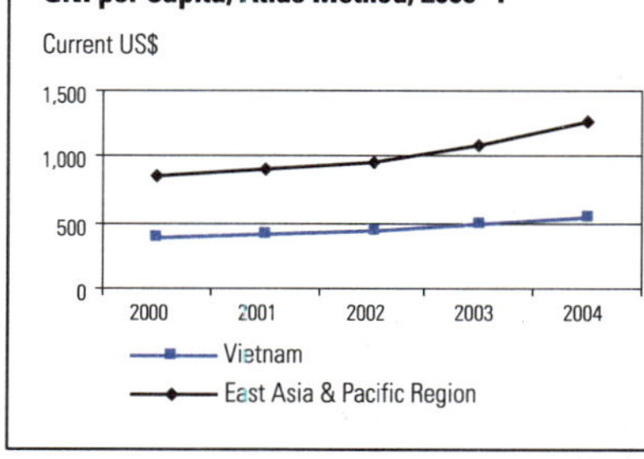

GNI per Capita, Atlas Method, 2000–4

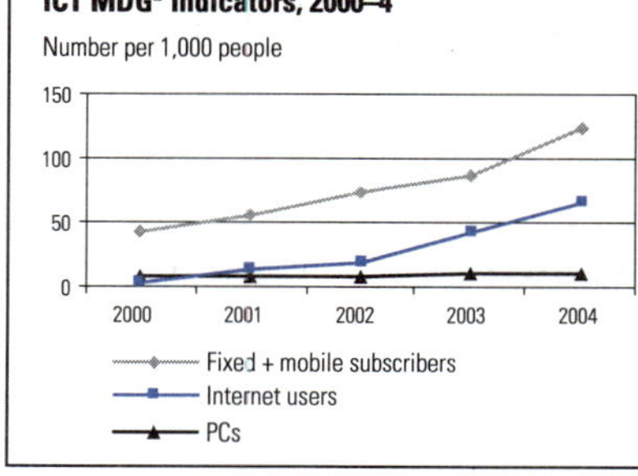

ICT MDG[b] Indicators, 2000–4

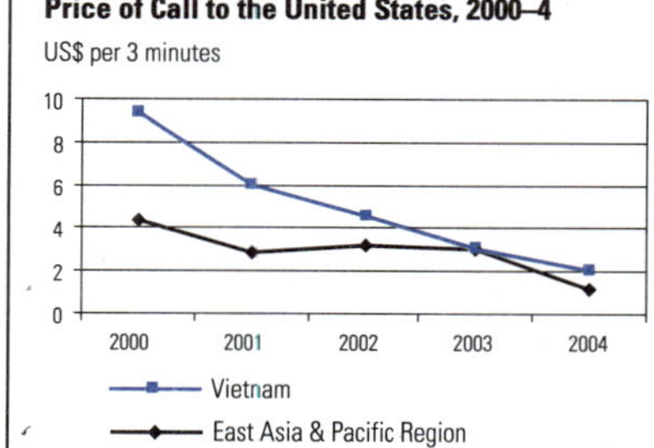

Price of Call to the United States, 2000–4

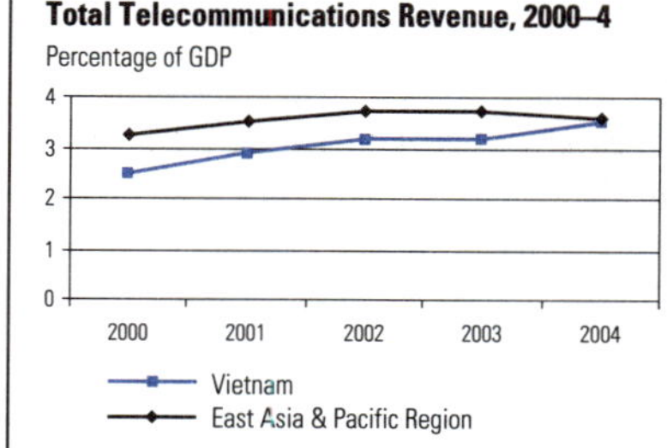

Total Telecommunications Revenue, 2000–4

Notes: Figures in italics are for years other than those specified. .. indicates data are not available. C = competition; GDP = gross domestic product; GNI = gross national income; ICT = information and communication technology; M = monopoly; MDG = Millennium Development Goal; P = partial competition; and PCs = personal computers.

a. Outgoing and incoming.

b. Millennium Development Goal indicators 47, 48a, and 48b.

Sources: Economic and social context: UNDP and World Bank; ICT sector structure: ITU, WEF; ICT sector performance: Global Insight/WITSA, ITU, Netcraft, UNDESA, UNPAN, and World Bank. Produced by the Global Information and Communication Technologies Department and the Development Economics Data Group. For complete information, see Definitions and Data Sources.

West Bank and Gaza

	West Bank and Gaza 2000	West Bank and Gaza 2004	Lower-middle-income group 2004	Middle East & North Africa Region 2004
Economic and social context				
Population, total (millions)	3	4	2,430	294
Urban population (% total population)	..	..	49	56
Poverty (% population below US$1 per day)	..	..	..	*2.4*
GNI per capita, Atlas method (current US$)	1,750	*1,120*	1,580	2,000
GDP growth, 1995–2000 and 2000–4 (%)	4.3	*–13.3*	5.7	4.5
Adult literacy rate (% ages 15 and over)	..	*92*	90	..
Primary, secondary, tertiary school enrollment (% gross)	78	*80*	*72*	*68*
ICT sector structure				
Separate telecommunications regulator	..	..		
Status of main fixed-line operator	..	..		
Level of competition: international long distance	..	..		
Level of competition: mobile	..	..		
Level of competition: Internet service provider	..	..		
Government prioritization of ICT (scale 1–7)	..	..	3.8	..
ICT sector performance				
Access				
Telephone main lines (per 1,000 people)	92	83	192	*118*
International voice traffic (minutes per person)[a]	28	*30*	..	*20*
Mobile subscribers (per 1,000 people)	59	303	255	*88*
Population covered by mobile telephony (%)	95	*95*	*76*	..
Internet users (per 1,000 people)	12	*43*	70	*47*
Personal computers (per 1,000 people)	..	*39*	38	*30*
Households with television (%)	85	*94*	*89*	*88*
Quality				
Telephone faults (per 100 main lines per year)	..	*97.0*	..	..
Broadband subscribers (per 1,000 people)	0.0	0.0	12.6	*0.2*
International Internet bandwidth (bits per person)	..	*12*	58	*15*
Affordability				
Price basket for fixed line (US$ per month, residential)	10.3	*9.4*	*5.5*	*4.9*
Price basket for mobile (US$ per month)	..	..	8.9	8.1
Price basket for Internet (US$ per month)	..	*25.4*	*25.3*	*24.5*
Price of call to United States (US$ per 3 minutes)	*1.11*	*1.03*	1.45	1.64
Institutional efficiency and sustainability				
Total telecommunications revenue (% GDP)	0.5	*0.6*	3.6	2.8
Total telephone subscribers per employee	263	..	*195*	..
Total telecommunications investment (% revenue)	32.3	..	*25.5*	*27.8*
ICT applications				
ICT expenditure (% GDP)	..	..	*5.1*	..
E-government readiness index (scale 0–1)	..	..	0.26	0.16
Secure Internet servers (per 1 million people)	..	..	1.6	0.6
Schools connected to the Internet (%)	..	..	..	..

GNI per Capita, Atlas Method, 2000–4

Current US$

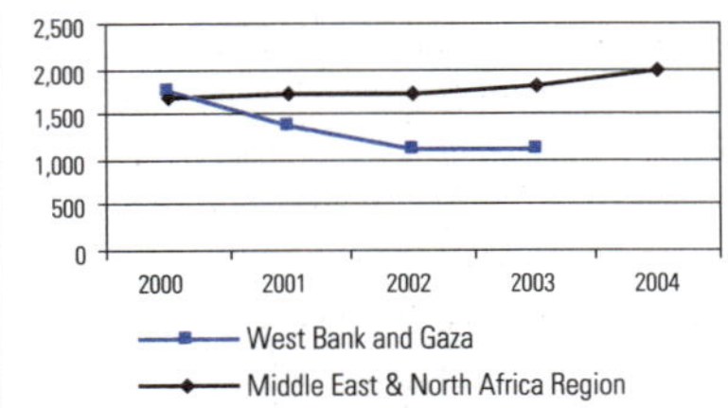

ICT MDG[b] Indicators, 2000–4

Number per 1,000 people

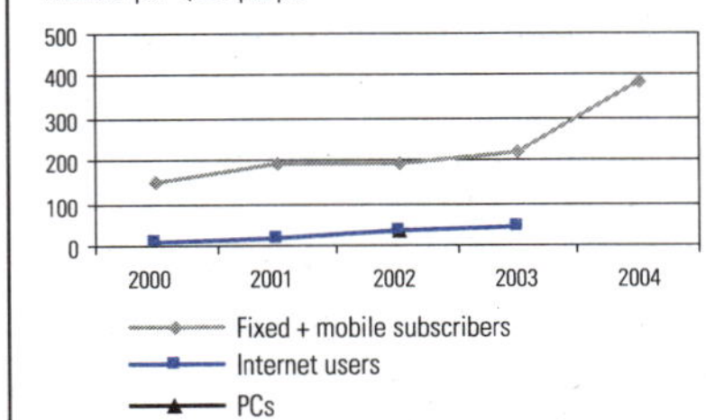

Price of Call to the United States, 2000–4

US$ per 3 minutes

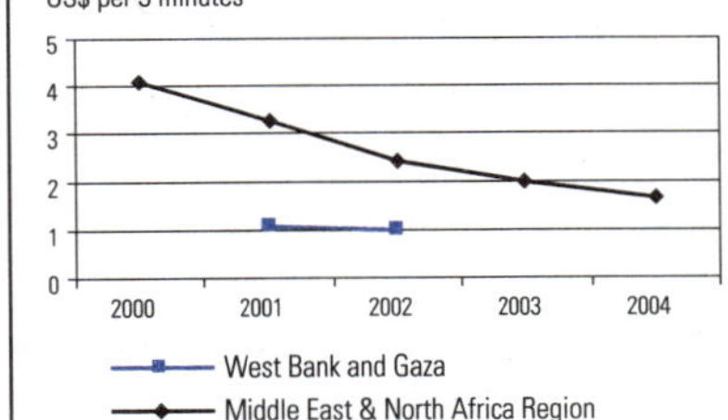

Total Telecommunications Revenue, 2000–4

Percentage of GDP

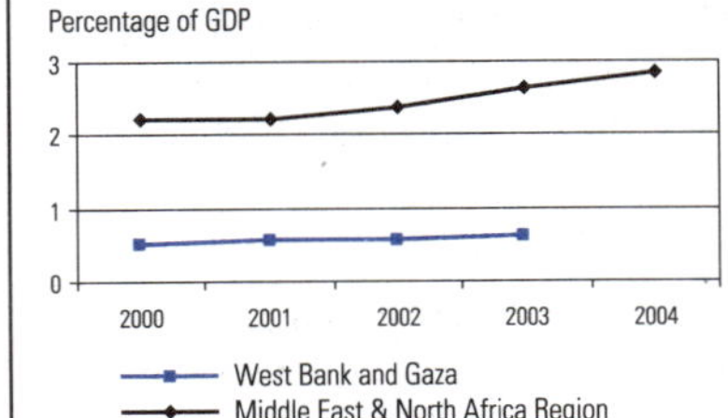

Notes: Figures in italics are for years other than those specified. .. indicates data are not available. C = competition; GDP = gross domestic product; GNI = gross national income; ICT = information and communication technology; M = monopoly; MDG = Millennium Development Goal; P = partial competition; and PCs = personal computers.

a. Outgoing and incoming.

b. Millennium Development Goal indicators 47, 48a, and 48b.

Sources: Economic and social context: UNDP and World Bank; ICT sector structure: ITU, WEF; ICT sector performance: Global Insight/WITSA, ITU, Netcraft, UNDESA, UNPAN, and World Bank. Produced by the Global Information and Communication Technologies Department and the Development Economics Data Group. For complete information, see Definitions and Data Sources.

Yemen, Republic of

	Yemen, Rep. of 2000	Yemen, Rep. of 2004	Low-income group 2004	Middle East & North Africa Region 2004
Economic and social context				
Population, total (millions)	18	20	2,338	294
Urban population (% total population)	25	26	31	56
Poverty (% population below US$1 per day)	*15.7*	..	..	*2.4*
GNI per capita, Atlas method (current US$)	420	570	510	2,000
GDP growth, 1995–2000 and 2000–4 (%)	5.6	3.6	5.4	4.5
Adult literacy rate (% ages 15 and over)	..	*49*	61	..
Primary, secondary, tertiary school enrollment (% gross)	53	*55*	*55*	*68*
ICT sector structure				
Separate telecommunications regulator	..	..		
Status of main fixed-line operator	*Public*	Public		
Level of competition: international long distance	*M*	M		
Level of competition: mobile	*C*	C		
Level of competition: Internet service provider	*M*	C		
Government prioritization of ICT (scale 1–7)	..	..	..	..
ICT sector performance				
Access				
Telephone main lines (per 1,000 people)	20	*36*	33	*118*
International voice traffic (minutes per person)[a]	8	*12*	*4*	*20*
Mobile subscribers (per 1,000 people)	2	*37*	48	*88*
Population covered by mobile telephony (%)	..	68	*43*	..
Internet users (per 1,000 people)	1	*5*	20	*47*
Personal computers (per 1,000 people)	2	*8*	*8*	*30*
Households with television (%)	43	*43*	*16*	*88*
Quality				
Telephone faults (per 100 main lines per year)	..	..	..	..
Broadband subscribers (per 1,000 people)	0.0	*0.0*	0.5	*0.2*
International Internet bandwidth (bits per person)	*0*	..	*3*	*15*
Affordability				
Price basket for fixed line (US$ per month, residential)	3.7	*3.0*	*6.6*	*4.9*
Price basket for mobile (US$ per month)	..	5.9	11.6	8.1
Price basket for Internet (US$ per month)	..	*30.8*	*45.5*	*24.5*
Price of call to United States (US$ per 3 minutes)	4.45	2.39	1.95	1.64
Institutional efficiency and sustainability				
Total telecommunications revenue (% GDP)	1.0	*1.5*	2.3	2.8
Total telephone subscribers per employee	72	*176*	*89*	..
Total telecommunications investment (% revenue)	51.1	*51.1*	*27.8*	*27.8*
ICT applications				
ICT expenditure (% GDP)	..	..	*4.1*	..
E-government readiness index (scale 0–1)	..	0.05	0.12	0.16
Secure Internet servers (per 1 million people)	..	*0.1*	0.3	0.6
Schools connected to the Internet (%)	..	..	..	..

GNI per Capita, Atlas Method, 2000–4
Current US$

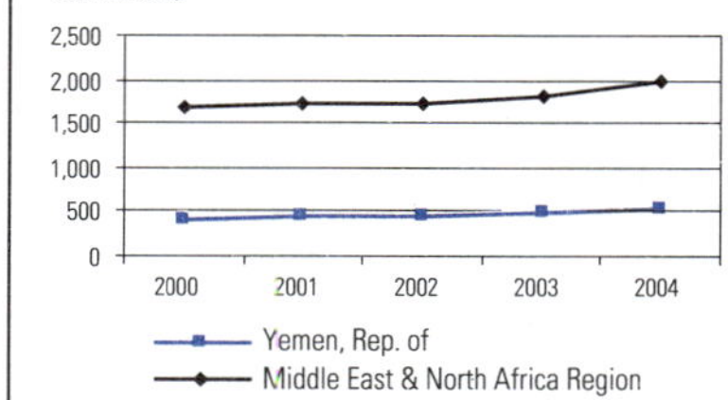

ICT MDG[b] Indicators, 2000–4
Number per 1,000 people

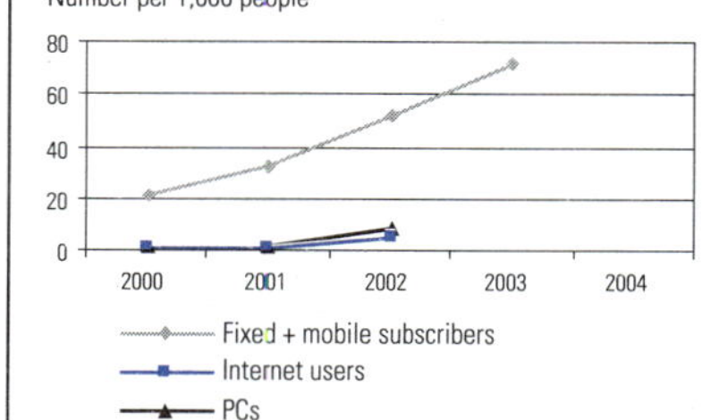

Price of Call to the United States, 2000–4
US$ per 3 minutes

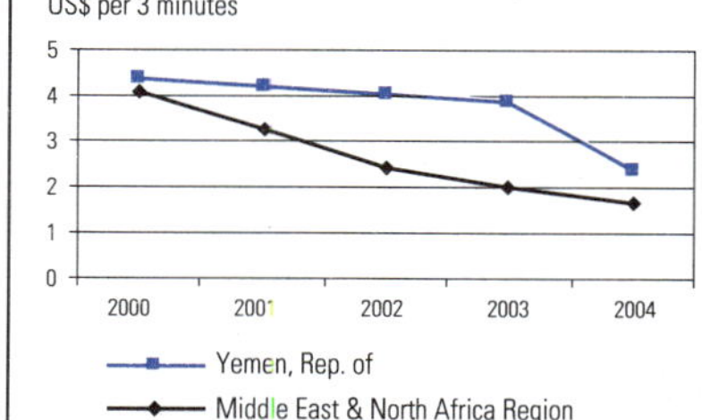

Total Telecommunications Revenue, 2000–4
Percentage of GDP

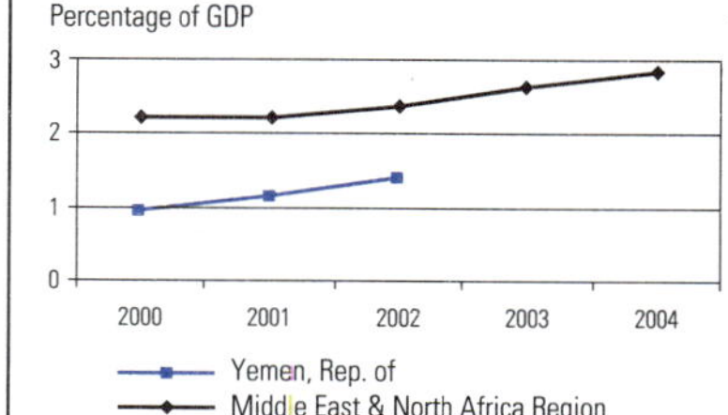

Notes: Figures in italics are for years other than those specified. .. indicates data are not available. C = competition; GDP = gross domestic product; GNI = gross national income; ICT = information and communication technology; M = monopoly; MDG = Millennium Development Goal; P = partial competition; and PCs = personal computers.

a. Outgoing and incoming.

b. Millennium Development Goal indicators 47, 48a, and 48b.

Sources: Economic and social context: UNDP and World Bank; ICT sector structure: ITU, WEF; ICT sector performance: Global Insight/WITSA, ITU, Netcraft, UNDESA, UNPAN, and World Bank. Produced by the Global Information and Communication Technologies Department and the Development Economics Data Group. For complete information, see Definitions and Data Sources.

Zambia

	Zambia 2000	Zambia 2004	Low-income group 2004	Sub-Saharan Africa Region 2004
Economic and social context				
Population, total (millions)	10	11	2,338	719
Urban population (% total population)	35	36	31	37
Poverty (% population below US$1 per day)	*63.7*	..	..	*46.4*
GNI per capita, Atlas method (current US$)	320	450	510	600
GDP growth, 1995–2000 and 2000–4 (%)	2.2	4.4	5.4	3.9
Adult literacy rate (% ages 15 and over)	*68*	..	61	59
Primary, secondary, tertiary school enrollment (% gross)	45	*48*	*55*	*52*
ICT sector structure				
Separate telecommunications regulator	Yes	Yes		
Status of main fixed-line operator	*Public*	Public		
Level of competition: international long distance	..	M		
Level of competition: mobile	*C*	P		
Level of competition: Internet service provider	*C*	P		
Government prioritization of ICT (scale 1–7)	..	3.7	..	*4.4*
ICT sector performance				
Access				
Telephone main lines (per 1,000 people)	8	8	33	17
International voice traffic (minutes per person)[a]	4	..	*4*	..
Mobile subscribers (per 1,000 people)	10	*23*	48	86
Population covered by mobile telephony (%)	51	..	*43*	..
Internet users (per 1,000 people)	2	5	20	15
Personal computers (per 1,000 people)	7	*9*	*8*	*12*
Households with television (%)	23	*26*	*16*	*15*
Quality				
Telephone faults (per 100 main lines per year)	90.9	*90.8*	..	..
Broadband subscribers (per 1,000 people)	0.0	*0.0*	0.5	*0.1*
International Internet bandwidth (bits per person)	0	*0*	*3*	*4*
Affordability				
Price basket for fixed line (US$ per month, residential)	4.6	*5.4*	*6.6*	*8.5*
Price basket for mobile (US$ per month)	..	13.1	11.6	13.5
Price basket for Internet (US$ per month)	..	*32.6*	*45.5*	*54.8*
Price of call to United States (US$ per 3 minutes)	2.57	*6.45*	1.95	2.43
Institutional efficiency and sustainability				
Total telecommunications revenue (% GDP)	2.0	..	2.3	5.0
Total telephone subscribers per employee	59	..	*89*	*144*
Total telecommunications investment (% revenue)	12.3	..	*27.8*	*27.8*
ICT applications				
ICT expenditure (% GDP)	..	..	*4.1*	..
E-government readiness index (scale 0–1)	..	0.00	0.12	0.11
Secure Internet servers (per 1 million people)	..	0.2	0.3	1.9
Schools connected to the Internet (%)	..	..	..	..

GNI per Capita, Atlas Method, 2000–4

Current US$

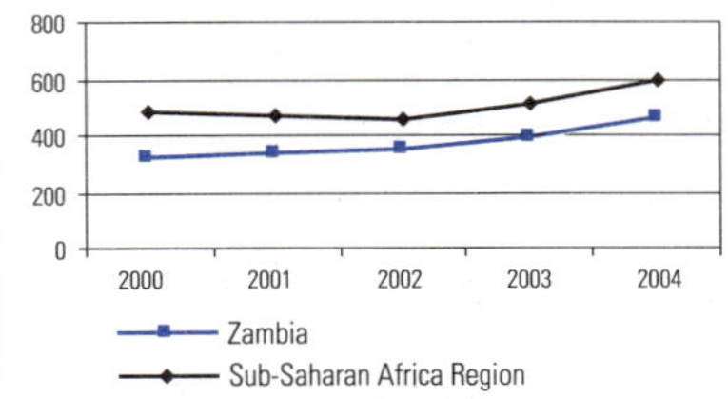

ICT MDG[b] Indicators, 2000–4

Number per 1,000 people

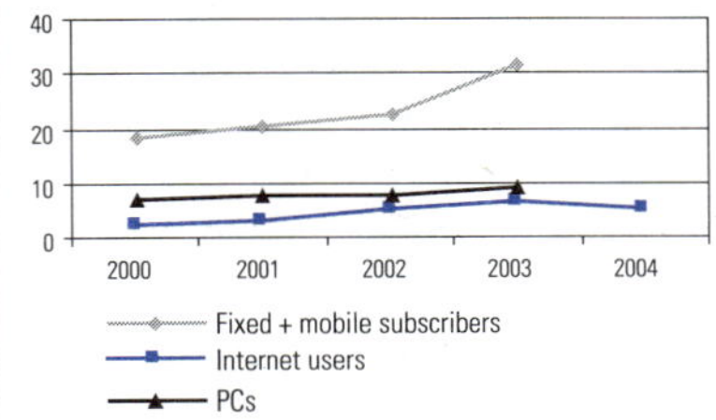

Price of Call to the United States, 2000–4

US$ per 3 minutes

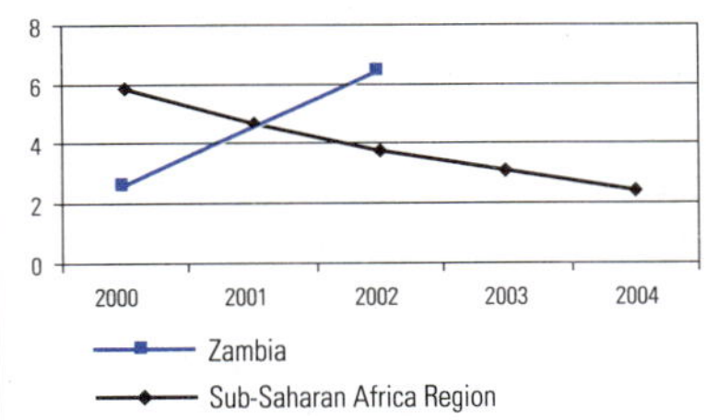

Total Telecommunications Revenue, 2000–4

Percentage of GDP

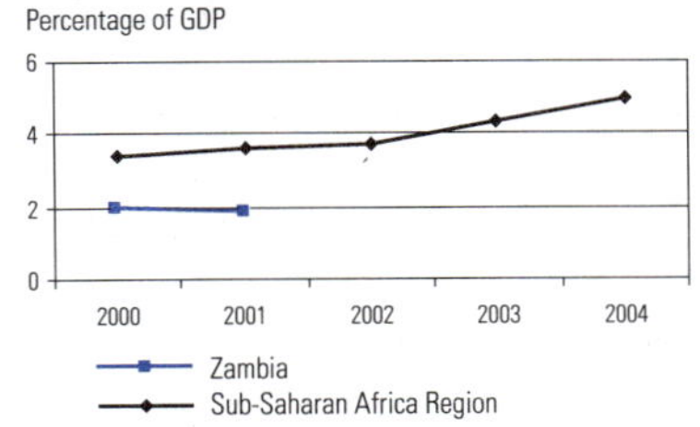

Notes: Figures in italics are for years other than those specified. .. indicates data are not available. C = competition; GDP = gross domestic product; GNI = gross national income; ICT = information and communication technology; M = monopoly; MDG = Millennium Development Goal; P = partial competition; and PCs = personal computers.

a. Outgoing and incoming.

b. Millennium Development Goal indicators 47, 48a, and 48b.

Sources: Economic and social context: UNDP and World Bank; ICT sector structure: ITU, WEF; ICT sector performance: Global Insight/WITSA, ITU, Netcraft, UNDESA, UNPAN, and World Bank. Produced by the Global Information and Communication Technologies Department and the Development Economics Data Group. For complete information, see Definitions and Data Sources.

Zimbabwe

	Zimbabwe 2000	Zimbabwe 2004	Low-income group 2004	Sub-Saharan Africa Region 2004
Economic and social context				
Population, total (millions)	13	13	2,338	719
Urban population (% total population)	34	35	31	37
Poverty (% population below US$1 per day)	*56.1*	..	..	*46.4*
GNI per capita, Atlas method (current US$)	440	..	510	600
GDP growth, 1995–2000 and 2000–4 (%)	1.9	*-7.0*	5.4	3.9
Adult literacy rate (% ages 15 and over)	..	*90*	61	59
Primary, secondary, tertiary school enrollment (% gross)	58	*52*	*55*	*52*
ICT sector structure				
Separate telecommunications regulator	No	Yes		
Status of main fixed-line operator	*Public*	Public		
Level of competition: international long distance	*M*	C		
Level of competition: mobile	*C*	C		
Level of competition: Internet service provider	*C*	C		
Government prioritization of ICT (scale 1–7)	..	3.1	..	*4.4*
ICT sector performance				
Access				
Telephone main lines (per 1,000 people)	20	*23*	33	17
International voice traffic (minutes per person)[a]	12	*13*	*4*	..
Mobile subscribers (per 1,000 people)	24	30	48	86
Population covered by mobile telephony (%)	..	..	*43*	..
Internet users (per 1,000 people)	4	30	20	15
Personal computers (per 1,000 people)	15	*47*	*8*	*12*
Households with television (%)	25	*26*	*16*	*15*
Quality				
Telephone faults (per 100 main lines per year)	..	..	..	..
Broadband subscribers (per 1,000 people)	*0.0*	*0.4*	0.5	*0.1*
International Internet bandwidth (bits per person)	1	*1*	*3*	*4*
Affordability				
Price basket for fixed line (US$ per month, residential)	6.3	*2.0*	*6.6*	*8.5*
Price basket for mobile (US$ per month)	..	17.8	11.6	13.5
Price basket for Internet (US$ per month)	..	*23.3*	*45.5*	*54.8*
Price of call to United States (US$ per 3 minutes)	4.36	..	1.95	2.43
Institutional efficiency and sustainability				
Total telecommunications revenue (% GDP)	2.5	..	2.3	5.0
Total telephone subscribers per employee	121	*140*	*89*	*144*
Total telecommunications investment (% revenue)	66.6	..	*27.8*	*27.8*
ICT applications				
ICT expenditure (% GDP)	4.2	*11.8*	*4.1*	..
E-government readiness index (scale 0–1)	..	0.02	0.12	0.11
Secure Internet servers (per 1 million people)	*0.1*	0.5	0.3	1.9
Schools connected to the Internet (%)	..	..	..	..

GNI per Capita, Atlas Method, 2000–4
Current US$

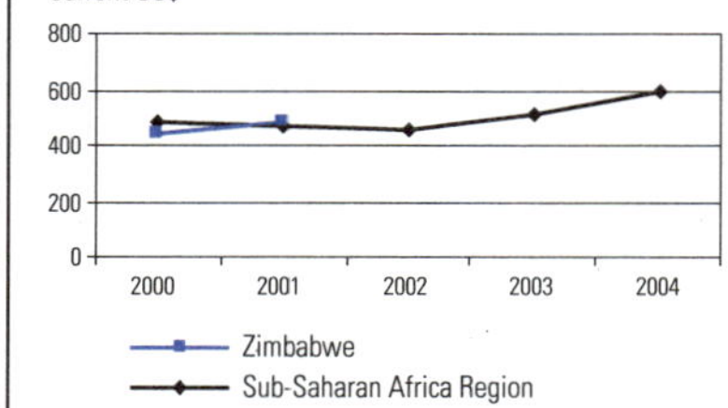

ICT MDG[b] Indicators, 2000–4
Number per 1,000 people

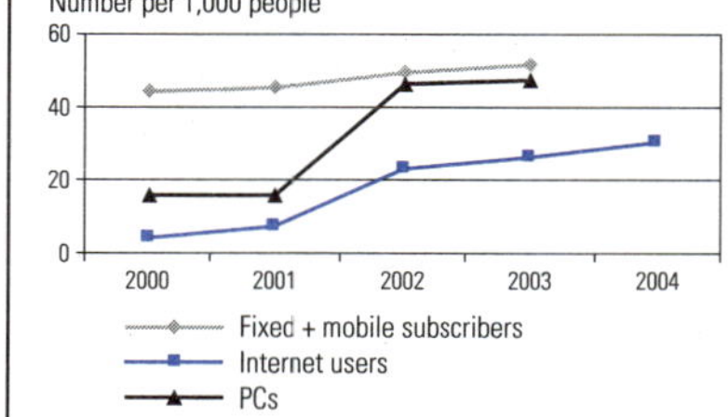

Price of Call to the United States, 2000–4
US$ per 3 minutes

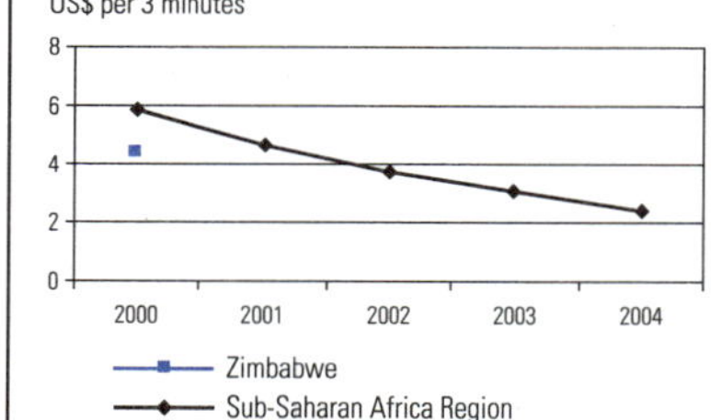

Total Telecommunications Revenue, 2000–4
Percentage of GDP

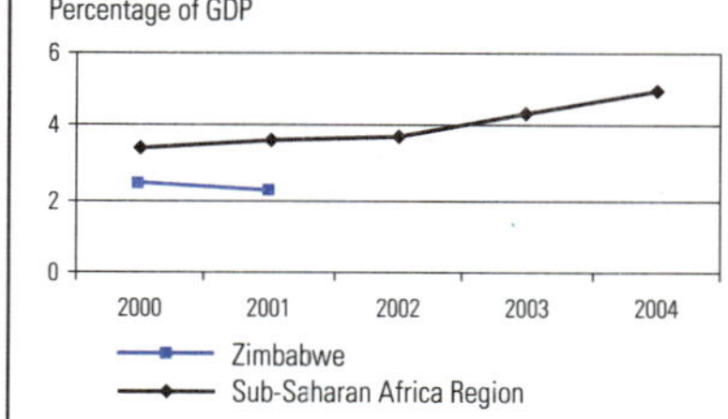

Notes: Figures in italics are for years other than those specified. .. indicates data are not available. C = competition; GDP = gross domestic product; GNI = gross national income; ICT = information and communication technology; M = monopoly; MDG = Millennium Development Goal; P = partial competition; and PCs = personal computers.

a. Outgoing and incoming.

b. Millennium Development Goal indicators 47, 48a, and 48b.

Sources: Economic and social context: UNDP and World Bank; ICT sector structure: ITU, WEF; ICT sector performance: Global Insight/WITSA, ITU, Netcraft, UNDESA, UNPAN, and World Bank. Produced by the Global Information and Communication Technologies Department and the Development Economics Data Group. For complete information, see Definitions Data and Sources.

Key ICT Indicators for Other Economies

	Population (thousands) 2004	GNI per capita, Atlas method (current US$) 2004	Telephone main lines (per 1,000 people) 2002–4[a]	Mobile subscribers (per 1,000 people) 2002–4[a]	Internet users (per 1,000 people) 2002–4[a]	Personal computers (per 1,000 people) 2002–4[a]
Afghanistan	..	..[b]	..	..	..	..
American Samoa	*57*	..[c]	..	..	..	..
Andorra	*66*	..[d]	683	786	152	..
Antigua and Barbuda	80	10,000	484	500	131	..
Aruba	*99*	..[d]	..	..	..	..
Bahamas, The	320	*14,920*	415	388	265	..
Bahrain	725	*12,410*	261	905	275	153
Barbados	272	*9,270*	495	517	370	104
Belize	283	3,940	122	221	128	132
Bermuda	64	..[d]	871	467	..	529
Bhutan	896	760	29	9	17	12
Brunei Darussalam	361	..[d]	252	..	..	77
Cape Verde	481	1,770	146	137	43	85
Cayman Islands	44	..[d]	..	..	..	..
Channel Islands	149	..[d]	..	..	..	..
Comoros	614	530	22	3	8	8
Cyprus	776	17,580	650	717	325	252
Djibouti	716	1,030	13	33	9	21
Dominica	71	3,650	334	132	176	112
Equatorial Guinea	506	..[c]	19	84	4	7
Faeroe Islands	48	..[d]	..	631	..	..
Fiji	848	2,690	122	132	66	49
French Polynesia	246	..[d]	218	375	146	292
Greenland	57	..[d]	448	522	442	..
Grenada	106	3,760	312	404	182	135
Guam	164	..[d]	..	..	314	..
Guyana	772	990	121	196	39	38
Haiti	8,592	390	16	17	10	..
Iceland	290	38,620	660	1,000	586	451
Iraq	25,261	..[e]	28	44	1	8
Isle of Man	77	..[d]	..	..	..	..
Kiribati	98	970	47	6	21	11
Korea, Democratic People's Republic of	22,745	..[b]	43	0	..	..
Liberia	3,449	110	2	15	1	..
Libya	5,674	4,450	135	18	29	24
Liechtenstein	34	..[d]	..	..	..	..
Luxembourg	450	56,230	805	1,222	460	598
Macao, China	449	..[d]	387	963	270	210
Maldives	300	2,510	105	378	97	100
Malta	401	12,250	511	744	302	254
Marshall Islands	60	2,370	78	10	25	57
Mayotte	172	..[c]	..	302	..	..
Micronesia, Federated States of	127	1,990	..	14	49	..
Monaco	33	..[d]	1,015	457	485	..
Netherlands Antilles	222	..[d]	..	..	..	..
New Caledonia	229	..[d]	236	363	267	..
Northern Mariana Islands	77	..[c]	..	..	..	..
Palau	20	6,870	..	..	204	..
Puerto Rico	3,929	..[d]	316	465	175	..
Qatar	637	..[d]	296	769	202	180

	Population (thousands) 2004	GNI per capita, Atlas method (current US$) 2004	Telephone main lines (per 1,000 people) 2002–4[a]	Mobile subscribers (per 1,000 people) 2002–4[a]	Internet users (per 1,000 people) 2002–4[a]	Personal computers (per 1,000 people) 2002–4[a]
Samoa	179	1,860	67	15	23	7
San Marino	28	..[d]	739	604	517	..
São Tomé and Principe	161	370	44	31	95	..
Seychelles	85	8,090	264	652	213	158
Solomon Islands	471	550	14	3	5	41
Somalia	9,938	..[b]	11	4	3	..
St. Kitts and Nevis	47	7,600	503	107	214	193
St. Lucia	164	4,310	309	826	..	151
St. Vincent and the Grenadines	108	3,650	250	91	64	128
Suriname	443	2,250	182	384	57	..
Timor-Leste	925	550	2	27	2	..
Tonga	102	1,830	111	33	29	20
Vanuatu	215	1,340	31	37	36	15
Virgin Islands (U.S.)	113	..[d]	..	..	272	..

Notes: Figures in italics are for years other than those specified. .. indicates data are not available. GNI = gross national income; ICT = information and communication technology.

a. Data are the latest year available in the period shown.
b. Estimated to be low income ($825 or less).
c. Estimated to be upper-middle income ($3,256–$10,066).
d. Estimated to be high income ($10,066 or more).
e. Estimated to be lower-middle income ($826–$3,255).

Sources: ITU and World Bank. Produced by the Global Information and Communication Technologies Department and the Development Economics Data Group. For complete information, see Definitions and Data Sources.

Definitions and Data Sources

This section provides definitions of the indicators used in the World Bank ICT At-a-Glance country tables. It also indicates the source of the data used.

Economic and Social Context

Total population is mid-year estimates based on the de facto definition of population, which counts all residents regardless of legal status or citizenship—except for refugees not permanently settled in the country of asylum, who are generally considered part of the population of their country of origin. (World Bank)

Urban population is the share of the total population living in areas defined as urban in each country. (United Nations)

Poverty is the percentage of the population living on less than $1.08 a day at 1993 international prices. As a result of revisions in PPP (purchasing power parity) exchange rates, poverty rates cannot be compared with poverty rates reported previously for individual countries. Data are from nationally representative primary household surveys conducted by national statistical offices or by private agencies under the supervision of government or international agencies and obtained from government statistical offices and World Bank Group country departments. (World Bank)

GNI per capita is the gross national income, converted to U.S. dollars using the World Bank Atlas method, divided by the mid-year population. GNI is the sum of value added by all resident producers plus any product taxes (less subsidies) not included in the valuation of output plus net receipts of primary income (compensation of employees and property income) from abroad. GNI, calculated in national currency, is usually converted to U.S. dollars at official exchange rates for comparisons across economies, although an alternative rate is used when the official exchange rate is judged to diverge by an exceptionally large margin from the rate actually applied in international transactions. To smooth fluctuations in prices and exchange rates, a special Atlas method of conversion is used by the World Bank. This applies a conversion factor that averages the country's exchange rates for a given year and the two preceding years, adjusted for the difference between the rate of inflation in the country and that in Japan, the United Kingdom, the United States, and the euro zone. (World Bank)

GDP growth is the average annual growth rate of gross domestic product at market prices based on constant local currency. Aggregates are based on constant 2000 U.S. dollars. GDP is the sum of gross value added by all resident producers in the economy plus any product taxes and minus any subsidies not included in the value of the products. It is calculated without making deductions for depreciation of fabricated assets or for depletion and degradation of natural resources. (World Bank, Organisation for Economic Co-operation and Development, United Nations)

Adult literacy rate is the percentage of people ages 15 and above who can, with understanding, read and write a short, simple statement about their everyday life. (United Nations Educational, Scientific, and Cultural Organization [UNESCO] Institute for Statistics)

Primary, secondary, tertiary school enrollment is the combined number of students enrolled in primary, secondary, and tertiary levels of education, regardless of age, as a percentage of the population of official school age for the three levels. (United Nations Development Programme)

ICT Sector Structure

Separate telecommunications regulator refers to whether a separate telecommunications regulator exists. (International Telecommunication Union)

Status of main fixed-line operator shows the status of the incumbent fixed-line operator. *Public* refers to a fully state-owned operator, *private* refers to a fully private operator, and *mixed* refers to a partially private operator. (International Telecommunication Union)

Level of competition: international long distance refers to the level of competition for international long distance telephone calls (M = monopoly, P = partial, C = full competition). (International Telecommunication Union)

Level of competition: mobile refers to the level of competition for digital cellular mobile services (M = monopoly, P = partial, C = full competition). (International Telecommunication Union)

Level of competition: Internet service provider refers to the level of competition for retail Internet access service (M = monopoly, P = partial, C = full competition). (International Telecommunication Union)

Government prioritization of ICT is based on replies to the following question: "Information and communication technologies (ICT) are an overall priority for the government" (1 = strongly disagree, 7 = strongly agree). (World Economic Forum)

ICT Sector Performance

Access

Telephone main lines is the number of telephone lines connecting a customer's equipment to the public switched telephone network, per 1,000 people. (International Telecommunication Union, World Bank)

International voice traffic is derived from the sum of international incoming and outgoing telephone traffic (in minutes) divided by total population. (International Telecommunication Union, World Bank)

Mobile subscribers is the number of subscribers to a public mobile telephone service using cellular technology, per 1,000 people. (International Telecommunication Union, World Bank)

Population covered by mobile telephony measures the percentage of people who are within range of a mobile cellular signal regardless of whether they are subscribers. (International Telecommunication Union, World Bank)

Internet users is the number of persons with access to the worldwide computer network, per 1,000 people. In some countries, surveys are carried out to measure Internet usage. These surveys show penetration as a percentage of a target age group that varies between countries. Therefore, this indicator divides the number of Internet users in the targeted age group by the total population. In countries where there are no regular surveys, data are estimated based on the number of subscribers. Therefore, the data are not strictly comparable. (International Telecommunication Union, World Bank)

Personal computers is the number of self-contained computers designed to be used by a single individual, per 1,000 people. (International Telecommunication Union, World Bank)

Households with a television is the percentage of households with a television set. Note that some countries report only the number of households with a color television set and therefore the true figure may be higher than reported. (International Telecommunication Union, World Bank)

Quality

Telephone faults is the number of reported telephone faults per year per 100 telephone main lines. (International Telecommunication Union, World Bank)

Broadband subscribers is the total number of broadband subscribers (that is, Digital Subscriber Line [DSL], cable modem, and other high-speed technologies), per 1,000 people. Because data are from national sources that have different definitions of broadband, the data are not strictly comparable. (International Telecommunication Union, World Bank)

International Internet bandwidth is the international Internet bandwidth (that is, contracted capacity of international connections between countries for transmitting Internet traffic) divided by the population. (International Telecommunication Union, World Bank)

Affordability

Price basket for fixed line is calculated based on a portion of the installation charge (one-fifth), monthly subscription charge and cost of local calls (15 peak and 15 off-peak calls of three minutes each). Data are compiled in the national currency and converted to U.S. dollars using the annual average exchange rate. (International Telecommunication Union, World Bank)

Price basket for mobile is calculated based on the pre-paid price for 25 calls per month spread over same mobile network, other mobile networks, and mobile to fixed calls and during peak, off-peak, and weekend times. The basket also includes the price of 30 text messages per month. Data are compiled in the national currency and converted to U.S. dollars using the annual average exchange rate. (World Bank)

Price basket for Internet is calculated based on the cheapest available tariff for accessing the Internet 20 hours a month (10 hours peak and 10 hours off-peak). The basket does not include the telephone line rental but does include telephone usage charges if applicable. Data are compiled in the national currency and converted to U.S. dollars using the annual average exchange rate. (International Telecommunication Union)

Price of call to United States is the cost of a three-minute, peak rate, fixed-line call from the country to the United States. Data are compiled in the national currency and converted to U.S. dollars using the annual average exchange rate. (International Telecommunication Union, World Bank)

Institutional efficiency and sustainability

Total telecommunication revenue is the amount of revenues from the provision of telecommunications services such as fixed, mobile, and data divided by GDP. (International Telecommunication Union, World Bank)

Total telephone subscribers per employee is the total number of telephone subscribers (fixed plus mobile) divided by total telecommunications employees. (International Telecommunication Union, World Bank)

Total telecom investment is the total telecommunications investment (capital expenditure) as a percentage of telecommunications revenue. (International Telecommunication Union, World Bank)

ICT Applications

ICT expenditure includes hardware (computers, storage devices, printers, and other peripherals); software (operating systems, programming tools, utilities, applications, and internal software development); services (information technology consulting, computer and network systems integration, Web hosting, data processing services, and other services); and communications services (voice and data communications services) and wired and wireless communications equipment. (Global Insight, Inc., published by World Information Technology and Services Alliance)

E-government readiness index is based on a five-stage model, ascending in nature, and building upon the previous level of sophistication of a government's online presence. These stages are Emerging, Enhanced, Interactive, Transactional, and Networked. Countries are scored on the basis of whether they provide specific products and services (1 = best). (United Nations Department of Economic and Social Affairs, United Nations Online Network in Public Administration and Finance)

Secure Internet servers is the number of servers using encryption technology for Internet transactions, per 1 million people. (Netcraft, http:// www.netcraft.com/)

Schools connected to the Internet is the percentage of primary and secondary schools in the country that have access to the Internet. (World Bank)

Contributors

Aref Adamali is a management consultant with the On The Frontier (OTF) Group. He is currently working in Afghanistan on the National Innovation and Competitiveness Project, developing growth strategies across a range of sectors from carpets to finance. Prior to joining OTF, he worked at the World Bank's Global Information and Communication Technologies Department (GICT) and with BearingPoint's nonprofit consulting practice. He has a master's degree from Columbia University's School of International and Public Affairs.

David A. Cieslikowski is a senior economist with the Development Economics Data Group of the World Bank. He has been the team coordinator of the Bank's flagship statistical publication, *World Development Indicators (WDI)*, and is the team's specialist on data related to the private sector, governance, transport, ICT, and science and technology. He developed, along with a Data Group colleague, the original ICT At-a-Glance table, and he collaborates with Bank staff on ICT-related statistical issues and the global Partnership on Measuring ICT for Development. He has an MA in International Relations from the University of Southern California.

George R. Clarke is a senior economist working for the Regional Program on Enterprise Development (RPED) in the Africa Private Sector Group at the World Bank. Before joining the RPED team in 2005, he was a member of the Competition Policy and Regulation team in the World Bank's Development Economics Research Group. His work has focused on privatization, competition, and regulation of infrastructure and on firm surveys. He was a core member of the team working on the *World Development Report 2005*. His research has been published in several books and academic journals including the *Journal of Law and Economics*, the *Journal of Public Economics*, and the *Journal of Development Economics*. He has a PhD in economics from the University of Rochester.

John Oliver Coffey is a consultant on a private sector development project in Manaus, Brazil. With the GICT Policy Division, he conducted research and analysis on the national ICT strategies of 40 countries. The findings of this study informed the *Analysis of National E-Strategies* report. He also contributed to the *E-Strategies Monitoring and Evaluation Toolkit* and studied e-government strategy implementations and the lessons to be learned from failed e-government initiatives. He has recently worked as an economist studying remittances and labor migration between Mexico and the United States, and as an investment analyst for a private equity investment firm in the Serbian telecommunications sector. He received a master's degree in international affairs from Columbia University's School of International and Public Affairs.

Pierre Guislain is the manager of the World Bank GICT Policy Division, where he is in charge of the Bank's ICT, telecommunications and postal sector strategy, projects and analytical portfolio. Prior to taking this position, he was based in Brussels, where he managed a joint program between the World Bank and the European Commission on Private Participation in Mediterranean Infrastructure. He has also worked in the West Africa Projects Department, the Legal Department, the Asia Technical Department, and the Private Sector Advisory Services Department at the World Bank. He has written and edited numerous publications on ICT, infrastructure sector reform, and privatization. He holds an MPA in economics and public policy from Princeton University and a graduate law degree from the University of Louvain.

Naomi Halewood is a consultant with the GICT Policy Division of the World Bank. She is currently working on a number of ICT projects in the East Asia and Pacific region and is part of the Monitoring & Evaluation (M&E) team, for which she performs data analysis and research. Prior to joining GICT, she worked in the Development Economics Data Group of the World Bank and for the Center for International Political Economy in New York. She has an MBA and an MA in international development from American University.

Kaoru Kimura is a consultant and team member of the Monitoring & Evaluation (M&E) team in the GICT Policy Division at the World Bank. Her work includes ICT data collection and analysis. Prior to joining GICT, Kaoru was at Nippon Telegraph and Telephone Corporation (NTT) Japan, where she worked on marketing, business consulting, and designing communication systems for multinational corporate clients. She has a master's degree in international development studies from the National Graduate Institute for Policy Studies (GRIPS) in Japan.

Bruno Lanvin is an adviser on e-strategies with the World Bank, based in Geneva (Switzerland). From June 2001 to December 2003, he was the manager of the Information for Development Program (infoDev). From 2000 to 2001, he was executive secretary of the Group of Eight (G-8) Digital Opportunity Task Force, and before that, he was head of Electronic Commerce in the United Nations Conference on Trade and Development (UNCTAD) in Geneva. He has occupied various senior positions in the UN system, including chief of cabinet of the director general of the United Nations in New York, head of Strategic Planning, and later chief of the SME Trade Competitiveness Unit of UNCTAD/SITE. He was the main drafter, team leader, and editor of *Building Confidence: Electronic Commerce and Development*, published in January 2000. In 2003 and in 2004, he co-authored the *Global Information Technology Report* (INSEAD, The World Economic Forum, infoDev). He holds an MBA from Ecole des Hautes Etudes Commerciales (HEC) in Paris and a PhD in economics from the University of Paris I (La Sorbonne).

Michael Minges is a senior market analyst at Telecommunications Management Group, Inc. (TMG), a telecommunications consulting firm. Before joining TMG, he worked as head of the Market, Economics and Finance Unit, Telecommunication Development Bureau at the International Telecommunication Union (ITU), where his functions included the analysis of telecommunications trends in developing nations, market studies of countries and regions, and regulatory and policy advice. While at the ITU, he launched the *World Telecommunication Development Report*, a principal industry publication, and designed the Digital Access Index, for measuring ICT progress. He also worked at the International Monetary Fund as an information technology specialist. He has an MBA from George Washington University.

William Prince is an information officer in the Development Economics Data Group of the World Bank, where he is responsible for production and content management of electronic data products, including the online and CD-ROM versions of *World Development Indicators* and *Global Development Finance*. He also provides overall data quality management and specialized data support for the Data Group's clients. Prior to joining the Bank, he was a statistical consultant for British Telecom. He has an MBA in Decision Analysis from Arizona State University.

Christine Zhen-Wei Qiang is a senior economist in the GICT Policy Division of the World Bank Group. Her main responsibilities include ICT policy and economics analysis as well as the design and management of ICT projects in Asia. She has published over 10 journal articles and book chapters on ICT for development, economic growth, and productivity. She holds a PhD in economics and an MSE in computer science and engineering from the Johns Hopkins University.

Zaid Safdar is a consultant in the GICT Policy Division of the World Bank. He specializes in economic, policy, and strategy aspects of telecommunications and ICT and has worked in the sector for seven years. Prior to joining the World Bank, he was a project manager at Lucent Technologies Bell Laboratories. He did product and business development for broadband, wireless, and Voice-over-IP segments of North American and East Asian markets. He has also worked in the Foreign Policy Department of the Brookings Institution. Zaid has an MA in international economics from the Johns Hopkins University Paul H. Nitze School of Advanced International Studies and an MS in engineering from the University of Pennsylvania.

Eric Swanson is program manager in the Development Economics Data Group, where he leads a team responsible for producing the *World Development Indicators* and monitoring the Millennium Development Goals. He works extensively with the UN and its specialized agencies to coordinate the compilation and dissemination of statistical information. He holds a PhD in economics from the State University of New York at Buffalo.

Björn Wellenius is an independent consultant on telecommunications policy, regulation, and economics, with experience in about 40 developing countries and transition economies. He advises the World Bank Group, law and consulting firms in Europe and North America, and governments and regulatory authorities. Until 1999 he was an adviser on telecommunications with the World Bank. His publications include five books on telecommunications and economic development as well as best practice notes and technical papers on universal service, rural infrastructure financing, competition policy, spectrum management, and sector reform. Before joining the World Bank, he was professor of telecommunications at Universidad de Chile. He has a PhD in physical sciences (telecommunications) from the University of Essex, England, and a degree in electronics and telecommunications engineering from Universidad de Chile.

Notes

Notes

Notes

Notes

Notes